Springer Complexity

Springer Complexity is an interdisciplinary program publishing the best research and academic-level teaching on both fundamental and applied aspects of complex systems—cutting across all traditional disciplines of the natural and life sciences, engineering, economics, medicine, neuroscience, social and computer science.

Complex Systems are systems that comprise many interacting parts with the ability to generate a new quality of macroscopic collective behavior the manifestations of which are the spontaneous formation of distinctive temporal, spatial or functional structures. Models of such systems can be successfully mapped onto quite diverse "real-life" situations like the climate, the coherent emission of light from lasers, chemical reaction-diffusion systems, biological cellular networks, the dynamics of stock markets and of the internet, earthquake statistics and prediction, freeway traffic, the human brain, or the formation of opinions in social systems, to name just some of the popular applications.

Although their scope and methodologies overlap somewhat, one can distinguish the following main concepts and tools: self-organization, nonlinear dynamics, synergetics, turbulence, dynamical systems, catastrophes, instabilities, stochastic processes, chaos, graphs and networks, cellular automata, adaptive systems, genetic algorithms and computational intelligence.

The three major book publication platforms of the Springer Complexity program are the monograph series "Understanding Complex Systems" focusing on the various applications of complexity, the "Springer Series in Synergetics", which is devoted to the quantitative theoretical and methodological foundations, and the "Springer Briefs in Complexity" which are concise and topical working reports, case studies, surveys, essays and lecture notes of relevance to the field. In addition to the books in these two core series, the program also incorporates individual titles ranging from textbooks to major reference works.

Indexed by SCOPUS, INSPEC, zbMATH, SCImago.

Series Editors

Understanding Complex Systems

Founding Editor: S. Kelso

More information about this series at http://www.springer.com/series/5394

Ihor Lubashevsky · Natalie Plavinska

Physics of the Human Temporality

Complex Present

Springer

Ihor Lubashevsky
University of Aizu
Aizu-Wakamatsu, Fukushima, Japan

Natalie Plavinska
University of Aizu
Aizu-Wakamatsu, Fukushima, Japan

ISSN 1860-0832 ISSN 1860-0840 (electronic)
Understanding Complex Systems
ISBN 978-3-030-82611-6 ISBN 978-3-030-82612-3 (eBook)
https://doi.org/10.1007/978-3-030-82612-3

This Springer imprint is published by the registered company Springer Nature Switzerland AG
The registered company address is: Gewerbestrasse 11, 6330 Cham, Switzerland

Preface

We have chosen the title of this book very carefully. It looks rather long but strictly reflects the problems posed in the book—Physics of the Human Temporality: Complex Present.

The word *physics* is essential because our mathematical constructions in their spirit are physical and turn to the main concepts developed in physics. Nowadays, there have emerged various branches of physics possessing individual paradigms, notions, and formalism. Newtonian mechanics, statistical physics, and quantum mechanics illustrate this fact. We hope that, in time, physics of the human mind also can become such a science.

It is necessary to accentuate that in no case we pretend to describe the physics of the human mind as a whole branch of science. In this book, we make the first step only and describe the temporal extension of the human mind in its own temporal dimension.

The term *time* is inextricably intertwined with the description and investigation of the inanimate world. Therefore, we introduce the term *temporality* to speak about the temporal dimension of the human mind and develop a novel concept of the temporal structure of the human mind.

The complex present is one of the basic structural elements of the human temporality. This element is very short—its duration is about 20 sec. Nevertheless, the complex present is of a multiscale structure and plays a crucial role in human behavior. In some sense, like one dewdrop reflects the whole world, the complex present reflects the basic properties of the temporal extension of the human mind. We may say that the present rather volumetric book focuses on describing how humans recognize reality here and now.

We consider that one of the characteristic features of the human mind is its temporal extension. For typical physical objects—particles, atoms, planets, etc.—only the present exists which may be conceived of as a point-like instant of time. In the human temporality, the past retained in memory, the imaginary future, and the present coexist and are closely intertwined affecting each other. Besides that, what humans perceive as the present also has a certain duration.

We have turned to the gist of phenomenology to describe the human temporality. It implies that there are two sources for our constructions. First, it is the consciousness as experienced from the first-person's point of view. Second, there are a variety of experimental data collected in psychology, neuropsychology, physiology, and other branches of cognitive science underlying our constructions. These premises enabled us to lay the foundation for the self-consistent theory of human temporality which comprises the past retained in memory, the imaginary future, and the perceived present. In this book, we consider only one of these parts, namely, the complex present.

For centuries, the notion of the now has been analyzed by many thinkers starting from the philosophers of ancient Greece. Already, St. Augustine noted the complexity and extension of the now. At the end of the nineteenth century, the philosopher and psychologist William James elaborated the concept of specious present to emphasize its extension in the human mind. At the beginning of the twentieth century, the philosopher Edmund Husserl proposed the account of the now having the tripartite structure comprising retention, primary impression, and protention.

In this book, we propose a novel concept of the present which is multicomponent and includes the connected past, the experiential now, and the connected future united by the possibility of common experience. According to experimental data, the experiential now is about 1-3 s, the connected past and future are about 10 s individually. The experiential now, in turn, consists of the immediate past, the immediate now, and the immediate future. The unity of the connected past, the experiential now, and the connected future called the *complex present* represents the minimal fragment of human temporality admitting a closed description.

The proposed mathematical formalism is aimed at the description of goal-oriented human actions and, thereby, it does not meet and cannot meet the paradigm of Newtonian mechanics. The analyzed systems are characterized by the four-component phase space comprising the spatial position, velocity, acceleration, and jerk of an object whose dynamics is governed by human actions. The Ostrogradsky formalism of systems with higher-order time-derivatives plays a crucial role in describing goal-oriented actions. The basic elementary entities of human goal-oriented actions—within our account—are space-time clouds which are fundamentally different from material points of Newtonian mechanics.

The novel concept of space-time clouds—the mental images of observed physical objects, subject's current actions on them, and anticipated actions—allows for the intrinsic uncertainty of human perception and anticipation. The description of the space-time clouds as distributed entities in the human temporality is based on the formalism of Hilbert spaces—the spaces with bilinear metric. In this sense, it is similar to the formalism of quantum mechanics but is dissipative in nature. The dynamics of space-time clouds admits the representation as a nonlocal random process in the time-temporality (2D-time) continuum.

The constructed mathematical formalism enables us to make the first step in studying physics of the human mind.

It is necessary to note that in developing a novel mathematical formalism for describing goal-oriented human actions, we have encountered difficulties in presenting our constructions. During developing our account, we discuss a wide variety of issues which are posed within different branches of philosophy, psychology, neuropsychology, physics, and that had to be taken into account in constructing the formalism and justifying the accepted premises and postulates. We combined these issues in the book following the logic of our mathematical constructions. As a result, discussion of these issues is distributed over different chapters, sections, and subsections.

These features are reflected in the style of writing the book. Chapter 1 is devoted to the history of the problem under consideration and the modern philosophical, psychological, and physical concepts underlying our constructions. Starting from Chap. 2, we present our constructions and justify them referring to the pertinent philosophical ideas and psychological facts. In the text, known and previously developed concepts are accompanied by the corresponding references. The main postulates of our account are singled out in the text using special format paragraphs. In the text, known and previously developed concepts are accompanied by the corresponding references. The main postulates of our account are singled out in the text using special format paragraphs. Each chapter and some sections are appended by conclusions where our premises and concepts are accumulated in condensed form for facilitating the text reading.

Aizu-Wakamatsu, Japan
February 2021

Ihor Lubashevsky
Natalie Plavinska

Contents

Toward Physics of the Human Mind

By the end of the twentieth century, it became evident that the notions and approaches developed in modern physics can be successfully used in describing a wide variety of complex phenomena in systems where the human factor is crucial. The emergence of such branches of physics as econophysics and sociophysics indicates the made advances (e.g., Chakrabarti et al. 2006, Helbing 2010, 2012, Galam 2012, Săvoiu 2013, Sen and Chakrabarti 2014, Slanina 2014, Abergel et al. 2017, 2019, for a review). In particular, the formalism of modern physics has been applied to modeling traffic flow (e.g., Helbing 2001, Kerner 2004, 2009, 2017, Treiber and Kesting 2013), pedestrian motion on overcrowded areas (e.g., Helbing and Johansson 2009, Schadschneider et al. 2009, Cristiani et al. 2014), opinion formation and voting dynamics (e.g., Castellano et al. 2009, Galam 2012), language evolution and cultural dynamics (e.g., Castellano et al. 2009, Heylighen and Chielens 2009, Port 2009, Kashima 2014), moral dynamics (Hegselmann 2009, Veit 2019), etc.

At the beginning of the twenty-first century, there has emerged a novel branch of physics that focuses directly on mental phenomena on their own. In particular, it poses a question about generalizing the notions and mathematical formalism of modern physics to allow for the properties unique to living beings, in particular, human beings. Perlovsky (2006) seems to be the first who coined the term *physics of the mind* with respect to this branch of physics. Nowadays, physics of the mind is taking first steps in development and closely intertwined with philosophy, psychology, neuropsychology, and cybernetics. In the present book, we confine our consideration to physics of the human mind separating this research direction from issues of animal mind (e.g., Wynne and Udell 2013, Spence et al. 2017, Andrews 2020) and ongoing debates about the essence (including its existence) of artificial mind (e.g., Penrose 1989, 1994, Franklin 1995, Chalmers 1996, Dennett 2005, 2017, Chalmers 2010, Hoya 2005, Stapp 2009, see also Pizzi 2008 for a review).

In constructing our account of the human mind, we turn to three sources.

Phenomenology of the human mind: As far as its philosophical aspects are concerned, physics of the human mind mainly rests upon *phenomenology* of the mind—the study of structures of consciousness as experienced from the first-person point of view (e.g., Smith 2013). In contrast to animals, humans are able to express their feelings, emotions, intentions, etc.. using well-developed language. Thereby we can analyze not only our own inner world, but also compare it with the inner worlds of others. In the present book, we develop physics of the human mind focusing mainly on human goal-oriented behavior.

In the early twentieth century, phenomenology became an individual branch of philosophy of mind due to the works of W. James (1842–1910), H.-L. Bergson (1859–1941), E. Husserl (1859–1938), M. Heidegger (1889–1976), J.-P. Sartre (1905–1980), M. Merleau-Ponty (1908–1961), and others. Below in this book, we will consider in detail the relationship between philosophy of the mind and the physics of the human mind.

The modern accounts of phenomenology (for introduction see, e.g., Thompson and Zahavi 2007, Gallagher and Zahavi 2012, and collection edited by Zahavi 2012), in particular, accounts of time perception (e.g., Dainton 2000, 2010, 2017a, and collection edited by Callender 2011, Arstila and Lloyd 2014, Phillips 2017) give us a hint about constructing the mathematical description of goal-oriented behavior and the underling basic principles. The physical theory of self-organization—synergetics developed by I. Prigogine (1917–2003), H. Haken (born 1927), W. Ebeling (born 1936) and applied also to modeling human behavior and the brain functioning (e.g., Kelso 2009a,b, 2014, Latash 2008b)—and phenomenology of the human mind have much in common. In particular, it is the possibility of describing the dynamics of complex hierarchical systems in terms of their upper-level properties. In our case, it implies that dealing with properties accessible to the mind, we can reconstruct in some way unconscious processes involved in analyzed phenomena.

Objective data about human behavior: In studying the human mind, we need to take into account not only conscious processes, but also unconscious ones which affect human actions and play the role of information sources. However, the person himself cannot report his unconscious reasons and motives for actions as well as unconscious details of processing perceived information. Physical measurements of human motor behavior, as well as various experiments, when the characteristics of subject's response like delay, intercity, *etc*., are recorded using physical tools provide physics of the human mind with essential details for verification of theoretical constructions. Here, we want to note some of the research directions that we refer to in developing our account of the human mind:

- The self-organization phenomena in *goal-oriented motor behavior* found by N. Bernstein (1896–1966) and investigated by his adherents including Scholz and Schöner (1999) proposed the novel concept of goal-oriented actions—the uncontrolled manifold hypothesis and Feldman (1966, 1986) put forward the equilibrium point hypothesis as a basic neural mechanism of action strategy implementation, see also Latash (2008a, 2012) for a review. The concept of unstable (uncontrolled) manifold was elaborated by Asai et al. (2009, 2013), Yoshikawa et al. (2016), Suzuki et al. (2016) for modeling human control.

- Theory of *unconscious (automatic) actions* including also effortless actions without attention and unconscious motivation (e.g., Sherman et al. 2014, Kihlstrom 2019, as well as collection edited by Morsella et al. 2009, Bruya 2010b, Ryan 2019 for a review). Automatic actions and their contribution to the efficiency of physical performance has been studied in detail within sport applications (e.g., Raab, Lobinger, Hoffmann, Pizzera and Laborde 2016, Raab, Wylleman, Seiler, Elbe and Hatzigeorgiadis 2016, Cappuccio 2019). Human intermittent control over unstable systems also illustrates the basic regularities of human cognition and its limit capacity (e.g., Loram et al. 2011, and collection edited by Ohira and Uzawa 2015).
- *Embodied cognition (embodied mind)* is the collection of theories appealing to the idea that cognition deeply depends on objects of the human body other than the brain. Without the involvement of the body in both sensing and acting, mental affairs would not exhibit the characteristics and properties they do (e.g., Wilson and Foglia 2015, Varela et al. 2016, Shapiro 2019a, as well collection edited by Shapiro 2014).

Objective data about cognitive states: Modern techniques such as fMRI and EEG make it possible to observe the brain functioning during mental operations. In this way, neuropsychologists could find out the regions of brain related to human inner cognitive states, as well as human actions, imagination, *etc*., and their dynamic properties. This research direction belongs to the modern branch of science called cognitive neuroscience (see, e.g., Baars and Gage 2010, Gazzaniga et al. 2019, for introduction to the analyzed issues). As a result, nowadays there have been many developed mathematical models for various cognitive processes turning explicitly or implicitly to the objective date of neuroscience; e.g., see collections edited by Busemeyer et al. (e.g., 2015), Yufik et al. (e.g., 2017), Batchelder et al. (e.g., 2017, 2018), Moustafa (e.g., 2018), and a recent review by Northoff et al. (2020) for relation between the brain regions and mental features like self, consciousness, emotion, *etc.*

Besides, we want to note a number of concepts and models proposed for analyzing cognitive phenomena described in terms attributed to the brain as a whole system, what are essential for our account.

- *Principle of minimizing free energy* for describing the brain functioning (Friston et al. 2006, Friston 2009, 2010).
- *Active inference* being the macro-level mechanism comprising loops of action and perception via which the free energy minimization is implemented (Friston et al. 2009, 2010, 2012).
- *Predictive coding paradigm* describing the adaptation of mental representations of physical objects and process via the comparison of top-down and bottom-up neural signals and governed by the free energy minimization (Friston and Kiebel 2009, Friston, FitzGerald, Rigoli, Schwartenbeck and Pezzulo 2017, Friston, Lin, Frith, Pezzulo, Hobson and Ondobaka 2017).

- Emotion emergence governed by free energy energy (Joffily and Coricelli 2013, Demekas et al. 2020, Michael 2020).
- Human goal-oriented behavior as a result of active inference (Botvinick and Toussaint 2012, Solway and Botvinick 2012, 2018).
- *Dynamic logic* united with the *knowledge instinct* and *aesthetic emotions* as a neural mechanism of predictive cording Perlovsky (2006, see also Perlovsky 2016, Schoeller et al. 2018 for a review).

The modern state of the art in modeling neural processes in the brain turning to the concept of free energy is presented in reviews by Friston (2018b), Ramstead et al. (2018, 2019), Badcock et al. (2019), Parr and Friston (2019), Smith et al. (2020).

The human mind and temporality: In the present book, we single out the human mind on its own as the main object of our investigation. However, in constructing our account, we turn to various concepts, models, and data obtained in psychology, neuroscience, physics of nonlinear processes, and philosophy.

We focus on the problem of mental time—the human temporality—which differs from time in physical reality in its basic properties. The problem of human temporality has common roots with a number of investigations put forward original concepts and models:

- The *operational architectonics* of brain and mind functioning proposed by Fingelkurts and Fingelkurts (2001, 2008), Fingelkurts et al. (2010, 2013).[1]
- The concept of a new branch of neuroscience—*spatiotemporal neuroscience* put recently forward by Northoff et al. (2020) which is actually based on the temporo-spatial theory of consciousness (Northoff and Huang 2017).[2]
- The *projective consciousness model* (Rudrauf et al. 2017, Williford et al. 2018) which assumes that consciousness is structured by a projective geometry governed by active inference.[3]

[1]Within this account of the mind-body problem, first, the spatial-temporal patterns of the external world via sensory modalities cause the emergence of a continuum of dynamics spatial-temporal patterns in the brain. Second, the latter patterns being the essence of the internal space-time of the brain are reflected in the phenomenal spatial-temporal patterns at the higher level of abstractness—phenomenal space-time the mind deals with (see Fingelkurts et al. 2020, 2019, for a review).

[2]The given approach posits that the brain possesses its own inner time and space which as "common currency" unites neural activity and mental features. Various dimensions of consciousness are grounded on this inner space-time (see also Northoff 2014, 2018). Moreover, conscious phenomena are characterized by considerable spatial and temporal extent (Dehaene et al. 2014, Koch et al. 2016, for a review), which endows the entities of this space-time with special properties. The irreducibility of space-time attributed to neuronal representations of real spatial temporal patterns is also emphasized by Buzsáki and Llinás (2017). Especially it concerns mental time travels—(re)experience of past episodes and episodes in the imagined future (e.g., Tulving 2002b, for a review).

[3]The model deeply integrates basic aspects of phenomenology such as intentionality, situated spatiality, temporal integration, self-awareness. In this context we want note the concept of *embodied self* (Newen 2018) implying the crucial role of the predictive coding process.

- The *consciousness state space model* (Berkovich-Ohana and Glicksohn 2014)—a phenomenological model for consciousness and selfhood which relates time, awareness, and emotion within one framework.[4]
- The concept of *inner time* with individual properties considered within general frameworks (e.g., Penrose 1989, 1994, Smythies 2003, Loewenstein 2013, as well as collection edited by Smythies and French 2018).

In describing temporal properties of human actions, we inevitably confront various theoretical problems related to goal-oriented aspects in the behavior. These theoretical problems are partially reflected in the collection edited by Levich (1995). Studying the behavior of animals, McFarland (1989) distinguishes goal-achieving, goal-seeking, and goal-directed types of actions, which implies the intrinsic temporal extension of such actions. In their review, Winning and Bechtel (2019) relate the goal-directedness to the emergence of time patterns representing the desired goal and its achievement. In this context, we want to point out the Anticipatory Systems Theory developed by Rosen (2012, 2nd edition, 1st edition published in 1985). According to the proposed concept, an anticipatory system is "a system containing a predictive model of itself and/or of its environment, which allows it to change state at an instant in accord with the model's predictions pertaining to a later instant" (Rosen 2012, p. 313). In this way, the anticipatory systems theory puts forward the teleological "anticipatory paradigm" which extends the classical "re-active paradigm" underlying the study of physical systems. For a review and extended discussion of anticipatory systems, a reader may be referred to Nadin (2010, 2012) as well as the collection edited by Nadin (2016).

In developing our account of human temporality, we face up the problem of mental representation fuzziness; this problem was accentuated by Perlovsky (2016). The mental representations cannot be described by notions dealing with point-like entities. In this context, we want to note the quantum approach to understanding and describing the mind where the uncertainty of quantum objects is an intrinsic property. Among a wide variety of investigations, we want to note the following:

- The quantum description of mind turning to the general principles (Stapp 2009, Loewenstein 2013, Atmanspacher 2020, e.g.,).
- The quantum models focus directly on uncertainty in human perception (e.g., Wang, Busemeyer, Atmanspacher and Pothos 2013, Bruza et al. 2015, Busemeyer et al. 2020).
- The applications of quantum theory to social and economic systems (e.g., Khrennikov 2010, Haven and Khrennikov 2013, Wendt 2015),
- The models focuses on the relationship between consciousness and quantum processing of information (Georgiev 2018, for a review),

[4]The model posits also that all phenomenological states fall into two categories of consciousness, core and extended. Core consciousness supports minimal selfhood that is short of temporal extension and its scope is the here and now. Extended consciousness supports narrative selfhood, which involves personal identity and continuity across time, as well as memory, imagination and conceptual thought.

- The quantum nature of top-down causation (Aharonov et al. 2018, Ellis 2018, Kent 2020),
- The quantum decision-making developed following different lines (Busemeyer and Bruza 2012, for a review), (Yukalov and Sornette 2008, 2009, 2010, 2011, also Yukalov and Sornette 2016 for a review), and (Khrennikov et al. 2018).

We hope that the present brief introduction will convince a reader that nowadays investigations of the human mind in the context of its relationship with reality is an actual problem not only of the philosophy and cognitive sciences, but also physics, in particular, theoretical physics. The presented bundle of concepts and models illustrates that up to now there is no universal approach to describing the human mind and tackling the mind-body problem. In this book, we want to propose an original mathematical account of the human temporality which is based on the advances in the noted branches of science.

Now to speak about particular applications of physics of the human mind is too early, however, the development of the theory of human temporality can give rise to novel concepts in theoretical physics. We expect that elucidating the properties of human temporality as an entity of the mind will reveal prospects in understanding how goal-oriented actions of individuals and their groups emerge, evolve, and are related to the accumulated experience. We will be able to deeply understand the hidden mechanisms governing social phenomena and cultural dynamics. Naturally, going in this way, it will be necessary to take into account the complexity of goal-oriented actions. This complexity includes a number of atemporal dimensions of the human temporality like emotions, motivations, meaning and norms, skill, conscious and unconscious aspects of human actions, social dimensions, etc., which can exhibit substantial variations on long-term scales. For a review of these factors, a reader may be referred, e.g., to Vallacher and Wegner (1987), DeShon and Gillespie (2005), Kozak et al. (2006), Liberman et al. (2007), Liberman and Trope (2014), Trautmann and van de Kuilen (2012), Adamopoulos (2012), Montaño and Kasprzyk (2015), Vandewalle et al. (2019), Hagger and Hamilton (2020), and references therein. Finally, we want to note that investigations of the human mind and the animal mind, their comparison should broaden the horizon of understanding ourselves and living beings.

References

Abergel, F., Aoyama, H., Chakrabarti, B. K., Chakraborti, A., Deo, N., Raina, D., & Vodenska, I. (Eds.). (2017). *Econophysics and Sociophysics: Recent Progress and Future Directions*. Springer International Publishing AG.

Abergel, F., Chakrabarti, B. K., Chakraborti, A., Deo, N., & Sharma, K. (Eds.). (2019). *New Perspectives and Challenges in Econophysics and Sociophysics*. Springer Nature Switzerland AG.

Adamopoulos, J. (2012). The emergence of social meaning: A theory of action construal. In K. T. örnblom, & A. Kazemi (Eds.). *Handbook of Social Resource Theory: Theoretical Extensions,*

Empirical Insights, and Social Applications (pp. 255–272). New York: Springer Science +Business Media.
Aharonov, Y., Cohen, E., & Tollaksen, J. (2018). Completely top–down hierarchical structure in quantum mechanics. *Proceedings of the National Academy of Sciences, 115*(46), 11730–11735.
Andrews, K. (2020). *The animal mind: An introduction to the philosophy of animal cognition* (2nd ed.) London: Routledge, the Taylor & Francis Group.
Arstila, V., & Lloyd, D. (Eds.). (2014). *Subjective time: The philosophy, psychology, and neuroscience of Temporality*. The MIT Press.
Asai, Y., Tasaka, Y., Nomura, K., Nomura, T., Casadio, M., & Morasso, P. (2009). A model of postural control in quiet standing: robust compensation of delay-induced instability using intermittent activation of feedback control. *PLoS One, 4*(7), e6169 (14 pages).
Asai, Y., Tateyama, S., & Nomura, T. (2013). Learning an Intermittent Control Strategy for Postural Balancing Using an EMG-Based Human-Computer Interface. *PLoS One, 8*(5), e62956 (19 pages).
Atmanspacher, H. (2020). Quantum approaches to consciousness. In E. N. Zalta (Ed.), *The Stanford Encyclopedia of Philosophy*, summer 2020 ed., Metaphysics Research Lab, Stanford University.
Baars, B. J., & Gage, N. M. (2010). *Cognition, brain, and consciousness: introduction to cognitive neuroscience* (2nd ed.), Academic Press. Elsevier Ltd.
Badcock, P. B., Friston, K. J., & Ramstead, M. J. (2019). The hierarchically mechanistic mind: A free-energy formulation of the human psyche. *Physics of Life Reviews, 31*, 104–121.
Batchelder, W. H., Colonius, H., Dzhafarov, E. N., & Myung, J. (Eds.). (2017). *New handbook of mathematical psychology:* (Vol. 1). Cambridge, UK: Cambridge University Press.
Berkovich-Ohana, A., & Glicksohn, J.(2014). The consciousness state space (css)—A unifying model for consciousness and self. *Frontiers in Psychology, 5*(341) (19 pages).
Botvinick, M., & Toussaint, M. (2012). Planning as inference. *Trends in Cognitive Sciences, 16* (10), 485–488.
Bruya, B. (Ed.). (2010). *Effortless attention: A new perspective in the cognitive science of attention and action*. The MIT Press.
Bruza, P. D., Wang, Z., & Busemeyer, J. R. (2015). Quantum cognition: A new theoretical approach to psychology. *Trends in Cognitive Sciences, 19*(7), 383–393.
Busemeyer, J. R., & Bruza, P. D. (2012). *Quantum models of cognition and decision*. Cambridge University Press.
Busemeyer, J. R., Kvam, P. D., & Pleskac, T. J. (2020). Comparison of markov versus quantum dynamical models of human decision making. *Wiley Interdisciplinary Reviews: Cognitive Science, 11*(4), e1526 (19 pages).
Busemeyer, J. R., Wang, Z., Townsend, J. T., & Eidels, A. (Eds.). (2015). *The oxford handbook of computational and mathematical psychology*. Oxford University Press.
Buzsáki, G., & Llin′as, R. (2017). Space and time in the brain. *Science, 358*(6362), 482–485.
Callender, C. (Ed.). (2011). *The oxford handbook of philosophy of time*. Oxford University Press.
Cappuccio, M. L. (Ed.). (2019). *Handbook of embodied cognition and sport psychology*. The MIT Press.
Castellano, C., Fortunato, S., & Loreto, V. (2009). Statistical physics of social dynamics. *Reviews of Modern Physics, 81*(2), 591–646.
Chakrabarti, B., Chakraborti, A., & Chatterjee, A. (2006). *Econophysics and sociophysics: Trends and perspectives*. Wiley-VCH Verlag GmbH & Co.
Chalmers, D. J. (1996). *The conscious mind*. Oxford University Press: Oxford, UK.
Chalmers, D. J. (2010). *The character of consciousness*. Oxford University Press.
Cristiani, E., Piccoli, B., & Tosin, A. (2014). *Multiscale modeling of pedestrian dynamics*. Springer International Publishing.
Dehaene, S., Charles, L., King, J.-R., & Marti, S. (2014). Toward a computational theory of conscious processing. *Current Opinion in Neurobiology, 25*, 76–84.

Demekas, D., Parr, T., & Friston, K. J. (2020). An investigation of the free energy principle for emotion recognition, *Frontiers in Computational Neuroscience, 14*(30) (17 pages).

Dennett, D. C. (2005). *Sweet dreams: Philosophical obstacles to a science of consciousness.* Cambridge, MA: The MIT Press.

Dennett, D. C. (2017). *From bacteria to bach and back: The evolution of minds.* W. W. Norton & Company.

DeShon, R. P., & Gillespie, J. Z. (2005). A motivated action theory account of goal orientation. *Journal of Applied Psychology, 90*(6), 1096–1127.

Ellis, G. F. R. (2018). Top-down causation and quantum physics. *Proceedings of the National Academy of Sciences, 115*(46), 11661–11663.

Feldman, A. G. (1966). Functional tuning of the nervous system with control of movement of maintenance of a steady posture of movement or maintenance of a steady posture: Ii. controllable parameters of the muscles. *Biophysics*, 11(3), 565–578.

Feldman, A. G. (1986). Once more on the equilibrium-point hypothesis (k model) for motor control. *Journal of Motor Behavior, 18*(1), 17–54.

Fingelkurts, A. A., & Fingelkurts, A. A. (2001). Operational architectonics of the human brain-biopotential field: Towards solving themind-brain problem. *Brain and Mind, 2*(3), 261– 296.

Fingelkurts, A. A., & Fingelkurts, A. A. (2008). Brain-mind operational architectonics imaging: Technical and methodological aspects. *The Open Neuroimaging Journal, 2*, 73–93.

Fingelkurts, A. A., Fingelkurts, A. A., & Neves, C. F. H. (2010). Natural world physical, brain operational, and mind phenomenal space–time. *Physics of Life Reviews, 7*(2), 195–249.

Fingelkurts, A. A., Fingelkurts, A. A., & Neves, C. F. (2013). Consciousness as a phenomenon in the operational architectonics of brain organization: Criticality and self-organization considerations. *Chaos, Solitons & Fractals, 55*, 13–31. Fingelkurts, A. A., Fingelkurts, A. A., Neves, C. F., & Kallio-Tamminen, T. (2019). Brain-mind operational architectonics: At the boundary between quantum physics and eastern metaphysics. *Physics of Life Reviews, 31*, 122–133.

Fingelkurts, A. A., Fingelkurts, A. A., & Kallio-Tamminen, T.: 2020, Selfhood triumvirate: From phenomenology to brain activity and back again, *Consciousness and Cognition,* 86, 103031.

Franklin, S. P. (1995). *Artificial Minds*. The MIT Press, Cambridge, MA.

Friston, K., Kilner, J., & Harrison, L. (2006). A free energy principle for the brain. *Journal of Physiology, Paris, 100*(1–3), 70–87.

Friston, K., & Kiebel, S. (2009). Predictive coding under the free-energy principle. *Philosophical Transactions of the Royal Society b: Biological Sciences, 364*(1521), 1211–1221.

Friston, K. J., Daunizeau, J., & Kiebel, S. J. (2009). Reinforcement Learning or Active Inference?, *PloS One*, *4*(7), e6421.

Friston, K. (2009). The free-energy principle: A rough guide to the brain? *Trends in Cognitive Sciences, 13*(7), 293–301.

Friston, K. J., Daunizeau, J., Kilner, J., & Kiebel, S. J. (2010). Action and behavior: A free-energy formulation. *Biological Cybernetics, 102*(3), 227–260.

Friston, K. (2010). The free-energy principle: A unified brain theory? *Nature Reviews Neuroscience, 11*(2), 127–138.

Friston, K., Breakspear, M., & Deco, G. (2012). Perception and self-organized instability. *Frontiers in Computational Neuroscience, 6*(44) (19 pages).

Friston, K. J., Lin, M., Frith, C. D., Pezzulo, G., Hobson, J. A., & Ondobaka, S. (2017). Active inference, curiosity and insight. *Neural Computation, 29*(10), 2633–2683.

Friston, K., FitzGerald, T., Rigoli, F., Schwartenbeck, P., & Pezzulo, G. (2017). Active inference: A process theory. *Neural Computation, 29*(1), 1–49.

Friston, K. (2018). Does predictive coding have a future? *Nature Neuroscience, 21*(8), 1019–1021.

Galam, S. (2012). *Sociophysics: A physicist's modeling of psycho-political phenomena.* New York, NY: Springer Science+Business Media.

Gallagher, S., & Zahavi, D. (2012). *The Phenomenological mind* (2nd ed.). Routledg:. Taylor & Francis Group.

Gazzaniga, M., Ivry, R. B., & Mangun, G. R. (2019). *Cognitive neuroscience: The biology of the mind* (5th ed.). W. W. Norton & Company Inc.

Georgiev, D. D. (2018). *Quantum information and consciousness: A gentle introduction.* CRC Press, Taylor & Francis Group.

Hagger, M. S., & Hamilton, K. (2020). Changing behavior using integrated theories. In M. S. Hagger, L. D. Cameron, K. Hamilton, N. Hankonen, & T. Lintunen (Eds.), *The handbook of behavior change* (pp. 208–224). Cambridge University Press.

Haven, E., & Khrennikov, A. (2013). *Quantum social science*. Cambridge University Press.

Hegselmann, R. (2009). Moral dynamics. In R. A. Meyers (ed.), *Encyclopedia of complexity and systems science* (pp. 5677–5692). New York, NY: SpringerScience+Buisiness Media, LLC.

Helbing, D. (2001). Traffic and related self-driven many-particle systems. *Reviews of Modern Physics, 73*, 1067–1141.

Helbing, D., & Johansson, A. (2009). Pedestrian, crowd and evacuation dynamics. In R. Meyers (Ed.), *Encyclopedia of complexity and systems science* (pp. 6476–6495). SpringerScience +Buisiness Media, LLC.

Helbing, D. (2010). *Quantitative sociodynamics: Stochastic methods and models of social interaction processes* (2nd ed.). Springer-Verlag.

Helbing, D. (Ed.). (2012). *Social self-organization: Agent-based simulations and experimentsto study emergent social behavior*. Springer-Verlag.

Heylighen, F., & Chielens, K. (2009) Evolution of culture, memetics. In R. A. Meyers (ed.), *Encyclopedia of complexity and systems science* (pp. 3205–3220). New York, NY: Springer-Science+Buisiness Media, LLC.

Hoya, T. (2005). *Artificial mind system: Kernel memory approach*. Springer-Verlag.

Joffily, M., & Coricelli, G. (2013). Emotional valence and the free-energy principle. *PLOS Computational Biology, 9*(6), e1003094 (14 pages).

Kelso, J. A. S. (2009a). Coordination dynamics. In R. A. Meyers (ed.), *Encyclopedia of complexity and systems science* (pp. 1537–1565). New York, NY: SpringerScience+Buisiness Media, LLC.

Kelso, J. A. S. (2009). Synergies: Atoms of brain and behavior. In D. Sternad (Ed.), *Progress in motor control: A multidisciplinary perspective* (pp. 83–91). Springer Science+Business Media, LLC.

Kelso, J. A. S. (2014). The dynamic brain in action: Coordinative structures, criticality, and coordination dynamics. In D. Plenz & E. Niebu (Eds.), *Criticality in neural systems* (pp. 67–104). Wiley-VCH Verlag GmbH & Co., KGaA.

Kerner, B. S. (2004). *The physics of traffic: Empirical freeway pattern features, engineering applications, and theory*. Springer-Verlag.

Kerner, B. S. (2009). *Introduction to modern traffic flow theory and control: The long road to three-phase traffic theory*. Springer-Verlag.

Kerner, B. S. (2017). *Breakdown in traffic networks: Fundamentals of transportation science*. Springer-Verlag.

Khrennikov, A. (2010). *Ubiquitous quantum structure: From psychology to finance*. Springer-Verlag.

Khrennikov, A., Basieva, I., Pothos, E. M., & Yamato, I. (2018). Quantum probability in decision making from quantum information representation of neuronal states. *Scientific Reports, 8*(1), 16225 (8 pages).

Kihlstrom, J. F. (2019). The motivational unconscious. *Social and Personality Psychology Compass, 13*(5), e12466 (18 pages).

Koch, C., Massimini, M., Boly, M., & Tononi, G. (2016). Neural correlates of consciousness: Progress and problems. *Nature Reviews Neuroscience, 17*(5), 307–321.

Kozak, M. N., Marsh, A. A., & Wegner, D. M. (2006). What do I think you're doing? Action identification and mind attribution. *Journal of Personality and Social Psychology, 90*(4), 543–555.

Latash, M. L. (2008a). *Neurophysiological basis of movement* (2nd ed.). Human Kinetics. Urbana, IL. Latash, M. L. (2008). *Synergy*. Oxford University Press.

Latash, M. L. (2012). *Fundamentals of motor control, academic press*. Elsevier Inc.

Levich, A. P. (Ed.) (1995). *On the way to understanding the time phenomenon: The constructions of time in natural science, part 1*. Singapore: World Scientific Publishing Co. Pte. Ltd.

Liberman, N., Trope, Y., McCrea, S. M., & Sherman, S. J. (2007). The effect of level of construal on the temporal distance of activity enactment. *Journal of Experimental Social Psychology, 43* (1), 143–149.

Liberman, N., & Trope, Y. (2014). Traversing psychological distance. *Trends in Cognitive Sciences, 18*(7), 364–369.

Loewenstein, W. R. (2013). *Physics in mind: A quantum view of the brain*. Basic Books.

Loram, I. D., Gollee, H., Lakie, M., & Gawthrop, P. J. (2011). Human control of an inverted pendulum: Is continuous control necessary? Is intermittent control effective? Is intermittent control physiological? *The Journal of Physiology, 589*(2), 307–324.

McFarland, D. (1989). *Problems of animal behaviour*. Harlow, Essex, UK: Longman Scientific & Technical.

Michael, M. T. (2020). Unconscious emotion and free-energy: A philosophical and neuroscientific exploration, *Frontiers in Psychology, 11*, Article 984 (14 pages).

Montaño, D. E., & Kasprzyk, D. (2015). Theory of reasoned action, theory of planned behavior, and the integrated behavioral model. In K. Glanz, B. K. Rimer, & K. Viswanath (Eds.), *Health behavior: Theory, research, and practice* (5th ed., pp. 95–124). Jossey-Bass.

Morsella, E., Bargh, J. A., & Gollwitzer, P. M. (Eds.). (2009). *Oxford handbook of human action*. New York, NY: Oxford University Press.

Moustafa, A. A. (Ed.) (2018). *Computational models of brain and behaviour*. Hoboken, NJ: John Wiley & Sons, Ltd.

Nadin, M. (2010). Anticipation: Annotated bibliography. *International Journal of General Systems, 39*(1), 35–133.

Nadin, M. (2012). The anticipatory profile. An attempt to describe anticipation as process. *International Journal of General Systems, 41*(1), 43–75.

Nadin, M. (Ed.). (2016). *Anticipation across disciplines*. Springer International Publishing AG.

Newen, A. (2018). The embodied self, the pattern theory of self, and the predictive mind, *Frontiers in Psychology, 9*, Article 2270 (14 pages).

Northoff, G. (2014). *Unlocking the brain, volume 2: Consciousness*. New York: Oxford University Press.

Northoff, G., & Huang, Z. (2017). How do the brain's time and space mediate consciousness and its different dimensions? temporo-spatial theory of consciousness (TTC). *Neuroscience & Biobehavioral Reviews, 80*, 630–645.

Northoff, G. (2018). *The spontaneous brain: From the mind-body to the world-brain problem*. The MIT Press.

Northoff, G., Wainio-Theberge, S., & Evers, K. (2020). Is temporo-spatial dynamics the "common currency" of brain and mind? in quest of "spatiotemporal neuroscience." *Physics of Life Reviews, 33*, 34–54.

Ohira, T., & Uzawa, T. (Eds.) (2015). *Mathematical approaches to biological systems: Networks, oscillations, and collective motions*. Tokyo: Springer Japan.

Parr, T., & Friston, K. J. (2019). Generalised free energy and active inference. *Biological Cybernetics, 113*(5–6), 495–513.

Penrose, R. (1989). *The emperor's new mind: Concerning computers, minds, and the laws of physics*. Oxford University Press.

Penrose, R. (1994). *Shadows of the mind: A search for the missing science of consciousness*. Oxford University Press.

Perlovsky, L. I. (2006). Toward physics of the mind: Concepts, emotions, consciousness, and symbols. *Physics of Life Reviews, 3*(1), 23–55.

Perlovsky, L. I. (2016). Physics of the mind. *Frontiers in Systems Neuroscience 10*, Article 84, 12 pages.

Phillips, I. (Ed.). (2017). *The Routledge handbook of philosophy of temporal experience*. Abingdon, Oxon: Routledge, Taylor & Francis Group.

Pizzi, R. M. R. (2008). Artificial mind. In F. Orsucci, & N. Sala (Eds.), *Reffiexing interfaces: The complex coevolution of information technology ecosystems* (pp. 83–94). Hershey, PA: Information Science Reference.

Port, R. (2009). Dynamics of language. In R. A. Meyers (Ed.), *Encyclopedia of Complexity and Systems Science* (pp. 2310–2323). New York, NY: SpringerScience+Buisiness Media, LLC.

Raab, M., Lobinger, B., Hoffmann, S., Pizzera, A., & Laborde, S. (Eds.). (2016). *Performance psychology: Perception, action, cognition, and emotion*. Academic Press. Elsevier Inc.

Ramstead, M. J. D., Badcock, P. B., & Friston, K. J. (2018). Answering schrödinger's question: A free-energy formulation. *Physics of Life Reviews, 24*, 1–16.

Ramstead, M. J., Constant, A., Badcock, P. B., & Friston, K. J. (2019). Variational ecology and the physics of sentient systems, *Physics of Life Reviews, 31*, 188–205. Special issue: Physics of Mind.

Rosen, R. (2012), *Anticipatory systems: Philosophical, mathematical, and methodological foundations* (2nd ed.). New York: Springer. With Contributions by Judith Rosen, John J. Kineman, and Mihai Nadin.

Rudrauf, D., Bennequin, D., Granic, I., Landini, G., Friston, K., & Williford, K. (2017). A mathematical model of embodied consciousness. *Journal of Theoretical Biology, 428*, 106–131.

Ryan, R. M. (Ed.). (2019). *The Oxford Handbook of Human Motivation* (2nd ed.). Oxford University Press.

Săvoiu, G. (2013). *Econophysics: background and applications in economics, finance, and sociophysics*. Academic Press.

Schadschneider, A., Klingsch, W., Klüpfel, H., Kretz, T., Rogsch, C., & Seyfried, A. (2009). Evacuation dynamics: Empirical results, modeling and applications. In R. Meyers (Ed.), *Encyclopedia of complexity and systems science* (pp. 3142–3176). SpringerScience+Buisiness Media LLC.

Schoeller, F., Perlovsky, L., & Arseniev, D. (2018). Physics of mind: Experimental confirmations of theoretical predictions. *Physics of Life Reviews, 25*, 45–68.

Scholz, J. P., & Schöner, G. (1999). The uncontrolled manifold concept: Identifying control variables for a functional task. *Experimental Brain Research, 126*(3), 289–306. *Science and human experience* (revised ed.) Cambridge, MA: The MIT Press. New foreword by Jon Kabat-Zinn and new introductions by Evan Thompson and Eleanor Rosch. Sen, P., & Chakrabarti, B. K. (2014). *Sociophysics: An Introduction*. Oxford University Press.

Shapiro, L. (Ed.). (2014). *The Routledge Handbook of Embodied Cognition*. Routledge. Taylor & Francis Group.

Shapiro, L. (2019). *Embodied cognition* (2nd ed.). Routledge. Taylor & Francis Group.

Sherman, J. W., Gawronski, B., & Trope, Y. (Eds.). (2014). *Dual-process theories of the social ind*. The Guilford Press.

Slanina, F. (2014). *Essentials of econophysics modelling*. Oxford University Press.

Smith, D. W. 2013. Phenomenology. In E. N. Zalta (Ed.), *The Stanford Encyclopedia of Philosophy* (summer 2018 ed.). Stanford, CA: Metaphysics Research Lab, Stanford University.

Smith, R., Schwartenbeck, P., Parr, T., & Friston, K. J. (2020). An active inference approach to modeling structure learning: Concept learning as an example case, *Frontiers in Computational Neuroscience 14*, Article 41, 24 pages.

Smythies, J. (2003). Space, time and consciousness. *Journal of Consciousness Studies, 10*(3), 47–56.

Smythies, J. R., & French, R. (Eds.). (2018). *Direct versus indirect realism: A neuro philosophical debate on consciousness*. Academic Press. Elsevier Inc.

Spence, C. E., Osman, M., & McElligott, A. G. (2017). Theory of animal mind: Human nature or experimental artefact? *Trends in Cognitive Sciences, 21*(5), 333–343.

Stapp, H. P. (2009). *Mind, Matter and Quantum Mechanics* (3rd ed.). Springer-Verlag.

Suzuki, Y., Morimoto, H., Kiyono, K., Morasso, P. G., & Nomura, T. (2016). Dynamic determinants of the uncontrolled manifold during human quiet stance. *Frontiers in Human Neuroscience,10*, Article 618, 20 pages.

Thompson, E., & Zahavi, D. (2007). Philosophical issues: Phenomenology. In P. D. Zelazo, M. Moscovitch, & E. Thompson (Eds.), *The Cambridge Handbook of Consciousness* (pp. 67–87). Cambridge University Press.

Trautmann, S. T., & van de Kuilen, G. (2012). Prospect theory or construal level theory? *Acta Psychologica, 139*(1), 254–260.

Treiber, M., & Kesting, A. (2013). *Traffic flow dynamics: Data*. Berlin: Springer-Verlag.

Tulving, E. (2002). Episodic memory: From mind to brain. *Annual Review of Psychology, 53*(1), 1–25.

Vallacher, R. R., & Wegner, D. M. (1987). What do people think they're doing? Action identification and human behavior. *Psychological Review, 94*(1), 3–15.

Vandewalle, D., Nerstad, C. G., & Dysvik, A. (2019). Goal orientation: A review of the miles traveled and the miles to go. *Annual Review of Organizational Psychology and Organizational Behavior, 6*(1), 115–144.

Varela, F. J., Thompson, E., & Rosch, E. (2016). *The embodied mind: Cognitive.*

Veit, W. (2019). Modeling morality. In A. Nepomuceno-Fernandez, L. Magnani, F. J. Salguero-Lamillar, C. Bares-Gomez, & M. Fontaine (Eds.), *Model-based reasoning in science and technology* (pp. 83–102). Cham, Switzerland: Springer Nature Switzerland AG.

Wang, Z., Busemeyer, J. R., Atmanspacher, H., & Pothos, E. M. (2013). The potential of using quantum theory to build models of cognition. *Topics in Cognitive Science, 5*(4), 672–688.

Wendt, A. (2015). *Quantum mind and social science: Unifying physical and social ontology*. Cambridge University Press.

Williford, K., Bennequin, D., Friston, K., & Rudrauf, D. (2018). The projective consciousness model and phenomenal selfhood. *Frontiers in Psychology, 9*, Article 2571, 18.

Wilson, R. A., & Foglia, L. (2015). Embodied cognition. In E. N. Zalta (Ed.), *The stanford encyclopedia of philosophy* (winter 2016 ed.). Metaphysics Research Lab, Stanford University.

Winning, J., & Bechtel, W. (2019). Being emergence vs. pattern emergence: Complexity, control and goal-directedness in biological systems. In S. Gibb, R. Hendry, & T. Lancaster (Eds.), *The routledge handbook of emergence* (pp. 134–144). London: Routledge, Taylor & Francis Group.

Wynne, C. D. L., & Udell, M. A. R. (2013). *Animal cognition* (2nd ed.). Palgrave Macmillan.

Yoshikawa, N., Suzuki, Y., Kiyono, K., & Nomura, T. (2016). Intermittent feedback-control strategy for stabilizing inverted pendulum on manually controlled cart as analogy to human stick balancing. *Frontiers in Computational Neuroscience, 10*, Article 34, 19.

Yufik, Y. M., Sengupta, B., & Friston, K. (Eds.). (2017). *Self-organization in the nervous system*. Lausanne: Frontiers Media SA, Frontiers in Systems Neuroscience.

Yukalov, V. I., & Sornette, D. (2008). Quantum decision theory as quantum theory of measurement. *Physics Letters A, 372*(46), 6867–6871.

Yukalov, V. I., & Sornette, D. (2009). Processing information in quantum decision theory. *Entropy, 11*(4), 1073–1120.

Yukalov, V. I., & Sornette, D. (2010). Mathematical structure of quantum decision theory. *Advances in Complex Systems, 13*(5), 659–698.

Yukalov, V. I., & Sornette, D. (2011). Decision theory with prospect interference and entanglement. *Theory and Decision, 70*(3), 283–328.

Yukalov, V. I., & Sornette, D. (2016). Quantum probability and quantum decision-making. *Philosophical Transactions of the Royal Society A: Mathematical, Physical and Engineering Sciences, 374*(2058), 20150100, 15.

Zahavi, D. (Ed.). (2012). *The oxford handbook of contemporary phenomenology*. Oxford University Press.

Part I
Problem of Time. Various Aspects

Chapter 1
Time From a Bird's Eye View

In the present chapter, first, we briefly touch on the main milestones in the history of philosophy of time from the days of ancient Greece up to the XIX century. Second, we pay special attention to the crucial concepts put forward in the philosophy of time and related branches of physics in the XX century except for the issues related to the Theories of Relativity. This separation reflects the paradigm shift that occurred in philosophy of time as well as physics at the border of the XIX–XX centuries. Starting from the works by Albert Einstein and Edmund Husserl devoted to the problems of time it has become clear that the properties of the subjective time as it exists in the mind and the objective time should and can be studied separately. The temporal connectivity of human cognition is of special interest for our further constructions, so we consider it individually.

In the present book we construct an account of the human temporal dimension treated as an individual object with its own properties. Therefore in discussing the problems of time posed in the XX century, we focus on the aspects of human time only. Nowadays, human time is usually understood as subjective time related to human perception of objective time; therefore, we also briefly discuss the psychophysical regularities and neural mechanisms of human recognition of time flow.

The problems to be considered in this chapter are needed to elucidate the notions and mathematical formalism required for constructing the theory of human temporality—the main subject-matter of the present book. In the present chapter, discussing classical accounts of time, we inevitably have to touch on the relationship between time and causality. In our account of temporality (Sect. 3.4), however, we prefer to categorize the relationship between events in terms of *intentional* relations, which does not require a predetermined direction of temporal order.

I. Lubashevsky and N. Plavinska, *Physics of the Human Temporality*, Understanding Complex Systems, https://doi.org/10.1007/978-3-030-82612-3_1

1.1 Philosophy of Time: Historical Milestones

During many centuries philosophers tried to gain insight into various challenging issues of time, and even their brief discussion goes far beyond the scope of this book. For such a review a reader may be referred, e.g., to Dowden (2001), Bardon (2013), Dyke and Bardon (2013), Markosian et al. (2014). In this section we focus only on issues that can impact upon the further constructions in describing the temporal properties of human mind. They are:

- Is time something real rather than an illusion of our mind?
- Could there be time without change or, as another facet of the same question, does time exist independently of events that occur in time?
- What is the structure of time, in particular, is the present a time component and does it exist?

Naturally these questions partly overlap and possible answers to them are not completely independent of one another. In the next sections we will discuss also these issues as well as others.

The modern philosophy of time is rooted in philosophy of ancient Greece (see, e.g., Bardon 2013). As the first thinker impacting upon our understanding of time we note Heraclitus (c. 535–c. 475 BC) who argued that reality is characterized by unending change, with nothing constant in the world. Nowadays Heraclitus's metaphor of the river of time in Plato's interpretation:

> Heraclitus, I believe, says that all things pass and nothing stays, and comparing existing things to the flow of a river, he says you could not step twice into the same river (Plato, *Cratylus* 402a = A6),

has become a component of our general culture.[1]

Parmenides (c. 515 BC) and his pupil Zeno of Elea (c. 490–c. 430 BC) shared the point of view that all visible changes are just illusion. In particular, the famous Zeno's paradox "The Arrow" is often referred to as the illustration of his arguments for the illusion of motion (for details see, e.g., Hoy 2013).

Plato (428/427–348/347 BC) developed the theory of Forms. He supposed that there is a world of perfect, immutable Forms which are reflected in the objects of material world. These objects undergo changes permanently and are accessible to our senses (e.g., Silverman 2014). For Plato only these eternal Forms are real and exist out of time; whereas the physical world and its time are illusions (e.g., Ariotii 1975).

In his *Physics* Aristotle (384–322 BC), Plato's pupil, put forward the account of time impacted upon the further development of philosophy of time for many centuries. Its basic ideas concern issues important also for our constructions, so let us discuss them in little more detail. There is a wide literature on Aristotle's approach to understanding time and a reader interested in its details may be referred, i.e., to

[1] The precise meaning of the Heraclitus's corresponding statement is discussed, e.g., by Graham (2015).

Coope (2005, 2009a), Falcon (2013) and references therein. We will not follow the internal logic of Aristotle's account and rearrange its crucial points in a manner elucidating its relationship with our account of human time.

Aristotle formulated several puzzles about time being the subject of debates for many centuries. One of these puzzles is:

> In the case of anything divisible, if it is, it is necessary that when it is, either all or some of its parts must exist. But of time, though it is divisible, some parts have been, some parts are to come, but no part is. The *now* is not a part. For the part measures and it is necessary that the whole is composed from the parts. But time is not thought to be composed out of *nows*. (*Physics*, 4.10: 218^{a} 3–8; *italic* is ours)

The pivot point of this puzzle is the statement that the *now*, i.e., the present instant is a point-like object and, thus, must be is indivisible and structureless. Comparing time with a continuous line and the now with a point on it, Aristotle draws a conclusion that the now cannot be a part of time. According to his logic points having no dimension cannot make up a continuous line. As we could say nowadays, Aristotle tacitly adopts the assumption that a complex object must be composed of a *countable* set of its basic elements. In this sense, any countable set of dimensionless instants cannot compose a stretching fragment of time having a finite duration. Removing the now from the constituent elements Aristotle inevitable faces up to the problem that in this case time does not exist.

Nowadays we could overcome this problem saying that time is the *uncountable* set of instants; the set of real numbers or a straight line as a set of points exemplifies such objects. However, turning to this alternative we meet other problems as to how time can be measured using dimensionless nows and how the properties of structureless nows can determine the properties of stretching parts of time. We have noted this puzzle and its relationship with dimensionless nows because in our account of human time the corresponding *human nows* have a structure and finite duration (see also, Lubashevsky 2017).

Another Aristotle's puzzle concerns the now itself and its change. He argues that we get into a logical contradiction independently of whether we consider the now to be always different or always the same. Moreover, dealing with dimensionless instants it is not clear how the current instant can simultaneously come into being and ceases to exist. Aristotle did not explain how to answer to his puzzles but gave an account of time copying with other aspects of time (Coope 2005, 2009a).

The gist of Aristotle's account is the proposition that time is *something of change* i.e., time is related to *change* but not *change*. Change is always associated with particular things and their changes can be faster or slow, whereas time is universal everywhere and in everything. In other words, within Aristotle's account change is ontologically prior to time (e.g., Coope 2005, 2009a, b; Falcon 2013). In constructing our account of human temporality we also consider that the human time arises due to the changes in our internal world. In this sense exactly changes have causal power with respect to human time.

In Aristotle's concept of time the description of mutual arrangement of instants plays an essential role. He put forward two accounts (Coope 2005). In one of them given in his *Metaphysics* Aristotle uses the *now* as the reference point and the *before* and the *after* are specified by their position relative to the *now*. This account may be regarded as a *static* description of temporal order. In the other account given in the *Physics*—which is the primary Aristotle's account of time—there is no special reference point and all instants as well as the corresponding changes are characterized in terms of *earlier* and *later* with respect to one another. In this *dynamic* account time gains properties of motion due to one and the same thing—the now—moving through time, being first earlier and then later (e.g., Hussey 1993, xliii–xliv). From our point of view both the two accounts are the precursors of the A- and B-theories of time developed at the border of XIX–XX centuries which will be discussed below.

We want to note that in *Physics* Aristotle posed also a question about the relation between time and the mind. He defends that time can exist only in a world in which there are ensouled beings able to recognize changes. His arguments and their interpretations are under debates up to now.

The next essential step in the philosophy of time with respect to the issues we discuss here was made by St. Augustine (354–430), a Christian theologian and philosopher. Augustine brought psychological, subjective aspects of time into Aristotle's account. Like Aristotle, he supposes that

- time is an infinitely divisible continuum,
- there would be no time if there were no change and no ensouled beings able to recognize change,
- time and change are distinct,
- change is ontologically prior to time.

A detailed review of these aspects can be found, e.g., in O'Daly (1981), Teske (1996, 2001), Knuuttila (2001).

However, unlike Aristotle, Augustine considers the present (the now in Aristotle's account) *to be a part of time* though it has no duration (*Confessions*, 11.15.20). In this way he faces up to another challenge about the *measurement* of time in the case when only one part of time—the present—is actual but has no duration. To overcome this problem Augustine proposes that time becomes measurable due to the "distension of the soul" which retains the past events in memory and anticipates future events. It is possible to say that Augustine defines time as a distension of the mind which holds on to the past while attending to the present in anticipation of the future (Teske 1996, 2001). This idea is reflected in the famous Augustine's example:

> Suppose I am about to recite a psalm which I know. Before I begin, my expectation is directed towards the whole. But when I have begun, the verses from it which I take into the past become the object of my memory. The life of this act of mine is stretched two ways, into my memory because of the words I have already said and into my expectation because of those which I am about to say. But my attention is on what is present: by that the future is transferred to become the past. As the action advances further and further, the shorter the expectation and the longer the memory, until all expectation is consumed, the

> entire action is finished, and it has passed into the memory. What occurs in the psalm as a whole occurs in its particular pieces and its individual syllables. The same is true of a longer action in which perhaps that psalm is a part. It is also valid of the entire life of an individual person, where all actions are parts of a whole, and of the total history of 'the sons of men' (Ps. 30: 20) where all human lives are but parts (*Confessions*, 11.28.38, Translation by Chadwick, 1991).

Although Augustine's subjective account of time is not a novelty in ancient philosophy, Augustine seems to be the first who clearly recognized that the question "What is time?" comprises two parts. One is "How is the consciousness of time's passage (by the human) possible?" The other is "What am I (the human) for whom time is at all in question?" (de Warren 2009). Trying to answer to this dual question Augustine draws the conclusion that the three parts of time—the past, present, and future—exist only in the present *memory* of the past, the present *intuition* of the present, and the present *expectation* of the future (*Confessions* 11.20.26). This proposition admits the interpretation as the *threefold present* existing in the human mind. Here we would like to attract attention to that Augustine actually introduces two components of time, one is the objective (physical) time when he speaks about the present, the other is the subjective time (existing in the human mind).

Following Ricoeur (1988, Sect. 1.1) in comparing the accounts proposed by Aristotle and Augustine, we see that both they accept the primacy of change over time.

As far as the relationship between the mind and time is concerned, Aristotle left this aspect vague. Augustine attempted to derive the extension and the measurement of time from the distension of the mind. In particular, the difference between Aristotle's and Augustine's accounts is reflected in the distinction between the notion of the *now* in Aristotle's sense and that of the *present* as it is understood by Augustine. The Aristotle' *now* is merely a moment where the mind makes a break in the continuity of movement; therefore such *nows* do not possess any special property. These breaks can be made anywhere and are used by the mind to distinguish two events as occurring before and after solely.

In Augustine's account of time the *present*, it does not matter which instant is chosen, is a point playing special role in the mind—the future and the past exist only in relation to a present and are characterized by different properties. Within Augustine's standpoint the temporal properties are evaluated by the mind from the present and the *earlier-later* relation is foreign to the notions of present, past, and future. We will return to these aspects of time description in discussing the modern A- and B-theories of time. Here we would like to emphasize that

> the problem of time cannot be attacked from a single side only, whether of the soul or of movement. The distension of the soul alone cannot produce the extension of time; the dynamism of movement alone cannot generate the dialectic of the threefold present. (Ricoeur 1988, p. 21)

Here we have payed special attention to Aristotle's and Augustin's standpoints on the nature of time because their accounts became the base for the following philosophical constructions. The questions they put forward are up to now are the subject of debates in various branches of philosophy and cognitive sciences. In our description of human temporality Augustin's understanding of time and its further development through centuries also takes a central place.

Concerning philosophy of time in Medieval Ages (for a detailed discussion of this issue see, e.g., Trifogli 2000; Porro 2001b), we note Henry of Ghent (c. 1217–1293), a scholastic philosopher at the University of Paris, and Giles of Rome (c. 1243–1316), an archbishop of Bourges. Their accounts of time admit actually the existence of two different components. The objective component—“sublunar time”—is related with the continuity of motion. The subjective component—“angelic time”—is a discrete succession of instants (e.g., Porro 2014; Lambertini 2014).

The concept of discrete time was considered in the Medieval philosophy within various aspects (Porro 2001a). In particular, the problem of introducing discrete time concerns the existence of a certain minimal time—an indivisible part of time having a finite extent. The introduction of minimal time endowing the present with a finite extent looks quite attractive because it opens a gate to solving Aristotle's puzzle about the existence of time and related Zeno's paradoxes. However, in this way new problems are met because the notion of a minimal part of a continuum seems to include two incompatible features: having an extent and being indivisible (Trifogli 2001, 2003, 2015). The possibility of discrete time and the corresponding philosophical and scientific problems are analyzed from the modern point of views by Van Bendegem (2011). In describing human temporality we will employ the notion of space-time cloud (Lubashevsky 2017) to allow for that the present in the mind has some extension in physical time but, being divisible, can possess a certain internal structure.

In Early Modern Philosophy (XVII–XVIII centuries) a whole bundle of new ideas and concepts about time were posed, let us briefly discuss some of them that contribute to our further constructions.

We will start from Dutch philosopher B. Spinoza (1632–1677), in particular, he scrutinized the notion of continuous motion in time whose instants have no extent. Spinoza draws the conclusion that time is nothing but a product of the imagination (e.g., Alexander 1921; Baugh 2010; Manning 2016). It does not exclude that duration is a real property of existing things and comes into being via the degree of existence (Baugh 2010). This concept is interesting for us because we turn to similar ideas to characterize the causal effects of the past retained in human memory and the imaginary future on the present in the mind.

An English philosopher and physician John Locke (1632–1704) focused his attention on another problem met within any account of time where only the point-like present exists. It concerns how we can perceive such time and recognize its passage. Locke himself did not discriminate between the objective and subjective components of time, nevertheless, his constructions opened a gate toward this. Nowadays he is regarded as a precursor of the phenomenological theory of human mind distinguishing between the outer and inner worlds of human beings, which will be discussed below in more detail. In our constructions of human temporality we also turn to Locke's idea about time arising via changes of mental states.

According Locke, time as some object must be attributed to the inner world of human beings as whole and is not reduced to our perception of particular events. He considered that there is an upper boundary of the speed with which mental states (ideas) can follow one another and our perception of time is based on the change of these states; therefore there should be some "instant" of the least duration. Actually this idea is the precursor of the notion of specious present[2]—time interval whose instants are perceived by human as simultaneous—playing essential role in modern concepts of subjective time.

The birth of Newtonian mechanics with its mathematical formalism impacted upon philosophy of time drastically. For a brief introduction to this issue and the contribution of Newton's predecessors including Italian polymath Galileo Galilei (1564–1642), Dutch mathematician and scientist Christiaan Huygens (1629–1695), and English Christian theologian and mathematician Isaac Barrow[3] (1630–1677) a reader may be referred, e.g., to DiSalle (2009), Rynasiewicz (2011), Schliesser (2013), Huggett and Hoefer (2015).

Sir Isaac Newton (1642–1726) in his *Principia*[4] laid the foundations of classical mechanics where he does not discriminate between objective and subjective time. In his picture of the world there is only physical time which together with space exists independently of all possible events, and is not affected by them. It should be noted that Newton himself introduced two categories of physical time; one is absolute, true, and mathematical, the other is relative, apparent, and common. However, he did not consider the problem of human perception of time, which was the reason for the appearance of many philosophical concepts accepting this point of view or rejecting it and introducing subjective time as an essential component of reality. In this sense the birth of Newtonian mechanics became a certain watershed in the further development of philosophy as well as physics.

German polymath and philosopher Gottfried Wilhelm (von) Leibniz (1646–1716) in his famous dispute with English philosopher Samuel Clarke (1675–1729) put forward a number of arguments against the Newtonian concept of time and space. Leibniz considers that time (as well as space) does not exists on its own. In these debates he employs the Principle of Sufficient Reason and the Principle of the Identity of Indiscernibles (for details see, e.g., Ballard 1960; Futch 2008; McDonough 2014). We want to pay attention to the former because it turns to the problem of causality which is essential in our constructions.

According to the Principle of Sufficient Reason every event must have a cause and always there is a reason why something goes so and not otherwise (e.g., Look 2013). This principle contradicts the Newtonian concept of time (and space) because all the instants of Newtonian absolute time are identical in properties. In this case

---

[2] The term *specious present* is considered to be coined by E. Robert Kelly (alias 'E. R. Clay') (1882). Then it was further developed by philosopher and psychologist William James (1890) (see, for details, Andersen and Grush 2009; Andersen 2014).

[3] The teacher of Isaac Newton.

[4] Mathematical Principles of Natural Philosophy, first publication in 1687 with further editions in 1713 and 1726.

Leibniz cannot find any reason for the world to be so as it is now rather than it was before or will be later. It should be emphasized that here Leibniz treats the world as a whole entity rather than just a possible configuration of its parts.

Leibniz put forward a causal theory of time where the structure of time and its anisotropy are considered to depend on the causal order. Moreover, Leibniz holds that the categories of *cause* and *effect* are logically prior to *change* and, in turn, *change* is logically prior to *time* (a detailed analysis of Leibniz's description of time is given, e.g., by Arthur 1985; Cover 1997; Futch 2002, 2008). Below we will return to causal theories of time to be discussed from the modern standpoint; here speaking about Leibniz's point of view it is worthy of noting that he also considered the possibility of complex structures of time which in modern terms can be characterized as branching time or multiple time streams (Newton-Smith 1980; Futch 2008).

Continuing our discussion of time problems we want to note two aspects of philosophical constructions by David Hume (1711–1776), a Scottish philosopher, historian, economist, and essayist.

Within Hume's Copy Principle (e.g., Kemp Smith 1941; Rosenberg 1993) all our ideas are copies of our impressions including all our sensations, passions, and emotions. All the ideas arising in the mind are derived from impressions and all the ideas resemble or correspond to impressions but not are identical to them because of the bounded capacity of human cognition. Here we would like to note that Hume's concept of relationship between complex impressions and ideas may be regarded as a certain kind independence of the human mind from physiological processes responsible for the formation of impressions. It allows us to ascribe individual dynamics and temporal dimension to the mind being the gist of our two-component description of human behavior (Lubashevsky 2017).

The second aspect of Hume's philosophy essential for our constructions is his famous attempt to combine within the cause-effect relationship the temporal asymmetry of a cause and its effect together with the simultaneity in their coexistence. According to Hume, for an event to cause another event both of them should exist in some way at one moment of time, thus, must be contiguous in time. Hume's concept of time contains the notion of minimal time interval, which opens a gate to overcoming this contradiction by saying that the effect of its cause arises just at the very next moment of time, although Hume himself denies that the effect and the cause may be separated by some time interval (Beauchamp 1974). It should be noted that in our account of human temporality finite intervals of subjective time also possess the status of existence, which endows the contemporaneity with complex temporal structure where even the effect may precede its cause. Moreover, if we want to deal with causal relationship between temporally distant events existing in the mind we need to turn to the idea of partial existence applicable to the inner world of human beings.

German philosopher Immanuel Kant (1724–1804), the central figure in early modern philosophy, put forward the famous doctrine of *transcendental idealism*. The gist of this doctrine is that in the mind we deal only the image of the world and the images of things rather than the world and things in it such as they exist on their own, independently of us, i.e., as things-in-themselves in Kant's terminology.

Let us discuss in more details some aspects of Kant's philosophy we share in our constructions.

First, for Kant (see, e.g., Cleve 1999; Dunlop 2009; Janiak 2016; Stang 2016) time and space are nothing but forms of *a priory* intuition—non-empirical, singular, immediate representations. Put differently, according to Kant, time is neither a thing nor a property of things; time is an attribute of the appearances of such things. Therefore in our account we will deal with the *human temporality* because no individual properties and own dynamics can be attributed to human time.

Second, for describing how external objects appear in our mind, Kant introduces the conceptual triad "*thing-in-itself* (physical object)–*appearance*–*representation* (metal image)." According to Kant, we have no access to the things-in-themselves, but their appearances—as empirical objects—are related to our sensory (in general sense) perception and, finally, to the mind as representations. In Kant's account, the appearance has an intermediate status since it is located between objective things and subjective representations. However, the ontological and epistemic status of appearance remaining uncertain in Kant's view is the subject of ongoing debates.

Nowadays, two rival interpretations of the appearance, the "two-world" view and the "two-aspect" view, take the central position in these debates (e.g., Stang 2016, for are view). The "two-world" interpretation regards things-in-themselves and appearances as two distinct entities, whereas the "two-aspect" interpretation takes this distinction as one between two aspects of the same thing. Both the interpretations pose challenges to our theoretical constructions. Within the "two-world" interpretation it is not clear how the two kinds of entities interact with each other. Within the "two-aspect" interpretation it is not clear why the two aspects are so different in properties, including their dynamics. Therefore, in our constructions, we will turn to the gist of novel accounts of appearance, namely, the "two-perspective" view by Robinson (1994), the account of Oberst (2015) which may be regarded as non-Berkelian phenomenalism, and the concept of the appearance "objective-objectual" status by Katrechko (2018). In particular, we will regard the appearance as an entity belonging to the body-in-the-self (Sect. 4.1). The interaction between the external objects, their appearances (the body-in-the-self), and the corresponding mental images (the mind-in-the-self) will be described via the human temporality within the concept of Blurring World (concluding section), which is related to the diachronic aspect of Robinson's account.

Georg Wilhelm Friedrich Hegel (1770–1831), a German philosopher and an important figure of German idealism, proposed the concept of *mathematical* and *spiritual* time. Mathematical (or external) time characterizes the motion of material objects in space. Spiritual time is the cumulative time in which we make progress, in which we both look backward and forward (e.g., Ferrarin 2001; Siep 2014). We want to emphasize that again the idea of two temporal dimensions is met in Hegel's philosophy.

On the border of the XIX–XX centuries a number of novel theories of time emerged. These theories have not only brought into focus new aspects of time but also changed our way of thinking about time. In this context we should note Bertrand Arthur William Russell (1872–1970), a British philosopher, logician, essayist and social critic; Henri Poincaré (1854–1912), a French mathematician, theoretical physicist, and philosopher of science; Alfred North Whitehead (1861–1947), a British mathematician, logician, and philosopher; Charlie Dunbar Broad (1887–1971), an English philosopher of science, epistemologist, and moral philosopher. A special attention should be paid to two Theories of Relativity proposed by Albert Einstein (1879–1955) in 1905 (Special Theory of Relativity) and 1915 (General Theory of Relativity). These theories (together with quantum mechanics) caused the paradigm shift in physics. In particular, it has been recognized that time and space are not independent and should be treated as just different dimensions of one space-time continuum whose choice depends on the external observer. As a result, the temporal order and the notion of simultaneity on their own lose their meaning. For an introduction to the modern theories of space-time continuum, in particular, microscopic properties of time as it is understood in the theory of Quantum Gravity a reader may be referred to Hawking and Mlodinow (2005), Hawking (1988), a philosophical discussion about these issues can be found, e.g., in Weinstein and Rickles (2015). It is necessary to emphasize that this concept of space-time continuum copes with the inanimate world.

The reality of time in various meanings and in its connection to the integrity of space-time continuum was one of the crucial issues discussed in philosophy of the XX century. Nowadays the majority of developed theories of time are divided into two big classes called the A-theories and the B-theories after J. E. McTaggart (1866–1925, English metaphysician, worked mainly at Trinity College, Cambridge). Naturally, there are theories that do not meet this classification, for example, a new class of theories of time called hybrid AB-theories is proposed by Baker (2007) as well as the the C-theories discriminating between the temporal order and the direction of time (Farr 2012, 2020a, b, for are view and modern interpretation).

McTaggart (1908) put forward two principally different ways of categorizing sequences of events according to their order in time.[5] Within the A-theories events are characterized by their temporal position "running from the far past through the near past to the present, and then from the present to the near future and the far future." In the B-theories any two events are characterized by their temporal positions relative to each other, i.e., one of them is earlier, the other is later. The watershed between these theories is how they answer to questions such as whether the temporal position of events or objects in the past, present, or future endows them with principally different *intrinsic* properties and, correspondingly, whether the present moment of time possesses some unique properties singling it out of other instants of time. In particular, A-theories are most suitable for tackling problems of human time. Below we will also turn to an A-theory for describing the subjective component and regard the imaginary future as well as the past retained in memory as real objects existing at the current present in the human mind. For a detailed discussion of the A- and B-

[5] Actually McTaggart put forward a more complex classification of temporal orders including also a category called C-sequences to overcome his conclusion about the unreality of time. This issue, however, is the subject of special philosophical research which is beyond the scope of this book; an interested reader may be referred, e.g., to modern reviews by Boccardi (2016), Tegtmeier (2016).

theories a reader may be referred, e.g, to Zimmerman (2005, 2010), Bigelow (2013), Markosian et al. (2014), Boccardi (2016, 2017).

The B-theories of time defended in classic works by such philosophers as Bertrand Russell (1915), Donald C. Williams (1951), John Jamieson Carswell Smart (1963, see also 2008), David Hugh Mellor (1981, 1998) are mainly related to the description of the inanimate world (Denhigh 1981; Dyke 2003; Smart 2008).

At the border of the XIX–XX centuries there have been emerged novel academic disciplines such as psychology and sociology dealing directly with the inner world of human beings. Philosophers also turned to the problems of human mind trying to elucidate its basic laws as they exist on their own, i.e., leaving aside the questions of their reducibility or irreducibility to physical reality. It has given rise to the new philosophical discipline called phenomenology with its particular application to describing consciousness, the phenomenological theory of mind (for an introduction see, e.g., Gallagher and Zahavi 2012; Smith 2013). Phenomenology has been employed in various guises for centuries, but it came into its own in the works of William James (1842–1910), an American philosopher and psychologist; Henri-Louis Bergson (1859–1941), a French philosopher; Edmund Gustav Albrecht Husserl (1859–1938) and Martin Heidegger (1889–1976), German philosophers; Jean-Paul Charles Aymard Sartre (1905–1980), a French philosopher, novelist, political activist, and literary critic; Maurice Merleau-Ponty (1908–1961), a French phenomenological philosopher, and others.

Among the basic ideas of phenomenology we want to note the concept of *specious present*—the finite time interval whose instants are perceived by humans as simultaneous—developed by William James (1890) and further elaborated by Charlie Dunbar Broad (1923).[6] Henri-Louis Bergson (1910/2001) also spoke about a complex structure of human cognition units where "several conscious states are organized into a whole." In our constructions, the specious present plays the role of human time present having an inner structure due to its finite duration.

Edmund Husserl formulated the basic ideas underlying modern phenomenology. We want to emphasize three important for us lines in his constructions. First, Husserl brought into phenomenology the notion of *intentionality* as a basic feature of human mind. Intentionality is a general notion describing, in particular, goal-oriented behavior of humans. In our constructions human goal-oriented behavior gives rise to that the imaginary future affects the present. Second, to describe in detail human time-consciousness, Husserl introduced three notions, *primal impression*, *retention*, and *protention* (see, e.g., Gallagher and Zahavi 2012) which deal with three slice of time: the nearest past, the present, and the nearest future. These tree slices exist at the present in the mind and endows human time with a certain complex inner structure. We share this concept and will employ it in describing the complex present. Third, for Husserl *phenomena* are complex objects comprising two complementary com-

[6] It should be noted that the first usage of the term *specious present* is attributed to E. Robert Kelly (alias 'E. R. Clay') and this notion in various guises appears in the works of James Ward (1843–1925), an English psychologist and philosopher, and Shadworth Hollway Hodgson (1832–1912), an English philosopher (Dainton 2017).

ponents: subjective acts of consciousness and the corresponding intentional objects (e.g., Smith 2013). Our idea of subjective and objective components in describing human behavior corresponds to this Husserl's concept.

Turning to the philosophy of Martin Heidegger we want to note the notion of *temporality* Heidegger developed in his famous work *Being and Time* (1927/1996, see also Heidegger 1985). In Heidegger's account of time this notion reflects the integrity of the past, present, and future of our existence in the world (e.g., Blattner 2005). In our book the temporality of human mind is the main subject-matter similar in some sense to this Heidegger's one.

In the XX century the advances in physics again aroused interest of philosophers in the existential issues of the past, present, and future and their relationship. Let us note some of these issues that can elucidate our further constructions. The first we want to discuss concerns the doctrine generically called *presentism* which posits that only the present exists, whereas neither the past nor the future exists. In this case, accepting the present to be a point-like instant of time, we inevitably face up to a number of logical problems already recognized by Aristotle and underlying the doubts about the reality of time. There have been proposed several approaches to overcoming these problems via endowing the present with some inner structure or finite duration. In particular, they involve the *degree presentism* proposed by Quentin Smith (2002), the *thick presentism* by Scott Hestevold (2008), the *compounded presentism* by Barry Dainton (2010), and the *priority presentism* by Sam Baron (2015).[7]

Here we want to pay special attention to the degree presentism implying that existence is not located at one instant—the present—but distributed over time. In other words, within this doctrine the reality may be conceived of as some cloud whose most dense part corresponds to the current *now*. We will use this idea in describing temporal properties of the subjective component. It should be emphasized that the degree presentism requires a two-dimensional structure of time in order to admit dynamics. For a system with distributed reality to change as the present now is shifted to another instant of time an additional dimension of time—meta-time—in which this change can take place is necessary. The idea of two-dimensional structure of time was put forward previously by John Jamieson Carswell Smart (1949, 1980) to deal with temporal properties of events embedded into the "river of time." For a thorough analysis of arguments for and against the introduction of two-dimensional structure of time a reader may be referred to Dainton (2010). Within our concept of objective and subjective components we also turn to this idea about two temporal dimensions, physical time and the human temporality.

The second issue we want to note is the bundle of theories of time turning to the alternative to presentism, the doctrine generically called *eternalism*. Eternalism admits the past and future to exist even though they are not currently present (e.g., Markosian et al. 2014). The appearance of this doctrine is partly stimulated by Einstein's Theories of Relativity, where the introduction of time as a particular coordinate of the space-time continuum can be done in many ways depending on the

[7] Various types of presentism and their relation to the mathematical formalism of Newtonian mechanics is discussed in more details in the book *Physics of the Human Mind* (2017).

state of external observer, for details a reader may be referred, e.g., to Craig (2001, 2000a, b), Balashov and Janssen (2003), Dainton (2010), Zimmerman (2011). Here we want to note the famous models of "moving spotlight" and "growing block" by Charlie Dunbar Broad (1923) (see also Adams 1986; Forrest 2004, 2008) and, especially, the *branching* model of time. One of the first philosophers formulated the idea of branching time in a precise manner was Henri Bergson (1910/2001) however its roots can be found in Aristotle's and Diodorus' philosophy (e.g., Øhrstrøm and Hasle 1995; Goranko and Galton 2015). In the XX century the idea of branching time became rather popular, in this connection we may note the short story "The Garden of Forking Paths" written in 1941 by Argentine writer and poet Jorge Luis Borges (1899–1986) where in artistic form he presents this concept of time. The branching time model was advanced, in particular, by Storrs McCall (1994), and the logic of branching time was developed by Arthur Norman Prior (1957), Saul Kripke (in his letter to Prior, 1958), Nuel Belnap (1992), and others (for details see, e.g., Øhrstrøm and Hasle 1995; Lowe 2013). In describing human time we can appeal to McCall's metaphor (1994) which considers the world to be like a tree: the past is the tree trunk, the tree branches represent possible futures, and the passage of time is conceived of as the trunk growth and the continuous disappearance of branches, one by one or the growth of new ones.

In summary, we want to single out the ideas of time that underlie our constructions:

- There are two temporal dimensions quite distinct from each other in the basic properties, physical and human time. In the book we focus our attention on human time where the description of *past*, *present*, and *future* play an essential role.
- The present of human time as well as all its instants possesses a complex internal structure and is characterized by a finite duration. The human present, the retained past, and the imaginary future partly overlap with one another rather than being strictly separated.
- Within human inner world the retained past and the imaginary future are real, have causal efficacy, and can change as physical time goes on and induces the "movement" of human present.
- Properties to be attributed to some event depend essentially on whether this event happened in the past or will happen in the future. Causal efficacy of this event gradually become weaker as the time distance between the present and analyzed instant increases.
- Cause is prior to time and the causal properties determine the structure of human time. The temporal order of events in the mind is fixed when these events are causally related to one another. The temporal arrangement of events in the imaginary future (and the retained past as well) can have branching structure. The temporal order of events belonging to different branches may change.
- The representations of physical objects in the mind are irreducible to these objects in their properties. The properties of representations can possess features not met in the inanimate world.

1.2 Personal Identity over Time

To illustrate the problems discussed in this section, we may ask what is common between 65-year professor Lubashevsky, writing this book, and the student Ihor just started to learn physics. They have different bodies and distinct life experiences. What is the reason to say that this student and the older man are the same person? Many centuries ago, similar questions were posed by philosophers of ancient Greece and up to now are under debate. The reason for discussing the identity problem in the context of the present book is due to personal identity over time endowing a human with temporal integrity. In other words, trying to understand the current behavior of a given person, his past retained in memory and imaginary future have to be taken into account.

In the present subsection we discuss the problem of identity over time from a rather general, philosophical standpoint. At the beginning of this subsection, we consider the temporal identity problem in the form applicable to animate and inanimate objects. Then this problem is confined to aspects essential for human nature. Finally, the modern concept of temporal parts characterizing personal identity over time is brought into focus.

This problem is interesting for us because, in studying goal-oriented actions, we will deal with temporal parts of action strategies as a basic element of their mathematical description (Sect. 6.2).

1.2.1 Ship-of-Theseus Problem

Many issues concerning the properties of time are tightly related to the questions about the identity of an object over time, which is often referred to as *diachronic identity*. The following legend rose to us due to Plutarch (c. CE 46–CE 120), a Greek historian, biographer, and essayist, is a paradigmatic example of the diachronic identity problem.

Ship of Theseus Paradox:

> The ship on which Theseus sailed with the youths and returned in safety, the thirty-oared galley, was preserved by the Athenians down to the time of Demetrius Phalereus. They took away the old timbers from time to time, and put new and sound ones in their places, so that the vessel became a standing illustration for the philosophers in the mooted question of growth, some declaring that it remained the same, others that it was not the same vessel. (Plutarch, I, XXIII (p. 49))

Actually the question posed by ancient Greek philosophers is how this ship can remain the same if all its parts have been entirely replaced, piece by piece.

The given example illustrates the main aspects discussed within the diachronic identity problem, namely, the meaning, criteria, and reasons of treating a *temporal* sequence of objects with different properties or constituent elements as time changes of the *same* object. In particular, it concerns composite systems whose parts are gained or lost as time goes on.

These issues are closely related to premises underlying the further description of human behavior whose elements cannot be addressed to a single instant of time and finite time intervals, maybe, with fuzzy boundaries have to be used as whole entities. Therefore, in the present subsection we discuss the problem of diachronic identity in detail focusing mainly on its particular version, *personal identity*, dealing with humans as conscious beings. The general description of diachronic identity—applicable to inanimate and animate objects—is beyond the scope of this book and a reader may be referred to reviews, e.g., by Oderberg (1993), Sider (2000), Pasnau (2011), Gallois (2016).

Aristotle proposed a solution to the diachronic identity problem within his account of motion. A moving object is a series of different entities attributed to different *nows* and these entities are united into something single via sharing the membership of temporal "before-and-after" order based on the *substantial form* persisting through time (e.g., Coope 2009a). In his *Metaphysics* Aristotle deploys the notion of substantial form which unifies some matter into a single object comprising two part. He treats soul and body as a special case of form and matter (e.g., Ainsworth 2016).

Aristotelian philosophical traditions dominated up to the 11th century, in particular, within the time of Greco-Arabic communication in metaphysics. For example, Avicenna (or Ibn Sīnā, c. 980–1037, a Persian polymath regarded as one of the most significant thinkers and writers of the Islamic Golden Age) claims that a human being is the same over time due to its soul enduring even as its body constantly changes because of ordinary biological processes of nutrition and growth (e.g., Pasnau 2011, Sect. 29.1).

This point of view was held and elaborated in more detail by Thomas Aquinas (1225–1274) and William of Ockham (also Occam, c. 1287–1347), ones of the most influential medieval scholastic philosophers and theologians. Ockham put forward two principles concerning possible changes in both the components of human being (Pasnau 2011). The *part-whole identity* thesis states that composite material substance is nothing but its various parts and, thus, when a thing gains or loses a part it is no longer the same thing. The *no-transfer* thesis postulates that substantial forms cannot transfer from one thing to another and, so when integral parts of a substance change, the substantial form must also change. The two theses give rise to a theological problem of human identity in the most pronounced case—after the death and the following Resurrection. If the body changes how the sole can be the same?

John Buridan (c. 1300–c. 1362), a French priest, philosopher, and scientific theorist, trying to resolve this theological puzzle singled out three kinds of diachronic identity (Pasnau 2011):

- the *total* identity of a thing that endures in virtue of the permanence of all its parts at once without any change, addition, or subtraction;

- the *successive* identity of a thing that consists in a continuous succession of its parts, one after another;
- the *partial* identity of a thing whose endurance is due to the permanence of some of its parts, whereas the other parts succeed one another.

Within Buridan's classification (*i*) the heavens exhibit the total identity over time, (*ii*) the successive identity is exemplified by the motion of physical objects, and (*iii*) the partial identity should be attributed to a human being whose body goes through a sequence of changes, whereas the intellectual soul remains the same. Broadly speaking, the successive identity of physical objects opens a gate to describing their dynamics using differential equations, whereas the partial identity of human personality underlies the introduction of temporality for modeling human behavior.

Having introduced several kinds of diachronic identity, Buridan actually demonstrated that this problem requires a sophisticated analysis of *temporal* interrelation between different parts of a given object taking into account its specific details. In particular, Nicole Oresme (c. 1320–1382), a French philosopher famous for his original ideas in economics, mathematics, and natural sciences, introduced the notion of *identity degree* emphasizing that the sameness may be fuzzy. He notes that we should distinguish between objects whose parts gradually replace one another and objects changing as a whole at every instant. For example, when its planks are gradually replaced one by one, the Ship of Theseus cannot be totally the same but admits the interpretation of its sameness with a certain degree until the last plank is replaced. Moreover, Thomas Hobbes (1588–1679), an English philosopher best known for his political thoughts, demonstrated that diachronic identity can be more ambiguous. By way of example, he modified the Ship-of-Theseus paradox. Let us assume that someone replaces the ship planks step by step using the same wood, collects all the discarded pieces, and builds a new ship with them. Finally we have two ships. Hobbes poses the question of whether both the two ships, only one of them, or none should be regarded as identical to the original ship of Theseus. This ambiguity can be overcome, as Hobbes supposes, if questions about diachronic identity are made specific with respect to aspects[8] brought into focus. In particular, the matter or the form of a given object may play the role of such aspects.

The given discussion of diachronic identity, turning to the Ship-of-Theseus paradox as a paradigmatic example, demonstrates that this problem seems to have no general solution. So it should be analyzed separately for various systems different in nature, which enables one to include into consideration the characteristic features a particular class of systems possess. For instant, the diachronic identity of

- the Ship of Theseus,
- a seed and the three grown from this seed,
- a person in his childhood and old age

[8] Hobbes actually uses the term *name* allowing interpretation in a very broad sense (e.g., Duncan 2017).

are actually quite distinct problems and different aspects can be involved in their solution. Below we will confine our consideration to the diachronic identity where the object in question is the human being.

1.2.2 Human Identity over Time

Locke seems to be the first who raised the personal identity issue as an individual problem on its own and formulated the approach to be called the phenomenological theory of mind in three centuries. Taking into account that humans have only limited insight, at least at his time, into the deep structure of both matter and mind, Locke proposed to use the grasp of ourselves as the reference point. In other words, he proposed to analyze human behavior looking at it through our own eyes, i.e., in the notions directly accessible in our mind. We would like to attract attention to this Locke's account of diachronic identity because in our further construction we will turn to this idea.

The concept of a person as the subject of consciousness who simultaneously recognizes himself as such a subject takes the central position in Locke's account, for details a reader may be addressed to Thiel (2011), Ainslie and Ware (2014), Uzgalis (2017). For Locke the basic mechanisms via which our consciousness emerges and develops are beyond our understanding. He also rejects the reducibility of human personality to issues concerning the soul or the body because they are just irrelevant to the individuality of a given person (Thiel 2011).

According to Locke, the consciousness has to be treated as the basic element in describing the mind turning, in particular, to introspection. It concerns also our ability to recognize our own identity through time. Exactly the consciousness constitutes the personality and personal identity—we would say, integrity over time—by unifying thoughts and actions. The unification of events happened at different time moments requires the memory for its implementation. However, the consciousness based on the memory cannot be equated to the memory; a person can forget some events and have false memories without losing the diachronic identity. Moreover, in explaining this diachronic identity the consciousness treated as a whole without specifying its particular aspects cannot be also referred to. Indeed, our consciousness of events changes as time goes on.

Trying to elucidate a particular aspect of consciousness that could be responsible for personal identity Locke puts forward the concept of *continuity of consciousness* as a mechanism giving rise to personal identity. Namely, a person at time t_1 and a person at time t_2 may be regarded as one subject if they are linked by continuity of consciousness (Thiel 2011). In this sense Locke constructs personal identity on arbitrary time scales via *local* diachronic identity. Thereby the consciousness of one person may be different at distant moments of time if there is some way to determine its sameness at the neighboring instants of time. To tackle the latter problem Locke notes that consciousness should locally involve not only the current instant of time but also the immediately preceding and following moments. Thereby, in Locke's account human consciousness acquires a complex temporal structure, the present also contains information about the past and future perceived from the present. Naturally this past-present-future complex gradually changes as time goes on.

Hume raises a doubt about Locke's account of personal identity based on the consciousness unity (e.g., Thiel 2011). Hume notes that when he turns to inner experience or introspection, each time he finds only a variety of distinct, particular perceptions, thoughts, feelings, etc. There is no experiential evidence of a simple soul, mind, or something else that remains the same through time. All we can find on the basis of introspection is that the mind is "nothing but a bundle or collection of different perceptions, which succeed each other with an inconceivable rapidity, and are in a perpetual flux and movement." (T 1.4.6.4)[9] This point of view has become known as the "bundle theory" of the self or the mind. For Hume the bundle structure of the mind does not provide any base for personal identity through time.

Although Hume accepts that in inner observation there can be found no evidence for ascribing identity to the self, we, nevertheless, have a "natural propension" (*ibid*) to believe in the simplicity and identity of our own selves. Trying to answer the question of what gives rise to this propension, Hume draws the distinction between the issue of *identity* and the issue of *simplicity* i.e., indivisibility. Hume supposes that objects with "perfect" diachronic identity which may be also simple must not change in time even partially. According to Hume, we ascribe diachronic identity to a subject whose successive perceptions are closely related to one another in some way, e.g., via their continuous sequence. It is a certain misinterpretation, "[t]he thought slides along the succession with equal facility, as if it consider'd only one object; and therefore confounds the succession with the identity." (T 1.4.2.34) Hume puts forward three relations—resemblance, contiguity, and causation—that give rise to our perception of the unity of ourselves. He attracts attention to the human memory playing a crucial role in the mind unification leading to personal identity.

> The memory [is] a faculty, by which we raise up the images of past perceptions[...] [T]he memory not only discovers the identity, but also contributes to its production, by producing the relation of resemblance among the perceptions. The case is the same whether we consider ourselves or others. (T 1.4.6.18)

For Hume it is causation via memory that is the main mechanism endows ourselves with integrity. In particular, he writes:

> As to *causation*; we may observe, that the true idea of the human mind, is to consider it as a system of different perceptions or different existences, which are link'd together by the relation of cause and effect, and mutually produce, destroy, influence, and modify each other. Our impressions give rise to their correspondent ideas; and these ideas in their turn produce other impressions.

[9] As previously, references in Hume's text prefaced by "T" are to David Hume, *A Treatise of Human Nature* (a critical edition), ed. David Fate Norton and Mary J. Norton (Oxford: Clarendon Press, 2007) Vol. 1.).

> One thought chases another, and draws after it a third, by which it is expell'd in its turn. (T 1.4.6.19)

In other words, according to Hume, memory combined with causation is actually the main source of personal identity. Moreover without memory we would have no notion of causation—chain of causes and effects constituting our personality (cf. T 1.4.6.20).

Leaping ahead, we would like to note that Hume's point of view on personal identity may be regarded as a certain base for developing a mathematical description of human temporality. The human temporality comprises many individual events retained in memory or expected to occur in the future and is not simple, i.e., indivisible entity. However, the cause-effect relations (in Hume's terms) or intentional relations (in our terms, Sect. 3.4) combining the events into a highly connected network makes it necessary to tackle the human temporality as a certain integral entity. Therefore, in explaining human behavior, the temporality cannot be divided into several parts without breaking essential intentional relations.

Thomas Reid (1710–1796), a Scottish philosopher, notes that memory is neither necessary nor sufficient for personal identity. He holds that it is impossible to account for personal identity in any terms other than itself. Thereby, in his account personal identity becomes simple and unanalyzable. Though memory is not the metaphysical ground of personal identity, it provides the first-personal evidence of personal identity. In this sense, Reid comes close to the future concept of phenomenology because his standpoint leads to the self as an individual entity with its own properties endowing our conscious perception of ourselves with integrity through time.

As the next step in developing the concept of personal identity, we want to note Kant's account based on his division of the reality into appearances (things-as-they-appear) and things-in-themselves; for a detailed introduction a reader may be referred to Cleve (1999), Dicker (2013), Ainslie and Ware (2014). In his account, Kant draws a fundamental distinction between the *self-as-it-appears* to our consciousness and the *self-as-it-is* on its own. For Kant, we have access only to the former self, whereas any knowledge about the latter self is principally impossible, which may be related to Reid's standpoint noted above. In Sect. 4.1 we turn to the concepts of the mind-in-the-self and the body-in-the-self, which overlap with the two kinds of the self in Kant's account.

Actually Kant regards cognition as a twofold process, first, our physical senses provide us with particular representations of perceived objects or events and, second, our mind is responsible for their combination (Dicker 2013). These representations forming a *temporally extended series* are the constituent components of human experience; Strawson (1966) calls this Kant's statement the temporality thesis: "human experience essentially shows temporal succession" (see also Brown 2006, p. 96). The given series of representations, however, are not merely collections of separate elements; the human mind combines them—using Kant's terminology, *synthesizes* them—into the ordered *manifold of representations* being something united into a certain entity. However, for the manifold of representations to emerge there should be some unity of consciousness, which leads Kant to the conclusion about the existence of personal identity. Strawson calls it the necessary unity of consciousness thesis: "there must be among our temporally extended series of experiences the unity that is needed for the possibility of self-consciousness" (*ibid*, p. 96).

Like Hume, Kant considers that we cannot find ourselves through empirical self-consciousness and direct introspection. However, unlike Hume, in his account Kant rejects the idea that diachronic identity of a human being is only specious. It poses a question about what unifies particular representations and endows their manifold with stable integrity. Kant's answer to this question is that the representations are unified by *being of* an object—a human being (Wolff 1963). Roughly speaking, a human with *one* consciousness, whose *endurance* is due to the memory, bears all the representation and integrates them in himself.

The integrity endows the manifold of representations with properties crucial for our further constructions. Leaping ahead, the manifold of representations admits an interpretation as a certain mental space, including its own temporal dimension, whose basic elements are all possible representations. The integrity of the given space is due to that all the relations between these representations are accessible to the mind. It enables us to consider the mental space to have some internal structure and its own properties attributed to it as a whole.

Returning to Kant's account let us briefly discuss the properties of the manifold of representations (for details see, e.g., Strawson 1966; Dicker 2004).

First, the representations recognized by the mind—the mental images of perceived objects and events—are related to one another in a non-arbitrary, rule-governed way. However, trying to elucidate these rules we face up to the dual nature of representations. It is necessary to be specific with respect to which aspect of these representations is under consideration, whether it is about how[10]

- the perceive external objects appear in the mind and how we operate with these mental images (the realm of the mind-in-the-self, Sect. 4.1),
- these images are formed by our sensory systems and then encoded in the neural networks (the realm of the body-in-the-self, Sect. 4.1).

Indeed humans are able to analyze their mental images, whereas the emergence of these images in the mind is implemented unconsciously. The rules accessible to the mind are applicable only to the representations as mental images. Therefore it is not clear how they can govern *completely* our perception of external objects which involves a number of unconscious physiological processes. Nevertheless if we accept the manifold of representations to emerge, then we should also admit that the properties of these mental images specify in some way, explicit or implicit one, the formation and evolution of the representations' manifold. Within Kant's account

---

[10] The consideration of representations from these two aspects overlaps with the modern accounts of Kant's appearances proposed by Robinson (1994), Oberst (2015), Katrechko (2018), see also Sect. 1.1.

this interaction may be characterized as a causal power of representations. Below we will turn to the gist of this concept in constructing our account of human temporality.

Second, although time on its own cannot be perceived directly, the enduring consciousness endows the representations' manifold with temporal structure admitting the introduction of two temporal orders. The first temporal order—subjective time—characterizes the order of how particular representations are apprehended. The second temporal order—objective time—arises via the interaction of the representations during the formation of their manifold. The two temporal orders may be different, which prompts us to regard the subjective time as an object possessing its own dynamics rather than been enslaved by physical time.

1.2.3 Temporal Parts

In the XX-century *personal identity* became an individual branch of philosophy involving a whole bundle of different specific problems (e.g., Hershenov 2012; Olson 2019, and collection of papers edited by Gasser and Stefan 2012). We will touch on only some of them that are related to the problem of persistence through time and are reflected in a question as to what is a reason to say that the past and future beings are "I."

Nowadays, there are singled out several classes of plausible answers to this question (e.g., Olson 2012, 2019). One of them, called *brute-physical views*, in most cases, implies that personal identity consists in the identity of the body. Following this line we return to the general problem of identity over time treating inanimate and animate objects from one standpoint. Another class, *called psychological-continuity views*, considers our persistence across time to be implemented via the appropriate links between our mental states. These links can underlie *direct* or *indirect* connections via human memory. In the latter case personal identity stems from the existence of a certain chain of overlapping memories subjected to the non-branching requirement (for a critical analysis of such an approach see Parfit 1984, 1995). Shoemaker (1984, 1997) proposed to generalize these psychological relations replacing memory with a more general notion of causal dependence, which, in particular, overcomes the problems of consciousness discontinuity, e.g., in sleeping.

The concept of *temporal parts* (see, e.g., Sider 2001; Olson 2007) also often referred to as "four-dimensionalism" has become popular in modern philosophy. The introduction of temporal parts is based on the idea that humans are extended not only through space but also through time. So any person exists as a *whole entity* at *any* moment of time, where the past, present, and future are just attributes of his structure. It should be noted that the turn to the concept of temporal parts may be closely related to the gist of Einstein's Theories of Relativity, where the division into spatial and temporal components of system position in the space-time continuum is not absolute and depends on the motion of observer (a detailed discussion of this issue can be found in Balashov 2008, 2011; Gilmore 2008; Hawley 2009). This concept plays an essential role in developing the theory of human temporality, so the remaining fragment of this section is devoted to its discussion in more detail.

The theory of persistence through time dealing with the concept of temporal parts is called *perdurantism*. It accepts that objects persist through time due to possessing successful temporal parts, so only part of persistent object—its current temporal part—is present at any instant of its existence (e.g., Lowe 2002, Chap. 3). As a rival theory, *endurantism* holds that any object persisting through time is *wholly present* at every instant of its existence, there is noting related to the given object beyond the present.

Leaping ahead, we want to note that within our two-component description of human behavior both the theories—perdurantism and endurantism—are employed as basic premises. The objective component is accepted to meet the endurance model of reality, whereas the subjective component is supposed to obey the perdurance principle with respect to its own temporal dimension.

Up to now the nature of temporal parts is rather far from being understood well (e.g., Hawley 2020). In particular, there is no widely accepted standpoint explaining what sort of entities temporal parts might be. The introduction of temporal parts was inspired by the notion of four-dimensional space-time continuum in Einstein's Theories of Relativity and the apparent existence of spatial parts for physical objects. Nevertheless, spatial parts are not an analogy to temporal parts and are drastically different in properties. Spatial objects possess composite structures and their spatial parts existing on their own can be related to the constituent components. The persistence of objects through time is not due to their possible composition. There are no evident reasons to claim that the parts representing the past, present, and future exist independently of one another.

Trying to elucidate the nature of temporal parts (e.g., Lowe 2002) put forward an approach that may be regarded as axiomatic, which we accept and employ in further constructions. Following logic of Lowe (2002, p. 51), processes are characterized by duration and extension over time is their intrinsic property.[11] Therefore temporal parts may be attributed to processes.[12] Moreover, Lowe (2002) supposes that temporal parts should be regarded as entities ontologically more fundamental than the persisting objects to which they belong.

Below we will use the human temporality as the basic object in describing the dynamics of subjective component which is not reduced at least causally to the collection of time instants in the person's past, present, and future. Actually in constructing the models for human behavior we will follow the strategy proposed by Lowe (2002)

---

[11] For a discussion of the philosophical and axiomatic background of intrinsic and extrinsic properties, a reader may be address, e.g., to Weatherson and Marshall (2018).

[12] For the introduction to ontology of processes a reader can be referred, e.g., to Seibt (2001, 2009, 2017), Campbell (2009). The relationship between process ontology and thick presentism was also discussed by Lubashevsky (2017).

for coping with the concept of temporal parts. Within this strategy, especially in dealing with goal-oriented actions, we will regard the human temporality as a certain object whose existence is justified by the explanatory success, utility, as well as the internal self-consistency and logical simplicity of the corresponding theories.

Summarizing the presented discussion of personal identity we want to emphasize that this notion implies:

- a certain integrity of the human mind over time such that the past, present, and future of a given person cannot be treated separately without loss of intentional relations,
- the introduction of an individual temporal dimension (human time) attributed to the mind and not necessarily coinciding with physical time,
- the possibility of introducing personal identity as a basic notion bearing but not reducible to various relations including intentional ones between the past, present, and future existing in the mind of a given person and changing as physical time goes on.

As a result, in a mathematical description of human behavior some of its *basic elements* should involve in their structure temporal intervals of human time. Correspondingly, the causal power can be attributed to these intervals as a whole rather than its instants separately.

1.3 Psychophysics: General Regularities

In the present section we want to discuss the mechanisms responsible for the general laws of psychophysics. In particular, we want to demonstrate that the main psychophysical laws can be represented in the universal form and, thereby, their mechanisms are based on the properties of the brain functioning rather than physiology of particular sense organs. In the following chapters we will turn to this universality for describing human response to external stimuli within the phenomenological approach dealing with the properties of the human mind.

1.3.1 Basic Psychophysical Laws & Fechner–Stevens Dilemma

The inner world of humans as an object admitting analytical investigation has attracted attention for many centuries starting from ancient Greek philosophy. An important step in this direction was made in the second half of 19th century when in the frameworks of psychology and physiology a new branch of science—psychophysics—appeared. The subject-matter of this science may be characterized as the *quantitative* description of the relationship between physical stimuli and the human sensations they produce.

Psychophysics is rooted in experimental psychology of the late 19th century, its birth is related to the publication of the famous book *Elements of Psychophysics* (1860) by Gustav Theodor Fechner (1801–1887), a German philosopher, physicist and experimental psychologist. In this book Fechner posed a number of questions about the interrelation between the intensity of various physical stimuli and human sensation of these stimuli (for introduction see, e.g., Gescheider 1997; Grondin 2016). Actually at its birth, psychophysics underwent bifurcation; Fechner himself drew a crucial distinction between *outer* and *inner* psychophysics, which differ in subject-matter (e.g., Scheerer 1992; Robinson 2010). According to Fechner, human sensation is implemented as a sequence of two transformations with own properties:

- *stimulus* $\rightarrow$ *psychophysical processes*,[13]
- *psychophysical processes* $\rightarrow$ *sensation*.

Outer psychophysics deals with the two transformations as a whole and studies the stimulus-sensation relationship based on direct observation and experimentation.

Inner psychophysics focuses on the latter transformation and has to draw inferences from the findings of outer psychophysics. Experimentation and quantification are rather problematic in inner psychophysics; for this reason inner psychophysics was rejected for a long time starting from Fechner's contemporaries. Nevertheless, Fechner himself considered that regularities found in outer psychophysics and being only approximate become strictly valid in inner psychophysics, which makes inner psychophysics important branch of science on its own. For the modern state of the art in inner psychophysics taking into account the advances in cognitive neuroscience and the mapping between mental quantities and the brain states a reader may be referred to Ritchie and Carlson (2016).

Based on the findings of his mentor Ernst Heinrich Weber (1795–1878), a German physician, Fechner formulated one of the basic psychophysical laws, *Weber's law*. Weber's law relates the perception threshold[14] ΔS of variations in the intensity S of some physical stimulus (e.g., sound or light) to its intensity S via the proportionality with a coefficient κ_w often referred to as the Weber fraction

$$\Delta S = \kappa_w S\,. \tag{1.1a}$$

Its later modification

$$\Delta S = \kappa_w S + \eta \tag{1.1b}$$

allows for this stimulus of low intensity via adding some constant η interpreted as the result of sensory noise.

[13] In modern terms psychophysical processes may be treated as the generation of neural signals by the corresponding sense organs.

[14] This perception threshold is also called the *just-noticeable difference*.

Naturally each sensory modality[15] is characterized by its own value of the Weber fraction, e.g., for visual length of lines, brightness, and loudness $\kappa_w = 0.04$, 0.08, and 0.1, respectively (Baird and Noma 1978).

Appealing to Weber's law, Fechner put forward the principle that an arithmetic series of mental magnitudes of perceived physical stimuli should correspond to a geometric series of their intensities (energies) (e.g., Gescheider 1997). In the framework of this principle the ratio $\Delta S/S$ is regarded as the induced increment in the sensory magnitude ΔM of the given physical stimulus, whence it follows that

$$M = \kappa_w \ln S \quad \text{for } S \gg \eta/\kappa_w. \tag{1.2}$$

This logarithmic relationship is often referred to as *Fechner's law*. It should be emphasized that the introduction of the additive term in Weber's law (1.1b) eliminates the formal logarithmic singularity of Fechner's law at $S \to 0$.

Fechner's description of the stimulus-sensation relationship was dominating for a hundred years until Stanley Smith Stevens (1906–1973), an American psychologist, put forward a new psychophysical law (Stevens 1957, 1961) called after him *Stevens' law*. When subjects are asked to *quantify* (numerically or, e.g., via squeezing a handgrip dynamometer) their sense of subjective intensity the $M(S)$-relationship is described more effectively by a power law

$$M = kS^{\beta_s}, \tag{1.3}$$

where the coefficient k and the exponent β_s are some constants determined by the physiological mechanisms of the corresponding sensory modality. The Stevens exponent β_s can be less or larger than or equal to 1, e.g., for visual length of lines, brightness, and loudness $\beta_s = 1.0$, 0.5, and 0.67, respectively, for electric shock $\beta_s = 3.5$ (e.g., Gescheider 1997).

The appearance of two different laws competing for the description of the stimulus-sensation relationship gave rise to long-term debates in psychophysics including the present time. These debates are focused on the possibility of reconciling Fechner's law and Steven's law. Both of them, on one hand, concern quantifying the inner sensation of external physical stimuli and, on the other hand, have functionally different forms. We call this issue the *Fechner–Stevens dilemma* for short. It should be noted that this dilemma can be also posed at a high level of describing the human mind in dealing with cultural aspects of human society (Dehaene et al. 2008; Cantlon et al. 2009).

To reconcile the two laws with each other MacKay (1963) almost immediately after Stevens' publication proposed a concept of the relationship between the inten-

[15] *Sensory modalities* are independent sensory systems such as vision, hearing, smell, taste, and touch. These traditional five senses were singled out by Aristotle. In 1904 the Austrian physiologist Max von Frey (1852–1932) identified the four component sensations of touch as heat, cold, pain, and pressure, pressure now being known to include sensations of texture and vibration. The sensory systems associated with the vestibular system, kinaesthesis, proprioception, and the magnetic sense are also sometimes regarded as sensory modalities (Colman 2015).

sity S of physical stimulus and the perceived magnitude M. A model for neural network functioning takes the central position in his concept. Namely, first, the frequencies of neuron operations caused by (*i*) the stimulus intensity and (*ii*) the brain internal activity are assumed to be related to S and M via Fechner's logarithmic law. Second, the stimulus-sensation comparison is supposed to be implemented via a linear relationship of the two frequencies. In this way MacKay comes directly to Stevens' law (1.3). Similar ideas were considered also by Treisman (1964), Ekman (1964), in addition see Ekman and Sjöberg (1965), Cook (1967) for their discussion. Nowadays this concept has developed into a branch of science that could be called neural psychophysics focusing its attention on neural mechanisms governing human inner perception of physical stimulus (for a review see, e.g., Johnson et al. 2002; Copelli et al. 2002; Billock and Tsou 2011; Adler et al. 2014, and references there in).

Another approach to tackling the Fechner–Stevens dilemma is to modify Weber's law in the vicinity of some special points accompanied by additional modifications in the understanding of how the internal sense of physical stimuli is formed with the increase in the stimulus intensity. The discussion of this approach was induced by publication of Krueger (1989) followed by a large number of commentaries published in the same journal; for a review see, e.g., Laming (1997), Norwich and Wong (1997), Zwislocki (2009).

The crucial point in the psychophysics development playing an essential role in our further constructions is related to the idea of converting the description of stimulus-sensation regularities from outer psychophysics to inner psychophysics posed by Gösta Ekman (1920–1971), a Swedish psychologist. Instead of dealing with the intensity of physical stimuli Ekman (1959) proposed to analyze the basic laws of human sensation in terms of sensory magnitudes. In particular, he demonstrated that the relationship between the variation of magnitude ΔM matching the just-noticeable difference ΔS of the corresponding physical stimulus and the sensory magnitude obeys the law

$$\Delta M = \kappa_e M \tag{1.4}$$

similar to Weber's law (1.1) in form. Relationship (1.4) is often referred to as *Ekman's law* and below the coefficient κ_e will be called the Ekman fraction. It should be noted that an idea similar to Ekman's law was put forward earlier by Franz Brentano (1838–1917), a German philosopher, psychologist, and priest whose work influenced also the development of phenomenology as a branch of philosophy (Brentano 1874/1995).

Keeping in mind our further constructions we want to emphasize that Ekman's law:

- presents a series obstacle to Fechner's logarithmic interpretation of the stimulus-sensation relationship based on Weber's law,
- is regarded as an immediate consequence of Weber's law and Stevens's law (e.g., Laming 1997).

As far as the latter item is concerned, the three laws are, in fact, not mutually independent. Weber's law admits much more reliable ways of verification in comparison with Ekman's law. Nevertheless, we will regard Ekman's law to be the basic one, whereas Weber's law to be a consequence because Ekman's law can be explicitly attributed to inner psychophysics directly dealing with the mind.

1.3.2 Universality of Inner Psychophysics

The basic psychophysical laws noted in the previous section except for Ekman's law describe the phenomena of outer psychophysics which deals with the sensory modalities as whole entities. Each sensory modality like vision, hearing, and smell is a complex sensory system comprising a sense organ (or specific receptors) and the corresponding neural network. The sense organ converts a physical stimulus affecting it into neural signals which, then, are proceeded by the neural network. The fact that the psychophysical laws are of the same *functional* form for all the modalities argues for the existence of a common basic mechanism underlying these laws. However, a particular implementation of this mechanism might be individual for each modality because the particular values of the Weber fraction κ_w and the Stevens exponent β_s are different for different modalities. We can conceive two possibilities of its implementation. Namely, the mechanism responsible for converting the intensity of neural signals generated by the corresponding sense organs into the sensory magnitudes of physical stimuli can be attributed either to

- the part of neural network belonging only to a given modality or
- the part of brain shared by the all the sensory modalities.

Naturally the conversion of physical stimuli into neural signal is individual for each modality, which in the latter cause explains the difference in particular values of the basic law parameters. To figure out which one of the two options is the case, additional arguments based on the available experimental data are necessary.

In this section we discuss the available evidence for the existence of a certain part of brain proceeding all the neural signals generated by different sensory modalities. In particular, the universality of inner psychophysic to be discussed below is, first, one of the arguments for the existence of this region. Second, the laws of inner psychophysics will be used further to justify assumptions about the properties of mental space—a particular implementation of the subjective component within the two-component approach.

Based on the experimental data, summarized, in particular, by Poulton (1967, 1968), the Canadian psychophysicist Robert Teghtsoonian (1932–2017) put forward a number of hypotheses about the universality of inner psychophysics. From our point of view these statements open a gate toward constructing a sophisticated mathematical theory of the mental processing of external stimuli that deals with properties attributed directly to the human mind. For this reason below we list them. According to Teghtsoonian (1971, 1973), Teghtsoonian and Teghtsoonian (1978):

1. The Ekman fraction takes approximately the same value equal to $\kappa_e = 0.03$ for the main sensory modalities, which assumes the strict relationship $\beta_s = \beta_s(\kappa_w)$ between the Weber fraction and the Stevens exponent.[16]
2. A wide variety of perceptual continua are characterized by a common maximal dynamic range,[17] R_M, of sensory magnitude with the estimate $\log R_M \sim 1.5$.
3. The maximal dynamic range R_S of stimulus intensity and the Stevens exponent β_s are related such that the product $\beta_s \cdot \log R_S = \log R_M$ is an approximately constant value.
4. In quantifying two stimuli within a given perceptual continuum the ratio r_S of their intensities and the ratio r_M of their sensory magnitudes are related via the expression $\log r_M = \beta_s \cdot \log r_S + k$, where the constant $k \sim 0.1$. Broadly speaking, humans, on the average, tend to constrict the subjective dynamic range in quantifying physical stimuli. In particular:
   - Within a fixed range of test stimuli the magnitude estimates are systematically biased towards the center of the tested range as it was figured out and studied previously by Hollingworth (1910) and Stevens and Greenbaum (1966). This bias called the *regression effect* causes an underestimation of large magnitudes and an overestimation of small ones.[18]
   - This bias towards the center becomes more pronounced as magnitudes increases, which is called the *range effect*.

The aforementioned properties enabled Teghtsoonian (1971, 1973, 2012) to posit a single central macrolevel mechanism responsible for all judgments of sensory magnitude. The similarity between human perception of physical stimuli based on our sense organs and the mental evaluation of abstract objects like numbers (Baird and Noma 1975; Noma and Baird 1975; Weissmann et al. 1975; Baird 1975a, b) is an additional argument for this mechanism. The same also concerns the fact that sensory modalities are not strictly independent of one another (e.g., Shimojo and Shams 2001, for a review); their interaction via this common mechanism of neural signal processing is quite expectable.

Also we have added the following items accumulating the results of other scholars to complete this description of the basic features of inner psychophysics.

---

[16] The analysis of a wider collection of experimental data by Laming (1997) demonstrates that the Ekman fraction obtained in the way used by Teghtsoonian (1971) can deviate substantially from the value $\kappa_e = 0.03$. It poses doubts about the existence of a strict relationship between Weber's law of *sensory discrimination* and Stevens's law of *sensory judgment* (Laming 1997). Below we argue for its existence and relate the scattering of calculated values of the Ekman fraction to the use of data obtained at the boundary of judgment possibility.

[17] Following, e.g., Teghtsoonian and Teghtsoonian (1997) the term *dynamic range* is understood here as the span (usually given in log-units) from the lowest to highest stimulus intensities or sensory magnitudes admitting a continuous grading.

[18] The underestimation and overestimations of long and short time intervals, respectively, was recognized even earlier by Karl von Vierordt (1818–1884), which is reflected in his famous book *The Experimental Study of the Time Sense* (English translation) (1868) (e.g., Lejeune and Wearden 2009, for details).

5. Judgments on perceived magnitudes are also affected by the recent history of evaluating physical stimuli belonging to the same perceptual continuum or, even, different one, which is often called the *sequential effects*. Their various aspects were thoroughly studied in the last century and for a recent review a reader may be referred, e.g., to Matthews and Stewart (2009a, b), Podlesek (2010), Zhao et al. (2017).
6. In particular, the sequential effects give rise to *hysteresis* in stimulus evaluation. It means that the sensory magnitude ascribed to a given stimulus by an observer can be different when the given stimulus is approaches from stimuli of higher or weaker intensity (Stevens 1957; Cross 1973).

The data accumulated in fMRI, EEG, and neuropsychological investigations also argue for the central mechanism. In particular Walsh and Bueti (Walsh 2003; Bueti and Walsh 2009) put forward a new paradigm about human judgment and evaluation of external stimuli called ATOM ("A Theory Of Magnitude"). The ATOM supposes that various dimensions of magnitude information are encoded by "common neural metrics" in the parietal cortex, which explains the emergence of common *neurocognitive* mechanisms governing human perception of various physical stimuli. Naturally, the situation can be more complex (Pinel et al. 2004). For a review of arguments for and against this universality a reader may be referred to Kadosh et al. (2008), Dormal and Pesenti (2012), Skagerlund et al. (2016). Hayes et al. (2014) generalized the ATOM by extending its scope onto memory, reasoning, and categorization traditionally treated as separate components of human cognition. The exemplar-based account of the relationship between categorization and recognition (Nosofsky et al. 2012) is also rather close to this paradigm.

1.3.3 Psychological Space

In the XX century inner psychophysic originally proposed by Fechner and rejected by his contemporaries was developed as an individual branch of science with its own problems and methods of analysis. In particular, nowadays the following premises are accepted and underlie particular investigations of various phenomena studied within inner psychophysics. These premises also play a crucial role in our further constructions.

Two-system implementation of stimulus perception: As originally posited by Fechner, the stimulus perception is considered to be implemented into two steps:

- the conversion of physical stimuli into neural signals by sense organs (perceptual subsystem) and
- the mental evaluation of these signals based on the central subsystem functioning.

After Fechner, Treisman (1964) seems to be the first who supposed the perceptual and central subsystems to be governed by their *own* laws. Naturally, the two subsystems mutually affect each other (see, e.g., Baird 1997, for are view and the discussion of similar concepts).

Memory effects: Petrov (2003), Petrov and Anderson (2005) accentuated the contribution of human memory to the mental evaluation of stimulus intensity. Namely, they emphasize that the neural image of a physical stimulus is proceeded by the central subsystem involving human memory, which allows for the effects of learning and causes changes in the state of this subsystem from trial to trial. Broadly speaking, their model introduces the *internal* dynamics into stimulus perception via formation and evolution of *anchors*[19] in the content-addressable memory.

Probabilistic properties of stimulus inner representation: Within the given premise it is posited that the uncertainty in evaluating sensory magnitudes is determined mainly by the functioning of the central subsystem. In this case noise in perceived physical stimuli as well as noise appearing in the stimulus transformation in into neural signals by sense organs is of minor importance. Petrov and Anderson (2003, 2005) allow for this effect endowing the relationship between the sensory magnitude M and the physical stimulus intensity S with an additive noise whose amplitude depends on the sensory magnitude. Their model is based on the interpretation of Weber's law (1.1) by Ekman (1959) where the just-noticeable differences ΔS is caused by perceptual noise. This noise additively enters the relationship $S \mapsto M$ with amplitude proportional to the mean value of sensory magnitude,

$$\langle (\delta M)^2 \rangle^{1/2} = \kappa_e \langle M \rangle \,. \tag{1.5}$$

The latter expression is the probabilistic interpretation of Ekman's law (1.4). Based on this relationship Petrov–Anderson model represents the relationship between the sensory magnitude and physical stimulus in the form

$$M = k S^{\beta_s} (1 + \kappa_e \xi) \,. \tag{1.6}$$

Here ξ is a noise of unit amplitude. Assuming the noise contribution to the sensory magnitude to be small and the just-noticeable differences ΔS to be specified by the amplitude of the random fluctuations in the sensory magnitude, Exp. (1.6) immediately gives rise to Weber's law (1.1) with $\kappa_w \sim \kappa_e / \beta_s$. The probabilistic concepts turning, in particular, to the notion of entropy have been used to model psychophysical regularities (e.g., Norwich 1993, 1987, 2005; Nizami 2009).

Psychological space: The concept of psychological space is rooted in the works of Louis Leon Thurstone (1887–1955), American psychophysicist pioneering in psychometrics, who formulated the *law of comparative judgment*. Later this law was converted then into the *law of categorical judgment* by Warren S. Torgerson

[19] In the given description the *anchors* are the associations between the magnitudes of neural images N (or even the intensity S of the original physical stimuli) and the corresponding elements of psychological space established in the previous trials. The use of anchors enhances the functioning of the central subsystem enabling it to skip some steps of its operations implemented previously.

(1924–1999), also American psychophysicist much contributed to psychometrics. A sophisticated mathematical description of psychological space appears for the first time in the works by Robert Duncan Luce (1925–2012), American mathematician and social scientist (e.g., Luce 1959, 1963; Luce and Galanter 1963; Luce and Suppes 1965, also Luce 2013 for a general discussion).

The gist of these constructions is the use of the probabilistic properties of human decision-making in categorizing various external stimuli for constructing the topology of the inner psychological space, whose points are various categories. Exactly the introduction of psychological space with its own properties and governing the final step in stimulus evaluation explains the universality of psychophysical laws for all the sensory modalities (e.g., Shepard 1957, 1958a, b, 1987; Nosofsky 1992). Among the recent publications this universality has been also discussed, e.g., by Dzhafarov and Colonius (2005a, b) and Nosofsky and Palmeri (2015), Winter et al. (2015a, b). In this connection we want to note the Bayesian approach to describing the interaction between the perceptual and central subsystems turning to the formalism of conditional probabilities (Petzschner and Glasauer 2011; Petzschner et al. 2015, for a review). The psychological space can be also introduced axiomatically (e.g., Narens 1996, 2002).

As far as psychological approach is concerned, the a German-American psychologist Kurt Lewin (1890—1947) put forward the concept of psychological force field describing psychological processes in topological spaces (Lewin 1935, 1936, 1938, 1948). Then the American psychologist George Kelly (1905–1967) formulated personal construct psychology, as complete theory of cognition, action, learning and intention, using geometry of psychological spaces as alternative to logic (Kelly 1991).

Naturally, the concept of psychological space is much wider and concerns many other aspects of mental phenomena, for details a ready may be referred to Fauconnier (1985), Eliot (1987), Gärdenfors (2000), Spivey (2007) and a review by Duch (2012).

The premises mentioned above argue for the efficiency of introducing some psychological space with its own properties and own dynamics for describing how physical stimuli are proceeded by the central neural subsystem and how their mental images emerge.

The given statement does not exclude an inevitable incompleteness of the description based on the phenomenological approach because there are some mental phenomena whose explanation requires knowledge about the underlying neurophysiological processes in the brain (e.g., Hurlburt and Schwitzgebel 2007; Schwitzgebel 2011; Duch 2011).

1.4 Psychophysics of Time Perception

Investigation of time perception actually started just after Fechner's works already discussed in Sect. 1.3.1, where we focused our attention on the general laws governing the human perception of external stimuli. Naturally, the time perception obeys these laws but also exhibits its own features to be discussed below.

The Austrian physicist and philosopher Ernst Mach (1838–1916) began experiments to analyze the human perception of time interval duration, in particular, with respect to Weber's law (1.1). Further, the German physiologist Karl von Vierordt (1818–1884) brought into focus the *time sense* as an inner phenomenon of human cognition, i.e., human direct experience of the lapse of time. One of his striking findings called now *Vierordt's law* describes the perceived duration depending on the actual duration of some event and the task complexity. Namely, it states that humans tend to overestimate short intervals and to underestimate long ones. For a detailed historical review of psychology and psychophysics of time a reader may be referred to Roeckelein (2000, 2008), Wearden (2016).

Below let us discuss in detail two issues about the time perception. First, we want to demonstrate that the perception of duration is not reduced only to the duration of physical time interval and also affected by the accompanying mental processes. Thereby the time perception has to be regarded as a phenomenon of inner psychophysics. Second, it is a nonlinear phenomena which poses a doubt about the possibility of treating the subjective time dimension as a collection of the corresponding time intervals.

Here, using the notion of subjective time, we imply the time intervals as they perceived and evaluated in the mind. In this context, the subjective time is opposed to objective time measured using a physical timer.

1.4.1 Time Perception as a Phenomenon of Inner Psychophysics

As far as mechanisms governing the human perception of time are concerned, up to now it is a challenging problem attracted much intention in the second half of the XX century. Since we have no special sense organ enabling us to experience the passage of time directly, there should be some inner processes that generate a sequence of events or a certain rhythm for perceiving time implicitly. As emphasized by the French psychologist Paul Fraisse (1911–1996), it is not time itself but what goes on in time that produces temporal effects and can be recognized by humans (Fraisse 1963, 1984).

Nowadays there are two rival approaches to explaining how humans perceive the passage of time. One of them may be characterized as *automatic*, the other is *cognitively controlled* (for a review see, e.g., van Rijn and Taatgen 2017, as well as Rammsayer 1999; Lewis and Miall 2003 for the corresponding neurophysiological evidence). Using the terms coined by Ivry and Hazeltine (1992) these approached may be referred to as *timing-with-a-timer* and *timing-without-a-timer*, respectively. Here the timer is understood as a certain pacemaker based on some cyclic unconscious process in the body (see also discussion in Block and Zakay 1996; Block 2003).

The French psychologist Marcel François and the American physiologist Hudson Hoagland seem to be the first who related the perception of time to some chemical processes in the body playing the role of internal clocks (François 1927; Hoagland 1933, 1935). However the theory of internal clocks in its modern understanding as timing-with-a-timer is based on two works by Creelman (1962) and Treisman (1963). These works provide an account of how an internal clock like a metronome pulse generator sends quantized ticks to an accumulator that counts these tics, enabling the cognitive comparison of time intervals and other timing tasks. The development of clock models of timing is reviewed, e.g., by Treisman (2013) and a unified description of possible neurophysiological mechanisms responsible for the pacemaker tick generation is presented by Allman et al. (2014). This concept of internal clocks has been also subjected to criticism for its fundamental limitations (e.g., Block 1990, 2003; van Rijn and Taatgen 2017).

Timing-with-a-timer: Nowadays there have been proposed several basic models that turn to Treisman's concept of internal pacemaker. Maybe, the most popular model is the *scalar timing* or *scalar expectancy theory* (SET) initially put forward by Gibbon (1977) for animals (see Gibbon and Church 1981, for quantitative details). Later it became clear that SET can be used also for describing human timing (see., e.g., Allan and Gibbon 1991; Wearden 1991a, b, 2002, as well as Allan 1998 for a review). This model comprises an internal clock as a pacemaker, memory effects, and decision-making processes. The pacemaker generates ticks, a stimulus enables the ticks to enter an accumulator, and the resulting accumulation of ticks representing the elapsed time is continuously sent to the working memory. When the perceived stimulus ends or a reinforcement happens, the final time value is stored in the reference (long-term) memory and can be used in comparison with other time intervals retained in the memory (a detailed description of SET can be found in Wearden 2016, Chap. 3).

The *behavioral theory of timing* (BeT) (Killeen and Fetterman 1988, 1993) focuses its attention mainly on animal timing. Like SET this model accepts the existence of some pacemaker whose ticks are accumulated and, then, retained in the memory. However, in contrast to SET, the behavioral theory assumes the rate of tick generation is not fixed but determined by the characteristic time interval between events of reinforcement sequence, e.g., feeding. In other words, this theory supposes that the external factor—the animal adaptation to the rate of reinforcement—determines the tick duration used in quantifying the perceived time.

Another example is the *learning to time* (LeT) model (Machado 1997; Machado and Keen 1999). Like BeT, it considers the animal timing to be governed by the adaptation and focuses its attention on the dynamics of reinforcement learning—the basic mechanism of this adaption. In contrast to SET, the given model assumes the accumulated ticks not to be retained directly in the reference memory but to be used in forming the sequence of intermediate behavioral states involved in the learning process.

In the *connectionist theory of timing* (Church and Broadbent 1991, 1990; Wearden and Doherty 1995) the internal clock is replaced by a set of oscillators different in frequency. In this case elapsed time is represented by the correlations between the states of these oscillators at the measured time.

For a review of models of timing based on the concept of pacemaker tick generation, a reader may be referred to Grondin (2010), Allman et al. (2014), Wearden (2016).

Timing-without-a-timer and integrative models: Strictly speaking, models dealing with cognitively controlled timing do not exclude the existence of some internal clock playing the role of pacemaker. They just focus their main attention on how an objective duration is subjectively perceived, retained in memory, and cognitively reproduced. In other words, they analyze the top-down influence of human cognition and memory system on time perception. In this sense the previously discussed timer-based models may be characterized as the bottom-up approach (cf. van Rijn 2016).

The *multiple timescale framework* (MTS) (Staddon and Higa 1999; Staddon et al. 2002) gives an example of models even not assuming explicitly the existence of any internal clock. According to MTS, temporal information is extracted from dynamics of the corresponding memory trace, its activation and the following decay (see also Marr 1999). For a detailed discussion of phenomena involving various cognitive processes which argue for the timing-without-a-timer models a ready may be referred to Block (2003). In particular, a subject can select some time interval available in his memory and relevant to a given particular task and, then, use it as a unit to quantify other intervals (see also Block and Zakay 2008; Block et al. 2010, 2016). In this context we also note the model by van Rijn and Taatgen (2008) for a bottom-level single clock combined with a nonlinear underlying time-scaling mechanism at the top of timing system. The given model allows for relatively accurate perception of overlapping time intervals when these intervals are analyzed within different sensory modalities.

There is a clear evidence for a multitude of timing mechanisms (e.g., Hass and Durstewitz 2014, 2016) including, in particular, the cumulative contribution of automatic and cognitively controlled systems of timing (Lewis and Miall 2003; Rammsayer and Ulrich 2005). It makes a generalized theory allowing for both the systems of timing rather attractive. It should be noted that the concept of two systems, one of which is fast and automatic, the other is slow and related to deliberate analysis is rather general and can be used for characterizing various mental processes including decision-making (for a review see, e.g., Barrouillet 2011; Kahneman 2011).

The integrative model of timing (Taatgen et al. 2007; van Rijn et al. 2014; van Rijn 2016; van Rijn and Taatgen 2017) may be regarded as such a generalization of timing description. It consists of several components that *together* explain how time is perceived, compared, and reproduced. This integrative model of timing, first, incorporates the theories of human memory, attention, and decision-making developed in psychology within the Adaptive Control of Thought-Rational (ACT-R) cognitive architecture (e.g., Anderson 2007). Second, an additional module is added to ACT-R which contains a timer and a means to read out and reset the timer. This timer is not independent of cognitive processes in its functioning. In particular, it generates ticks with a rate gradually decreasing with time while accumulating the ticks quantifying the perceived duration of analyzed events. It should be noted that the idea to combine the pacemaker-accumulator model of timing with the concept of clock characterized by nonlinear internal timescales was discussed previously by Church and Deluty (1977) as well as Staddon and Higa (1999) in order to archive Weber's law.

Timing and psychophysical processes: The inclusion of cognitive processes in any account of time perception necessarily poses questions about the influence of such factors as attention, mental load, emotions on timing as well as the relative contribution of conscious and unconscious processes into human timing. Since the second half of XX century there has been a great deal of theoretical and experimental investigations aimed at elucidating these issues and a vast literature representing them from different standpoints is available (an annotated bibliography can be found, e.g., in Roeckelein 2000, 2008). Nevertheless, up to now, there is no one concept satisfying the majority of scholars and the available experimental data are rather intricate or even contradictory. For a review of these problems a reader may be referred to Block and Zakay (2001), Zakay and Block (2004), Brown (2008), Wearden (2016), Zakay (2016). In this section we note only some of them essential for our further constructions.

Concerning the roles of conscious and unconscious factors in human timing the problem of estimating their relative contributions becomes pronounced in comparing the experimental results obtained within two general paradigms. In the *prospective* paradigm, subjects are informed that they will be asked to judge the duration of analyzed intervals and their attention is consciously directed to the passage of time. In contrast, within the *retrospective* paradigm subjects are not aware that they will be required to judge time. These subjects are asked for an unexpected time judgment only after the analyzed events have ended. Therefore prospective judgments involve essential attentiveness to time, whereas retrospective judgments are based on either temporal information previously retained in the memory or unconsciously accumulate during experiments. Duration judgments in the prospective and retrospective paradigms reveal the differences between them in governing mechanisms.

Conscious Factors in Human Timing

The *attentional-gate model* developed by Zakay and Block (1995, 1996, 1997), Block and Zakay (1996) for prospective judgments puts forward the concept of attentional resources as its basic premise (Zakay 1989; Zakay and Block 1996). This concept postulates the existence of a limited pool of attentional resources shared by all cognitive processes (Kahneman 1973; Norman and Bobrow 1975a; Navon and Gopher 1979; Navon 1984). The effects of attention on human timing is widely investigated also within the *dual-task* experiments when participants simultaneously performed temporal and nontemporal tasks (Brown 1997, 2008, for a review). It should be noted that Thomas and Weaver (1975) seem to be the first who described the interplay between temporal and nontemporal information processing.

Conscious factors in their influence on human timing are not reduced to the competition for attentional resources only, they can also intensify cognitive processes as a whole. In particular, the model for timing-with-a-timer by Treisman et al. (1990,

1992) takes into account that arousal enhances the clock pace. Burle and Casini (2001) emphasize that activation of human perception via arousal and distribution of attention over various mental tasks affect time judgment via different mechanisms. A detailed analysis of possible generalizations of timing models allowing for the effects of attention and activation can be found in Lejeune (1998).

Other cognitive factors also affect human timing. For example, the appearance of unexpected event can give rise to the oddball effect. Namely, in a series of stimuli being of the same objective duration, a low probability stimulus (oddball) lasts subjectively longer than a typical stimulus (Rose and Summers 1995; Tse et al. 2004; Pariyadath and Eagleman 2007, including Eagleman 2008; Birngruber et al. 2014 for a review). Emotions also contribute to time perception as demonstrated in a variety of experiments since the second half of XX century. In particular, Langer et al. (1961) analyzed the effect of potential danger on time judgments, Angrilli et al. (1997) tested how time judgment is affected by the degree of the pleasantness or unpleasantness of shown pictures. In the last decades much of researches devoted to a tight coupling between the emotionality of stimuli and their apparent duration were conducted by Sylvie Droit-Volet and her colleagues (e.g., Droit-Volet et al. 2013, 2015; Fayolle et al. 2015; Droit-Volet 2016; Droit-Volet and Gil 2016). A detailed discussion of time perception and emotions can be found in Droit-Volet (2014) and Wearden (2016, Chap. 5). Droit-Volet and Meck (2007) also demonstrated that emotional arousal and valence can cause both increases and decreases in subjective time.

The memory integrity is also essential in human timing endowing it with complex properties, in particular, a possible interference between different time scales. This has been demonstrated directly by Taatgen and van Rijn (2011) studying the correlations between perceived durations of relatively short and long stimuli analyzed within one experiment.

Unconscious Factors in Human Timing

The temporal and nontemporal aspects of analyzed events can be intertwined. So the structure of these events can provided access to their temporal information even in cases when attention is mainly directed to nontemporal features of stimuli. It is the gist of the *contextual-change model* (Block and Reed 1978; Block 1982, 1985, see, also, 1990; 1992) developed to describe duration judgments within retrospective paradigm (for a detailed discussion see, e.g., Block and Gruber 2014). This model assumes that duration judgments are based on contextual information available in memory. The regularity of temporal organization of analyzed stimuli makes retrospective duration judgments rather accurate (Boltz 1995, 1998, 2005; Boltz et al. 1998).

Although retrospective timing is usually considered to be based on unconscious perception of time, cognitive processes indirectly contribute to such time judgments (e.g., Block and Zakay 2001). In particular, it has been demonstrated that distribution of attention between several nontemporal tasks (Brown 1985, 1997; Brown and Stubbs 1992, also Brown 2008 for a review) and emotions (Droit-Volet 2016) can play a significant role in retrospective timing.

Whether it is possible to describe prospective and retrospective timing from a common standpoint on mechanisms governing the corresponding time perception is the subject of debates. For a discussion of arguments for and against the common description of the two types of timing a reader may be referred, e.g., to Wearden (2016), Addyman et al. (2016).

1.4.2 Psychophysical Laws of Time Perception

In this subsection we return to Weber's law and Steven's law already discussed from the general point of view in Sect. 1.3.1. Here we consider their particular implementation for the time perception. As noted previously there is no sense organ enabling us to feel directly the passage of time. So some internal physiological processes playing the role of inner timer have to be involved in quantifying the duration of events. Therefore the general concept of two-transformations implementing human sensation (Sect. 1.3.1) should be modified for the time perception as follows:

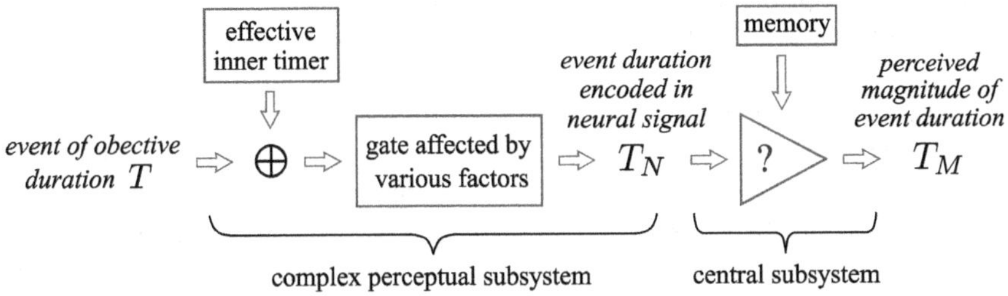

Here the perceptual subsystem comprising an effective inner timer and a possible gate controlling the timer tick accumulation and being affected by various psychophysical factors (attention, emotions, etc.) generates some neural signal. The temporal component of this signal structure bears the information about the external event duration T quantified in timer ticks T_N, i.e., it performs the first transformation $T \to T_N$. At the next step the central subsystem converts the generated neural signal into a mental image whose properties specify the perceived magnitude of event duration T_M. The mechanisms of the second transformation $T_N \to T_M$—forming the subject-matter of inner psychophysics—are rather far from being understood well.

In this subsection we want to present arguments for that the perceived magnitude T_M and the information about the event duration T_N encoded by the perceptual subsystem should be different in property. Thereby they have to be governed by different mechanisms. Actually we will try to justify that the available experimental data exclude the possibility of identifying the values T_N and T_M.

There are several experimental paradigms used to analyze Weber's law in the case of time perception. Their examples are experiments on (*i*) verbal estimation of event duration, (*ii*) reproducing target time intervals, (*iii*) discriminating two time intervals with respect to the relations "shorter," "equal," and "longer." The analyzed time intervals can be just a time gap between two events with or without auditory or/and visual stimuli filling this interval, the duration of a physical stimulus or their sequence, some time scales as objects retained in memory. Besides, the task of time evaluation can be accompanied by other non-temporal tasks requiring a certain redistribution of subject's attention. For a detailed review of these investigations a reader may be referred to Rammsayer and Grondin (2000), Grondin (2003, 2014), Wearden (2016).

In the context of timing, the found variability ΔT of time estimates increases linearly with the duration T of the interval to be timed:

$$\Delta T = \kappa_T T \,. \tag{1.7}$$

Here κ_T is the Weber fraction for the time perception. Equality (1.7) is the implementation of Weber's law (1.1a) for timing and the coefficient κ_T should be constant at least approximately.

The available experimental data argue for a more complex relationship between ΔT and T, e.g., plausible generalizations proposed by Getty (1975), Killeen and Weiss (1987) assume a considerable increase in the Weber fraction $\kappa_T(T)$ in the region of short time intervals. Moreover, Bizo et al. (2006) based on experiments on timing by pigeons and also referring data for humans (Getty 1975) found that the $\kappa_T(T)$-dependence is U-shaped, which poses a doubt about the applicability of Weber's law to the time perception within a wide range of scales. For the sub-second time scales the Weber fraction remains constant with high accuracy, however, is affected essentially by particular conditions of timing. These results enabled (Merchant et al. 2008) to suggest the existence of a partially overlapping *distributed* mechanism underlying the ability to quantify time in different contexts. Summarizing the experimental data for the Weber fraction in time perception, Grondin (2014) argues for the existence of multiple timing systems. Moreover, keeping in mind numerous dual-process interpretations (Evans 2011, 2008, 2003) used in cognitive psychology, Grondin supposes that these systems can be discriminated in properties turning to a generic dichotomy between systematic/analytic, implicit/explicit, unconscious/conscious systems, to name only a few. The timing mechanisms governing the transformations $T \to T_N$ and $T_N \to T_M$ are characterized by this dichotomy and must possess quite distinct properties.

We want to emphasize that the popular scalar expectancy theory (SET) discussed in Sect. 1.4.1 describes the cumulative contributions of perceptual and central subsystems to timing. As a results it bears the characteristic features which can be attributed to both of them individually. Indeed, on one hand, SET describes human perception of time turning to the timing-with-a-timer concept. On the other hand, it employs the scaling idea that a person to evaluate the duration of an event selects a certain time interval as a unit for some reasons, which is not reduced to pure physiology (for details see Wearden et al. 1997; Wearden 2016).

If Weber's law actually concerns the perception threshold (fuzzy threshold) in discriminating time intervals, Stevens' law for the time perception (cf. Exp. 1.3)

$$T_M = k\, T^{\beta_T} \,. \tag{1.8}$$

describes the scaling relationship between the objective duration T and its mental evaluation T_M. There are generalizations of this power-law relationship, in particular, the parallel-clock model (Eisler 1975, see also Eisler 2003) takes into account that time is unstoppable. Therefore, two time registers have to be used, e.g., in reproducing stimulus duration. One accumulates ticks of inner timer from the moment when an analyzed stimulus appears to the end of its reproduction; the other accumulates the ticks only during the stimulus reproduction. Then the ratio of objective durations is converted into the ratio of perceived durations, which is characterized by the exponent β_T. As before, this tick accumulation may be assume to be governed by mechanisms of neurophysiological level, whereas the ratio conversion involves consciousness.

As in the case of Weber's law complex structure of effective perceptual subsystem affected also by various psychological factors is responsible for significant scattering in the experimental data for the Stevens exponent β_T. Eisler (1976), Allan (1983) reviewed many experimental data-sets and demonstrated that the average value $\langle\beta_T\rangle$ of the Stevens exponent is less but rather close to $\beta_T = 1$, however, for different individuals and particular experimental conditions the value of β_T can show considerable scattering in the region [0.5:1.3]. For a recent discussion of this issue a reader may be referred to García-Pérez (2014), Grondin and Laflamme (2015).

In this context another regularity of time perception is worthy of noting; its the law of *stimulus bisection*. Generally within the classical bisection task subjects are asked to classify a trial stimulus with respect to its proximity to one of two given stimuli from the same perceptual continuum. Speaking about the time perception this task matches the *time bisection problem* concerning the stimulus whose duration T_M^m is perceived as the mean value of the perceived durations T_{M1} and T_{M2} of other two stimuli. The time bisection problem was analyzed with respect to various aspects, e.g., by Wearden (1991b), Allan and Gibbon (1991), Wearden and Ferrara (1995), Kopec and Brody (2010); for a detailed review a reader may be addressed to Levy et al. (2015), Wearden (2016).

A meta-analysis of classification data (Kopec and Brody 2010) reports that the bisection point lies near the geometric mean for ratios of referent duration about 2 or less and lies near the arithmetic mean for ratios of 4 or greater. As far as a theoretical explanation of the geometric-mean-law is concerned, Luce (1961), also Luce and Galanter (1963) seems to be the first who supposed that the given law stems from the fact that in mind we compare the ratios of stimulus durations rather than their absolute values. Thereby the time bisection may be regarded to be governed completely by the regularities of inner psychophysics.

In the present subsection we have demonstrated that time perception is governed by mechanisms of two types. The mechanisms of the first type are responsible for encoding the information about the duration of perceived stimuli in the neurophysiological signals. Nowadays, these mechanisms are more or less understood and the accumulation of ticks generated by inner timers during analyzed events is one of the dominant points of view on how the temporal information is encoded. The other type mechanisms govern processing this information in the brain. However, how this information is processed at the level of the central neural network is poorly understood up to now. It makes describing the second type mechanisms within the phenomenological approach attractive and it is actually incorporated, explicitly or implicitly, in the developed models for time perception.

The gist of phenomenological approach is as follows. Processing neural signals bearing the information about external stimuli, the brain creates their representations as individual objects with their own properties. These representations have physiological and mental aspects. Whether the mental aspects are reducible to the physiological aspects is a matter of long-term debates, nevertheless, it is widely accepted that they are tightly interconnected. Metaphorically speaking, the mental aspects of such a representation—the mental image of perceived stimulus—and its physiological aspects—the underlying neurophysiological state of the brain—are two facets of one object. The properties of one facet must be reflected in the properties of the other facet and vice verse. Thereby analyzing one of its faces we may hope to elucidate the properties of this object as a whole entity.

Mental images are accessible to the mind. So studying the stimulus perception from the first-person point of view—the gist of phenomenology—can open a gate to understanding the basic regularities of how these images emerge and evolve in the mind as well as how we operate with them. Mathematical implementation of this approach is the main subject-matter of the present book.

1.5 Conclusion

In this chapter we discussed various aspects of time problem which highlight the complexity of questions we face up to when trying to penetrate deeper into the properties of time and related phenomena. The presented analysis of particular issues being in focus during the past centuries makes it apparent that up to now we do not clearly discriminate between objective and subjective time in their basic properties. Therefore, from out point of view, as the first step we need to recognize that there are two temporal dimensions which should be considered separately. They are:

- objective time, i.e., time of the inanimate world admitting precise measurements and
- human temporal dimension, i.e., human inner time which is not accessible to the external observer.

Because the notion of subjective (human) time with point-like instants is ill-defined, in the present book we introduce the notion of human temporality as the temporal dimension of the mind together with its temporal integrity. Below we will discuss this term in more detail including its use in the classical framework of phenomenology.

We also draw the conclusion that human temporality is related but not reduced to objective time. Human temporality possesses its own properties, for example, may be characterized by non-stationary rate of flow, a branching structure, dependence on external conditions, etc. As a result, in order to describe temporal phenomena in the mind and human actions a novel mathematical formalism in required. This formalism has to deal with notions and objects that cannot be introduced within the classical scope of physics.

Three aspects discussed in this chapter should be emphasized individually because they impact on our further constructions.

- What is the present (the *now*—something treated as the current instant of time) is a challenging problem. It is clear, at least, that the *now* in physical reality and the *now* in the mind are quite distinct entities and deserve individual consideration. Speaking about the inner world of human being its now is a complex entity with its internal structure and own properties sensitive to what is going on in the external world.
- The human now is of finite extent and overlaps with the past and the future in the mind. At least, in describing human actions all the three temporal parts should be integrated in some way.
- There are two basic concepts of embedding in time flow, they are represented by the A-theories and B-theories of time. The A-type relations (*before-now-after*) are met in mental phenomena and seem to be inapplicable to the physical world. The B-type relations (*earlier-later*) are typical for the physical world and are likely to be met in the inner world of human being too.

References

Adams, R. M. (1986). Time and thisness. *Midwest Studies in Philosophy, 11*(1), 315–329.

Addyman, C., French, R. M., & Thomas, E. (2016). Computational models of interval timing. *Current Opinion in Behavioral Sciences,8*, 140–146. Time in perception and action.

Adler, M., Mayo, A., & Alon, U. (2014). Logarithmic and power law input-output relations in sensory systems with fold-change detection. *PLOS Computational Biology,10*(8), e1003781 (14 pp).

Ainslie, D., & Ware, O. (2014). Consciousness and personal identity. In A. Garrett (Ed.), *The Routledge handbook of eighteenth century philosophy* (pp. 245–264). London: Routledge Taylor & Francis Group.

Ainsworth, T. (2016). Form vs. matter. In E. N. Zalta (Ed.), *The Stanford encyclopedia of philosophy* (Spring 2016 ed.). Stanford: Metaphysics Research Lab, Stanford University.

Alexander, S. (1921). *Spinoza and time*. London: Geoge Allen & Unwin Ltd.

Allan, L. G. (1983). Magnitude estimation of temporal intervals. *Perception & Psychophysics, 33*(1), 29–42.

Allan, L. G. (1998). The influence of the scalar timing model on human timing research. *Behavioural Processes, 44*(2), 101–117.

Allan, L. G., & Gibbon, J. (1991). Human bisection at the geometric mean. *Learning and Motivation, 22*(1), 39–58.

Allman, M. J., Teki, S., Griffiths, T. D., & Meck, W. H. (2014). Properties of the internal clock: First- and second-order principles of subjective time. *Annual Review of Psychology, 65*(1), 743–771.

Andersen, H. (2014). The development of the "specious present" and James's views on temporal experience. In V. Arstila & D. Lloyd (Eds.), *Subjective time: The philosophy, psychology, and neuroscience of temporality* (pp. 25–42). Cambridge: The MIT Press.

Andersen, H., & Grush, R. (2009). A brief history of time-consciousness: Historical precursors to James and Husserl. *Journal of the History of Philosophy, 47*(2), 277–307.
Anderson, J. R. (2007). *How can the human mind occur in the physical universe?* New York, NY: Oxford University Press.
Angrilli, A., Cherubini, P., Pavese, A., & Manfredini, S. (1997). The influence of affective factors on time perception. *Perception & Psychophysics, 59*(6), 972–982.
Anonymous (E. Robert Kelly). (1882). *The alternative: A study in psychology*. London: Macmillan and Co.
Ariotii, P. E. (1975). The concept of time in western antiquity. In T. Fraser & N. Lawrence (Eds.), *The Study of Time II, Proceedings of the Second Conference of the International Society for the Study of Time Lake Yamanaka—Japan* (pp. 69–80). Berlin: Springer.
Arthur, R. T. W. (1985). Leibniz's theory of time. In K. Okruhlik & J. R. Brown (Eds.), *The natural philosophy of Leibniz* (pp. 263–313). Dordrecht: D. Reidel Publishing Company.
Augustine, S. (1991). *Saint Augustine confessions (Oxford world's classics)*. New York: Oxford University Press Inc. Translated with an Introduction and notes by Henry Chadwick.
Baird, J. C. (1975a). Psychophysical study of numbers: IV. Generalized preferred state theory. *Psychological Research,38*(2), 175–187.
Baird, J. C. (1975b). Psychophysical study of numbers: V. Preferred state theory of matching functions. *Psychological Research,38*(2), 189–207.
Baird, J. C. (1997). *Sensation and judgment: Complementarity theory of psychophysics*. Mahwah, NJ: Erlbaum.
Baird, J. C., & Noma, E. (1975). Psychophysical study of numbers: I. Generation of numerical responses. *Psychological Research, 37*(4), 281–297.
Baird, J. C., & Noma, E. (1978). *Fundamentals of scaling and psychophysics*. New York: Wiley.
Baker, L. R. (2007). *The metaphysics of everyday life: An essay in practical realism*. Cambridge, UK: Cambridge University Press.
Balashov, Y. (2008). Persistence and multilocation in spacetime. In D. Dieks (Ed.), *The ontology of spacetime II* (pp. 59–81). Amsterdam: Elsevier B.V.
Balashov, Y. (2011). Persistence. In C. Callender (Ed.), *The Oxford handbook of philosophy of time* (pp. 13–40). Oxford: Oxford University Press.
Balashov, Y., & Janssen, M. (2003). Presentism and relativity. *British Journal for the Philosophy of Science, 54*(2), 327–346.
Ballard, K. E. (1960). Leibniz's theory of space and time. *Journal of the History of Ideas, 21*(1), 49–65.
Bardon, A. (2013). *A brief history of the philosophy of time*. New York: Oxford University Press.
Baron, S. (2015). The priority of the now. *Pacific Philosophical Quarterly, 96*(3), 325–348.
Barrouillet, P. (2011). Dual-process theories and cognitive development: Advances and challenges. *Developmental Review,31*(2–3), 79–85. Special issue: Dual-process theories of cognitive development.
Baugh, B. (2010). Time, duration and eternity in Spinoza. *Comparative and Continental Philosophy,2*(2), 211–233.
Beauchamp, T. L. (1974). Hume on causal contiguity and succession. *Dialogue, 13*(2), 271–282.
Belnap, N. (1992). Branching space-time. *Synthese, 92*(3), 385–434.
Bergson, H. (1910/2001). *Time and free will: An essay of the immediate data of consciousness*. New York: Dover Publications, Inc. Translation by F. L. Pogson, M.A.
Bigelow, J. (2013). The emergence of a new family of theories of time. In H. Dyke & A. Bardon (Eds.), *A companion to the philosophy of time* (pp. 151–166). Chichester, West Sussex, UK: Wiley-Blackwell, Wiley.
Billock, V. A., & Tsou, B. H. (2011). To honor Fechner and obey Stevens: Relationships between psychophysical and neural nonlinearities. *Psychological Bulletin, 137*(1), 1–18.
Birngruber, T., Schröter, H., & Ulrich, R. (2014). Duration perception of visual and auditory oddball stimuli: Does judgment task modulate the temporal oddball effect? *Attention, Perception, & Psychophysics, 76*(3), 814–828.

Bizo, L. A., Chu, J. Y. M., Sanabria, F., & Killeen, P. R. (2006). The failure of Weber's law in time perception and production. *Behavioural Processes,71*(2), 201–210. Interval timing: The current status.

Blattner, W. (2005). Temporality. In H. L. Dreyfus & M. A. Wrathall (Eds.), *A companion to Heidegger (Blackwell companions to philosophy)* (pp. 311–324). Oxford: Blackwell Publishing Ltd.

Block, R. A. (1982). Temporal judgments and contextual change. *Journal of Experimental Psychology: Learning, Memory, and Cognition, 8*(6), 530–544.

Block, R. A. (1985). Contextual coding in memory: Studies of remembered duration. In J. A. Michon & J. L. Jackson (Eds.), *Time, mind, and behavior* (pp. 169–178). Berlin: Springer.

Block, R. A. (1990). Models of psychological time. In R. A. Block (Ed.), *Cognitive models of psychological time* (pp. 1–35). Hillsdale, NJ: Lawrence Erlbaum Associates, Inc.

Block, R. A. (1992). Prospective and retrospective duration judgment: The role of information processing and memory. In F. Macar, V. Pouthas, & W. J. Friedman (Eds.), *Time, action and cognition: Towards bridging the gap* (pp. 141–152). Dordrecht: Springer-Science+Business Media, B.V.

Block, R. A. (2003). Psychological timing without a timer: The roles of attention and memory. In H. Helfrich (Ed.), *Time and mind II: Information processing perspectives* (pp. 41–60). Cambridge, MA: Hogrefe & Huber.

Block, R. A., & Gruber, R. P. (2014). Time perception, attention, and memory: A selective review. *Acta Psychologica,149*(Supplement C), 129–133. Including special section articles of Temporal processing within and across senses - Part-2.

Block, R. A., & Reed, M. A. (1978). Remembered duration: Evidence for a contextual-change hypothesis. *Journal of Experimental Psychology: Human Learning and Memory, 4*(6), 656–665.

Block, R. A., & Zakay, D. (1996). Models of psychological time revisited. In H. Helfrich (Ed.), *Time and mind* (pp. 171–195). Cambridge, MA: Hogrefe & Huber.

Block, R. A., & Zakay, D. (2001). Retrospective and prospective timing: Memory, attention, and consciousness. In C. Hoerl & T. McCormack (Eds.), *Time and memory: Issues in philosophy and psychology* (pp. 59–76). Oxford, UK: Oxford University Press.

Block, R. A., & Zakay, D. (2008). Timing and remembering the past, the present, and the future. In S. Grondin (Ed.), *Psychology of time* (pp. 368–394). Bingley: Emerald Group Publishing Ltd.

Block, R. A., Hancock, P. A., & Zakay, D. (2010). How cognitive load affects duration judgments: A meta-analytic review. *Acta Psychologica, 134*(3), 330–343.

Block, R. A., Hancock, P. A., & Zakay, D. (2016). Physical load affects duration judgments: A meta-analytic review. *Acta Psychologica,165*(Supplement C), 43–47.

Boccardi, E. (2016). Recent trends in the philosophy of time: An introduction to time and reality I. *Manuscrito: Revista Internacional de Filosofía,39*(4), 5–34.

Boccardi, E. (2017). The passage of time and its enemies: An introduction to time and reality II. *Manuscrito: Revista Internacional de Filosofía,40*(1), 5–41.

Boltz, M. G. (1995). Effects of event structure on retrospective duration judgments. *Perception & Psychophysics, 57*(7), 1080–1096.

Boltz, M. G. (1998). The processing of temporal and nontemporal information in the remembering of event durations and musical structure. *Journal of Experimental Psychology: Human Perception and Performance, 24*(4), 1087–1104.

Boltz, M. G. (2005). Duration judgments of naturalistic events in the auditory and visual modalities. *Perception & Psychophysics, 67*(8), 1362–1375.

Boltz, M. G., Kupperman, C., & Dunne, J. (1998). The role of learning in remembered duration. *Memory & Cognition, 26*(5), 903–921.

Brentano, F. (1874/1995). *Psychology from an empirical standpoint*. London: Routledge. Translated by Antos C. Rancurello, D.B. Terrell and Linda L. McAlister.

Broad, C. D. (1923). *Scientific thought*. London: Routledge & Kegan Paul.

Brown, C. (2006). *Peter Strawson*. Stocksfield: Acumen Publishing Ltd.

Brown, S. W. (1985). Time perception and attention: The effects of prospective versus retrospective paradigms and task demands on perceived duration. *Perception & Psychophysics, 38*(2), 115–124.
Brown, S. W. (1997). Attentional resources in timing: Interference effects in concurrent temporal and nontemporal working memory tasks. *Perception & Psychophysics, 59*(7), 1118–1140.
Brown, S. W. (2008). Time and attention: Review of the literature. In S. Grondin (Ed.), *Psychology of time* (pp. 111–138). Bingley: Emerald Group Publishing Ltd.
Brown, S. W., & Stubbs, D. A. (1992). Attention and interference in prospective and retrospective timing. *Perception, 21*(4), 545–557.
Bueti, D., & Walsh, V. (2009). The parietal cortex and the representation of time, space, number and other magnitudes. *Philosophical Transactions of the Royal Society B: Biological Sciences, 364*(1525), 1831–1840.
Burle, B., & Casini, L. (2001). Dissociation between activation and attention effects in time estimation: Implications for internal clock models. *Journal of Experimental Psychology: Human Perception and Performance, 27*(1), 195–205.
Campbell, R. (2009). A process-based model for an interactive ontology. *Synthese, 166*(3), 453–477.
Cantlon, J. F., Cordes, S., Libertus, M. E., & Brannon, E. M. (2009). Comment on "Log or linear? Distinct intuitions of the number scale in Western and Amazonian indigene cultures". *Science,323*(5910), 38–38.
Church, R. M., & Broadbent, H. A. (1990). Alternative representations of time, number, and rate. *Cognition,37*(1), 55–81. Special issue Animal Cognition.
Church, R. M., & Broadbent, H. A. (1991). A connectionist model of timing. In M. L. Commons, S. Grossberg, & J. E. R. Staddon (Eds.), *Neural network models of conditioning and action: Quantitative analyses of behavior* (pp. 225–240). Hillsdale, NJ: Lawrence Erlbaum Associates.
Church, R. M., & Deluty, M. Z. (1977). Bisection of temporal intervals. *Journal of Experimental Psychology: Animal Behavior Processes, 3*(3), 216–228.
Cleve, J. V. (1999). *Problems from Kant*. New York: Oxford University Press Inc.
Colman, A. M. (2015). *A dictionary ofpsychology*. Oxford reference (4th ed.). Oxford, UK: Oxford University Press.
Cook, M. L. (1967). The power law as a special case of Fechner's law. *Perceptual and Motor Skills, 25*(1), 51–52.
Coope, U. (2005). *Time for aristotle: Physics IV.10–14*. Oxford, UK: Clarendon Press, Oxford University Press.
Coope, U. (2009a). Aristotle: Time and change. In R. L. Poidevin, P. Simons, A. McGonigal, & R. P. Cameron (Eds.), *The Routledge companion to metaphysics* (pp. 39–47). London: Routledge Taylor & Francis Group.
Coope, U. (2009b). Change and its relation to actuality and potentiality. In G. Anagnostopoulos (Ed.), *A companion to aristotle* (pp. 277–291). Chichester, West Sussex: Wiley-Blackwell.
Copelli, M., Roque, A. C., Oliveira, R. F., & Kinouchi, O. (2002). Physics of psychophysics: Stevens and Weber-Fechner laws are transfer functions of excitable media. *Physical Review E,65*(6), 060901 (R, 4 pp).
Cover, J. A. (1997). Non-basic time and reductive strategies: Leibniz's theory of time. *Studies in History and Philosophy of Science Part A, 28*(2), 289–318.
Craig, W. L. (2000a). *The tensed theory of time: A critical examination*. Dordrecht: Springer Science+Business Media B.V.
Craig, W. L. (2000b). *The tenseless theory of time: A critical examination*. Dordrecht: Springer Science+Business Media B.V.
Craig, W. L. (2001). *Time and the metaphysics of relativity*. Dordrecht: Springer Science+Business Media B.V.
Creelman, C. D. (1962). Human discrimination of auditory duration. *The Journal of the Acoustical Society of America, 34*(5), 582–593.
Cross, D. V. (1973). Sequential dependencies and regression in psychophysical judgments. *Perception & Psychophysics, 14*(3), 547–552.
Dainton, B. (2010). *Time and space* (2nd ed.). Durham: Acumen.

Dainton, B. (2017). Temporal consciousness. In E. N. Zalta (Ed.), *The Stanford encyclopedia of philosophy* (Winter 2018 ed.). Stanford: Metaphysics Research Lab, Stanford University.

de Warren, N. (2009). *Husserl and the promise of time: Subjectivity in transcendental phenomenology*. Cambridge: Cambridge University Press.

Dehaene, S., Izard, V., Spelke, E., & Pica, P. (2008). Log or linear? Distinct intuitions of the number scale in Western and Amazonian indigene cultures. *Science, 320*(5880), 1217–1220.

Denhigh, K. G. (1981). *Three concepts of time*. Berlin: Springer.

Dicker, G. (2004). *Kant's theory of knowledge: An analytical introduction*. New York: Oxford University Press.

Dicker, G. (2013). Transcendental arguments and temporal experience. In H. Dyke & A. Bardon (Eds.), *A companion to the philosophy of time* (pp. 410–431). Chichester, West Sussex, UK: Wiley-Blackwell, Wiley.

DiSalle, R. (2009). Space and time: Inertial frames. In E. N. Zalta (Ed.), *The Stanford encyclopedia of philosophy* (Winter 2016 ed.). Stanford: Metaphysics Research Lab, Stanford University.

Dormal, V., & Pesenti, M. (2012). Processing magnitudes within the parietal cortex. In A. Costa & E. Villalba (Eds.), *Horizons in neuroscience research* (Vol. 8, pp. 107–140). New York: Nova Science Publishers Inc.

Dowden, B. (2001). Time. In J. Fieser & B. Dowden (Eds.), *The internet encyclopedia of philosophy*. With ongoing update.

Droit-Volet, S. (2014). What emotions tell us about time. In V. Arstila & D. Lloyd (Eds.), *Subjective time: The philosophy, psychology, and neuroscience of temporality* (pp. 477–506). Cambridge: The MIT Press.

Droit-Volet, S. (2016). Emotion and implicit timing. *PLOS ONE,11*(7), e0158474 (15 pp).

Droit-Volet, S., & Gil, S. (2016). The emotional body and time perception. *Cognition and Emotion,30*(4), 687–699. PMID: 25817441.

Droit-Volet, S., & Meck, W. H. (2007). How emotions colour our perception of time. *Trends in Cognitive Sciences, 11*(12), 504–513.

Droit-Volet, S., Fayolle, S., Lamotte, M., & Gil, S. (2013). Time, emotion and the embodiment of timing. *Timing & Time Perception,1*(1), 99–126.

Droit-Volet, S., Wearden, J. H., & Zélanti, P. S. (2015). Cognitive abilities required in time judgment depending on the temporal tasks used: A comparison of children and adults. *The Quarterly Journal of Experimental Psychology, 68*(11), 2216–2242.

Duch, W. (2011). Neurodynamics and the mind. In *Proceedings of the International Joint Conference on Neural Networks* (pp. 3227–3234). San Jose, CA: IEEE Press.

Duch, W. (2012). Mind-brain relations, geometric perspective and neurophenomenology. *American Philosophical Association Newsletter, 12*(1), 1–7.

Duncan, S. (2017). Thomas Hobbes. In E. N. Zalta (Ed.), *The Stanford encyclopedia of philosophy* (Summer 2017 ed.). Stanford: Metaphysics Research Lab, Stanford University.

Dunlop, K. (2009). "The unity of time's measure": Kant's reply to Locke. *Philosophers' Imprint, 9*(4), 1–31.

Dyke, H. (2003). Temporal language and temporal reality. *The Philosophical Quarterly, 53*(212), 380–391.

Dyke, H., & Bardon, A. (Eds.). (2013). *A companion to the philosophy of time*. Chichester, West Sussex, UK: Wiley-Blackwell, Wiley.

Dzhafarov, E. N., & Colonius, H. (2005a). Psychophysics without physics: A purely psychological theory of Fechnerian scaling in continuous stimulus spaces. *Journal of Mathematical Psychology,49*(1), 1–50.

Dzhafarov, E. N., & Colonius, H. (2005b). Psychophysics without physics: Extension of Fechnerian scaling from continuous to discrete and discrete-continuous stimulus spaces. *Journal of Mathematical Psychology,49*(2), 125–141.

Eagleman, D. M. (2008). Human time perception and its illusions. *Current Opinion in Neurobiology,18*(2), 131–136. Cognitive neuroscience.

Eisler, H. (1975). Subjective duration and psychophysics. *Psychological Review, 82*(6), 429–450.

Eisler, H. (1976). Experiments on subjective duration 1868–1975: A collection of power function exponents. *Psychological Bulletin, 83*(6), 1154–1171.

Eisler, H. (2003). The parallel-clock model: A tool for quantification of experienced duration. In R. Buccheri, M. Saniga, & W. M. Stuckey (Eds.), *The nature of time: Geometry, physics and perception* (pp. 19–26). Dordrecht: Springer Science+Business Media.

Ekman, G. (1959). Weber's law and related functions. *The Journal of Psychology, 47*(2), 343–352.

Ekman, G. (1964). Is the power law a special case of Fechner's law? *Perceptual and Motor Skills, 19*(3), 730.

Ekman, G., & Sjöberg, L. (1965). Scaling. *Annual Review of Psychology, 16*(1), 451–474.

Eliot, J. (1987). *Models of psychological space: Psychometric, developmental, and experimental approaches (with contributions by Heinrich Stumpf)*. New York: Springer.

Evans, J. S. B. T. (2003). In two minds: Dual-process accounts of reasoning. *Trends in Cognitive Sciences, 7*(10), 454–459.

Evans, J. S. B. T. (2008). Dual-processing accounts of reasoning, judgment, and social cognition. *Annual Review of Psychology, 59*, 255–278.

Evans, J. S. B. T. (2011). Dual-process theories of reasoning: Contemporary issues and developmental applications. *Developmental Review,31*(2–3), 86–102. Special issue: Dual-process theories of cognitive development.

Falcon, A. (2013). Aristotle on time and change. In H. Dyke & A. Bardon (Eds.), *A companion to the philosophy of time* (pp. 47–58). Chichester, West Sussex, UK: Wiley-Blackwell, Wiley.

Farr, M. (2012). On a- and b-theoretic elements of branching spacetimes. *Synthese, 188*(1), 85–116.

Farr, M. (2020a). C-theories of time: On the adirectionality of time. *Philosophy Compass*. To be published.

Farr, M. (2020b). Causation and time reversal. *The British Journal for the Philosophy of Science,71*(1), 177–204. Published online 23 June 2018.

Fauconnier, G. (1985). *Mental spaces: Aspects of meaning construction in natural language*. Cambridge, Massachusetts: The MIT Press.

Fayolle, S., Gil, S., & Droit-Volet, S. (2015). Fear and time: Fear speeds up the internal clock. *Behavioural Processes,120*(Supplement C), 135–140.

Ferrarin, A. (2001). *Hegel and Aristotle*. Cambridge, UK: Cambridge University Press.

Forrest, P. (2004). The real but dead past: A reply to Braddon-Mitchell. *Analysis, 64*(4), 358–362.

Forrest, P. (2008). General facts, physical necessity, and the metaphysics of time. In D. W. Zimmerman (Ed.), *Oxford studies in metaphysics* (Vol. 2, pp. 137–152). New York: Oxford University Press.

Fraisse, P. (1963). *The psychology of time*. New York: Harper & Row. (Originally published in French, *Psychologie du temps*, Paris, 1957).

Fraisse, P. (1984). Perception and estimation of time. *Annual Review of Psychology, 35*(1), 1–37.

François, M. (1927). Contribution à l'étude du sens du Temps. La température interne comme facteur de variation de l'appréciation subjective des durées. *L'année psychologique, 28*(1), 186–204.

Futch, M. J. (2002). Supervenience and (non-modal) reductionism in Leibniz's philosophy of time. *Studies in History and Philosophy of Science Part A, 33*(4), 793–810.

Futch, M. J. (2008). *Leibniz's metaphysics of time and space*. New York: Springer Science + Business Media B.V.

Gallagher, S., & Zahavi, D. (2012). *The phenomenological mind* (2nd ed.). London: Routledge, Taylor & Francis Group.

Gallois, A. (2016). Identity over time. In E. N. Zalta (Ed.), *The Stanford encyclopedia of philosophy* (Winter 2016 ed.). Stanford: Metaphysics Research Lab, Stanford University.

García-Pérez, M. A. (2014). Does time ever fly or slow down? The difficult interpretation of psychophysical data on time perception. *Frontiers in Human Neuroscience,8*, Article 415 (19 pp).

Gärdenfors, P. (2000). *Conceptual spaces: The geometry of thought*. Cambridge, Massachusetts: The MIT Press.

Gasser, G., & Stefan, M. (Eds.). (2012). *Personal identity: Simple or complex?* Cambridge, UK: Cambridge University Press.

Gescheider, G. A. (1997). *Psychophysics: The fundamentals* (3rd ed.). Mahwah, NJ: Lawrence Erlbaum Associates.
Getty, D. J. (1975). Discrimination of short temporal intervals: A comparison of two models. *Perception & Psychophysics, 18*(1), 1–8.
Gibbon, J. (1977). Scalar expectancy theory and Weber's law in animal timing. *Psychological Review, 84*(3), 279–325.
Gibbon, J., & Church, R. M. (1981). Time left: Linear versus logarithmic subjective time. *Journal of Experimental Psychology: Animal Behavior Processes, 7*(2), 87–108.
Gilmore, C. (2008). Persistence and location in relativistic spacetime. *Philosophy Compass, 3*(6), 1224–1254.
Goranko, V., & Galton, A. (2015). Temporal logic. In E. N. Zalta (Ed.), *The Stanford encyclopedia of philosophy* (Winter 2015 ed.). Stanford: Metaphysics Research Lab, Stanford University.
Graham, D. W. (2015). Heraclitus. In E. N. Zalta (Ed.), *The Stanford encyclopedia of philosophy* (Fall 2015 ed.). Stanford: Metaphysics Research Lab, Stanford University.
Grondin, S. (2003). Studying psychological time with Weber's law. In R. Buccheri, M. Saniga, & W. M. Stuckey (Eds.), *Proceedings of the NATO Advanced Research Workshop on the Nature of Time: Geometry, Physics and Perception* (pp. 33–41). Dordrecht: Springer-Science+Business Media, B.V.
Grondin, S. (2010). Timing and time perception: A review of recent behavioral and neuroscience findings and theoretical directions. *Attention, Perception, & Psychophysics, 72*(3), 561–582.
Grondin, S. (2014). About the (non)scalar property for time perception. In H. Merchant & V. de Lafuente (Eds.), *Neurobiology of interval timing* (pp. 17–32). New York: Springer Science+Business Media.
Grondin, S. (2016). *Psychology of perception*. Switzerland: Springer International Publishing.
Grondin, S., & Laflamme, V. (2015). Stevens's law for time: A direct comparison of prospective and retrospective judgments. *Attention, Perception, & Psychophysics, 77*(4), 1044–1051.
Hass, J., & Durstewitz, D. (2014). Neurocomputational models of time perception. In H. Merchant & V. de Lafuente (Eds.), *Neurobiology of interval timing*. Advances in experimental medicine and biology (Vol. 829, pp. 49–71). New York, NY: Springer Science+Business Media.
Hass, J., & Durstewitz, D. (2016). Time at the center, or time at the side? Assessing current models of time perception. *Current Opinion in Behavioral Sciences,8*(Supplement C), 238–244. Time in perception and action.
Hawking, S. (1988). *A brief history of time: From big bang to black holes*. London: Bantam Press.
Hawking, S., & Mlodinow, L. (2005). *A briefer history of time*. London: Bantam Press.
Hawley, K. (2009). Metaphysics and relativity. In R. L. Poidevin, P. Simons, A. McGonigal, & R. P. Cameron (Eds.), *The Routledge companion to metaphysics* (pp. 507–516). London: Routledge Taylor & Francis Group.
Hawley, K. (2020). Temporal parts. In E. N. Zalta (Ed.), *The Stanford encyclopedia of philosophy* (Summer 2020 ed.). Stanford, CA: Metaphysics Research Lab, Stanford University. https://plato.stanford.edu/entries/temporal-parts/.
Hayes, B. K., Heit, E., & Rotello, C. M. (2014). Memory, reasoning, and categorization: Parallels and common mechanisms. *Frontiers in Psychology,5*, Article 529 (9 pp).
Heidegger, M. (1927/1996). *Being and time*. Albany: State University of New York. A translation of Sein und Zeit by Joan Stambaugh.
Heidegger, M. (1985). *History of the concept of time*. Bloomington: Indiana University Press. Translated by Theodore Kisiel.
Hershenov, D. B. (2012). Personal identity. In N. A. Manson & R. W. Barnard (Eds.), *The continuum companion to metaphysics* (pp. 198–222). New York: Continuum International Publishing Group.
Hestevold, H. S. (2008). Presentism: Through thick and thin. *Pacific Philosophical Quarterly, 89*(3), 325–347.
Hoagland, H. (1933). The physiological control of judgments of duration: Evidence for a chemical clock. *The Journal of General Psychology, 9*(2), 267–287.

Hoagland, H. (1935). *Pacemakers in relation to aspects of behavior*. New York: The Macmillan Company.
Hollingworth, H. L. (1910). The central tendency of judgment. *The Journal of Philosophy, Psychology and Scientific Methods, 7*(17), 461–469.
Hoy, R. C. (2013). Heraclitus and Parmenides. In H. Dyke & A. Bardon (Eds.), *A companion to the philosophy of time* (pp. 9–29). Chichester, West Sussex, UK: Wiley-Blackwell, Wiley.
Huggett, N., & Hoefer, C. (2015). Absolute and relational theories of space and motion. In E. N. Zalta (Ed.), *The Stanford encyclopedia of philosophy* (Spring 2018 ed.). Stanford: Metaphysics Research Lab, Stanford University.
Hurlburt, R. T., & Schwitzgebel, E. (2007). *Describing inner experience? Proponent meets skeptic*. Cambridge, Massachusetts: The MIT Press.
Hussey, E. (1993). *Aristotle physics books Ill and IV: Translated with introduction and notes*. Oxford: Oxford University Press.
Ivry, R. B., & Hazeltine, R. E. (1992). Introduction: Models of timing-with-a-timer. In F. Macar, V. Pouthas, & W. J. Friedman (Eds.), *Time, action and cognition: Towards bridging the gap* (pp. 183–189). Dordrecht: Springer-Science+Business Media, B.V.
James, W. (1890). *The principles of psychology* (Vol. 1). New York: Henry Holt and Company.
Janiak, A. (2016). Kant's views on space and time. In E. N. Zalta (Ed.), *The Stanford encyclopedia of philosophy* (Winter 2016 ed.). Stanford: Metaphysics Research Lab, Stanford University.
Johnson, K. O., Hsiao, S. S., & Yoshioka, T. (2002). Review: Neural coding and the basic law of psychophysics. *The Neuroscientist, 8*(2), 111–121.
Kadosh, R. C., Lammertyn, J., & Izard, V. (2008). Are numbers special? An overview of chronometric, neuroimaging, developmental and comparative studies of magnitude representation. *Progress in Neurobiology, 84*(2), 132–147.
Kahneman, D. (1973). *Attention and effort*. Englewood Cliffs, NJ: Prentice-Hall.
Kahneman, D. (2011). *Thinking, fast and slow*. New York: Farrar, Straus and Giroux.
Katrechko, S. (2018). Kant's appearance as an objectual representation. *International Journal of Philosophy, 2108*(7), 44–59.
Kelly, G. A. (1991). *The psychology of personal constructs. Volume I: A theory of personality*. London: Routledge. Taylor & Francis Group. Originally published in 1955.
Kemp Smith, N. (1941). *The philosophy of David Hume: A critical study of its origins and central doctrines*. London: MacMillan.
Killeen, P. R., & Fetterman, J. G. (1988). A behavioral theory of timing. *Psychological Review, 95*(2), 274–295.
Killeen, P. R., & Fetterman, J. G. (1993). The behavioral theory of timing: Transition analyses. *Journal of the Experimental Analysis of Behavior, 59*(2), 411–422.
Killeen, P. R., & Weiss, N. A. (1987). Optimal timing and the Weber function. *Psychological Review, 94*(4), 455–468.
Knuuttila, S. (2001). Time and creation in Augustine. In E. Stump & N. Kretzmann (Eds.), *The Cambridge companion to Augustine* (pp. 103–115). Cambridge, UK: Cambridge University Press.
Kopec, C. D., & Brody, C. D. (2010). Human performance on the temporal bisection task. *Brain and Cognition, 74*(3), 262–272.
Krueger, L. E. (1989). Reconciling Fechner and Stevens: Toward a unified psychophysical law. *Behavioral and Brain Sciences, 12*(2), 251–267.
Lambertini, R. (2014). Giles of Rome. In E. N. Zalta (Ed.), *The Stanford encyclopedia of philosophy* (Winter 2014 ed.). Stanford: Metaphysics Research Lab, Stanford University.
Laming, D. (1997). *The measurement of sensation*. New York: Oxford University Press.
Langer, J., Wapner, S., & Werner, H. (1961). The effect of danger upon the experience of time. *The American Journal of Psychology, 74*(1), 94–97.
Lejeune, H. (1998). Switching or gating? The attentional challenge in cognitive models of psychological time. *Behavioural Processes, 44*(2), 127–145.
Lejeune, H., & Wearden, J. H. (2009). Vierordt's the experimental study of the time sense (1868) and its legacy. *European Journal of Cognitive Psychology, 21*(6), 941–960.

Levy, J. M., Namboodiri, V. M. K., & Hussain Shuler, M. G. (2015). Memory bias in the temporal bisection point. *Frontiers in Integrative Neuroscience,9*, Article 44 (11 pp).
Lewin, K. (1935). *A dynamic theory of personality: Selected papers*. New York: McGraw-Hill Book Company, Inc. Translated by Donald K. Adams and Karl E. Zener.
Lewin, K. (1936). *Principles of topological psychology*. New York: McGraw-Hill Book Company, Inc. Translated by Fritz Heider and Grace M. Heider.
Lewin, K. (1938). *The conceptual representation and the measurement of psychological forces*. Durham, NC: Duke University Press.
Lewin, K. (1948). *Resolving social conflicts: Selected papers on group dynamics*. New York: Harper & Brothers Publishes. Edited by Gertrud Weiss Lewin, Foreword by Gordon W. Allport.
Lewis, P. A., & Miall, R. C. (2003). Distinct systems for automatic and cognitively controlled time measurement: Evidence from neuroimaging. *Current Opinion in Neurobiology, 13*(2), 250–255.
Look, B. C. (2013). Gottfried Wilhelm Leibniz. In E. N. Zalta (Ed.), *The Stanford encyclopedia of philosophy* (Summer 2017 ed.). Stanford: Metaphysics Research Lab, Stanford University.
Lowe, E. J. (2002). *A survey of metaphysics*. Oxford: Oxford University Press.
Lowe, E. J. (2013). Branching time and temporal unity. In F. Correia & A. Iacona (Eds.), *Around the tree: Semantic and metaphysical issues concerning branching and the open future* (pp. 73–80). Dordrecht: Springer Science+Business Media.
Lubashevsky, I. (2017). *Physics of the human mind*. Cham: Springer International Publishing AG.
Luce, R. D. (1959). On the possible psychophysical laws. *Psychological Review, 66*(2), 81–95.
Luce, R. D. (1961). A choice theory analysis of similarity judgments. *Psychometrika, 26*(2), 151–163.
Luce, R. D. (1963). Detection and recognition. In D. Luce, R. R. Bush, & E. Galanter (Eds.), *Handbook of mathematical psychology* (Vol. 1, pp. 103–189). New York: Wiley.
Luce, R. D. (2013). Analogs in Luce's global psychophysical theory of Stevens's psychophysical regression effect. *The American Journal of Psychology, 126*(1), 45–52.
Luce, R. D., & Galanter, E. (1963). Psychophysical scaling. In D. Luce, R. R. Bush, & E. Galanter (Eds.), *Handbook of mathematical psychology* (Vol. 1, pp. 245–307). New York: Wiley.
Luce, R. D., & Suppes, P. (1965). Preference, utility, and subjective probability. In D. Luce, R. R. Bush, & E. Galanter (Eds.), *Handbook of mathematical psychology* (Vol. 3, pp. 249–410). New York: Wiley.
Machado, A. (1997). Learning the temporal dynamics of behavior. *Psychological Review, 104*(2), 241–265.
Machado, A., & Keen, R. (1999). Learning to time (LET) or scalar expectancy theory (SET)? A critical test of two models of timing. *Psychological Science, 10*(3), 285–290.
MacKay, D. M. (1963). Psychophysics of perceived intensity: A theoretical basis for Fechner's and Stevens' laws. *Science, 139*(3560), 1213–1216.
Manning, R. (2016). Spinoza's physical theory. In E. N. Zalta (Ed.), *The Stanford encyclopedia of philosophy* (Winter 2016 ed.). Stanford: Metaphysics Research Lab, Stanford University.
Markosian, N., Sullivan, M., & Emery, N. (2014). Time. In E. N. Zalta (Ed.), *The Stanford encyclopedia of philosophy* (Fall 2016 ed.). Stanford: Metaphysics Research Lab, Stanford University.
Marr, J. (1999). The whirligig of time: Some thoughts on Staddon and Higa. *Journal of the Experimental Analysis of Behavior, 71*(2), 281–284.
Matthews, W. J., & Stewart, N. (2009a). Psychophysics and the judgment of price: Judging complex objects on a non-physical dimension elicits sequential effects like those in perceptual tasks. *Judgment and Decision Making,4*(1), 64–81.
Matthews, W. J., & Stewart, N. (2009b). The effect of interstimulus interval on sequential effects in absolute identification. *The Quarterly Journal of Experimental Psychology,62*(10), 2014–2029.
McCall, S. (1994). *Model of the universe: Space-time, probability, and decision*. New York: Oxford University Press.
McDonough, J. K. (2014). Leibniz's philosophy of physics. In E. N. Zalta (Ed.), *The Stanford encyclopedia of philosophy* (Spring 2014 ed.). Stanford: Metaphysics Research Lab, Stanford University.

McTaggart, J. E. (1908). The unreality of time. *Mind, New Series, 17*(68), 457–474.
Mellor, D. H. (1981). *Real time*. New York: Cambridge University Press.
Mellor, D. H. (1998). *Real time II*. London: Routledge.
Merchant, H., Zarco, W., & Prado, L. (2008). Do we have a common mechanism for measuring time in the hundreds of millisecond range? Evidence from multiple-interval timing tasks. *Journal of Neurophysiology, 99*(2), 939–949.
Narens, L. (1996). A theory of ratio magnitude estimation. *Journal of Mathematical Psychology, 40*(2), 109–129.
Narens, L. (2002). *Theories of meaningfulness*. Mahwah, NJ: Lawrence Erlbaum Associates.
Navon, D. (1984). Resources—A theoretical soup stone? *Psychological Review, 91*(2), 216–234.
Navon, D., & Gopher, D. (1979). On the economy of the human-processing system. *Psychological Review, 86*(3), 214–255.
Newton-Smith, W. (1980). *The structure of time*. London: Routledge & Kegan Paul.
Nizami, L. (2009). A computational test of the information-theory based entropy theory of perception: Does it actually generate the Stevens and Weber-Fechner laws of sensation? In S. I. Ao, L. Gelman, D. W. L. Hukins, A. Hunter, & A. M. Korsunsky (Eds.), *Proceedings of the World Congress on Engineering. Vol. 2. WCE 2009, July 1–3, 2009, London, U.K.* (pp. 1853–1858). Newswood Limited. Retrieved 1 Feb 2017.
Noma, E., & Baird, J. C. (1975). Psychophysical study of numbers: II. Theoretical models of number generation. *Psychological Research, 38*(1), 81–95.
Norman, D. A., & Bobrow, D. G. (1975a). On data-limited and resource-limited processes. *Cognitive Psychology, 7*(1), 44–64.
Norwich, K. H. (1987). On the theory of Weber fractions. *Perception & Psychophysics, 42*(3), 286–298.
Norwich, K. H. (1993). *Information, sensation and perception*. San Diego, California: Academic.
Norwich, K. H. (2005). Physical entropy and the senses. *Acta Biotheoretica, 53*(3), 167–180.
Norwich, K. H., & Wong, W. (1997). Unification of psychophysical phenomena: The complete form of Fechner's law. *Perception & Psychophysics, 59*(6), 929–940.
Nosofsky, R. M. (1992). Similarity scaling and cognitive process models. *Annual Review of Psychology, 43*(1), 25–53.
Nosofsky, R. M., & Palmeri, T. J. (2015). An exemplar-based random-walk model of categorization and recognition. In J. R. Busemeyer, Z. Wang, J. T. Townsend, & A. Eidels (Eds.), *The Oxford handbook of computational and mathematical psychology* (pp. 142–164). New York: Oxford University Press.
Nosofsky, R. M., Little, D. R., & James, T. W. (2012). Activation in the neural network responsible for categorization and recognition reflects parameter changes. *Proceedings of the National Academy of Sciences of the United States of America, 109*(1), 333–338.
Oberst, M. (2015). Two worlds and two aspects: On Kant's distinction between things in themselves and appearances. *Kantian Review, 20*(1), 53–75.
O'Daly, G. J. P. (1981). Augustine on the measurement of time: Some comparisons with Aristotelian and Stoic texts. In H. J. Blumenthal & R. A. Markus (Eds.), *Neoplatonism and early Christian thought: Essays in honour of A.H. Armstrong* (pp. 171–179). London: Variorum Press.
Oderberg, D. S. (1993). *The metaphysics of identity over time*. London: The Macmillan Press LTD.
Øhrstrøm, P., & Hasle, P. F. V. (1995). *Temporal logic: From ancient ideas to artificial intelligence*. Dordrecht: Kluwer Academic Publishers.
Olson, E. T. (2007). *What are we? A study in personal ontology*. Oxford: Oxford University Press.
Olson, E. T. (2012). In search of the simple view. In G. Gasser & M. Stefan (Eds.), *Personal identity: Simple or complex?* (pp. 44–62). Cambridge, UK: Cambridge University Press.
Olson, E. T. (2019). Personal identity. In E. N. Zalta (Ed.), *The Stanford encyclopedia of philosophy* (Spring 2021 ed.). Stanford: Metaphysics Research Lab, Stanford University.
Parfit, D. (1984). *Reasons and persons*. New York: Oxford University Press.

Parfit, D. (1995). The unimportance of identity. In H. Harris (Ed.), *Identity: Essays based on Herbert Spencer lectures given in the University of Oxford* (pp. 13–45). New York: Oxford University Press.

Pariyadath, V., & Eagleman, D. (2007). The effect of predictability on subjective duration. *PLOS ONE,2*(11), e1264 (6 pp).

Pasnau, R. (2011). *Metaphysical themes: 1274–1671*. New York: Oxford University Press.

Petrov, A. A. (2003). Additive or multiplicative perceptual noise? Two equivalent forms of the ANCHOR model. In R. Alterman & D. Kirsch (Eds.), *Proceedings of the 25th Annual Cognitive Science Society* (pp. 922–927). Hillsdale, NJ: Erlbaum.

Petrov, A. A., & Anderson, J. R. (2005). The dynamics of scaling: A memory-based anchor model of category rating and absolute identification. *Psychological Review, 112*(2), 383–416.

Petzschner, F. H., & Glasauer, S. (2011). Iterative Bayesian estimation as an explanation for range and regression effects: A study on human path integration. *Journal of Neuroscience, 31*(47), 17220–17229.

Petzschner, F. H., Glasauer, S., & Stephan, K. E. (2015). A Bayesian perspective on magnitude estimation. *Trends in Cognitive Sciences, 19*(5), 285–293.

Pinel, P., Piazza, M., Bihan, D. L., & Dehaene, S. (2004). Distributed and overlapping cerebral representations of number, size, and luminance during comparative judgments. *Neuron, 41*(6), 983–993.

Plutarch. (1914). *Plutarch's lives* (Vol. I). London: Harvard University Press. With an English translation by Bernadotte Perrin.

Podlesek, A. (2010). Sequential effects are not trivial context effects in psychophysical research. *Review of Psychology, 17*(1), 39–42.

Porro, P. (2001a). Angelic measures: Aevum and discrete time. In P. Porro (Ed.), *The medieval concept of time: Studies on the scholastic debate and its reception in early modern philosophy* (pp. 131–160). Leiden: Koninklÿke Brill NV.

Porro, P. (Ed.). (2001b). *The medieval concept of time: Studies on the scholastic debate and its reception in early modern philosophy*. Leiden: Koninklÿke Brill NV.

Porro, P. (2014). Henry of Ghent. In E. N. Zalta (Ed.), *The Stanford encyclopedia of philosophy* (Fall 2014 ed.). Stanford: Metaphysics Research Lab, Stanford University.

Poulton, E. C. (1967). Population norms of top sensory magnitudes and S. S. Stevens' exponents. *Perception & Psychophysics,2*(7), 312–316.

Poulton, E. C. (1968). The new psychophysics: Six models for magnitude estimation. *Psychological Bulletin, 69*(1), 1–19.

Prior, A. N. (1957). *Time and modality*. Oxford: Clarendon Press.

Rammsayer, T. H. (1999). Neuropharmacological evidence for different timing mechanisms in humans. *The Quarterly Journal of Experimental Psychology Section B,52*(3), 273–286. PMID: 10467900.

Rammsayer, T. H., & Grondin, S. (2000). Psychophysics of human timing. In R. Miller (Ed.), *Time and the brain* (pp. 181–194). Amsterdam: Harwood Academic Publishers.

Rammsayer, T., & Ulrich, R. (2005). No evidence for qualitative differences in the processing of short and long temporal intervals. *Acta Psychologica, 120*(2), 141–171.

Ricoeur, P. (1988). *Time and narrative* (Vol. 3). Chicago: The University of Chicago Press. Translated by Kathleen McLaughlin and David Pellauer.

Ritchie, J. B., & Carlson, T. A. (2016). Neural decoding and "Inner" psychophysics: A distance-to-bound approach for linking mind, brain, and behavior. *Frontiers in Neuroscience,10*, Article 190 (8 pp).

Robinson, D. K. (2010). Fechner's "Inner psychophysics". *History of Psychology, 13*(4), 424–433.

Robinson, H. (1994). Two perspectives on Kant's appearances and things in themselves. *Journal of the History of Philosophy, 32*(3), 411–441.

Roeckelein, J. E. (2000). *The concept of time in psychology: A resource book and annotated bibliography*. London: Greenwood Press.

Roeckelein, J. E. (2008). History of conceptions and accounts of time and early time perception research. In S. Grondin (Ed.), *Psychology of time* (pp. 1–50). Bingley: Emerald Group Publishing Ltd.

Rose, D., & Summers, J. (1995). Duration illusions in a train of visual stimuli. *Perception, 24*(10), 1177–1187.

Rosenberg, A. (1993). Hume and the philosophy of science. In D. F. Norton (Ed.), *The Cambridge companion to Hume* (1st ed., pp. 64–89). New York: Cambridge University Press.

Russell, B. (1915). On the experience of time. *The Monist, 25*(2), 212–233.

Rynasiewicz, R. (2011). Newton's views on space, time, and motion. In E. N. Zalta (Ed.), *The Stanford encyclopedia of philosophy* (Summer 2014 ed.). Stanford: Metaphysics Research Lab, Stanford University.

Scheerer, E. (1992). Fechner's inner psychophysics: Its historical fate and present status. In H.-G. Geissler, S. W. Link, & J. T. Townsend (Eds.), *Cognition, information processing, and psychophysics: Basic issues* (pp. 3–21). Mahwah: Lawrence Erlbaum Associates, Inc.

Schliesser, E. (2013). Newton's philosophy of time. In H. Dyke & A. Bardon (Eds.), *A companion to the philosophy of time* (pp. 87–101). Chichester, West Sussex, UK: Wiley-Blackwell, Wiley.

Schwitzgebel, E. (2011). *Perplexities of consciousness*. Cambridge, Massachusetts: The MIT Press.

Seibt, J. (2001). Formal process ontology. In N. Guarino, B. Smith, & C. Welty (Eds.), *Proceedings of the International Conference on Formal Ontology in Information Systems-Volume 2001* (pp. 333–345). New York: ACM.

Seibt, J. (2009). Forms of emergent interaction in general process theory. *Synthese, 166*(3), 479–512.

Seibt, J. (2017). Process philosophy. In E. N. Zalta (Ed.), *The Stanford encyclopedia of philosophy* (Summer 2020 ed.). Stanford: Metaphysics Research Lab, Stanford University.

Shepard, R. N. (1957). Stimulus and response generalization: A stochastic model relating generalization to distance in psychological space. *Psychometrika, 22*(4), 325–345.

Shepard, R. N. (1958a). Stimulus and response generalization: Deduction of the generalization gradient from a trace model. *Psychological Review,65*(4), 242–256.

Shepard, R. N. (1958b). Stimulus and response generalization: Tests of a model relating generalization to distance in psychological space. *Journal of Experimental Psychology,55*(6), 509–523.

Shepard, R. N. (1987). Toward a universal law of generalization for psychological science. *Science, 237*(4820), 1317–1323.

Shimojo, S., & Shams, L. (2001). Sensory modalities are not separate modalities: Plasticity and interactions. *Current Opinion in Neurobiology, 11*(4), 505–509.

Shoemaker, S. (1984). Personal identity: A materialist's account. In S. Shoemaker & R. Swinburne (Eds.), *Personal identity: Great debates in philosophy*. Faith and the future (pp. 67–132). Oxford, UK: Basil Blackwell Publisher Ltd.

Shoemaker, S. (1997). Self and substance. *Noûs,31*(s11: Supplement: Philosophical Perspectives, 11, Mind, Causation, and World), 283–304.

Sider, T. (2000). Recent work on identity over time. *Philosophical Books, 41*(2), 81–89.

Sider, T. (2001). *Four dimensionalism: An ontology of persistence and time*. New York: Oxford University Press.

Siep, L. (2014). *Hegel's phenomenology of spirit*. Cambridge, UK: Cambridge University Press. Translated by Daniel Smyth.

Silverman, A. (2014). Plato's middle period metaphysics and epistemology. In E. N. Zalta (Ed.), *The Stanford encyclopedia of philosophy* (Fall 2014 ed.). Stanford: Metaphysics Research Lab, Stanford University.

Skagerlund, K., Karlsson, T., & Träff, U. (2016). Magnitude processing in the brain: An fMRI study of time, space, and numerosity as a shared cortical system. *Frontiers in Human Neuroscience,10*, Article 500 (12 pp).

Smart, J. J. C. (1949). The river of time. *Mind, 58*(232), 483–494.

Smart, J. J. C. (1963). *Philosophy and scientific realism*. New York: Routledge.

Smart, J. J. C. (1980). Time and becoming. In P. Van Inwagen (Ed.), *Time and cause: Essays presented to Richard Taylor* (pp. 3–16). Dordrecht: Springer Science+Business Media B.V.

Smart, J. J. C. (2008). The tenseless theory of time. In T. Sider, J. Hawthorne, & D. W. Zimmerman (Eds.), *Contemporary debates in metaphysics* (pp. 226–238). Oxford, UK: Blackwell Publishing Ltd.

Smith, D. W. (2013). Phenomenology. In E. N. Zalta (Ed.), *The Stanford encyclopedia of philosophy* (Summer 2018 ed.). Stanford, CA: Metaphysics Research Lab, Stanford University. https://plato.stanford.edu/archives/sum2018/entries/phenomenology/.

Smith, Q. (2002). Time and degrees of existence: A theory of 'degree presentism'. In C. Callender (Ed.), *Time, reality & experience*. Royal institute of philosophy supplement (Vol. 50, pp. 119–136). Cambridge, UK: Cambridge University Press.

Spivey, M. (2007). *The continuity of mind*. New York: Oxford University Press, Inc.

Staddon, J. E. R., & Higa, J. J. (1999). Time and memory: Towards a pacemaker-free theory of interval timing. *Journal of the Experimental Analysis of Behavior,71*(2), 215–251.

Staddon, J. E. R., Chelaru, I. M., & Higa, J. J. (2002). A tuned-trace theory of interval-timing dynamics. *Journal of the Experimental Analysis of Behavior,77*(1), 105–124.

Stang, N. F. (2016). Kant's transcendental idealism. In E. N. Zalta (Ed.), *The Stanford encyclopedia of philosophy* (Spring 2016 ed.). Stanford: Metaphysics Research Lab, Stanford University.

Stevens, S. S. (1957). On the psychophysical law. *Psychological Review, 64*(3), 153–181.

Stevens, S. S. (1961). To honor Fechner and repeal his law. *Science, 133*(3446), 80–86.

Stevens, S. S., & Greenbaum, H. B. (1966). Regression effect in psychophysical judgment. *Perception & Psychophysics, 1*(5), 439–446.

Strawson, P. F. (1966). *The bounds of sense: An essay on Kant's critique of pure reason*. London: Methuen.

Taatgen, N., & van Rijn, H. (2011). Traces of times past: Representations of temporal intervals in memory. *Memory & Cognition, 39*(8), 1546–1560.

Taatgen, N. A., Van Rijn, H., & Anderson, J. (2007). An integrated theory of prospective time interval estimation: The role of cognition, attention, and learning. *Psychological Review, 114*(3), 577–598.

Teghtsoonian, R. (1971). On the exponents in Stevens' law and the constant in Ekman's law. *Psychological Review, 78*(1), 71–80.

Teghtsoonian, R. (1973). Range effects in psychophysical scaling and a revision of Stevens' law. *The American Journal of Psychology, 86*(1), 3–27.

Teghtsoonian, R. (2012). The standard model for perceived magnitude: A framework for (almost) everything known about it. *The American Journal of Psychology, 125*(2), 165–174.

Teghtsoonian, R., & Teghtsoonian, M. (1978). Range and regression effects in magnitude scaling. *Perception & Psychophysics, 24*(4), 305–314.

Teghtsoonian, R., & Teghtsoonian, M. (1997). Range of acceptable stimulus intensities: An estimator of dynamic range for intensive perceptual continua. *Perception & Psychophysics, 59*(5), 721–728.

Tegtmeier, E. (2016). Time and order. *Manuscrito: Revista Internacional de Filosofía,39*(4), 157–168.

Teske, R. (2001). Augustine's philosophy of memory. In E. Stump & N. Kretzmann (Eds.), *The Cambridge companion to Augustine* (pp. 148–158). Cambridge, UK: Cambridge University Press.

Teske, R. J. (1996). *Paradoxes of time in Saint Augustine*. Milwaukee, Wisconsin: Marquette University Press.

Thiel, U. (2011). *The early modern subject: Self-consciousness and personal identity from Descartes to Hume*. New York: Oxford University Press.

Thomas, E. A. C., & Weaver, W. B. (1975). Cognitive processing and time perception. *Perception & Psychophysics, 17*(4), 363–367.

Treisman, M. (1963). Temporal discrimination and the indifference interval: Implications for a model of the "internal clock". *Psychological Monographs: General and Applied, 77*(13), 1–31.

Treisman, M. (1964). Sensory scaling and the psychophysical law. *Quarterly Journal of Experimental Psychology, 16*(1), 11–22.

Treisman, M. (2013). The information-processing model of timing (Treisman, 1963): Its sources and further development. *Timing & Time Perception, 1*(2), 131–158.
Treisman, M., Faulkner, A., Naish, P. L. N., & Brogan, D. (1990). The internal clock: Evidence for a temporal oscillator underlying time perception with some estimates of its characteristic frequency. *Perception, 19*(6), 705–742.
Treisman, M., Faulkner, A., & Naish, P. L. N. (1992). On the relation between time perception and the timing of motor action: Evidence for a temporal oscillator controlling the timing of movement. *The Quarterly Journal of Experimental Psychology Section A, 45*(2), 235–263.
Trifogli, C. (2000). *Oxford physics in the thirteenth century (ca. 1250-1270): Motion, infinity, place, and time*. Koninklÿke Brill NV.
Trifogli, C. (2001). Averroes's doctrine of time and its reception in the scholastic debate. In P. Porro (Ed.), *The medieval concept of time: Studies on the scholastic debate and its reception in early modern philosophy* (pp. 57–82). Leiden: Koninklÿke Brill NV.
Trifogli, C. (2003). Thomas Wylton against minimal times. *Early Science and Medicine, 8*(4), 404–417.
Trifogli, C. (2015). The reception of Averroes' view on motion in the Latin west. In P. J. Bakker (Ed.), *Averroes' natural philosophy and its reception in the Latin west* (pp. 127–140). Leuven: Leuven University Press.
Tse, P. U., Intriligator, J., Rivest, J., & Cavanagh, P. (2004). Attention and the subjective expansion of time. *Perception & Psychophysics, 66*(7), 1171–1189.
Uzgalis, W. (2017). John Locke. In E. N. Zalta (Ed.), *The Stanford encyclopedia of philosophy* (Summer 2017 ed.). Stanford: Metaphysics Research Lab, Stanford University.
Van Bendegem, J. P. (2011). The possibility of discrete time. In C. Callender (Ed.), *The Oxford handbook of philosophy of time* (pp. 145–162). Oxford: Oxford University Press.
van Rijn, H. (2016). Accounting for memory mechanisms in interval timing: A review. *Current Opinion in Behavioral Sciences,8*, 245–249. Time in perception and action.
van Rijn, H., & Taatgen, N. A. (2008). Timing of multiple overlapping intervals: How many clocks do we have? *Acta Psychologica, 129*(3), 365–375.
van Rijn, H., & Taatgen, N. A. (2017). The integrative account of psychological time. In S. E. F. Chipman (Ed.), *The Oxford handbook of cognitive science (Oxford handbooks)* (pp. 151–168). New York: Oxford University Press.
van Rijn, H., Gu, B.-M., & Meck, W. H. (2014). Dedicated clock/timing-circuit theories of time perception and timed performance. In H. Merchant & V. de Lafuente (Eds.), *Neurobiology of interval timing*. Advances in experimental medicine and biology (Vol. 829, pp. 75–99). New York, NY: Springer Science+Business Media.
Walsh, V. (2003). A theory of magnitude: Common cortical metrics of time, space and quantity. *Trends in Cognitive Sciences, 7*(11), 483–488.
Wearden, J. (2016). *The psychology of time perception*. London: Palgrave Macmillan.
Wearden, J. H. (1991a). Do humans possess an internal clock with scalar timing properties? *Learning and Motivation,22*(1), 59–83.
Wearden, J. H. (1991b). Human performance on an analogue of an interval bisection task. *The Quarterly Journal of Experimental Psychology Section B,43*(1), 59–81.
Wearden, J. H. (2002). Traveling in time: A time-left analogue for humans. *Journal of Experimental Psychology: Animal Behavior Processes, 28*(2), 200–208.
Wearden, J. H., & Doherty, M. F. (1995). Exploring and developing a connectionist model of animal timing: Peak procedure and fixed-interval simulations. *Journal of Experimental Psychology: Animal Behavior Processes, 21*(2), 99–115.
Wearden, J. H., & Ferrara, A. (1995). Stimulus spacing effects in temporal bisection by humans. *The Quarterly Journal of Experimental Psychology Section B, 48*(4), 289–310.
Wearden, J. H., Denovan, L., & Haworth, R. (1997). Scalar timing in temporal generalization in humans with longer stimulus durations. *Journal of Experimental Psychology: Animal Behavior Processes, 23*(4), 502–511.

Weatherson, B., & Marshall, D. (2018). Intrinsic vs. extrinsic properties. In E. N. Zalta (Ed.), *The Stanford encyclopedia of philosophy* (Spring 2018 ed.). Stanford: Metaphysics Research Lab, Stanford University.

Weinstein, S., & Rickles, D. (2015). Quantum gravity. In E. N. Zalta (Ed.), *The Stanford encyclopedia of philosophy* (Winter 2016 ed.). Stanford: Metaphysics Research Lab, Stanford University.

Weissmann, S. M., Hollingsworth, S. R., & Baird, J. C. (1975). Psychophysical study of numbers III. Methodological applications. *Psychological Research, 38*(1), 97–115.

Williams, D. C. (1951). The Myth of passage. *Journal of Philosophy, 48*(15), 457–472.

Winter, B., Matlock, T., Shaki, S., & Fischer, M. H. (2015a). Mental number space in three dimensions. *Neuroscience & Biobehavioral Reviews,57*, 209–219.

Winter, B., Marghetis, T., & Matlock, T. (2015b). Of magnitudes and metaphors: Explaining cognitive interactions between space, time, and number. *Cortex,64,* 209–224.

Wolff, R. P. (1963). *Kant's theory of mental activity: A commentary on the transcendental analytic of the critique of pure reason*. Cambridge, Massachusetts: Harvard University Press.

Zakay, D. (1989). Subjective time and attentional resource allocation: An integrated model of time estimation. In I. Levin & D. Zakay (Eds.), *Time and human cognition: A life-span perspective* (pp. 365–397). Amsterdam: Elsevier Science Publishers B.V.

Zakay, D. (2016). Psychological time. In B. Mölder, V. Arstila, & P. Øhrstrøm (Eds.), *Philosophy and psychology of time* (pp. 53–66). Cham: Springer International Publishing.

Zakay, D., & Block, R. A. (1995). An attentional-gate model of prospective time estimation. In V. De Keyser, G. D'Ydewalle, & A. Vandierendonck (Eds.), *Time and the dynamic control of behavior* (pp. 167–178). Liège, Belgium: Universite Liège.

Zakay, D., & Block, R. A. (1996). The role of attention in time estimation processes. In M. A. Pastor & J. Artieda (Eds.), *Time, internal clocks and movements* (pp. 143–164). Amsterdam: Elsevier Science B.V.

Zakay, D., & Block, R. A. (1997). Temporal cognition. *Current Directions in Psychological Science, 6*(1), 12–16.

Zakay, D., & Block, R. A. (2004). Prospective and retrospective duration judgments: An executive-control perspective. *Acta Neurobiologiae Experimentalis, 64*(3), 319–328.

Zhao, H., Andersson, B., Guo, B., & Xin, T. (2017). Sequential effects in essay ratings: Evidence of assimilation effects using cross-classified models. *Frontiers in Psychology,8*, Article 933 (10 pp).

Zimmerman, D. (2010). The A-theory of time, presentism, and open theism. In M. Y. Stewart (Ed.), *Science and religion in dialogue* (Vol. II, pp. 789–809). Oxford, UK: Blackwell Publishing Ltd.

Zimmerman, D. W. (2005). The A-theory of time, the B-theory of time, and 'Taking tense seriously'. *Dialectica, 59*(4), 401–457.

Zimmerman, D. W. (2011). Presentism and the space-time manifold. In C. Callender (Ed.), *The Oxford handbook of philosophy of time* (pp. 163–244). Oxford: Oxford University Press.

Zwislocki, J. J. (2009). *Sensory neuroscience: Four laws of psychophysics*. New York: Springer-Science+Buisiness Media, LLC.

Chapter 2
Temporal Structure of Now from a Close-Up View

By the end of the XX century, the problem of human now attracted much attention, and at the beginning of the XXI century, a number of its comprehensive accounts have been proposed. The accounts to be discussed below can be combined under the term *temporal experience* or *temporal consciousness*. In the present chapter we want to analyze them in detail.

During this analysis, in parallel, we will formulate some of the main premises making up the gist of our account of *human temporality* and describe the basic elements of individual temporal dimension attributed to the mind. We intend to focus special attention on issues related to the temporal structures of the human now. In particular, the following issues are discussed in detail:

- The structure of a single unit of temporal experience.
- How these units are combined into the stream of experiences.
- The available experimental data elucidating the details and particular time scales characterizing the experiential now.

The concept of experiential now to be proposed has to:

- explain in detail what we understand under the term the human now and justifies the accepted premises,
- elucidate the basic elements of the human temporality and their properties that should be allowed for in our further constructions,
- introduce the basic physical notions to be employed in mathematical description of human perception on the temporal scales of human now.

2.1 Specious Present: Short Comments

The notion of present is crucial for many philosophical and mathematical constructions. Its various interpretations are related to a number of paradoxes in physics and

I. Lubashevsky and N. Plavinska, *Physics of the Human Temporality*,
Understanding Complex Systems, https://doi.org/10.1007/978-3-030-82612-3_2

philosophy of time; we discussed some of them in Sect. 1.1. The doctrine of presentism in its naïve form accepts that only the present exists, neither the future nor the past exist. The interpretation of the present as a structureless point-like object gives rise to many problems in the concept of presentism. Some of them can be overcome by turning to the theory of *thick* presentism (introduced by Hestevold 2008) treating the present as a time interval with infinitely short duration. We discussed thick presentism with the main attention focused on its application to the mathematical premises of Newtonian mechanism previously (Lubashevsky 2017). An important feature of thick presentism is the possibility of equipping the present with some inner structure. We have noted this because in our constructions we will turn to temporal structure of human now as it is perceived by humans and appears in the mind.

In spite of its long-term history, the problems of human now is still the matter of ongoing debated and there is no one common term reflecting all the facets of the human now. For this reason various authors, focusing on different aspects of how humans perceive the present, introduced different terms referring to the same object of investigation. Moreover, in literature these terms used often interchangeably, in particular, it concerns the terms the *experiential now*, *experiential present*, and *specious present*.

The notion of specious present is considered to be coined by E. R. Clay,[1] however its main development is related to works by the American philosopher and psychologist William James (1842–1910). A historical review concerning the precursors of James' specious present, including the British philosophers Shadworth Hollway Hodgson (1832–1912), Thomas Reid (1710–1796), Dugald Stewart (1753–1828), Thomas Brown (1778–1820), and Sir William Hamilton (1788–1856), can be found in publications by Andersen and Grush (2009), Andersen (2014). Introduced the notion of specious present, James emphasizes that to make sense of our temporal experience we need to distinguish the strict (or mathematical) present from the experiential (or specious) present; whereas the first is indeed durationless, the second possesses a brief duration, sufficient to accommodate the change and persistence we find in our immediate experience (Dainton 2017a).

Just after James put forward the concept of specious present, the German psychologist and philosopher William Stern (1871–1938) posed a question about the structure of this temporal object—*presence-time* in Stern's notations—including the contribution of just-passed and nearly-expected contents (see, e.g., Dainton 2017b).

In the XX century the concept of specious present was generalized within the account of internal time consciousness by the German philosopher Edmund Husserl (1859—1938) who established the school of phenomenology.[2] The central pivot of Husserl's account is the tripartite structure of the perceived now, according to which our experience of temporal moments consists of three components: *retention*, *pro-*

[1] This name is know from the *Principles of Psychology* (1980) by William James, where he enigmatically and misleadingly refers to "E. R. Clay" as a person anonymously published a book introducing the notion of specious present. However, there is no known person such as "E. R. Clay," which is discussed in detail, e.g., by Andersen and Grush (2009), Andersen (2014).

[2] The basic concepts of phenomenology and their relation to the problem analyzed within the present book are discussed in more details in Sect. 3.2.

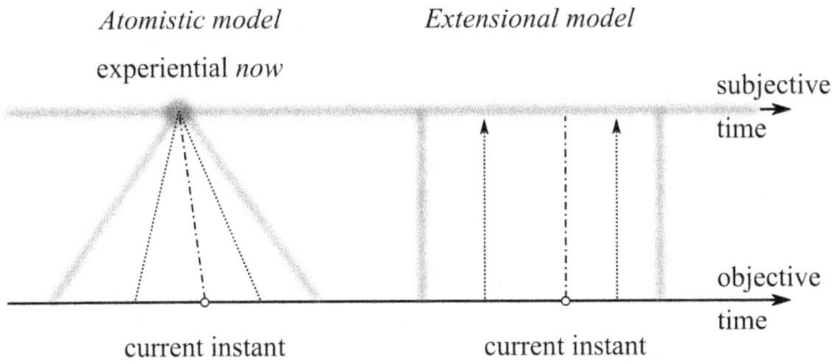

Fig. 2.1 Qualitative illustration of the atomistic and extensional models of temporal experiences. Blurred lines imply that the corresponding temporal boundaries and points can be fuzzy

tention, and *primal impression*. According to Husserl (1991), in our consciousness we simultaneously hold (*i*) contents experienced as just past, (*ii*) events anticipated to happen just now, and (*iii*) impression of what the present is (see for details Gallagher 1998; Gallagher and Zahavi 2012). These components are integrated into one entity—the human now. The mutual conversion of these components endows us with perception of continuous change in time. In other words, the perceived now includes its short history and its possible short future, with this tripartite structure being an inherent property of any conscious experience (Lloyd 2004, 2012).

In the given chapter we want to focus the main attention on the interpretation of the basic properties of the experiential now and below we will use exactly this term to avoid possible misunderstanding.

2.2 Atomism and Extensionalism

Here we return to the modern accounts of human time developed by the beginning of the XXI century. This section as well as Sects. 2.3 and 2.4 concerns what is that we perceive as the present, i.e., the state of the world around us at the current moment of time as it is experienced and then recognized. There are two main approaches to characterizing this present—*atomism* and *extensionalism*. It is necessary to note beforehand that the term atomism or extensionalism refers to a certain bundle of particular accounts (for a review see, e.g., Chuard 2011, 2017; Lee 2014b, c; Dainton 2010, 2017a; Phillips 2014b). Figure 2.1 qualitatively illustrates the two approaches to the experiential now: atomism and extensionalism.

For completeness, we also note the account of temporal experiences generically called the *retentional models*. As standardly conceived of, retentional *now* has no extension in objective time. Nevertheless, it is a complex object comprising immediate experience combined with a collection of representations of the recent past. All these representations are packed into the current momentary experience (for details see, e.g., Dainton 2017a).

2.3 Atomistic Accounts of Experiential Now

In the present section we will confine our consideration to atomistic accounts. Although our theory of human temporality in its spirit is closer to the extensionalist paradigm we will discuss atomistic accounts in detail because the ideas posited by modern versions of atomism also contribute to our further constructions.

2.3.1 Atomism

Historically, the Atomistic Theory considered the current experience of even temporally extended events strictly instantaneous (Broad 1923; Husserl 1991). Its recent versions at least remain neutral about the instantaneity of such experiences. Nevertheless they accept that the temporal properties of current experience do not reflect in the right way the temporal properties of perceived events. In particular, the temporal order of events can be lost in their experience.

Tye (2003) put forward an atomistic account based on the *unity of experience*—the thesis that our sensory modalities *together* give rise to the atomic experience of now bearing also some information about events' non-simultaneity. From Tye's account we take the idea that *within atomic experience* humans are able to perceive the order of event appearance. In other words, the temporal order without perceiving a time lap between two non-simultaneous events can be a component of the atomic experience. We will return to this issue below in discussing the corresponding psychophysical data. The general arguments for the claim that an experience of succession is not a succession of experiences are discussed by Hoerl (2013).

Le Poidevin (2007) accepts the atomic experience to bear the awareness of motion and the passage of time. He comes to conclusion that:

> we could hardly both perceive two events as present and perceive one as occurring before the other. What we experience as present is not divisible in this way. No part of what is perceived as present is simultaneously perceived as past and future. One way of understanding James's specious present, then, is to take it as involving a mixture of perception, memory, and anticipation. The term 'cognised present' certainly suggests more than raw perception (p. 81).

We want to emphasis two aspects in Poidevin' account. First, the experiential now is of finite duration and involves perception, memory, and anticipation. From our point of view, the experiential now is the basic "brick" of human temporality.

Second, it is the possibility of treating the experiential now as a certain linear space whose elements are experienced now. Namely, we may accept that experience does indeed give us the impression of change and succession, but no very clear sense that this belongs to a particular moment. We are aware of it, and that is all. Describing succession as phenomenologically present does not tie it down to a moment. Allied to this is the thought, urged by Dennett (1991), that there is no definite moment at which a piece of information becomes conscious (p. 92).

2.3.2 *Extended Atomism*

Lee (2014a, b, c) in his account of *extended atomism* emphasizes that atomic experiences need not be instantaneous events.[3] Therefore Lee has to introduce the notion of experience atomism turning to another criterion, namely, he writes

> [a]tomic experiences are atomic in the sense that they do not contain shorter experiences as temporal parts. If an atomic experience is realized by an extended process (like 40hz neural firing), then this process may have shorter *physical events* as proper temporal parts, even if it doesn't have *experiences* as proper temporal parts; for example, these shorter physical events may not be sufficient for any experiences to exist. To be clear, if an experiential property instantiation occurs fundamentally over a short interval in this way, we can say that it has shorter experiential temporal parts in a *derivative* sense (Lee 2014b, pp. 152–153).

Below we will turn to a similar idea of the atomic experience with hidden temporal structure.

One of the cornerstones of atomistic accounts is the question of how atomic experiences can form a continuum endowing our consciousness with perception of smooth time flow. Lee proposes a solution to this problem. Let us discuss this solution in more detail because it opens a gate for our constructions of human temporality.

Within Lee's account atomic experiences are not isolated elements but overlap in time. In this manner, a series of atomic experiences can form a certain continuum rather than a discrete set. Lee (2014a, see also 2014c) puts forward the concept of *trace integration* as a mechanism responsible for the relationship between the continuum of atomic experiences and the continuum of the perceived events. The gist of this mechanism is that the brain integrates signals received from sensor organs and generates the representation of perceived events in the present atomic experience. In the brain the processing of information about even a point-like event is a complex processes *extended* in time. Thereby two neighboring atomic experiences may bear the information about the same event, which is the implementation of their overlapping.

We want to note that an atomistic account similar to Lee's extended atomism was proposed earlier by Grush (2005, 2007, 2008). Grush himself does not classify his

---

[3] The idea that atomic experience may have a finite duration was also discussed by Broad (1925).

account as atomism. He introduces the term *the trajectory estimation model* to refer to the description of the relationship between temporal experiences and the time pattern of perceived events in objective time. Actually he prefers to speak about his account as the specious present doctrine interpreted withing the trajectory estimation model (Grush 2007). However in the framework of this book we may consider his account a certain version of atomism. Each atomic experience—*perceptual act* in Grush's terminology—aggregates the information about non-simultaneous events within some finite interval of objective time. This aggregation is characterized by properties we want to discuss in more detail.

First, Grush's model explicitly states that the atomic experience is determined by events happened just now, in the nearest past, and expected to occur in the immediate future.

Second, two objectively non-simultaneous events contributing to a given atomic experience are not perceived as simultaneous. Following Grush's logic this atomic experience is not a simple algebraic sum of the perceived events. Instead, Grush writes:

> [t]he specious present doctrine, on the interpretation [he offers], is that the content of perception is a temporal interval, and this means that it is an interval within which relations of succession, before and after, and so forth, obtain. None of this is captured in a time-lapse photo (2007, p. 31).

The given Grush's claim prompted us to attribute an extended phase space to the atomic experience of a moving object where the its *perceived* position and velocity are *functionally independent* quantities.

Wiese (2017) generalized Grush's account to emphasizes that the trajectory estimation model has to deal with different types of representations in aggregating the information about perceived events. Remembering and imagining involves *conceptual* representations, whereas experiencing events involves *perceptual* representations. The mechanisms governing the two types of representations are different and, in particular, characterized by different time scales. To allow for this feature Wiese proposes the *hierarchical trajectory estimation model* introducing the fast and slow variable into the mathematical description developed by Grush (2004, 2005, 2006). In this case the fast variables can be enslaved by the slow ones or exhibit fast variations when the slow variables practically do not change. This idea underlies the introduction of shot-term strategies of behavior specified by slow variables in our further constructions.

In his account Wiese elso emphasizes that the events, even point-like ones, need not to be represented as having determinate temporal boundaries, he calls them fuzzy boundaries. In our notations, it is related to the idea that point-like instants of objective time convert into some temporal clouds being the basic elements of human temporality.

There is also an account of atomistic type called the *snapshot conception* implying that the perception of events is reduced to instantaneous slices—*snapshots*. In this case, the stream of consciousness is composed of continuous succession of still snapshots, which is called *the cinematic model* (for a review see Chuard 2011, 2017). Here we have noted this conception for completeness only.

2.3.3 Atomic Experience as Structural Element of Human Temporal Dimension

The heart of atomistic accounts considered in the previous sections is the notion of *atomic experience*—the representation of the human now which is assumed to be indivisible in the mind. This notion is relatively new and its understanding is a matter of lively debate, in particular, each account ascribes to it some individual properties.

In our account, which is not atomistic, the notion of atomic experience as an element of consciousness also plays a crucial role. It does not represent the experiential now completely but is its essential part. This part, on the one hand, represents some piece of reality of finite duration and, on the other hand, is indivisible in the mind. For the latter reason atomic experience cannot be embedded into objective time explicitly. Thereby, we propose to regard it as an elementary "brick" of the temporal dimension of the mind (human temporal dimension) possessing its own properties.

Based on extended atomism discussed above, we summarize the properties to be attributed to atomic experience within our account of human temporality (Fig. 2.2).

- The atomic experience (or, more strictly, atomic temporal experience) is a basic *constituent* element of the temporal dimension attributed to the mind. On the one hand, the atomic experience represents human perception of the temporal arrangement of events or objects as they are embedded in the real space-time continuum. On the other hand, it is indivisible in the mind. This indivisibility implies that there are no temporal parts of the atomic experience that can be experientially or consciously singled out from it as some constituents.

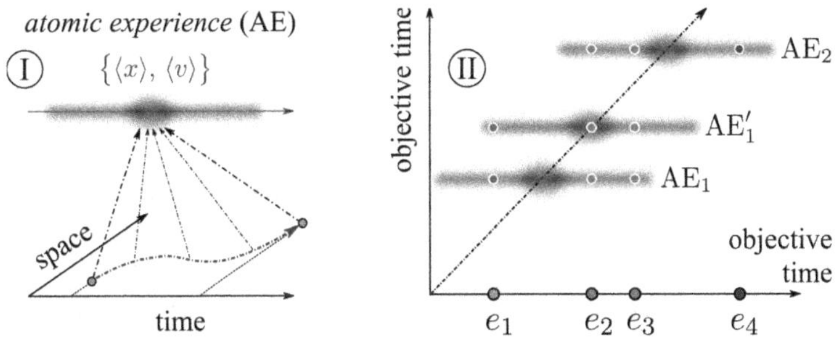

Fig. 2.2 Illustration of the atomic experience as an elementary object of the mental temporal experience which aggregates information about reality within some finite interval of objective time (I), and the overlapping of neighboring instantiations of atomic experience via perceived events $\{e_i\}$ (II)

- The atomic experience corresponds to a finite interval of objective time with fuzzy boundaries. It aggregates in some way the information about perceived events such that
  - the non-temporal properties of these events are averaged,
  - certain pieces of information about the event temporal arrangement can be also saved and attributed to the atomic experience as a whole.

  For example, the atomic experience of an observed object moving in space bears the information about its perceived position and the mean velocity given to the mind as functionally independent quantities. Figure 2.2(I) illustrates the relationship between the atomic experience content and the properties of perceived events.
- The individual contribution of perceived events as well as their collection changes with objective time. Because the present of objective time as an individual object is beyond human cognitive abilities only the change in the atomic experience is able to make us aware of the passage of time. In this sense, the change is prior to time in the mind.
- As objective time goes on:
  - An instantiation of atomic experience, as an aggregate of perceived events making up a certain set, continues to exist for some (objective) time. During this time the atomic experience undergoes quantitative change due to variations in the individual contributions of the same events.
  - Then its content changes *qualitatively* and a new instantiation of atomic experience emerges.

  However, the neighboring atomic experiences are not independent, they *overlap* with each other. This overlapping means that there are common events contributing to the content of both the atomic experiences (Fig. 2.2II). The overlapping of atomics experiences gives rise to the temporal experience continuum and the experience of time flow smoothness.
- There are two types of variables characterizing the state of a given atomic experience that admit the interpretation as fast and slow variables.
  - *Fast* variables describe the inner details of how perceived events are embedded into a given instantiation of atomic experience. They are "invisible" for the mind and specify the interaction between the atomic experience as a mental object and the corresponding neural processes in the brain. The fast variables may be used in describing the dynamics of atomic experience but have to be included into a certain automatic component of mental states lying beyond the human cognition.
  - *Slow* variables describing the given instantiation of atomic experience as a whole. They cannot exhibit essential variations on temporal scales about the characteristic duration of atomic experience and, so, can describe the interrelations between neighboring atomic experiences. The slow variables are accessible to the mind.

2.4 Extensionalist Accounts of Experiential Now

As noted at Sect. 2.2, there is another approach to describing temporal experiences, it is *extensionalism* (Fig. 2.1). The extensionalist doctrine claims that experiences of events must (*i*) themselves exhibit the temporal properties of these events and (*ii*) contain different temporal parts (Rashbrook-Cooper 2017). In other words, extensionalism rejects the indivisibility of experiential now and its reducibility to the atomic experience. Within the extensionalist paradigm temporal experiences are *processes* and their temporal properties mirror the temporal properties of perceived events.

Alexius Meinong (1853–1920), an Austrian philosopher, and Ludwig Wilhelm Stern (1871–1938), a German psychologist and philosopher, are the precursors of extensionalism. Nowadays several variants of extensionalism have been proposed by Mundle (1966), Foster (1979, 1982, 1991), Dainton (2000, 2008b, 2017a), Phillips (2010, 2014a, b), Hoerl (2013), Rashbrook (2013a, b, c), Soteriou (2010, 2013).

A benchmark for elucidating common and individual aspects of various extensionalist accounts is how they regard the famous statement of William James given in *The Principles of Psychology*:

> a succession of feelings, in and of itself, is not a feeling of succession. (James 1890, Vol. I, p. 629)

Other versions of this statement can be found, e.g., in (Hoerl 2013).

Hoerl (2013) and Rashbrook (2013a) presented a number of arguments leading to the conclusion that an experience of succession requires a succession of experiences. It, in particular, prompted them to accept extensionalism for describing human perception of time. Naturally, the relationship between the experience of succession and the succession of experience is not necessarily of a simple form. In the general case, an experience of succession is not merely the corresponding succession of experiences. The accounts to be discussed below propose different philosophical models for the temporal structure of experience of successive events.

2.4.1 The Naïve Extensionalism

In order to illustrate the most common aspects of extensionalism let us consider one of its models—the Naïve theory of temporal experiences. It should be noted that in spite of its simplicity this model is highly intuitive and capable of withstanding the central line of extensionalism criticism (Rashbrook 2013a, c; Phillips 2014a, b).

Answering to the question of the relation between the temporal structure of experience and the temporal structure of the objects of experience, the Naïve extensionalism accepts that our experience inherits the temporal structure of the events (Phillips 2014b). The temporal structure of the world is just reproduced one-to-one in the stream of consciousness (Fig. 2.3). Our account includes the main ideas of two rival philosophical approaches to understanding the human now, namely, extended atomism and extensionalism.

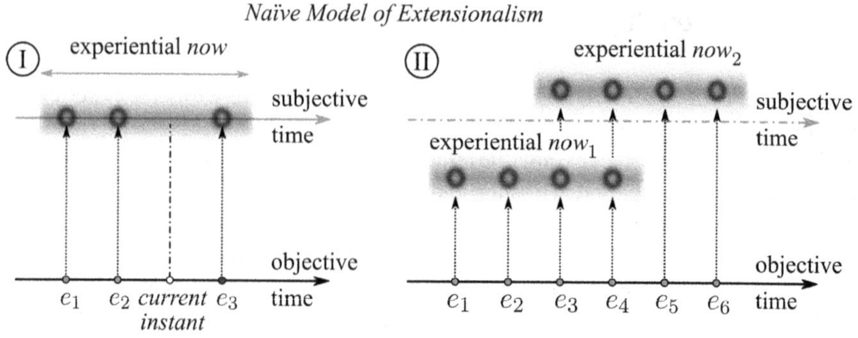

Fig. 2.3 The temporal structure of experiential now within the Naïve model of extensionalism (I) and the corresponding mechanism by which the neighboring instantiations of experiential now overlap with each other, endowing human consciousness with continuity (II)

Here we meet two fundamental notions actually each extensionalist account deals with. They are *synchronic unity* and *diachronic unity* of consciousness. Within the problem under consideration, the synchronic unity concerns events—simultaneous or non-simultaneous in objective time—that are perceived as coincidental. The diachronic unity concerns non-simultaneous events that are related to the experiential now.

The concepts of synchronic unity and diachronic unity are particular implementations of a more general concept analyzed in philosophy of mind—the unity of consciousness; for introduction to this issue a reader may be referred, e.g., to Dainton (2000), Tye (2003). It is worthy of noting that the roots of diachronic unity as the concept of predictive processing can be traced back to Kant's and Locke's philosophy (see, e.g., Swanson 2016), the account developed by Hermann von Helmholtz for the relationship between sensory and mental images (see Friston 2003; Adams et al. 2013; Seth 2014, for are view), is one of the key ideas in Husserl's phenomenology, and has attracted much attention in the XX-century (e.g., Bayne 2010; Brook and Raymont 2017; Dainton 2017a).

We want to emphasize that the Naïve model of extensionalism—as well as all its other variants—first, considers the experiential now to be extended in the temporal dimension of the mind. This extension means the experiential now to be divisible into temporal parts that consciously and experientially are discriminated and admit the classification in terms of before-after interrelations. The possibility of singling out a sequence of non-simultaneous experiences which contribute to the experiential now is the essence of diachronic unity. Figure 2.3(I) illustrates the experiential now to possess a finite duration in subjective time and to contain separated elements distributed along the axis of subjective time.

Second, within the Naïve extensionalism synchronic unity actually does not exist or, speaking more strictly, only events being synchronic in the external world are synchronic in human perception. Thereby the Naïve extensionalism inevitably faces up to a number of puzzles related to the problem of simultaneity in objective time with point-like instants.[4]

The Naïve extensionalism provides an intuitively logical explanation for the continuity of consciousness. Turning to the bottom-up approach (Broad 1923, 1938; Foster 1991; Dainton 2000, see also Husserl 1991) , which is illustrated in Fig. 2.3(II), we see that there are some common events contributing to both the neighboring instantiations of the experiential now. It relates different experiential nows to one another and endows our consciousness with continuity—we would say homogeneity—in going through the temporal sequence of events $\{e_i\}$.

2.4.2 *Composite Extensionalism*

In what follow we will discuss the following two issues:

- the temporal structure of experiential now and
- the continuity of consciousness.

We want to present our understanding of the experiential now taking into account the modern philosophical constructions of the British philosophers Barry Dainton and Michael Tye (see Dainton 2000, 2017a; Tye 2003, for are view). We accept the experiential now to have a complex temporal structure comprising different elements which, in turn, are of complex structure. It is reflected in the title of the present section.

In this section we introduce two new notions: *synchronic unit* and *diachronic unit.* These notions play a crucial role in describing temporal aspects of human perception, so, at first, we want to elucidate the reasons for their introduction.

As far as objective (physical) time is concerned, its very basic element is the point-like instant and other temporal structures may be conceived of as sets of time instants. To cope with human temporal dimension we also need its own basic element (or elements) for constructing more complex structures as some sets of such elements. We call this very basic element of human temporal dimension the *synchronic unit.* Leaping ahead, we note that the temporal structure of experiential now may be treated as a fusion of three synchronic units, for this reason it will be called the *diachronic unit.*[5] Thereby in our constructions the diachronic unit is not strictly elementary.

However, before passing to the detailed description of synchronic and diachronic units we want to return to the notion of experiential now and explain what exactly we understand under this term.

The experiential now is the aggregate of all the mental images of events that are experienced (*sensorily* perceived) right now. These events are understood as physical objects being at a given state at a given instant of objective time. Having a color, having a form, having a texture, etc. exemplify such events. The experiential now is

---

[4] We already discussed this issue and related ones in Sect. 1.1.

[5] The term *diachronic* is understood as an aspect involving in itself different moments of time and treating them as something with internal integrity.

a generic term referring to all features and aspects characterizing mental and sensory images of events experienced right now.

Speaking about the temporal structure of experiential now we should discriminate two issues:

- how the structure of experiential now is embedded into objective time;
- what is the structure of mental (subjective) time in the scope of experiential now.

The first issue concerns the temporal structure of neural signals' processing in the brain that relates the properties of perceived physical objects to the properties attributed to their mental images in the mind. This processing of neural signals is mainly unconscious and only the final result becomes accessible to the mind. Thereby the properties of perceived objects are attributed to their mental images at the level of experiential now. Even for a single event the integration of neural signals is a complex nonlinear process in objective time which involves a multitude of time instants contributing to the final result. However, the temporal structure of this process should be rather stable because of its unconsciousness, which allows us to suppose it to possess a certain universality. The given universality does not exclude that the mind, in principle, can affect the unconscious processing of neural signals; its effect is just implicit. In particular, conscious redistribution of attention between different mental tasks is reflected in the unconscious processing of perceived events but, broadly speaking, it cannot not affect the inner details of how this processing is implemented. The universality of processing neural signals at unconscious level enables us to regard the temporal structure of experiential now as a certain entity with internal integrity.

The second issue is about whether the temporal structure of experiential now is divisible in the mind. Accepting, in general, the extensionalist paradigm, we consider the experiential now to be mentally divisible and to contain some temporal parts accessible to the mind, namely, the *immediate past*, the *immediate now*, and the *immediate future* experienced right now. The terms the immediate past, the immediate now, and the immediate future are also generic names referring to all the features and aspects of the corresponding mental images.

Each of these temporal parts comprises the mental images of events not only experienced as simultaneous but also *cognitively* perceived as simultaneous. Therefore in the aspect of its temporal structure each of them represents the very basic element of human temporal dimension; we call it the *synchronic unit*. This element is not anymore divisible in the mind. Naturally, the "embedding" of the synchronic unit into objective time does not map it at a point-like instant, the synchronic unit is of finite duration in objective time. The immediate past, the immediate now, and the immediate future are some analogy to the atomic experience of extended atomism.

The mental invisibility allows us to treat the synchronic unit of immediate past, the synchronic unit of immediate now, and the synchronic unit of immediate future as three indivisible elements of human temporal dimension. We call their aggregate the *diachronic unit*. It should be emphasized that the diachronic unit is not a simple "mechanistic" sum of its constituent synchronic units; it is their fusion, which endows the diachronic unit also with properties not exhibited by its synchronic units

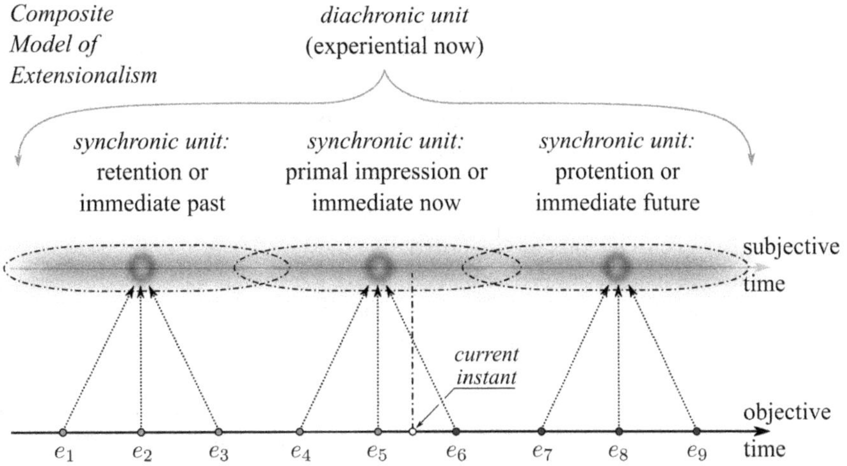

Fig. 2.4 The temporal structure of experiential now within the composite model of extensionalism. The notion of *retention*, *primal impression*, and *protention* have been introduced by Husserl, we will use them along with the *immediate past*, the *immediate now*, and the *immediate future* as synonyms

separately. The diachronic unit is always one and represents the temporal structure of experiential now.

As illustrated above we posits that experiential now is a complex mental entity which is reflected in its multicomponent temporal structure. For this reason we call our account of experiential now *composite extensionalism*. The following two sections will elucidate this concept.

2.4.3 Temporal Structure of Experiential Now

The composite extensionalism accepts the experiential now to possess the temporal structure shown in Fig. 2.4. Its basic constituent elements are the synchronic units and the diachronic unit uniting them. The diachronic unit shown in this figure is assumed to comprise three synchronic units representing the immediate past, the immediate now, and the immediate future.

The synchronic unit generally comprises events being non-simultaneous in objective time. Therefore, on the one hand, it is the indivisible primitive of human temporal dimension. On the other hand, at the level of physical reality it is a complex object. Following Rashbrook (2013c) and rephrasing James' statement we may say that the introduction of the synchronic unit illustrates that

> an experience of conjunction, in and of itself, is not a conjunction of experiences.

From our point of view, the synchronic unit coincides with the atomic experience discussed above within the extended atomism except for the reducibility of the experiential now to the synchronic unit.

As far as the properties, in particular, causal power of the synchronic unit are concerned, there are two rival approaches to their description—the bottom-up and top-down paradigms (see Rashbrook 2013c, for a detailed discussion).

Bottom-up theorists like Dainton accept that all the properties of a given instantiation of the synchronic unit are inherited directly from the properties of the corresponding events making up this unit and playing the role of its supervenient base. Thereby that what we can do with the synchronic unit in the mind is to operate logically with its properties and interrelations between its instantiations (Dainton 2000).

The top-down approach represented by the accounts of Tye (2003), Bayne and Chalmers (2003), Bayne (2005, 2010) is based on the claim that the synchronic unit is a basic element of human temporal dimension and exists in the mind on its own. Thereby in mental operations we can cope with the synchronic unit as an object possessing its own properties including causal power ascribed to it a whole.

Within the two-component approach (Lubashevsky 2017) accepting the subjective and objective components to possess their own properties, the top-down and bottom-up approaches are reconciled. Namely, on the one hand, the basic features of the synchronic unit are determined by the particular properties of the corresponding events. On the other hand, in the mind we operate with the synchronic unit as a basic element of our consciousness, which affects external objects via our actions. In other words, the synchronic unit possesses two facets. One of them reflects how it *aggregates* the properties of perceived events or objects including temporal ones, where the brain acts as an intermediary between the external world and the mind. The other facet reflects how the synchronic unit appears in the mind with properties attributed to it as a whole. This facet can bear some temporal information about the perceived events and objects, for example, their temporal order, but this information is averaged and fragmentary. It does not enable the mind to reconstruct—experientially or cognitively—the individual details of events arrangement in reality.

The diachronic unit aggregates in itself not only events perceived as happening right now but also events recognized as ones belonging to the past or the expected future but *experienced* at the current instant of time. In this sense, the immediate past and the immediate future co-exist with the immediate now and may be regarded as temporal parts of experiential now. We want to emphasize that although the immediate past and the immediate future are experienced now, we recognize that there is some time lag between the events belonging to them and the events perceived as currently presented. The possibility of introducing the given tripartite structure as some mental object with its own properties is the essence of the concept of diachronic unity and accepted within Dainton's (2000) account as a basic premise. Dainton also uses the term *co-consciousness* for the diachronic unity.

> Diachronic co-consciousness is what binds together the adjacent phases of a stream of consciousness, and so is responsible for the very existence of

> the stream as a temporally extended whole...[i.e.,] the experienced unity of streams of consciousness from moment to moment...One reason [to accept this notion] is the fact that change and persistence are directly experienced. Change and persistence both take time. How could we directly experience change and persistence unless experience itself encompasses a temporal interval? (Dainton 2000, pp. 113–114)

The tripartite structure—the succession of the immediate past, the immediate now, and the immediate future experienced *together*—forms the diachronic unit of human temporal dimension (Fig. 2.4). Exactly this diachronic unit admits the interpretation as experiential now. It should be noted that neither the immediate past nor the immediate future is merely collections of separated events. In our mind, the immediate past and the immediate future are indivisible mental objects of human temporal dimension—the synchronic units. The introduction of the notion of the diachronic unit allows us to say, following James, that

> an experience of succession, in and of itself, is not a *simple* succession of experiences.

Dealing with the relationship between the own properties of the diachronic unit, on the one hand, and the individual properties of events composing it, on the other hand, we again face the dilemma of accepting the top-down or bottom-up paradigm as the main premise. We suppose that turning to the two-component description, as previously, opens a gate to resolving this dilemma.

2.4.4 Continuity of Consciousness

The aspects of composite extensionalism discussed above provide a plausible answer to the problem of how we can feel the present moment and the change in the environment as simultaneous and mutually independent experiences. This answer turns to the idea that the experiential now includes the synchronic and diachronic units integrating the information about perceived events over a finite interval of objective time. As far as the experience of the passage of time is concerned, this problem requires additional discussion, which is the subject of the present subsection.

The passage of time tacitly implies that it is some continuous process and its experience is also a continuous process, which is reflected in the term *continuity of consciousness*. At first glance, it seems that the conception of experiential now composed of the synchronic units as indivisible elements of human temporal dimension and the continuity of consciousness are in a contradiction to each other. Indeed, dealing with synchronic units, the concept of discrete changes following one another step-wise as time goes on looks more naturally than the continuity. We already faced up to this problem in the discussion of extended atomism, where its solution was discussed (Sect. 2.3.2). The gist of this solution—the overlapping of neighboring instantiations of atomic experience—holds true in the case of composite extensionalism too. However, its implementation in the case of composite extensionalism is not merely reduced to the previous constructions. It also elucidates aspects important for our theory of temporality, for example, illustrates that this theory cannot meet the paradigm of Newtonian mechanics. For this reason let us discuss the problem of continuity of consciousness in the framework of composite extensionalism in more detail.

Following the logic of Dainton (2000, 2010, 2017a) we will start from a *pulse theory* adopted, e.g., by Whitehead (1929/1978), Sprigge (1984). The given theory accepts that the stream of consciousness is partitioned into *pulses* of experience which individually come into being as wholes rather than part by part. These pulses are of finite duration, which enables the persistence and the change to be experienced simultaneously. In this sense such pulses may be categorized as synchronic units.

The pulse theory faces up to the problem of how individual pulses, especially neighboring in time but appearing and disappearing one after another, are interrelated. One of the promising accounts of extensionalism called the *overlap model* much advocated by Foster (1991, 1982, 1979) endows the pulses of experience with fuzzy temporal boundaries (for a detailed review with own arguments for this model see Dainton 2000, 2017a). Thereby these pulses can overlap. As a result, the immediate experience related to the current pulse has also partial access to experiential details of neighboring pulses in the immediate past and the immediate future. It enables the continuous stream of consciousness to emerge. Besides, because the inter-pulse and intra-pulse relations must be different in properties and, thus, can be discriminated in our consciousness, the immediate experience is able to possess the temporal parts—retention, primal impression, and protention.

Figure 2.5 illustrates our interpretation of the overlap model. The persistence and the change are assumed to be experienced *directly*, i.e., they appear in our mind as independent, mutually irreducible aspects of our perception. It is possible only in the case when the *now* in our mind is not a point-like instant of objective time and there are at least two different channels through which the information about the external world gets the mind even within one sensory modality.

Thereby, the moment when we recognize that some event has occurred (or will occur right now) corresponds to a certain finite interval of objective time. Let us denote the characteristic duration of this interval by τ_a. All the events inside this interval are perceived as simultaneous and we are not able to order them *consciously* in objective time. Figure 2.5 illustrates this; for example, the time distance between two event e_4 and e_5 is much less than τ_a and, so, we cannot perceive that the two events are separated by some *finite time lag*. However, we do recognize that the event e_5 follows the event e_4 as a non-temporal fact if the time lag between them is not too short.[6]

Thereby, for our consciousness a certain neighborhood of the current instant t_0 of objective time whose thickness is about τ_a has no temporal structure, it exists as a whole structureless object. It is an implementation of synchronic unit that may be referred to as the immediate now (Fig. 2.5). To describe this in mathematical terms

[6] The relationship between the perception of event temporal arrangement and neurophysiological processes in the brain will be discussed in Sect. 2.5.

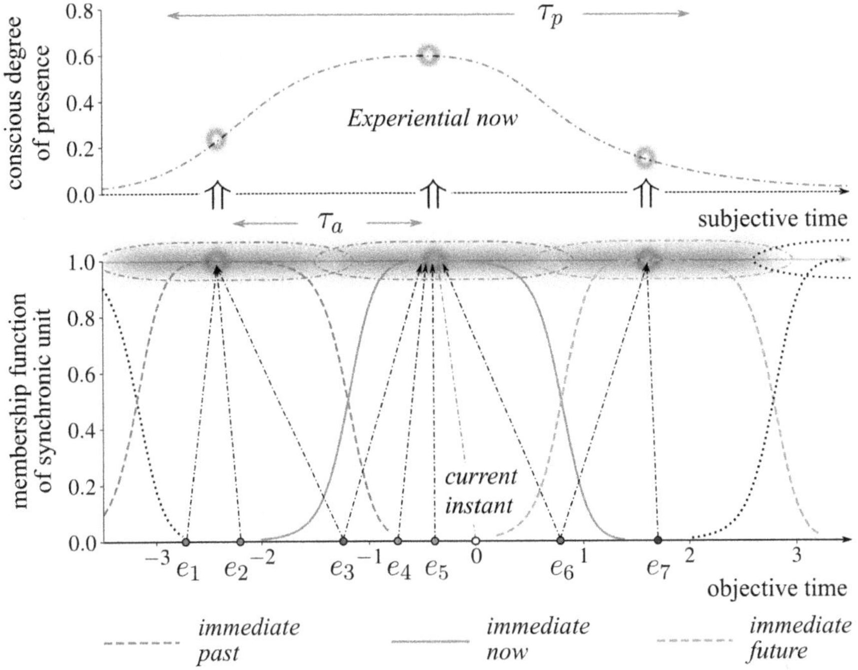

Fig. 2.5 Illustration of the extensional overlap model with fuzzy synchronic units of temporal experiences and the implementation of the retention-impression-protention overlapping at the level of individual events

we can use the notion of the membership function

$$m_n(t, t_0) = \begin{cases} 1, & \text{if } |t - t_0| \ll \tau_0\,, \\ 0, & \text{if } |t - t_0| \gg \tau_0\,. \end{cases} \tag{2.1}$$

as a measure that a given event e happened (or to happen) at time t belongs to the immediate now in our consciousness. For $|t - t_0| \sim \tau_0$ the membership function $m_n(t, t_0)$ gradually decreases from 1 to 0 as the time distance $|t - t_0|$ between the two moments increases. Actually this membership function bears all the *temporal* information about the events of the immediate now. Within a single synchronic unit the information that its events are not simultaneous is processed in another way and attributed to the synchronic unit as a *non-temporal* property.

When the time lag between events happened in the past or expected to occur in the nearest future and the current instant t_0 exceeds τ_a substantially, e.g., the events e_1 and e_7, we are able to order them in time *consciously*. In other words, we recognize these events to be separated from the present by a certain *finite* time interval. In this case our short-term memory and internal (neurophysiological) models responsible for unconscious short-term prediction must underlie our perception of temporal order.

Naturally, the time order of two distant events mutually separated by a short time lap, e.g., e_1 and e_2 in Fig. 2.5, is beyond our cognition. It enables us to introduce the synchronic unit of immediate past and the synchronic unit of immediate future—the pulses of retention or protention. To quantify the belonging of events to these synchronic units we can introduce the membership functions of the immediate past and the immediate future in a similar way:

$$m_p(t, t_0) = \begin{cases} 1, & \text{if } t_0 - t \gg \tau_0\,, \\ 0, & \text{otherwise}\,, \end{cases} \qquad m_f(t, t_0) = \begin{cases} 1, & \text{if } t - t_0 \gg \tau_0\,, \\ 0, & \text{if otherwise}\,. \end{cases} \tag{2.2}$$

These three functions may be assumed to obey the equality

$$m_p(t, t_0) + m_n(t, t_0) + m_f(t, t_0) = 1\,, \tag{2.3}$$

however other rules of their superposition are also feasible.

It should be emphasized that the unity of short-term memory and sensory memory endows our perception with sensory aspects and enables us to experience the immediate past and the immediate future as coming into being right now (e.g., Cowan 2008, 2015; Dharani 2015).

The events separated from the current instant by time lags about τ_a take a special place in the extensionalist account of temporal experiences, in Fig. 2.5 the events e_3 and e_6 illustrate it. In this case such events may be equally attributed to the immediate now and the immediate past or the immediate future. Exactly this feature allows us to speak about the pulse overlapping. In terms of the membership functions it means that

$$m_p(t, t_0) \sim m_n(t, t_0) \quad \text{or} \quad m_n(t, t_0) \sim m_f(t, t_0)\,. \tag{2.4}$$

The overlapping of these synchronic units endows them with direct interconnection and justifies the introduction of diachronic unit possessing three temporal parts. On the one hand, the diachronic unit integrating these three parts may be treated as a whole—the experience now. On the other hand, these parts bear temporal properties accessible individually for our consciousness. In this way the experiential now becomes a complex object with several interrelated parts and different time scales forming, maybe, some hierarchy. In mathematical terms the diachronic unity enables us to introduce the notion of conscious degree of presence F_p, F_n, F_f to quantify the individual contributions of the immediate past, the immediate now, and the immediate future to the diachronic unit, respectively (Fig. 2.5). How the conscious degree of presence and the corresponding membership functions are interrelated will be clarified in developing the theory of human temporality.

The overlapping between the immediate past, the immediate now, and the immediate future endows the stream of experience with continuity. Indeed, even if we cannot perceive the current instant of objective time as some object on its own, we do recognize how some events happened just now *gradually* become objects of our memory with experiential trace remaining in our *now*. This continuity brings into focus the concept of stream of consciousness as a certain process.

2.4.5 Tripartite Structure of Experiential Now as a Gateway to Its Multi-dimensional Dynamics

In this subsection we summarize the discussion about composite extensionalism. First of all, we want to underline that, from our point of view, there is no irreconcilable contradiction between extended atomism and composite extensionalism. Both of them deal with the synchronic units as the basic "bricks" of human temporal dimension. The main difference is that extended atomism accepts experiential now to be just one synchronic unit—atomic experience, whereas composite extensionalism accepts experiential now to comprise three such units—the immediate past, the immediate now, and the immediate future.

As illustrated below, this tripartite structure of experiential now grants our mind conscious and experiential access to the a number of basic properties characterizing the temporal arrangement of perceived events. In this sense this tripartite structure opens a gateway toward multi-dimensional dynamics of mental images of event succession.

The presented analysis of extensionalism allows us to state that the stream of consciousness based on the tripartite temporal structure of experiential now—the diachronic unit—is a continuous process implying the following:

- The immediate past, the immediate now, and the immediate future are synchronic units—the indivisible elements of human temporal dimension recognized as whole entities with own properties. Due to these synchronic units being extended in objective time the properties attributed to them *individually* bear the information about the averaged characteristics of the event time arrangement. In particular the mean position and the mean velocity of an observed moving object can be attributed to these units as features currently experienced independently of each other.
- The immediate past, the immediate now, and the immediate future are united in our current experience and, in this way, form another element of human temporal dimension—the diachronic unit. The diachronic unit contains temporal parts—the three synchronic units—with their individual properties. Nevertheless, it is also perceived as a whole entity—experiential now—with some properties attributed to it directly. These own properties of experiential now can be derivatives of the individual properties of the immediate past, the immediate now, and the immediate future. Speaking about an observed moving object the tripartite structure grants our consciousness a direct access to experiencing the object acceleration and jerk (time derivative of acceleration). In this way the phase space of the corresponding mental image may contain the spatial position of moving object, its velocity, acceleration,

and jerk.[7] For comparison, we note that within Newtonian mechanics the phase space of physical object consists of only the spacial position and velocity. Thereby the dynamics of mental images does not fit the paradigm of Newtonian mechanics and requires an individual mathematical formalism.

- The immediate past, the immediate now, and the immediate future overlap with one another, which reflects their fuzzy nature. This overlapping is implemented at the level of individual events when one event may be attributed to both the neighboring synchronic units. The overlapping of these temporal parts of experiential now enables them to continuously and gradually convert into one another, which endows our cognition with the feeling of passage of time.
- In mathematical terms the relationship between the given synchronic units (the level of mind) and the constituent events (the level of physical reality) can be described using the notion of membership functions, Eqs. (2.1) and (2.2), and similar ones aggregating other properties of perceived events. In these terms, the continuity of consciousness is represented by the *perceived gradual* changes of the membership functions

$$m_p(t, t_0),\ m_n(t, t_0),\ m_f(t, t_0)$$

as *objective* time t_0 goes on although objective time t_0 on its own is not perceived. In other words, the mind can perceive time flow through the change of the three synchronic units. In this sense the change must be prior with respect to time in the mind.

We want to emphasize that the latter statement is not in a contradiction with the psychophysical conception of time perception turning to the accumulation of internal clock ticks (Sect. 1.4). These ticks are just perceived events.

2.5 Time Scales of Temporal Experience: Neurophysiological Data and Models

The statement that the experiential now has a temporal structure with complex properties is one of the cornerstones in our constructions of human temporality. In the previous sections we presented a general description of this structure.

In the present section we want to justify the accepted assumptions and for this purpose turn to available neurophysiological data. These data concern how we perceive various properties of temporal arrangement of non-simultaneous events at different time scales.

---

[7] In particular, we have met the necessity of employing this 4D-phase space for describing human actions in the car-following process based our experiments with car-driving simulator (Lubashevsky 2017, Sect. 7.6).

Keeping in mind the account of Pöppel (1997, 2009) we discern four levels in the integration of event properties over objective time in the formation of experiential now:

- The first level corresponds to the fuzzy boundary separating (*i*) the region of time scales where the information about event arrangement is completely lost in experiential now from (*ii*) the region where this information, at least partly, is aggregated in its properties. This level is discussed in Sect. 2.5.1 devoted to the lower boundary of temporal order perception.
- The second level corresponds to the region where the information about event arrangement is unconsciously integrated into certain temporal entities. These entities allow for the possibility of extracting particular properties of event arrangement *inside* the corresponding intervals from the results of integration. The given level is discussed in Sect. 2.5.2 devoted to the scales related to the partial perception of event temporal arrangement.
- The third level corresponds to the conscious integration of perceived events into a coherent global temporal structure on scales about a few seconds. This structure representing directly the experiential now comprises temporally discriminated units of the immediate past, the immediate now, and the immediate future. At this level we meet the intra-relations between these units as whole entities as well as the inter-relations between them and the experiential now which admits interpretation in terms of the bottom-up and up-down causation. In Sect. 2.6 we discuss the available neurophysiological data arguing for the noted properties of the third level.
- The fourth level, noted here for completeness, corresponds to complex present to be discussed in Sect. 3.1 where the experiential now taking the central position is one of the three constituent elements.

Naturally, these levels partly overlap and in the intermediate zones the experience of event succession exhibits some features of both the neighboring levels. A sophisticated review of the corresponding experimental data and their discussion from different standpoints can be found in Wittmann (2011, 2015), Elliott and Giersch (2016).

2.5.1 Lower Boundary of Temporal Order Perception

According to neurophysiological data our sense organs respond to external stimuli at different rates. The eyes detect changes much more quickly than the tongue detects changes of test.

In this subsection we want to estimate the lower boundary of time scales contributing to the relationship between the experiential now and current changes. Below this boundary all the details of event temporal arrangement are completely lost for the human mind. The desired estimates may be extracted from the neurophysiological data for the time thresholds below which humans cannot perceive any aspect of before-after relation. On shorter time scales events non-simultaneous in objective time are experienced as strictly simultaneous.

The given threshold is usually estimated turning to psychophysical experiments assessing the perception of non-simultaneity of two events. Unfortunately there can be given no precise estimate of this threshold because the sensory modalities are quite distinct from each other in response delay. The auditory and visual modalities are the quickest ones, their thresholds are about a few milliseconds. The other sensory modalities, e.g., test and tactile sensations have higher thresholds getting some tens in milliseconds with respect to non-simultaneity (e.g., Vatakis and Spence 2007; van Wassenhove 2009). Inter-modal stimulation leads to even higher thresholds, however, perception of stimuli whose processing involves several modalities simultaneously is a spacial case and lies outside the scope of the present book.

Empirical evidence shows that temporal order thresholds across modalities may be estimated on the average as 20–60 ms.[8] For a detailed description of the noted data a reader may be referred to Fink et al. (2006), Wittmann (2011), Babkoff and Fostick (2013), Elliott and Giersch (2016).

The noted neurophysiological data allow us to suppose that the temporal structure of experiential now cannot contain inhomogeneities whose scales are less than a certain value τ_0. The particular values of τ_0 can change essentially for different sensory modalities, however $\tau_0 \lesssim$ 20–60 ms may be adopted as a characteristic estimate of τ_0.

As far as mathematical aspects are concerned, it turns out that the time scales characterizing the other levels of event integration within the experiential now are much large than 20–60 ms, which will enable us to regard τ_0 as a small parameter. In particular, in describing the contribution of point-like events to the experiential now we may average the moment of their appearance over time intervals of duration about τ_0.

2.5.2 *Partial Perception of Event Temporal Arrangement*

In Sect. 2.4 we introduced the notion of synchronic unit which plays a crucial role in our constructions. The synchronic unit is the basic element of human temporal dimension; other elements of this dimension are composed of synchronic units.

[8] This averaged estimate of perception threshold can be explained turning to the temporal binding by gamma-band synchrony hypothesis (Singer 1993; Singer and Gray 1995; Gray 1999). This binding hypothesis states that neurons in anatomically distinct regions can synchronously encode different features of an external stimulus by firing together in a single gamma cycle (see for a review Engel et al. 2001; Buzsáki and Draguhn 2004; Ahmed and Cash 2013). In humans the gamma type neural oscillations are characterized by the 25–100 Hz frequency-band, which corresponds to time scales about 10–40 ms (Gold 1999; Hughes 2008, for are veiw). The relationship between gamma oscillations and cognitive phenomena was also analyzed by Joliot et al. (1994), Başar-Eroglu et al. (1996), Fries et al. (2007), Ehm et al. (2011).

For example, the diachronic unit consists of three synchronic units representing the immediate past, the immediate now, and the immediate future forming together the experiential now.

Within our account the following properties are attributed to the synchronic unit:

1. It can be conceived of as a *blurred instant* of human temporal dimension. The blurriness is the basic feature distinguishing human time from objective time. In Newtonian mechanics the instants of objective (physical) time are point-like elements; also for this reason theory of human temporality cannot meet the paradigm of Newtonian mechanics.
2. As well as the point-like instant is indivisible in physical reality, the synchronic unit (blurred instant) is indivisible in the mind.
3. Due to their blurriness synchronic units can mutually overlap.
4. As objective time goes on, the synchronic units of immediate past, now, and future continuously convert into one another. It endows us with the sense of the continuous passage of time.
5. The synchronic unit is extended in objective time. Thereby the synchronic unit bears partial information about the temporal arrangement of perceived events. However the corresponding properties of event temporal arrangement are attributed to the synchronic unit as a whole. This feature, in particular, enables us to *perceive directly* not only the position of objects but also their motion.

In the present subsection we focus the main attention on Property 5 and discuss the available neurophysiological data and phenomena justifying it and elucidating its details.

Event Temporal Arrangement and Its Perception

Leaping ahead, we note that the synchronic unit corresponds to time scales from 20–60 ms up to 300–500 ms. These rather wide estimates of the lower and upper boundaries are caused by that the sensory systems (visual modality, auditory modality, taste modality, etc.) significantly differ from each other in the rate of their response to changes. The lower boundary is determined by the perception threshold τ_0 of time order discussed in Sect. 2.5.1. The upper boundary $\tau_s \sim$ 300–500 ms actually separates the synchronic and diachronic units in time scales.

We want to emphasize that the upper boundary of time scales of synchronic unit also plays the role of the watershed between conscious and unconscious perception of event temporal arrangement. The matter is that the temporal processing of intervals longer than 300–500 ms is cognitively mediated, whereas fast and parallel temporal processing of shorter intervals is not accessible to cognitive control (Michon 1985), for a review see also Buonomano et al. (2009), Spencer et al. (2009), Rammsayer and Troche (2014), White (2020). It is worthy of noting that as early as 1889 Hugo Münsterberg (1863–1916), a German-American psychologist, put forward a hypothesis that time scales less than one third of a second are perceived directly, whereas temporal information on longer time scales has to be reconstructed by higher mental processes.

Below in the present subsection we argue that the human sensory systems integrate events over time scales about τ_s. However this integration is not just simple averaging on τ_s, the temporal structure of integration is rater complex. The complexity of this structure enables the sensory systems to extract partial properties of the event temporal arrangement on intermediate scales including scales much less than τ_s from the result of integration. In our consciousness these extracted properties are attributed to the synchronic unit as a whole entity.

Starting from the second half of the XX-century, there has been a great deal of studying human perception of subsecond intervals and nowadays a vast literature concerning this issue is available, a reader may be referred to recent reviews by Grondin (2010), Block and Grondin (2014), Wearden (2016), Grondin et al. (2018). Here we will confine our discussion only to some neurophysiological data illustrating directly the issues under consideration.

Non-simultaneity and temporal order of events: To illustrate the complex structure of information processing on scales about τ_s, at first, we want to note the results of experiments concerning the perception of:

- the event non-simultaneity which is often referred to as the perception of the stimulus onset asynchrony in psychological literature and
- the temporal order or the temporal order judgment.

The two aspects of event perception can be discriminated experimentally because there are situations when being aware that two events did not occur simultaneously, we still are unable to tell which one of them occurred first (e.g., Wittmann 2011).

As experimentally demonstrated, humans are able to compare the duration of signals on scales of order 50 ms or even shorter (e.g., Rammsayer and Grondin 2000; Rammsayer and Troche 2014; Rammsayer et al. 2015). The lower temporal boundaries characterizing the perception of stimulus onset asynchrony and the temporal order judgments are also of similar scales (for a review see, e.g., Weiß and Scharlau 2011; Yarrow et al. 2016). As should be expected, these time scales are of order of the threshold $\tau_0 \sim 20$–60 ms characterizing the human ability to perceived time order in principle (Sect. 2.5.1).

The given estimates for the perception of stimulus onset asynchrony and time order judgments seem to be in direct contradiction to the statement about the integration of events on scales $\tau_s \gg \tau_0$. However, as follows from the experiments to be described below, we do not recognize such short time intervals between events immediately when these events occurred in the reality. A sensory system, first, integrates events of scales τ_s and, only then, short-term characteristics are extracted from the result and become accessible to the mind.

Perception integrity on scales of τ_s: To illustrate the existence of this integrity we turn to two type phenomena.

First, it is several illusions found by the Nakajima group in the auditory perception of two successive intervals whose total duration is about 300–400 ms (for a review see ten Hoopen et al. 2008). Their core is the *time shrinking* phenomenon (Nakajima et al. 1991),[9] which in the bilateral case is reflected in the underestimation of longer intervals and the overestimation of shorter intervals (Sasaki et al. 1998; Miyauchi and Nakajima 2005). The integrity of this phenomenon is justified by the fact that it undergoes step-wise breakdown when the interval difference exceeds a threshold about 100 ms (ten Hoopen et al. 2006). In the latter case the duration of each interval is evaluated separately and no considerable underestimation or overestimation is detected. The time-shrinking is also observed within the visual modality and exhibits similar properties (Nagaike et al. 2016, and references therein for a review).

Second, it is phenomena investigated by Scharnowski et al. (2009) and Pilz et al. (2013) which argue not only for the perception integrity but also for the *multi-channel* information processing on scales of τ_s. The authors analyzed perception of two successive stimuli when after a certain time interval mental processing of information is disturbed in some way. As demonstrated, for the delay time in disturbance onset about 45–90 ms only the processing of the first stimulus is affected and, as a result, the second stimulus becomes dominant. For the delay time about 95–420 ms the situation is opposite; the second stimulus is depressed and the first one dominates. It is only after delay time about 400–500 ms the perception of both the stimuli is not disturbed (for a review see Elliott and Giersch 2016). These results admit the interpretation as the parallel processing of first and second stimuli separately and the final fusion determining their cumulative perception.

It is worthy of noting that the idea about the perception integrity is the gist of the *temporal profile model* by Stelmach and Herdman (1991) and the *model of temporal order perception* based on the concepts proposed by Sternberg and Knoll (1973) and then elaborated by Schneider and Bavelier (2003); for a review a reader may be referred to García-Pérez and Alcalá-Quintana (2018). Besides, both the models suppose that different aspects of information about event temporal arrangement are processed separately via different particular mechanisms—*independent channels* in terms of Sternberg and Knoll (1973). The idea about multi-channel information processing before decision making is also supported, e.g., by Bruno and Cicchini (2016).

The upper integration limits: The discussed phenomena illustrate not only the complex temporal structure of event integration hidden for our consciousness but also justify the estimate of its upper temporal boundary $\tau_s \sim 300$–500 ms accepted above.

The given estimate is also justified by the paradigmatic experiments of tapping a table with our fingers. When the inter-tap interval is substantially shorter than 300 ms this repetitive movement is too fast for us to be able to have a representation of individual finger taps. Only when the movement slows down and the inter-tap interval gets 300 ms individuated finger taps as following one another are experienced (e.g., Peters 1989; Wittmann et al. 2001; Wittmann 2016). The duration about 300 ms seems to be a typical threshold in perceiving the time order in a sequence of events

[9] The time-shrinking phenomenon in auditory time perception has been first reported by Nakajima and ten Hoopen in (1988) in Japanese.

via the visual and auditory modalities (for a review see Warren and Obusek 1972; Ulbrich et al. 2009). In experiments involving more complex tasks or slower sensory modalities the duration of integration can be longer. For a detailed description of tapping experiments a reader may be referred to Repp (2005, 2006), Repp and Su (2013).

Neurological Mechanisms of Timing on Scales of τ_s

In the present present part of this subsection we touch some basic features of neurological mechanisms underlying the perception integrity discussed above. These mechanisms have to be characterized as distributed, i.e., attributed to the brain as a whole (Mauk and Buonomano 2004; Paton and Buonomano 2018; Bakhurin et al. 2017). This feature, in particular, is one of the arguments for the applicability of phenomenological approach to describing temporal experiences. Naturally there are certain correlations between mental operations with time and different parts of the brain (see, e.g., Merchant et al. 2013; Aghdaee et al. 2014; Gupta 2014, for are view).

As far as particular neurophysiological mechanisms of human timing on scales in question are concerned, all of them may be divided into two big classes. One class comprises the 'dedicated' mechanisms assuming human timing to be implemented via specialized neurological processes belonging to the level of the brain as a whole. For a detailed description of dedicated models a reader may be referred, e.g., to Mauk and Buonomano (2004), Ivry and Schlerf (2008), Goel and Buonomano (2014), Hass and Durstewitz (2014), van Rijn et al. (2014).

The other class—the 'intrinsic' type models—accepts human timing to be based on local and common properties of neural circuits (for a review see Mauk and Buonomano 2004; Ivry and Schlerf 2008; Goel and Buonomano 2014; Hass and Durstewitz 2014). Below we will confine our discussion to the intrinsic models because in the last decade there have been accumulate clear evidence for their significant role on scales under consideration (Spencer et al. 2009; Goel and Buonomano 2014, for are veiw).

The *state-dependent network* models (Buonomano and Mauk 1994; Mauk and Donegan 1997; Buonomano and Merzenich 1995; Buonomano 2000; Medina and Mauk 2000) are actually the heart of the 'intrinsic' account. Its gist is that a wide variety of timing tasks as well as many other cognitive processes must be implemented in the brain based on some common mechanism not specialized in performing a particular operation.

The notions of *active* and *passive* states of neural networks together with the *recurrent connectivity* between these states take the central place in such models (for a review see, e.g., Mauk and Buonomano 2004; Karmarkar and Buonomano 2007; Buonomano and Maass 2009; Buonomano 2014). Similar notions underlie our further constructions of human behavior within complex present, so let us briefly discuss their particular implementation within the state-dependent network models.

An *active* state of neural network is specified by a particular arrangement of neurons that are firing at a given time moment. Inactive neurons, however, are also able to contribute to the brain functioning because even a silent network can bear a memory of what happened in the recent past. In this sense the not-firing neurons in hidden states are also active but their activity is not "visible" for the downstream neurons (Buonomano 2014). So the dynamics of neural networks is determined by the cumulative contribution of active and hidden states together with external stimuli.

Even for fixed external stimuli neural networks change their states continuously and in the space of their possible states the neural networks form highly complex temporal patterns. The gist of state-dependent network models is the assumption that an external stimulus influencing the network dynamics gives rise to the emergence of certain cyclic variations in the network states. These variations bear the information about the stimulus. Put another way, these networks possess complex properties of returning to their initial states (or going in their close proximity)—the intrinsic *recurrent connectivity*—and external stimuli just select certain sets of this recurrence. The given feature endows such neural networks with the capability for decoding various time properties of events practically in the same way (e.g., Paton and Buonomano 2018; Goudar and Buonomano 2018).

In our interpretation, the state-dependent network models posit the existence of some cooperative variables such that, on the one hand, change slow on time scales $\tau \gg \tau_s$. On the other hand, they govern the variations in the state of neural networks within time intervals about or shorter than τ_s.

We will also turn to the idea of two states in human actions—active and passive ones—where actually both of them are active in the mind but the latter is hidden for the external observer. Besides, humans response to current events will be related to the existence of slow variables governing fast variations in the states of the immediate past, the immediate now, and the immediate future as well as correlations between them.

In *summary* to this subsection, the presented discussion argues for the introduction of synchronic unit as the elementary entity of human temporal dimension possessing the following properties:

- The synchronic unit emerges in the mind as an indivisible (so having no temporal parts) time entity responsible for unconscious integration of event temporal arrangement. Its indivisibility in the mind is due to the unconsciousness of this integration and the underlying neurological mechanisms may be categorized as distributed, intrinsic ones.
- The processing of the corresponding information can be implemented via different independent channels with own properties.
- The time scales of integration are bounded from above by a value $\tau_s \sim 300$–500 ms, which may be treated as the characteristic dimension of synchronic unit in objective time. After integration on scales of τ_s the result becomes accessible to the mind.
- The synchronic unit possesses a certain complex time structure hidden for our consciousness characterized by the following features:

- It just averages events on scales about $\tau_0 \sim 20$–60 ms such that all the information about their temporal arrangement on smaller scales is completely lost.
- On time scales from τ_0 to τ_s ($\tau_0 \lesssim \tau \lesssim \tau_s$) the synchronic unit is able to extract partially properties of event temporal arrangement from the result of integration. These properties are attributed to the synchronic unit as a whole entity and become accessible to the mind. Speaking about the mathematical representation of this integration, the given capability of partial extraction enables us to suppose that the corresponding kernel of integration may have a multi-scale form individual for each channel.
- In describing the synchronic unit the inequality $\tau_0 \ll \tau_s$ may be accepted beforehand.

2.6 Conscious Level of Experiential Now

In the previous subsections we considered the structure of diachronic unit that comprises three synchronic units and the characteristics of the synchronic unit as an indivisible entity. Here discuss the contents of the diachronic unit as well as that of its constituent elements. We will turn to the corresponding neurophysiological data and phenomena to justify our constructions.

Figure 2.6 illustrates the gist of our account of how the information about perceived events becomes accessible to the mind where the diachronic unit plays the role of converter. In brief, via sensory organs the subject perceives events that after a certain delay become accessible to the mind. Because this delay is distributed in objective time rather than characterized by a fixed value, what the subject perceives as the present is also distributed over objective time and is characterized by a certain duration $\tau_s \lesssim 0.5$ s. Above we have introduced the concept of *immediate now* in order to refer to this element of consciousness and discussed its characteristic properties.

Events can be *experienced* as present ones even when the corresponding physical stimuli *currently* do no affect directly the sense organs. It is due to the sensory memory and inner neurophysiological models we gain during our life for short-term, unconscious anticipation of events in the nearest future. As will be clarified below, such experience of events along with the conscious experience of the immediate now belongs to the lower level of consciousness, in other words, it is not reducible to more elementary fragments of consciousness. Besides, the immediate past, the immediate now, and the immediate future are closely interrelated and cannot be studied separately without losing some their basic features. Indeed, the content of the immediate now quickly becomes the content of sense memory, both of them unconsciously generate the immediate future, which in turn becomes the immediate now due to the passage of objective time.[10] It is a reason for treating them as the constituent components of one entity—the diachronic unit—with the tripartite structural integrity and

[10] This issue is discussed in detail in Sect. 5.6 and underlies the principle of self-consistency of human perception on different time scales.

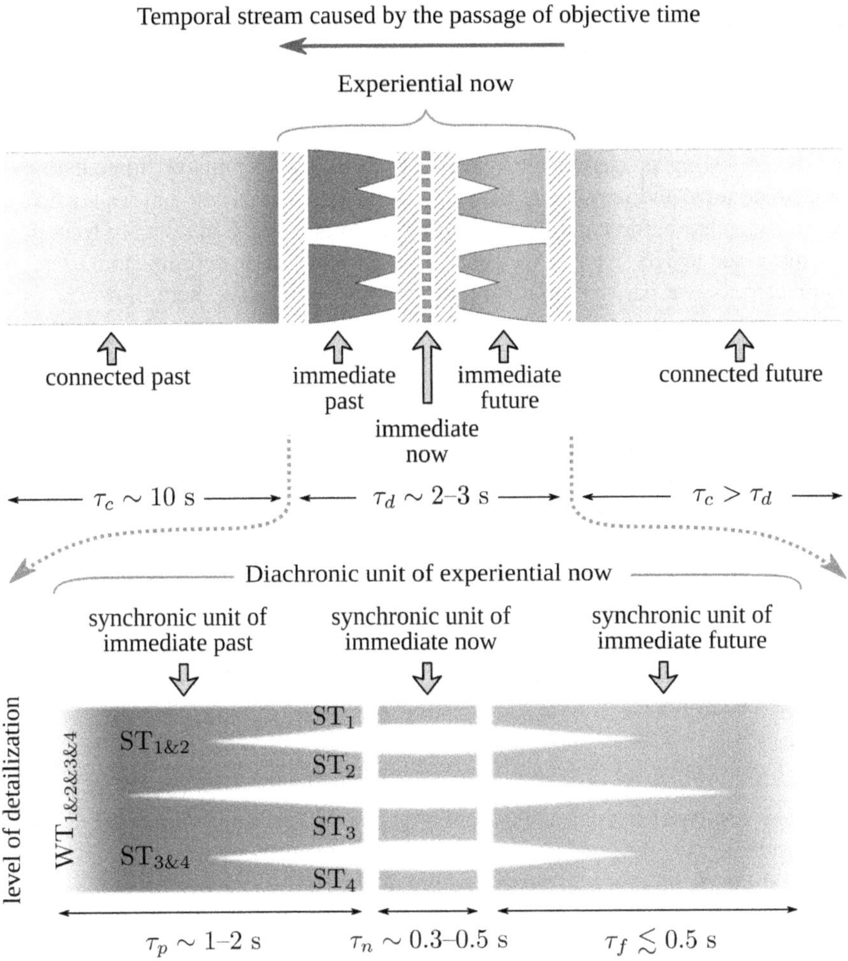

Fig. 2.6 Structure and contents of experiential now. At the neurophysiological level the immediate past is based on short-term encoding and short-term consolidation. Through the two processes traces of sensory memory (S-traces) that are initially created just after the integration of perceived information within the immediate now (here ST_1–ST_4) are gradually converted into more stable traces of working memory (W-traces), here $WT_{1\&2\&3\&4}$

own properties. It should be noted that due to the tripartite structure the diachronic unit is able to describe intentional, goal-oriented behavior on short-term scales (cf. Tanida and Pöppel 2006; Pöppel 2009; Pöppel and Bao 2014; White 2017).

2.6.1 *The Diachronic Unit and Its Constituent Elements: Basic Properties*

To be determined in discussing particular phenomena, at the beginning we want to list the following properties attributed to the synchronic units of immediate past, immediate now, and immediate future as well as the diachronic unit as a whole. It should be noted beforehand that some of these properties were already formulated and discussed in Sect. 2.5.2 as the generic properties of the synchronic unit. Here, for completeness, we partly reproduce the previous statements where, however, other aspects are accentuated.

First, concerning the synchronic unit of immediate now (labeled with N):

N.1: The unconscious processing of information about events perceived as having occurred just now must be completed. In other words, the properties of events happened within the time horizon $\tau_n \sim 0.3$–$0.5\,\text{s}$ ($\tau_n \sim \tau_s$) are integrated in some way and become accessible to the mind. Below the value τ_n will be referred to as the characteristic duration of immediate now. This piece of information is retained in memory (sensory memory) as a certain trace.

N.2: In the mind these time-averaged properties are attributed to the unit of immediate now and regarded as the characteristics of the current state of environment. The mind can operate only with these properties in analyzing the current state of environment.

Second, concerning the synchronic unit of immediate past (labeled with P):

P.1: The results of unconscious integration at previous time moments within the horizon $\tau_p \sim 1$–$2\,\text{s}$ ($\tau_p \sim \tau_d$) are saved in some way in memory. It implies that the memory traces generated previously within the horizon of 1–2 s fuse together to form the consolidated unit of immediate past. The time scale τ_p may be regarded as the duration of immediate past in objective time. The information about events is aggregated into the properties of the immediate past and attributed to it as a whole.

P.2: The information retained in the unit of immediate past, i.e., its content remains active. In other words, the properties of immediate past

- are freely accessible to the mind, i.e., accessible without any retrieval threshold;
- can be involved into the unconscious processing of current events, i.e., affect the content of the immediate now unit;
- are able to change due to the interaction between the synchrony units making up the present diachronic unit—the experiential now.

Third, concerning the synchronic unit of immediate future (labeled with F):
The feasibility of introducing the synchronic unit of immediate future as a individual constitutive element of human temporal dimension is due to the following properties.

F.1: There are neurological mechanisms endowing us with the ability to experience events of the recognized nearest future as ones having come into being already. For this reason the immediate future may be regarded as a component of the experiential now. The characteristic duration τ_f of the immediate now regarded as the time horizon within which experiencing future events is possible can be estimated as $\tau_f \lesssim 0.5\,\text{s}$.

F.2: The brain functioning in processing sensory signals is such that:

(*a*) the information accumulated within the immediate now and the immediate past is used for the event prediction on scales about τ_f;

(*b*) this short-term prediction is involved into higher-order cognitive processes in a way enabling humans (*i*) to predict the dynamics of observed objects on much longer scales and (*ii*) to choose quasi-optimal actions in responding to the environment.

The given properties endows the immediate future with the anticipation efficiency.

F.3: The details of information processing within the immediate future are inaccessible to the mind and, so, the functioning of the immediate future is unconscious. Only the final result of this processing becomes accessible to the mind as the integral properties of perceived events attributed to the immediate future as a whole. For this reason, the immediate future admits the interpretation as a basic (indivisible) component of human temporal dimension—the synchronic unit.

Fourth, concerning the diachronic unit (labeled with D):

D.1: The properties attributed to the diachronic unit as a whole:

- differ from the individual properties of the synchronic unit of immediate past, the synchronic unit of immediate now, and the synchronic unit of immediate future;
- aggregate the individual properties of the three constituent units. This aggregation is implemented consciously in the mind[11];
- can be used by the mind as a certain conscious channel of information about the current state of environment and its dynamics;
- have the power of top-down causation. It is due to the mind being able to affect the immediate past, the immediate now, and the immediate future within the experiential now as some external factor.

D.2: The interrelations between the diachronic unit and its constituent synchronic units govern complex processes of event integration on multiple scales. Therefore the hierarchical temporal structure of experiential now is not strictly fixed, e.g., the duration of its various elements may change depending on particular situations. This hierarchical structure reflecting the past-present-future

[11] The conscious level of this aggregation, however, may be regarded as the very basic one for the human mind. In some sense, this aggregation is implemented at the boundary between unconscious and conscious processes.

integrity of experiential now is determined by the integration of perceived events into a coherent global temporal pattern on scales about $\tau_d \sim 2$–3 s.

Below we want to discuss neurophysiological phenomena justifying the adoption of the aforementioned properties.

2.6.2 *Properties of Immediate Now*

The corresponding phenomena arguing for the properties N.1 and N.2 were already discussed in Sect. 2.5.2. They are the time shrinking illusion and the masking effect characterizing the unconscious integration of events on scales $\tau \lesssim \tau_n \sim 0.3$–$0.5$ s. Only the cumulative results of this integration become accessible to the mind. The stability of the observed *integral* phenomena with respect to human individuality and various accompanying effects enables us to regard them as phenomena reflecting the basic features of human perception and being natural and necessary rather than abnormal (ten Hoopen et al. 2008). In Fig. 2.6 the immediate now is represented by the central fragment colored in orange.

2.6.3 *Properties of Immediate Past*

In our account the immediate past is characterized by several aspects studied individually in cognitive sciences, which is reflected in the structure of discussing neurophysiological data and phenomena underlying P.1 & P.2.

Experiential presentness of immediate past: The term immediate past as a constituent component of the experiential now describes perception of events that are currently perceived as present, even if the subject recognizes that they have occurred and are already in the past. This experiential presentness of the immediate past becomes possible due to the sensory memory. In a simplified form, the short-term retaining may be represented by the diagram (cf. Roediger et al. 2017; Cowan 2017b)

$$\begin{aligned}&\text{perception} \rightarrow \text{sensory memory} \rightarrow \text{fragile memory}\\&\qquad\rightarrow \text{(short-term) working memory}\end{aligned} \tag{2.5}$$

where the upper line represents the conversion of information implemented directly within the immediate past.

Sensory memory retains the initial short-lived details of perceived events; how things looked, sounded, smelled, etc. (e.g., Cowan 2008, 2015). Summarizing the previous studies, including the seminal investigations by Sperling (1960), Cowan (1988) argues that there are two stages of sensory memory where the short stage—a literal sensory afterimage—is characterized by the duration of a few hundred milliseconds. Speaking about the interaction between the immediate past and the immediate now, the short stage of sensory memory may be regarded as the result of event integration just completed within the immediate now and transferred to the immediate past. In Fig. 2.6 it is a relatively short part of the immediate past adjacent to the immediate now and colored with dark orange.

The following stage of sensory memory is substantially longer than the first one; according to Cowan it lasts for seconds. It is essential that within this stage the afterimage before finally being fixed in the working memory possesses properties including both sensory features such as color, loudness, smell, etc. as well as more abstract features such as shapes or meanings. Besides, Cowan (2001) suggested that the capacity of this memory stage is limited to three to five items when each item is an integrated unit; for a review see also Cowan (2015).

Referring to visual modality it has been recently supposed that the transition from sensory memory to working memory is implemented via an intermediate stage—fragile memory (for details and review see Sligte et al. 2010; Vandenbroucke et al. 2011; Pinto et al. 2013). This intermediate stage is characterized by several features:

- its capacity is less than the capacity of sensory memory and higher than that of working memory (e.g., Sligte et al. 2010);
- its lifetime is of seconds (e.g., Vandenbroucke et al. 2011);
- it can be easily overwritten by new stimulation and its dependence on attention is rather weak (Vandenbroucke et al. 2011; Pinto et al. 2013).

The fragile memory seems may be related to second stage of sensory memory discussed above.

The noted properties of the sensory memory and the fragile memory:

- substantiate the possibility of introducing the concept of the immediate past as the component of experiential now that belongs to the past but is experienced in the human now;
- justify the accepted estimate $\tau_p \sim 1$–2 s for the duration of the immediate past and its adaptability to particular conditions of event integration;
- determine the number of independent variables we may use for describing the properties of the immediate past as a whole.

Integrity of immediate past: At the neurophysiological level the immediate past is the result of fusion of sensory memory traces (S-traces). Figure 2.6 illustrates this fusion by the region with color transition from dark orange to blue. S-traces are created when the information about perceived events accumulated within the immediate now is passed to the sensory memory. The transmission of this information and its initial encoding are thought to be fast and occur in parallel. As a result, such traces bear many details about perceived events. However, S-traces are lost quickly if they are not attended and converted into more stable traces of working memory, W-traces (e.g., Cowan 1984). W-traces do not bear sensory information, so the conversion of S-traces into W-traces determines the boundary between the immediate past and the connected past domain (Fig. 2.6).

The fusion of S-traces is implemented via two processes called short-term encoding and short-term consolidation. Short-term consolidation governs the dynamics of short-term encoding (but not its content) as well as endows W-traces with durability and robustness. For the two processes to begin attention must be focused on the corresponding events or objects and they last until the completion or until attention is shifted toward another object. For details and review a reader may be referred, e.g., to Jolicœur and Dell'Acqua (1998), Nieuwenstein and Wyble (2014), Ricker and Cowan (2014), Ricker (2015), Bayliss et al. (2015), Ricker and Hardman (2017), Ricker et al. (2018), Langerock et al. (2018). In our account we consider the two processes the main mechanisms determining the dynamics of immediate past, with short-term consolidation endowing the immediate past with integrity.[12]

Short-term consolidation is characterized by a certain autonomy reflected, first, in its independence of encoding (Jolicœur and Dell'Acqua 1998; Nieuwenstein and Wyble 2014; Ricker and Cowan 2014). Second, having started, it can be completed even when the stimulus is no longer physically present (e.g., Bayliss et al. 2015; Ricker and Hardman 2017; Ricker et al. 2018).

Short-term consolidation is a substantially nonlinear process. In particular, the running consolidation process can capture attention and prevent the use of attention for other purposes. It gives rise to the phenomena called *attentional blink*, at least, is one of its main explanations (e.g., Wyble et al. 2009, 2011; Lagroix et al. 2012). The attentional blink is a marked impairment in identifying the second of two targets presented in close succession (Raymond et al. 1992; Shapiro et al. 1997); for a recent review on the attentional blink a reader may be referred to Dux and Marois (2009), Martens and Wyble (2010).

The nature, especially the time course of short-term consolidation is rather far from being understood well (Ricker and Hardman 2017; Langerock et al. 2018). It concerns, in particular, whether a sequence of S-traces is converted into the corresponding W-traces as one-to-one, many-to-one, or in a more sophisticated scenario depending on their similarity and complexity. In this context we want to note two neural-type models, the eSTST model (Wyble et al. 2009; Martens and Wyble 2010; Wyble et al. 2011; Nieuwenstein and Wyble 2014) and the ST^2 model (Bowman and Wyble 2007; Craston et al. 2009). They have common roots and allow for parallel consolidation governed by the competition for attention between traces or its sharing, the trace interference, and the dynamic interaction between the amount of attention and memory traces. The possibility of parallel consolidation reflected in complex properties of attentional blink has been demonstrated experimentally (e.g., Mance et al. 2012; Rideaux et al. 2015). There are, naturally, other models for the trace consolidation, e.g., the boost-and-bounce model (Olivers et al. 2007; Olivers and Meeter 2008) turning to the idea of attention activation-inhibition; for their discussion and comparison a reader may be referred to Lagroix et al. (2012), Ricker (2015).

---

[12] In the cited publications short-term consolidation is analyzed mainly for visual working memory or cases where visual modality plays the governing role. However, this phenomenon observed also within other modalities, in particular, for auditory working memory (e.g., Shen and Alain 2011).

Short-term consolidation is affected by similarity of traces. In particular, in subsequent visual input of targets and masks the result shows more interference from similar than from dissimilar masks. Similar masks require more time to consolidate and elicit lower overall performance than did dissimilar masks (Vogel et al. 2006; Blalock 2013). Besides grouping by similarity in working memory is beneficial for its performance (Peterson and Berryhill 2013, see for are view also Li et al. 2018).

Taking into account of the aforesaid about the possibility of parallel consolidation and the similarity effects we have adopted the following parallel, many-to-one scenario. The initial S-traces are divided in a few groups—different channel of short-term encoding—that are processed separately. Within one channel all the active S-traces are merged into one W-trace. It should be noted that consolidation even within one channel is not necessarily related to one property attributed to the created W-trace. Results reported by Stevanovski and Jolicœur (2011) argue for this idea, namely, they demonstrated that equivalent attentional resources can be necessary for consolidation of single- and multi-feature items. The given fact may be also explained turning to the gist of boost-and-bounce model when at the initiation of parallel consolidation process the required attention is determined by the cumulative demand of all the channels. For a discussion of argument for and against the memory models dealing with multi-feature objects or single-feature traces a reader may be referred to Langerock et al. (2018).

The data and phenomena presented below argue for the following scenario:

- For a single item the consolidation processes seems to be completed between 0.5 and 1 s, however, it can be longer for tasks with heavy demands (Ricker et al. 2018; Langerock et al. 2018, for are view). In particular, Bayliss et al. (2015) and De Schrijver and Barrouillet (2017) provided data for the upper estimate of consolidation duration as 2 s. The individual life of an S-trace just after its creation and before consolidation affects it essentially seems to last for a few hundred milliseconds or even shorter (cf. Vogel et al. 2006). So several S-traces may be involved simultaneously in short-term consolidation.
- The capacity of working memory seems to be limited to three to five items when each item is an integrated unit (e.g., Cowan 2015). So the fusion of several S-traces representing the same object at different time moments is necessary for memorizing other items (see Vogel et al. 2006 and Langerock et al. 2018 for a discussion of this aspect).
- In remembering color patterns—which exemplifies short-term consolidation of information about stimuli *continuous* in nature—two types of W-traces were found (Hardman et al. 2017). They are a single item bearing information about color continuous properties and about two items (on the average) with the corresponding categorical information. So several S-traces especially concerning graduate variations in a property of observed object must fuse together.
- Longer encoding/consolidation times result in better memory performance, which suggests a substantial advantage for remembering features presented in working memory via an integrative way (for a review see Langerock et al. 2018). The

integration has also to affect memory precision; the corresponding experimental evidence is discussed, e.g., by Bae and Flombaum (2013).

- Within short-term memory the time representation is nonlinear, with its compression approximately obeying the logarithmic law (for recent investigations and review see, e.g., Singh and Howard 2017; Singh et al. 2018; Howard 2018). It implies that a sequence of S-traces as a whole has to undergo essential transformation during consolidation.
- W-traces must constantly change to represent dynamic environment adequately. This change can be implemented via either updating or resetting. In the former case some part of existing W-trace is modified, whereas in the latter case actually a new W-trace replaces the previous one. The transition between the two modes is determined by the break of object-to-representation mapping (Balaban et al. 2018). For the reset to become possible the initially created S-traces should fuse to form a certain integral entity. This fusion also should endow the mechanism of active forgetting of unnecessary information via trace interference with efficiency; for discussion of plausible mechanics of forgetting on short-term scale a reader may be referred to Lemaire and Portrat (2018), Rhodes and Cowan (2018), Morey and Cowan (2018), Lewis-Peacock et al. (2018)

The aforementioned facts enable us to accept that the short-term consolidation is the neurophysiological base of the immediate past. As a result, short-term consolidation determines the boundary between the immediate past and the connected past domain (Fig. 2.6) and endows the immediate past with:

- integrity and indivisibility at the level of consciousness provided only S-traces at the final stage of consolidation become accessible to the mind (this issue will be justified and discussed below in more details);
- some temporal structure hidden from the mind whose dynamics is:
  - autonomous in the sense that it is not affected *explicitly* by cognition and, thereby, admits description within the classical formalism of physics,
  - nonlinear and sensitive to external circumstances via distribution of attention enabling also the mind to affect *implicitly* the dynamics of immediate past as a whole[13];
- short-term encoding that:
  - is governed by short-term consolidation,
  - is based on the information about perceived events accumulated within the immediate now which, thereby, plays the role of initial conditions for the encoding process,

[13] Depending on specific external conditions this distribution of attention should make one neurophysiological factors dominant and depress others at the level of cognition. It may explain the existence of *ceteris paribus* laws—the observed regularities in human behavior in spite of many features being outside the direct control—even without reference to the cooperative interaction between the members of some community (for details see Lubashevsky 2017).

- gives rise to the averaging of information about perceived events on time scales about $\tau_p \sim 1$–$2\,\mathrm{s}$ which is borne by the S-traces at the final stage of consolidation.

The immediate past as an element of the basic level of consciousness: Here we start discussion about the relationship of conscious and unconscious in the mind. We want to elucidate the reasons for supposing that the properties attributed to the immediate past as a whole belong to the lower level of consciousness. It means that these properties are accessible to the mind as just given to it rather than admitting derivation from more basic features. Below we will return to this problem turning to the notion of diachronic unit.

In our account of the experiential now the properties of immediate past are directly related to the transitions when S-traces are converted into W-traces, after the short-term consolidation being completed, and, then, they become accessible to the mind. To justify this let us turn to the key features of the modern accounts of working memory mainly those that pose also relevant questions about consciousness (for a review see Hassin, Bargh, Engell and McCulloch 2009; Velichkovsky 2017; Sergent 2018).

※ Generally neurophysiological accounts of working memory, e.g., Baddeley's multicomponent model (Baddeley 2012, for are view), are based on the assumption that working memory operates on consciously represented information. Baddeley's model suggests a hierarchical organization of working memory—the central executive and several enslaved systems for transient storage of modality-specific information. In our account we relate the enslaved systems to the immediate past of the corresponding sensory modalities. Conscious accessibility to these systems is intuitively evident, we are just aware of perceiving external objects (Velichkovsky 2017). This type accessibility can be described within the concept of *phenomenal consciousness* (Block 1995, for are view see also Block 2007a; Overgaard 2018). In Sect. 3.2.5 we will present arguments for treating phenomenal consciousness as the lower level of consciousness, which advocates for the made assertion about the immediate past. The activation-based models (Cowan 1999; Engle 2002; Oberauer 2002a) put forward the question about the role of attention in working memory. These models, maybe tacitly, accept that consciousness can be equated with the content of the focus of attention (Velichkovsky 2017).

※ The relationship between consciousness, attention, and working memory was considered in detail by Lamme (2003) for the visual domain. In particular, he singles out three possible states of perceived information: unconscious, unattended conscious, and attended conscious.[14] Speaking about consciousness he draws a distinction between visual sensory memory[15] and visual working memory. This distinction

[14] A more detailed classification of such memory states taking into account the cognition-induced top-down modulation has be proposed by Jacob et al. (2015).

[15] Actually in analyzing consciousness (awareness) Lamme (2003) deals with the second stage of visual sensory memory—fragile visual memory (cf. Velichkovsky 2017). Fragile memory was already discussed above in the part devoted to the experiential presentness of immediate past.

can be illustrated turning to our everyday experience; we are aware of a detailed visual scene when looking at it but it quickly reduces to a few focal objects when we close eyes. In Lamme's account sensory memory (actually visual fragile memory) is associated with phenomenal consciousness, whereas working memory is associated with *access consciousness*. In Sect. 3.2.5 we will discuss the concept access consciousness in more detail. Here, leaping ahead, we note that access consciousness admits interpretation as a higher level of consciousness; its objects can be consciously analyzed and operated with. As will be clarified further, the properties of diachronic unit attributed to it as a whole are characterized by access consciousness.

※ The relationship between conscious awareness, attention, and working memory is intricate. It turned out that working memory can be involved without awareness into processing perceived information as well as operate on nonconscious representations. Moreover, the process through which the contents of working memory reach awareness is subject to various factors (i.e. attention, intention, motivation, heuristics), which may lead to dissociation between the contents of working memory and the contents of subjective conscious experience (Soto and Silvanto 2014).

Speaking about particular phenomena we want to note the following. Investigations by Soto et al. (2011), Soto and Silvanto (2014), Dutta et al. (2014) provide the evidence of working memory being able to maintain unconscious items. Employing the attentional blink effects Bergström and Eriksson (2014, 2015), Pincham et al. (2016) found arguments for the existence of non-conscious processing of information in visual working memory. The concept of implicit working memory—a system for operative storage and information processing whose functioning is not consciously accessible—has been elaborated by Hassin (2005), Hassin, Bargh, Engell and McCulloch (2009) to cope with mental operations lying outside of conscious awareness using a generalized framework.[16] These non-conscious operations include, in particular,

- maintenance of information on short-term scales,
- context-relevant updating of information,
- goal-relevant computations involving active representations,[17]
- learning and preference formation.

※ As far as the relation between explicit (conscious) and implicit (unconscious) memory systems is concerned, Ji et al. (2017) have presented arguments for mutually independence of the two systems in the sense of that none of them enslaves the other, these memory systems play individual roles in processing information of perceived events. Naturally they interact with each other and this interaction between conscious and unconscious components of working memory is multifactorial (Baars

[16] A closely related notion of implicit short-term memory was elaborated in parallel by McKone (1995, 1998), McKone and Dennis (2000) and Maljkovic and Nakayama (1994, 2000), Maljkovic and Martini (2005).

[17] A detailed discussion of non-conscious goal pursuit including the governing neurophysiological mechanisms can be found in Eitam et al. (2008), Hassin et al. (2008), Hassin, Bargh and Zimerman (2009).

and Franklin 2003). In this context we want to note an original model for this interaction proposed by Jacobs and Silvanto (2015); they suggest that the mind does not operate on actual memory traces but rather requires new representations to be created for the conscious domain. Plausible neurophysiological mechanisms and processes implementing implicit working memory are analyzed, in particular, by Dutta et al. (2014), see also Velichkovsky (2017) for a review.

※ The existence of two complimentary components of working memory has enabled Sergent (2018) to propose two scenarios for the timing of conscious access.

Scenario 1 (*delay*): Conscious processing occurs during the build-up of sensory representations within sensory cortices. Thereby conscious perception should be time-locked to the moment of sensory input.

Scenario 2 (*desynchronization*): Conscious processing occurs later within broadcasting of sensory information to the global network. Thereby, conscious perception is time-locked to this broadcasting event, not necessarily to sensory input. It explains, in particular, the following phenomena:

- retro-perception, i.e., perceiving the past when conscious perception is triggered by retrospective attention (Sergent et al. 2013; Thibault et al. 2016; Xia et al. 2016);
- the universality in conscious processing of different types of stimuli which is due to a common governing mechanism functioning at the level of the brain as a whole[18];
- a flexible interaction between unconscious and conscious processes depending on external conditions, intentions, and goals.

The facts and phenomena noted above enable us to accept that:

- There are properties of perceived events that become accessible to the mind just after the consolidation of S-traces bearing the corresponding information is completed and these S-traces are converted into initial W-traces. These properties
    - can be attributed to the immediate past as a whole consciously indivisible entity,
    - admit categorization in terms of phenomenal consciousness—the lower level of consciousness (see Sect. 3.2.5).
- There are properties that after the consolidation of corresponding S-traces are processed within implicit (unconscious) working memory for a certain time before becoming accessible to the mind. The time moment of their appearance in our cognition is not recognized with an accuracy required for embedding it precisely in the stream of conscious representations. Therefore the later properties:
    - can be taken as just given to the mind and merged with the former properties together within one integral characteristic of the immediate past as well as attributed also to the immediate now. Besides, via the internal mechanisms of

[18] We have already discussed this issue in the context of inner psychophysics, Sect. 1.3.2, and in Sect. 3.2.5 will return to it again in discussing the concept of global workspace speaking about the diachronic unit as a integral entity.

anticipation (see discussion below) they also can be "passed" to the immediate future.
- o can be raised up to the level of diachronic unit provided they remain a focus. In this case they should be attributed to the diachronic unit as a whole and the status of their conscious accessibility becomes higher and corresponds to access consciousness (see Sect. 3.2.5).

- Due to the ability of attention to control short-term consolidation as well as the interaction between implicit and explicit memory systems, the state of diachronic unit is able to affect the noted channels of processing the information about perceived events. In particular, it endows the diachronic units with top-down causal power to influence its constituent parts—the immediate past, the immediate now, and the immediate future. Therefore the own properties of diachronic unit have to contribute to the set of basic quantities that must be taken into consideration within a phenomenological theory of the experiential now.

2.6.4 Properties of Immediate Future

At first, we want to note the main premises underlying the modern concept of neurological mechanisms governing human perception and short-term anticipation. In our opinion, they are closely related to the properties of immediate future. We mean the following three basic ideas:

- the *internal models* for motor control, *forward* and *inverse* ones, via which the central nervous system internally simulates the behavior of the motor system in planning, control and learning,
- the *predictive coding* paradigm which deals with the functional description of brain processes underlying human anticipation,
- the *cognitive penetrability* of perceptual experience, which concerns whether higher-level cognitive states (concepts, intentions, expectations) are able to affect perceptual experience understood as a continous process.[19]

For reviews of these concepts as well as neurological and cognitive processes underlying the human ability to predict the future and to pursue a goal a reader may be referred to Ishikawa et al. (2016), Newen et al. (2017a), Friston (2018b), Arstila (2018) and several special issues edited by Schröger et al. (2015), Newen et al. (2017b), Raftopoulos and Lupyan (2018).

Figure 2.7 illustrates the gist of our account of how the synchronic unit of immediate future is related to the underlying mechanisms of human anticipation including conscious and unconscious processes to be discussed below.

---

[19] In the given context the discussion of the cognitive penetrability problem is focused actually on the *cognitive impenetrability* of the early stage of perception which is related to the properties of immediate future. We will return to the cognitive penetrability problem again within the context of the diachronic unit.

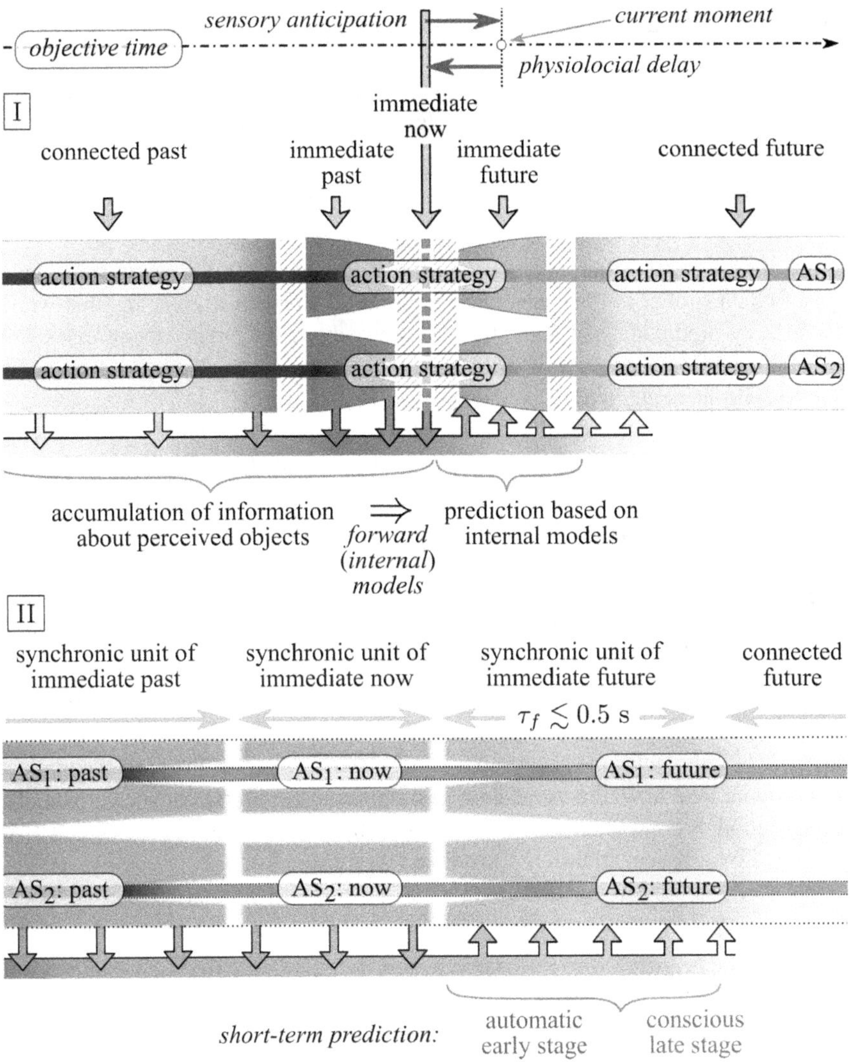

Fig. 2.7 Structure and contents of the immediate future. Based on information about perceived objects accumulated within the immediate now and the immediate past, forward internal models generate neural signals similar to neural signals to be generated by expected physical stimuli. This conversion is unconscious (early stage, sensory anticipation) and only the final result (late stage) becomes accessible to the mind and is compared with sensory predictions generated via inverse internal models by currently used mental model of the environment. This mental model is modified if only the discrepancy between the sensory anticipation and mental model prediction is found, which gives rise to the formation of action strategies—temporal sequences of intentional actions. Sensory anticipation also partly compensates the physiological delay in sensory signal processing

Internal models and experiencing future events: Humans are able not only to anticipate future events and respond to them in advance but also to see, hear something before these events happen, i.e., our sensory system can also anticipate future events. It is due to the existence of neurophysiological mechanisms based on

- the sensory information accumulated right now and
- the embodied knowledge of the corresponding regularities accumulated previously, i.e., emerged via learning.

These mechanisms—often called *internal models* (Cisek 2009)—are able to generate sensory data similar to ones generated by external physical stimuli. In other words, the sensory prediction of future states of the body or the environment arises from mimicking their respective dynamics through the use of internal models (for a review see, e.g., Bubic et al. 2010).

There are two varieties of internal models, *forward* and *inverse* models (e.g., Wolpert et al. 1998; O'Reilly et al. 2008; Ishikawa et al. 2016, for are view).

> Forward models capture the forward or causal relationship between inputs to the system, such as the arm, and the outputs. A forward dynamic model of the arm, for example, predicts the next state (e.g. position and velocity) given the current state and the motor command. In contrast, inverse models invert the system by providing the motor command that will cause a desired change in state. They are, therefore, well suited to act as controllers as they can provide the motor command necessary to achieve some desired state transition (Wolpert et al. 1998, p. 338, leftcolumn).

Figure 2.7(I) illustrates the accumulation of information about the perceived objects in the immediate now, the immediate past and, maybe, in the connected past. The forward models convert this information into sensory signals imitating signals to be generated by physical stimuli in the nearest future.

We want to emphasis that the neurological mechanisms underlying internal models

- function at the level of brain as a whole and
- "bring" the nearest future into our current now.

Therefore, first, sensory anticipation is a common phenomenon concerning all the main sensory modalities rather than proprioception[20] only.

Second, the internal models enable the subject to perceive future events on short time scales as events already come into the existence. In particular, various illusions like Kanisza's figures (Kanizsa 1955) justify the possibility of dissociation of percepts from sensory input (for a review see, e.g., Pearson and Westbrook 2015; Aggelopoulos 2015).[21]

Summarizing the aforesaid we may state that

---

[20] The concept of internal models was originally developed for the human motor behavior (e.g., Miall and Wolpert 1996; Wolpert et al. 1998). So proprioception—the sense of self-movement and body position—is the paradigmatic example where sensory anticipation is crucial.

[21] Meta-stable phantom percepts—the perception of a sensory experience in the absence of a physical stimulus—may be also explained turning to the interaction between bottom-up compensation and top-down updating of the model with some failure in one or both mechanisms, resulting in a

(i) the existence of neurological processes enabling short-term sensory anticipation and
(ii) the possibility of percept-stimulus dissociation

justify the introduction of the immediate future as a fragment of human temporal dimension. This fragment comprises events that are expected to occur in the nearest future and already experienced as having come into being (Property F.1). It is a reason to regard the immediate future as a basic constituent component of the experiential now and its mathematical model.

To estimate the duration τ_f of immediate future let us turn to the available neurophysiological data. Nowadays it is widely acknowledged that the cerebellum plays the role of movement sensor (e.g., Ishikawa et al. 2016; Therrien and Bastian 2019). However its role as a certain sensor seems to be much wider (Guell et al. 2018; Schmahmann 2019). Therefore the delay time in the exchange of information between the cerebellum and cerebral cortex may be used to estimate the time horizon of unconscious prediction and, thereby, the duration of immediate future.

Referring to the available experimental data (e.g., Cao et al. 2017; Marek et al. 2018) we may estimate this duration as 100–380 ms. A similar estimate is also obtained based on the general analysis of plausible computational mechanisms of sensorimotor control (Franklin and Wolpert 2011), which justifies the estimate $\tau_f \lesssim 0.5$ s (Property F.1 and Fig. 2.7(II)).

Sensory anticipation and cognitive prediction: Sensory anticipation is involved into complex cognitive processes enabling humans to predict efficiently the dynamics of environment on scales larger than the duration of experiential now. Nowadays this process is described within the *predictive coding* paradigm. So let us note its basic premises to elucidate the corresponding role of sensory anticipation and, thereby, the properties of the immediate future.

Dealing with visual perception, Rao and Ballard (1999) seem to be the first who elaborated the concept of predictive coding. In its most general formulation, this concept

> understands the brain as a hierarchically structured prediction-testing machine. In particular, higher-order cognitive states serve as a basis for top-down predictions of lower-order states. On the lower-order state the prediction error is computed and reported bottom-up to the higher-levels in the cognitive architecture. In response to this bottom-up prediction error the brain then revises its higher-order cognitive state in such a way that it reduces or even minimizes the prediction error. Thus, [predicting coding] replaces the standard picture of bottom-up processing from lower order-states (the inputs to the senses being the lowest state) upwards to higher-order states by a more dynamic interplay between higher-order and lower-order cognitive states. (Newen et al. 2017a, p. 3)

---

constant prediction-error. A reader may be referred to a review Mohan and Vanneste (2017) for theoretical models and various sensory modalities.

For a detailed description of the predictive coding paradigm a reader may be referred to Rao and Ballard (1999), Lee and Mumford (2003), Friston (2005), Friston and Stephan (2007), Friston et al. (2009), Friston (2010), Brown et al. (2011), and also a review by Clark (2013). Here we want to emphasis two of its premises crucial for our discussion[22]:

- The collection of signals generated by the sensory system and the set of higher-level models of environment are equipollent in their contribution to the prediction processes.
- Only the "prediction errors" resulting from the comparison of the sensory data with the prediction of the currently used model propagate in the bottom-up direction and can cause the model modification.[23] Therefore when there is no remarkable discrepancy between the sensory signals and the model prediction, the model remains unchanged.

The noted premises of the predictive coding paradigm allow us to state the following:

- Without sensory anticipation our cognition of events would be inevitably delayed at least for the time required for the initial integration of neural signals—the core function of immediate now. Sensory anticipation—the core function of immediate future—can partly compensate this physiological delay in the information processing and, thereby, keep the perceived events and their sensory images synchronized in objective time. It is illustrated in Fig. 2.7(I) as the temporal shift to the left caused by the physiological delay and the counterbalancing shift to the right due to sensory anticipation. For this type compensation to be feasible, the characteristic time scales of the immediate now and the immediate future have to be of the same order. The latter seems to be the case, at least, the above estimates for the duration of the immediate now and the immediate future, $\tau_n \sim \tau_f \lesssim 0.5\,\text{s}$, argue for it.
- Sensory anticipation implies a certain dissociation between sensory signals and percepts. It concerns events expected to occur in the nearest future based on our sensory and cognitive awareness about the current state of environment. So the immediate past, the immediate now, and the immediate future bear some pieces of information about the same—the current state of environment—but at different time moments. The mind dealing with all of them can extract richer information about the environment than if it would rely upon only one of them. We will return to this issue in the discussion about the diachronic unit in order to justify the use

[22] Other postulates and the core notions of the predictive coding paradigm such as

- the *Bayesian brain* (e.g., Friston 2003, 2009; Clark 2013; Hohwy 2013),
- the *active inference* process (e.g., Friston et al. 2009, 2010, 2013, 2014, 2016),
- the *principle of free-energy minimization* developed by Friston (2003, 2009, 2010), Friston and Stephan (2007), Feldman and Friston (2010) for describing human perception and behavior

will be discussed below within a relevant context.

[23] In this sense, the predictive coding paradigm is rooted in the account of Hermann von Helmholtz (1821–1894), a German physician and physicist, who first seized on the idea of the brain as a hypothesis tester (for details see, e.g., Hohwy 2013).

of extended phase space with higher-order time derivatives in our constructions. Here we have noted this aspect to emphasize that the synchronic unit of immediate future is the essential component of human cognition.

In addition to the aforesaid we want to focus attention on another aspect of predictive coding paradigm. Namely, in its framework lower-order sensory signals and higher-order internal models of *human actions* are treated as complementary, equipollent components. Their interaction is such that the current action minimizes the discrepancy between prediction and sensation.

To elaborate this idea and its mathematical implementation we put forward the notion of *action strategy*.[24] Due to expected changes in a dynamic environment, not only the parameters of a particular action model may vary in time. Pursuing some goal a person may intend to implement a sequence of different actions step-by-step. The term *action strategy* is introduced to refer to such an action or their sequence as a single whole. Figure 2.7 depicts action strategies as straight narrow stripes passing through all the shown temporal fragments from the connected past to the connected future.[25]

Below we will return to discussing the properties of action strategies and the predictive coding paradigm reformulated in special terms to describe the diachronic unit embedded into the connected past and the connected future. Here finalizing the justification of F.2 property, we note two general characteristic features of action strategies:

- An action strategy may be maintained unchanged on scales much longer than the duration of experiential now until the necessity of its correction becomes evident.
- An event of correcting the higher-order models caused by a mismatch between the expected and perceived dynamics of environment admits the interpretation as a transition between possible action strategies.

In this way we actually have introduced a new basic component—the action strategy—for describing relationship between the sensory evaluation of environment states and goal-oriented intentional actions. The sensory anticipation being the essence of immediate future contribution to human cognition is a crucial component of this interaction.

Cognitive impenetrability of immediate future: Cognitive penetrability as the *direct* top-down effect of higher-level cognitive states on sensory perception was first introduced by Pylyshyn (1980, see also Fodor and Pylyshyn 1981). Up to now its existence is debated; arguments in favor of and against it are reviewed and analyzed in detail, e.g., by Firestone and Scholl (2016, see also the following critical comments), Lammers et al. (2017), Newen and Vetter (2017), O'Callaghan et al. (2017), Arstila (2018), Cecchi (2018) as well as Raftopoulos in his monograph (2019).

---

[24] Below we will use the terms *action strategy* and *strategy of actions* interchangeably.

[25] For a discussion of action strategies as cognitive phenomenon and the underlying neurological mechanisms a reader may be referred to Overgaard and Mogensen (2014), Mogensen and Overgaard (2017, 2018) and references therein. The notions of the connected past and the connected future are elaborated in Sect. 3.1 as temporal entities related but not belonging to the experiential now.

A challenging problem besetting discussions about cognitive penetrability is that perception is not a homogeneous undifferentiated process. For example, visual perception consists of two stages affected differently by cognition, namely, early vision—affected *indirectly*—and late vision—affected *directly* (e.g., Raftopoulos 2019). The existence of two stages quite distinct in properties with respect to the dependence on cognitive top-down phenomena should be the case for other sensory modalities. In particular, it is so for auditory modality (e.g., Pressnitzer and Hupé 2006; Moore and Gockel 2012; Kaya and Elhilali 2017) and olfactory modality (e.g., Keller 2011; Mori et al. 2013).

Now we want to focus attention on the early stage of perception. Following the account of Raftopoulos (2009, 2014, 2015, 2017, 2019) we accept that the early stage of perception is not *directly* affected by cognition.[26] It is our final argument for regarding the temporal fragment of immediate future as the synchronic unit.

Indeed, according, e.g., to Raftopoulos the information processing during the early stage of perception does not use cognition as an informational resource and, so, has to be cognitively impenetrable. Naturally, for example, cognitively driven attention can prepare the sensory system to process some items faster and more efficiently. However such effects are *indirect*; within the early stage the information processing does not change cognitive states (see also Gross 2017). In other words, cognition can determine some structural or functional parameters of processing perceived information but these parameters cannot change during the early stage.

The forward internal models discussed above do not require the contribution of mental processes for their implementation. Thereby sensory anticipation based on the bottom-up neural processes

- should be categorized as the early stage of perception and
- must be unconscious with respect to particular details of how the perceived information is integrated.

In contrast to the early stage of perception, the late stage is cognitively penetrable and the mind has a direct access to it (Raftopoulos 2011, 2019). Therefore only the integral properties characterizing the perceived events and attributed to the early stage as a whole entity have to be accessible to the mind. It matches exactly the notion of synchronic unit and justifies the introduction the synchronic unit of immediate future.

The features discussed above demonstrate that the immediate future as the synchronic (i.e., indivisible) unit of human temporal dimension:

- does exist as an individual component of experiential now whose duration τ_f admits the estimate $\tau_f \lesssim 0.5\,\mathrm{s}$,
- represents sensory anticipation as the main neural process underlying it,
- is involved substantially into human prediction of future events on time scales larger than the duration of experiential now,

[26] For a detail comparison of various accounts of cognitive penetrability a reader may be referred to Raftopoulos (2019).

- is cognitively impenetrable and provides the mind with access only to the results of sensory anticipation averaged on time scales about τ_f.

2.6.5 Mental Images and Their Experience

Before passing directly to the fusion of the immediate past, the immediate now, and the immediate future into the diachronic unit, we need to discuss a certain basic feature distinguishing physical objects from their mental images. This feature underlies the predictive coding and is the gist of the concept of mathematical eidoi taking the central position in describing human actions (Sects. 3.3.3 and 3.3.4). Broadly speaking, in this subsection, we want to discuss the properties of perceived objects that, on the one hand, are extracted from sensory signals. On the other hand, they do not require the sensory input to be currently present (see, e.g., Colombo 2012; Thomas 2014; Pitt 2020 for an introductory discussion). In this sense they are applicable to characterizing (*i*) the immediate past, the immediate now, the immediate future, (*ii*) the experiential now as an individual entity.

As note above predictive coding paradigm deals with:

- sensory images of perceived objects, i.e., lower-order sensory signals generated by the corresponding external stimuli,
- higher-order cognitive states representing these objects in the mind in some way, and
- the interaction between sensory images and cognitive states governed by internal models responsible for the conversion of sensory signals into the "characteristics" of higher-order cognitive states or vice versa and, thus, making their comparison possible.

It is essential that this conversion is implemented hierarchically, where the levels of hierarchy are determined by the details of how cognition is involved into the conversion. It endows human behavior with high variability and gives rise to a variety of problems devoted to the relationship between sensation and the corresponding mental representations.

In the given context at least two entirely distinct issues concerning physical objects and their mental images may be singled out. One of them is about how a certain generic, mental type-object representing a whole class of physical objects emerges for the first time. Broadly speaking, this issue is about how our categories and notions emerge, which is an individual challenging problem; its discussion is outside the scope of the present book.

The other issue—the subject-matter of the present subsection—concerns the transformation of properties we already attributed to mental images within some categorization. The posed question is how these properties change when a perceived object gets out of our physical contact, which is illustrated in Fig. 2.8. Its upper fragment illustrates the analyzed properties of experiential now.

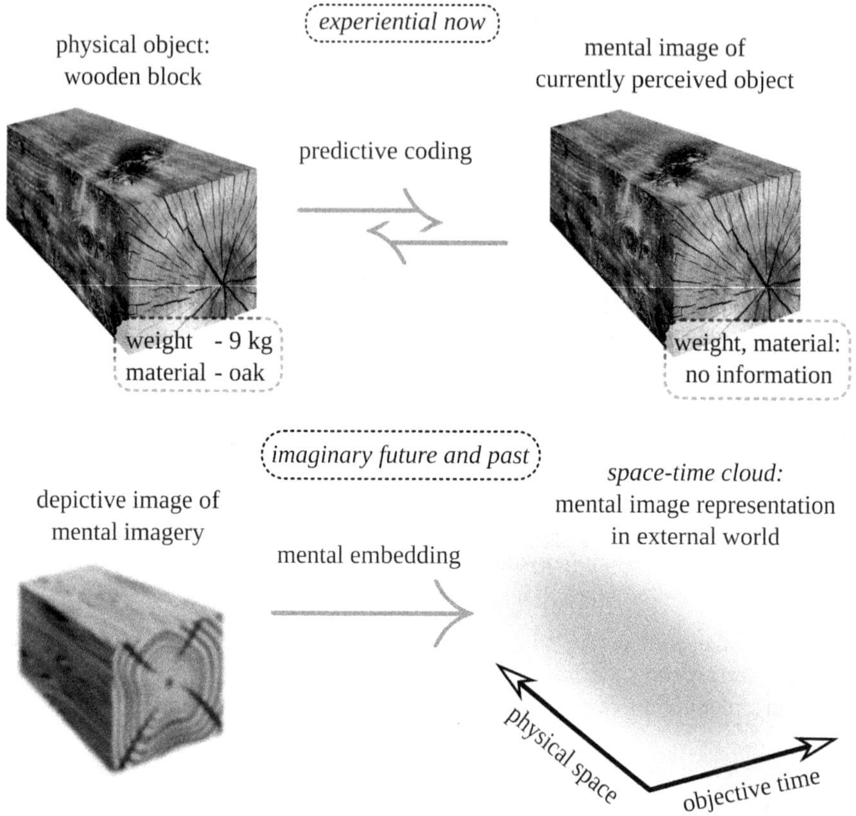

Fig. 2.8 Illustration of how the properties of mental images change when a perceived physical object gets out of the direct contact with the external observer. The upper part (the experiential now) represents the relationship between the perceived physical object—a wooden block—and its mental image formed within the direct physical contact of the observer with this object. The object categorization and the detailed properties attributed to the object image in the mind are governed via neural mechanisms described by predictive coding paradigm. The lower fragment (imaginary future and past) illustrates, first, the properties of physical object that are irreversibly lost in its mental image after the retaining in and retrieval from memory, here it is the cracked wood texture. Second, the others, like the space-time position of imaginary object in future or past remain accessible to the mind but their accuracy drop substantially. The notion of space-time cloud describing the uncertainty of the mental embedding of an image into the external world in illustrated here as the embedding of wooden block in the space-time continuum

When looking at some physical object—a wooden block in the given case—and, maybe, touching it we, first, have its sensory image including, e.g., the object color pattern, the visible details of its geometric form, the tactile sensation, etc.[27] Second, we have a mental image of this object which:

- categorizes the object as a representative of a certain class of physical objects—in the shown case it is a wooden building piece of some construction having the form of cuboid,
- bears a detailed set of properties characterizing the physical properties of the perceived object—here, in particular, the size of block, at least, approximately, the state of cracked wood and the detailed pattern of crack network.

The adequate mapping of the object properties onto the image properties is due to the neural mechanisms and learning process forming the gist of the predictive coding. Naturally the properties of a perceived physical object on its own and the properties attributed to its mental image in the mind are not the same. First, even we directly recognize, e.g., the size of object our perception is bounded, which is reflected in Weber's law (1.1) discussed in Sect. 1.3.1. Second, there are object properties remaining outside our cognition, in Fig. 2.8 they are the block weight and its particular wood material.

The lower part of Fig. 2.8 illustrates the wooden block image in the imaginary future and past. When the perceived object gets out of our view, its image is retained in memory and can be retrieved from it for mental operations with the imaginary object in estimating the previous events or predicting future ones. In this case it is reasonable to single out, at least, two aspects of the image description. First, it is the descriptive characteristic of the object as the member of a certain class, in Fig. 2.8 it is some cuboid-like building block. In particular situations to be modeled further this feature will be assumed not to change. Therefore the state of an object perceived right now and its imaginary states in future or past, in principle, should admit a description dealing with the same collection of variables.

The second aspect concerns the details and the precision of particular properties for images retrieved from memory. As illustrated in Fig. 2.8 (lower fragment) some properties of physical objects can be irreversibly lost. For example, turning to memory the crack pattern or wood texture of the shown wooden block cannot be recalled and may be formally reconstructed based only on the general ideas about the structure of old wooden blocks. The other properties remain accessible to the mind but their accuracy usually drops substantially.

We want to pay a special attention to this accuracy drop because it impacts on our further mathematical constructions. After a certain time when the perceived object—the wooden block—got out of view, the observer can just not remember the exact place occupied by the block previously. In this case he may claim that the block was "somewhere over there." The same also concerns moments of time. For example, if the given wooden block is moving, based on visual information the observer is able to estimate the time moment when the block will reach a certain special position only

[27] Speaking about a sensory image we mean sensation rather than categorization, e.g., sensory image of the red color is the redness as a sensational quality of "being red" rather than a component of color pallet.

roughly. In this case his estimate like "about one minute" may imply 30 s or 2 min within the same level of confidence.

Therefore in Fig. 2.8 we illustrate the mental embedding of the image into the external world in terms of some fuzzy region in the *space-time* continuum. The spatial dimension of this region can be much larger than the object size, which together with uncertainly in temporal dimension has prompted us to introduce the notion of space-time cloud (Lubashevsky 2017) to characterize such properties of mental images.

The notion of space-time cloud makes it evident that the difference between physical objects and their mental images in properties

- is not so strong that physical objects and their mental images cannot be compared with each other in principle,
- is rather significant and, as a result, the description of mental image dynamics requires an individual formalism which is outside the scope of the classical paradigm of physics.

First, in mathematical terms:

- physical objects may be conceived of as material points of the external world, at least, approximately, i.e., they admit the mathematical representation as point-like objects whose states are completely determined at the current instant of time[28];
- as far as spatio-temporal properties are concerned with, the mental images of objects may be conceived of as fuzzy regions in the same phase[29] space which are characterized by finite spatial and temporal dimensions.

Thereby, the main difference between the physical objects and their mental images is more quantitative rather than qualitative, it is just the different degree of localization in the same space. So a physical object and its mental image may be compared as two space-time clouds, one of them is just a δ-function or something similar, the other is finitely distributed in the space-time continuum.

Second, a space-time cloud—a fuzzy region in the space-time continuum—is attributed to a given mental image as a single entity indivisible into smaller parts without lost of causality.[30] Besides, its spacial and temporal dimensions—*spatial & temporal* uncertainty in characterizing the position of imaginary object—may

---

[28] The concept of material point is justified for classical physics; quantum physics treats physical particles as some kind of spatial clouds—wave functions—but in temporal dimensions they are not extended.

[29] Here the phase space of a system in question is understood as space whose points represent all the possible states of this system at a given instant of time. In other words, the collection of phase variables specifies the properties that can be attributed to the given system at a single time moment. Generally, no particular properties of the system dynamics are attributed to the phase space.

[30] In the present book, we consider human actions on physical objects which move in space as whole entities without their division into components characterized by individual dynamics. For this reason, to avoid unnecessary over-complication in particular mathematical constructions, we regard the analyzed physical objects as material points. The stated indivisibility of the space-time cloud is the case for physical objects of this type only. For complex physical objects, e.g., a fan with rotating blades, a more sophisticated structure of multicomponent space-time clouds is required.

depend on the details of system dynamics and human actions. This feature endows mental images with properties atypical for physical objects and the dynamics that cannot be described explicitly within the classical paradigm of physicals.

The noted aspects argue for treating mental images and the corresponding physical objects as closely intertwined complementary elements equipollent in describing human actions.

In the discussion about the mental images of physical objects one crucial assumption has been tacitly accepted. Namely, a mental image retained in and then retrieved from memory is accepted to possess a picture-like format including the space-time embedding of the corresponding object into the environment. This format enables the mind to deal with such images as individual single entities which bear properties resembling the properties of the corresponding images of objects currently perceived. The remaining part of the present subsection is devoted to justifying this proposition.

Broadly speaking, there are two principally different ways of how mental images can be represented in the mind. One is the descriptive format, i.e., the propositional, language-like, symbolic representation. It implies that in the mind we deal with the corresponding pieces of information as some individual and mutually independent components. The other is the picture-like representation; using the terminology of Kosslyn (2005) we will call it depictive format not confining it to the visual modality only. Exactly the depictive representation must be quasi-perceptual due to its plausible conversion into sensory signals via the internal models. This conversion brings the mental image of physical object on its own into the experiential now. Put differently, it enables us to experience not only the currently presented objects but also imaginary ones.

The issue about quasi-perceptual experience of mental imagery and its forms is rooted in the philosophy of ancient Greece, for an introductory review a reader may be referred to Thomas (2014). In the late 20th century the related debates got a new level due to the novel techniques of observing the brain functioning directly (e.g., Pearson and Kosslyn 2015). In particular, it has been demonstrated that the vast majority of brain areas activated during visual perception are also activated during visual mental imagery (Kosslyn et al. 1997, 2006; Ganis et al. 2004, for a meta-analysis of these data see Winlove et al. 2018). Mental imagery is also able to shape low-level sensory representations (e.g., Pearson et al. 2008). Naturally the visual modality is not unique in the given aspect, mental imagery can deal with events involving several sensory modalities, which justifies the notion of *multimodal mental imagery* (Nanay 2018b). Moreover, the descriptive and depictive formats being different are not mutually exclusive. In particular, absorbing reading of a fictional narrative often induces the occurrence of highly vivid images in the mind of reader (Scarry 2001, for are view see, e.g., Brosch 2018).

As a certain mathematical paradigm in describing the relationship between a mental image and the sensory perception of the corresponding physical object, Shepard (1978) put forward the principle of functional equivalence between perception and imagination. Rejecting a strict one-to-one relationship ("first-order" isomorphism) between the images and objects, he turns to

> a more abstract or "second-order" isomorphism in which the functional relations among objects as imagined must to some degree mirror the functional relations among those same objects as actually perceived. Thus, to the extent that mental images can substitute for perceptual images, subjects should be able to answer questions about objects as well when those objects are merely imagined as when they are directly perceived ...For in this way, we can obtain information about global, holistic relations without having to make any assumptions in advance about what local, particularistic features may be important in any one object. ...[T]his paradigm has the advantage that we do not limit our investigation to the concretely externalizable component of mental images. The subjects may well base their judgments on the inherently internal, more abstract components as well. (Shepard 1978, p. 131)

The principle of functional equivalence between perception and imagination was verified within various experimental paradigms, for a review see, e.g., Shepard and Cooper (1982), Kosslyn (1980), Thomas (2014) as well as Cooper and Shepard (1984), Cooper (1990), Borst and Kosslyn (2008), Galdo-Alvarez et al. (2016). Finke (1980) argues that the equivalence in imagery and perception is based on the activation of complex information-processing mechanisms involving many levels of sensory modalities. It should endow mental images with complex properties simultaneously including unimodal, multimodal, or even amodal aspects (for a detailed discussion see Finke 1989).

The principle of functional equivalence underlies the possibility of describing mental representations of object motion in the past or the future based on the terms characterizing its sensory perception.

As far as the particular nature of the representation of mental images is concerned, it is still a matter of ongoing debate discussing, in particular, the pictorial/analog, propositional/descriptional, and enactive/motor accounts of imagery (e.g., Thomas 2014; Palmiero et al. 2019; Pitt 2020, for are view). Speaking about the spatial properties of imagery, we want to note two main theoretic frameworks of visual and spatial knowledge processing proposed in cognitive science. They are the theory of mental imagery (Finke 1989; Kosslyn 1994; Kosslyn et al. 2006) and the mental model theory (Johnson-Laird 1989, 1998; Tversky 1993); their comparison is given, in particular, by Sima et al. (2013). Both the theories draw a distinction between spatial representations of mental images (with amodal properties, e.g., spatial arrangement) and visual mental images (with modality dependent properties, e.g., color, texture). In our account we focus on spatio-temporal aspects of mental images and, what is essential for the models we develop below, the noted frameworks suppose the spatial representation of mental images to be analog, at least, approximately (see Thomas 2014; Palmiero et al. 2019, for a detailed discussion). This type representation admits the description of mental transformations—mental changes in image spacial representations—in terms of continuous variables. In particular, the concept of mental rotation (Shepard and Metzler 1971) justified by the accumulated experimental data (e.g., Shepard and Cooper 1982; Shepard and Metzler 1988; Cooper 1990; Thomas 2014) turns to continuous transformations of mental images. Besides, as demonstrated in these experiments, the larger the rotation angle, the longer the reaction time required for subjects to compare two visualized figures via mentally rotating one of them. This effect enables us to introduce the notion of rotational velocity to quantify mental rotation and suppose that effort of mental rotation should depend also on the rotational velocity.

Summarizing the presented discussion of mental image, the analog nature of the depictive format of mental images prompts us to introduce imaginary trajectories for describing object motion in the past or the future.

It should be also noted that the depictive representation of an imaginary object and the mental image of this object when it is currently perceived should be intrinsically related via working memory; for a detailed discussion of this issue a reader may be referred to Kosslyn (2007), Moulton and Kosslyn (2009), Pearson et al. (2015). In particular, Pearson et al. (2015) state that:

> activity patterns in the visual cortex are not merely similar across visual mental imagery and perception: activity patterns encode a common set of visual representations. …[the available data] lend further support to the conceptualization of visual mental imagery as a *weak* or *noisy* form of top-down perception that can in some cases take the place of bottom-up perception. (p. 593)…[M]ental images seem to behave much like *weak* versions of externally trigged perceptual representations. (p. 599 *italic* is ours)

This weakness of mental images should be reflected in the decrease of their vividness and accuracy as the time lag between the current moment and the moment in question increases, other conditions being equal. It has to be a direct consequence of mental imagery being based the working memory. In particular, experiments by Hyun and Luck (2007) on the mental rotation of some figure coupled with a mental search task in a certain alternate sequence provide evidence for the given proposition. Similar investigations of image vividness (D'Angiulli and Reeves 2002; D'Angiulli et al. 2013) also argue for this time behavior.

Concluding this discussion about mental images of physical objects we want to emphasize the following.

- According to predictive coding paradigm, our conscious operations with physical objects are based on their mental images. In modeling human actions these images admit interpretation as single entities with own properties and individual dynamics. It implies that in describing human actions physical object and their mental images have to be treated as complementary, equipollent elements.
- The mental images of currently perceived objects and imaginary objects (e.g., as elements of imaginary future or past) are characterized by similar properties. Their difference is more quantitative than qualitative. We are able to experience imaginary events, thereby:
  - The imaginary dynamics of observed objects in the immediate past and the immediate future is directly experienced and so these synchronic units have to be included in the experiential now.

  - Being partially experienced at present, the dynamics of imaginary objects retained in working memory and expected in the imaginary future can be included in the description of human goal-oriented actions. In particular, this feature underlies the concept of the complex present (Sect. 3.1).

- The mental images may be attributed with the same properties that are exhibited by the corresponding physical objects provided the physical properties specified by continuous variables are blurred. This blurriness (fuzziness) allows for the intrinsic decrease in the accuracy of mental imagery concerning, in particular, the corresponding continues variable as well as objective time. The concept of space-time clouds illustrates this fuzziness with respect to mental images of material points—physical objects conceived of as located at a given point of the space-time continuum.
- The motion of an observed object in the past as it is retained in memory or the anticipated motion in the future can be described in terms of imaginary continuous trajectories or their multitudes of object motion in physical space.

2.6.6 Properties of Diachronic Unit

We have introduced the notion of diachronic unit as the fragment of human temporal dimension comprising the three synchronic units—the immediate past, the immediate now, and the immediate future. In this sense the diachronic unit is not elementary; its structure is accessible to the mind. Thereby, speaking about the content of diachronic unit, we deal with mental images of perceived objects that integrate or emerge from the mental images created within the immediate past, the immediate now, and the immediate future. So the properties of mental images attributed to the level of diachronic unit may be conceived of as the properties of perceived physical objects converted into the properties of mental images as it has been discussed in the previous subsection. In particular, in describing the content of diachronic unit we may turn to mathematical terms like the spatial positions and velocities of observed physical objects provided the points of space-time continuum are understood as space-time clouds.

The diachronic unit is not just an "arithmetic" sum of its constituent elements, i.e., the diachronic unit can possess own properties that its constituent synchronic units do not have. To illustrate this we turn to the premises accepted by Marr in developing his famous theory of vision (Marr 1982) and compare them with the main ideas of the synergy of diachronic unit. A reader may be referred to Glennerster (2007), Bermúdez (2014) for a modern review and critical discussion of Marr's theory of vision and to Aitchison and Lengyel (2017) for discussing the theory premises from the standpoint of predictive coding paradigm.

For describing how humans perceive three-dimensional objects Marr (1982) put forward the concept of three-stage process of visual perception:

- At the first stage, a two-dimensional initial sketch of sensory information as it is perceived by the eye's retina is formed.
- At the second stage, a pseudo-three-dimensional picture of objects is created based on the signs of depth; at this stage, not all the characteristics of three dimensions are reflected.
- At the final third stage, a full-fledged three-dimensional picture of objects showing their orientation in space and mutually arrangement is created.

Now let us discuss the three stages in relation to our scenario of diachronic unit emergence to elucidate its key points.

The first stage of vision ends with the creation of a two-dimensional sketch (*primal sketch*) of the visual scene as it is represented on the eye's retina. The primal sketch comprises the elementary components of visual pictures like edges, regions, etc.

Here we want to draw a certain analogy between the primal sketch and the individual content of synchronic units which becomes accessible to the mind when the unconscious processing of sensory information is completed. As illustrated in Fig. 2.9 (lower fragment) these unconscious processes imply the accumulation of information about perceived objects together with sensory anticipation—unconscious prediction of the objects' future states based on internal models. By way of example, Fig. 2.9 illustrates the situation when the perceived object is a moving physical body and the synchronic units of immediate past, immediate now, and immediate future bear the information about its spatial position and the motion velocity $\{x, v\}$. Due to the tripartite structure of experiential now these quantities are discriminated in the mind as ones characterizing the body motion with some delay, just now, and one step ahead, $\{x_{-\tau_p}, v_{-\tau_p}\}$, $\{x_n, v_n\}$, $\{x_{+\tau_f}, v_{+\tau_f}\}$, respectively. To avoid possible misunderstanding, we want to emphasize that the quantities $\{x, v\}$ are the properties attributed to the mental image of moving body rather than the real position and velocity of this physical body; their difference was discussed in the previous subsection.

Turning to the analogy between the primal sketch and the synchronic unit we may consider the quantities $\{x, v\}$ the elementary components of the motion perception belonging to the lower level of consciousness. The latter feature implies that the velocity at which the body moves in space is given to the mind directly rather than is a results of mental "differentiation."

The second stage is characterized by the creation of a pseudo-three-dimensional picture ($2^1/_2$*D-sketch*) of objects bearing rough information about the depth of its details. The properties of surface fragments perceived at the first stage (e.g., their levels of luminous intensity) enable the mind to order the visible elements in depth. However, this arrangement is rather "mechanistic," the initially invisible (occluded or otherwise hidden from our view) fragments of objects are absent and, thereby, the connectedness of visible elements can remain unrecognized. In this case the observed three-dimensional objects cannot be reconstructed from the visually perceived elements.

Returning to the diachronic unit synergy we draw an analogy between the $2^1/_2D$-sketch and the transition from the tripartite structure of the three synchronic units

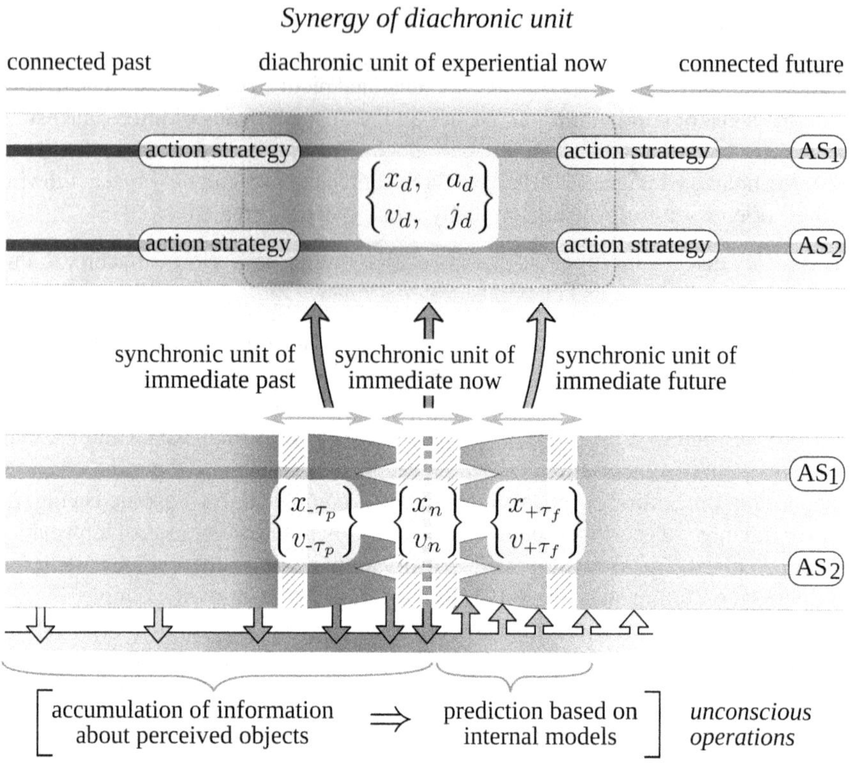

Fig. 2.9 Schematic representation of how the mental image of perceived physical objects form at the level of diachronic unit from the corresponding mental images already emerged at the level of the synchronic units of immediate past, now, and future. By way of example, the figure's lower fragment illustrates the content of the synchronic units of immediate past, now, and future stemming from the accumulation of sensory information about a moving object combined with sensory anticipation of its further dynamics. The upper fragment illustrates the diachronic unit emergence in combing the three synchronic units into a single whole. The properties attributed to the emergent image are of two types. First, it is the properties just aggregating the properties of images at the level of synchronic units. Here it is the position x_d and the velocity v_d of perceived moving body as the characteristics of mental images. The other type involves emergent properties that the constituent synchronic units separately do not possess. Here the emergent properties are presented by the body acceleration a_d and jerk j_d. The upper fragment illustrates also the accepted assumption that only the experiential now as a single whole contributes to the strategies of actions in their quality evaluation. So this contribution admits a description in terms of the variables $\{x_d, v_d, a_d, j_d\}$ determining the structure of the mental phase space in motion perception

to the diachronic unit (Fig. 2.9). Here the mind combines them together within the inherited temporal order. This, however, does not lead to the emergence of novel properties that can be attributed individually to the level of diachronic unit.

The third stage is the final phase of vision when a full-fledged three-dimensional picture (3D-*model*) of the visual scene with volumetric objects and their mutual arrangement is created. It is due to the ability of the mind to reconstruct the invisible fragments of object surfaces based on purely mental operations without a contribution of visual modality. Thereby these 3D-images of observed objects bear properties of two types; the properties reflecting the sensory details of visual modality as well as the properties reflecting features of higher-order cognitive processes (e.g., Raftopoulos 2011, 2019). In particular, such mental images emerge

> through the synergy of bottom-up processing transmitting information registered at the lower levels or prediction errors, and top-down processing transmitting information relevant to the testing of hypotheses concerning the probable causes of the input, and in so far as the processes constructing these hypotheses are informed by high-level knowledge about worldly objects, visual perception unifies cognition and late vision; these two become intertwined. (Raftopoulos 2019, p. 267)

This means that sensation and cognition are equipollent components of human vision and the emergence of 3D-images obeys predictive coding paradigm.

In cognitive psychology such mental processes of image reconstruction are also called *amodal* perception or *amodal* completion (e.g., Bahrick 2010, see also Bahrick and Lickliter 2010). These terms are introduced to emphasize that creating mental images of complex objects when not all their elements are accessible to human senses can be implemented mentally and is common to all the sensory modalities (auditory, vision, tactile, etc.) For a detailed discussion of amodal perception a reader may be referred also to Tse (1999), Singh (2004), Nanay (2010, 2018a), Briscoe (2011).

Within the used analogy, we relate the 3D-model creation to the emergence of the diachronic unit as a single whole. There can be singled out two types of properties attributed to mental images belonging to the level of diachronic unit.

The *first type* properties just aggregate the corresponding properties of images at the level of synchronic units. Turning to the example of body motion perception, this aggregation admits interpretation as the mapping

$$\begin{Bmatrix} x_{-\tau_p} \\ v_{-\tau_p} \end{Bmatrix} \oplus \begin{Bmatrix} x_n \\ v_n \end{Bmatrix} \oplus \begin{Bmatrix} x_{+\tau_f} \\ v_{+\tau_f} \end{Bmatrix} \Longrightarrow \begin{Bmatrix} x_d \\ v_d \end{Bmatrix}, \tag{2.6}$$

where the resulting values of the body position and the motion velocity $\{x_d, v_d\}$ are attributed to the level of diachronic unit. Here the sign $\oplus$ of unification is used symbolically because within this cognitive "averaging" e.g., the individual weights ascribed to the synchronic units can be different and change depending on a particular situation.

Although the quantities $\{x_d, v_d\}$ characterize the mental image of moving body in the scope of diachronic unit, mapping (2.6) does not represent the emergence of novel properties. Indeed, the difference in the quantities $\{x, v\}$ characterizing the mental images at the levels of synchronic units and diachronic unit is more quantitative rather than qualitative. They correspond to the same components of sensory perception—"being at some place" and "moving at some speed," respectively. We may draw an analogy between conversion (2.6) and the incorporation of visible elements into volumetric images during the conversion of 2½*D*-sketch into 3D-model.

The *second type* properties admit categorization as emergent ones. The synergy of bottom-up and top-down processes endows the diachronic unit with own properties irreducible to the properties of its constituent synchronic units taken separately. In terms proposed by Briscoe (2011), these emergent properties arise through the top-down "cognitive" completion (*C-completion*).

Turning again to the example of motion perception, the emergent properties attributed to the mental image of moving object at the level of diachronic unit can be illustrated by the mappings

$$v_{+\tau_f} \ominus v_{-\tau_p} \circledcirc v_n \Longrightarrow a_d\,, \tag{2.7a}$$

$$v_{+\tau_f} \oplus v_{-\tau_p} \ominus v_n \Longrightarrow j_d\,. \tag{2.7b}$$

As previously, here the symbols $\oplus$, $\ominus$, and $\circledcirc$ are used schematically to exhibit the general structure of calculating the rates of first and second order at which a given variable changes. Mappings (2.7) represent the higher-level cognitive functions endowing the mental image at the level of diachronic unit with characteristics showing (*i*) how fast the velocity of moving object changes in time—the acceleration a_d—as well as (*ii*) how fast this rate in its turn also changes in time—the jerk j_d.

These image properties correspond to sensory information that is lost at the level of synchronic units and *cognitively* reconstructed at the level of diachronic unit. We consider this reconstruction a direct analogy to amodal perception and assume both of them to be based on the same neural mechanisms. Although such emergent properties attributed to the level of diachronic unit are of cognitive nature, they must become the components of depictive images of mental imagery. It enables us to *experience* them directly as some properties perceived via additional channels of the corresponding sensory modality. In particular, it is a reason for including the acceleration and jerk of observed moving object in the list of variables specifying the state of the experiential now as a whole entity (Fig. 2.9, upper fragment). Here we should clarify the relation between the diachronic unit and the experiential now. The diachronic unit represents an element of human temporal dimension, whereas the experiential now is how we perceive the external wold and ourselves in it within the diachronic unit.

Speaking about the images of diachronic unit level, the difference between their properties inherited from the level of synchronic units and their properties emerging via synergy of diachronic unit should be reflected in time scales of their dynamics. Experimental data concerning visual and tactile perception of acceleration and deceleration (e.g., Gori et al. 2013) argue for this statement. In particular, Gori et al. estimate the integration time required to make a judgment about acceleration/deceleration as a value about 1 s. The global-motion integration stage can even be of a longer duration around 3 s with an intermediate stage of duration around 1 s (Burr and Santoro 2001). We want to remind that the duration of early vision—the duration of the immediate now or the immediate future—is characterized by time scales about 0.2–0.3 s. For a detailed review of temporal properties characterizing such event integration a reader may be referred to White (2017).

Emerging via cognitive cohesion of its synchronic units, the diachronic unit should be sensitive to mental states affecting the sensory process as a whole. This dependence is illustrated, e.g., by attentional modulation of speed-change perception (e.g., Yang et al. 2018), reflecting the substantial contribution of cognitive processes to perceiving changes in motion. Moreover there are, at least, two qualitatively different attention systems affecting visual perception (Bao and Pöppel 2007). So the fusion of the immediate past, the immediate now, and the immediate future into the diachronic unit seems to be a complex process implemented via several "channels."

As far as the relationship between the diachronic unit and the action strategies is concerned with, action strategies characterize intentional goal-oriented actions on time scales much longer than 1 s. So the action strategy cannot change essentially on time scales about 1 s except for moments when a subject makes decision on its modification. Therefore the contribution of the experiential now to the mental evaluation of a given action strategy should be determined completely by the perception of system state in the scope of the diachronic unit as a single whole. Put differently, in evaluating the quality of a given action strategy, the particular details of images at the level of synchronic units are of minor importance if their cumulative contribution to the level of diachronic unit does not change. It implies that in modeling intentional goal-oriented actions the experiential now admits a description based on the properties of mental images belonging to the level of diachronic unit. Speaking about the motion perception, it means that the mental phase space of the experiential now has to involve, at least, the four variables $\{x_d, v_d, a_d, j_d\}$,[31] which demonstrates that the corresponding formalism cannot obey the paradigm of Newtonian mechanics in the general case. Figure 2.9 (upper fragment) illustrates this depicting the action strategies AS_1 and AS_2 connected with the region of diachronic unit containing the phase variables of the experiential now.

We want to emphasize that the made statement about the experiential now and goal-oriented actions does not exclude top-down effects of action strategies on the internal features of diachronic unit synergy; they are just implicit. Changes in mental evaluation of possible action strategies in their quality gives rise to reevaluation of the system current state within the scope of diachronic unit as a single whole.

[31] The idea that human perception of motion is characterized by higher order time derivatives, in particular, the acceleration a of the perceived moving object and its jerk $j = da/dt$ is not new. For example, the acceleration of car motion may be included in the list of phase variables determining the driver behavior (for a general review including of models and empirical data see, e.g., Kerner 2009, 2017; Treiber and Kesting 2013). In particular, driving simulator experiments on car-following demonstrated that the acceleration and jerk must be regarded as additional independent phase variables for modeling driver actions (Lubashevsky 2017; Lubashevsky and Morimura 2019). The jerk as a phase variable may be taken into account in describing sensorimotor control (for a review see, e.g., Todorov 2004; Liu and Todorov 2007; Biess et al. 2007, 2011; Biess 2013). The description of mental processes based on free energy also assumes that the acceleration, jerk, and other higher-order time derivatives may be included in the list of phase variables (e.g., Friston 2008; Friston et al. 2008, 2009, 2010; Buckley et al. 2017), in particular, it concerns emotion modeling (Joffily and Coricelli 2013; Friston et al. 2018).

Because C-completion is crucial in the diachronic unit emergence, the induced cognitive processes can get the level of synchronic units. As a result, e.g., the attention redistribution between the immediate past, the immediate now, and the immediate future, is able to affect the unconscious processing of sensory information as well as the bottom-up conversion of image properties similar to (2.6) and (2.7).

Summarizing the presented discussion about the properties of diachronic unit, we accept that:

- The diachronic unit as an individual element of human temporal dimension with duration about $\tau_d \sim 2$–3 s represents the experiential now as a whole entity with internal tripartite structure.
- The diachronic unit
    - on the one side, emerges via the fusion of three synchronic units—the immediate past, the immediate now, and the immediate future,
    - on the other side, is irreducible to its constituent synchronic units due to cognitive processes similar to C-completion contributing to its synergy.
- The diachronic unit bears the properties caused by both
    - the bottom-up aggregation of sensory data, which reflects its ability of sensory perception, and
    - the top-down cognitive processing of the individual sensory information borne by the constituent synchronic units, which enables the extraction of additional details about the perceived events.
- Due to the cognitive processing of cumulative sensory information at the level of diachronic unit accompanied by the accepted model of observed phenomena, the diachronic unit bears
    - the properties inherited from the constituent synchronic units and averaged on temporal scales about its duration τ_d as well as
    - the own properties stemming from the comparison of the constituent synchronic units in their individual properties but not reducible to them; these properties characterize how the individual properties of the constituent synchronic units change in time.
- The own properties attributed to the diachronic unit and generated via independent channels (one or several) combined with the properties borne individually by the constituent synchronic units make up the complete description of the experiential now—the phase space of its states.
- The tripartite structure of diachronic unit makes it plausible that its own properties can characterize the first and second order time derivatives of the individual properties of the constituent synchronic units.

Generally speaking, the diachronic unit supervenes upon the synchronic units of immediate past, immediate now, and immediate future but is not reducible to them in properties. Figure 2.9 illustrates this supervenience.

As far as the conscious aspects of diachronic unit, i.e., the experiential now are concerned, their detailed discussion is postponed to the next section. Here, leaping ahead, we note that their categorization depends on the understanding of consciousness and up to now there is no stable concept satisfying the majority of researches. Keeping in mind the later version of Raftopoulos's account (2019), we suppose that the properties inherited from the three synchronic units are just given to the mind, i.e., they belong to the base level of consciousness. The own properties of diachronic unit are derived from them, which involves some cognitive processes. So these properties of the experiential now have to belong to a higher level of consciousness. However, the mind seems to have no efficient access to the *details* of the transitions between the two levels (cf. Raftopoulos 2019). Therefore, we treat the properties of the two types as the equipollent components of the experiential now states. In particular, the motion state of observed object at the level of diachronic unit is described in terms of its position, velocity, acceleration, and jerk—the dimensions of the four-component phase space. The difference between the properties of the experiential now—inherited from the constituent synchronic units and its own—reflect in their sensitivity to intentions in human actions, in particular, to allocated attention. In other words, the sensitivity of the own properties must be much higher than that of the inherited properties.

2.7 Conclusion

We have considered modern accounts of human now and analyzed in detail extended atomism and extensionalism—two rival accounts. In this Chapter we have started to develop a theory of human temporality and proposed our account of experiential now combining the main ideas of extended atomism and extensionalism. Experience has been put forward as the main criterion for specifying what is now in the mind. Namely, everything that even has been perceived or will be perceived but is felt as experienced now belongs to the experiential now.

We have accepted the idea that the experiential now of finite extension in time possesses hierarchical temporal structure. Turning to the available psychological and neurological data, this view has been justified and the underlying neural mechanisms were discussed. Mainly these mechanisms are responsible for the unconscious processing of perceived information and only the final integral results become accessible to the mind. Predictive coding paradigm has been accepted as the gist of this information processing, which allows for top-down and bottom-up interaction between neural signals generated by sensory modalities and human cognition.

As the basic features of the proposed account of experiential now we note the following.

- There are three constituent temporal parts of the experiential now represented in human temporal dimension as the corresponding synchronic units:
  - the immediate past

  - the immediate now,
  - the immediate future.

  The introduction of these temporal parts is partly inspired by Husserl's tripartite structure of human now including retention, primary impression, and protention.
- The three temporal parts are recognized in the mind as individual entities separated in time, on the one hand, and having no conscious temporal structure, on the other hand. This conscious structurelessness implies that:

  - The mind is unable to separate events belonging to one of these parts by finite time lags, which is emphasized in the term *synchronic unit* used to refer to the corresponding temporal part.
  - Nevertheless, the details of event arrangement inside one synchronic unit can be extracted, at least, partly, via the unconscious integration of perceived information. These details become accessible to the mind at the level of the corresponding synchronic unit and attributed to this unit as properties just given to the mind. The mind is unable to reconstruct directly the hidden relationship between these properties.

  It is necessary to note that the direct unconscious integration of perceived information is implemented within immediate now, short-term encoding and memory consolidation governs the immediate past and endow it with integrity, whereas internal models of automatic anticipation give rise to the immediate future.
- In human temporal dimension the synchronic units of the immediate past, immediate now, and the immediate future are the basic indivisible elements bearing physical properties of perceived events contained in them.
- The union of the immediate past, immediate now, and the immediate future and their partial fusion in the mind is the essence of the experiential now having the conscious tripartite structure. Experiential now is based on its constituent components but is not reduced to them in properties. The integral duration of experiential now is estimated as 2–3 s.
- In human temporal dimension, the experiential now is represented by the diachronic unit having the temporal tripartite structure. In spite of its structural divisibility the diachronic unit is regarded as a certain fundamental fragment bearing own properties integrating but not reducible to the properties of its constituent synchronic units.
- Dealing with perception of moving objects, it is accepted that at the level of the experiential now four quantities are attributed to the diachronic unit as a single entity. They are the spacial position x of perceived object, its velocity v, acceleration a, and jerk j characterizing the object motion on scales about 2–3 s. These quantities are just given to the mind without recognizable relationship existing in physical reality. Moreover the quantitative evaluation of the perceived motion properties is characterized by, maybe, significant uncertainty intrinsic to the perception system.
- The first pair of the motion properties, $\{x, v\}$, is actually the result of information processing within the synchronic units of the immediate past, immediate now, and

the immediate future, whereas the second pair $\{a, j\}$ is due to the conscious (semi-conscious) processing of information within the diachronic unit. For this reason mental conditions, e.g., attention redistribution should mainly affect the second pair of motion characteristics.
- Neural mechanisms responsible for perception, memorizing, and imagination seem to be closely intertwined in the brain which is reflected in the mind. Therefore the given description of motion perception should be applicable to motion imagination. It argues for introducing the extended phase space $\{x, v, a, j\}$ in the mathematical description of the mental image of moving objects—real or imaginary ones.

As noted above properties characterizing perception of system dynamics are just given to the mind as mutually unrelated quantities with significant, intrinsic uncertainty. Therefore the mathematical description of human actions—especially intentional, goal-oriented ones—cannot be confined to the experiential now only.

The mental images of currently perceived objects and imaginary objects can be experienced in a similar way, in this sense, their difference is more quantitative than qualitative. This similarity underlies the introduction of the concept of action strategy as a single entity combing related actions located also outside experiential in the past and the future.

References

Adams, R. A., Shipp, S., & Friston, K. J. (2013). Predictions not commands: Active inference in the motor system. *Brain Structure and Function, 218*(3), 611–643.

Aggelopoulos, N. C. (2015). Perceptual inference. *Neuroscience & Biobehavioral Reviews, 55*, 375–392.

Aghdaee, S. M., Battelli, L., & Assad, J. A. (2014). Relative timing: From behaviour to neurons. *Philosophical Transactions of the Royal Society of London B: Biological Sciences, 369*(1637), 20120472 (11 pages).

Ahmed, O. J., & Cash, S. S. (2013). Finding synchrony in the desynchronized EEG: The history and interpretation of gamma rhythms. *Frontiers in Integrative Neuroscience, 7*, Article 58 (7 pages).

Aitchison, L., & Lengyel, M. (2017). With or without you: Predictive coding and Bayesian inference in the brain. *Current Opinion in Neurobiology, 46*, 219–227. Special Issue: Computational Neuroscience.

Andersen, H. (2014). The development of the "specious present" and James's views on temporal experience. In V. Arstila & D. Lloyd (Eds.), *Subjective time: The philosophy, psychology, and neuroscience of temporality* (pp. 25–42). Cambridge: The MIT Press.

Andersen, H., & Grush, R. (2009). A brief history of time-consciousness: Historical precursors to James and Husserl. *Journal of the History of Philosophy, 47*(2), 277–307.

Arstila, V. (2018). When is cognitive penetration a plausible explanation? *Consciousness and Cognition, 59*, 78–86.

Baars, B. J., & Franklin, S. (2003). How conscious experience and working memory interact. *Trends in Cognitive Sciences, 7*(4), 166–172.

Babkoff, H., & Fostick, L. (2013). The role of tone duration in dichotic temporal order judgment. *Attention, Perception, & Psychophysics, 75*(4), 654–660.

Baddeley, A. (2012). Working memory: Theories, models, and controversies. *Annual Review of Psychology, 63*(1), 1–29.

Bae, G. Y., & Flombaum, J. I. (2013). Two items remembered as precisely as one: How integral features can improve visual working memory. *Psychological Science, 24*(10), 2038–2047.

Bahrick, L. E. (2010). Amodal perception. In E. B. Goldstein (Ed.), *Encyclopedia of perception* (Vol. 1, pp. 44–46). Los Angeles: SAGE Publishers Inc.

Bahrick, L. E., & Lickliter, R. (2010). Perceptual development: Intermodal perception. In E. B. Goldstein (Ed.), *Encyclopedia ofperception* (Vol. 2, pp. 753–756). Los Angeles: SAGE Publishers Inc.

Bakhurin, K. I., Goudar, V., Shobe, J. L., Claar, L. D., Buonomano, D. V., & Masmanidis, S. C. (2017). Differential encoding of time by prefrontal and striatal network dynamics. *Journal of Neuroscience, 37*(4), 854–870.

Balaban, H., Drew, T., & Luria, R. (2018). Delineating resetting and updating in visual working memory based on the object-to-representation correspondence. *Neuropsychologia, 113,* 85–94.

Bao, Y., & Pöppel, E. (2007). Two spatially separated attention systems in the visual field: Evidence from inhibition of return. *Cognitive Processing, 8*(1), 37–44.

Başar-Eroglu, C., Strüber, D., Schürmann, M., Stadler, M., & Başar, E. (1996). Gamma-band responses in the brain: A short review of psychophysiological correlates and functional significance. *International Journal of Psychophysiology, 24*(1), 101–112. New Advances in EEG and cognition.

Bayliss, D. M., Bogdanovs, D. M., & Jarrold, C. (2015). Consolidating working memory: Distinguishing the effects of consolidation, rehearsal and attentional refreshing in a working memory span task. *Journal of Memory and Language, 81,* 34–50.

Bayne, T. (2005). Divided brains and unified phenomenology: A review essay on michael tye's consciousness and persons. *Philosophical Psychology, 18*(4), 495–512.

Bayne, T. (2010). *The unity of consciousness*. Oxford, UK: Oxford University Press.

Bayne, T. J., & Chalmers, D. J. (2003). What is the unity of consciousness? In A. Cleeremans (Ed.), *The unity of consciousness: Binding, integration, and dissociation* (pp. 23–58). Oxford, UK: Oxford University Press.

Bergström, F., & Eriksson, J. (2014). Maintenance of non-consciously presented information engages the prefrontal cortex. *Frontiers in Human Neuroscience, 8,* Article 938 (10 pages).

Bergström, F., & Eriksson, J. (2015). The conjunction of non-consciously perceived object identity and spatial position can be retained during a visual short-term memory task. *Frontiers in Psychology, 6,* Article 1470 (19 pages).

Bermúdez, J. L. (2014). *Cognitive science: An introduction to the science of the mind* (2nd ed.). New York, NY: Cambridge University Press.

Biess, A. (2013). Shaping of arm configuration space by prescription of non-Euclidean metrics with applications to human motor control. *Physical Review E - Statistical, Nonlinear, and Soft Matter Physics, 87*(1), 012729 (15 pages).

Biess, A., Flash, T., & Liebermann, D. G. (2011). Riemannian geometric approach to human arm dynamics, movement optimization, and invariance. *Physical Review E - Statistical, Nonlinear, and Soft Matter Physics, 83*(3), 031927 (11 pages).

Biess, A., Liebermann, D. G., & Flash, T. (2007). A computational model for redundant human three-dimensional pointing movements: Integration of independent spatial and temporal motor plans simplifies movement dynamics. *Journal of Neuroscience, 27*(48), 13045–13064.

Blalock, L. D. (2013). Mask similarity impacts short-term consolidation in visual working memory. *Psychonomic Bulletin & Review, 20*(6), 1290–1295.

Block, R. A., & Grondin, S. (2014). Timing and time perception: A selective review and commentary on recent reviews. *Frontiers in Psychology, 5,* Article 648 (3 pages).

Block, N. (1995). On a confusion about a function of consciousness. *Behavioral and Brain Sciences, 18*(2), 227–247.

Block, N. (2007). Consciousness, accessibility, and the mesh between psychology and neuroscience. *Behavioral and Brain Sciences, 30*(5–6), 481–499.

Borst, G., & Kosslyn, S. M. (2008). Visual mental imagery and visual perception: Structural equivalence revealed by scanning processes. *Memory & Cognition, 36*(4), 849–862.

Bowman, H., & Wyble, B. (2007). The simultaneous type, serial token model of temporal attention and working memory. *Psychological Review, 114*(1), 38–70.
Briscoe, R. E. (2011). Mental imagery and the varieties of amodal perception. *Pacific Philosophical Quarterly, 92*(2), 153–173.
Broad, C. D. (1923). *Scientific thought*. London: Routledge & Kegan Paul.
Broad, C. D. (1925). *The mind and its place in nature*. London: Routledge & Kegan Paul.
Broad, C. D. (1938). *An examination of McTaggart's philosophy* (Vol. II, Part I). Cambridge: Cambridge University Press.
Brook, A., & Raymont, P. (2017). The unity of consciousness. In E. N. Zalta (Ed.), *The stanford encyclopedia of philosophy*, summer (2017th ed.). Metaphysics Research Lab, Stanford University.
Brosch, R. (2018). What we 'see' when we read: Visualization and vividness in reading fictional narratives. *Cortex, 105*, 135–143. Special Issue: The Eye's Mind - visual imagination, neuroscience and the humanities.
Brown, H., Friston, K., & Bestmann, S. (2011). Active inference, attention, and motor preparation. *Frontiers in Psychology, 2*, Article 218 (10 pages).
Bruno, A., & Cicchini, G. M. (2016). Multiple channels of visual time perception. *Current Opinion in Behavioral Sciences, 8*, 131–139. Time in perception and action.
Bubic, A., Von Cramon, D. Y., & Schubotz, R. (2010). Prediction, cognition and the brain. *Frontiers in Human Neuroscience, 4*, 25 (15 pages).
Buckley, C. L., Kim, C. S., McGregor, S., & Seth, A. K. (2017). The free energy principle for action and perception: A mathematical review. *Journal of Mathematical Psychology, 81*, 55–79.
Buonomano, D. V. (2000). Decoding temporal information: A model based on short-term synaptic plasticity. *Journal of Neuroscience, 20*(3), 1129–1141.
Buonomano, D. V. (2014). Neural dynamics based timing in the subsecond to seconds range. In H. Merchant & V. de Lafuente (Eds.), *Neurobiology of interval timing* (pp. 101–117). New York, NY: Springer Science+Business Media.
Buonomano, D. V., Bramen, J., & Khodadadifar, M. (2009). Influence of the interstimulus interval on temporal processing and learning: Testing the state-dependent network model. *Philosophical Transactions of the Royal Society of London B: Biological Sciences, 364*(1525), 1865–1873.
Buonomano, D. V., & Maass, W. (2009). State-dependent computations: Spatiotemporal processing in cortical networks. *Nature Reviews Neuroscience, 10*, 113.
Buonomano, D. V., & Mauk, M. D. (1994). Neural network model of the cerebellum: Temporal discrimination and the timing of motor responses. *Neural Computation, 6*(1), 38–55.
Buonomano, D. V., & Merzenich, M. M. (1995). Temporal information transformed into a spatial code by a neural network with realistic properties. *Science, 267*(5200), 1028–1030.
Burr, D. C., & Santoro, L. (2001). Temporal integration of optic flow, measured by contrast and coherence thresholds. *Vision Research, 41*(15), 1891–1899.
Buzsáki, G., & Draguhn, A. (2004). Neuronal oscillations in cortical networks. *Science, 304*(5679), 1926–1929.
Cao, L., Veniero, D., Thut, G., & Gross, J. (2017). Role of the cerebellum in adaptation to delayed action effects. *Current Biology, 27*(16), 2442–2451.e1–e3.
Cecchi, A. S. (2018). Cognitive penetration of early vision in face perception. *Consciousness and Cognition, 63*, 254–266.
Chuard, P. (2011). Temporal experiences and their parts. *Philosophers' Imprint, 11*(11), 1–28.
Chuard, P. (2017). The snapshot conception of temporal experiences. In I. Phillips (Ed.), *The Routledge handbook of philosophy of temporal experience* (pp. 121–132). Abingdon, Oxon: Routledge, Taylor & Francis Group.
Cisek, P. (2009). Internal models. In M. D. Binder, N. Hirorawa, & U. Windhorst (Eds.), *Encyclopedia of neuroscience* (pp. 2009–2012). Berlin: Springer-Verlag GmbH.
Clark, A. (2013). Whatever next? predictive brains, situated agents, and the future of cognitive science. *Behavioral and Brain Sciences, 36*(03), 181–204.

Colombo, B. (2012). Mental imagery. In N. M. Seel (Ed.), *Encyclopedia of the sciences of learning* (pp. 2187–2191). LLC, London: Springer Science+Business Media.

Cooper, L. A. (1990). Manipulation of visual information. In J. I. Elkind, S. K. Card, J. Hochberg, & B. M. Huey (Eds.), *Human performance models for computer-aided engineering* (pp. 144–158). San Diego, CA: Academic Press Inc.

Cooper, L. A., & Shepard, R. N. (1984). Turning something over in the mind. *Scientific American, 251*(6), 106–114.

Cowan, N. (2008). Sensory memory. In J. H. E. Byrne, & H. L. V. Roediger III (Eds.), *Learning and memory: A comprehensive reference* (Vol. 2, pp. 23–32). Cognitive psychology of memory. Oxford: Academic Press.

Cowan, N. (2017b). Working memory: The information you are now thinking of. In J. H. e. Byrne, & J. T. v. Wixted (Eds.), *Learning and memory: A comprehensive reference* (2nd ed., Vol. 2, pp. 147–161). Cognitive psycholoogy of memory. Amsterdam: Elsevier Ltd.

Cowan, N. (1984). On short and long auditory stores. *Psychological Bulletin, 96*(2), 341–370.

Cowan, N. (1988). Evolving conceptions of memory storage, selective attention, and their mutual constraints within the human information-processing system. *Psychological Bulletin, 104*(2), 163–191.

Cowan, N. (1999). An embedded-processes model of working memory. In A. Miyake & P. Shah (Eds.), *Models of working memory: Mechanisms of active maintenance and executive control* (pp. 62–101). Cambridge: Cambridge University Press.

Cowan, N. (2001). The magical number 4 in short-term memory: A reconsideration of mental storage capacity. *Behavioral and Brain Sciences, 24*(1), 87–114.

Cowan, N. (2015). Sensational memorability: Working memory for things we see, hear, feel, or somehow sense. In P. Jolicoeur, C. Lefebvre, & J. Martinez-Trujillo (Eds.), *Mechanisms of sensory working memory: Attention and performance XXV* (pp. 5–22). San Diego: Academic Press.

Craston, P., Wyble, B., Chennu, S., & Bowman, H. (2009). The attentional blink reveals serial working memory encoding: Evidence from virtual and human event-related potentials. *Journal of Cognitive Neuroscience, 21*(3), 550–566.

Dainton, B. (2006). *Stream of consciousness: Unity and continuity in conscious experience* (6th ed.). London: Routledge, Taylor & Francis Group. Rev. ed. 2006.

Dainton, B. (2008). Sensing change. Philosophical. *Issues, 18*(1), 362–384.

Dainton, B. (2010). *Time and space* (2nd ed.). Durham: Acumen.

Dainton, B. (2017a). Temporal consciousness. In E. N. Zalta (Ed.), *The stanford encyclopedia of philosophy*, winter (2018th ed.). Metaphysics Research Lab, Stanford University.

Dainton, B. (2017b). William Stern's "Psychische Präsenzzeit." In I. Phillips (Ed.), *The Routledge handbook of philosophy of temporal experience* (pp. 107–118). Routledge, Taylor & Francis Group: Abingdon, Oxon.

D'Angiulli, A., & Reeves, A. (2002). Generating visual mental images: Latency and vividness are inversely related. *Memory & Cognition, 30*(8), 1179–1188.

D'Angiulli, A., Runge, M., Faulkner, A., Zakizadeh, J., Chan, A., & Morcos, S. (2013). Vividness of visual imagery and incidental recall of verbal cues, when phenomenological availability reflects long-term memory accessibility. *Frontiers in Psychology, 4*, Article 1 (18 pages).

De Schrijver, S., & Barrouillet, P. (2017). Consolidation and restoration of memory traces in working memory. *Psychonomic Bulletin & Review, 24*(5), 1651–1657.

Dennett, D. C. (1991). *Consciousness explained.* New York, NY: Back Bay Books, Little, Brown and Company.

Dharani, K. (2015). *The biology of thought: A neuronal mechanism in the generation of thought–a new molecular model* (pp. 53–74) Chap. 3, Memory. San Diego: Academic Press.

Dutta, A., Shah, K., Silvanto, J., & Soto, D. (2014). Neural basis of non-conscious visual working memory. *NeuroImage, 91*, 336–343.

Dux, P. E., & Marois, R. (2009). The attentional blink: A review of data and theory. *Attention, Perception & Psychophysics, 71*(8), 1683–1700.

Ehm, W., Bach, M., & Kornmeier, J. (2011). Ambiguous figures and binding: EEG frequency modulations during multistable perception. *Psychophysiology, 48*(4), 547–558.
Eitam, B., Hassin, R. R., & Schul, Y. (2008). Nonconscious goal pursuit in novel environments: The case of implicit learning. *Psychological Science, 19*(3), 261–267.
Elliott, M. A., & Giersch, A. (2016). What happens in a moment. *Frontiers in Psychology, 6*, Article 1905 (7 pages).
Engel, A. K., Fries, P., & Singer, W. (2001). Dynamic predictions: Oscillations and synchrony in top-down processing. *Nature Reviews Neuroscience, 2*(10), 704–716.
Engle, R. W. (2002). Working memory capacity as executive attention. *Current Directions in Psychological Science, 11*(1), 19–23.
Feldman, H., & Friston, K. J. (2010). Attention, uncertainty, and free-energy. *Frontiers in Human Neuroscience, 4*, Article 215 (23 pages).
Fink, M., Ulbrich, P., Churan, J., & Wittmann, M. (2006). Stimulus-dependent processing of temporal order. *Behavioural Processes, 71*(2), 344–352. Interval timing: The current status.
Finke, R. (1989). *Principles of mental imagery*, A Bradford book. Cambridge, MA: The MIT Press.
Finke, R. A. (1980). Levels of equivalence in imagery and perception. *Psychological Review, 87*(2), 113–132.
Firestone, C., & Scholl, B. J. (2016). Cognition does not affect perception: Evaluating the evidence for "top-down" effects. *Behavioral and Brain Sciences, 39*, e229.
Fodor, J. A., & Pylyshyn, Z. W. (1981). How direct is visual perception?: Some reflections on Gibson's "ecological approach". *Cognition, 9*(2), 139–196.
Foster, J. (1979). In *Self-defence*.In G. F. Macdonald (Ed.), *Perception and identity: Essays presented to A. J. Ayer with his replies to them* (pp. 161–185). London: The Macmillan Press Ltd.
Foster, J. (1982). *The case for idealism*. London: Routledge & Kegan Paul.
Foster, J. (1991). *The immaterial self: A defence of the Cartesian dualist conception of the mind*. London: Routledge.
Franklin, D. W., & Wolpert, D. M. (2011). Computational mechanisms of sensorimotor control. *Neuron, 72*(3), 425–442.
Fries, P., Nikolić, D., & Singer, W. (2007). The gamma cycle. *Trends in Neurosciences, 30*(7), 309–316. July INMED/TINS special issue–Physiogenic and pathogenic oscillations: The beauty and the beast.
Friston, K. (2003). Learning and inference in the brain. *Neural Networks, 16*(9), 1325–1352. Neuroinformatics.
Friston, K. (2005). A theory of cortical responses. *Philosophical Transactions of the Royal Society B: Biological Sciences, 360*(1456), 815–836.
Friston, K. (2008). Hierarchical models in the brain. *PLOS Computational Biology, 4*(11), e100021 (24 pages).
Friston, K. (2009). The free-energy principle: A rough guide to the brain? *Trends in Cognitive Sciences, 13*(7), 293–301.
Friston, K. (2010). The free-energy principle: A unified brain theory? *Nature Reviews Neuroscience, 11*(2), 127–138.
Friston, K. (2018). Does predictive coding have a future? *Nature Neuroscience, 21*(8), 1019–1021.
Friston, K. J., Daunizeau, J., & Kiebel, S. J. (2009). Reinforcement learning or active inference?. *PloS ONE, 4*(7), e6421.
Friston, K., Schwartenbeck, P., Fitzgerald, T., Moutoussis, M., Behrens, T., & Dolan, R. J. (2013). The anatomy of choice: Active inference and agency. *Frontiers in Human Neuroscience, 7*(598), Article 598, 1–18.
Friston, K. J., Daunizeau, J., Kilner, J., & Kiebel, S. J. (2010). Action and behavior: A free-energy formulation. *Biological Cybernetics, 102*(3), 227–260.
Friston, K., FitzGerald, T., Rigoli, F., Schwartenbeck, P., O'Doherty, J., & Pezzulo, G. (2016). Active inference and learning. *Neuroscience & Biobehavioral Reviews, 68*, 862–879.

Friston, K. J., Joffily, M., Barrett, L. F., & Seth, A. K. (2018). Active inference and emotions. In A. S. Fox, R. C. Lapate, A. J. Shackman, & R. J. Davidson (Eds.), *The nature of emotions: Fundamental questions* (2nd ed., pp. 28–33). New York, NY: Oxford University Press.
Friston, K., Schwartenbeck, P., FitzGerald, T., Moutoussis, M., Behrens, T., & Dolan, R. J. (2014). The anatomy of choice: Dopamine and decision-making. *Philosophical Transactions of the Royal Society B: Biological Sciences, 369*(1655), 20130481.
Friston, K. J., & Stephan, K. E. (2007). Free-energy and the brain. *Synthese, 159*(3), 417–458.
Friston, K. J., Trujillo-Barreto, N., & Daunizeau, J. (2008). Dem: A variational treatment of dynamic systems. *NeuroImage, 41*(3), 849–885.
Galdo-Alvarez, S., Bonilla, F. M., González-Villar, A. J., & de-la Peña, M. T. C. (2016). Functional equivalence of imagined vs. real performance of an inhibitory task: An EEG/ERP study. *Frontiers in Human Neuroscience, 10*, Article 467.
Gallagher, S. (1998). *The inordinance of time*. Ivanston, Illinois: Northwestern University Press.
Gallagher, S., & Zahavi, D. (2012). *The phenomenological mind* (2nd ed.). London: Routledge, Taylor & Francis Group.
Ganis, G., Thompson, W. L., & Kosslyn, S. M. (2004). Brain areas underlying visual mental imagery and visual perception: An fMRI study. *Cognitive Brain Research, 20*(2), 226–241.
García-Pérez, M. A., & Alcalá-Quintana, R. (2018). Perceived temporal order and simultaneity: Beyond psychometric functions. In A. Vatakis, F. Balcı, D. M. Luca, & A. Correa (Eds.), *Timing and time perception: Procedures, measures, and applications* (pp. 263–294). Leiden: Brill.
Glennerster, A. (2007). Marr's vision: Twenty-five years on. *Current Biology, 17*(11), R397–R399.
Goel, A., & Buonomano, D. V. (2014). Timing as an intrinsic property of neural networks: evidence from in vivo and in vitro experiments. *Philosophical Transactions of the Royal Society of London B: Biological Sciences*, *369*(1637), 20120460 (12 pages).
Gold, I. (1999). Does 40-Hz oscillation play a role in visual consciousness? *Consciousness and Cognition, 8*(2), 186–195.
Gori, M., Sciutti, A., Jacono, M., Sandini, G., Morrone, C., & Burr, D. C. (2013). Long integration time for accelerating and decelerating visual, tactile and visual-tactile stimuli. *Multisensory Research, 26*(1–2), 53–68.
Goudar, V., & Buonomano, D. V. (2018). Encoding sensory and motor patterns as time-invariant trajectories in recurrent neural networks. *eLife*, *7*, e31134 (28 pages).
Gray, C. M. (1999). The temporal correlation hypothesis of visual feature integration. *Neuron, 24*(1), 31–47.
Grondin, S., Hasuo, E., Kuroda, T., & Nakajima, Y. (2018). Auditory time perception. In R. A. Bader (Ed.), *Springer handbook of systematic musicology* (pp. 423–440). Germany: Springer-Verlag GmbH. Chap. 21, S. Koelschy (Ed.), Part C: Music psychology-physiolog.
Grondin, S. (2010). Timing and time perception: A review of recent behavioral and neuroscience findings and theoretical directions. *Attention, Perception, & Psychophysics, 72*(3), 561–582.
Gross, S. (2017). Cognitive penetration and attention. *Frontiers in Psychology*, *8*, Article 221 (12 pages).
Grush, R. (2004). The emulation theory of representation: Motor control, imagery, and perception. *Behavioral and Brain Sciences, 27*(3), 377–396.
Grush, R. (2005). Internal models and the construction of time: Generalizing from state estimation to trajectory estimation to address temporal features of perception, including temporal illusions. *Journal of Neural Engineering, 2*(3), S209–S218.
Grush, R. (2006). How to, and how not to, bridge computational cognitive neuroscience and Husserlian phenomenology of time consciousness. *Synthese, 153*(3), 417–450.
Grush, R. (2007). Time and experience. In T. Müller (Ed.), *Philosophie der zeit: Neue analytische ansätze* (pp. 27–44). Frankfurt a. M.: Klostermann.
Grush, R. (2008). Temporal representation and dynamics. *New Ideas in Psychology, 26*(2), 146–157. Dynamics and Psychology.
Guell, X., Gabrieli, J. D. E., & Schmahmann, J. D. (2018). Embodied cognition and the cerebellum: Perspectives from the dysmetria of thought and the universal cerebellar transform theories. *Cortex*,

100, 140–148. Embodiment disrupted: Tapping into movement disorders through syntax and action semantics.
Gupta, D. S. (2014). Processing of sub- and supra-second intervals in the primate brain results from the calibration of neuronal oscillators via sensory, motor, and feedback processes. *Frontiers in Psychology*, *5*, Aritcle 816 (16 pages).
Hardman, K. O., Vergauwe, E., & Ricker, T. J. (2017). Categorical working memory representations are used in delayed estimation of continuous colors. *Journal of Experimental Psychology. Human Perception and Performance*, *43*(1), 30–54.
Hass, J., & Durstewitz, D. (2014). Neurocomputational models of time perception. In H. Merchant & V. de Lafuente (Eds.), *Neurobiology of interval timing, Advances in experimental medicine and biology* (Vol. 829, pp. 49–71). NY, New York: Springer Science+Business Media.
Hassin, R. R. (2005). Nonconscious control and implicit working memory. Oxford series in social cognition and social neuroscience. In R. R. Hassin, J. S. Uleman, & J. A. Bargh (Eds.), *The new unconscious* (pp. 196–222). New York, NY: Oxford University Press Inc.
Hassin, R. R., Aarts, H., Eitam, B., Custers, R., & Kleiman, T. (2008). Non-conscious goal pursuit and the effortful control of behavior. In E. Morsella, J. A. Bargh, & P. M. Gollwitzer (Eds.), *Oxford handbook of human action* (pp. 549–566). New York, NY: Oxford University Press Inc.
Hassin, R. R., Bargh, J. A., Engell, A. D., & McCulloch, K. C. (2009). Implicit working memory. *Consciousness and Cognition, 18*(3), 665–678.
Hassin, R. R., Bargh, J. A., & Zimerman, S. (2009). Automatic and flexible: The case of nonconscious goal pursuit. *Social Cognition, 27*(1), 20–36.
Hestevold, H. S. (2008). Presentism: Through thick and thin. *Pacific Philosophical Quarterly, 89*(3), 325–347.
Hoerl, C. (2013). A succession of feelings, in and of itself, is not a feeling of succession. *Mind, 122*(486), 373–417.
Hohwy, J. (2013). *The predictive mind*. Oxford, UK: Oxford University Press.
Howard, M. W. (2018). Memory as perception of the past: Compressed time in mind and brain. *Trends in Cognitive Sciences, 22*(2), 124–136.
Hughes, J. R. (2008). Gamma, fast, and ultrafast waves of the brain: Their relationships with epilepsy and behavior. *Epilepsy & Behavior, 13*(1), 25–31.
Husserl, E. (1991). *On the phenomenology of the consciousness of internal time (1893–1917)* (J. B. Brough, Trans.). Dordrecht: Kluwer Academic Publishers.
Hyun, J.-S., & Luck, S. J. (2007). Visual working memory as the substrate for mental rotation. *Psychonomic Bulletin & Review, 14*(1), 154–158.
Ishikawa, T., Tomatsu, S., Izawa, J., & Kakei, S. (2016). The cerebro-cerebellum: Could it be loci of forward models?. *Neuroscience Research, 104*, 72–79. Body representation in the brain.
Ivry, R. B., & Schlerf, J. E. (2008). Dedicated and intrinsic models of time perception. *Trends in Cognitive Sciences, 12*(7), 273–280.
Jacob, J., Jacobs, C., & Silvanto, J. (2015). Attention, working memory, and phenomenal experience of wm content: memory levels determined by different types of top-down modulation. *Frontiers in Psychology*, *6*, Article 1603 (7 pages).
Jacobs, C., & Silvanto, J. (2015). How is working memory content consciously experienced? the 'conscious copy' model of WM introspection. *Neuroscience & Biobehavioral Reviews, 55*, 510–519.
James, W. (1890). *The principles of psychology* (Vol. 1). New York: Henry Holt and Company.
Ji, E., Lee, K. M., & Kim, M.-S. (2017). Independent operation of implicit working memory under cognitive load. *Consciousness and Cognition, 55*, 214–222.
Joffily, M., & Coricelli, G. (2013). Emotional valence and the free-energy principle. *PLOS Computational Biology*, *9*(6), e1003094 (14 pages).
Johnson-Laird, P. N. (1989). Mental models. A Bradford book. In M. I. Posner (Ed.), *The foundations of cognitive science* (pp. 469–499). Cambridge, MA: The MIT Press.
Johnson-Laird, P. N. (1998). Imagery, visualization, and thinking. In J. Hochberg (Ed.), *Perception and cognition at century's end* (pp. 441–467). San Diego: Academic Press.

Jolicœur, P., & Dell'Acqua, R. (1998). The demonstration of short-term consolidation. *Cognitive Psychology, 36*(2), 138–202.

Joliot, M., Ribary, U., & Llinás, R. (1994). Human oscillatory brain activity near 40 Hz coexists with cognitive temporal binding. *Proceedings of the National Academy of Sciences of the United States of America, 91*(24), 11748–11751.

Kanizsa, G. (1955). Margini quasi-percettivi in campi con stimolazione omogenea. *Rivista di Psicologia, 49*(1), 7–30. English translation: Quasi-perceptual margins in homogenously stimulated fields. In S. Petry, & G. E. Meyer (Eds.), (1987). *The perception of illusory contours* (pp. 40–49). New York: Springer.

Karmarkar, U. R., & Buonomano, D. V. (2007). Timing in the absence of clocks: Encoding time in neural network states. *Neuron, 53*(3), 427–438.

Kaya, E. M., & Elhilali, M. (2017). Modelling auditory attention. *Philosophical Transactions of the Royal Society B: Biological Sciences, 372*(1714), 20160101.

Keller, A. (2011). Attention and olfactory consciousness. *Frontiers in Psychology, 2*, Article 380 (13 pages).

Kerner, B. S. (2009). *Introduction to modern traffic flow theory and control: The long road to three-phase traffic theory*. Berlin: Springer.

Kerner, B. S. (2017). *Breakdown in traffic networks: Fundamentals of transportation science*. Berlin: Springer.

Kosslyn, S. (1994). *Image and brain: The resolution of the imagery debate*. Cambridge, MA: The MIT Press.

Kosslyn, S. M. (1980). *Image and mind*. Cambridge, MA: Harvard University Press. [Revised edition is published in 1986].

Kosslyn, S. M. (2005). Mental images and the brain. *Cognitive Neuropsychology, 22*(3–4), 333–347.

Kosslyn, S. M. (2007). Remembering images. In M. A. Gluck, J. R. Anderson, & S. M. Kosslyn (Eds.), *Memory and mind: A Festschrift for Gordon H. Bower* (Chap. 7, pp. 93–109). New York: Lawrence Erlbaum Associates, Taylor & Francis Group.

Kosslyn, S. M., Thompson, W. L., & Alpert, N. M. (1997). Neural systems shared by visual imagery and visual perception: A positron emission tomography study. *NeuroImage, 6*(4), 320–334.

Kosslyn, S. M., Thompson, W. L., & Ganis, G. (2006). *The case for mental imagery*. New York, NY: Oxford University Press.

Lagroix, H. E. P., Spalek, T. M., Wyble, B., Jannati, A., & Di Lollo, V. (2012). The root cause of the attentional blink: First-target processing or disruption of input control? *Attention, Perception, & Psychophysics, 74*(8), 1606–1622.

Lamme, V. A. F. (2003). Why visual attention and awareness are different. *Trends in Cognitive Sciences, 7*(1), 12–18.

Lammers, N. A., de Haan, E. H., Pinto, Y. (2017). No evidence of narrowly defined cognitive penetrability in unambiguous vision. *Frontiers in Psychology, 8*, Article 852 (6 pages).

Langerock, N., Vergauwe, E., Dirix, N., & Barrouillet, P. (2018). Is memory better for objects than for separate single features? the temporal hypothesis. *Journal of Experimental Psychology: Learning, Memory, and Cognition, 44*(6), 898–917.

Le Poidevin, R. (2007). *The images of time: An essay on temporal representation*. New York: Oxford University Press Inc.

Lee, G. (2014). Extensionalism, atomism, and continuity. In L. N. Oaklander (Ed.), *Debates in the metaphysics of time* (pp. 149–173). Bloomsbury Academic.

Lee, G. (2014). Temporal experience and the temporal structure of experience. *Philosophers' Imprint, 14*(3), 1–21.

Lee, G. (2014). Experiences and their parts. In D. Bennett & C. Hill (Eds.), *Sensory integration and the unity of consciousness* (pp. 287–321). Cambridge, MA: The MIT Press.

Lee, T. S., & Mumford, D. (2003). Hierarchical Bayesian inference in the visual cortex. *Journal of the Optical Society of America A, 20*(7), 1434–1448.

Lemaire, B., & Portrat, S. (2018). A computational model of working memory integrating time-based decay and interference. *Frontiers in Psychology, 9*, Article 416 (16 pages).

Lewis-Peacock, J. A., Kessler, Y., & Oberauer, K. (2018). The removal of information from working memory. *Annals of the New York Academy of Sciences, 1424*(1), 33–44.
Li, J., Qian, J., & Liang, F. (2018). Evidence for the beneficial effect of perceptual grouping on visual working memory: an empirical study on illusory contour and a meta-analytic study. *Scientific Reports*, *8*(1), 13864 (20 pages).
Liu, D., & Todorov, E. (2007). Evidence for the flexible sensorimotor strategies predicted by optimal feedback control. *Journal of Neuroscience, 27*(35), 9354–9368.
Lloyd, D. (2004). *Radiant cool: A novel theory of consciousness*. Cambridge, MA: The MIT Press.
Lloyd, D. (2012). Neural correlates of temporality: Default mode variability and temporal awareness. *Consciousness and Cognition, 21*(2), 695–703.
Lubashevsky, I. (2017). *Physics of the human mind*. AG, Cham: Springer International Publishing.
Lubashevsky, I., & Morimura, K. (2019). Physics of mind and car-following problem. In B. S. Kerner (Ed.), *Complex dynamics of traffic management, Encyclopedia of complexity and systems science series* (pp. 559–592). New York, NY: Springer Science+Business Media, LLC.
Maljkovic, V., & Nakayama, K. (1994). Priming of pop-out: I. role of features. *Memory & Cognition*, *22*(6), 657–672.
Maljkovic, V., & Nakayama, K. (2000). Priming of popout: III. a short-term implicit memory system beneficial for rapid target selection. *Visual Cognition*, *7*(5), 571–595.
Maljkovic, V., & Martini, P. (2005). Short-term memory for scenes with affective content. *Journal of Vision, 5,* 215–229.
Mance, I., Becker, M. W., & Liu, T. (2012). Parallel consolidation of simple features into visual short-term memory. *Journal of Experimental Psychology: Human Perception and Performance, 38*(2), 429–438.
Marek, S., Siegel, J. S., Gordon, E. M., Raut, R. V., Gratton, C., Newbold, D. J., Ortega, M., Laumann, T. O., Adeyemo, B., Miller, D. B., Zheng, A., Lopez, K. C., Berg, J. J., Coalson, R. S., Nguyen, A. L., Dierker, D., Van, A. N., Hoyt, C. R., McDermott, K. B., Norris, S. A., Shimony, J. S., Snyder, A. Z., Nelson, S. M., Barch, D. M., Schlaggar, B. L., Raichle, M. E., Petersen, S. E., Greene, D. J., & Dosenbach, N. U. F. (2018). Spatial and temporal organization of the individual human cerebellum. *Neuron, 100*(4), 977–993.e1–e7.
Marr, D. (1982). *Vision: A computational investigation into the human representation and processing of visual information*, W. H. Freeman and Company, San Francisco. In 2010 the MIT press re-published the book with a foreword from Shimon Ullmann and an afterword from Tomaso Poggio under ISBN 9780262514620.
Martens, S., & Wyble, B. (2010). The attentional blink: Past, present, and future of a blind spot in perceptual awareness. *Neuroscience and Biobehavioral Reviews, 34*(6), 947–957.
Mauk, M. D., & Buonomano, D. V. (2004). The neural basis of temporal processing. *Annual Review of Neuroscience, 27*(1), 307–340.
Mauk, M. D., & Donegan, N. H. (1997). A model of Pavlovian eyelid conditioning based on the synaptic organization of the cerebellum. *Learning & Memory, 4*(1), 130–158.
McKone, E. (1995). Short-term implicit memory for words and nonwords. *Journal of Experimental Psychology: Learning, Memory, and Cognition, 21*(5), 1108–1126.
McKone, E. (1998). The decay of short-term implicit memory: Unpacking lag. *Memory & Cognition, 26*(6), 1173–1186.
McKone, E., & Dennis, C. (2000). Short-term implicit memory: Visual, auditory, and cross-modality priming. *Psychonomic Bulletin & Review, 7*(2), 341–346.
Medina, J. F., & Mauk, M. D. (2000). Computer simulation of cerebellar information processing. *Nature Neuroscience, 3,* 1205–1211.
Merchant, H., Harrington, D. L., & Meck, W. H. (2013). Neural basis of the perception and estimation of time. *Annual Review of Neuroscience, 36*(1), 313–336.
Miall, R., & Wolpert, D. M. (1996). Forward models for physiological motor control. *Neural Networks*, *9*(8), 1265–1279. Four Major Hypotheses in Neuroscience.
Michon, J. A. (1985). The compleat time experiencer. In J. A. Michon & J. L. Jackson (Eds.), *Time, Mind, and Behavior* (pp. 20–52). Berlin: Springer.

Miyauchi, R., & Nakajima, T. (2005). Bilateral assimilation of two neighboring empty time intervals. *Music Perception: An Interdisciplinary Journal, 22*(3), 411–424.
Mogensen, J., & Overgaard, M. (2017). Reorganization of the connectivity between elementary functions - a model relating conscious states to neural connections. *Frontiers in Psychology, 8*, Article 625 (21 pages).
Mogensen, J., & Overgaard, M. (2018). Reorganization of the connectivity between elementary functions as a common mechanism of phenomenal consciousness and working memory: From functions to strategies. *Philosophical Transactions of the Royal Society of London B: Biological Sciences, 373*(1755), 20170346 (11 pages).
Mohan, A., & Vanneste, S. (2017). Adaptive and maladaptive neural compensatory consequences of sensory deprivation-from a phantom percept perspective. *Progress in Neurobiology, 153,* 1–17.
Moore, B. C. J., & Gockel, H. E. (2012). Properties of auditory stream formation. *Philosophical Transactions of the Royal Society B: Biological Sciences, 367*(1591), 919–931.
Morey, C. C., & Cowan, N. (2018). Can we distinguish three maintenance processes in working memory? *Annals of the New York Academy of Sciences, 1424*(1), 45–51.
Mori, K., Manabe, H., Narikiyo, K., & Onisawa, N. (2013). Olfactory consciousness and gamma oscillation couplings across the olfactory bulb, olfactory cortex, and orbitofrontal cortex. *Frontiers in Psychology*, *4*, Article 743 (13 pages).
Moulton, S. T., & Kosslyn, S. M. (2009). Imagining predictions: Mental imagery as mental emulation. *Philosophical Transactions of the Royal Society B: Biological Sciences, 364*(1521), 1273–1280.
Mundle, C. W. K. (1966). Augustine's pervasive error concerning time. *Philosophy, 41*(156), 165–168.
Münsterberg, H. (1889). *Beiträge zur experimentellen Psychologie: Heft 2*. Mohr, Freiburg: Akademische Verlagsbuchhandlung von J. C. B.
Nagaike, A., Mitsudo, T., Nakajima, Y., Ogata, K., Yamasaki, T., Goto, Y., & Tobimatsu, S. (2016). 'Time-shrinking perception' in the visual system: A psychophysical and high-density ERP study. *Experimental Brain Research, 234*(11), 3279–3290.
Nakajima, Y., & ten Hoopen, G. (1988). The effect of preceding time intervals on duration perception. *Proceedings of the Autumn Meeting of Acoustical Society of Japan*, 381–382. https://ci.nii.ac.jp/naid/10003864756/en/.
Nakajima, Y., ten Hoopen, G., & Van Der Wilk, R. (1991). A new illusion of time perception. *Music Perception: An Interdisciplinary Journal, 8*(4), 431–448.
Nanay, B. (2010). Perception and imagination: Amodal perception as mental imagery. *Philosophical Studies, 150*(2), 239–254.
Nanay, B. (2018). Multimodal mental imagery. *Cortex*, *105*, 125–134. Special Issue: The Eye's Mind - visual imagination, neuroscience and the humanities.
Nanay, B. (2018a). The importance of amodal completion in everyday perception. *i-Perception*, *9*(4), Article 2041669518788887 (16 pages).
Newen, A., & Vetter, P. (2017). Why cognitive penetration of our perceptual experience is still the most plausible account. *Consciousness and Cognition*, *47*, 26–37. Cognitive Penetration and Predictive Coding.
Newen, A., Marchi, F., & Brössel, P. (2017a). Introduction—cognitive penetration and predictive coding. Pushing the debate forward with the recent achievements of cognitive science. *Consciousness and Cognition*, *47*, 1–5. Cognitive Penetration and Predictive Coding.
Newen, A., Marchi, F., & Brössel, P. (Eds.). (2017b). *Cognitive Penetration and Predictive Coding*, *47*. Consciousness and Cognition, Special Issue, 112 pages.
Nieuwenstein, M., & Wyble, B. (2014). Beyond a mask and against the bottleneck: Retroactive dual-task interference during working memory consolidation of a masked visual target. *Journal of Experimental Psychology: General, 143*(3), 1409–1427.
Oberauer, K. (2002). Access to information in working memory: Exploring the focus of attention. *Journal of Experimental Psychology: Learning, Memory, and Cognition, 28*(3), 411–421.

O'Callaghan, C., Kveraga, K., Shine, J. M., Adams, R. B., & Bar, M. (2017). Predictions penetrate perception: Converging insights from brain, behaviour and disorder. *Consciousness and Cognition, 47*, 63–74. Cognitive Penetration and Predictive Coding.

Olivers, C. N. L., & Meeter, M. (2008). A boost and bounce theory of temporal attention. *Psychological Review, 115*(4), 836–863.

Olivers, C. N. L., van der Stigchel, S., & Hulleman, J. (2007). Spreading the sparing: Against a limited-capacity account of the attentional blink. *Psychological Research, 71*(2), 126–139.

O'Reilly, J. X., Mesulam, M. M., & Nobre, A. C. (2008). The cerebellum predicts the timing of perceptual events. *The Journal of Neuroscience, 28*(9), 2252–2260.

Overgaard, M. (2018). Phenomenal consciousness and cognitive access. *Philosophical Transactions of the Royal Society of London B: Biological Sciences, 373*(1755), 20170353 (6 pages).

Overgaard, M., & Mogensen, J. (2014). Visual perception from the perspective of a representational, non-reductionistic, level-dependent account of perception and conscious awareness. *Philosophical Transactions of the Royal Society B: Biological Sciences, 369*(1641), Article 20130209 (12 pages).

Palmiero, M., Piccardi, L., Giancola, M., Nori, R., D'Amico, S., & Olivetti Belardinelli, M. (2019). The format of mental imagery: From a critical review to an integrated embodied representation approach. *Cognitive Processing, 20*(3), 277–289.

Paton, J. J., & Buonomano, D. V. (2018). The neural basis of timing: Distributed mechanisms for diverse functions. *Neuron, 98*(4), 687–705.

Pearson, J., & Kosslyn, S. M. (2015). The heterogeneity of mental representation: Ending the imagery debate. *Proceedings of the National Academy of Sciences, 112*(33), 10089–10092.

Pearson, J., Clifford, C. W. G., & Tong, F. (2008). The functional impact of mental imagery on conscious perception. *Current Biology, 18*(13), 982–986.

Pearson, J., Naselaris, T., Holmes, E. A., & Kosslyn, S. M. (2015). Mental imagery: Functional mechanisms and clinical applications. *Trends in Cognitive Sciences, 19*(10), 590–602.

Pearson, J., & Westbrook, F. (2015). Phantom perception: Voluntary and involuntary nonretinal vision. *Trends in Cognitive Sciences, 19*(5), 278–284.

Peters, M. (1989). The relationship between variability of intertap intervals and interval duration. *Psychological Research, 51*(1), 38–42.

Peterson, D. J., & Berryhill, M. E. (2013). The gestalt principle of similarity benefits visual working memory. *Psychonomic Bulletin & Review, 20*(6), 1282–1289.

Phillips, I. (2010). Perceiving temporal properties. *European Journal of Philosophy, 18*(2), 176–202.

Phillips, I. (2014). The temporal structure of experience. In V. Arstila & D. Lloyd (Eds.), *Subjective time: The philosophy, psychology, and neuroscience of temporality* (pp. 139–158). Cambridge: The MIT Press.

Phillips, I. (2014). Experience of and in time. *Philosophy Compass, 9*(2), 131–144.

Pilz, K. S., Zimmermann, C., Scholz, J., & Herzog, M. H. (2013). Long-lasting visual integration of form, motion, and color as revealed by visual masking. *Journal of Vision, 13*(10), 12 (11 pages).

Pincham, H. L., Bowman, H., & Szucs, D. (2016). The experiential blink: Mapping the cost of working memory encoding onto conscious perception in the attentional blink. *Cortex, 81*, 35–49.

Pinto, Y., Sligte, I. G., Shapiro, K. L., & Lamme, V. A. F. (2013). Fragile visual short-term memory is an object-based and location-specific store. *Psychonomic Bulletin & Review, 20*(4), 732–739.

Pitt, D. (2020). Mental representation. In E. N. Zalta (Ed.), *The Stanford encyclopedia of philosophy*, Spring 2020 ed. Metaphysics Research Lab, Stanford University.

Pöppel, E. (1997). A hierarchical model of temporal perception. *Trends in Cognitive Sciences, 1*(2), 56–61.

Pöppel, E. (2009). Pre-semantically defined temporal windows for cognitive processing. *Philosophical Transactions of the Royal Society B: Biological Sciences, 364*(1525), 1887–1896.

Pöppel, E., & Bao, Y. (2014). Temporal windows as a bridge from objective to subjective time. In V. Arstila & D. Lloyd (Eds.), *Subjective time the philosophy, psychology, and neuroscience of temporality* (pp. 241–261). Cambridge: The MIT Press.

Pressnitzer, D., & Hupé, J.-M. (2006). Temporal dynamics of auditory and visual bistability reveal common principles of perceptual organization. *Current Biology, 16*(13), 1351–1357.
Pylyshyn, Z. W. (1980). Computation and cognition: Issues in the foundations of cognitive science. *Behavioral and Brain Sciences, 3*(1), 111–132.
Raftopoulos, A. (2009). *Cognition and perception: How do psychology and neural science inform philosophy?*, A Bradford Book. London, England: The MIT Press.
Raftopoulos, A. (2011). Late vision: Processes and epistemic status. *Frontiers in Psychology, 2*, Article 382 (12 pages).
Raftopoulos, A. (2014). The cognitive impenetrability of the content of early vision is a necessary and sufficient condition for purely nonconceptual content. *Philosophical Psychology, 27*(5), 601–620.
Raftopoulos, A. (2015). The cognitive impenetrability of perception and theory-ladenness. *Journal for General Philosophy of Science, 46*(1), 87–103.
Raftopoulos, A. (2017). Pre-cueing, the epistemic role of early vision, and the cognitive impenetrability of early vision. *Frontiers in Psychology, 8*, Article 1156 (12 pages).
Raftopoulos, A., & Lupyan, G. (Eds.). (2018). *Pre-cueing Effects on Perception and Cognitive Penetrability*, Frontiers Media, Frontiers in Psychology, Lausanne.
Raftopoulos, A. (2019). *Cognitive penetrability and the epistemic role of perception*. Cham, Switzerland: Palgrave Macmillan.
Rammsayer, T. H., Borter, N., & Troche, S. J. (2015). Visual-auditory differences in duration discrimination of intervals in the subsecond and second range. *Frontiers in Psychology, 6*, Article 1626 (7 pages).
Rammsayer, T. H., & Grondin, S. (2000). Psychophysics of human timing. In R. Miller (Ed.), *Time and the brain* (pp. 181–194). Amsterdam: Harwood Academic Publishers.
Rammsayer, T. H., & Troche, S. J. (2014). Elucidating the internal structure of psychophysical timing performance in the sub-second and second range by utilizing confirmatory factor analysis. In H. Merchant & V. de Lafuente (Eds.), *Neurobiology of interval timing* (pp. 33–47). New York, NY: Springer Science+Business Media.
Rao, R. P. N., & Ballard, D. H. (1999). Predictive coding in the visual cortex: A functional interpretation of some extra-classical receptive-field effects. *Nature Neuroscience, 2*(1), 79–87.
Rashbrook, O. (2013). An appearance of succession requires a succession of appearances. *Philosophy and Phenomenological Research, 87*(3), 584–610.
Rashbrook, O. (2013). The continuity of consciousness. *European Journal of Philosophy, 21*(4), 611–640.
Rashbrook, O. (2013). Diachronic and synchronic unity. *Philosophical Studies, 164*(2), 465–484.
Rashbrook-Cooper, O. (2017). Atomism, extensionalism and temporal presence. In I. Phillips (Ed.), *The Routledge handbook of philosophy of temporal experience* (pp. 133–145). Abingdon, Oxon: Routledge, Taylor & Francis Group.
Raymond, J. E., Shapiro, K. L., & Arnell, K. M. (1992). Temporary suppression of visual processing in an rsvp task: An attentional blink? *Journal of Experimental Psychology: Human Perception and Performance, 18*(3), 849–860.
Repp, B. H. (2005). Sensorimotor synchronization: A review of the tapping literature. *Psychonomic Bulletin & Review, 12*(6), 969–992.
Repp, B. H. (2006). Rate limits of sensorimotor synchronization. *Advances in Cognitive Psychology,* 2(2–3), 163–181.
Repp, B. H., & Su, Y.-H. (2013). Sensorimotor synchronization: A review of recent research (2006–2012). *Psychonomic Bulletin & Review, 20*(3), 403–452.
Rhodes, S., & Cowan, N. (2018). Attention in working memory: Attention is needed but it yearns to be free. *Annals of the New York Academy of Sciences, 1424*(1), 52–63.
Ricker, T. J., & Cowan, N. (2014). Differences between presentation methods in working memory procedures: A matter of working memory consolidation. *Journal of experimental psychology. Learning, Memory, and Cognition,40*(2), 417–428.

Ricker, T. J. (2015). The role of short-term consolidation in memory persistence. *AIMS Neuroscience, 2*(4), 259–279.
Ricker, T. J., & Hardman, K. O. (2017). The nature of short-term consolidation in visual working memory. *Journal of Experimental Psychology: General, 146*(11), 1551–1573.
Ricker, T. J., Nieuwenstein, M. R., Bayliss, D. M., & Barrouillet, P. (2018). Working memory consolidation: Insights from studies on attention and working memory. *Annals of the New York Academy of Sciences, 1424*(1), 8–18.
Rideaux, R., Apthorp, D., & Edwards, M. (2015). Evidence for parallel consolidation of motion direction and orientation into visual short-term memory. *Journal of Vision, 15*(2), Article 17 (12 pages).
Roediger, H. L., Zaromb, F. M., & Lin, W. (2017). A typology of memory terms. In J. H. e. i. c. Byrne, & R. v. e. Menzel (Eds.), *Learning and memory: A comprehensive reference* (2nd ed., Vol. 1, pp. 7–19). Learning theory and behavior. Amsterda: Elsevier Ltd.
Sasaki, T., Nakajima, Y., & ten Hoopen, G. (1998). Categorical rhythm perception as a result of unilateral assimilation in time-shrinking. *Music Perception: An Interdisciplinary Journal, 16*(2), 201–222.
Scarry, E. (2001). *Dreaming by the book*. Princeton, NJ: Princeton University Press.
Scharnowski, F., Rüter, J., Jolij, J., Hermens, F., Kammer, T., & Herzog, M. H. (2009). Long-lasting modulation of feature integration by transcranial magnetic stimulation. *Journal of Vision, 9*(6), 1 (10 pages).
Schmahmann, J. D. (2019). The cerebellum and cognition. *Neuroscience Letters, 688*, 62–75. The Cerebellum in Health and Disease.
Schneider, K. A., & Bavelier, D. (2003). Components of visual prior entry. *Cognitive Psychology, 47*(4), 333–366.
Schröger, E., Kotz, S., & SanMiguel, I. (Eds.). (2015). *Predictive and Attentive Processing in Perception and Action, 1626*. Brain Research, Special Issue, 280 pages.
Sergent, C. (2018). The offline stream of conscious representations. *Philosophical Transactions of the Royal Society of London B: Biological Sciences, 373*(1755), 20170349 (13 pages).
Sergent, C., Wyart, V., Babo-Rebelo, M., Cohen, L., Naccache, L., & Tallon-Baudry, C. (2013). Cueing attention after the stimulus is gone can retrospectively trigger conscious perception. *Current Biology, 23*(2), 150–155.
Seth, A. K. (2014). A predictive processing theory of sensorimotor contingencies: Explaining the puzzle of perceptual presence and its absence in synesthesia. *Cognitive Neuroscience, 5*(2), 97–118.
Shapiro, K. L., Raymond, J. E., & Arnell, K. M. (1997). The attentional blink. *Trends in Cognitive Sciences, 1*(8), 291–296.
Shen, D., & Alain, C. (2011). Temporal attention facilitates short-term consolidation during a rapid serial auditory presentation task. *Experimental Brain Research, 215*(3–4), 285–292.
Shepard, R. N., & Cooper, L. A. (1982). *Mental images and their transformations*, A Bradford book. Cambridge, MA: The MIT Press.
Shepard, R. N. (1978). The mental image. *American Psychologist, 33*(2), 125–137.
Shepard, R. N., & Metzler, J. (1971). Mental rotation of three-dimensional objects. *Science, 171*(3972), 701–703.
Shepard, S., & Metzler, D. (1988). Mental rotation: Effects of dimensionality of objects and type of task. *Journal of Experimental Psychology: Human Perception and Performance, 14*(1), 3–11.
Sima, J. F., Schultheis, H., & Barkowsky, T. (2013). Differences between spatial and visual mental representations. *Frontiers in Psychology, 4*, Article 240.
Singer, W. (1993). Synchronization of cortical activity and its putative role in information processing and learning. *Annual Review of Physiology, 55*(1), 349–374.
Singer, W., & Gray, C. M. (1995). Visual feature integration and the temporal correlation hypothesis. *Annual Review of Neuroscience, 18*(1), 555–586.
Singh, I., & Howard, M. W. (2017). Recency order judgments in short term memory: Replication and extension of hacker (1980). *bioRxiv*.

Singh, I., Tiganj, Z., & Howard, M. W. (2018). Is working memory stored along a logarithmic timeline? converging evidence from neuroscience, behavior and models. *Neurobiology of Learning and Memory, 153*, Part A, 104–110. MCCS 2018: Time and Memory.

Singh, M. (2004). Modal and amodal completion generate different shapes. *Psychological Science, 15*(7), 454–459.

Sligte, I., Vandenbroucke, A., Scholte, H. S., & Lamme, V. (2010). Detailed sensory memory, sloppy working memory. *Frontiers in Psychology, 1*, Article 175 (10 pages).

Soteriou, M. (2010). Perceiving events. *Philosophical Explorations, 13*(3), 223–241.

Soteriou, M. (2013). *The mind's construction: The ontology of mind and mental action*. Oxford, UK: Oxford University Press.

Soto, D., Mäntylä, T., & Silvanto, J. (2011). Working memory without consciousness. *Current Biology, 21*(22), R912–R913.

Soto, D., & Silvanto, J. (2014). Reappraising the relationship between working memory and conscious awareness. *Trends in Cognitive Sciences, 18*(10), 520–525.

Spencer, R. M. C., Karmarkar, U., & Ivry, R. B. (2009). Evaluating dedicated and intrinsic models of temporal encoding by varying context. *Philosophical Transactions of the Royal Society of London B: Biological Sciences, 364*(1525), 1853–1863.

Sperling, G. (1960). The information available in brief visual presentations. Psychological Monographs: General and Applied, 74(11). *Whole No., 498*, 1–29.

Sprigge, T. L. S. (1984). *The vindication of absolute idealism*. Edinburgh: Edinburgh University Press.

Stelmach, L. B., & Herdman, C. M. (1991). Directed attention and perception of temporal order. *Journal of Experimental Psychology: Human Perception and Performance, 17*(2), 539–550.

Sternberg, S., & Knoll, R. L. (1973). The perception of temporal order: Fundamental issues and a general model. In S. Kornblum (Ed.), *Attention and performance: IV* (pp. 629–685). New York: Academic Press.

Stevanovski, B., & Jolicœur, P. (2011). Consolidation of multifeature items in visual working memory: Central capacity requirements for visual consolidation. *Attention, Perception, & Psychophysics, 73*(4), 1108–1119.

Swanson, L. R. (2016). The predictive processing paradigm has roots in kant. *Frontiers in Systems Neuroscience, 10*, Article 79 (13 pages).

Tanida, K., & Pöppel, E. (2006). A hierarchical model of operational anticipation windows in driving an automobile. *Cognitive Processing, 7*(4), 275–287.

ten Hoopen, G., Miyauchi, R., & Nakajima, Y. (2008). Time-based illusions and the auditory mode. In S. Grondin (Ed.), *Psychology of time* (pp. 139–187). Bingley: Emerald Group Publishing Ltd.

ten Hoopen, G., Sasaki, T., Nakajima, Y., Remijn, G., Massier, B., Rhebergen, K. S., & Holleman, W. (2006). Time-shrinking and categorical temporal ratio perception: Evidence for a 1:1 temporal category. *Music Perception: An Interdisciplinary Journal, 24*(1), 1–22.

Therrien, A. S., & Bastian, A. J. (2019). The cerebellum as a movement sensor. *Neuroscience Letters, 688*, 37–40. The Cerebellum in Health and Disease.

Thibault, L., van den Berg, R., Cavanagh, P., & Sergent, C. (2016). Retrospective attention gates discrete conscious access to past sensory stimuli. *PLoS ONE, 11*(2), e0148504 (13 pages).

Thomas, N. J. T. (2014). Mental imagery. In E. N. Zalta (Ed.), *The Stanford encyclopedia of philosophy*, Summer 2019 ed. Metaphysics Research Lab, Stanford University.

Todorov, E. (2004). Optimality principles in sensorimotor control. *Nature Neuroscience, 7*(9), 907–915.

Treiber, M., & Kesting, A. (2013). *Traffic flow dynamics: Data models and simulation*. Berlin: Springer.

Tse, P. U. (1999). Volume completion. *Cognitive Psychology, 39*(1), 37–68.

Tversky, B. (1993). Cognitive maps, cognitive collages, and spatial mental models. In A. U. Frank, & I. Campari (Eds.), *Spatial information theory: A theoretical basis for GIS. COSIT 1993* (Vol. 716, pp. 14–24). Lecture notes in computer science. Berlin: Springer.

Tye, M. (2003). *Consciousness and persons: Unity and identity*. Cambridge, MA: The MIT Press.

Ulbrich, P., Churan, J., Fink, M., & Wittmann, M. (2009). Perception of temporal order: The effects of age, sex, and cognitive factors. *Aging, Neuropsychology, and Cognition, 16*(2), 183–202.

van Rijn, H., Gu, B.-M., & Meck, W. H. (2014). Dedicated clock/timing-circuit theories of time perception and timed performance. In H. Merchant & V. de Lafuente (Eds.), *Neurobiology of interval timing, Advances in experimental medicine and biology* (Vol. 829, pp. 75–99). NY, New York: Springer Science+Business Media.

van Wassenhove, V. (2009). Minding time in an amodal representational space. *Philosophical Transactions of the Royal Society of London B: Biological Sciences, 364*(1525), 1815–1830.

Vandenbroucke, A. R. E., Sligte, I. G., & Lamme, V. A. F. (2011). Manipulations of attention dissociate fragile visual short-term memory from visual working memory. *Neuropsychologia, 49*(6), 1559–1568. Attention and Short-Term Memory.

Vatakis, A., & Spence, C. (2007). Crossmodal binding: Evaluating the "unity assumption" using audiovisual speech stimuli. *Perception & Psychophysics, 69*(5), 744–756.

Velichkovsky, B. B. (2017). Consciousness and working memory: Current trends and research perspectives. *Consciousness and Cognition, 55,* 35–45.

Vogel, E. K., Woodman, G. F., & Luck, S. J. (2006). The time course of consolidation in visual working memory. *Journal of Experimental Psychology: Human Perception and Performance, 32*(6), 1436–1451.

Warren, R. M., & Obusek, C. J. (1972). Identification of temporal order within auditory sequences. *Perception & Psychophysics, 12*(1), 86–90.

Wearden, J. (2016). *The psychology of time perception.* London: Palgrave Macmillan.

Weiß, K., & Scharlau, I. (2011). Simultaneity and temporal order perception: Different sides of the same coin? evidence from a visual prior-entry study. *Quarterly Journal of Experimental Psychology, 64*(2), 394–416.

White, P. A. (2017). The three-second "subjective present": A critical review and a new proposal. *Psychological Bulletin, 143*(7), 735–756.

White, P. A. (2020). The perceived present: What is it, and what is it there for? *Psychonomic Bulletin & Review, 27*(4), 583–601.

Whitehead, A. N. (1929/1978). In D. R. Griffin, & D. W. Sherburne (Eds.), *Process and reality: An essay in cosmology*, corrected edn. New York: The Free Press, A Division of Macmillan Publishing Co., Inc.

Wiese, W. (2017). Predictive processing and the phenomenology of time consciousness: A hierarchical extension of Rick Grush's trajectory estimation model. In T. K. Metzinger, & W. Wiese (Eds.), *Philosophy and predictive processing*. MIND Group, Frankfurt am Main, chapter 26, p. (21 pages).

Winlove, C. I. P., Milton, F., Ranson, J., Fulford, J., MacKisack, M., Macpherson, F., & Zeman, A. (2018). The neural correlates of visual imagery: A co-ordinate-based meta-analysis. *Cortex, 105,* 4–25. Special Issue: The Eye's Mind - visual imagination, neuroscience and the humanities.

Wittmann, M. (2011). Moments in time,.*Frontiers in Integrative Neuroscience, 5,* Article 66 (9 pages).

Wittmann, M. (2015). Modulations of the experience of self and time. *Consciousness and Cognition, 38,* 172–181.

Wittmann, M. (2016). The duration of presence. In B. Mölder, V. Arstila, & P. Øhrstrøm (Eds.), *Philosophy and psychology of time* (pp. 101–113). Cham: Springer International Publishing.

Wittmann, M., von Steinbüchel, N., & Szelag, E. (2001). Hemispheric specialisation for self-paced motor sequences. *Cognitive Brain Research, 10*(3), 341–344.

Wolpert, D. M., Miall, R. C., & Kawato, M. (1998). Internal models in the cerebellum. *Trends in Cognitive Sciences, 2*(9), 338–347.

Wyble, B., Bowman, H., & Nieuwenstein, M. (2009). The attentional blink provides episodic distinctiveness: Sparing at a cost. *Journal of Experimental Psychology: Human Perception and Performance, 35*(3), 787–807.

Wyble, B., Potter, M. C., Bowman, H., & Nieuwenstein, M. (2011). Attentional episodes in visual perception. *Journal of Experimental Psychology: General, 140*(3), 488–505.

Xia, Y., Morimoto, Y., & Noguchi, Y. (2016). Retrospective triggering of conscious perception by an interstimulus interaction. *Journal of Vision, 16*(7), 3.
Yang, T., Strasburger, H., Pöppel, E., & Bao, Y. (2018). Attentional modulation of speed-change perception in the perifoveal and near-peripheral visual field. *PLOS ONE, 13*(8), e0203024 (17 pages).
Yarrow, K., Martin, S. E., Di Costa, S., Solomon, J. A., & Arnold, D. H. (2016). A roving dual-presentation simultaneity-judgment task to estimate the point of subjective simultaneity. *Frontiers in Psychology*, *7*, Article 416 (19 pages).

Chapter 3
Human Temporality: Qualitative Description

The present chapter is devoted to qualitative description of the human temporality and the complex present as its constituent component. Speaking about the complex present, special attention will be focused on action strategies representing intentional, goal-oriented actions and reflecting their essence.

The concept of space-time clouds will be elaborated in detail in order to describe mental representation of perceived physical objects as well as intentional actions in mathematical terms.[1] This concept has to allow for intrinsic uncertainty in human perception of physical objects and quantitative evaluation of properties attributed to the corresponding mental images.

At the final step, the human temporality will be introduced as a specific mathematical object describing human behavior and comprising past-oriented mental time travels, the complex present, and future-oriented mental time travels. It will be elucidated that the complex present can be regarded as the minimal fragment of human temporal dimension within which intentional, goal-oriented actions can admit a closed mathematical description. Besides, focusing on the complex present we will formulate the basic concepts underlying the tcorresponding mathematical formalism to be elaborated in the next Part.

3.1 Complex Present: Outside Experiential Now

As demonstrated in the previous chapter, the experiential now is a complex individual fragment of human temporal cognition with its own properties. Being the central part, the experiential now is closely intertwined with other two crucial parts: the connected

[1] The concept of space-time clouds has been introduced in the previous book (Lubashevsky 2017) and also partly discussed in Sect. 2.6.5 in the context of mental image description.

I. Lubashevsky and N. Plavinska, *Physics of the Human Temporality*,
Understanding Complex Systems, https://doi.org/10.1007/978-3-030-82612-3_3

past and the connected future. Leaping ahead, we posit that the connected past, the experiential now, and the connected future together constitute the *complex present* which may be treated as the minimal fragment of intentional, goal-oriented actions admitting a closed description.[2] So we meet again the tripartite temporal structure at a new level. The number of such levels is also three.

Each of these levels is characterized by its own mechanism determining the event connectedness:

- The synchronic unit corresponds to events cognitively perceived as simultaneous and, thereby, simultaneously experienced (cognition).
- The experiential now comprises events experienced as simultaneous but admitting some temporal order perceived explicitly (sensory experience).
- The complex present comprises events (and also actions) perceived as intentionally simultaneous and experientially connected (intentionality combined with sensory experience).

The noted connectedness criteria are ordered according to their strength such that each following level contains the preceding one as a component.

The present section is mainly devoted to elucidating two issues:

– What is the mechanism via which the connected past, the experiential now, and the connected future are tightly related to one another such that a mathematical description of the human temporality requires the consideration of the complex present as a whole entity with internal integrity.
– What temporal scales characterize the connected past and the connected future. Put differently, what are the minimal time scales within which human intentional, goal-oriented behavior admits a closed description.

For the first step, we want to explain why the experiential now on its own, i.e., treated as a separated entity admits no mathematical description able to predict how the states of experiential now change in time flow.

The experiential now comprises the mental images of currently perceived objects. The properties attributed to these images in the mind are the ultimate result of neural signal processing which, for the most part, is unconscious. Thereby, the properties of mental images are merely given to the mind rather than stem from mental analytic manipulations. In this sense the image properties are mutually independent and just characterize different aspects of perceived physical objects. For example, in observing the motion of a physical object, we perceive its position in space, the speed of motion (i.e. how fast or slow the object is moving) and, maybe, the acceleration and jerk. However, it is no more than the sensory perception of the moving object, we do not cognitively relate its motion characteristics to one another via the corresponding time derivatives.

Therefore, to relate various properties of mental images to one another we have to go outside the experiential now. Put differently, we need (*i*) to *cognitively* reconstruct

[2] In the previous chapter we only introduced the terms the *connected past* and the *connected future*. In the present chapter we elaborate these notions in detail.

the states of perceived environment in the nearest past which are saved in working memory and (*ii*) to imagine its plausible states expected in the nearest future. These states are directly related to the experiential now but are not currently experienced as events having occurred right now. In other words, for the cognitive reconstruction of physical relationships between the perceived properties to become feasible, the contribution of the past retained in memory and, maybe, the imaginary future is necessary. Turning to the example of a moving object, to relate the perceived speed of the object to its spatial positions, the subject needs to compare the object positions on time scales exceeding substantially the duration of experiential now. Only in this case, uncertainty in perceiving space-time scales can be ignored in evaluating the speed as the time derivative of the spatial position.

3.1.1 Goal-Oriented Behavior and Action Strategies

Intentional, goal-oriented behaviour is a special kind of human activity where action strategies[3]

- determine the content of the connected past and the connected future,
- unite the connected past, the experiential now, and the connected future into the complex present as a certain entity with intrinsic integrity.

Here we want to note the characteristic properties of action strategies to elucidate their role in the complex present. Each action strategy related to pursuing a certain goal is a sequence of actions that individually belong to one of the following types:

1. actions implemented in the connected past and retained in working memory or imagined to have been implemented,
2. actions being currently under implementation or, speaking more strictly, experienced as such,
3. actions imagined to be implemented in the connected future.

The actions of one strategy are intentionally or/and logically linked in the mind. These actions are aimed at maintaining the current state of environment or changing it at a certain instant of time.

In the present book, we confine our consideration to short-term phenomena in human intentional, goal-oriented behavior. For this reason, in describing the complex present we suppose that:

- The complex present involves only one goal. All the action strategies under consideration are aimed at pursuing this common goal.
- Action strategies do not merge or split in the complex present, they admit interpretation as different, separated mental entities.
- Each strategy is homogeneous in structure.

---

[3] The notion of action strategy has been introduced in Sect. 2.6.4 turning to predictive coding paradigm.

- The individual structure of action strategies including their past and future temporal parts does not change within the scope of complex present.
- Reevaluating the priority of one or another action strategy is a continuous process in the experiential now.
- The choice of an action strategy for its further use is implemented within the experiential now. The choice of a new strategy can be conceived of as a transition between different strategies in the experiential now.
- Each imaginary action belonging to one of the strategies bears some degree of sensory experience inherited from the experiential now for the given type of actions. This feature endows the complex present as a whole entity with the aspect of sensory perception. In other words, all the actions in the complex present—currently executed, retained in working memory, or imaginary ones—are sensorily experienced by the subject. Naturally, the sensory perception of retained or imaginary actions is weaker in strength in comparison with actions belonging to the experiential now.[4]

Leaping ahead, we may say that action strategies within the complex present are very simple examples of the human temporality. In the general case, (*i*) the human temporality involves a variety of different goals, (*ii*) the corresponding action strategies may merge or split, forming a complex network of intentional links, and (*iii*) this mental network of past and future actions admits reconstruction as a whole, which is reflected in rearranging actions, the corresponding events, as well as goals among different intentional links or/and their reevaluation.

Figure 3.1 illustrates the complex present comprising its three constituent components: the connected past, the experiential now, and the connected future. The horizontal, narrow stripes AS_1–AS_3 illustrate action strategies going through the connected past, the experiential now, and the connected future and uniting them into a single whole. These strategies individually possess a fixed action structure that does not change withing the complex present, which is noted with symbols $\bigcirc$, $\square$, and $\triangle$, respectively, shown at both the sides of each stripe.

The choice of a new strategy is illustrated as the transition $AS_2 \Rightarrow AS_2$ representing the change of the currently used strategy AS_2 for the strategy AS_1 within the experiential now. This transition as a continuous process is of finite duration in objective time but has to be located within the experiential now and, thus, must be experienced as instantaneous. In Fig. 3.1 is depicted as the overlap between two fragments of the stripes AS_2 and AS_1 corresponding to their state of being selected for implementation.

The properties attributed to mental images within the experiential now, here, $\{x_d, v_d, a_d, j_d\}$ are given to the mind as mutually independent quantities. The relationship between these quantities can be reestablished based on the information contained in the connected past or the connected future via memory or imagination.

[4] As discussed in Sect. 2.6.5, there are two formats each mental image possesses—descriptive and picture-like (depictive) ones. Exactly the depictive format endows us with the ability to sensorily perceive imaginary actions, objects, or events, but this sensory perception is weak.

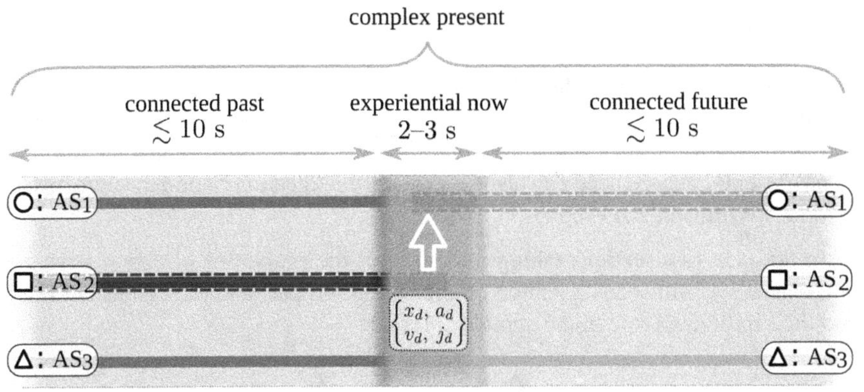

Fig. 3.1 Schematic representation of the complex present and its three constituent components—the connected past, the experiential now, the connected future. The horizontal stripes AS_1–AS_3 going through the connected past, the experiential now, and the connected future illustrate individual strategies of actions linked by intentional relations within one strategy. The strategies under consideration do not merge or split in the mind as well as do not change their action structure; the latter is illustrated by individual symbols ○, □, and △, at the both sides of each stripe. The upward arrow inside the experiential now illustrates an intentional change of currently used strategy (AS_2) for a new one (AS_1)

It should be noted that the three-level structure of event connectedness stated at the beginning of Sect. 3.1 and partly illustrated in Fig. 3.1 is reminiscent of James's description of presence in terms of different phenomena (James 1890, Chap. XIV). It involves the uniform transition from momentary appearances of thoughts, perceptual experiences, up to the short-term memory-related fading out of mental content. A similar concept of *mental presence* is put forward by Wittmann (2011, 2016). Below we cite it in more details to emphasize the governing role of working memory accepted also in our account of complex present; its concerns also the general features characterizing the temporal structure of mental presence and the complex present.

> Whereas the *experienced moment* forms an elementary unit, a temporally unified percept, *mental presence* involves the experience of a perceiving and feeling agent ("my self") within a window of extended presence, a phenomenon that is based on working memory function...An *experienced moment* happens now, for a short but extended moment. *Mental presence* encloses a sequence of such moments for the representation of a unified experience of presence...
>
> *Mental presence* is bound to the ability of maintaining mental representations in an active state for a certain period of time...*[M]ental presence* is related to the fact that once attended objects slowly phase out of experience over time; that is, the phenomenally experienced sliding window of *mental presence* co-occurs with the constant loss of memory contents...*Mental presence* is a temporal platform of multiple seconds within which an individual is aware of herself and the environment, where sensory-motor perception, cog-

> nition, and emotion are interconnected features of representation leading to phenomenal experience...
>
> Within the realm of the *mental presence* an individual can be considered fully operational as she tracks current conditions, compares them with memories of past incidences, and makes plans for the future. The past and the future as presently experienced can also be explicitly judged in relation to subjective time. (Wittmann 2011, pp. 5–7)

In the next two sections taking into account the properties of action strategies noted here, we will focus attention on the connected past and the connected future and their individual role in the complex present.

3.1.2 Connected Past

Before passing directly to the discussion of the connected past we need a short digression to clarify the terminology to be used below.

> **Brief digression:** In describing human actions in the connected past as well as the connected future we should clearly discriminate between their two aspects:
>
> – how these actions are implemented in reality,
> – how these actions are memorized and then represented in the mind or just imagined as certain hypothetical events.
>
> Keeping in mind the former, we will use the terms *physical object* and *real actions* described how the state of a physical object changes in objective time under human actions in physical reality. Besides, not to overload the text we will use just *object* and *actions* when it is evident that a physical object and real actions with them are under consideration.
>
> If the latter is under consideration, in order to refer to the mental image of a physical object we may use the term *mental object*. In this case the mental representation of a real action will be referred to as an *action representation*, the mental image of a hypothetical action will be called an *imaginary action*, and the term *mental image of action* or just *action image* will be used as a generic name of the both.

As noted in the previous section, each action strategy is homogeneous, which means that all its actions (*i*) are intentionally aimed at reaching or pursuing one common goal and (*ii*) are of the same kind. Within the connected past such actions may be imaginary or retained in memory as previously implemented; in any case these actions concern the same system comprising a fixed collection of physical objects.

Each of these actions embedded into objective time is characterized by that a certain mental object (as well as the corresponding physical object)

1. changed or could change its current state for a new one, or

2. kept or could keep its current state unchanged

at a given instant of time as a *consequence* of human intentional, goal-oriented actions. In the given context we may consider the states of mental objects to be unchanged if (*i*) no actions were taken to govern the dynamics of the corresponding physical objects or (*ii*) actions were implemented according to a certain initially fixed plan. In should be noted that a state of mental objects categorized as unchanged does not exclude time variations in their properties under these conditions.

In the description of the connected past we will mainly use the term *event* with respect to the analyzed human actions. In this way we focus on what and when it happened or hypothetically could happen with mental objects. Put differently, speaking about actions as events we concentrate at their embedding into the external physical world; treating the actions as elements of strategies we focus on their mutual intentional/logic interrelationship. It should be emphasized that if even a given event corresponds to a real human action, its characteristics reflect how it was retained in memory rather than what exactly happened and when it occurred in objective time. Therefore in describing human actions as they are perceived from the first-person point of view, all the events are actually mental entities and belong to the scope of the human mind.

Now we are ready to specify the notion of connected past. The *connected past* is the set of all possible events that

– already happened, retained and kept in working memory or
– imagined as happened in the past

and directly reflect human actions aimed at reaching the goal under consideration. Action strategies or, speaking more strictly, their temporal parts belonging to the past may be conceived of as strata of the connected past.

We want to note that the notion of connected past is not reduced to the collection of real events memorized and in this way saved in the mind even within their distortions caused by the bounded capacity of human cognition. Human imagination endows the connected past with properties not met in physical systems even with memory effects.[5] In particular, an action strategy that has been implemented in reality only in part, in the mind may contribute to its mental evaluation and the choice preference as a whole entity.

In our account of human temporality, the role of the connected past is the accumulation of information about the adequacy of a given action strategy with respect to reaching the desired goal. Speaking more strictly, this information concerns the preference of a given action strategy as a quantity (or set of quantities) attributed to the strategy as a whole entity and quantifying its choice among the other possible action strategies within the experiential now. Indeed, we cannot change the past but we can gain experience of how our actions are reflected in the external world and learn how to make them more efficient in reaching a goal. The introduction of the

---

[5] In describing physical systems with complex dynamics, the reduction of phase space can give rise to *effective* memory effects, which, particular, is reflected in the modern formalism of equations with fractional derivatives (see, e.g., Uchaikin 2013a, b).

connected past is justified by that the individual contribution of an event to the priority of some action strategy can be determined, at least partly, by the properties of the given event on its own. Put differently, the individual contribution of a given event to the preference of a given strategy is determined by two factors:

- how the given event is related to the motion state of physical objects; it is an aspect of action strategy;
- what are the outcome and cost of the corresponding action as an event characterizing of its embedding into reality; it is an aspect of the connected past.

The connected past as a constituent component of the complex present is characterized by its direct connectedness with the desired goal. It enables us to turn to psychological data concerning human conversation and dialogues. A pause in a conversation about 6 s may be felt as disturbingly long (Wittmann 2011), in particular, it can mean the change of conversation subject. The matter is that when two events are separated by an interval remarkably longer than 3 s, the experience of "emptiness" occurs, the events are not perceived to be bound together, and the time lag can attract the attention (Wackermann 2007). The duration of working memory—a crucial component of the complex present—when its information is not actively updated or rehearsed can be estimated from below as about 10–15 s (Goldstein 2015, for details see also Cowan 2015, 2017b; Simmering 2016).

For these reasons, we regard the connected past duration as the temporal horizon within which the memorized events are kept in working memory in the connection with currently experienced events and estimate it as a temporal scale about 10 s (Fig. 3.1).

Speaking about the connected past as a component of human temporal dimension, the temporal order of action images within one action strategy is essential. In this context it is worthy of noting Bernstein's concept of *engrams* as precursors of movements storied in memory which he proposed in 1935. Engrams reflect topology of planned movements rather than their temporal metrics. According to Bernstein, when the initial part of an engram has been implemented in a particular case, the implementation of the other parts should be also initiated in the appropriate time moments.

We ascribe individual causal power to action strategies as whole entities and assume that the action sequence is their intrinsic properties. However, the time moments of their implementation attributed to the action images—mental representations of real actions as well as imaginary actions—reflect particular details in the dynamics of the complex present. Moreover, we suppose that the particular temporal metrics of action strategy, i.e., the particular duration of objective time intervals between the real actions is not crucial. In other words, its contribution to the priority of action strategy is of minor importance. As evidence for the latter, we note that within short-term memory, the time representation is nonlinear, with its compression approximately obeying the logarithmic law (for recent investigations and review see, e.g., Singh and Howard 2017; Singh et al. 2018; Howard 2018). Thereby the temporal structure of the connected past as a component of human temporal dimension is not crucial and may be approximated in a way chosen for convenience reasons. However, the details of its embedding in the external world is essential, which is not in contradiction with the previous statement. The matter is that the outcomes and costs of real actions are attributed as individual properties of their mental representations. It is an additional argument for using the term the complex present in describing such intentional, goal-oriented actions.

3.1.3 Connected Future

The connected future is the third constituent component of the complex present and can be specified by analogy with the connected past. Namely, let us consider all the imaginary actions that

- admit interpretation as future events with respect to the experiential now in its embedding into object time,
- belong to possible action strategies related to reaching or pursuing the desired goal and where the state of experiential now or, speaking more strictly, its mental representation plays the role of initial conditions.

Then the *connected future* is the set of all the imaginary events representing these actions with respect to their particular content—expected imaginary actions with system objects at expected moments of time.

The role of the connected future is predictive estimation of a given action strategy with respect to (*i*) its efficiency in reaching (pursuing) the desired goal and (*ii*) the corresponding cost of its implementation.

Predictive estimation of possible action strategies is the pivot point in describing human choice of some optimal strategy for its further implementation in the scope of experiential now. Here the optimality of action strategy is understood in the general manner, which implies its evaluation turning to various criteria, including the rational estimate of strategy efficiency, emotional factors, cognitive biases, etc. Moreover, optimization of action strategies may be determined by the cumulative contribution from several channels mutually independent in the mind.

As in the case of the connected past, the real temporal pattern of action strategies seems not to be so significant in comparison with the temporal order of imaginary actions that are expected to be implemented one-by-one in the connected future. In this sense, the connected future as an element of human temporal dimension bears the information about the temporal order of events connected to one another via intentional or/and logical relations rather than the information about time intervals. It emphasizes once more that the human temporal dimension and the human temporality cannot be reduced to an ordered 1D-set of durationless basic elements similar to the time axis. Moreover, accepting the gist of Bernstein's engrams we may conceive of an action strategy within the connected future as a sequence of imaginary actions whose temporal order is fixed, whereas the moments of their initiation are not determined. The moments when the expected actions are initiated in reality belong to the experiential now and reflect the current state of physical objects as it is represented in the mind.

We want to emphasize that this statement concerning the minor role of temporal structure characterizing action images within one action strategy holds only in the scope of complex present. In describing mental time travels into the past retained in memory and into imaginary future the time lag between a given event and the complex present affects such phenomena substantially.

As will be elucidated in the next two sections, the connected future and the connected past have common roots—working memory. Moreover, their properties are so close to each other that we can treat in the same way imaginary events belonging to the connected past and the connected future. Therefore, we suppose that the duration of connected future should be about the duration of connected past. So, turning to the details discussed in Sect. 3.1.2, the duration of connected future must be estimated as a time scale of order of 10 s (Fig. 3.1).

Finalizing this discussion about the connected feature, we want to note two examples illustrating intentional, goal-oriented actions and particular problems to be analyzed further.

First, in our live almost all of us tried to walk with a cup full of coffee or tea carrying it, e.g., from the kitchen to the dining room. It requires permanent balancing of the cup not to spill the liquid. The goal of these actions is to hold the cup as steady as possible minimizing its random movements. In this case an action strategy may be conceived of as a sequence of intentional actions aimed at depressing the cup unwanted movements when their amplitude becomes remarkable. Such actions may be separated by the fragments of "no-action" behavior when liquid fluctuations in the cup are not significant and no correction is necessary. The given example illustrates intentional actions aimed at pursuing a goal, which is a permanent process rather than a finite sequence of steps.

Depending on various conditions, e.g., the level of coffee in the cup or the walking speed, the carrying without spill can be feasible but not a simple task. In this case the task implementation requires special attention and cognitive control. In particular,

- the time horizon of accumulating the information about the current style of working and its motor-control (the connected past),
- the time horizon in anticipating coffee spill possibility (the connected future)

should must be a little longer that the time interval during which the amplitude of coffee oscillations in the cup becomes critical. As demonstrated by Mayer and Krechetnikov (2012), under similar conditions the mean time to the coffee spilling is about 5 s or a little bit longer, in space it corresponds to of 7–10 steps. The noted time estimates fit the expected duration of connected past and the connected future.

Second, almost all of us in our childhood met a situation when our school teacher asked us a question and we tried to guess the right answer because we did not know it with certainty. Just during answering or immediately after it we recognized that it would be better to choose another version and asked the teacher to allow us to change our answer. Nowadays similar situations are studied within the problems usually referred to as *changes of mind in decision-making* or *confidence and evidence in decision-making* and there is a vast literature about them. We will return to these issues latter in the present book, here, by way of example, we want to note publications by Resulaj et al. (2009), Van den Berg et al. (2016) which support ideas similar to our understanding of the change of mind and its mathematical description. The given example illustrates multi-stage decision-making where the action strategies are aimed at reaching a certain (final) goal and it can be done via a finite number of stages. In this case, at a certain moment of time a person choices an action strategy for implementation but, then, before reaching the final goal can change his mind and select another strategy.

3.1.4 Remembering and Imaginary

Remembering, imaginary, and future thinking are not independent cognitive phenomena and seem to be rooted in common neural mechanisms. The idea about a close relationship between memory and imagination is not new; ancient and medieval scholars considered memory to be intrinsically linked with various forms of imagination and future-oriented thinking, in particular, the formation of intentions, plans and expectations, as well as the pursuit of goals (e.g., Black 2000; De Brigard 2017; O'Callaghan 2017). At the end of XX-century this idea had a revival based on new neurological data. Ingvar (1985) in his essay 'Memory of the future' summarized evidence that the frontal/prefrontal cortex handles the temporal organization of behavior and cognition, and that the same structures house the action programs or plans for future behavior and cognition. Tulving (1985) put forward the concept of *episodic memory* as the basic memory element enabling humans to mentally re-experience the past and pre-experience the future. According to Tulving (1972 see also Tulving 2002, 2005), episodic memory is responsible for receiving and storing information about temporally dated events, i.e., about the *what*, the *where*, and the *when*. Exactly such events, memorized or imaginary ones, make up the temporal structure of the complex present.

In 2007, three concurrent publications by Addis et al. (2007), Hassabis et al. (2007), and Szpunar et al. (2007) proposed a novel paradigm concerning a common neural mechanism of episodic foresight and episodic memory. Within this paradigm, the hippocampus is treated as a hub in a crucial neural network governing memory as well as imagination, which is now commonly accepted by most researchers. This hypothesis, nevertheless, still requires further research for its irrefutable verification (Miloyan et al. 2019).

Nowadays the relationship between memory and imagination has become a rapidly growing topic of interdisciplinary research and even a brief review of its main results is beyond the capacity of our discussion. As a certain resume of this research, we want to cite (Devitt and Addis 2016, p. 107):

> Neuropsychological and neuroimaging studies reveal a striking overlap in the neural networks recruited during memory and imagination, suggesting an inter-

> play between these two processes. It has been hypothesized that episodic memory contributes the necessary ingredients for the simulation of novel events, while semantic memory provides an organizational framework upon which an imagined event is constructed. However, while it is clear that imagination draws heavily on memory, this relationship appears to be a bidirectional one.

For details a reader may be referred, e.g., to reviews by Schacter et al. (2012), Mullally and Maguire (2014), Cheng et al. (2016), Szpunar and Schacter (2018), Addis (2018), paper collections edited by Bar (2011), Michaelian et al. (2016), as well as the paper collection edited by Broadway et al. (2015) for the relationship between time perception and imagination. In particular, as a certain generalization we note the conclusion by Addis (2018, p. 64) that

> the brain's default mode network underpins the remembering, imagining and perception of event representations via a common simulation process that is constructive and iterative, involving the interaction of pre-existing knowledge and reinstated perceptual content with incoming experiential information to create and refine online mental simulations of experience.

Following Marchetti (2014), Breeden et al. (2016), Dere et al. (2018) we note that in order to reconstruct past events and to construct imaginary future scenarios in an accurate way, at least, three major elements are required.

※ First, it is a certain level of autonoetic consciousness—the awareness of our own existence in time; put it differently, the ability to mentally place ourselves in the past, in the future and, to thus, be able to examine our own thoughts (for details see, e.g., Arzy et al. 2008; Nyberg et al. 2010; Vandekerckhove et al. 2014; Abraham and Bubic 2015). Exactly autonoetic consciousness enables us to speak about actions strategies as whole entities existing in the mind and possessing individual causal power.

※ Second, it is attention; attention ensures the selection of basic elements or pieces of information (see also Marchetti 2010; Ganeri 2017). Below in constructing the psychological/physiological cost of action strategy implementation we will return to the discussion of attention.

※ Third, it is working memory of a sufficiently high capacity and processing power needed for reconstructing past events and constructing imagined future scenarios as detailed as possible (Breeden et al. 2016). This requirement on working memory becomes more demanding when past events are used for the preparation for similar events in the future. For example it is the case in predicting plausible consequences of future actions through accumulating the information about various alternative options for similar actions (Marchetti 2014; Breeden et al. 2016). In particular, it a typical situation when the goal of actions is to keep some physical system near its unstable equilibrium—one of example to be analyzed in detail future.

3.1.5 Complex Present and Working Memory

Here we want to focus on the concept of working memory. We already discussed its properties previously in the context of the immediate past (Sect. 2.6.3) and the connected past (Sect. 3.1.2). However, its contribution is not reduced to memorizing the past events; as explained in the previous subsection, neural mechanisms governing memory processes are also responsible for imagination. In particular, the connected future is based on these mechanisms. Thereby the connected past and the connected future should exhibit common features and the complex present as a whole should be rooted in the mechanisms of working memory.

As noted by Baddeley (2012), the term *working memory* evolved from the earlier concept of short-term memory and the two are still on occasion used interchangeably. Nowadays, however, the term short-term memory is used mainly to refer to the simple temporary storage of information, in contrast to working memory, which implies a combination of storage and mental manipulation.[6] In our account of complex present we treat the transition from short-term memory to working memory as a certain fuzzy boundary between immediate past and the connected past.

Mental manipulations with past, present, and future require working memory which ensures that events—representing real events or purely imaginary ones—interconnected with one another in the mind are maintained active during their processing and assembling. However, acknowledging the crucial role of working memory in such mental manipulations we face up to a challenge caused by the low capacity of working memory. According to Cowan (2005, 2015) the capacity of working memory seems to be limited to three to five chunks—different elements of memory holding integrated episodes in the case under consideration. Therefore, when multiple stimuli or events have to be processed, they should interfere with one another in competing for the information processing in the brain.

A solution to this problem has been proposed by Baddeley (2000) added the *episodic buffer* to the multicomponent model of working memory (see also Baddeley 2007). The episodic buffer

> is assumed to hold integrated episodes or chunks in a multidimensional code. In doing so, it acts as a buffer store, not only between the components of [working memory], but also linking [working memory] to perception and [long-term memory]. It is able to do this because it can hold multidimensional representations, but like most buffer stores it has a limited capacity. (Baddeley 2000, p. 15, right column)

Put differently, episodic buffer is dedicated to linking information across domains to form integrated units of visual, spatial, and verbal information within a sequence of episodes ordered chronologically. Baddeley made the further assumption that retrieval from the buffer occurred through conscious awareness. In this way, like

[6] For a detailed discussion of various accounts of working memory and its properties considered from different standpoints, including short-term memory and long-term memory as well as the plausible reconciling of different accounts, a reader may be referred to Cowan (2015, 2017a, b), Adams et al. (2018).

Baars (1988, 1997) he considers consciousness to serve as a mechanism for binding related episodes into mental objects perceived with a certain integrity; in our account this object is an action strategy. We want to emphasize that the noted multicomponent model operates with real objects and events as they are retained in memory; imagination is not taken into account.

The multicomponent model of working memory has been extended by Breeden et al. (2016) and Dere et al. (2018) in order to allow for mental manipulations with imaginary objects too.

> The mental time travel platform serves as a supervisory system that receives input from the declarative memory systems (episodic and semantic memory) and integrates information from declarative memory sources into a coherent past event or imagined scenario. The mental time travel system maintains these scenarios for two operative workbenches, the experience reconstruction system and the scenario simulation system. The experience reconstruction system serves as an operative workbench where past events can be re-experienced/relived or used to deliberate about alternative response options and outcomes in similar situations. It is also used to extract information about own behavior, preferences and aversions, personality, self-perception and identity. The scenario simulation system is an operative workbench that is used to construct anticipated or imaginary future scenarios to act out different response alternatives and to predict their differential outcomes/consequences. The scenario simulation system is needed for the planning of actions in the future. (Dere et al. 2018, p. 4)

We consider working memory, as it is represented by extended multicomponent model, to be the neurological mechanism underlying the basic feature of the complex present and endowing it with integrity as a certain element of human temporal dimension.

From our point of view, the extended multicomponent model by Breeden et al. (2016), Dere et al. (2018) explains how the problem of working memory limit capacity is overcame in dealing with action strategies comprising individually many elements. The main idea is that an action strategy is just one chunk containing actions as episodes linking to each other explicitly. In this case the limit capacity problem comes up again in evaluating the preference of a given action strategy with respect to other possible strategies. The matter is that in choosing an action strategy for implementation in a some optimal way we need to perform in parallel two mental operations:

- to keep in the mind the properties of strategies characterizing their preference and,
- to update permanently these properties as time goes on.

Both of them require much attention and so should interfere with one another, which seems to have to prevent the two operation from their efficient implementation.

This problem is actually a particular case of a more general mystery of human cognition. Namely, it is the ability to regulate, coordinate, and efficiently implement a sequence of many actions and thoughts in accordance with internally maintained behavioral goals. In other words, it is the high capacity of cognitive control. The solution to this challenge proposed in the last two decades in studying working memory underlies our description of strategy dynamics, so let us consider it in more details.

In this context we want to note, first, the *executive-attention theory* of working memory elaborated by Engle (2002), Engle and Kane (2003), Kane et al. (2007). The gist of this theory is the the synergy of "attentional" and "memorial" processes in maintaining and recovering the access to information that is relevant to ongoing tasks and in blocking the access to task-irrelevant information. In this way the conflict between goal-maintenance and reactive response is resolved.

Second, it is the *dual-component model* of working memory proposed by Unsworth and Engle (2007a, b) which may be regarded as some development of the attention-memory synergy. It posits the existence of two components of working memory entirely distinct from each other in properties. Namely, the given model represents working memory as a combination of two components: a dynamic attention component—*primary memory*—and a probabilistic cue-dependent search component—*secondary memory* (see, e.g., Unsworth et al. 2012; Shipstead et al. 2014, for are view and detailed analysis). Unsworth and Engle (2007a) describe the working memory components as follows:

- Primary memory serves to maintain a number of separate representations active for ongoing processing by means of the continued allocation of attention. As noted before, the number of such mental elements—the limit capacity of working memory—is from three to five. So when more than this limit number of items should be presented to reflect, e.g., incoming new information, some items currently contained in primary memory are displaced from it. If a new task distracting attention from the previous activity or interfering with it has to be implemented, all the items in primary memory are displaced through disengagement of attention from them. The removed items are placed into secondary memory and must be recalled from it.
- Secondary memory contains items displaced from primary memory. If these items are necessary again for mental operations they must be retrieved from secondary memory through a probabilistic cue-dependent search process. Retrieval from secondary memory is a competitive process, therefore, some cues may be more efficient than others depending on the current situation and the tasks under implementation. The effectiveness of used cues depends on the similarity between the contextual elements at recall and the contextual elements at encoding. The more dissimilar the two contexts, the lower the probability of recall. In particular, items presented close to the recall period are likely to share more contextual elements than those presented much earlier.

Third, it is the *dual-mechanism theory* of control developed by Braver et al. (2005, 2007, for a review see also Braver 2012). It is devoted to another aspect of mental operations based on working memory. The central hypothesis of this theory is that cognitive control is implemented within two distinct operating modes: *proactive control* and *reactive control*.

> – The proactive control mode can be conceptualized as a form of 'early selection' in which goal-relevant information is actively maintained in a sustained manner, before the occurrence of cognitively demanding events, to optimally bias attention, perception and action systems in a goal-driven manner (Miller and Cohen 2001).
> – [I]n reactive control, attention is recruited as a 'late correction' mechanism that is mobilized only as needed, in a just-in-time manner, such as after a high interference event is detected (Jacoby et al. 1999).
>
> Thus, proactive control relies upon the anticipation and prevention of interference before it occurs, whereas reactive control relies upon the detection and resolution of interference after its onset. (Braver 2012, p. 106, itemizing is ours,the style of references has been also changed.)

It is necessary to emphasize that the implementation of proactive control creates a high *cognitive* demand because the top-down goal maintenance activity should facilitate the efficient processing of expected upcoming events. It involves sustained active maintenance of goals, which results in less conflict experienced when unexpected events occur. By contrast, reactive control relies upon detection of unexpected events and drives bottom-up reactivation of task goals. In this control mode task goals are not actively maintained and an occurring event may give rise to an inappropriate response with higher probability in comparison with proactive control.

Each of the two modes has its own advantages and disadvantages (see Braver 2012, p. 108). In particular, as far as proactive control is concerned:

pros: - Plans and their implementation can be continually adjusted to facilitate successful completion of the desired goal.
- The continuous maintenance of internal goals makes cognitive processing less sensitive to other unexpected but potentially relevant sources of bottom-up information.

cons: - Strong resource consumption is required for continuous goal maintenance. The low capacity of working memory reflected in limitations of goal representation in the focus of attention (e.g., Cowan 2001; McElree 2001; Oberauer 2002b) hinders maintenance of other information that could be held in working memory.
- The engagement of proactive control is dependent on the availability of contextual cues that are strong and reliable enough to trigger goal activation and maintenance in advance of the time when those goals are needed.

With respect to reactive control they are:

pros: - Other tasks and goals can be carried out efficiently between events when the main goal representations are activated at the time in which they are needed and then the memory resources are freed up.
- Due to reactive control being stimulus driven and transient it does not require contextual cues. It reduces demands on attentional resources of working memory substantially.

cons: - Repeated reactivation of the goal is required, which reflects in greater dependence on the trigger events themselves. If possible trigger events are insufficiently discriminative they will not drive the desired reactivation.
- Because of stimulus dependent reactivation reactive control is much more vulnerable to transient attentional capture or orienting effects that can disrupt the ability to trigger goal reactivation when necessary.

From the comparison of the given pros and cons it is seen that the proactive control and reactive control may be regarded as complementary parts of intentional, goal-oriented actions using the resources of working memory for remembering and anticipation.

The possibility of switching between proactive control and reactive control leads to the idea that weak differences between two tasks can induce significant changes in an cognitive control strategy an individual prefers depending on situational factors. For a detailed discussion of this issue including experimental evidence a reader may be addressed to Braver (2012), Redick (2014), Wiemers and Redick (2018).

The dual-component model of working memory and the dual-mechanism theory of cognitive control prompt us to accept the following scenario of how action strategies are selected for implementation and, then, are implemented. This scenario consists of three items.

- When an action strategy has been selected for implementation, it is placed in primary memory as its single element being in the focal point of attention. In this case the concept of top-down proactive control describes the strategy implementation aimed at reaching or pursuing the desired goal. The strategy efficiency is permanently estimated based on its predictions and current sensory experience of physical objects in the environment. Until the discrepancy between the predictions and experience is not significant all the other action strategies are not taken into account and the current strategy is implemented according to its original plan. Therefore this stage may be referred to as the *passive phase* of human behavior. Its duration can be of order of the complex present duration.
- When the discrepancy between the strategy predictions and sensory experience becomes essential and the necessity of correcting current actions is evident, the current strategy is displaced from primary memory and a new action strategy is searched in secondary memory. This cue-dependent search process is based on
  - the information characterizing the mental representation of the current state of environment and
  - the information about the efficiency of action strategies accumulated previously during their implementation and retained in memory.

  This search process admits interpretation as bottom-up reactive control; its duration can be compared with the duration of experiential now and is typically much shorter than the duration of complex present. Thereby the choice of new action strategy may be categorized as an *action point*[7] within the *active phase* of human behavior

---

[7] Our understanding of action points is rooted in the concept of action points put forward by Todosiev (1963), Todosiev and Barbosa (1963/64) for describing driver behavior.

to emphasize the difference in duration between proactive control and reactive control in the given case (Fig. 3.1).

- The cue in searching an action strategy appropriate to the current conditions is the degree of preference characterizing the individual properties of all the possible strategies quantitatively. This preference degree, on the one side, reflects the information about a given action strategy accumulated previously under different conditions. On the other side, it reflects also forgetting this information which becomes pronounced with the growth of time lag between the current moment and the moments when the strategy was implemented previously. In the simplest case the memory about the strategy properties can fade on scales about the complex present duration.

Below we will call this scenario *intermittent dynamics of human actions* or just human *intermittent actions* (behavior) for short to emphasize the permanent switching between the proactive implementation of choosing action strategies and reactive choice of a new strategy for implementation. Also we will the term human *piecewise-passive control* to emphasize the difference of the two modes of human actions in their duration.

The proposed concept of human intermittent actions within the complex present gives rise to the idea of *inertia of proactive control* or *proactive inertia* for short. When a certain action strategy is under implementation the limit capacity of primary memory does not allow us to analyze the properties of other possible strategies in parallel. It is illustrated in Fig. 3.2. During successful implementation of the strategy AS_1 other action strategies, e.g., the strategy AS_2, are left beyond our attention. As soon as we recognize that the efficiency of strategy AS_1 has dropped essentially and come close to the region where strategy implementation is inefficient we:

- halt the implementation of AS_1 (in Fig. 3.2 this moment is depicted as transition from blue to orange),
- focus our attention on other possible strategies retrieved from second memory and select a new action strategy (AS_2) that is most efficient under the changed conditions (in Fig. 3.2 this process corresponds to the time interval t_{ap} and depicted in orange)
- follow the selected strategy AS_2 while it is efficient.

The change of action strategies is not implemented immediately at the moment when the strategies AS_1 and AS_2 become equivalent in efficiency. In Fig. 3.2 this moment is noted as t_{se}. The delay Δ between the moments t_{se} and t_{ap} can be compatible with the duration of the complex present.

We regard this delay as the effect of *inertia of proactive control* (*proactive inertia*) because this delay admits interpretation as a certain bias towards the current choice. The considered inertia effect is caused by the limit capacity of working memory. In our following constructions we will turn to the idea of multi-channels responsible for the mutually independent processing of information to allow for the proactive inertia.

It should be noted that similar effects including the classical status quo bias (Samuelson and Zeckhauser 1988) are met within various psychological contexts

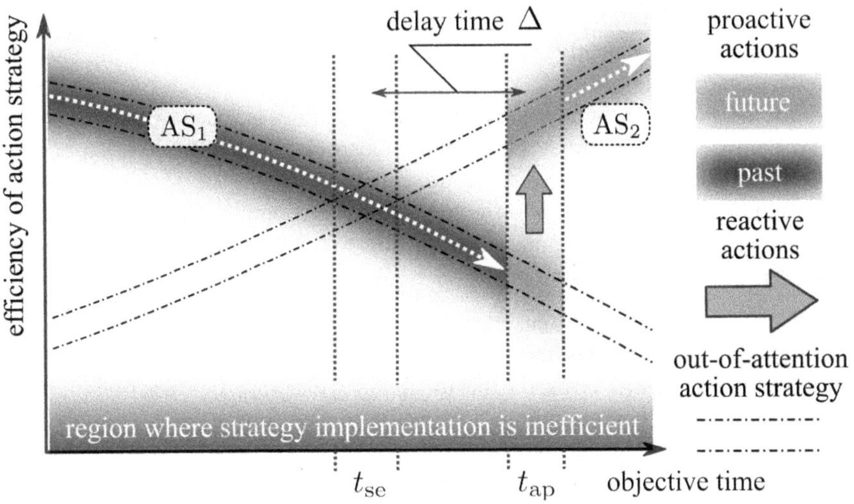

Fig. 3.2 Illustration of the proactive control inertia caused by the limit capacity of working memory. The horizontal axis is objective time, t_{se} denotes the time of strategy equivalence, t_{ap} denotes the time of action point. The vertical axis represents the efficiency of action strategy. The horizontal darken layer exhibits the region where strategy implementation is inefficient. The region above this layer corresponds to strategies acceptable in principle. The implementation of the shown two strategies AS_1 and AS_2 exhibits the characteristic features of intentional, goal-oriented actions within the complex present. Blurred blue lines depict the fragments of action strategies being currently under implementation. Yellow arrows illustrate the direction of event flow. A strategy left beyond our attention is not accessible to the mind, which is illustrated with dashed-dotted lines. The region where the two strategies become equivalent in efficiency is depicted as t_{se}. The effect of proactive control inertia is reflected in the delay time Δ between the time of strategy equilibrium t_{se} and the time of action point t_{ap}. The action point represents the change of strategies; in this case the properties of all the possible strategies are accessible to the mind, which is shown in orange. The newly selected strategy AS_2 remains in focus whereas the strategy AS_1 is removed from attention

and based in different mechanisms, for their review a reader may be referred to Akaishi et al. (2014), Alós-Ferrer et al. (2016), Jung and Dorner (2018), Jung et al. (2018).

Concluding this discussion about the relationship between the complex present and working memory we want to draw the following conclusions.

- The combination of remembering and imagination endows the complex present with integrity and allows us to regard the complex present as an individual fragment of human temporal dimension.
- Remembering and imagination united within the complex present are represented by a collection of action strategies implementing intentional actions aimed at one common goal. In this sense, action strategies are elements of the human temporality confined to the complex present. Their dynamics admits the description on terms of human intermittent actions.

- The mind operates with action strategies as whole entities bearing their own properties which are permanently updated when the current action has been implemented in the experiential now.
- The complex present in its content and temporal scales is based on working memory. Using available estimates of working memory duration the duration of complex present can be estimated as a time scale of 10–20 s.
- In modeling intentional, goal-oriented actions the complex present is the minimal fragment of human temporal dimension which admits closed mathematical description. In this description:
  - the connected past is responsible for accumulating the information about system state for further actions, put differently, for learning;
  - the experiential now is responsible for updating the properties of action strategies (in addition to their selection for further implementation);
  - the connected future is responsible for specifying the choice of action sequences and their plausible temporal patterns.

3.2 Phenomenology: Basic Concepts

In creating our account of human temporality we turn to the gist of phenomenology as the philosophical approach to studying mental phenomena and the structures of consciousness as experienced from the first-person point of view (e.g., Smith 2013, for introduction). In this section we want to discuss the basic concepts of phenomenology and argue for the possibility of their application to developing mathematical models for intentional, goal-oriented behavior of humans.

3.2.1 Philosophical Background

Speaking about the human temporality as a fundamental component of time-consciousness and phenomenological description of the mind developed in the XX-century, it is necessary to note as milestones the accounts proposed notably by Edmund Husserl, Martin Heidegger, Jean-Paul Sartre, Maurice Merleau-Ponty, and many others.

In their accounts Husserl's theory of time-consciousness holds a central position where temporality is addressed to the most fundamental temporal structure of the stream of consciousness (Husserl 1991, Collected Works, IV). In particular, there can be singled out several basic premises in Husserl's phenomenological account of mind, including those often referred to as *intentionality*, *phenomenological reduction*, *self-consciousness*, *embodiment*, and *temporality* itself. The noted philosophical notions underlie our constructions, therefore we want to discuss them in more detail. The goal of this discussion is to clarify the reasons we turned to them in the mathematical description of human actions.

3.2.2 *Intentionality and Phenomenological Reduction*

Husserl's notion of *intentionality* is defined as openness toward otherness, put differently, according to Husserl, consciousness is not self-enclosed. All the mental objects are related—explicitly or implicitly—to objects of the external world; all the diverse forms of consciousness—a perception, an expectation, a thought, a fantasy, etc.—are characterized by the intending of an object.[8] Husserl's phenomenological methodology also accepts a stronger proposition called *phenomenological reduction* (e.g., Thompson and Zahavi 2007, for details).

In our account of intentional, goal-oriented actions we also accept the concept of phenomenological reduction but in a weaker form to be called the *effective phenomenological reduction*. In physics the description of various complex systems including continuum media often turns to the gist of phenomenology and is also known under the name *effective theories*; we already discussed this issue previously concerning the mathematical description of the mind (Lubashevsky 2017). The main idea of such phenomenological theories is to single out fast and slow variables such that slow variables enslave fast ones. In this case, the mathematical description of the corresponding system can be confined to a certain collection of state variables containing only slow ones. Concerning the relation between mental images and the corresponding physical objects we suppose that cognitive states and unconscious processes in the brain underlying these states are inextricably intertwined and can be described together turning to the characteristics of mental images.

We accept the gist of the *non-representationalist* interpretation of intentionality (defended by Zahavi 2003) which enables us to endow the mental images with individual properties and dynamics irreducible to that of physical objects. Here the irreducibility is understood as the fact that mental images are not enslaved by the corresponding physical objects.

Let us now formulate the version of phenomenological reduction underlying our further mathematical constructions.

Effective phenomenological reduction:

1. Mental images and mental operations with them can be described in terms of quantities (physical variables) characterizing the corresponding objects in the external world. In particular:

[8] There are mental states like various moods (e.g., anxiety, depression, and boredom) that are not object-directed. However, it does not exclude such states from the scope of phenomenology; intentionality is not confined to object-directedness, it is about openness and moods are due to real events in the external world.

- the description of mental images employing the physical variables of the corresponding images is closed in the sense that there are no other type variables (*i*) required for describing mental images and (*ii*) not enslaved by these physical variables;
- the given closeness of mathematical vocabulary does not imply that mathematical models of mental images are necessarily reduced to that of the corresponding physical objects in property.

2. Mental images exhibit their own dynamics irreducible to the dynamics of the corresponding physical objects, which implies that:

   - the dynamics of mental images is not enslaved by the dynamics of the corresponding physical objects;
   - the laws governing the dynamics of mental images are not necessarily reduced to the laws governing the dynamics of the corresponding physical objects.

To elucidate the given propositions we want to note the following.

Proposition 1*:* For each mental image, memorized or purely imaginary one, there is, in principle, some physical object or a group of objects that exhibits, at least, qualitatively the properties attributed to this image. In other words, a mental image is a *qualitative* representative of a physical object or several objects combined in some way which may be realistic or purely hypothetical. In the latter case the image may represent also objects that cannot be assembled from their details in reality like impossible cube or other impossible objects invented by M. C. Escher (see,. e.g., Bool et al. 1992, for picture collection). We share the point of view that we just cannot imagine objects that in no way—as a whole or in parts—are not reflected in reality. The given assumption does not exclude development of novel notions or invention of novel technical tools, they can emerge step-by-step via generalization and reconsideration of already available notions or tools.

This type reducibility enables us to speak about the embedding of mental images into physical reality and treat the obtained images of mental images as mental objects described in physical terms like spatial position, time moment, the speed of motion, etc. Although such physical variables are assumed to make up the vocabulary sufficient for describing the properties of mental images, it does not mean that the mathematical model of physical objects and the models of their mental images are similar in properties. For example, for physical objects admitting the description in terms of material points their images should match some space-time clouds reflecting the uncertainty of human perception. Using Dainton's terminology (Dainton 2008a, Sect. 2.2) we may say that Proposition 1 implies that the *phenomenal space* of embedded mental images coincides with the physical space or, speaking more strictly, with its certain subspace. Contrarily, mathematical representations of physical objects and their mental images embedded in the space are entirely distinct objects.

Proposition 2*:* Irreducibility of the dynamics of mental images to the dynamics of the corresponding objects affected by human actions is one of the main arguments for the two component description of human actions (Lubashevsky 2017). This description implies that, first, the objective component—the external world—and the subjective component—its mental image—possess their own dynamics. Although mental images emerge in the mind due to the existence of some physical objects they do not necessarily need the explicit presence of the corresponding physical objects. Mental images on their own may be engaged in human actions and evolve as time goes on. Second, human actions aimed at governing the dynamics of physical objects are based on their mental images, these mental images in turn change their properties also due to changes in the environment. It couples the dynamics of objective and subjective components with each other endows systems affected by human actions with complex properties not met in physical systems.

3.2.3 Self-Consciousness, Agency, and Ownership

In contemporary phenomenology there is a bundle of closely intertwined notions—*phenomenal consciousness*, *self-consciousness*, *self-awareness* are among them—which also admit different individual interpretations (Gallagher and Zahavi 2012, Chap. 3). These notions concern mental states that have a subjective and experiential character.

> [A] subjective character, a certain phenomenal quality, correspond[s] to what it is like to live through or undergo that state. This is what makes the mental state in question phenomenally conscious. (Thompson and Zahavi 2007, pp. 75, right column)

As far as experiential quality is concerned, each of us has direct access only to our own experience, which is often called the *first-personal givenness* of experience. It leads to the issue of *self-awareness* and make the experience *subjective*. The first-personal givenness of own experience is a basic feature of phenomenal consciousness. Therefore a conscious experience,

> in addition to being of or about its intentional object is prereflectively manifest to itself...[T]ransitive phenomenal consciousness (consciousness-of) is also intransitive self-consciousness (see Kriegel 2007). Intransitive self-consciousness is a primitive form of self-consciousness in the sense that (*i*) it does not require any subsequent act of reflection or introspection, but occurs simultaneously with awareness of the object; (*ii*) does not consist in forming a belief or making a judgement; and (*iii*) is passive in the sense of being spontaneous and involuntary. (Thompson and Zahavi 2007, pp. 76, right column, style of citation has been changed.)

These components belonging to the very basic level of phenomenological consciousness prompt us to assume that the mental image of a physical or imaginary object is given to the mind as a whole entity. In this case the individual properties of even some parts of this image are attributed to the image as a whole.

Unlike physical objects existing objectively, there is an owner for mental images. Only the owner—the self—can modify a mental image changing its parts in the mind, whereas the parts of a physical object can exhibit their own dynamics.

As arguments for this point of view, we turn to another notion of phenomenology, it is *agency* or *self-agency*. Generally speaking, an agent is a being with the capacity to act, and the term *agency* denotes the exercise or manifestation of this capacity (Schlosser 2015). Nowadays the concept of agency is widely debated from various aspects which, however, are rather far from the problems analyzed in the present book except for the following. Gallagher (2000a, b) singles out two senses of agency:

- *Sense of agency:* The sense that I am the one who is causing or generating an action. For example, the sense that I am the one who is causing something to move, or that I am the one who is generating a certain thought in my stream of consciousness.
- *Sense of ownership:* The sense that I am the one who is undergoing an experience. For example, the sense that my body is moving regardless of whether the movement is voluntary or involuntary. (Gallagher 2000a, p. 15, *italic* and itemizing is ours)

Both the two senses should be categorized as just given to the mind, i.e., as pre-reflective senses. Put differently, when we experience something our experience is immediately and implicitly given to us as *our* experience (Gallagher 2007, 2012a, see also Gallagher and Zahavi 2012). When we implement some intentional action or generate a certain thought, our attention is focused on the results of this action or thought rather than the physiological details of how this action is implemented or how the though appears in the mind.

It want to emphasize that Jean-Paul Sartre, probably the best-known defender of the phenomenological theory of self-consciousness, considered pre-reflective self-consciousness to constitute the consciousness of something at the very basic level. Citing his famous book *Being and Nothingness* (Sartre 1956) where, in particular, he analyzes ontological aspects of consciousness, it is possible to say that experience does not simply exist, it exists in such a way that it is implicitly self-given (for details see, e.g., Zahavi 2005).

The first-personal givenness of experience with pre-reflective experience of agency and ownership allow us to hold the following proposition.

Pre-reflective ownership of mental images: In intentional actions involving mental operations with images of real physical as well as purely imaginary objects, these images are processed as whole entities. None of their parts exhibits its own dynamics caused by this part itself. All the possible changes in these mental images occur via the top-down effect of the self which is the *owner* of all the images accessible via working memory, i.e., currently being in the active state admitting direct access to the mind. This ownership of mental images by the self is characterized by that:

- Only the initial and final states of a mental image undergoing transformation are accessible to the mind, the intermediate phase of this transformation is implemented unconsciously and, so, its details are inaccessible to the mind. Therefore, if an image transformation is (*i*) related to sharp changes in the image state and (*ii*) implemented on time scales of the experiential now, then for the mind it looks like a step-wise transition between possible states.

- The choice of states engaged into image transitions as well as characteristic properties of these transitions can directly reflect
  - the state of a given image as a whole entity taken in isolation,
  - the state of affairs in the system of all mental images currently accessible to the mind.

As will be seen below, the principle of pre-reflective ownership of mental images enables us, in particular, to use non-Hermitian operators to describe transformations of space-time clouds and their dynamics admitting step-wise transitions.

3.2.4 Embodiment and Phenomenology: Reconciliation

Embodied cognition is a novel research direction in philosophy of mind and cognitive sciences that accepts cognition to be deeply depend upon features of the human body; for a review of the current state of the art in embodied cognition a reader may be referred to Wilson and Foglia (2015), Shapiro (2019a, b) and collections edited by Borghi and Pecher (2012), Shapiro (2014). The gist of embodied cognition hypothesis is the claim that the format of cognitive representations, at least partly, is modality-specific rather than purely abstract or amodal. Thereby, embodiment proposes a direct explanation of how the brain—the organ isolated from the environment—can generate images closely representing external objects. Nowadays there have been proposed a wide range of embodied cognition concepts endowing the physical body with a weaker or stronger contribution to cognition and, finally, the mind. In particular, the concepts met at the opposite poles may be characterized as follows:

> Weak embodiment is the view that concepts are not represented *only* by sensory/motor processes, but are also represented at an abstract or modality-independent level …In contrast to weak embodiment, radical or strong embodiment posits that (all aspects of all) concepts are represented in a modality-specific format. …[T]he strongest form that the embodied cognition hypothesis could possibly take: there is no abstract, modality-neutral conceptual content, but rather only information represented in modality-specific input and output systems. (Mahon 2015, p. 423, left column)

For a critical discussion of arguments for and against the weak and strong embodiment a reader may be referred, e.g., to Chemero (2013), Wilson and Golonka (2013), Kiverstein and Miller (2015), Beckes et al. (2015), Mahon (2015), Goldinger et al. (2016), Shapiro (2019a).

As far as phenomenology is concerned, the idea of embodiment can be found in Husserl's early works (for details see, e.g., Gallagher 2014). In particular, according to Husserl, human action properties are not cognitive additions to perception but are implicit, intrinsic elements belonging to the intentional structure of perception. These ideas were elaborated in detail by Merleau-Ponty (1945/2005). In his *Phenomenology of Perception* Merleau-Ponty introduced the notion of *body schema* to emphasize that the body is a perceiver and perception intrinsically involves both sensory and motor processes.

> [M]y whole body for me is not an assemblage of organs juxtaposed in space. I am in undivided possession of it and we know where each of my limbs is through a body image [schema] in which all are included. (Merleau-Ponty 1945/2005, pp. 112–113)

We want to point out that Merleau-Ponty taking over Goldstein (1931/1971) draws distinction between *concrete* movement, e.g., grasping, and *abstract* movement, e.g., pointing. Grasping seems to be more basic (it survives certain pathologies where pointing does not) and characterizes our normal motor intentionality in its non-representational, non-conceptual form (e.g., Kelly 2003).

In our construction we turn to the concept of embodiment formulated as:

> *Embodiment Thesis:* Many features of cognition are embodied in that they are deeply dependent upon characteristics of the physical body of an agent, such that the agent's beyond-the-brain body plays a significant causal role, or a physically constitutive role, in that agent's cognitive processing. (Wilson and Foglia 2015, Sect. 3)

We accept the understanding of the embodiment thesis as a certain intermediate interpretation of embodied cognition—*grounding-by-interaction* proposed by Mahon and Caramazza (2008). Their account combines (*i*) the view that mental images and concepts are, at some level, *abstract* and *symbolic* with (*ii*) the idea that sensory and motor information may *instantiate* online conceptual processing. It is essential that the activation of the motor system involved in this processing is fast (its delay is about 200 ms) and automatic, i.e., unconscious.

However, before passing directly to particular aspects of embodied cognition that underlie our mathematical constructions, it is necessary to clarify a certain issue seeming to hamper applying phenomenology to mathematical description of human behavior. The matter is that the gist of embodied cognition especially in its radical form seems to exclude the possibility of constructing a closed mathematical description of human actions in terms of the first-person point of view. In particular, it prompted Varela et al. (1991, Chap. 2) to pose a question about the breakdown of phenomenology. Indeed phenomenology, on its own, is limited to the analysis of conscious phenomena and, strictly speaking, is not able to penetrate beyond our conscious experience of how things appear in the mind.

A plausible solution to this problem can be found in the *enactive* account of radical embodied cognition proposed by Varela et al. (1991, 2016). For an extensive contemporary statement of the enactive approach to the mind a reader may be referred also to Thompson (2007), Gallagher (2017a). The concept of *autonomy* introduced by Varela (1979) based on Maturana and Varela's (1980)[9] theory of autopoiesis takes the central position in the enactive approach. There are two constituent premises of autonomy, *operational closure* and *precariousness* (see, e.g., Di Paolo and Thompson 2014).

Operational closure implies that a system exhibiting the given property can be conceived of as a network of different processes which:

- mutually depend on one another and affect one another,
- may depend on some external processes and affect some other external processes but without feed-back effects.

Thereby the dynamics of operationally closed system is determined by the state of inner processes and the state of environment playing the role external conditions. It should be emphasized that an operationally closed system is not isolated from dependencies and interactions.

Precariousness implies that without the interaction between these inner processes any inner process necessarily stop or run down. Put differently, it is not possible for a precarious process in an operationally closed network to exist solely on its own and the effect of environment.

Describing the resulting properties of a precarious, operationally closed system, Di Paolo and Thompson write that

> it sustains itself in time partially due to the activity of its own constituent processes. Moreover, because these processes are precarious, the system is always decaying. The "natural tendency" for each constituent process is to stop, a fate the activity of the other processes prevents. The network is constructed on a double negation. The impermanence of each individual process tends to affect the network negatively if sustained unchecked for a sufficient time. It is only the effect of other processes that curb these negative tendencies... Thus, a precarious, operationally closed system is inherently restless, and in order to sustain itself despite its intrinsic tendencies towards internal imbalance, it requires energy, matter, and relations with the outside world. (Di Paolo and Thompson 2014, p. 72)

Autonomy of such systems also gives rise to their *adaptivity*, i.e., the ability to regulate their inner processes in relation to external conditions (Di Paolo 2005). This adaptivity should endow an autonomous system with a certain kind of stability. After its transient stage has finished the phase portrait of inner processes' dynamics must become stable and be determined by the state of environment. Put differently,

- the nonlinear interaction between the constituent inner processes causes the instability of the system steady state and endows the process ensemble with complex dynamics;
- individual intrinsic dissipation of the inner processes necessitates the permanent external input of energy and matter for the developed time variations to sustain,

[9] The cited monograph of Maturana and Varela is the second edition, the first edition is published in 1972.

– the combination of the external input (energy and mass supply) with internal dissipation causes the formation of stable (metastable) attractor of the system dynamics whose properties are affected by the state of environment.

The characteristic properties of this attractor may be used to categorize the dynamics of precarious, operationally closed system as its certain state.

Returning now to the relationship between phenomenology and enactive embodied cognition we see that autonomous neurophysiological processes and mental (conscious) states can be related in the same way.

Namely, on small time scales neurophysiological processes should exhibit complex dynamics whose features are not accessible to the mind and must be characterized as unconscious. These neurophysiological processes may spread over the whole body thereby the details of the body arrangement in physical space have to become the intrinsic characteristics of neurophysiological processes. The mental states just cannot change remarkably on such time scales. It is quite natural to estimate these scales from above by the characteristic duration of the immediate past, the immediate now, and the immediate future.

On larger time scales about or longer than the duration of experiential now the averaged properties of neurophysiological processes, first, become accessible to the mind. Second, the neurophysiological processes are enslaved by the mental states in the sense that if a mental state changes than the dynamics of neurophysiological processes follows this change as a result of their unconscious adaptation.

These features justify the effective phenomenological reduction (Sect. 3.2.2) even in the case when the mind does not have *direct* access to the dynamics of neurophysiological processes, i.e., the course of how neurophysiological states follow one another in their dynamics is unconscious. Through step-by-step learning we acquire (*i*) the knowledge of how neurophysiological processes with mostly unconscious dynamics are reflected in our cognition and (*ii*) the skill in inducing or responding to the appearance of these processes. Naturally it is not necessarily one-to-one correlation between the phenomenological representation of neurophysiological processes and the processes themselves, only the integral results of unconscious processing of perceived information can become accessible to the mind. Nevertheless, via the trial-and-error learning operations within the person-level dimension—remembering, imagining, deciding, dreaming, etc.—and sensorimotor actions become interrelated at least under the standard conditions (cf. Gallagher 2014).

The embodiment of cognitive processes comprising conscious and unconscious components enables us to pose two questions about the relation between the properties of mental images of perceived external objects and the own properties of these objects. In particular, it is how the physical characteristics of object motion in the real space are reflected in the local characteristics of the mental image dynamics. This issue is crucial for our further constructions, so let us discuss it in more details.

3.2.5 Forms of Consciousness

In his pioneering works (1988, 1997, 2002, 2005), Bernard Baars put forward the general concept of *global workspace* to emphasize the integrative nature of consciousness. The integrity of consciousness implies the existence of many neural networks comparable in size with the brain and processing external information in parallel—the idea having a long history (e.g., Newell 1994). According to Baars' account of consciousness, unconscious processes take place in local modules distributed across the brain and, we would add, also across the body. The consciousness as an aspect of the mind involves *some* access to the information borne by all these elements and the ability to broadcast this information on a global brain/body scales. Baars and Franklin (2007) and Baars et al. (2013) generalized the concept of global workspace to including the dynamics of coordination and neural signal propagation over different task-related networks as well as the interaction with the unconscious cognitive system. The latter is understood as a human system responsible for *processing* the perception, memory, learning, etc. *implicitly*, i.e., without being aware of it and, thus, not readily available to introspection (Rudman and Spencer 2007, see also Bargh and Morsella 2008; Brownstein 2018). For a detailed description of the global workspace concept and its neural basis a reader may be referred to the reviews by Dehaene and Changeux (2011), Raffone et al. (2014). Block (1995), Block (2007a) turned to the notions of *access consciousness* and *phenomenal consciousness* to describe the relationship between consciousness and underlying neural phenomena. Following this way Dehaene et al. (1998, see Dehaene et al. 2006; Dehaene and Changeux 2011 for a review) put forward the concept of *global neuronal workspace* focusing on how sensory information gets selected to be broadcast throughout the cortex. For details and discussion of similar concepts a reader may be referred to the review by Raffone et al. (2014).

We have noted these versions of global workspace as some theoretical arguments for the possibility of describing mental images turning to general frameworks rather than notions applicable only to specific sensory modalities. By way of illustration, we note the model of working memory proposed by Jacobs and Silvanto (2015). It states that working memory introspection does not operate on the actual memory trace but rather requires a new representation to be created for the conscious domain. This conscious representation exists in addition and in parallel to the actual memory representation.

Nowadays, global workspace and its various aspects are a subject of ongoing debates, see, e.g., an introductory to a special journal issue by Fazekas and Overgaard (2018) and the following issue's papers. Here we want to focus on the notions of phenomenal consciousness and access consciousness, other notions and concepts will be touched on in their relation to the two ones.

Let us, at first, elucidate the meaning of phenomenal consciousness and access consciousness, P-consciousness and A-consciousness, for short. It should be noted beforehand that these notions are fundamental and cannot be defined strictly in a noncircular way (Block 1995). So the following is just their explanation rather than definition. Referring to Block (1995):

> P-consciousness is experience. P-consciousness properties are experiential ones. P-conscious states are experiential, that is, a state is P-conscious if it has experiential properties. The totality of the experiential properties of a state are "what it is like" (Nagel 1974) to have it. Moving from synonyms to examples, we have P-conscious states when we see, hear, smell, taste, and have pains. P-conscious properties include the experiential properties of sensations, feelings, and perceptions, …includ[ing] thoughts, desires, and emotions. [p. 230]
>
> A state is access-conscious (A-conscious) if, in virtue of one's having the state, a representation of its content is (1) inferentially promiscuous (Stich 1978), that is, poised for use as a premise in reasoning, (2) poised for rational control of action, and (3) poised for rational control of speech. [p. 231]

The implementation of P- and A-consciousness as elements of the global neural workspace is analyzed, e.g., by Raffone and Pantani (2010).

For Block the characteristic feature of A-consciousness is the *reportability* of the corresponding information, which implies this information to be a component of the working memory content (Block 2011, see also discussion by Carruthers 2017a). The reportability criterion, however, posed other challenges to science of cognition (see, e.g., Naccache 2018; Phillips 2018; Stazicker 2018) to be the subject of future research. Below we also propose an interpretation of reportability criterion for constructing a phase space of mental images.

Another characteristics of the relationship between P- and A-consciousness is the relative richness of P-consciousness with respect to A-consciousness based on working memory with limited capacity (Block 2007b, 2014). Put differently, the contents of P-conscious experience *overflow* what people are able to report. As a result the overflow effect is an argument for the existence of P-consciousness without access, i.e., a mental content currently experienced yet not accessed for further report. These kinds of consciousness and their properties have been much debated during the last decades; for details and arguments for and against these ideas a reader may be referred to recent reviews by Bronfman et al. (2014), Brown (2014), Raffone et al. (2014), Windey and Cleeremans (2015), Gross (2018), Lamme (2018), Overgaard (2018), Usher et al. (2018), Ward (2018).

The overflow effect allows us to describe mental images of external objects even at the vary basic level of their perception in terms of a few phase variables to be used in higher-level mental operations.

The watershed between the two kinds of consciousness or, in some sense, the interface between unconscious-conscious is up to know rather far from being understood well. As a result, there are various plausible interpretations of these notions proposing different answers to questions about P- and A-consciousness and their interrelations (e.g., Overgaard 2018).

Below we propose our account of the two types of consciousness applicable to describing the states of synchronic and diachronic units. It should be noted that the notions of P- and A-consciousness introduced above are not necessarily identical to the notions we will use: experienced consciousness (E-consciousness) and consciousness accessible to the mind (AM-consciousness). Following the logic of Sect. 2.4.2 we will use the criterion of *analyzability* to discriminate between E-consciousness and AM-consciousness. Here the analyzability of a given property or phenomenon related in the mind to some conscious image is understood as follows.

Criterion of analyzability: We call a property attributed to a mental image of a currently perceived physical object (or even purely imaginary object) *consciously analyzable* (or just *analyzable*) if a subject is consciously able:

(*i*) to decompose this image into its constituent components,
(*ii*) to recognize their individual properties at least at the experiential level, and
(*iii*) to report (to recognize) how the given property of this image attributed to it as a whole is reconstructed from the individual properties of its constituent components.

In other words, a given instantiation of consciousness is analyzable if it admits a representation via the *analysis-synthesis* strategy.[10]

We call a properties of a mental image *consciously unanalyzable* (or just an *unanalyzable*) when through experiential perception a subject is

(*i*) aware of the existence of the given property as a characteristic of this image, in particular, he is able to report on the property existence,
(*ii*) not able to explain *turning to his experiential perception* how he draws the conclusion about reported information.

Below we will refer to consciousness of analyzable properties of mental images as to AM-consciousness implying their accessibility to the mind with respect to the analysis of these properties, i.e., some reduction to their constituent components. Consciousness of *unanalyzable* properties will be regarded as E-consciousness to emphasize that the awareness of such properties is just given to the mind and cannot be reconstructed based on more basic features.

3.3 Space-Time Clouds

The concept of space-time cloud is the third cornerstone in our description of intentional, goal-oriented actions in addition to the notions of the complex present and the human temporality. The present section is devoted to elucidating this concept and underling premises.

---

[10] Previously (Lubashevsky 2017) we turned to the analysis-synthesis strategy as the criterion for categorizing the physical within the context of mind-body problem as well as the strong emergence.

3.3.1 Phase Space of Mental Images

There are, at least, two reasons for us to turn to categorizing conscious phenomena in terms of E-consciousness and AM-consciousness.

First, in these terms the synchronic units and the diachronic unit become objects of E-consciousness and AM-consciousness, respectively. In this case, the introduction of synchronic and diachronic units, on the one side, and the division of conscious phenomena in two classes of E- and AM-consciousness, on the other side, become self-consistent. Indeed, the synchronic units as well as consciously unanalyzable instantiations of E-consciousness are based on the unconscious integration of information about the experienced events on scales of order of 500 ms. Some particular details characterizing the temporal arrangement of these events become accessible to the mind as properties attributed to the synchronic unit as a whole entity. It is quite natural to suppose that this processing of perceived information should be implemented within some local brain aria specific for each sensory modality and take into account the contribution of neural networks distributed over the whole body.

The diachronic unit as well as the analyzable instantiations of AM-consciousness imply the essential contribution of conscious processing of information about perceived events at the level of global workspace. The corresponding governing mechanism must be common for all the sensory modalities and operate at the level of brain as a whole. Thereby, within the given interpretation of E- and AM-consciousness the distinction between the neural mechanisms governing the two kinds of consciousness is rather evident. Whether other approaches to categorizing consciousness into similar types leads to clearly distinct governing mechanisms is an open problem (see, e.g., Peremen and Lamy 2014).

Second, the given interpretation of E- and AM-consciousness is able to propose some plausible solutions to problems currently debated. In particular, it concerns (*i*) the phenomenal overflow (Block 2007b, 2011, 2014, see also D'Aloisio-Montilla 2017; Odegaard et al. 2018 for discussion), (*ii*) the partial awareness (e.g., Kouider and Dupoux 2004; Windey et al. 2014), and (*iii*) the gradual change of conscious experience implying the possibility of endowing consciousness with a certain *degree* (Sergent and Dehaene 2004; Overgaard et al. 2006; Nieuwenhuis and De Kleijn 2011; Windey et al. 2014; Elliott et al. 2016; Ward 2018; Lohse and Overgaard 2017).

In our account the transition from E-consciousness to AM-consciousness corresponds to the consolidation of sensory memory and the conversion of sensory memory traces rich in properties into traces of working memory substantially limited in capacity (Sect. 2.6.3). This transition is not strictly fixed, in particular, it depends on attention admitting conscious control. As a result, high-level mental states are able to affect, at least partly, which properties of sensorily perceived objects will be inherited in working memory and the degree of their conversion. It enables complex interrelation between E-consciousness and AM-consciousness based on top-down and bottom-up causal processes.

As noted above E-consciousness introduced using the analyzability criterion characterizes the properties of the synchronic unit of immediate past, immediate now, and immediate future. Strictly speaking, the corresponding properties of diachronic unit have to be classified as entities of AM-consciousness. However, they are the simplest ones among other properties of AM-conscious type because the diachronic unit is the direct aggregation of its three synchronic units. In this sense,

- the immediate past, the immediate now, and the immediate future from the side of unconscious processing of information and
- the experiential now from the side of conscious processing of information

form the basic boundary of most detailed description of perceived events accessible to the mind. In principle, the details of how the experiential now aggregates the properties of its constituent units can be intentionally controlled. It, however, requires that special attention be focused on these details because such intentional actions correspond to the limit capacity of cognitive control. For these reasons characteristic properties attributed to mental images of physical objects based on the experiential now will be regarded as the *phase variables* specifying their current state. The following proposition elucidates the given statement.

Phase space of mental images: Let us consider the mental image of some physical object whose position in space, shape, color, surface luminous intensity, etc. are consciously perceived and attributed to the image as its properties $\{x_i\}$ to be further retained in memory. The latter implies that attention is consciously focused on these properties. Time change of these physical properties is unconsciously processed via sensory modalities and the characteristic rates $\{v_i\}$ at which the variables $\{x_i\}$ change in time become accessible to the mind at the level of E-consciousness. Within the experiential now aggregating the immediate past, immediate now, and immediate future the properties $\{x_i, v_i\}$, first, are averaged and, second, are used to produce new quantities $\{a_i, j_i\}$ characterizing the second and third order rates at which the perceived physical properties change in time.[11] After that the properties $\{x_i, v_i, a_i, j_i\}$ are attributed to the image as a whole and characterize its state within the experiential now. They may be categorized as entities belonging to the lower level of AM-consciousness—the most detailed description of image dynamics accessible to the mind within time scales of the diachronic unit. AM-conscious perception of the variables $\{x_i, v_i, a_i, j_i\}$ is most essentially reflected in that the degree of attention focused on time difference in $\{x_i, v_i\}$ for the immediate past, immediate now, immediate future can accentuate or diminish the role of $\{a_i, j_i\}$.

In our further mathematical constructions we will

- refer to the quantities $\{x_i, v_i, a_i, j_i\}$ as the *phase variables* of a given mental image and
- assume that the *phase space*

$$\mathbb{R}_{[4]} = \Big[\{x_i, v_i, a_i, j_i\}\Big]$$

---

[11] In special cases like car driving (e.g., Lubashevsky and Morimura 2019; Lubashevsky 2017) there can be particular mechanisms enabling direct perception of $\{a_i, j_i\}$—the second and third order rates at which the properties of physical objects change in time.

can be used for completely specifying the current state of this mental image and its dynamics *within the experiential now.*

It is necessary to emphasize that:

- It is embodiment that enables the mind to operate with the image of a physical object in terms of variables characterizing the physical properties of the given object.
- The phase variables $\{x_i, v_i, a_i, j_i\}$ characterize the state of mental image within the diachronic unit—a basic time fragment of human temporal dimension which is of finite duration—rather than at a given point-like instant of objective time.
- Although the variables $\{v_i, a_i, j_i\}$ characterizing first, second, and third order rates at which the physical quantities of perceived object (reflected in $\{x_i\}$) change in time, these variables are not related to $\{x_i\}$ via the corresponding time derivatives. This relationship is restored on time scales exceeding substantially the duration of diachronic unit within the complex present.
- The image of physical object cannot be represented as a point in the phase space $\mathbb{R}_{[4]}$, it is a certain finite region with fuzzy boundaries—the spatial component of the space-time cloud describing the mental image of perceived physical object on space-time scales of the experiential now.

3.3.2 *Effective Trialism of Phenomenological Description*

Speaking about mental images of physical objects, we have to draw a distinction between their two aspects. First one concerns such images as just mental entities we deal with symbolically, in particular, using the corresponding names in describing visible objects around us. In this case, e.g., speaking about just a volleyball ball we imply a ball with certain properties such as size, weight, material, etc., i.e., the properties attributed to it on its own. Such properties, broadly speaking, are the same for all the objects of a given type and the corresponding mental image actually represents the whole class of these objects. However, how this ball is located in space with respect to other objects is not an issue involved in consideration.

Second aspect concerns the details characterizing only the given particular object individually. Here, to be specific, we focus on how its spatial location is represented in the mind. For example, a volleyball player must be able to estimate the position of the visible ball with respect to the net and its current velocity to select adequate actions. For a detailed discussion of these and related issues a reader may be referred to, e.g., Thomas (2014).

The present section is mainly confined to the second aspect, namely, how mental images of physical objects are embedded into the mental image of physical space-time structure.

The concept of embodied cognition accentuates the role of the unconscious processing of perceived information based on particular sensory modalities involving

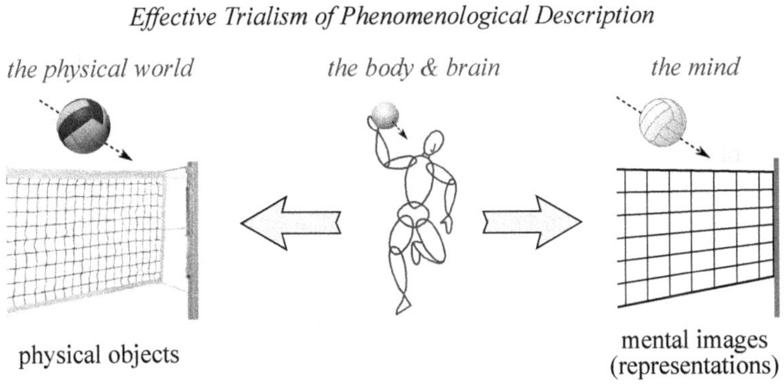

Fig. 3.3 Illustration of effective trialism reflecting the gist of embodied cognition

in their functioning the brain as well as the body. As discussed in Sect. 3.2.4 we turn to the interpretation of embodiment by Mahon and Caramazza (2008) taking an intermediate position between its weak and strong versions. In this case the phenomenological description of mental images and their dynamics has to deal with three complementary components—the physical (external) world, the neural network spreading over the whole body and responsible for unconscious processing of perceived information (the body-&-brain), and the mind. Each of these components is characterized by its own properties and partly exhibits individual dynamics. It is the mutual interaction between the three component that endows humans with complex behavior. To emphasize this aspect we introduce the term *effective trialism of phenomenological description* or just *effective trialism* (Fig. 3.3). In this sense the body-&-brain responsible for unconscious processing of information and the mind comprising various mental phenomena are two constituent components of human nature, at least, withing phenomenological approach.

It should be noted that the idea of trialism is not new. In particular, Cottingham (1985) proposed Cartesian trialism as a generalization of the mind-body dualism of Rene Descartes. In addition to the two substances of mind and body, Cottingham introduces a third constituent component—Sensation—belonging to the union of mind and body. Generalizing Cottingham's account of trialism, Njikeh (2019) replaces the component Sensation by the component called *Submind* comprising memories, sensation, emotions and reflexes. Kon and Miller (2015) put forward a novel methodology able to generalize the available accounts of temporal experiences and to reconcile bottom-up and top-down approaches. They introduce a three-level hierarchical architecture of the human *now* attributing its levels own properties of how we perceive the events. The bottom level represents objective time with point-like instants and physical events. The top level represents the temporal structure of subject's conscious experiences. The intermediate level represents the supervenience base of temporal experiences rooted in the corresponding neurophysiological processes in the brain and, maybe, the body.

Figure 3.3 illustrates the introduced concept of effective trialism. The three complementary components—the physical world, the body-&-brain, and the mind—are supposed to equipollently contribute to the formation and dynamics of mental images of physical object currently perceived, the images retained in and then retrieved from memory, or purely imaginary images.

As noted above, in this trio the body-&-brain component endows us with the ability to understand the spatio-temporal organization of external world and to decode adequately the sensory information about the mutual arrangement of perceived physical objects. As a result via the trial-and-error learning some mental image of the spatio-temporal structure of external world emerges in the mind. In particular, we acquire the ability to manipulate with mental images of physical objects in the imaginary space in a similar way as the corresponding physical objects can be manipulated (moved, turned, deformed, etc.) in reality (cf. Clark 2015). Mental rotation whose investigations were stimulated by Shepard and Metzler (1971, 1988), Vandenberg and Kuse (1978) illustrates such mental manipulations, for a general discussion of mental rotation and related issues a reader may be referred to Guillot et al. (2012), Thomas (2014).

The concept of embodied cognition proposes a plausible solution to the issue of how the mind possess the information about spatial structure of the external world and manage complex actions of the body. Embodied cognition is the mechanism via which the mental image of the space-time structure of external world emerges in the mind and reflect the characteristics properties of the real space-time structure. Moreover, for describing the mental image of this structure the collection of notions applicable to characterizing the real space-time structure of external world should be sufficient.

Summarizing this discussion of embodiment we accept the following statement.

Physical closure of mental-mental embedding: Mental operations with images of currently perceived physical objects or imaginary objects with respect to their mutual arrangement in the mental image of external world admits the complete description turning to the basic notions used for describing the physical properties of real objects in the external world. Put differently, in planning our actions aimed at reaching a certain goal and estimating possible changes in the external world caused by our actions we:

- operate with the mental images of real physical objects or imaginary ones,
- embed the mental images of these objects into the mental image of the external world and, so,
- create some imaginary spatio-temporal arrangement of object images which can changing with time,
- describe this spatio-temporal arrangement and its dynamics in terms that can be applied to characterize the external world.

The physical closure of mental-mental embedding means that this description is sufficient in principle, i.e., all the required notions and mathematical formalism can be constructed employing the basic notions of physical space-time continuum.

Here we do not imply that the mathematical models of physical objects are applicable to describing their mental images. The situation is just opposite as will be elucidated below. The given proposition means, in particular, that the mental-mental embedding can be described using the notion of spatial coordinates and their time derivatives of various order. Returning to the phenomenological description of human actions, the effective trialism actually proposes the two-component description of human actions dealing with:

- the external physical world (objective component),
- the inner human world as a set of mental images and their interrelations (subjective component),
- the possibility of describing mental images employing the basic notion like spatial coordinates and the instant rates of their changes in the real of physical world.

The latter feature is due to embodiment, which opens a gate to describing the interaction of the two components via actions and perceiving the results of actions.

3.3.3 Space-Time Cloud as Mathematical Eidos of Mental Image

In order to speak about mental-mental embedding we need a special term different from the terms "image" and "representation." These notions are widely used in conventional and scientific language and have many different aspects, which leads to confusion in our mathematical constructions. A new term is needed to refer to entities of mental-mental embedding in quantifying the spatio-temporal aspects of image arrangement rather than picture-like representations of physical objects perceived via sensory modalities.

For this reason we want to turn to the term *eidos* (pl. *eidoi*) which goes back to Plato's philosophy where eidoi are not sensibly perceptible entities, they are accessible only via the mind. We will use the term *mathematical eidos* for dealing with mathematical objects that describe the basic spatio-temporal properties the metal images exhibit in embedding into the mental image of space-time continuum. It is necessary to emphasize that a mathematical eidos is not a mental image of a certain physical object as it is perceived via sensory modalities. It represents how we mentally *evaluate* quantitatively or semi-quantitatively the spatio-temporal relations of a given object—currently observed or purely imaginary one—with respect to the surrounding objects. In this sense the mathematical eidos should be categorized as the conceptual type mental representation which has no experiential ("what-it's-like") features (see, e.g., Pitt 2020; Siegel 2016, for discussion of mental operations with such objects and related philosophycal issues).

Our account implies that the notion of mathematical eidos includes in itself two particular entities: *material point* and *space-time cloud*. Below we want to elucidate the common and distinct features of these entities which are illustrated in Fig. 3.4.

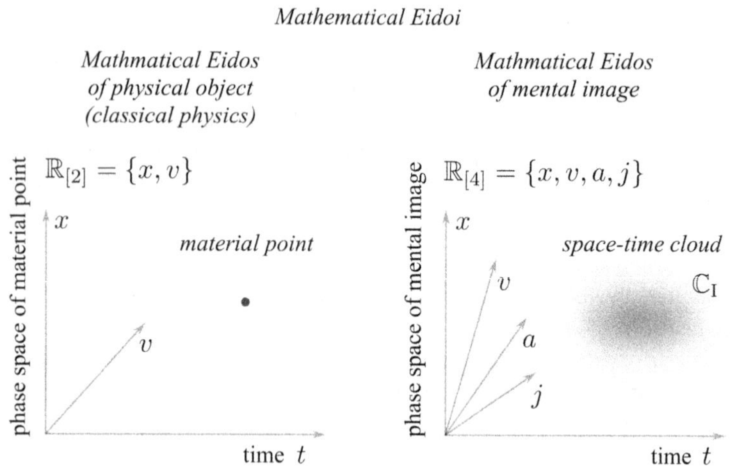

Fig. 3.4 The concept of mathematical eidoi introduced to describe the embedding of mental images into the physical space or, speaking more strictly, into the mental image of the physical space-time continuum. The result of embedding is the space-time cloud of object mental image $\mathbb{C}_\mathrm{I}$

It should be pointed out that in the previous sections the notion of space-time cloud was noted in different contexts but here this notion is specified more precisely.

Let us, at first, list the properties uniting the mathematical eidoi of physical objects and mental images into one category as well as differentiate them.

- *Common feature:* Both the mathematical eidoi can be described in terms of the same collection of variables including:
  - the spatial coordinates and temporal variable representing the dimensions of the physical space-time continuum;
  - the instant velocity, acceleration, etc., i.e., various order rates characterizing the temporal variations in the spatial coordinates of a point formally fixed in some manner;
  - similar quantities and their temporal rates that are attributed to a given physical object in describing, e.g., its shape, weight, surface luminance, etc.
- *Distinct features:* Although the material point and the space-time cloud are entities admitting a description within one language they are entirely distinct from each other in the following:

  1. *Localization degree*, i.e., certainty/uncertainty in particular magnitudes[12] of quantities characterizing the corresponding entity.

[12] Here a magnitude is understood as a number assigned to a quantity characterizing some property of an entity, so, this quantity can be compared with other similar quantities. For example, it is the particular volume of an object or the length of a vector.

- *The material point* has no dimensions. Namely, first, its position in space, velocity, etc. as well as physical properties attributed to it (e.g., the mass and electric charge) take particular magnitudes. Second, what is the essence of temporal localization, the state of the physical object the material point represents is completely determined at any point-like instant of time.
- *The space-time cloud* has finite spatial and temporal dimensions. Namely, first, the mental quantitative estimate of perceived object properties (including, the spatial localization, the velocity of motion, etc. as well as other physical properties attributed to the perceived object in the mind) bears some uncertainty intrinsic to human cognition. Second, the simultaneity of two events is also characterized by some uncertainty.

2. *Influence of subjectivity*, i.e., the dependence on or independence from human mental states.

   - The properties of *the material point* representing some physical object do not depend on external observer.
   - The properties of *the space-time cloud* depend substantially on the mental states of observer. Moreover, different types of consciousness can contributed to this dependence individually.

3. *The structure of phase space*, put differently, the number of independent dimensions characterizing the current state of affairs for a given entity[13] (Fig. 3.4).

   - For *the material point* the current state of the corresponding physical object considered within Newtonian mechanics is completely specified[14] as a point of the two-component phase space $\mathbb{R}_{[2]} = \{x, v\}$ comprising the spatial position (x) and instant velocity (v) of the given material point.
   - For *the space-time cloud* in order to describe the current state of affairs reduced to the analysis of its spatial characteristics generally four-component phase space is required. Dealing with the mental-mental embedding of a point-like physical object these space $\mathbb{R}_{[4]} = \{x, v, a, j\}$ comprises the spatial position (x), instant velocity (v), acceleration (a), and jerk (j) of the perceived object.[15]

As comments on the listed features, we want to note that, first, the common feature stems directly from the principle of physical closure of mental-mental embedding (Sect. 3.3.2). In other words, due to embodiment our imagination is deeply rooted

[13] The meaning of currentness for the space-time cloud which is not localized in the temporal dimension is clarified below.

[14] If the other properties of a given physical object represented by the material point are fixed, i.e., do not change with time in independently, at least, partly, of the spatial position of material point and its velocity.

[15] In fact there various representations of the given four-component phase space $\mathbb{R}_{[4]}$ where, e.g., the jerk j may be replaced by another variable describing human intentional actions in more direct way. This case met in describing driver behavior within the car-following scenario has been analyzed by Lubashevsky (2017), Lubashevsky and Morimura (2019).

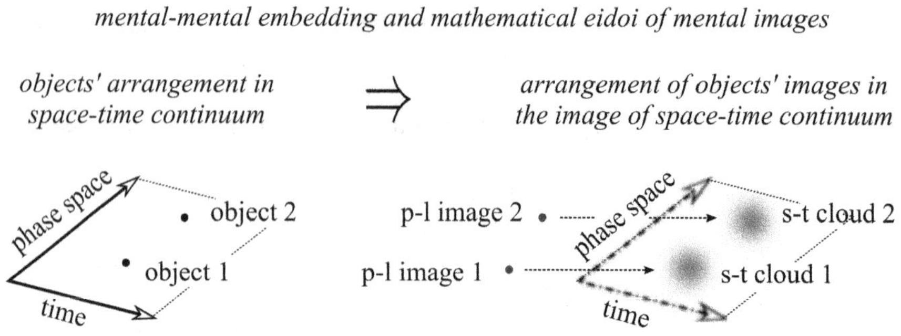

Fig. 3.5 Illustration of mental-mental embedding via formation of picture-like representations (p-l images) of physical objects in the mind and their mental embedding in the image of space-time continuum as space-time clouds (s-t clouds)

in physical reality. As should be emphasized, this type similarity does not lead to the conclusion that the space-time cloud is just a set (uncountable set) of material points. The matter is that the material point is the mathematical eidos of a certain physical object, so, the material point bears some properties of this object. The space-time cloud does formally contain a set (fuzzy set) of points belonging to the physical space-time continuum. However no physical properties can be attributed to these points individually. The space-time cloud as an *integral entity* bears all the physical properties perceived via sensory modalities or just attributed to some imaginary object. Put differently, the space-time cloud is the basic mathematical element representing a given image in the external world which cannot be divided into smaller parts bearing separately some *physical* properties.

Second, turning to the distinct features of space-time cloud, it is necessary to emphasize once more that the uncertainty in quantitative description of the corresponding image concerns the physical variables, in particular, objective time. These variables characterize the mental embedding of image into the external world rather then the image as it arises in the mind, which is illustrated in Fig. 3.5.

In the case shown in Fig. 3.5, two distant objects are clearly perceived and there is no spatial overlapping between their images in the mind. The situation changes drastically if we need to make some decision based on our mental estimate of the distance between them, e.g., respond in some way when the two visible objects come close to each other. Because there are no spatial scales fixed and clearly perceived in the mind, the visible distance between these objects has to be compared with some imaginary spatial scale. This comparison (*i*) is always approximate and its uncertainty is often rather high except for special cases and (*ii*) depends substantially on the current mental state, in particular, on attention focused on this issue. This uncertainty underlies the introduction of the space-time cloud with the blur depending on the subject mental states.

It is necessary to emphasize that there are two different factors contributing to the uncertainty of space-time cloud. The first one is caused by physiology of sensory modalities. The second factor is related to the deliberate, purely mental analysis of perceived distance. In this case the uncertainty is not related to any visible phenomenon and must be attributed directly to the image mental-mental embedding.

Actually the same can be said with respect to the uncertainty in objective time. Simultaneity of two events in the framework of the experiential now may be treated as the uncertainty caused by the physiology of sensory modalities. Two events clear perceived to have occurred at different instants of time, nevertheless, can be regarded as simultaneous if the time difference in their occurrence does not contribute to analyzed phenomena or, for some reason, such events have to be categorized as simultaneous within a certain context. Naturally, the latter type uncertainty can be attributed only to the space-time cloud and depends essentially on the subject mental state. Exactly in this sense the currentness of states of affairs is understood in describing the phase space of the mental-mental embedding.

Brief digression: We want to note the conception of fuzziness put forward by Snyder (1947) in order to specify the position of a physical object in space-time continuum. Based on this conception Doplicher et al. (1995) have argued for revising the notion of space-time at short length scales and given strong arguments in favor of a noncommutative geometry in developing, e.g., a theory of quantum gravity, for a review of this idea a reader may be referred to Madore (1997), Sidharth (2005), Scholtz (2018). In our account we prefer to speak about the fuzziness of mathematical eidos rather than the uncertainty of space-time metric because the fuzziness of mental-mental embedding is caused by mental operations with images rather than formal points of the space-time continuum. Besides, we want to note that the space-time cloud should not be treated as a standard object of fuzzy sets; it is not a fuzzy member-function representing an image in space. Indeed, for example, the comparison of two entities of mental-mental embedding must be based on the overlapping of the corresponding space-time clouds, which is closer to the gist of quantum mechanics rather then the formalism of fuzzy linear spaces. Nevertheless the relationship between the notion of space-time cloud and the conception of fuzziness may be a matter of further investigations because of a plausibly common aspects of quantum mechanics and the conception of fuzziness (see, e.g., Schmitt et al. 2009; Pykacz 2013 for characteristic examples).

Now we want to discuss the dimensional features of space-time clouds. In the previous sections we touched upon this issue, in particular, in Sect. 3.3.1 we introduced the four-component phase space $\mathbb{R}_{[4]} = \{x, v, a, j\}$ of mental images rooted in the properties of diachronic unit (Sect. 2.6.6). Nevertheless some additional comments are required to elucidate the structure of space-time cloud meeting the completeness criterion.

It would seem, at first glance, that the space-time cloud is composed of two dimensions, spatial and temporal ones; however, the situation is more intricate. The space-time cloud has to represent the uncertainties in quantitative evaluation of all the properties attributed in the mind to the image of physical object. Among these properties we note, first, the pair of the spatial position x of perceived (or imaginary) object and its velocity v. The two properties are just given to the mind via the unconscious processing of information. In this sense the object position and velocity are entities of E-consciousness and, thereby, should be treated as mutually independent quantities *within* the experiential now. This independence can be extended up to the scales characterizing the temporal uncertainty of space-time cloud including the deliberate evaluation of time difference within which the states of perceived object may or have to be treated as simultaneous. Indeed, let for some reasons we do not intend to discriminate the states of perceived object according their temporal order within a certain time interval or just cannot do this. In this case, it seems rather doubtful that we will do this with respect to time variations in the object position to find the the velocity of its motion.

The second pair of variables comprises the acceleration a and jerk j of perceived object. Whereas its spatial position x and velocity v are quantities appearing in the mind via the unconscious processing of information within the synchronic unit, the acceleration a and jerk j become accessible to the mind via the information processing within the diachronic unit (Sect. 2.6.6). The information processing at the level of the diachronic unit is caused by its tripartite structure and, strictly speaking, should be categorized as the lower level of AM-consciousness. However, the conscious processing of information within the diachronic unit requires high concentration on the difference in the motion velocities perceived within the immediate past, the immediate now, and the immediate future. Thereby via embodiment, including the learning process, this information processing can be implemented in a manner similar to unconscious processes. Namely, only the final result—the perceived value of acceleration and, maybe, jerk—become accessible to the mind.[16] For this reason, we have included the acceleration and jerk into the list of mutually independent phase variables and treat them as the quantities of E-consciousness (Sect. 3.2.5). Naturally, the contribution of the acceleration and jerk to how a perceived object is represented in the mind has to be much more sensitive to attention in comparison with that of position and velocity.

Therefore, in describing the spatial dimensions of space-time cloud, we have to take into account at least four components. They are the spatial position x of perceived objects, its velocity v, acceleration a, and jerk j (Fig. 3.4) characterized as quantities of E-consciousness. In this sense, the four quantities may be treated as mutually independent phase variables of the complete four-component phase space $\mathbb{R}_{[4]} = \{x, v, a, j\}$. As a result the mental-mental embedding has to be described in terms of space-time clouds possessing the structure $\mathbb{R}_{[4]} \times t$. Each dimension of this structure bears the individual information about uncertainty in the mental evaluation of the corresponding phase variable.

It is worthy of noting once more that the four phase variables are just perceived and their quantitative interrelation are not recognized within the diachronic unit. Only on scales of the complex present their interrelation is restored.

---

[16] There can be other mechanisms governing E-consciousness that provide us with the direct sense of acceleration, e.g., it is inertia in car-driving. Besides, in car-driving the jerk is implicitly related to the car pedal position and its time variation, which also admit interpretation as a embodied mechanism responsible for the sensation of jerk (Lubashevsky 2017; Lubashevsky and Morimura 2019).

To summarize the aforesaid we may posit the following for the mental image of physical object currently perceived, retained in and then retrieved from memory or purely imaginary one.

Space-time cloud of mental image: Mental evaluation of spacial and temporal properties characterizing the observed arrangement of physical objects in the space-time continuum deals with the mathematical eidos—the mental embedding of object image into the mental image of the physical space-time continuum. We call this type mathematical eidos *space-time cloud of object mental image* $\mathbb{C}_{\mathrm{I}}$ to accentuate that the given mathematical eidos:

- may be conceived of as some finite size region with blurred (fuzzy) boundaries belonging to the space that combines in itself the extended phase space $\mathbb{R}_{[4]}$ with temporal dimension, $\mathbb{R}_{[4]} \times t$,
- may possess a certain internal structure which can change with time depending on changes in (*i*) the external world and (*ii*) subject mental state;
- play the role of some entity distributed in a certain region of the space $\mathbb{R}_{[4]} \times t$ and representing the related physical object in this space. In particular, in inferring the similarity of two observed physical objects in their properties determined by the mental-mental embedding, the overlapping between the corresponding eidoi (space-time clouds) can be used as a quantitative measure.[17]

The space-time cloud is one of the basic elements in describing intentional, goal-oriented actions. The efficiency or adequacy of these actions is estimated via observing their particular results causing or reflecting changes in environment. The results of these actions can be *quantified* (but not just perceived) only via time variations in the properties of the corresponding space-time clouds. In this sense the space-time cloud plays the same role as the material point in Newtonian mechanics.

3.3.4 Space-Time Cloud of Action Strategy

The notion of *space-time cloud of action strategy* (or just *action strategy cloud* for short) takes the central position in our constructions. So the purpose of the present section is to elucidate this notion on its own and in relation to the space-time clouds of objects involved in human actions belonging to the given action strategy.

There are various action strategies in the mind and each of them admits a multitude of individual implementations in physical reality. Let us, at first, discuss the properties of one implementation of a given action strategy.

---

[17] It should be noted that even in a case when the degree of cloud overlapping is high and thus two observed physical objects must be categorized as equivalent in properties determined by mental-mental embedding, their picture-like mental images can be entirely distinct from each other.

Implementation of Action Strategy

A particular implementation AS_p of any action strategy $\mathcal{AS}$ in the external world (physical reality) may be conceived of as a certain temporally ordered sequence of states which the physical system under consideration passes step-by-step. Namely,

$$\mathrm{AS}_p = \bigcup_{\alpha \in \{\alpha\}_t} S_\alpha \,, \tag{3.1}$$

where the indices $\{\alpha\}$ are ordered such that the corresponding states S_α be ordered in objective time. In this representation several aspects should be explained in more details.

1. As far as the states of physical system are concerned, we note the following:

- The physical system in issue is assumed to combine all the objects either involved in human actions or able to influence them. In this case, the system state S_α at a given instant of objective time t_α represents the corresponding physical states of all the objects making up the system.

2. Representation (3.1) deals with the physical system states. These states determine the implementation of actions forming the action strategy $\mathcal{AS}$. This feature is the reason to regard Exp. (3.1) as action strategy implementation. Namely:

- The particular implementation of any action is determined completely by the transition between two system states which form a pair of neighboring elements in sequence (3.1).

By way of example, Fig. 3.6 illustrates a sequence of states representing some action strategy. These state admit a description dealing with the spatial coordinates of physical objects x and their velocities $v = dx/dt$ as the phase variables. Besides, the time dependence of spatial coordinates $x(t)$ is smooth. Thereby when it is fixed as the parametric representation of the corresponding action strategy $\mathrm{AS}_p(t)$, the phase states of the physical system (as well as the implementation of the corresponding actions) are completely determined. Indeed, in this case, for physical objects the sequence of their states $\mathrm{AS}_p(t) = \{x,\ v = dx/dt\}_t$ is determined completely by the time dependence $\{x(t)\}$. The same concerns the actions' implementation which should admit the parametrization $\{x, dx/dt, d^2x/dt^2, \ldots\}_t$. Put differently, for physical systems illustrated in Fig. 3.6 the smooth path $\{x(t)\}$ of system motion is the only one independent characteristic of action strategy implementation. All the other characteristics of such a system, e.g. $\{v = dx/dt\}_t$, are the derivatives of the path $\{x(t)\}$.

3. Any action strategy is evaluated from the standpoint of the experiential now and every one of its possible implementations contains temporal parts belonging to the connected past, the experiential now, and the connected future. It is shown in Fig. 3.6 using different colors for the temporal parts. Thereby:

- Any physical implementation of action strategy, i.e., sequence (3.1) contains phase states:

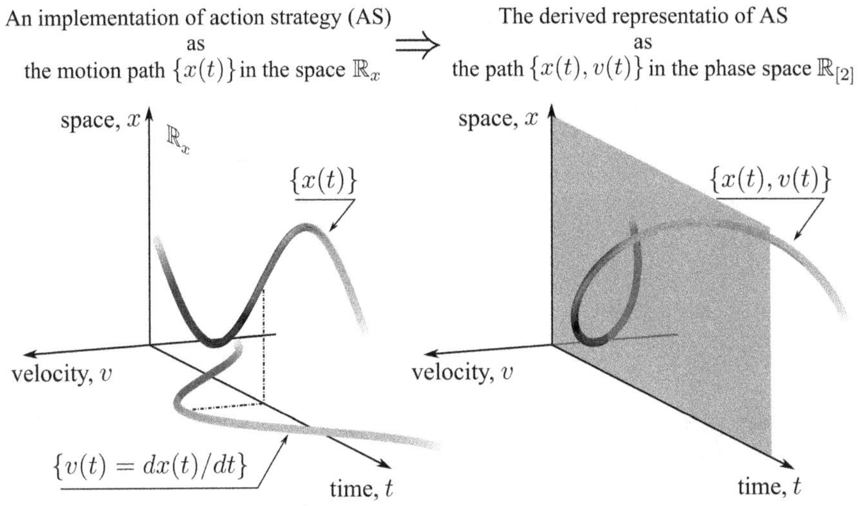

Fig. 3.6 One of possible implementations of a given action strategy in physical reality. *Left fragment:* The action strategy admitting representation as the path $\{x(t)\}$ of physical system motion in the space $\mathbb{R}_x$. *Right fragment:* the trajectory $\{x(t), v(t)\}$ of system motion in the phase space $\mathbb{R}_{[2]} = \{x, v\}$ which is the derivative of the path $\{x(t)\}$

- that the system occupied in the connected past (which is retained in memory) or just imagined to be occupied (the corresponding actions are also either really taken in the connected past or purely imaginary ones);
- comprising the state where the system is currently located and the states to those the system can transfer from the current one within the experiential now;[18]
- that the system is anticipate (or imagined) to pass through in the connected future.

The given temporal structure of action strategy implementation enables us to deal with different implementations of one action strategy as a collection of entities existing in the mind simultaneously and, so, available for the choice between them.

Multitude of Action Strategy Implementations

Any action strategy $\mathcal{AS}$ admits a multitude $\{\mathrm{AS}_p\}$ of particular implementations in physical reality. Figure 3.7 illustrates this multitude as a bundle of implementations close to one another in property.

[18] These transitions between different phase states can be implemented in reality only due to the finite duration of experiential now. In the mind, where all the events within the experiential now may be treated as simultaneous, the given transitions take the step-wise form. In this case for their description an extended phase space is required, e.g., $\mathbb{R}_{[4]}$, where the jerk—the phase variable representing the highest order time derivative of system spatial position—is supposed to be the control parameter affected directly by human actions and so is allowed to exhibit step-wise variations.

Multitude of implementations of one action strategy

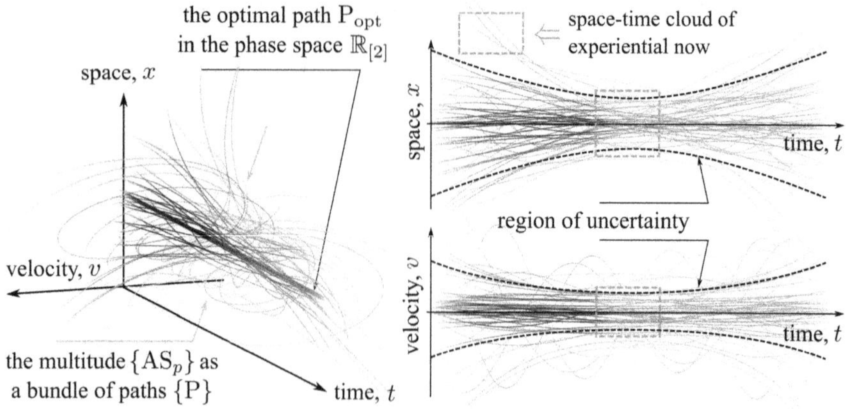

Fig. 3.7 Illustration of implementation multitude $\{AS_p\}$ for an action strategy $\mathcal{AS}$ aimed at governing the motion of a physical system admitting a description of its phase states in terms of continuum state variables comprising the position x of the system in space and its instant velocity v. *Left fragment:* The implementation multitude $\{AS_p\}$ as a bundle of paths P each of them depicts possible system motion in the space-time continuum $\mathbb{R}_{[2]} \times t$. *Right fragment:* The projection of the path bundle $\{P\}$ onto the planes $\{x, t\}$ and $\{v, t\}$, which visualizes the dependence of the bundle xv-dimensions on the time lag between the experiential now and time moments of the connected past and the connected future

In the multitude $\{AS_p\}$ of implementations there can be a certain optimal version AS_{opt} or a collection of such versions $\{AS_{opt}\}$ singled out from the other implementations using some criterion of optimality. However, the mind is unable to disentangle these optimal versions from other similar implementations. Moreover the mind cannot unambiguously and accurately trace the relationship between the past and the future for each of the possible implementations. Let us elucidate these propositions turning to Fig. 3.7.

By way of example, we consider a physical system admitting a description of its phase states in terms of the spacial position x and the instant velocity v—the continuous state variables. In this case, by virtue of Exp. (3.1), for any strategy $\mathcal{AS}$ of actions aimed at governing the dynamics of this system its possible implementations can be represented as paths $\{P\}$ of system motion in the phase space $\mathbb{R}_{[2]} = \{x, v\}$. The left fragment in Fig. 3.7 depicts the implementation multitude $\{AS_p\}$ of action strategy $\mathcal{AS}$ as the bundle of paths $\{P\}$ in $\mathbb{R}_{[2]} \times t$. Each one of these paths contains three temporal parts corresponding to the connected past (P^-, shown in blue), the experiential now (P^0, shown in orange), and the connected future (P^+, shown in green), respectively. The three parts are merged together and treated as a single entity $P = \{P^-, P^0, P^+\} = \{x(t), v(t)\}_{t_{CP}}$ in evaluating the action strategy $\mathcal{AS}$ based on its possible implementations. Here the subscript t_{CP} denotes that the analyzed time interval corresponds to the complex present.

In the given case, all the paths of the bundle {P} are similar in property, which is the reason to consider them equivalent[19] and to regard in the mind the bundle {P} as a single entity. The latter seems to be not just a reasonable manner but the only possible way of cooping with the implementation multitude $\{AS_p\}$ in the mind. Indeed, the mind is unable to discriminate between two similar paths reliably because of spatial and temporal uncertainty in identifying points of the space-time continuum. Moreover, this uncertainty becomes more pronounced when the point identification is based on memory or imagination, which is the case when the temporal parts P^- and P^+ are involved in consideration (Fig. 3.7). For the same reason, as clear seen in the right fragment of Fig. 3.7, in dealing with the bundle of paths {P} for any element P of them the unambiguous uniting $\{P^-, P^0, P^+\}$ of its temporal parts is not feasible.

In the scope of present discussion we may confine it, without loss of generality, to the case when there is only one optimal implementation AS_{opt} of the strategy $\mathcal{AS}$. For the system shown in Fig. 3.7 it is the optimal path P_{opt} "invisible" to the mind in the sense that the mind cannot single it out from the bundle {P} of all the possible paths representing the action strategy $\mathcal{AS}$. Nevertheless, the optimal path P_{opt} arises in the mind in a certain hidden way. In categorizing a given path $P = \{x(t), v(t)\}_{t_{CP}}$ of system motion as a possible implementation of the strategy $\mathcal{AS}$ we implicitly compare it with the optimal path $P_{opt} = \{x_{opt}(t), v_{opt}(t)\}_{t_{CP}}$. Namely, the closer the paths P, P_{opt} to each other, the higher the probability of accepting the path P as an implementation of $\mathcal{AS}$. However, even in the case $P = P_{opt}$ this probability has to be less than one because of the mind inability to discriminate two similar paths precisely. Thereby this comparison procedure, on the one hand, singles out the optimal path P_{op} from the other possible implementations of strategy $\mathcal{AS}$, on the other hand, remains it hidden from the mind.

Summarizing the aforesaid about the action strategy implementations we want to underling that:

- Every action strategy $\mathcal{AS}$ as a mental entity admits a multitude $\{AS_p\}$ of possible implementations in physical reality which are treated as equivalent. However, the mind is unable (*i*) to discriminate individual implementations between one another unambiguously and (*ii*) to match reliably the temporal parts of the same implementation. For this reason the only possible way to cope with the implementation

[19] As clear from this description the equivalence of two paths P^1 and P^2 is understood as some fuzzy relation, which is also reflected in the boundary blurring of action strategy clouds. In this sense the same fuzzy relation is the intrinsic property of path equivalence and the structure of action strategy clouds. Its specific description will be elucidated in the further constructions; here we confine the discussion to the general gist of fuzziness conception. Besides, it should be noted that in the classic theory of fuzzy sets the notion of fuzzy equivalence has been also introduced and analyzed from various standpoints (e.g., Zadeh 1971; Chakraborty and Das 1983a, b; Murali 1989; Ovchinnikov 1991; Ćirić et al. 2007; Li and He 2014); for a review a reader may be referred to Ovchinnikov (2000), Boixader et al. (2000). However, whether the formalism based on space-time clouds with their overlapping as the measure of object equivalence can be reduced to the fuzzy set formalism turning to the notion of member functions requires an individual investigation.

multitude $\{AS_p\}$ of action strategy $\mathcal{AS}$ in the mind is to regard it as a certain single entity with properties attributed to it as a whole.

- Thereby, a description of action strategies cannot be based directly on the properties of their implementations treated separately. Put differently, the properties of action strategies are based on the individual *physical* properties of their implementations but are not some arithmetic sum of them.
- For every action strategy $\mathcal{AS}$ there can be a collection of its optimal implementations $\{AS_{opt}\}$ which, however, are hidden from the mind. Therefore the optimal implementations $\{AS_{opt}\}$ of action strategy $\mathcal{AS}$ may be involved in mental operations only implicitly.

If two implementations differ essentially from each other, then they are not equivalent and represent different action strategies. If some implementation is not too different from the bundle of implementations of one strategy and at the same time it is not too different from the bundle of implementations of another strategy, then it can be assigned to both the strategies, which underlies the mechanism of the overlapping between action strategies.

Because we cannot operate with individual implementations but can do this with their groups, we turn to the concept of space-time clouds.

Space-Time Cloud as the Mathematical Eidos of Action Strategy

In Sect. 3.3.3 we discussed the space-time cloud of a mental image representing some physical object in the mind. This type space-time clouds have initial sources being physical objects which we see in reality or imagine in the mind. In the latter case an imaginary object inherits its properties from real objects similar to it.

The space-time cloud of action strategy is of another type. The source of action strategy cloud is the motion of some physical object in space and time whose temporal parts belonging to the past and the future are memorized or/and anticipated. Put differently, the source of action strategy cloud is a path of motion of physical object, which without human memory and the ability to anticipate cannot be treated as an object with own properties.

To simplify understanding the following constructions we want, beforehand, to emphasize that action strategies to be analyzed and their space-time clouds are considered within the complex present. In this case the individual structure of action strategies is relatively simple, it is a sequence of actions following one another, i.e., with a regular temporal order (Sect. 3.1). Leaping ahead, we note that the space-time cloud of some action strategy can be conceived of as its optimal implementation(s) with the cloud of uncertainty related to human perception and evaluation of surrounding objects and their motion.

The temporal structure of action strategies and the resulting temporal structure of their clouds are inherited from the complex present. Both the structures contain temporal parts matching the connected past, the experiential now, and the connected future, in Fig. 3.8 they are depicted in blue, orange, and green, respectively. These temporal parts have many common properties. So to elucidate some property of the action strategy cloud characterizing it as a whole, we may focus on its temporal part where this property is most pronounced.

To be specific we accept the following definition.

Space-time cloud of action strategy: The space-time cloud $\mathbb{C}_{\mathrm{AS}}$ of an action strategy $\mathcal{AS}$ is a mathematical eidos representing how the given action strategy is embedded into the mental image of the physical space-time continuum. In the case under consideration this mental-mental embedding is confined to the complex present. Here the action strategy $\mathcal{AS}$ is understood as the multitude of its implementations in physical reality regarded as equivalent.

To discuss particular properties of the action strategy cloud, it is necessary, at first, to clarify the dimensions of space-time continuum that are required to specify the action strategy embedding. As elucidated in Sects. 2.5, 2.6 and also summarized in Sect. 2.7, phase variables of four types are required to describe human perception of physical object motion in the general case. They are the spacial position x, instant velocity $v = dx/dt$, acceleration $a = d^2x/dt^2$, and jerk $j = d^3x/dt^3$. Within the experiential now these motion characteristics are evaluated via sensory modalities as mutually independent quantities. Then the resulting magnitudes are attributed to the mental image of perceived object. Therefore, the extended phase space $\mathbb{R}_{[4]} = \{x, v, a, j\}$ is required for the mental-mental embedding of action strategies within the experiential now.

The state of object motion mentally evaluated in this way can be retained in memory and used further for evaluating events anticipated in the future or imagined to occur previously. It becomes feasible due to embodied cognition based on the memorizing and learning of physical actions and their mental evaluation. Naturally, the degree of uncertainty in evaluating the imaginary states of object motion should be higher than this uncertainty within the experiential now. It is reflected in the shape of the action strategy cloud, whose characteristic width should increase with the time lag between a given instant of time and the experiential now (Fig. 3.8).

Therefore for action strategies (as well as for the images of physical objects) the spatial component of the mental-mental embedding has to include the extended phase space $\mathbb{R}_{[4]}$. This phase space with objective time t form the desired description $\mathbb{R}_{[4]} \times t$ of space-time continuum (Fig. 3.8).

Here, once more, we want to emphasize the difference between the spatial position x of a physical object, its velocity v, acceleration a, and jerk j as physical quantities and their magnitudes perceived via sensory modalities and retained in memory. As physical quantities they are closely related to one another, in particular, the velocity $v = dx/dt$, acceleration $a = d^2/dt^2$, and jerk $j = d^3/dt^3$ are the derivatives of the object motion path $\{x(t)\}$ in the space $\mathbb{R}_x$. The mental image of the corresponding object is characterized by these state variables as independent quantities. This independence means that the perceived magnitudes of the given quantities are just given to the mind via E-consciousness. Moreover, the perceived magnitudes are

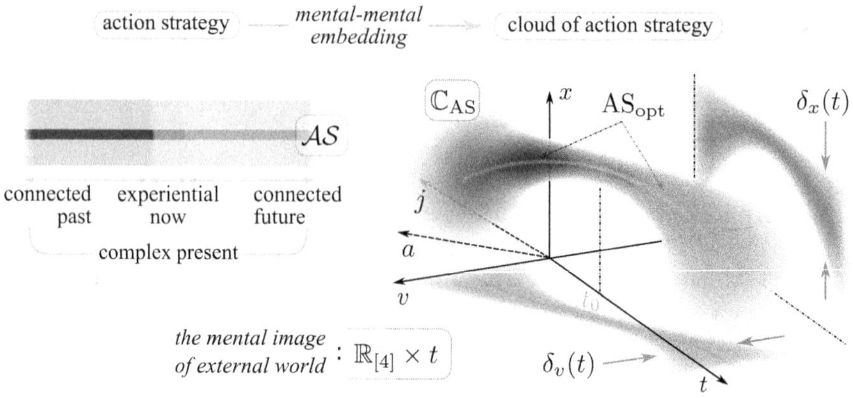

Fig. 3.8 The space-time cloud $\mathbb{C}_{\text{AS}}$ of an action strategy $\mathcal{AS}$ as the mathematical eidos, i.e., the mental-mental embedding of a given action strategy as a mental entity into the mental image $\mathbb{R}_{[4]} \times t$ of the space-time continuum for a system illustrated in Fig. 3.6. The space-time continuum is represented by the dimensions comprising the spacial coordinates x, the instant velocity v of motion, the acceleration a, and jerk j, which make up the extended phase space $\mathbb{R}_{[4]} = \{x, v, a, j\}$, together with objective time t. The projections of the space-time cloud $\mathbb{C}_{\text{AS}}$ onto the planes $\{x, t\}$ and $\{v, t\}$ are shown in gray, AS_{opt} is the optimal implementation of the cloud $\mathbb{C}_{\text{AS}}$ inaccessible to the mind

characterized by substantial uncertainty. Thereby, the state of object motion as it is experienced is represented in the space-time continuum $\mathbb{R}_{[4]} \times t$ as a certain fuzzy region. It is necessary to point out that the dimensions of the extended phase space $\mathbb{R}_{[4]} = \{x, v, a, j\}$ match the perceived magnitudes of the corresponding characteristics of real motion rather than their specific values. The latter are actually hidden from the mind.

There is another difference between the perceived magnitudes and the corresponding physical values that is reflected in the non-linearity of psychophysical laws (Sects. 1.3 and 1.4). A detailed description of this issue, however, is outside the scope of the present book.

The aforesaid gives rise to the representation of action strategy cloud illustrated in Fig. 3.8. Its construction comprises several steps:

- For every element AS_p in the implementation multitude $\{\text{AS}_p\}$ of action strategy $\mathcal{AS}$—it is represented by sequence (3.1) of physical object states:

- Some state is selected and its mental image is embedded into the space-time continuum $\mathbb{R}_{[4]} \times t$. This mental-mental embedding maps the physical object in the given state onto some fuzzy region in $\mathbb{R}_{[4]} \times t$, which is described in details in Sect. 3.3.3.
- Then the regions obtained in this way for all the object states from the implementation AS_p are merged together, forming the individual space-time cloud of AS_p.

- o Finally, all the individual clouds of action strategy implementations from the multitude $\{\mathrm{AS}_p\}$ are merged together, forming the space-time cloud $\mathbb{C}_{\mathrm{AS}}$ of the action strategy $\mathcal{AS}$. This merging is due to that there is no reason to draw a distinction between different implementations whose contribution to the action strategy $\mathcal{AS}$ is equivalent.

The space-time cloud $\mathbb{C}_{\mathrm{AS}}$ obtained in this way is regarded as the mental-mental embedding of the action strategy $\mathcal{AS}$ into the mental space-time continuum $\mathbb{R}_{[4]} \times t$. In other words, $\mathbb{C}_{\mathrm{AS}}$ is the mathematical eidos of the action strategy $\mathcal{AS}$.

By virtue of employed procedure the space-time cloud $\mathbb{C}_{\mathrm{AS}}$ has no fine structure reflecting particular details characterizing an individual implementation AS_p of the action strategy $\mathcal{AS}$. Any path within the region $\mathbb{C}_{\mathrm{AS}}$ may be treated as a particular implementation of the action strategy $\mathcal{AS}$. The latter does not imply that such paths can contain fragments bearing sharp changes in the system dynamics not caused by external factors or particular intentions of subjects governing the system dynamics. Due to the perceived magnitudes of velocity, acceleration, and jerk such sharp variations—actually step-wise variations on the scales of the experiential now—cannot belong to $\mathbb{C}_{\mathrm{AS}}$ as a region of $\mathbb{R}_{[4]} \times t$.

Summarizing this discussion we can characterize the action strategy cloud as follows.

- Within quantitative description of action strategy $\mathcal{AS}$ its implementation in physical reality is mentally represented in the form of some region $\mathbb{C}_{\mathrm{AS}}$ with blurred boundaries in the space-time continuum $\mathbb{R}_{[4]} \times t$. This region is referred to as the space-time cloud $\mathbb{C}_{\mathrm{AS}}$ of action strategy $\mathcal{AS}$. It has three constituent temporal parts matching the connected past, the experiential now, and the connected future (Fig. 3.8).
- The action strategy cloud $\mathbb{C}_{\mathrm{AS}}$ has no fine structure at the micro-level whose scales could represent the difference between equivalent implementations of $\mathcal{AS}$. The temporal structure attributed to $\mathbb{C}_{\mathrm{AS}}$ as a whole entity reflects:
  - o the general difference between its temporal parts matching the connected past and the connected future, on the one side, and the experiential now, on the other side. In particular, for a fixed instant of time t the dimensions of the corresponding cross-section should depend essentially on the time lag between the given instant of time t and the experiential now. These dimensions reflect the uncertainty in the mental evaluation of perceived physical properties; in Fig. 3.8 it is $\delta_x(t)$ and $\delta_v(t)$.
  - o the contribution of several factors to the uncertainty in the mental evaluation of physical properties characterizing action strategies. In particular:

- the uncertainty characterizing the perception of physical objects, which is reflected in the properties of space-time cloud of mental images;
- variations in the corresponding properties of object motion caused by difference among the particular implementations of action strategy $\mathcal{AS}$; due to the equivalence of these implementations the induced variations may be treated as some kind of uncertainty.

Whence it follows that the space-time cloud of action strategy is not reduced to a certain collection of space-time clouds representing mental-mental embedding of object images.

- As far as the temporal part of $\mathbb{C}_{\mathrm{AS}}$ belonging to the connected past is concerned with, the uncertainty in the evaluation of perceived properties depends also on:

- whether the physical system passed through the corresponding phase state and this event has been retained in memory or
- its is a purely imaginary event included in consideration for the continuity of the action strategy.

In other words, the temporal structure of $\mathbb{C}_{\mathrm{AS}}$ bears also the information about the action strategy realization as a collection of actions taken in reality and purely imaginary actions.

- In selecting an action strategy for implementation or/and evaluating its efficiency, the mind operates with its space-time cloud rather than its individual implementation in physical reality. Therefore in describing the properties of action strategy $\mathcal{AS}$ accessible to the mind via AM-consciousness the space-time cloud $\mathbb{C}_{\mathrm{AS}}$ must be the main mathematical element representing $\mathcal{AS}$ in the space-time continuum.

The listed features of action strategy cloud are the main reason to speak about the action strategy cloud as a certain type of mathematical eidos. For example, let two particular sequences of system states, AS_p^1 and AS_p^2, (see Exp. 3.1) are regarded as the specific implementations of two action strategies $\mathcal{AS}^1$ and $\mathcal{AS}^2$. Then whether the two action strategies are different or identical should be determined based on the analysis of the corresponding space-time clouds. Namely, at the first step, two space-time clouds $\mathbb{C}_{\mathrm{AS}}^1$ and $\mathbb{C}_{\mathrm{AS}}^2$ have to be constructed by including all the plausible sequences of actions that are equivalent to AS_p^1, AS_p^2, respectively. In other words, the two sequences of states should be embedded into the space $\mathbb{R}_{[4]} \times t$. At the second step, the mutual overlapping between the clouds $\mathbb{C}_{\mathrm{AS}}^1$, $\mathbb{C}_{\mathrm{AS}}^2$ can serve as a measure in quantifying the proximity of the strategies $\mathcal{AS}^1$, $\mathcal{AS}^2$ to each other.

Structure of Action Strategy Cloud: Kernel

In the proposed qualitative description of intentional, goal-oriented actions an action strategy $\mathcal{AS}$ comprising a multitude $\{\mathrm{AS}_p\}$ of possible implementations in reality is represented as some cloud in the space-time continuum $\mathbb{R}_{[4]} \times t$. In the multitude $\{\mathrm{AS}_p\}$ of implementations there are implementations that can be regarded as the best, i.e., optimal ones. However, the amount of such implementations is relatively small and the mind is unable to single out them from the multitude $\{\mathrm{AS}_p\}$. All the other implementations are acceptable and dominating in amount. So the mind mainly deals with them in evaluating the properties of the action strategy $\mathcal{AS}$. Exactly these implementations determine the uncertainty in strategy evaluation and are responsible for the cloud formation. Put differently, the acceptable implementations endows the action strategy cloud with volumetric properties in the space $\mathbb{R}_{[4]} \times t$, whereas the optimal ones determine the cloud position on the average.

In this context we meet two mechanisms via which the space-time clouds of action strategies acquire their volumetric properties.

First, it is the mechanism of *E-embedding* based on the unconscious processing of information about the state of observed physical objects. Only the final results of this processing become accessible to the mind as just given ones. Previously (Sect. 3.2.5) we introduced the notion of E-consciousness to categorize this mechanism. We will use the term *E-embedding* to refer to space-time cloud emergence via mental-mental embedding characterized by *perception-caused uncertainty*. In particular, the space-time cloud $\mathbb{C}_{\mathrm{I}}$ of a physical object image I arises exactly via E-type embedding (Sect. 3.3.3).

Second, it is the mechanism of *AM-embedding* based on conscious processing of E-conscious information about the dynamics of observed objects with evaluating their states (*i*) in the connected past, (*ii*) experienced right now, and (*iii*) in the connected future. We have categorized this inferential estimate of observed object dynamics in terms of AM-consciousness (Sect. 3.2.5). Within this mechanism the volumetric properties of a strategy space-time cloud $\mathbb{C}_{\mathrm{AS}}$ are due to the *acceptability* of different implementations of the action strategy $\mathcal{AS}$. Thereby a given implementation AS of the action strategy $\mathcal{AS}$ is treated as acceptable based on the AM-conscious evaluation of its efficiency. In this case, the subject may recognize the difference between AS and the optimal implementation $\mathrm{AS}_{\mathrm{opt}}$. We will use the term *AM-embedding* to refer to the given mechanism of mental-mental embedding via which strategy space-time clouds emerge and are characterized by *evaluation-caused uncertainty*

The localization of an action strategy cloud $\mathbb{C}_{\mathrm{AS}}$ in the space-time continuum $\mathbb{R}_{[4]} \times t$ is determined by the AM-embedding. Indeed, the cognitive evaluation of system dynamics within AM-consciousness is governed by requirements imposed on subject's action for effectiveness of goal-oriented behavior. It implies that the perception-caused uncertainty, at least, has to enable the required cognitive control over system dynamics to be feasible.

Because the perception-caused uncertainty originates from E-consciousness the redistribution of attention is the main channel to control it. Thereby the efficiency of goal-oriented actions imposes an requirement on the degree of attention focused on the given intentional actions. The minimal degree of attention is related to the limit case when the perception-caused uncertainty and the evaluation-caused uncertainty become equivalent. For a higher degree of attention the perception-caused uncertainty is lower than the evaluation-caused uncertainty.

In comparing a sequence of some actions with an action strategy $\mathcal{AS}$ the mind deals with the given sequence of actions embedded in the mental image of environment. The volumetric properties of this embedding are determined by the E-embedding with the perception-caused uncertainty. Therefore for describing such comparison mathematically we need to single out **a special part of the action strategy cloud** $\mathbb{C}_{\mathrm{AS}}$ to be called the *kernel* $\mathbb{K}_{\mathrm{AS}}$ of action strategy $\mathcal{AS}$ or the *action strategy kernel* for short.

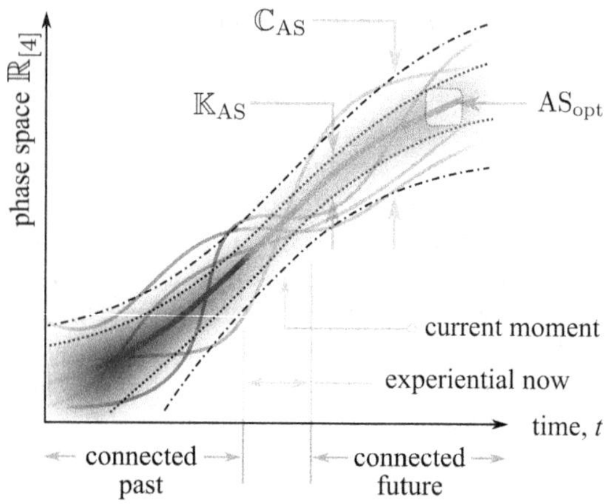

Fig. 3.9 Illustration of the action strategy kernel $\mathbb{K}_{AS}$ and the action strategy cloud $\mathbb{C}_{AS}$ as the mathematical eidoi of the action strategy optimal implementation AS_{opt} characterized by the perception-caused uncertainty (E-embedding) and the evaluation-caused uncertainty (AM-embedding)

The action strategy kernel $\mathbb{K}_{AS}$ is introduced as follows. Without loss of generality we may confine our consideration to the situation when there is only one optimal implementation AS_{opt} of the action strategy $\mathcal{AS}$ in its representative multitude $\{AS_p\}$, in particular,

$$AS_{opt} = \{x(t), v(t), a(t), j(t)\}_{opt} \tag{3.2}$$

in the space-time continuum $\mathbb{R}_{[4]} \times t$. In this case the kernel $\mathbb{K}_{AS}$ is the space-time cloud arising from mental-mental embedding of AS_{opt} into the mental image of environment. In other words, the kernel $\mathbb{K}_{AS}$ of the action strategy $\mathcal{AS}$ is the mathematical eidos of its optimal implementation AS_{opt} characterized by perception-caused uncertainty.

Finalizing our discussion about action strategy clouds it is worthy of noting that the notions of action strategy cloud $\mathbb{C}_{AS}$ and the action strategy kernel $\mathbb{K}_{AS}$ as mathematical eidoi are closely intertwined. We may say that

- the action strategy cloud $\mathbb{C}_{AS}$ is the mathematical eidos of AS_{opt} in its mental-mental embedding characterized by the *evaluation-based uncertainty* (AM-embedding),
- the action strategy kernel $\mathbb{K}_{AS}$ is the mathematical eidos of AS_{opt} in its mental-mental embedding characterized by the *perception-based uncertainty* (E-embedding),

which is illustrated in Fig. 3.9.

For the mind the optimal implementation AS_{opt} of the action strategy $\mathcal{AS}$ is not accessible. By contrast, the action strategy cloud $\mathbb{C}_{AS}$ and the action strategy kernel $\mathbb{K}_{AS}$ are directly accessible and, thereby, may be included as the basic elements in the mathematical description of goal-oriented actions. Within this description what properties have to be attributed to action strategy clouds can be elucidated by turning the description of subject's actions from the standpoint of external observer.

Then based on the obtained results and turning to the notion of attention degree, the properties of action strategy kernels can be also elucidated.

Besides, as noted previously, within the experiential now the state variables $\{x, v, a, j\}$ of the extended phase space $\mathbb{R}_4$ are treated as independent quantities accessible to the mind via E-consciousness. On scales of the complex present the situation becomes more intricate due to the corresponding contribution of AM-consciousness. In particular, if total variations in the quantities $\{x, v, a, j\}$ on temporal scales of the complex present exceed essentially the perception uncertainty, then the relationship of these variables

$$v = \frac{dx}{dt}, \qquad a = \frac{dv}{dt} = \frac{d^2x}{dt^2}, \qquad j = \frac{da}{dt} = \frac{d^2v}{dt^2} = \frac{d^3x}{dt^3}$$

can be restored, i.e., become accessible to the mind, at least, approximately. In this case, we can turn to the gist of Lagrange mechanics with higher order time derivatives to describe the properties of action strategy optimal implementation and then the corresponding properties of action strategy kernel and action strategy cloud.

Space-Time Cloud of Object Motion

We have considered two types of space-time clouds:

- the space-time cloud of the image representing a certain mental image of a physical object focused on the object current state,
- the space-time cloud of action strategy representing a certain multitude of object motion trajectories within the complex present leading to a desired goal.

Here we want to return to the problem of describing human perception of a physical object focusing on its motion within the complex present. It is important because the description of goal-oriented actions has to cope with two related issues, how humans perceive and evaluate motion of observed objects within some time interval and the corresponding planned actions.

As in the case of perceiving the current state of a physical object, we can recognize its current position, its motion trajectory in the nearest past, and the anticipated motion in the nearest future rather accurately. However, within quantitative or semi-quantitative evaluation of these characteristics with respect to the surrounding objects, we again meet the problem of mental-mental embedding of the observed object motion into the space-time continuum $\mathbb{R}_{[4]} \times t$.

As in the case of action strategies, the mental-mental embedding of object motion poses a number of questions about:

- the multitude of trajectories equivalent to the observed one in properties playing an essential role of subject's actions,
- the uncertainty in evaluating the current state of observed object,
- the variations in trajectory characteristics caused by the individual difference between equivalent motion trajectories.

Repeating actually one-to-one the previous constructions and arguments devoted to the analysis of action strategies, we come to the notion of space-time cloud object motion introduced as follows.

Space-time cloud of object motion image: *The space-time cloud* $\mathbb{C}_{\mathrm{MT}}$ *of the mental image of object motion trajectory* MT (or *motion trajectory cloud* for short) is a mathematical eidos representing how the trajectory of observed object is embedded into the mental image of the physical space-time continuum. The cloud $\mathbb{C}_{\mathrm{MT}}$ is similar to the action strategy kernel in properties, namely, it is the mathematical eidos of the motion trajectory MT in its mental-mental embedding determined by the E-embedding and characterized by the perception-based uncertainty.

In the present section we have discussed two types of space-time clouds having similar structures:

- the space-time cloud $\mathbb{C}_{\mathrm{AS}}$ of action strategy $\mathcal{AS}$ and its kernel $\mathbb{K}_{\mathrm{AS}}$,
- the space-time cloud $\mathbb{C}_{\mathrm{MT}}$ of object motion trajectory MT in the external world.

These clouds are the images of mental-mental embedding that represent how the mind maps a trajectory of some object motion within the complex present into the space-time continuum $\mathbb{R}_{[4]} \times t_{\mathrm{CP}}$ in its quantitative evaluation. For the space-time cloud $\mathbb{C}_{\mathrm{AS}}$ of action strategy $\mathcal{AS}$ as well as its kernel $\mathbb{K}_{\mathrm{AS}}$, this trajectory is a purely imaginary one supposed to match a hypothetical trajectory meeting a certain criterion of optimality or some collection of such trajectories. For the space-time cloud $\mathbb{C}_{\mathrm{MT}}$ of motion trajectory MT it is the motion trajectory of observed object. The volumetric property of these clouds—their blurring around the corresponding motion trajectory is determined by the cumulative contribution of two mechanisms:

1. E-embedding characterized by the perception-caused uncertainty. It represents the unconscious processing of perceived information whose final results become accessible to the mind as just given ones. In the case of perceiving and evaluating the states of a physical object at some instant of time, E-embedding is the mechanism responsible for the emergence of space-time cloud $\mathbb{C}_{\mathrm{I}}$ of the object image I. Concerning

   - the observed motion trajectory MT of some physical object or
   - the imaginary motion trajectory $\mathrm{AS}_{\mathrm{opt}}$ being the optimal implementation of some action strategy $\mathcal{AS}$

   E-embedding gives rise to the space time cloud $\mathbb{C}_{\mathrm{MT}}$ of MT and the kernel $\mathbb{K}_{\mathrm{AS}}$ of $\mathcal{AS}$, respectively. The emergence of $\mathbb{C}_{\mathrm{MT}}$ and $\mathbb{K}_{\mathrm{AS}}$ is due to that the subject cannot recognize the difference between motion trajectories being in a close proximity to MT and $\mathrm{AS}_{\mathrm{opt}}$, respectively.
2. AM-embedding characterized by the evaluation-caused uncertainty. The gist of this mechanism is the categorization of various implementations {AS} of one action strategy $\mathcal{AS}$ as equivalent within the complex present. This categorization

does not exclude that the subject may recognize the difference between the elements of the multitude {AS}. AM-embedding is the mechanism responsible for the emergence of action strategy clouds exemplified by $\mathbb{C}_{\mathrm{AS}}$. The kernel of $\mathbb{K}_{\mathrm{AS}}$ of action strategy $\mathcal{AS}$ is always a part of its space-time cloud $\mathbb{C}_{\mathrm{AS}}$ because the cognitive indistinguishability of two motion trajectories inevitably leads to their acceptability as equivalent.

In conclusion, speaking about the cumulative contribution of the two mechanisms to the mental-mental embedding of action strategy, we want to note the following.

- E-embedding with the perception-based uncertainty is due to unconscious processing of perceived information. So it can be governed by the mind only indirectly focusing or defocusing attention on or from the implementation of current action strategy $\mathcal{AS}$.
- AM-embedding with the evaluation-based uncertainty is governed by the mind in the framework of the analysis-synthesis strategy (Sect. 3.2.5) consciously and intentionally operating with E-conscious information directly. In this sense AM-embedding "envelops" E-embedding, where E-conscious information provided by the sensory modalities plays the role of the initial sources.
- If the degree of attention devoted to the action strategy $\mathcal{AS}$ is higher than a certain threshold, the kernel by $\mathbb{K}_{\mathrm{AS}}$ of the action strategy $\mathcal{AS}$ is contained inside its cloud $\mathbb{C}_{\mathrm{AS}}$ but does not coincide with $\mathbb{C}_{\mathrm{AS}}$. In other words, in the space-time continuum $\mathbb{R}_{[4]} \times t$ the inclusion $\mathbb{K}_{\mathrm{AS}} \subset \mathbb{C}_{\mathrm{AS}}$ holds. In this case, it is possible to categorized intentional, goal-oriented actions as *strictly rational*.
- If the degree of attention devoted to the action strategy $\mathcal{AS}$ is below this threshold, the perception-based uncertainty also plays the role of the evaluation-based uncertainty. Under this condition the kernel $\mathbb{K}_{\mathrm{AS}}$ and the space-time cloud $\mathbb{C}_{\mathrm{AS}}$ coincide with each other as the regions in the space-time continuum $\mathbb{R}_{[4]} \times t$. In this case, various cognitive biases (e.g., status quo bias) and human traits (like novelty seeking) can affect essentially intentional, goal-oriented actions in contrast to the strictly rational behavior.

Space-time clouds of dynamical type extend through the complex present and, thereby, they possess three temporal parts corresponding to the connect past, the experiential now, and the connected future. Uncertainty in mental evaluation of object properties depends essentially whether it is based on sensory modalities or mental operations. Therefore the cloud blurring corresponding to the connected past and the connected future can be much stronger than that of the experiential now.

In further theoretical construction we will treat the space-time clouds of dynamic type as one of the main elements in the mathematical description of intentional, goal-oriented actions. Physical properties characterizing human perception of all the main aspects of object motion and goal-oriented actions will be attributed to these clouds as whole entities. Thereby the dynamic type space-time clouds may be regarded as the mathematical edoi of object motion in the complex present. In addition, we want to note that:

- The possibility of introducing a certain optimality criterion for action strategies opens a gate toward using the formalism of Lagrange mechanics with higher-order time-derivatives.
- The comparison of goal-oriented actions and perceived dynamics of object motion based on the cloud overlapping prompts us to turn to some basic ideas posed within quantum mechanics. However, their particular realization in describing goal-oriented actions can be essentially different because of the existence of the "owner" of space-time clouds—the self (Sect. 3.2.3).
- Due to the rational aspects of AM-embedding the properties of action strategy clouds should admit a mathematical description dealing with the characteristics of physical object motion accessible to the external observer. It also opens a gate to describing E-embedding by introducing an additional characteristics like the attention degree, cognitive biases and human traits.

3.4 Human Temporality

The main element of objective or subjective time is a point-like instant having no duration. For this reason we turn to the conception of *temporality* taking the central position in the presented investigation and reflected in the title of this book "Physics of the Human Temporality: Complex Present." The main element of temporality is the space-time cloud, i.e., an element with intrinsic extension in time and, correspondingly, in space.

Another reason to turn to the temporality as an individual entity of investigation is that in our life we permanently meet the necessity of reconstructing past events, anticipating future events, and creating various imaginary scenarios. Thus, the memorized past and the imaginary future exist in the mind, endowing it with properties anomalous for the inanimate world. Exactly the temporality is the conception required for describing these properties.

3.4.1 Temporality: General Aspects

Husserl (1991, Collected Works, IV) offered phenomenology of the intrinsic temporality of experience setting aside assumptions about human time as objective or measurable by the clock. As the basic element of temporality, Husserl put forward a certain time interval of finite duration which somehow contains all three temporal parts—the tripartite structure of retention, primal impression, and protention. Each of these components never appears in isolation; in this sense their tripartite structure is a indivisible component of intrinsic temporality.

Heidegger actually raised this tripartite element of temporality up to the general level of existence attributing it to the notion of *being* as an intrinsic aspect. Namely, in the book *Being and Time* (1927/1996) he introduces the distinctive *unity* of human existence *Dasein*[20] referring to the experience of *being* that is peculiar to humans. In the context of the human temporality we want to emphasize temporal structure of Dasein as the *unit* combing in itself past, present, and future. In particular, the term *being-ahead-of-itself-being-already-in-the-world* implies that in embedding the indivisible unit of human consciousness, all the temporal relations are engaged simultaneously (Blattner 2005).

Following the gist of embodiment Varela et al. (1991, see also new edition 2016) proposed a naturalized account of intrinsic temporality that integrates phenomenological and neurophysiological elements. For the further development of this approach including also the enactivist account of time consciousness a reader may be referred to Thompson (2007), Clark (2008), Gallagher (2011a, 2016, 2017a, b).

Describing the dynamical properties of intrinsic temporality underlying Husserl's tripartite structure and their relationship with intentional actions Gallagher (2011a, 2016, 2017a) singles out three time scales

- The *elementary* scale (varying between 10 and 100 ms)
- The *integrative* scale (varying from 0.5 to 3 s)
- The *narrative* scale involving memory, deliberation, and planning (on scales larger as well as much large than 3 s).

The first two types of temporal scales we discussed previously in the context of the experiential now. Here we note some details about the narrative scales; following Gallagher:

> The narrative scale is meant to capture longer time periods that scale to complex actions and cognitive processes that may involve recollection, planning, intention formation, and so on. Further distinctions could be made (one could think of developmental and evolutionary timescales, for example)…(e.g., 2017a, p. 9)

We single out two type of narrative scales, one is related to the complex present discussed in Sect. 3.1, the other type of narrative scales concerns intrinsic temporal properties of mental time travels. Mental time travels deserve an own investigation, so below we will touch them only briefly in aspects related to the complex present and long-term phenomena.

Here we want to note the relation between narrative scales and a special type of intentional actions. According to Pacherie (2000, 2006, 2007) there can be single out three main stages in goal-oriented actions: the formation of future-directed intentions (F-intentions), present-directed intentions (P-intentions), and motor intentions (M-intentions).[21]

---

[20] The German word *Dasein* means "being there" and emphasizes the fact of existence in a particular place and at a particular moment of time. The use of Dasein as a special term referring to human existence actually emphasizes holistic aspects of human beings (e.g., Käufer 2005; Haugeland 2013).

[21] Within the development of the given description of intentional actions Pacherie (2008) started to use the term D-intentions instead of F-intentions to emphasize that distal intentions play no role in the guidance and control of actions once started, for a review see Mylopoulos and Pacherie (2019).

- *F-intentions* are formed before the action and represent the whole action as a unit. They are usually detached from the situation of action and specify types of actions rather than tokens. Their content is therefore conceptual and descriptive. F-intentions are also, as Bratman (1987) points out, subject to distinctive normative pressures for consistency and coherence: in particular, they should be means-end coherent, consistent with the agent's beliefs and consistent with other intentions he or she may have.
- *P-intentions* serve to implement action plans inherited from F-intentions. They anchor the action plan both in time and in the situation of action and thus effect a transformation of the descriptive contents of the action plan into perceptual-actional contents constrained by the present spatial as well as non-spatial characteristics of the agent, the target of the action, and the surrounding context.
- [*M-intention*, the] final stage in action-specification involves the transformation of the perceptual-actional contents of P-intentions into sensorimotor representations (M-intentions) through a precise specification of the spatial and temporal characteristics of the constituent elements of the selected motor program. (Pacherie 2007, *italic* and itemizing (-) is our)

The three types of intentional actions are involved in complex hierarchical interaction where each of them plays a role complementary to and partly independent of the others (for a detail discussion see Pacherie 2008).

It should be noted that Gallagher (2011a, 2016) relates F(D)-intentions to the narrative scales, whereas P-intentions are considered to correspond to the integrative scale. However, according to Pacherie's classification of intentions, F(D)-intentions are detached from the particular conditions of current situation and the leading role of P-intentions is implementing the action plan inherited from F(D)-intentions. Therefore, we attribute P-intentions to the complex present whereas F(D)-intentions are supposed to be related to mental time travels; the latter aspect will be elucidated in the next section as the fundamental feature of the human temporality on long-term scales. In this case, M-intentions based on the direct perception-action relationship must be attributed to the experiential now.

Exactly the hierarchical structure of F(D)-, P-, M-intentions with the transition from F(D)-intentions to P-intentions attributed to the boundary of the complex present is one of the fundamental reasons enabling us to consider the description of human actions within the complex present to admit a closed mathematical description of their dynamics. Similar conclusion about the possible closeness of the experiential now with respect to its mathematical description does not hold. It is caused by that the basic mechanisms governing evaluation of the experiential now deal with temporal scales much larger than the characteristic duration of experiential now (Sect. 2.6.4 concerning predictive coding paradigm).

In the next section we will discuss in more details our concept of long-term temporality based on mental time travels; here we want to briefly discuss the gist of mental time travels in their relation to intrinsic temporality met in literature. In particular Hutto and McGivern (2016, pp. 162–163) note:

> [m]ental time travel apparently underwrites our ability to reflect upon, evaluate, and appraise our past actions and to make decisions…in light of those actions when planning and motivating future actions…Mental time travel provides an emotionally hot yet reflective way of connecting our pasts, presents, and possible futures—a way that increases our chances of making better choices about and, thus, of acting better in the present and future.

Narratives play crucial role in mental time travels being a some of glue joining various events in our past and future into structures with a certain integrity. Narratives can provide explanations of explicit or implicit motives and reasons for someone actions and, thereby, enable us to understand them with a wider nexus. To emphasize this aspect, Goldie introduces the notion of narrative as follows.

> A narrative is a representation of events which is shaped, organized, and coloured, presenting those events, and the people involved in them, from a certain perspective or perspectives, and thereby giving narrative structure—coherence, meaningfulness, and evaluative and emotional import—to what is narrated. (Goldie 2012, p. 8)

Dealing with long-terms scales the narrative structures seem to be the most general form representing how events—real or purely imaginary ones—are interconnected in the mind.

Imagination as a mental phenomenon including its relationship with memory has been the subject of investigations for centuries starting from Ancient Greek philosophy. Mental time travels and temporality are also widely considered from this point of view; a reader can find a thorough discussion on these issue, e.g., in monograph by Michaelian (2016b) and collections edited by Michaelian et al. (2016), Bernecker and Michaelian (2017), Kind (2017). In particular, Klein et al. (2002, for a review see Klein and Steindam 2016) posit that intrinsic temporality is complex in nature. Namely, they proposed a relationship between different types of memory and types of temporality that can be seen instead of a single, unitary form of temporal experience. In particular, they hypothesize that intrinsic temporality can be partitioned into sub-types.

In this context we want to point out issues posed by De Brigard and Gessell (2016) in discussing the neural mechanisms of mental time travels. They focus on distinction between *intentional objects* and *intentional contents*. As noted in Sect. 3.2.2, mental states are intentional in nature because they are *about* something. *Intentional objects* are those things that mental states are about, they could be actual objects as well as some imaginary things. The latter may be exemplified by mathematical eidoi when they are regarded as type-representatives of mental-mental embedding. In this context mathematical eidoi are symbolic entities rather than some mental images. For a discussion of such objects of thought a reader may be referred to Crane (2013). How an intentional object is specifically represented in the mind is the *intentional content* of the corresponding mental state (e.g., Crane 2009, 2013). To cope with the intentional contents of mental time travels, De Brigard and Gessell turn to *representationalism* dealing with *representations* carrying the information about intentional objects as they are represented in the mind. Within the given context, the crucial point of representationalism is that representations acquire the status of individual entities possessing their own properties.[22]

It enables De Brigard and Gessell (2016) to draw a distinction between:

- *intentional content* which reflects in some way but does not coincide with the states of the corresponding real objects,
- *representational vehicle* understood as some neural mechanisms underlying the implementation of mental representations in the brain.

De Brigard and Gessell emphasize that the representative vehicle of mental time travels should be characterized by temporal dimension—temporal structure—in addition to physical dimensions such as size, locality, etc. generally attributed to representational vehicles (e.g., Clark 2008, 2015; Rowlands 2010).

In this way De Brigard and Gessell put forward the *dynamics-structure view* focusing on two sources of temporal features of mental time travels. First, it is the intentional contents *about* time with the underlying neural mechanisms combined with the *constructive episodic simulation hypothesis* (Schacter and Addis 2007a, b; Schacter et al. 2007). This hypothesis states that past and future thinking share overlapping cognitive and neural substrates (for a review see, e.g., Schacter et al. 2012; Conway et al. 2016). The second source stems from the temporal structure of representational vehicle which argues for the *scene construction hypothesis* (Hassabis and Maguire 2007). The scene construction—the process of mentally generating and maintaining a complex and coherent scene or event—is regarded as a basic cognitive mechanism which involves retrieving and integrating disparate details stored in modality-specific regions of the brain (for a review see, e.g., Mullally and Maguire 2014; Szpunar et al. 2016).

Plausible common neural mechanisms underlying the two hypotheses and providing a more general framework for relationship between memory, imagination, goal-oriented behavior, etc. are discussed e.g., by Hassabis and Maguire (2007), Schacter et al. (2012), Schacter and Madore (2016), Addis (2018). Referring, e.g., to Addis (2018) it is possible to suppose that there should be a neural network that

> underpins the remembering, imagining and perception of event representations via a common simulation process that is constructive and iterative, involving the interaction of pre-existing knowledge and reinstated perceptual content with incoming experiential information to create and refine online mental simulations of experience. [p. 64]

Mental time travels and temporality have common roots, namely, they imply a certain integrity of imaginary past and imaginary future at the level of brain. So temporality as a constituent component of diachronic mental phenomena should posses a complex structure. In particular,

---

[22] The particular forms of relationship between the content of representations and their objects is a matter of ongoing debates and for details and rival approaches a reader may be referred to Pitt (2020), Lycan (2015), Siegel (2016).

> [i]ntrinsic temporality includes more than lived or phenomenological time; it includes a temporal structuring that shapes action and experience, but might not be experienced as such. (Gallagher 2016, p. 206)

The aforesaid enables us to accept the following.

Characteristic features of the human temporality: Temporality as a constituent component of mental phenomena:

1. combines in some way the memorized or imaginary past, the imaginary future, the experiential now, and goal-oriented actions such that their union is able to maintain a certain integrity;
2. possesses complex temporal structure whose elements are
   - characterized by different time scales and the corresponding own properties reflected in different aspects of mental phenomena,
   - involved in close hierarchical interaction admitting top-down and bottom-up causation;
3. implies that episodic memory (i.e., the human capacity to remember past events) and episodic imagination (i.e., the human ability to imaging possible events) share common neural mechanism;
4. is rooted in the human capacity to generate and maintain a complex and coherent scene of objects in the mind.

Feature 1 is a argument for treating the human temporality is an individual entity with some properties ascribed to it as a whole.

Feature 2 justifies the idea of attributing to temporality some temporal structure as a certain constituent element.

Feature 3 enables us to suppose that the same collection of quantities (phase variables) used for describing mental images of real objects at first perceived via sensory modalities and then retained in memory can be employed for describing mental images of purely imaginary objects. We will refer to the given statement as the *phase space commonality of memory and imagination*.

Feature 4 may be regarded as a reason for introducing the notion of mental-mental embedding dealing with the embedding of mental images of external objects into the the mental image of physical space-time continuum.

3.4.2 *Long-Term Temporality: Qualitative Description*

In the previous sections we briefly discussed the notion of intrinsic temporality posed in philosophy of mind and, partly, in psychology. Here we intend to put forward our own account of temporality and will use the term the *human temporality* to emphasize its relation to the human mind. Naturally this account of human temporality inherits the general aspects of temporality noted previously. Nevertheless, the necessity of proposing a new account is due to that we need a special notion of the human temporality suitable to use in developing a mathematical formalism of goal-oriented human actions.

We employ the *human temporality* as a general term to refer to all possible *temporal* aspects of how we perceive or imagine the future and the past as well as their plausible relation to the experiential now including out actions. Put differently, the human temporality comprises the imaginary future and past, events retained in memory, as well as the experiential now embedded in time stream with intentional actions as one complex entity with internal integrity. The human temporality includes past-oriented mental time travels, the complex present, and future-oriented mental time travels united in the mind; the present book is devoted to the fragment of temporality matching the complex present. The temporality related to past-oriented and future-oriented mental time travels requires individual investigation. In the present section we intend to elucidate the general aspects of the human temporality.

As noted above the notion of temporality is contrasted with the notion of time. Speaking about *time* we—explicitly or implicitly—imply a point-like instant or a certain uncountable set instants of time. However, the mind does not operate with time instants, all the temporal fragments accessible to it in divisible or indivisible forms are, first, extended in time and, thus, are of finite duration. Second, these temporal fragments do not have fixed boundaries; in this sense they are fuzzy. Therefore, our account of human temporality does not contain elements that might be interpreted as point-like instants of time.

Now let us discuss the conception of the human temporality in a more specific way, which is illustrated in Fig. 3.10. Leaping ahead we can note that there are two main constituent parts of temporality: (*i*) *human temporal dimension* composed of all the possible conscious events and (*ii*) the *structure of intentions* uniting these events together.

Here we use the term *conscious event* or just *event* for short to denote an event in the conventional sense—i.e., that something happens, changes/not changes, appears/disappears in the external/inner world—which, in addition, has been recognized and fixed in memory.[23] There are several constituent components of such conscious events. Below we present their rather general classification.

First, it is an intentional object—real or imaginary—and what happens with it. When an intentional object is real, the corresponding event actually concerns its mental image rather than the object on its own. Moreover, in this case, what has happened with the real object is represented in the conscious event as what has happened with its mental image in the mental image of the external world. Some temporal aspect is always attributed to the given component—explicitly or implicitly. This aspect may be specified in terms of a certain temporal category like "in future," "in past," "soon," "now." A particular time (not to point-like instant of time), e.g., "on June, 2, 1988" can be used to mark an event. Beside, its temporal aspect also

[23] For a discussion of events from the general standpoint a reader may be referred to Casati and Varzi (2014).

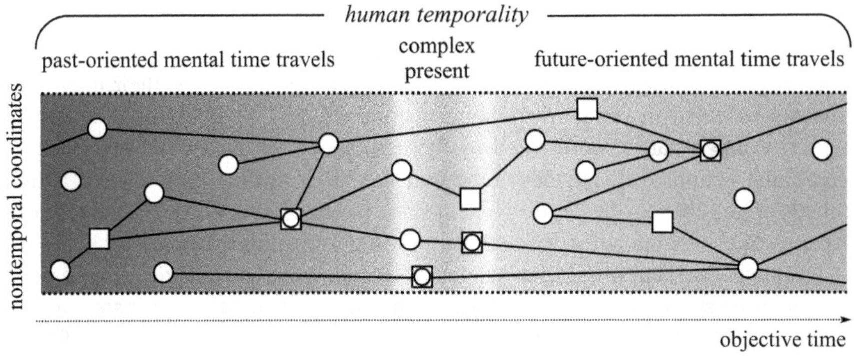

Fig. 3.10 Illustration of the human temporality as the complete set of conscious events happened in the past and retained in memory as well as purely imaginary events (already imagined or imaginable) with all the possible connections between them. Circles, squares, and their unions illustrate events representing only intentional objects, intentional actions, or both the components, respectively. The shown axis of objective time and the orthogonal direction labels as "nontemporal coordinates" illustrate that these events are *mentally* embedded in the real space-time continuum in some way; the details of this embedding, including the accuracy may change as time goes on. The solid lines connecting circles represent relations of various types which admit bidirectional effects of past-on-future as well as future-on-past in the mind

may be of indeterminate type such as "once" or "one day." The intentional object as the component of a given conscious event bears the information about its state and, maybe, dynamics within the spatio-temporal scales characterizing the given event as a whole.

Second, it is subject's intentional action.[24] Within the given component the self plays the role of intentional object. In particular, the specific details of a given action—real or imaginary—and, maybe, the specific intentions underlying exactly this action are somehow fixed in memory. A certain temporal aspect is also attributed to the given component. The intentional action as the component of a given conscious event bears the information about its individual properties including the physiological and mental "cost" of action, accuracy, etc. and their relation to the action efficiency.

Intentional objects and intentional actions are, in some sense, individual components of conscious events. Namely, a given event may contain one of them or both and in the latter case the corresponding intentional object and intentional action just complement each other.

It should be noted that the possibility of an event being about conscious action only is not in contradiction to the gist of intentionality. Indeed, although every intentional action, by definition, implies the existence of a certain intentional object which this action is aimed at, the corresponding intentional object may be the component of another event. Figure 3.10 depicts events representing only intentional objects or

[24] For a general discussion of the notion of human actions a reader may be addressed to Wilson and Shpall (2012).

intentional actions in the form of open circles or squares, respectively, the composite shapes illustrate events containing both the components.

Formally speaking, *the third component* of conscious events is their intentional contents reflected in the corresponding mental states. The intentional content of a given event is partly represented by the corresponding intentional object or/and intentional action. Both of them represent mental images of objects and actions directly accessible to the mind. Nonetheless, we have singled out the intentional contents of events into an individual class for the following reasons.

An intentional action and what happens to an intentional object are always embedded into time stream in some way. Even if several events of indeterminate temporal aspect and related to one another in some way are currently in the focus of attention they are temporally ordered in the mind. By contrast, the mental state bearing the intentional information about all these events integrates them into some common entity and has no *own* temporal aspect. In this sense the mental state may be regarded as a some kind "representational vehicle" (which however is of mental rather than neurophysiological nature) that carries all the aspects of intentional contents from event to event without any time delay. In the worst case when a given mental state is currently activated all the corresponding events should be either in the primary or secondary memory (see Sect. 3.1.5) and the retrieval items from the secondary memory is a relatively quick process. Moreover, speaking about goal-oriented behavior particular intentional actions and particular changes in the state of intentional objects can be individually evaluated only at the level integrating the corresponding events into one entity. Besides, appraising different ways of getting one goal with respect to their preference in possible choice can be implemented also at this level only. No properties evaluating the individual (partial) contribution of intentional actions and intentional objects within events to getting the desired goal can be attributed to them in the general case.

For these reasons, we treat the mental state carrying the intentional contents of all the events involved in implementing one goal or related to one another somehow else as a certain kind connection possessing a certain temporal structure. Below we will call this structure of events or its connected branch the *temporal structure of intentions* or just *intentional structure* for short. Figure 3.10 depicts these connections as solid lines. Within the given approach intentional structure becomes an entity of the inter-event level—*macro-level*. Therefore below the term *conscious event* will be understood as an entity of *micro-level*—the level representing particular intentional actions and particular details of what happening to an intentional object.

The properties characterizing goal-oriented behavior and admitting its quantitative description are attributed to intentional structures as the basic elements. It does not mean that nothing can be finally ascribed to individual intentional actions and the state of intentional objects following the top-down constructions. It is just not clear whether in the general case the resulting micro-level description is independent of the macro-level event structure.

In this sense we again meet predictive coding paradigm.[25] The macro-level intentional structures are the basic elements bearing the properties of goal-oriented behavior, whereas the micro-level of individual states of intentional actions and intentional objects reflects physical and neurophysiological constraints. Experience accumulated during the implementation of intentional actions with real objects has to endow the relationship of macro- and micro-levels with a certain self-consistency. It is also worthy of noting that within Pacherie's terminology (Sect. 3.4.1) intentional structure corresponds to P- and/or F(D)-intentions, whereas individual events may be categorized as F-intentions.

Now we can formulate the desired account of human temporality which focuses on intentional, in particular, goal-oriented behavior.

Notion of the human temporality: The human temporality is the general entity representing temporal connectivity of the human mind and intentional, in particular, goal-oriented actions extended in time. The human temporality comes into being via the merging of the following constituent components.

- *Human temporal dimension* as a container of all the possible events that are assigned or can be assigned in the mind to the past, future, experiential now, or even indeterminate temporal category. Human temporal dimension includes:
  - events experienced right now, i.e., belonging to the experiential now,
  - events happened in the past and retained in memory,
  - imaginary events of any temporal category,
  - events that are not currently presented in the mind but can arise in future.
- *Intentional network* containing all the possible temporal structures of intentions connecting, i.e., integrating the events of human temporal dimension into entities of one intentional content. It is the intentional network that bears the properties of intentional, in particular, goal-oriented human behavior. There are different types of intentional structures including, in particular, ones based on
  - inter-event causal-like relations of macro-level,
  - emotion-induced actions;
  - inter-subject interaction, i.e., collective effects.

We understand that the present account of human temporality can be generalized and extended for describing mental time travels and other mental long-term phenomena. The given description of the human temporality allows for the main aspects that should be taken into account for the analysis of the complex present.

[25] We already discussed the gist of predictive coding paradigm in Sect. 2.6.

3.4.3 Long-Term Temporality: Structural Elements

In this section we focus on two issues: particular properties of human temporal dimension and intentional network. The attention is mainly focused on the latter issue because it is the international network that governs goal-oriented human behavior being non-local in time.

Human temporal dimension is just a container for events allocated in it according to some temporal order. No particular properties like possessing fixed spatial dimensions and the time axis are attributed to it on its own. Only particular events bear such properties individually.

• There is the only one feature attributed to human temporal dimension individually, it is the existence of the owner—the self. Therefore, our account of human temporality is specified from the standpoint attached to the experiential now. On the one hand, it enables us to employ temporal categories like "in past," "in future," "now" in the mathematical analysis of goal-oriented behavior. For this reason, the given description of the human temporality has to be categorized as the A-type theory of temporal properties of mind.[26] On the other hand, within this approach we face up to a challenging problem that the event classification based on temporal categories like "in future" and "in past" is not stable and inevitably changes as physical time goes on. This problem is overcome within the concept of two-dimensional time to be developed below. Two dimensional time combines in itself the human temporality and physical time as individual complementary dimensions.

• The last class of events of human temporal dimension—events that are not currently presented in the mind but can arise in future—has been introduced in order to avoid the necessity of extending human temporal dimension as objective time goes on and new, unaccepted events come into play. It does not give rise to any problem in modeling intentional actions because such events are not currently included into intentional network and, thus, cannot affect human actions being currently under consideration.

Intentional network: Several premises of modern theory of remembering and imagining underlie the proposed concept of intentional network. Their detailed discussion, however, is outside the scope of the present book.

Our understanding of temporality includes three elements: past-oriented mental time travels, the complex present, and future-oriented mental time travels. Although the present book is devoted to the complex present, we need to briefly discus also mental time travels, whose detailed description is, naturally, an individual investigation. So here we confine myself to rather general comments about plausible mechanisms of remembering and their relationship with past-oriented and future-oriented mental time travels. It suffices to elucidate the gist of the human temporality.

During the last decades there has been taken a new step toward understanding the general mechanisms governing remembering, for a detailed discussion of the made

[26] It should be emphasized that within the classical paradigm of physics the dynamics of physical objects is described by the B-type theories (Sect. 1.1).

advances a reader may be addressed to Michaelian (2016a, b), Michaelian and Sutton (2017), Michaelian and Robin (2018). The causal theory of memory by Martin and Deutscher (1966) has undergone significant development in several directions. The gist of this theory is that remembering an episode necessarily involves the *transmission* of content from experience to retrieval which is causally implemented via a *memory trace*. Among the directions of causal theory generalizations we want to note the following concepts being essential for our constructions.

• The content of retrieved representations is, at least partly, produced, i.e., generated at the time of retrieval rather than being derived, i.e., transmitted from the content of the corresponding experience (e.g. Michaelian and Robin 2018). It opens a gate to the idea that

> the process of remembering the past is executed by the same cognitive system as the process of imagining the future [and the past] ... Both imagining the future and remembering the past draw on stored content originating in experience of past events. Imagining a future event does not, of course, draw on content originating in the experience of the particular event imagined. By the same token, the mental time travel framework suggests that remembering a past event does not necessarily draw on content originating in experience of the particular event remembered. (Michaelian and Robin 2018, p. 27)

For this reason in our account of human temporality we do not make a sharp distinction between events corresponding to real objects and actions, on the one side, and imaginary objects and actions, on the other side, in combing them together within the intentional network. Among the theories supporting this idea we want to note:

- The simulation theory proposed by Shanton and Goldman (2010). Based on the findings in developmental psychology and cognitive neuroscience, they turn to the simulational account of mind reading which involves the imitation, copying, or reexperience.
- The account of memory by De Brigard (2014) where he offers a picture of memory as an integral part of a larger system. He claims that remembering is a particular operation of this cognitive system that permits the flexible recombination of possible past events in the service of constructing mental simulations of possible future events.
- The simulation theory developed by Michaelian (2016a, b) treating episodic memory and episodic future thought as processes carried out by a common episodic construction system. Michaelian supposes that there is no difference in kind between cases in which one imagines an actual past event on the basis of one's experience and cases in which one imagines a past event on another basis. According to the simulation theory of memory, the difference between remembering real past events and imagining some past events is determined by the reliability of the imaginative process. In this case, remembering is the complex interplay of (*i*) an

accurate representation of the event, (*ii*) confabulation[27] resulting in an inaccurate representation, and (*iii*) misremembering.
- The functionalist theory by Fernández (2018) on which a mental state–called memory—plays a memory-type functional role that may be initiated rather than is necessarily caused by the past experience.

The noted accounts of memory are usually called *postcausal theories of remembering* to emphasize that a *causal* connection between experienced events and their mental representations based on memory traces is not necessary. Here the causality is understood in its classic meaning. These theories consider the remembering to depend on what happens when the subject remembers, rather than on whether there is some causal relationship between the retrieved representation and experiential representation (e.g., Michaelian and Robin 2018).

Accepting the direct relationship between imagination and remembering we face up to the necessity of developing the notion of causality to cope with mental phenomena where imagination plays the role of cause. In particular, Perrin (2018) proposed *procedural causal theory* assuming this type causation to be of complex structure. This theory posits that the causal connection of remembering occurs not between the retrieved memory and the earlier experience but between the reconstructive process producing the memory and the constructive process producing the experience. In this connection we also want to note the review by Walker and Gopnik (2013) as an introductory to the relation between the imagination and causal cognition with relevance to recent developments in computational theories of human learning.

In our constructions we will turn to the notion of causality between intentional structures taken at different instants of objective time, which requires the introduction of two-dimensional time.

• The direct transmission of content from experience to retrieval—one of the premises accepted by the classical causal theory of memory—is generalized within the conception of *distributed traces* Sutton (1998, see also Michaelian and Sutton 2017). If Martin and Deutscher's theory understands memory traces as local, individually stored entities with explicit content, then the distributed traces may be conceived of as many entities which superpositionally represent some content. The idea of distributed traces is the gist of *connectionist* account. It deals with superpositional connectivity of memory traces which are blended, i.e., not laid down independently once and for all. They are reconstructed rather than reproduced (for a review, e.g., Robins 2016).

The *multiple trace theory* developed by Nadel and Moscovitch (1997), Moscovitch and Nadel (1998), Nadel et al. (2000) for episodic memory consolidation represents a neural mechanism of memory turning to the gist of distributed traces.[28] According to

[27] Confabulation is a memory error defined as the production of fabricated, distorted, or misinterpreted memories about oneself or the world, without the conscious intention to deceive.

[28] The key idea of multiple trace theory is also discussed by Hintzman and Block (1971), Hintzman (1986, 1988) within a simulation model of episodic memory. This model assumes that each experience produces a separate memory trace and that knowledge of abstract concepts is derived from the pool of episodic traces at the time of retrieval. A retrieval cue contacts all traces simultaneously,

this theory each time an episodic memory is retrieved it is reencoded, which leads to the formation of multiple traces mediated by ensembles of hippocampal-neocortical neurons (Moscovitch et al. 2006). For details and a review of multiple trace theory a reader may be referred, e.g., to Nadel et al. (2007); Winocur and Moscovitch (2011); Moscovitch et al. (2016); Sekeres et al. (2018).

It is necessary to emphasize that memory formation is a highly dynamic process. In particular, memory traces are transformed over time in a number of ways including (*i*) the selective strengthening of only some, but not all, traces, (*ii*) the integration of new information into existing knowledge stores, and (*iii*) the establishment of new linkages within existing knowledge stores (e.g., Nadel et al. 2012; Kroes and Fernández 2012; Genzel and Wixted 2017).

This dynamics concerns not only episodic memory traces. According to the trace transformation theory developed by Moscovitch (2007), Winocur and Moscovitch (2011)

> in the intact brain, the detailed, hippocampus-dependent version of the memory, and the generalized or semanticized cortical version of the memory, can co-exist. The original memory and the transformed memory representations are in dynamic flux or interaction, and the conditions at retrieval influence which version of the memory is expressed. (Sekeres et al. 2017, p. 24)

The conception of distributed traces and the underlying neural mechanisms noted above argue for ascribing the intentional content directly to the intentional structure comprising a number of events somehow related to one another. Thereby, the intentional network bears all the possible intentional contents whereas an individual event of this network represents how a particular intention is implemented in reality or an imaginary world.

• The structure of intentional network implies that a number of conscious events—representing perception as well as actions—can be united by the same intentional content. In this case such events must belong to one connected branch of intentional network. So two events belonging to different components of intentional network must be independent with respect to intentions and goal-oriented actions. Whether a connected branch of intentional network may bear more than one intentional content deserves an individual investigation. However, dealing with the complex present it is possible to confine our consideration to the case when the only one type of intentions is involved into goal-oriented actions and suppose that all the events of a connected branch are related via a given intentional content.

The details of mathematical formalism employed for describing the human temporality depend essentially on the basic type of intentional contents, namely, whether inter-event causal relations, emotions, or inter-subject interactions determine how events as micro-level entities are united by the stream of intentional content. In the present book we confine myself to causal-type relations of intentional network. The description of other types (emotion-based intentions and intentions determined by

activating each according to its similarity to the cue, and the information retrieved from memory reflects the summed content of all activated traces responding in parallel.

inter-subject relations) require individual investigation which can use the formalism to be developed here as a certain background.

The notion of causal relations based on remembering and imagination is up to now a challenging problem being the matter of on-going debates. For a detailed discussion of memory-based causality a reader may be addressed, e.g., to Debus (2017), Michaelian and Robin (2018). As far as ontological issues of mental causation including causality of imagination are concerned with, the corresponding discussion of arguments for and against can be found in the collection edited by Gibb et al. (2013) and reviews by Robb and Heil (2018) and Walker and Gopnik (2013). Accepting the conception of distributed traces we face up to the problem of memory-based causality when remembering is reconstructive rather than reproductive (e.g., Michaelian and Robin 2018).

In our account of human temporality mental causality is understood as diachronic relations between

- conscious events and intentional structures taken at one instant of objective time and
- the corresponding events and intentional structures taken at another instant of objective time

that are accessible to the mind.

These diachronic relations

- can be represented as the "because-of" statement as well as
- meet the criterion of analyzability (Sect. 3.2.4), i.e., are accessible via the analysis-synthesis strategy.

This type mental causality implies that, first, (*i*) intentional structures involving past events retained in memory and currently retrieved from it, (*ii*) imaginary events assigned to the past and the future and currently accessible to the mind determine human actions in the future.

Second, this type relationship is not localized in the experiential now. The matter is that in making decision about actions in the future, the subject turns to the evaluation of plausible action strategies based on the accumulated experience. In this sense mental causality deals with dynamical properties of the human temporality that continuously accumulate subject's experience at all the previous moments of time. It should be noted that learning can convert this experience based on cognitive analysis into embodied perception of plausible strategies of actions.

Third, the analyzability of such causal relations implies that in conscious evaluation of possible action strategies, graduate changes in the properties of the corresponding events are reflected in the graduate changes in the strategies' preference. Moreover, events can be analyzed, at first, separately and, then, analyzed together taking into account their arrangement in a given action strategy. In mathematical terms, this statement can be regarded as a deliberate analysis of possible realizations of action strategies. Their quantitative evaluation does not depends drastically on whether a given strategy is actual, purely imaginary, or contains real and imaginary components together. In particular, it means that such causal relations may be categorized as entities of AM-consciousness.

The effect of emotions on the evaluation of action strategies is quite distinct from that of analyzable causality; emotions are just given to the mind. Put differently, emotion-induced relations between events must be categorized as entities of E-consciousness and their mathematical description requires individual formalism.

Speaking about relations of causal type where inter-subject interactions play a crucial role we meet another problem also requiring individual mathematical formalism. The matter is that inter-subject interaction endows the mind with holistic properties (e.g., Dainton 2000; Thompson 2007). Such properties attributed to the mind of one person depend essentially on the properties of other persons with respect not only their particular realizations (token-properties) but also their qualitative features, e.g., the existence (type-properties).[29]

• The concept of mental time travels implies that there should be no qualitative difference between episodic memory and episodic future thought, which is the gist of *continuism* put foreword by Perrin (2016). According to continuism any difference between past events retained in memory and purely imaginary ones is merely quantitative. On the contrary, *discontinuism*—the traditional view—maintains that there are qualitative differences between episodic memory and episodic future thought, naturally, they may possess various quantitative similarities. In particular, Debus (2014, 2018) introduces the notions of *R-memories* and *S-imaginations* for recollective memories of past events and sensory imaginations of future events, respectively, as mental occurrences of two different kinds. The *embeddedness* of R-memories and S-imaginations into a context of relevant thoughts endows human behavior with a certain self-consistency. A detailed analysis of arguments for and against continuism and discontinuism is presented in the review by Perrin and Michaelian (2017).

Our account of human temporality partly inherits the basic ideas of the two formally rival conceptions. First, continuism is accepted in describing the most general properties of events. Namely, it enables us to describe the properties of past events retained in memory and purely imaginary events using the same collection of phase variables. In other words, these events are treated as some kind objects admitting the embedding into a common phase space.

Second, within the complex present the connected past and the connected future exhibit essential asymmetry in their *functional* role. The past events—R-memories—contribute to learning, i.e., accumulating the information about the properties of surrounding objects and plausible actions. The expected future events—S-imaginations—contribute to the choice of possible action strategies whose evaluation is based on the information accumulated previously. In this sense the complex present must be categorized as an entity with temporal structure of the A-type, i.e., where the *before-now-after* relations take the central position in event description. In contrast, mental time travels are likely to be entities for which intentional structure is of the

[29] Phenomena of this type should be categorized as holistic, they are typical for human nature. Nevertheless holistic problems can be met in systems of another nature, e.g., holistic problems related to physical systems are considered by Healey (2016).

B-type, i.e., where there are only relations of the *earlier-later* type. Naturally the detailed categorization of mental time travels according to the role of the present requires individual investigation and the existence of some hybrid AB-form is quite plausible.

3.4.4 Temporality: Realm of Complex Present

In the previous section we discussed the concept of human temporality pointing out its general aspects concerning human temporal dimension and intentional network. The present book, however, focuses mainly on intentional, goal-directed actions whose analysis can be confined to the complex present comprising the connected past, the experiential now, and the connected future. The particular features of the three constituent components play a significant role in describing such actions, so we need a more detailed account of human temporality allowing for these feature. Therefore, the present subsection is devoted to elucidating the human temporality in the realm of complex present with cumulative duration about 10–20 s. The corresponding properties of the human temporality to be discussed below underlying the mathematical formalism required for describing this type goal-oriented actions.

For the possibility of confining the intentional structure of human actions to the complex present it is required that:

- the system current state can be evaluated based on actions within the connected past,
- short-term anticipation within the connected future can provide the information enabling the subject to govern the system dynamics efficiently.

Under these conditions, in detailing the concept of the human temporality presented in the previous section the following has to be accepted.

- Intentional behavior on scales of the complex present should be confined to achieving only one goal common for all the actions under consideration.
- All the other actions not directly involved in achieving this goal has to be ignored or left out of the question.[30]
- Thereby, dealing with intentional, goal-oriented actions confined to the complex present only one type of intentionality may be taken into account and the intentional content is about achieving the given goal only.

---

[30] Strictly speaking, there are many factors hidden from our attention during particular actions or deliberately ignored that, nevertheless, *implicitly* affect the actions being the focus of attention. So we need some additional arguments and evidence for the possibility of removing them form the list of factors taken into account directly. The collection of *ceteris paribus laws* and the notion of *nomological machine* (Cartwright 1989, 1999; Pemberton and Cartwright 2014) can be referred to in justifying this approach; we discussed this issue in the previous book (Lubashevsky 2017).

Besides, dealing with inter-event causal-like intentions we may consider only one intentional object (maybe, combining in itself several strongly connected objects) and its environment being the focus of attention within analyzed actions.

Thereby the description of the human temporality within the complex present can be represented in the form that (*i*) reduces the general description of the human temporality (Sect. 3.4.2) to the case of the complex present and (*ii*) bears also some features intrinsic to the complex present only. The following is the proposed account of human temporality confined to the complex present.

The structure of the human temporality within the complex present: In the realm of complex present the human temporality comprises the following component.

- *Human temporal dimension* reduced to the connected past, the experiential now, and the connected future, i.e., the *complex present*. It plays the role of the container of all the possible events related to achieving a given goal and admitting the following classification:
  - Events experienced right now, i.e., belonging to the experiential now. These events determine the collection of plausible action strategies between which the subject can *currently* select a strategy for implementation. Put differently, the content of the experiential now specifies the collection of acceptable actions currently available for the subject.
  - Events happened in the past and whose information is aggregated in evaluating the action strategies and ordering them according to the choice preference, at least, partly. Learning based on the previous experience can convert this evaluation into the embodied form with a significant contribution of E-consciousness.
  - Imaginary events belonging to the connected past that represent (*i*) plausible actions available for choice previously but not selected because another action was taken and (*ii*) the imaginary changes in the state of external objects corresponding to these actions.
  - Imaginary events expected to occur in the connected future within plausible implementation of the action strategies leading to the desired goal.
- *Intentional network* reduced to the collection of possible strategies of reaching the desired goal under various conditions. In the realm of complex present, these action strategies are the basic elements of the intentional network. In the case under consideration, the intentional relations are supposed to be of the causal type understood as inter-event relations emerging in the mind within AM-consciousness.
- The following *additional properties* are attributed to the human temporality in the realm of complex present.
  - Imaginary events make up imaginary fragments of action strategies. These fragments together with the fragments implemented previously (or being under the current implementation) make up the collection of all the plausible action strategies.
  - Within one action strategy all the involved events are temporally ordered.

- Based on the previous experience (due to learning) imaginary fragments of action strategies belonging to the connected past can bear properties similar to the properties of really implemented fragments. The difference between these properties may be quantitative rather than qualitative, which underlies the evaluation of a given action strategy as a whole.
- The priority—represented as a certain quantity—in selecting the given strategy is attributed to the strategy as a whole entity.
- In the general case, an action strategy links the connected past to the connected future going through the experiential now.
- Different strategies may have common fragments characterized by "points" of splitting, merging, and intersecting. These "branching points" are determined by events representing plausible actions that can be taken within the experiential now. Typically an action strategy can have only a few "branching points."

3.5 Conclusion

Human Temporality

In the present chapter we have elaborated the general concept of the human temporality which includes three temporal parts closely intertwined, they are:

- past-oriented mental time travels,
- the complex present,
- future-oriented mental time travels.

Each of the three parts admits an individual consideration and a closed mathematical description within certain frameworks.

Turning to the functional role as another criterion of categorization, we have singled out two constituent components of the human temporality. They are:

- *Human temporal dimension*: Human temporal dimension is regarded as a container for all the possible events that have been or can be recognized and fixed somehow in the mind. These conscious events bear particular information about the situational aspects—including temporal and spatial ones—characterizing mental images of physical objects—perceived or imaginary. The mind operates with conscious events as the elements of the very basic level of the human temporality.
- *Intentional network*: Intentional network represents all the possible temporal structures of intentions. These temporal structures integrate the involved conscious events into entities with one intentional content. The intentional network describes, in particular, goal-oriented actions and bears their properties characterizing desired goals and possible ways of achieving them.

Complex Present

The complex present is a fragment of the human temporality characterized, as a whole, by time scales of order of 10 s and having special properties. First, in its turn the complex present possesses the tripartite temporal structure comprising

- the connected past,
- the experiential now,
- the connected future

playing the role of containers for events that are either experienced as occurring right now or experienced as directly involved into current actions.

Second, within the complex present the intentional network is reduced to the collection of all action strategies aimed at achieving one common goal. The corresponding human actions are characterized by one type intentional relations.

In the complex present the action strategy—the temporality ordered sequence of conscious events determining a particular way of achieving a desired goal—is the main element describing human goal-oriented behavior. So all the properties characterizing the goal-oriented actions—including the preference in selecting a particular action strategy among all the possible ones—are attributed to the action strategy as a whole entity.

Generally any action strategy links the connected past with the connected future going through the experiential now. Thereby each action strategy reflects in its properties the individual contribution of the three parts of the complex present.

- The connected past is responsible for accumulating the information about the efficiency of action strategies based on previously taken actions. In this sense the connected past is involved in the process of learning.
- In the experiential now the action strategy being currently under implementation can be changed for another one. The action strategy change within the experiential now is the essence of human behavior dynamics.
- The connected future together with the experiential now determines the collection of action strategies currently available for achieving the desired goal.

Mental-Mental Embedding

The mind operates with different types of mental images:

- images of physical objects,
- images of trajectories representing object dynamics,
- images of action strategies,
- the image of the present environment including the image of physical space-time continuum.

The dynamic relationship between the mental images and the external world is established and governed by the neural mechanisms combined within the predictive coding paradigm.

Speaking about a mental image of object, object motion trajectory, or action strategy, we draw a distinction between:

- the mental image on itself, i.e., as it is just given to the mind and
- mental-mental embedding of this image as quantitative evaluation of image properties in the mind.

We have introduced the notion of mental-mental embedding to describe quantitative evaluation of image properties, in particular, how this mental image is embedded into the mental image of environment including the mental image of space-time continuum.

Mental evaluation of observed states of surrounding objects and the results of human actions in response to changes in the environment is based on the quantitative estimation of object arrangement as it is represented in the mind. Therefore, we regard the entities generated by mental-mental embedding of images of physical objects, motion trajectories, or action strategies as the main elements entering the desired mathematical formalism of intentional, goal-oriented actions.

Space-Time Clouds

We have introduced the notion of space-time clouds to describe the entities generated by mental-mental embedding. The term space-time cloud reflects that in evaluating the observed states of physical objects including their position in space and time uncertainty plays an essential role.

Space-time clouds may be conceived of as some regions in the space-time continuum with blurred boundaries. This blurring is caused by that in mental-mental embedding different points of the environment may be treated as equivalent with a certain degree and this degree can changes gradually.

To categorize various aspects of space-time clouds with respect to the levels of consciousness we have singled out two specific mechanisms of mental-mental embedding:

- E-embedding (characterized by perception-based uncertainty) is determined by the unconscious processing of perceived information. Only the final results of this processing become accessible to the mind. The mind can affect the properties of E-embedding mainly via focusing/defocusing attention on/from the subject's corresponding actions.
- AM-embedding (characterized by evaluation-based uncertainty) is determined by the conscious processing of E-conscious information accessible to the mind in details and admitting the rational analysis of obtained results.

The perception-based uncertainty is always lower than or coincides with the evaluation-based uncertainty. Their identity is attained when the degree of attention drops below a certain threshold. In this case, various cognitive biases and human traits affect goal-oriented actions essentially. In the case, when the attention degree is above threshold *rational criteria* start to play dominant role in governing such actions.

We have introduced three types of space-time clouds.

- *The space-time cloud of mental image of physical object* $\mathbb{C}_{\mathrm{I}}$:
  For a physical object the current state of its dynamics is described by the extended phase space $\mathbb{R}_{[4]} = \{x, v, a, j\}$ comprising the spatial position x of this object, its velocity v, acceleration a, and jerk j. The introduction of the extended phase space $\mathbb{R}_{[4]}$ reflects the properties of the experiential now discussed in Sect. 2.6.6. Therefore $\mathbb{C}_{\mathrm{I}}$ may be conceived of as a certain fuzzy region in the space-time continuum $\mathbb{R}_{[4]} \times t$.
  The temporal extend of $\mathbb{C}_{\mathrm{I}}$ is about the characteristic duration of experiential now, $\tau_d \sim$2–3 s. The characteristic dimensions of $\mathbb{C}_{\mathrm{I}}$ in the extended phase space $\mathbb{R}_{[4]}$ are determined by the perception uncertainty based on mental operations with observed objects (naturally together with their physical dimensions). Thereby this type space-time clouds should be categorized as a result of E-embedding.
- *The space-time cloud of motion trajectory* $\mathbb{C}_{\mathrm{MT}}$:
  The given type space-time cloud represents the E-embedding of a motion trajectory. Thereby $\mathbb{C}_{\mathrm{MT}}$ may be also conceived of as a certain fuzzy region in the space-time continuum $\mathbb{R}_{[4]} \times t$ and is characterized by the perception-based uncertainty.
  The temporal extension of $\mathbb{C}_{\mathrm{MT}}$, however, is about the characteristic duration of complex present, $\tau_{cp} \gtrsim$10 s. The dimensions of $\mathbb{C}_{\mathrm{MT}}$ in the extended phase space $\mathbb{R}_{[4]}$ are determined by the cumulative contribution of two factors. First, it is the perception uncertainty as in the case of $\mathbb{C}_{\mathrm{I}}$. Second, in mental-mental embedding all the possible trajectories equivalent to the given one are identified. Thereby the dimensions of $\mathbb{C}_{\mathrm{MT}}$ in $\mathbb{R}_{[4]}$ reflect also the individual difference in the properties of equivalent trajectories. The original trajectory determines in some sense the center position of its space-time cloud.
- *The space-time cloud of action strategy* $\mathbb{C}_{\mathrm{AS}}$:
  Any action strategy $\mathcal{AS}$ includes a special type motion trajectory—the motion trajectory $\mathrm{AS}_{\mathrm{opt}}$ representing the optimal implementation of the action strategy $\mathcal{AS}$. So the space-time cloud $\mathbb{C}_{\mathrm{AS}}$ of action strategy $\mathcal{AS}$ inherits some properties of motion trajectory cloud.
  Special features of action strategy cloud $\mathbb{C}_{\mathrm{AS}}$ are due to its emergence via AM-embedding of $\mathrm{AS}_{\mathrm{opt}}$ characterized by evaluation-based uncertainty.

In our account of goal-oriented actions, the space-time clouds are main mathematical elements describing human perception of intentional physical objects as they are embedded into the external world and the subject's actions with them.

Kernel of Action Strategy

For any action strategy $\mathcal{AS}$ there can be singled out a special part of its space-time cloud $\mathbb{C}_{\mathrm{AS}}$ called the kernel $\mathbb{K}_{\mathrm{AS}}$. We have defined the kernel $\mathbb{K}_{\mathrm{AS}}$ as the mathematical eidos of the optimal implementation $\mathrm{AS}_{\mathrm{opt}}$ of the action strategy $\mathcal{AS}$ emerging via E-embedding. Thereby the kernel $\mathbb{K}_{\mathrm{AS}}$ is characterized by the perception-based uncertainty.

When the degree of attention focused on the action strategy $\mathcal{AS}$ exceeds the threshold determining the equivalence of the perception-based uncertainty and the evaluation-based uncertainty and the kernel $\mathbb{K}_{\mathrm{AS}}$ is located inside the action strategy cloud, $\mathbb{K}_{\mathrm{AS}} \subset \mathbb{C}_{\mathrm{AS}}$.

When the attention degree drops below this threshold the kernel $\mathbb{K}_{\mathrm{AS}}$ and the space time cloud $\mathbb{C}_{\mathrm{AS}}$ of action strategy $\mathcal{AS}$ coincide with each other as the regions of $\mathbb{R}_{[4]} \times t$.

The introduction of action strategy kernel is crucial for our further construction because:

- The comparison of an action strategy $\mathcal{AS}$ (a mental entity) and a certain observed trajectory MT of intentional object motion (an object of the external world) in various aspects inevitably turns to the three elements: the space-time cloud $\mathbb{C}_{\mathrm{AS}}$, the kernel $\mathbb{K}_{\mathrm{AS}}$, and the motion trajectory cloud $\mathbb{C}_{\mathrm{MT}}$.
- The introduction the attention degree whose variations determine
  - the transitions $(\mathbb{K}_{\mathrm{AS}} \subset \mathbb{C}_{\mathrm{AS}}) \Longleftrightarrow (\mathbb{K}_{\mathrm{AS}} = \mathbb{C}_{\mathrm{AS}})$ inducing
  - the transition between the rational behavior and the behavior affected by irrational factors

  grants us the ability to deal with the *own inner* dynamics of the mind whose properties are not reducible to the properties of images representing the perceived external objects in the mind.

References

Abraham, A., & Bubic, A. (2015). Semantic memory as the root of imagination. *Frontiers in Psychology, 6*, Article 325 (5 pages).

Adams, E. J., Nguyen, A. T., & Cowan, N. (2018). Theories of working memory: Differences in definition, degree of modularity, role of attention, and purpose. *Language, Speech, and Hearing Services in Schools, 49*(3), 340–355.

Addis, D. R. (2018). Are episodic memories special? on the sameness of remembered and imagined event simulation. *Journal of the Royal Society of New Zealand, 48*(2–3), 64–88.

Addis, D. R., Wong, A. T., & Schacter, D. L. (2007). Remembering the past and imagining the future: Common and distinct neural substrates during event construction and elaboration. *Neuropsychologia, 45*(7), 1363–1377.

Akaishi, R., Umeda, K., Nagase, A., & Sakai, K. (2014). Autonomous mechanism of internal choice estimate underlies decision inertia. *Neuron, 81*(1), 195–206.

Alós-Ferrer, C., Hügelschäfer, S., & Li, J. (2016). Inertia and decision making. *Frontiers in Psychology, 7*, Article 169 (9 pages).

Arzy, S., Molnar-Szakacs, I., & Blanke, O. (2008). Self in time: Imagined self-location influences neural activity related to mental time travel. *Journal of Neuroscience, 28*(25), 6502–6507.

Baars, B. J. (2005). Global workspace theory of consciousness: Toward a cognitive neuroscience of human experience. In S. Laureys (Ed.), *The boundaries of consciousness: Neurobiology and neuropathology*, Vol. 150 of *Progress in brain research* (pp. 45–53). Elsevier.

Baars, B. J., & Franklin, S. (2007). An architectural model of conscious and unconscious brain functions: Global workspace theory and IDA. *Neural Networks, 20*(9), 955–961. Brain and Consciousness.

Baars, B. J., Franklin, S., & Ramsoy, T. Z. (2013). Global workspace dynamics: Cortical "binding and propagation" enables conscious contents. *Frontiers in Psychology, 4*, Article 200 (22 pages).

Baars, B. J. (1988). *A cognitive theory of consciousness*. New York, NY: Oxford University Press Inc.

Baars, B. J. (1997). *In the theater of consciousness: The workspace of the mind*. New York, NY: Oxford University Press Inc.

Baars, B. J. (2002). The conscious access hypothesis: Origins and recent evidence. *Trends in Cognitive Sciences, 6*(1), 47–52.

Baddeley, A. D. (2007). *Working memory, thought, and action*. Oxford University Press.

Baddeley, A. (2000). The episodic buffer: A new component of working memory? *Trends in Cognitive Sciences, 4*(11), 417–423.

Baddeley, A. (2012). Working memory: Theories, models, and controversies. *Annual Review of Psychology, 63*(1), 1–29.

Bar, M. (Ed.). (2011). *Predictions in the brain: Using our past to generate a future*. New York, NY: Oxford University Press.

Bargh, J. A., & Morsella, E. (2008). The unconscious mind. *Perspectives on Psychological Science, 3*(1), 73–79.

Beckes, L., IJzerman, H., & Tops, M. (2015). Toward a radically embodied neuroscience of attachment and relationships. *Frontiers in Human Neuroscience, 9*, Article 266 (11 pages).

Bernecker, S., & Michaelian, K. (Eds.). (2017). *The Routledge handbook of philosophy of memory*. London: Routledge, Taylor & Francis Group.

Bernstein, N. A. (1935). The problem of interrelation between coordination and localization. *Archives of Biological Science, 38*, 1–35 (in Russian).

Black, D. L. (2000). Imagination and estimation: Arabic paradigms and western transformations. *Topoi, 19*(1), 59–75.

Blattner, W. (2005). Temporality. In H. L. Dreyfus & M. A. Wrathall (Eds.), *A companion to Heidegger (Blackwell companions to philosophy)* (pp. 311–324). Oxford: Blackwell Publishing Ltd.

Block, N. (1995). On a confusion about a function of consciousness. *Behavioral and Brain Sciences, 18*(2), 227–247.

Block, N. (2007). Consciousness, accessibility, and the mesh between psychology and neuroscience. *Behavioral and Brain Sciences, 30*(5–6), 481–499.

Block, N. (2007). Overflow, access, and attention. *Behavioral and Brain Sciences, 30*(5–6), 530–548.

Block, N. (2011). Perceptual consciousness overflows cognitive access. *Trends in Cognitive Sciences, 15*(12), 567–575.

Block, N. (2014). Rich conscious perception outside focal attention. *Trends in Cognitive Sciences, 18*(9), 445–447.

Boixader, D., Jacas, J., & Recasens, J. (2000). Fuzzy equivalence relations: Advanced material. In D. Dubois & H. Prade (Eds.), *Fundamentals of fuzzy sets* (pp. 261–290). New York: Springer Science+Business Media, LLC.

Bool, F. H., Kist, J. R., Locher, J. L., & Wierda, F. (1992). *M. C. Escher: His life and complete graphic work (with a fully illustrated catalogue)*. New York, NY: Abradale Press/Harry N. Abrams, Inc. Edited by J.L. Locher, Translated from the Dutch by Tony Langham and Plym Peters.
Borghi, A. M., & Pecher, D. (Eds.). (2012). *Embodied and grounded cognition*. Frontiers in Psychology, Frontiers Media SA.
Bratman, M. E. (1987). *Intention, plans, and practical reason*. Cambridge, MA: Cambridge University Press.
Braver, T. S., Satpute, A. B., Rush, B. K., Racine, C. A., & Barch, D. M. (2005). Context processing and context maintenance in healthy aging and early stage dementia of the alzheimer's type. *Psychology and Aging, 20*(1), 33–46.
Braver, T. S. (2012). The variable nature of cognitive control: A dual mechanisms framework. *Trends in Cognitive Sciences, 16*(2), 106–113.
Braver, T. S., Gray, J. R., & Burgess, G. Y. (2007). Explaining the many varieties of working memory variation: Dual mechanisms of cognitive control. In A. R. A. Conway, C. Jarrold, M. J. Kane, A. Miyake, & J. N. Towse (Eds.), *Variation in working memory* (pp. 76–106). Oxford: Oxford University Press.
Breeden, P., Dere, D., Zlomuzica, A., & Dere, E. (2016). The mental time travel continuum: On the architecture, capacity, versatility and extension of the mental bridge into the past and future. *Reviews in the Neurosciences, 27*(4), 421–434.
Broadway, J. M., Zedelius, C. M., Schooler, J. W., & Grondin, S. (Eds.). (2015). *The long and short of mental time travel-self-projection over time-scales large and small, frontiers in psychology*. Lausanne, SA: Frontiers Media.
Bronfman, Z. Z., Brezis, N., Jacobson, H., & Usher, M. (2014). We see more than we can report: "cost free" color phenomenality outside focal attention. *Psychological Science, 25*(7), 1394–1403.
Brown, R. (2014). Consciousness doesn't overflow cognition. *Frontiers in Psychology, 5*, Article 1399 (3 pages).
Brownstein, M. (2018). *The implicit mind: Cognitive architecture, the self, and ethics*. New York, NY: Oxford University Press.
Carruthers, P. (2017). Block's overflow argument. *Pacific Philosophical Quarterly, 98*(S1), 65–70.
Cartwright, N. (1989). *Nature's capacities and their measurement*. New York: Oxford University Press Inc.
Cartwright, N. (1999). *The dappled world: A study of the boundaries of science*. Cambridge: Cambridge University Press.
Casati, R., & Varzi, A. (2014). Events. In E. N. Zalta (Ed.), *The Stanford encyclopedia of philosophy* (winter 2015 ed.). Metaphysics Research Lab, Stanford University.
Chakraborty, M. K., & Das, M. (1983a). On fuzzy equivalence: I. *Fuzzy Sets and Systems, 11*(1), 185–193.
Chakraborty, M. K., & Das, M. (1983b). On fuzzy equivalence: II. *Fuzzy Sets and Systems, 11*(1), 299–307.
Chemero, A. (2013). Radical embodied cognitive science. *Review of General Psychology, 17*(2), 145–150.
Cheng, S., Werning, M., & Suddendorf, T. (2016). Dissociating memory traces and scenario construction in mental time travel. *Neuroscience & Biobehavioral Reviews, 60*, 82–89.
Ćirić, M., Ignjatović, J., & Bogdanović, S. (2007). Fuzzy equivalence relations and their equivalence classes. *Fuzzy Sets and Systems, 158*(12), 1295–1313.
Clark, A. (2008). *Supersizing the mind: Embodiment, action, and cognitive extension*. Oxford, NY: Oxford University Press.
Clark, A. (2015). *Surfing uncertainty: Prediction, action, and the embodied mind*. Oxford, NY: Oxford University Press.
Conway, M. A., Loveday, C., & Cole, S. N. (2016). The remembering-imagining system. *Memory Studies, 9*(3), 256–265.
Cottingham, J. (1985). Cartesian trialism. *Mind (New Series), 94*(374), 218–230.

Cowan, N. (2005). *Working memory capacity*, Essays in Cognitive Psychology. New York, NY: Psychology Press, Taylor & Francis Group.

Cowan, N. (2017a). The many faces of working memory and short-term storage. *Psychonomic Bulletin & Review, 24*(4), 1158–1170.

Cowan, N. (2017b). Working memory: The information you are now thinking of. In J. H. e. Byrne, J. T. v. Wixted (Eds.), *Learning and memory: A comprehensive reference* (2nd ed., Vol. 2, pp. 147–161) Cognitive Psycholoogy of Memory. Amsterdam: Elsevier Ltd.

Cowan, N. (2001). The magical number 4 in short-term memory: A reconsideration of mental storage capacity. *Behavioral and Brain Sciences, 24*(1), 87–114.

Cowan, N. (2015). Sensational memorability: Working memory for things we see, hear, feel, or somehow sense. In P. Jolicoeur, C. Lefebvre, & J. Martinez-Trujillo (Eds.), *Mechanisms of sensory working memory: Attention and performance XXV* (pp. 5–22). San Diego: Academic Press.

Crane, T. (2009). Is perception a propositional attitude? *The Philosophical Quarterly, 59*(236), 452–469.

Crane, T. (2013). *The objects of thought*. Oxford, UK: Oxford University Press.

Dainton, B. (2000). *Stream of consciousness: Unity and continuity in conscious experience* (6th ed.). London: Routledge, Taylor & Francis Group. revised edition, 2006.

Dainton, B. (2008). *The phenomenal self*. New York: Oxford University Press Inc.

D'Aloisio-Montilla, N. (2017). Imagery and overflow: We see more than we report. *Philosophical Psychology, 30*(5), 545–570.

De Brigard, F. (2017). Memory and imagination. In S. Bernecker & K. Michaelian (Eds.), *The Routledge handbook of philosophy of memory* (pp. 127–140). London: Routledge, Taylor & Francis Group.

De Brigard, F. (2014). Is memory for remembering? recollection as a form of episodic hypothetical thinking. *Synthese, 191*(2), 155–185.

De Brigard, F., & Gessell, B. S. (2016). Time is not of the essence: Understanding the neural correlates of mental time travel. In K. Michaelian, S. B. Klein, & K. K. Szpunar (Eds.), *Seeing the future: Theoretical perspectives on future-oriented mental time travel* (pp. 153–179). New York, NY: Oxford University Press.

Debus, D. (2017). Memory causation. In S. Bernecker & K. Michaelian (Eds.), *The Routledge handbook of philosophy of memory* (pp. 63–75). London: Routledge, Taylor & Francis Group.

Debus, D. (2014). 'mental time travel': Remembering the past, imagining the future, and the particularity of events. *Review of Philosophy and Psychology, 5*(3), 333–350.

Debus, D. (2018). Memory, imagination, and narrative. In F. MacPherson & F. Dorsch (Eds.), *Perceptual imagination and perceptual memory* (pp. 72–95). Oxford, UK: Oxford University Press.

Dehaene, S., & Changeux, J.-P. (2011). Experimental and theoretical approaches to conscious processing. *Neuron, 70*(2), 200–227.

Dehaene, S., Changeux, J.-P., Naccache, L., Sackur, J., & Sergent, C. (2006). Conscious, preconscious, and subliminal processing: A testable taxonomy. *Trends in Cognitive Sciences, 10*(5), 204–211.

Dehaene, S., Kerszberg, M., & Changeux, J.-P. (1998). A neuronal model of a global workspace in effortful cognitive tasks. *Proceedings of the National Academy of Sciences, 95*(24), 14529–14534.

Dere, E., Dere, D., de Souza Silva, M. A., Huston, J. P., & Zlomuzica, A. (2018). Fellow travellers: Working memory and mental time travel in rodents. *Behavioural Brain Research, 352*, 2–7. Special Issue: Animal Model of the Year 2036: Novel Perspectives in Behavioral Neuroscience.

Devitt, A. L., & Addis, D. R. (2016). Bidirectional interactions between memory and imagination. In K. Michaelian, S. B. Klein & K. K. Szpunar (Eds.), *Seeing the future: Theoretical perspectives on future-oriented mental time travel* (chapter 5, pp. 93–115). New York, NY: Oxford University Press.

Di Paolo, E., & Thompson, E. (2014). The enactive approach. In L. Shapiro (Ed.), *The Routledge handbook of embodied cognition* (pp. 68–71). London: Routledge, Taylor & Francis Group.

Di Paolo, E. A. (2005). Autopoiesis, adaptivity, teleology, agency. *Phenomenology and the Cognitive Sciences, 4*(4), 429–452.

Doplicher, S., Fredenhagen, K., & Roberts, J. E. (1995). The quantum structure of spacetime at the planck scale and quantum fields. *Communications in Mathematical Physics, 172*(1), 187–220.

Elliott, J. C., Baird, B., & Giesbrecht, B. (2016). Consciousness isn't all-or-none: Evidence for partial awareness during the attentional blink. *Consciousness and Cognition, 40,* 79–85.

Engle, R. W., & Kane, M. J. (2003). Executive attention, working memory capacity, and a two-factor theory of cognitive control. In B. H. Ross (Ed.), *The psychology of learning and motivation: Advances in research and theory* (Vol. 44, pp. 145–199). San Diego, CA: Elsevier Academic Press.

Engle, R. W. (2002). Working memory capacity as executive attention. *Current Directions in Psychological Science, 11*(1), 19–23.

Fazekas, P., & Overgaard, M. (2018). Perceptual consciousness and cognitive access: An introduction. *Philosophical Transactions of the Royal Society B: Biological Sciences, 373*(1755), 20170340 (4 pages).

Fernández, J. (2018). The functional character of memory. In K. Michaelian, D. Debus & D. Perrin (Eds.), *New directions in the philosophy of memory* (pp. 53–72). New York, NY: Routledge, Taylor & Francis Group.

Gallagher, S. (2000a). Philosophical conceptions of the self: Implications for cognitive science. *Trends in Cognitive Sciences, 4*(1), 14–21.

Gallagher, S. (2000b). Self-reference and schizophrenia: A cognitive model of immunity to error through misidentification. In D. Zahavi (Ed.), *Exploring the self: Philosophical and psychopathological perspectives on self-experience* (pp. 203–239). Amsterdam: John Benjamins Publishing Co.

Gallagher, S. (2014). Phenomenology and embodied cognition. In L. Shapiro (Ed.), *The Routledge handbook of embodied cognition* (pp. 9–18). London: Routledge, Taylor & Francis Group.

Gallagher, S. (2016). Timing is not everything: The intrinsic temporality of action. In R. Altshuler & M. J. Sigrist (Eds.), *Time and the philosophy of action* (pp. 205–221). New York, NY: Routledge, Taylor& Francis Group.

Gallagher, S. (2017a). *Enactivist interventions: Rethinking the mind.* Oxford, UK: Oxford University Press.

Gallagher, S. (2017b). The past, present and future of time-consciousness: From Husserl to Varela and beyond. *Constructivist Foundations, 13*(1), 91–97.

Gallagher, S., & Zahavi, D. (2012). *The phenomenological mind* (2nd ed.). London: Routledge, Taylor & Francis Group.

Gallagher, S. (2007). The natural philosophy of agency. *Philosophy Compass, 2*(2), 347–357.

Gallagher, S. (2011). Time in action. In C. Callender (Ed.), *The Oxford handbook of philosophy of time* (pp. 420–438). Oxford: Oxford University Press.

Gallagher, S. (2012). Multiple aspects in the sense of agency. *New Ideas in Psychology, 30*(1), 15–31.

Ganeri, J. (2017). Mental time travel and attention. *Australasian Philosophical Review, 1*(4), 353–373.

Genzel, L., & Wixted, J. T. (2017). Cellular and systems consolidation of declarative memory. In N. Axmacher & B. Rasch (Eds.), *Cognitive neuroscience of memory consolidation* (pp. 3–16). Cham, Switzerland: Springer International Publishing.

Gibb, S. C., Lowe, E. J., & Ingthorsson, R. D. (Eds.). (2013). *Mental causation and ontology.* Oxford, UK: Oxford University Press.

Goldie, P. (2012). *The mess inside: Narrative, emotion, & the mind.* Oxford, UK: Oxford University Press.

Goldinger, S. D., Papesh, M. H., Barnhart, A. S., Hansen, W. A., & Hout, M. C. (2016). The poverty of embodied cognition. *Psychonomic Bulletin & Review, 23*(4), 959–978.

Goldstein, K. (1931/1971). Über zeigen und greifen. In A. Gurwitsch, E. M. G. Haudek & W. E. Haudek (Eds.), *Selected Papers/Ausgewählte Schriften* (pp. 263–281). Netherlands, Dordrecht: Springer.

Goldstein, E. B. (2015). *Cognitive psychology: Connecting mind, research, and everyday experience* (4th ed.). Stamford, USA, Wadsworth: Cengage Learning.

Gross, S. (2018). Perceptual consciousness and cognitive access from the perspective of capacity-unlimited working memory. *Philosophical Transactions of the Royal Society of London B: Biological Sciences, 373*(1755), 20170343 (11 pages).

Guillot, A., Hoyek, N., & Collet, C. (2012). Mental rotation and functional learning. In N. M. Seel (Ed.), *Encyclopedia of the sciences of learning* (pp. 2222–2223). Boston, MA: Springer Science+Business Media, LLC.

Hassabis, D., Kumaran, D., Vann, S. D., & Maguire, E. A. (2007). Patients with hippocampal amnesia cannot imagine new experiences. *Proceedings of the National Academy of Sciences, 104*(5), 1726–1731.

Hassabis, D., & Maguire, E. A. (2007). Deconstructing episodic memory with construction. *Trends in Cognitive Sciences, 11*(7), 299–306.

Haugeland, J. (2013). *Dasein disclosed.* Cambridge, MA: Harvard University Press. Edited by Joseph Rouse.

Healey, R. (2016). Holism and nonseparability in physics. In E. N. Zalta (Ed.), *The Stanford encyclopedia of philosophy* (spring 2016 ed.). Metaphysics Research Lab, Stanford University.

Heidegger, M. (1927/1996). *Being and time.* State University of New York, Albany. A Translation of Sein und Zeit by Joan Stambaugh.

Hintzman, D. L. (1986). "schema abstraction" in a multiple-trace memory model. *Psychological Review, 93*(4), 411–428.

Hintzman, D. L. (1988). Judgments of frequency and recognition memory in a multiple-trace memory model. *Psychological Review, 95*(4), 528–551.

Hintzman, D. L., & Block, R. A. (1971). Repetition and memory: Evidence for a multiple-trace hypothesis. *Journal of Experimental Psychology, 88*(3), 297–306.

Howard, M. W. (2018). Memory as perception of the past: Compressed time in mind and brain. *Trends in Cognitive Sciences, 22*(2), 124–136.

Husserl, E. (1991). *On the phenomenology of the consciousness of internal time (1893–1917).* Translated by Brough, John Barnett. Dordrecht: Kluwer Academic Publishers.

Hutto, D. D., & McGivern, P. (2016). Updating the story of mental time travel: Narrating and engaging with our possible pasts and futures. In R. Altshuler & M. J. Sigrist (Eds.), *Time and the philosophy of action* (pp. 141–157). New York, NY: Routledge, Taylor& Francis Group.

Ingvar, D. H. (1985). Memory of the future: An essay on the temporal organization of conscious awareness. *Human Neurobiology, 4*(3), 127–136.

Jacobs, C., & Silvanto, J. (2015). How is working memory content consciously experienced? the 'conscious copy' model of WM introspection. *Neuroscience & Biobehavioral Reviews, 55,* 510–519.

Jacoby, L. L., Kelley, C. M., & McElree, B. D. (1999). The role of cognitive control: Early selection versus late correction. In S. Chaiken & Y. Trope (Eds.), *Dual-process theories in social psychology* (pp. 383–400). New York, NY: The Guilford Press.

James, W. (1890). *The principles of psychology* (Vol. 1). New York: Henry Holt and Company.

Jung, D., & Dorner, V. (2018). Decision inertia and arousal: Using neurois to analyze bio-physiological correlates of decision inertia in a dual-choice paradigm. In F. D. Davis, R. Riedl, J. vom Brocke, P.-M. Léger & A. B. Randolph (Eds.), *Information systems and neuroscience (Lecture notes in information systems and organisation, Vol. 25)* (pp. 159–166). Cham: Springer International Publishing AG.

Jung, D., Stäbler, J., & Weinhardt, C. (2018). *Investigating cognitive foundations of inertia in decision-making: Discussion paper heikamaxy 2018.* Technical report, Karlsruher Institut für Technologie (KIT).

Kane, M. J., Conway, A. R. A., Hambrick, D. Z., & Engle, R. W. (2007). Variation in working memory capacity as variation in executive attention and control. In A. R. A. Conway, C. Jarrold, M. J. Kane, A. Miyake, & J. N. Towse (Eds.), *Variation in working memory* (pp. 21–48). Oxford: Oxford University Press.

Käufer, S. (2005). Logic. In H. L. Dreyfus & M. A. Wrathall (Eds.), *A companion to Heidegger (Blackwell companions to philosophy)* (pp. 141–155). Oxford, UK: Blackwell Publishing Ltd.
Kelly, S. D. (2003). Merleau-ponty on the body. In M. A. Proudfoot (Ed.), *The philosophy of body* (pp. 62–76). Malden, MA: Wiley-Blackwell.
Kind, A. (Ed.). (2017). *The Routledge handbook of philosophy of imagination*. London: Routledge, The Taylor & Francis Group.
Kiverstein, J., & Miller, M. (2015). The embodied brain: Towards a radical embodied cognitive neuroscience. *Frontiers in Human Neuroscience, 9*, Article 237 (11 pages).
Klein, S. B., Loftus, J. & Kihlstrom, J. F. (2002). Memory and temporal experience: The effects of episodic memory loss on an amnesic patient's ability to remember the past and imagine the future. *Social Cognition, 20*(5), 353–379.
Klein, S., & Steindam, C. (2016). The role of subjective temporality in future-oriented mental time travel. In K. Michaelian, S. Klein, & K. Szpunar (Eds.), *Seeing the future: Theoretical perspectives on future-oriented mental time travel* (pp. 135–152). Oxford: Oxford University Press Inc.
Kon, M., & Miller, K. (2015). Temporal experience: Models, methodology and empirical evidence. *Topoi, 34*(1), 201–216.
Kouider, S., & Dupoux, E. (2004). Partial awareness creates the "illusion" of subliminal semantic priming. *Psychological Science, 15*(2), 75–81.
Kriegel, U. (2007). Philosophical theories of consciousness: Contemporary western perspectives. In P. D. Zelazo, M. Moscovitch, & E. Thompson (Eds.), *The Cambridge handbook of consciousness* (pp. 35–66). Cambridge, UK: Cambridge University Press.
Kroes, M. C. W., & Fernández, G. (2012). Dynamic neural systems enable adaptive, flexible memories. *Neuroscience & Biobehavioral Reviews, 36*(7), 1646–1666. Memory Formation.
Lamme, V. A. F. (2018). Challenges for theories of consciousness: Seeing or knowing, the missing ingredient and how to deal with panpsychism. *Philosophical Transactions of the Royal Society of London B: Biological Sciences, 373*(1755), 20170344 (12 pages).
Li, B., & He, W. (2014). The structures of intuitionistic fuzzy equivalence relations. *Information Sciences, 278*, 883–899.
Lohse, M., & Overgaard, M. (2017). Emotional priming depends on the degree of conscious experience. *Neuropsychologia*. on-line.
Lubashevsky, I., & Morimura, K. (2019). Physics of mind and car-following problem. In B. S. Kerner (Ed.), *Complex dynamics of traffic management, encyclopedia of complexity and systems science series* (pp. 559–592). New York, NY: Springer Science+Business Media, LLC.
Lubashevsky, I. (2017). *Physics of the human mind*. Cham: Springer International Publishing AG.
Lycan, W. (2015). Representational theories of consciousness. In E. N. Zalta (Ed.), *The Stanford encyclopedia of philosophy* (summer 2015 ed.). Metaphysics Research Lab, Stanford University.
Madore, J. (1997). Classical gravity on fuzzy space-time. *Nuclear Physics B - Proceedings Supplements, 56*(3), 183–190. Theory of Elementary Particles.
Mahon, B. Z., & Caramazza, A. (2008). A critical look at the embodied cognition hypothesis and a new proposal for grounding conceptual content. *Journal of Physiology-Paris, 102*(1), 59–70. Special issue: Links and Interactions Between Language and Motor Systems in the Brain.
Mahon, B. Z. (2015). What is embodied about cognition? *Language, Cognition and Neuroscience, 30*(4), 420–429.
Marchetti, G. (2014). Attention and working memory: Two basic mechanisms for constructing temporal experiences. *Frontiers in Psychology, 5*, Article 880 (15 pages).
Marchetti, G. (2010). *Consciousness, attention and meaning*. New York: Nova Science Publisher Inc.
Martin, C. B., & Deutscher, M. (1966). Remembering. *The Philosophical Review, 75*(2), 161–196.
Maturana, H. R., & Varela, F. J. (1980). *Autopoiesis and cognition: The realization of the living* (2nd ed.). Dordrecht: D. Reidel Publishing Co.
Mayer, H. C., & Krechetnikov, R. (2012). Walking with coffee: Why does it spill? *Physical Review E, 85*(4), 046117 (7 pages).

McElree, B. (2001). Working memory and focal attention. *Journal of Experimental Psychology: Learning, Memory, and Cognition, 27*(3), 817–835.

Merleau-Ponty, M. (1945/2005). *Phenomenology of perception*. Paris: Gallimard. Translated by Colin Smith, edition published in the Taylor and Francis e-Library, 2005.

Michaelian, K. (2016a). Confabulating, misremembering, relearning: The simulation theory of memory and unsuccessful remembering. *Frontiers in Psychology, 7*, Article 1857 (13 pages).

Michaelian, K. (2016b). *Mental time travel: Episodic memory and our knowledge of the personal past*. Cambridge: The MIT Press.

Michaelian, K., & Robin, S. K. (2018). Beyond the causal theory? fifty years after Martin and Deutscher. In K. Michaelian, D. Debus & D. Perrin (Eds.), *New directions in the philosophy of memory* (pp. 13–32). New York, NY: Routledge, Taylor & Francis Group.

Michaelian, K., & Sutton, J. (2017). Memory. In E. N. Zalta (Ed.), *The Stanford encyclopedia of philosophy* (summer 2017 ed.). Metaphysics Research Lab, Stanford University.

Michaelian, K., Klein, S. B., & Szpunar, K. K. (Eds.). (2016). *Seeing the future: Theoretical perspectives on future-oriented mental time travel*. New York, NY: Oxford University Press.

Miller, E. K., & Cohen, J. D. (2001). An integrative theory of prefrontal cortex function. *Annual Review of Neuroscience, 24*(1), 167–202 PMID: 11283309.

Miloyan, B., McFarlane, K. A., & Suddendorf, T. (2019). Measuring mental time travel: Is the hippocampus really critical for episodic memory and episodic foresight?, *Cortex*. Available online 13 February 2019.

Moscovitch, M. (2007). Memory: Why the engram is elusive. In H. L. Roediger III, Y. Dudai & S. M. Fitzpatrick (Eds.), *Science of memory: Concepts* (pp. 17–21). Oxford University Press.

Moscovitch, M., Nadel, L., Winocur, G., Gilboa, A., & Rosenbaum, R. S. (2006). The cognitive neuroscience of remote episodic, semantic and spatial memory. *Current Opinion in Neurobiology, 16*(2), 179–190. Cognitive Neuroscience (Special Issue).

Moscovitch, M., Cabeza, R., Winocur, G., & Nadel, L. (2016). Episodic memory and beyond: The hippocampus and neocortex in transformation. *Annual Review of Psychology, 67*(1), 105–134.

Moscovitch, M., & Nadel, L. (1998). Consolidation and the hippocampal complex revisited: In defense of the multiple-trace model. *Current Opinion in Neurobiology, 8*(2), 297–300.

Mullally, S. L., & Maguire, E. A. (2014). Memory, imagination, and predicting the future: A common brain mechanism? *The Neuroscientist, 20*(3), 220–234.

Murali, V. (1989). Fuzzy equivalence relations. *Fuzzy Sets and Systems, 30*(2), 155–163.

Mylopoulos, M., & Pacherie, E. (2019). Intentions: The dynamic hierarchical model revisited. *Wiley Interdisciplinary Reviews: Cognitive Science, 10*(2), e1481.

Naccache, L. (2018). Why and how access consciousness can account for phenomenal consciousness. *Philosophical Transactions of the Royal Society of London B: Biological Sciences, 373*(1755), 20170357 (9 pages).

Nadel, L., Hupbach, A., Gomez, R., & Newman-Smith, K. (2012). Memory formation, consolidation and transformation. *Neuroscience & Biobehavioral Reviews, 36*(7), 1640–1645. Memory Formation (Special Issue).

Nadel, L., & Moscovitch, M. (1997). Memory consolidation, retrograde amnesia and the hippocampal complex. *Current Opinion in Neurobiology, 7*(2), 217–227.

Nadel, L., Samsonovich, A., Ryan, L., & Moscovitch, M. (2000). Multiple trace theory of human memory: Computational, neuroimaging, and neuropsychological results. *Hippocampus, 10*(4), 352–368.

Nadel, L., Winocur, G., Ryan, L., & Moscovitch, M. (2007). Systems consolidation and hippocampus: Two views. *Debates in Neuroscience, 1*(2), 55–66.

Nagel, T. (1974). What is it like to be a bat? *The Philosophical Review, 83*(4), 435–450.

Newell, A. (1994). *Unified theories of cognition (The William James Lectures, 1987)*. Cambridge, MA: Harvard University Press.

Nieuwenhuis, S., & De Kleijn, R. (2011). Consciousness of targets during the attentional blink: A gradual or all-or-none dimension? *Attention, Perception, & Psychophysics, 73*(2), 364–373.

Njikeh, K. D. (2019). Derician trialism: The concept of human composition into the mind, submind and body substances/components. *International Journal of Philosophy, 7*(1), 17–19.
Nyberg, L., Kim, A. S. N., Habib, R., Levine, B., & Tulving, E. (2010). Consciousness of subjective time in the brain. *Proceedings of the National Academy of Sciences, 107*(51), 22356–22359.
Oberauer, K. (2002). Access to information in working memory: Exploring the focus of attention. *Journal of Experimental Psychology: Learning, Memory, and Cognition, 28*(3), 411–421.
O'Callaghan, J. (2017). Thomas aquinas. In S. Bernecker & K. Michaelian (Eds.), *The Routledge handbook of philosophy of memory* (pp. 461–469). New York, NY: Routledge, Taylor & Francis Group.
Odegaard, B., Chang, M. Y., Lau, H. & Cheung, S.-H. (2018). Inflation versus filling-in: Why we feel we see more than we actually do in peripheral vision. *Philosophical Transactions of the Royal Society of London B: Biological Sciences, 373*(1755), 20170345 (10 pages).
Ovchinnikov, S. (1991). Similarity relations, fuzzy partitions, and fuzzy orderings. *Fuzzy Sets and Systems, 40*(1), 107–126.
Ovchinnikov, S. (2000). An introduction to fuzzy relations. In D. Dubois & H. Prade (Eds.), *Fundamentals of fuzzy sets* (pp. 233–259). New York: Springer Science+Business Media, LLC.
Overgaard, M. (2018). Phenomenal consciousness and cognitive access. *Philosophical Transactions of the Royal Society of London B: Biological Sciences, 373*(1755), 20170353 (6 pages).
Overgaard, M., Rote, J., Mouridsen, K., & Ramsøy, T. Z. (2006). Is conscious perception gradual or dichotomous? a comparison of report methodologies during a visual task. *Consciousness and Cognition, 15*(4), 700–708. Special Issue on Introspection.
Pacherie, E. (2000). The content of intentions. *Mind & Language, 15*(4), 400–432.
Pacherie, E. (2006). Toward a dynamic theory of intentions. In S. Pockett, W. P. Banks, & S. Gallagher (Eds.), *Does consciousness cause behavior?* (pp. 145–167). Cambridge, MA: The MIT Press.
Pacherie, E. (2007). The sense of control and the sense of agency. *Journal Psyche, 13*(1), 1–30.
Pacherie, E. (2008). The phenomenology of action: A conceptual framework. *Cognition, 107*(1), 179–217.
Pemberton, J., & Cartwright, N. (2014). Ceteris paribus laws need machines to generate them. *Erkenntnis, 79*(10), 1745–1758.
Peremen, Z., & Lamy, D. (2014). Do conscious perception and unconscious processing rely on independent mechanisms? a meta-contrast study. *Consciousness and Cognition, 24*, 22–32.
Perrin, D. (2016). Asymmetries in subjective time In K. Michaelian, S. B. Klein & K. K. Szpunar (Eds.), *Seeing the future: Theoretical perspectives on future-oriented mental time travel* (chapter 3, pp. 39–61). New York, NY: Oxford University Press.
Perrin, D. (2018). A case for procedural causality in episodic recollection. In K. Michaelian, D. Debus & D. Perrin (Eds.), *New directions in the philosophy of memory* (pp. 33–51). New York, NY: Routledge, Taylor & Francis Group.
Perrin, D., & Michaelian, K. (2017). Memory as mental time travel. In S. Bernecker & K. Michaelian (Eds.), *The Routledge handbook of philosophy of memory* (pp. 228–239). London: Routledge, Taylor & Francis Group.
Phillips, I. (2018). The methodological puzzle of phenomenal consciousness. *Philosophical Transactions of the Royal Society of London B: Biological Sciences, 373*(1755), 20170347 (9 pages).
Pitt, D. (2020). Mental representation. In E. N. Zalta (Ed.), *The Stanford encyclopedia of philosophy* (Spring 2020 ed.). Metaphysics Research Lab, Stanford University.
Pykacz, J. (2013). Fuzzy sets in foundations of quantum mechanics. In R. Seising, E. Trillas, C. Moraga & S. Termini (Eds.), *On fuzziness: A homage to Lotfi A. Zadeh* (Vol. 2, pp. 553–557). Berlin: Springer.
Raffone, A., Srinivasan, N., & van Leeuwen, C. (2014). The interplay of attention and consciousness in visual search, attentional blink and working memory consolidation. *Philosophical Transactions of the Royal Society of London. Series B, Biological Sciences, 369*(1641), 20130215 (16 pages).
Raffone, A., & Pantani, M. (2010). A global workspace model for phenomenal and access consciousness. *Consciousness and Cognition, 19*(2), 580–596.

Redick, T. S. (2014). Cognitive control in context: Working memory capacity and proactive control. *Acta Psychologica, 145,* 1–9.

Resulaj, A., Kiani, R., Wolpert, D. M., & Shadlen, M. N. (2009). Changes of mind in decision-making. *Nature, 461,* 263–266.

Robb, D., & Heil, J. (2018). Mental causation. In E. N. Zalta (Ed.), *The Stanford encyclopedia of philosophy* (summer 2019 ed.). Metaphysics Research Lab, Stanford University.

Robins, S. (2016). Representing the past: Memory traces and the causal theory of memory. *Philosophical Studies, 173*(11), 2993–3013.

Rowlands, M. (2010). *The new science of the mind: From extended mind to embodied phenomenology*. Cambridge, MA: The MIT Press.

Rudman, L. A., & Spencer, S. J. (2007). The implicit self. *Self and Identity, 6*(2–3), 97–100.

Samuelson, W., & Zeckhauser, R. (1988). Status quo bias in decision making. *Journal of Risk and Uncertainty, 1*(1), 7–59.

Sartre, J.-P.S. (1956). *Being and nothingness: A phenomenological essay on ontology*. New York: Washington Square Press. Translated by Hazel E. Barnes.

Schacter, D. L., & Addis, D. R. (2007a). The cognitive neuroscience of constructive memory: Remembering the past and imagining the future. *Philosophical Transactions of the Royal Society B: Biological Sciences, 362*(1481), 773–786.

Schacter, D. L., & Addis, D. R. (2007b). Constructive memory: The ghosts of past and future. *Nature, 445,* 27.

Schacter, D. L., Addis, D. R., & Buckner, R. L. (2007). Remembering the past to imagine the future: The prospective brain. *Nature Reviews Neuroscience, 8*(9), 657–661.

Schacter, D. L., Addis, D. R., Hassabis, D., Martin, V. C., Spreng, R. N., & Szpunar, K. K. (2012). The future of memory: Remembering, imagining, and the brain. *Neuron, 76*(4), 677–694.

Schacter, D. L., & Madore, K. P. (2016). Remembering the past and imagining the future: Identifying and enhancing the contribution of episodic memory. *Memory Studies, 9*(3), 245–255.

Schlosser, M. (2015). Agency. In E. N. Zalta (Ed.), *The Stanford encyclopedia of philosophy* (fall 2015 ed.). Metaphysics Research Lab, Stanford University.

Schmitt, I., Nürnberger, A., & Lehrack, S. (2009). On the relation between fuzzy and quantum logic. In R. Seising (Ed.), *Views on fuzzy sets and systems from different perspectives: Philosophy and logic, criticisms and applications* (pp. 417–438). Berlin: Springer.

Scholtz, F. G. (2018). Classical dynamics on three-dimensional fuzzy space. *Physical Review D, 98*(10), 104058 (18 pages).

Sekeres, M. J., Winocur, G. and Moscovitch, M. (2018). The hippocampus and related neocortical structures in memory transformation. *Neuroscience Letters, 680,* 39–53. New Perspectives on the Hippocampus and Memory (Special Issue).

Sekeres, M. J., Moscovitch, M., & Winocur, G. (2017). Mechanisms of memory consolidation and transformation. In N. Axmacher & B. Rasch (Eds.), *Cognitive neuroscience of memory consolidation* (pp. 17–44). Cham, Switzerland: Springer International Publishing.

Sergent, C., & Dehaene, S. (2004). Is consciousness a gradual phenomenon?: Evidence for an all-or-none bifurcation during the attentional blink. *Psychological Science, 15*(11), 720–728.

Shanton, K., & Goldman, A. (2010). Simulation theory. *Wiley Interdisciplinary Reviews: Cognitive Science, 1*(4), 527–538.

Shapiro, L. (2019). *Embodied cognition* (2nd ed.). London: Routledge, Taylor & Francis Group.

Shapiro, L. (Ed.). (2014). *The Routledge handbook of embodied cognition*. London: Routledge, Taylor & Francis Group.

Shapiro, L. A. (2019). Flesh matters: The body in cognition. *Mind & Language, 34*(1), 3–20.

Shepard, R. N., & Metzler, J. (1971). Mental rotation of three-dimensional objects. *Science, 171*(3972), 701–703.

Shepard, S., & Metzler, D. (1988). Mental rotation: Effects of dimensionality of objects and type of task. *Journal of Experimental Psychology: Human Perception and Performance, 14*(1), 3–11.

Shipstead, Z., Lindsey, D. R., Marshall, R. L., & Engle, R. W. (2014). The mechanisms of working memory capacity: Primary memory, secondary memory, and attention control. *Journal of Memory and Language, 72*, 116–141.

Sidharth, B. G. (2005). *The universe of fluctuations: The architecture of spacetime and the universe*. Dordrecht, The Netherlands: Springer. (see also arXiv:physics/0311040).

Siegel, S. (2016). The contents of perception. In E. N. Zalta (Ed.), *The Stanford encyclopedia of philosophy* (winter 2016 ed.). Metaphysics Research Lab, Stanford University.

Simmering, V. R. (2016). *Working memory capacity in context: Modeling dynamic processes of behavior, memory, and development*. Monographs of the Society for Research in Child Development. With commentaries by Nelson Cowan.

Singh, I., & Howard, M. W. (2017). Recency order judgments in short term memory: Replication and extension of hacker (1980), *bioRxiv*.

Singh, I., Tiganj, Z., & Howard, M. W. (2018). Is working memory stored along a logarithmic timeline? converging evidence from neuroscience, behavior and models. *Neurobiology of Learning and Memory, 153, Part A*, 104–110. MCCS 2018: Time and Memory.

Smith, D. W. (2013). Phenomenology. In E. N. Zalta (Ed.), *The Stanford encyclopedia of philosophy* (summer 2018 ed.). Stanford, CA: Metaphysics Research Lab, Stanford University. https://plato.stanford.edu/archives/sum2018/entries/phenomenology/.

Snyder, H. S. (1947). Quantized space-time. *Physical Review, 71*(1), 38–41.

Stazicker, J. (2018). Partial report is the wrong paradigm. *Philosophical Transactions of the Royal Society of London B: Biological Sciences, 373*(1755), 20170350 (9 pages).

Stich, S. P. (1978). Beliefs and subdoxastic states. *Philosophy of Science, 45*(4), 499–518.

Sutton, J. (1998). *Philosophy and memory traces: Descartes to connectionism*. Cambridge, UK: Cambridge University Press.

Szpunar, K. K. & Schacter, D. L. (2018). Memory and future imagining. In J. T. E.-i.-C. Wixted, E. A. Phelps & L. V. E. Davachi (Eds.), *Stevens' handbook of experimental psychology and cognitive neuroscience. Volume 1: Learning & memory* (4th ed., chapter 5, pp. 145–168). Wiley: New York.

Szpunar, K. K., Spreng, R. N., & Schacter, D. L. (2016). Toward a taxonomy of future thinking. In K. Michaelian, S. B. Klein, & K. K. Szpunar (Eds.), *Seeing the future: Theoretical perspectives on future-oriented mental time travel* (pp. 21–35). New York, NY: Oxford University Press.

Szpunar, K. K., Watson, J. M., & McDermott, K. B. (2007). Neural substrates of envisioning the future. *Proceedings of the National Academy of Sciences, 104*(2), 642–647.

Thomas, N. J. T. (2014). Mental imagery. In E. N. Zalta (Ed.), *The Stanford encyclopedia of philosophy* (summer 2019 ed.). Metaphysics Research Lab, Stanford University.

Thompson, E. (2007). *Mind in life: Biology, phenomenology, and the sciences of mind*. Cambridge, MA: The Belknap Press of Harvard University Press.

Thompson, E., & Zahavi, D. (2007). Philosophical issues: Phenomenology. In P. D. Zelazo, M. Moscovitch, & E. Thompson (Eds.), *The Cambridge handbook of consciousness* (pp. 67–87). Cambridge, UK: Cambridge University Press.

Todosiev, E. P. (1963). *The action point model of the driver-vehicle system*, PhD thesis, The Ohio State University. (Ph.D. Dissertation, Ohio State University, 1963).

Todosiev, E. P., & Barbosa, L. C. (1963/64). A proposed model for the driver–vehicle system. *Traffic Engineering, 34*, 17—20.

Tulving, E. (1985). Memory and consciousness. *Canadian Psychology/Psychologie canadienne, 26*(1), 1–12.

Tulving, E. (2002). Chronesthesia: Conscious awareness of subjective time. In D. T. Stuss & R. T. Knight (Eds.), *Principles of Frontal Lobe function* (chapter 20, pp. 311–325). New York, NY: Oxford University Press.

Tulving, E. (2005). Episodic memory and autonoesis: Uniquely human?. In H. S. Terrace & J. Metcalfe (Eds.), *The missing link in cognition: Origins of self-reflective consciousness* (chapter 1, pp. 3–56). New York, NY: Oxford University Press.

Tulving, E. (1972). Episodic and semantic memory. In E. Tulving & W. Donaldson (Eds.), *Organization of memory* (pp. 381–402). Cambridge: Academic Press.

Uchaikin, V. V. (2013a). *Fractional derivatives for physicists and engineers: Volume I. Background and theory*. Beijing: Higher Education Press; Berlin: Springer.

Uchaikin, V. V. (2013b). *Fractional derivatives for physicists and engineers: Volume II: Applications*. Beijing: Higher Education Press; Berlin: Springer.

Unsworth, N., & Engle, R. W. (2007). The nature of individual differences in working memory capacity: Active maintenance in primary memory and controlled search from secondary memory. *Psychological Review, 114*(1), 104–132.

Unsworth, N., & Engle, R. W. (2007). On the division of short-term and working memory: An examination of simple and complex span and their relation to higher order abilities. *Psychological Bulletin, 133*(6), 1038–1066.

Unsworth, N., Redick, T. S., Spillers, G. J., & Brewer, G. A. (2012). Variation in working memory capacity and cognitive control: Goal maintenance and microadjustments of control. *Quarterly Journal of Experimental Psychology, 65*(2), 326–355.

Usher, M., Bronfman, Z. Z., Talmor, S., Jacobson, H., & Eitam, B. (2018). Consciousness without report: Insights from summary statistics and inattention 'blindness'. *Philosophical Transactions of the Royal Society of London B: Biological Sciences, 373*(1755), 20170354 (11 pages).

Van den Berg, R., Zylberberg, A., Kiani, R., Shadlen, M. N., & Wolpert, D. M. (2016). Confidence is the bridge between multi-stage decisions. *Current Biology, 26*(23), 3157–3168.

Vandekerckhove, M., Bulnes, L. C., & Panksepp, J. (2014). The emergence of primary anoetic consciousness in episodic memory. *Frontiers in Behavioral Neuroscience, 7*, Article 210 (8 pages).

Vandenberg, S. G., & Kuse, A. R. (1978). Mental rotations, a group test of three-dimensional spatial visualization. *Perceptual and Motor Skills, 47*(2), 599–604.

Varela, F. J., Thompson, E., & Rosch, E. (2016). *The embodied mind: Cognitive science and human experience* (Revized edition ed.). Cambridge, MA: The MIT Press. New foreword by Jon Kabat-Zinn and new introductions by Evan Thompson and Eleanor Rosch.

Varela, F. J. (1979). *Principles of biological autonomy*. New York: Elsevier North Holland.

Varela, F. J., Thompson, E., & Rosch, E. (1991). *The embodied mind: Cognitive science and human experience*. Cambridge, MA: The MIT Press.

Wackermann, J. (2007). Inner and outer horizons of time experience. *The Spanish Journal of Psychology, 10*(1), 20–32.

Walker, C. M., & Gopnik, A. (2013). Causality and imagination. In M. Taylor (Ed.), *The Oxford handbook of the development of imagination* (pp. 342–358). New York, NY: Oxford University Press.

Ward, E. J. (2018). Downgraded phenomenology: How conscious overflow lost its richness. *Philosophical Transactions of the Royal Society of London B: Biological Sciences, 373*(1755), 20170355 (6 pages).

Wiemers, E. A., & Redick, T. S. (2018). Working memory capacity and intra-individual variability of proactive control. *Acta Psychologica, 182*, 21–31.

Wilson, A. D. & Golonka, S. (2013). Embodied cognition is not what you think it is. *Frontiers in Psychology, 4*, Article 58 (13 pages).

Wilson, R. A., & Foglia, L. (2015). Embodied cognition. In E. N. Zalta (Ed.), *The Stanford encyclopedia of philosophy* (winter 2016 ed.). Metaphysics Research Lab, Stanford University.

Wilson, G., & Shpall, S. (2012). Action. In E. N. Zalta (Ed.), *The Stanford encyclopedia of philosophy* (winter 2016 ed.). Metaphysics Research Lab, Stanford University.

Windey, B., Vermeiren, A., Atas, A., & Cleeremans, A. (2014). The graded and dichotomous nature of visual awareness. *Philosophical Transactions of the Royal Society of London B: Biological Sciences, 369*(1641), 20130282 (11 pages).

Windey, B., & Cleeremans, A. (2015). Consciousness as a graded and an all-or-none phenomenon: A conceptual analysis. *Consciousness and Cognition, 35*, 185–191.

Winocur, G., & Moscovitch, M. (2011). Memory transformation and systems consolidation. *Journal of the International Neuropsychological Society, 17*(5), 766–780.

Wittmann, M. (2011). Moments in time. *Frontiers in Integrative Neuroscience, 5*, Article 66 (9 pages).
Wittmann, M. (2016). The duration of presence. In B. Mölder, V. Arstila, & P. Øhrstrøm (Eds.), *Philosophy and psychology of time* (pp. 101–113). Cham: Springer International Publishing.
Zadeh, L. A. (1971). Similarity relations and fuzzy orderings. *Information Sciences, 3*(2), 177–200.
Zahavi, D. (2003). *Husserl's phenomenology*. Stanford, CA: Stanford University Press.
Zahavi, D. (2005). *Subjectivity and selfhood: Investigating the first-person perspective*. Cambridge, MA: The MIT Press.

Chapter 4
Interaction of Human Temporality and External World

We have introduced the human temporality as an individual component of the mind, which implies that it bears its own properties and possesses individual dynamics. In the framework of employed phenomenological approach, the human temporality reflects the external world in properties but is not reduced to it. The state of the human temporality and its changes can be specified in terms of properties attributed to it directly.

Having introduced the human temporality, we can pose a question about the interaction between the human temporality and the external world. The human temporality can change due to the changes in the environment and its physical objects. These changes in the external world become accessible to the mind via the properties attributed to the corresponding mental images being the components of the human temporality. Therefore, confining our consideration to the human temporality only, we cannot describe changes in the external world in terms of properties attributed to the physical objects. So to describe the dynamics of the human temporality, we need some being that belongs to the external world and, at the same time, to the inner world. We will call this being *the self*.

The purpose of the given chapter is to elucidate the notion of self understood in this way and to justify the basic features of mathematical formalism underlying our further constructions.

4.1 The Self as Bridge Between External World and Human Temporality

Before starting our discussion of the self we want to clarify the general aspects concerning the description of dynamics of the human temporality.

I. Lubashevsky and N. Plavinska, *Physics of the Human Temporality*,
Understanding Complex Systems, https://doi.org/10.1007/978-3-030-82612-3_4

4.1.1 *Dynamics of Human Temporality: General Representation*

Within phenomenological approach to describing human intentional, goal-oriented actions we meet two equipollent temporal dimensions. The first one is physical time of objective reality. A discussion of physical time properties is outside the scope of the present book.[1] So, in dealing with objective time, we will confine our consideration to the classical formalism of physics, in particular, to the laws of Newtonian mechanics and the principle of temporal locality. This principle means that all the quantities entering the physical laws should be taken at the current moment of time. Within Newtonian mechanics the instant of time referred to as the current moment continuously moves along the axis of objective time at a fixed rate.

The second one is the temporal dimension of the human mind bearing intentional network, i.e., the human temporality. The intentional network includes temporally distributed conscious events and is regarded as a given integral entity at each instant of objective time. Concerning human temporal dimensions on its own, the experiential now together with the complex present are treated as the main reference points, at least, in dealing with human goal-oriented behavior on time scales compatible with the duration of complex present. Therefore in describing their properties, all conscious events, at first, have to be categorized according to their arrangement with respect to the complex present. In turn, on scales of the complex present, the experiential now plays the same role. It should be noted that the given choice of reference point in the human temporality is caused directly by its A-type structure.

Therefore, to cope with the dynamics of the human temporality, we turn to its two-component representation illustrated in Fig. 4.1. At each moment of objective time the human temporality consists of human temporal dimension with past, present, and future conscious events embedded into it and connected into the intentional network. All these components of the human temporality are accessible to the mind and may be used as reasons for actions at the next moments of objective time. In this sense, any conscious event independently of whether its is associated with the past, present, or future in the mind, exists in the inner world and possesses causal power.

At the next moment of objective time, the human temporality varies as a whole, i.e., all the temporal parts of intentional network can undergo transformations. These transformations are synchronized through complex present taking the central position in human temporal dimension at any instant of objective time (Fig. 4.1).

There can be singled out several channels via which the transformation of the human temporality takes place. In particular, they are:

- Time flow in objective reality such that complex present, or speaking more strictly, the experiential now is related to different moments of objective time. It causes the drift of intentional network as a whole from the future to the past. In Fig. 4.1 it is illustrated by the 1-2-3 fragment drifting from the right to the left as objective time goes on. This channel represents the conversion of the future part of the human

[1] We have touched the properties of objective time and the related problems in Sect. 1.1.

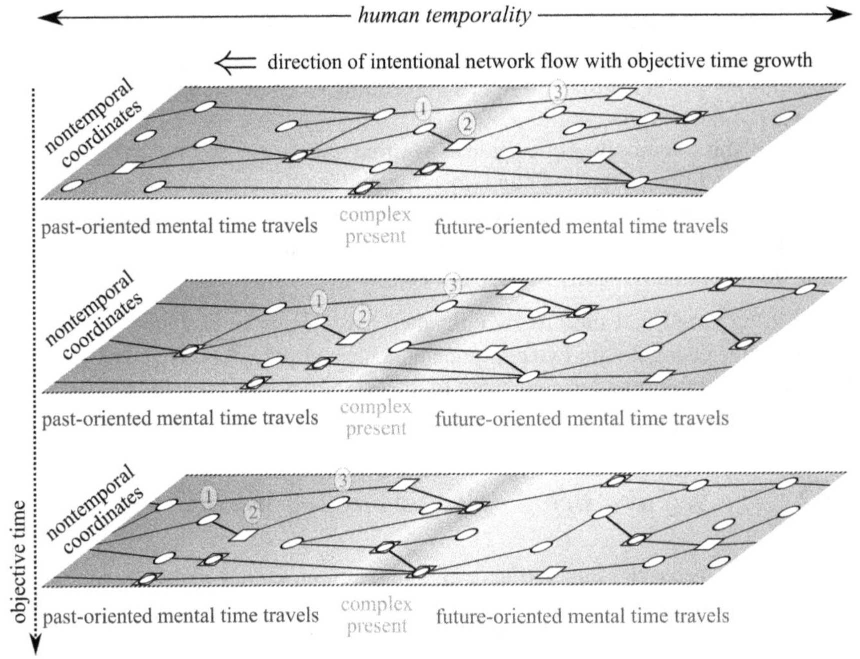

Fig. 4.1 Illustration of the human temporality dynamics as objective time goes on. Origin of human temporal dimension is attached to complex present so the growth of objective time causes the drift of intentional network as a whole from the right to left, fragment 1-2-3 visualizes this drift

temporality into its past part via going through the complex present, i.e., through the experiential now.

- Internal neural mechanisms governing the unconscious processing of previously perceived information like implicit working memory (Sect. 2.6.3) and forgetting. Strictly speaking these mechanisms have to be categorized as physiological, i.e., belonging to physical reality. Nevertheless, they should not depend substantially on changes in the environment and its physical objects. It reasons us to suppose that these processes have to be reflected in the regularities of the human temporality existing on their own and admitting a description remaining inside the realm of human temporality. Through this channel various fragments of the human temporality acquire the ability to change independently of its other fragments, at least, in part.
- Conscious processing of perceived information about changes in the environment and its physical objects. This information processing is accessible to the mind and its result is represented in consciousness as changes in the properties of corresponding mental images. The given channel of temporality transformation reflects real events in the external world but does not enable the description of the human temporality dynamics to go beyond the realm of the mind. The corresponding changes in the human temporality are treated in the mind as caused by the real

external events but the specific details of how these events have been implemented in the external world can be hidden from the mind. Put differently, this channel represents changes in the mental-mental embedding of object images into the mental images of the external world.

- Mechanisms of embodied cognition converting intentions existing in the mind into particular actions of the body and, then, perceiving the results of these action via sensory modalities. Actually it is a bridge between intentions existing in the mind and their particular implementations in objective reality. The details of this bridge functioning remain mainly outside our cognition.

It should be noted that the last two channels are closely intertwined and may be treated as one channel with two components. One component is related to the mind, the other via embodiment is connected to the external world.

4.1.2 The Self as a Being Belonging to Two Worlds

In order to describe the interaction of the human temporality and objective reality we turn to the notion of *the self*—one of the basic ideas accepted in philosophy of mind and psychology. The use of the self in different contexts has a long history. In particular, following the line of Western philosophy some analogy to the self is met in the treatises of Plato and Aristotle. The self was in the focus of the early modern philosophy (e.g., Barresi and Martin 2011; Thiel 2011). The modern notion of the self contains many facets, e.g., the sense of ownership of body parts ('this is my arm'), the sense of agency ('this is my action'), the sense of authorship of thoughts ('this is my thought') as well as the transtemporal unification of self-related information within mental time travels are among them (Synofzik et al. 2008b).

Nowadays, a broad spectrum of various accounts of the self has emerged. At one of the opposite poles, we meet the no-self concept accepting the conscious self to be just the content of a model created by the brain (Metzinger 2003, 2009, 2011). At the opposite pole, we meet the claim that only the narrative self—primarily an abstract entity—has to be presupposed (Schechtman 2007, 2011; Hardcastle 2008). Among the available accounts of the self we turn to the concept of the self generally called *the embodied self* (e.g., Cassam 2011; Newen 2018). This self is able to link intentional actions as conscious events initially existing in the mind with their bodily implementations in reality. The embodied self intertwines components of AM-consciousness, E-consciousness, as well as unconscious bodily awareness of how implement desired actions and other elements of embodied cognition. In particular, according to Newen (2018) the concept of embodied self posits that:

> self-consciousness results from an integration of an embodied, basic affective flow with an intentional object (the self as agent or as center of perception, imagination or thought) where this integration remains anchored in an embodied self. [PDF-version, p. 2, right column]

In parallel Gallagher (2013), Gallagher and Daly (2018), on the one side, and Newen et al. (2015), Newen (2018), on the other side, put forward a novel account of the embodied self—*the pattern theory of the self*. This account focuses on the variety and flexibility of self-conscious phenomena.

Within Newen's account the self is the embodied human being. The pattern theory of the self emphasizes its constitutive aspect—the integrative pattern of characteristic features which is anchored in the body and which determines the body as the anchoring unit for self-conscious experiences (Newen 2018). Gallagher's account of the self focuses on the dynamical aspect of its constitution (cf. also Buhrmann and Di Paolo 2017) as well as that

> the self (as an agent) is more "in-the-world" than "in the brain." ...[A] self is constituted as a pattern of a sufficient number of characteristic factors, including embodied, experiential, affective, behavioral, intersubjective, psychological/cognitive, reflective, narrative, extended, and normative factors...a set of components dynamically interrelated in a pattern...This means that an intervention that affects one factor will involve modulations in the other factors. Adjustments in one aspect, above a certain threshold, will lead via dynamical interactions to changes in others. (Gallagher and Daly 2018, PDV-version, p. 3, right column)

The pattern theory of the self raises the predictive coding paradigm[2] to the level of human being as a whole entity. Namely, first, perceived information about the external world is processed through all the levels of the hierarchical constituent structure of the self. Then, the predictive states are compared with the actual states and, if necessary, are corrected at each level from the basic levels of sensory modalities up the level of cognitive states. Finally, the constituent structure of the self (self-model) can change. In particular, according to Newen (2018) the pattern theory of the self implies that

> [w]e have to presuppose two parallel hierarchies of processing, one based on prior hypotheses for the external world, and one based on prior hypotheses for the self. The self-related information is integrated into the self-model, and in certain situations into the contextually relevant part of the self-model which is the working self. The self is a product of the same sensory information which is used by the cognitive system to construct and modify hypotheses about the external world as well as hypotheses about an internal self. [PDV-version, p. 11, right column, see also Fig. 4 therein]

In this context it is worthy of noting that the mental-mental embedding and the corresponding mathematical eidoi discussed in Sect. 3.3.3 obey the given generalization of predictive coding paradigm. Indeed, the concept of mental-mental embedding supposes that in *evaluating* the states of environment the mind operates with mental images of external physical objects embedded in the mental image of the external world. Therefore, the interaction with reality can cause not only correction of particular properties attributed to the object images but also can lead to a modification of the external world representation in the mind as well as change our way of thinking about the nature of external objects and, so, the basics of their representation.

In the context of the present section want to note the modern dualistic accounts of the mind-body problem—*non-Cartesian substance dualism*—proposed by Lowe (e.g., 2008, 2010, 2014, 2018, a review of philosophy of Lowe can be found in

---

[2] We have discussed the gist of the predictive coding paradigm in Sect. 2.6.4 and the corresponding citations are given there too. For a general review of the concept of predictive processing a reader may be referred to Friston (2010), Hohwy (2013, 2018), Clark (2015).

collection edited by Carruth et al. 2018). Lowe accepts the existence of physical and mental substances. The physical substance bears all the physical properties that can be attributed to *the body*. The mental substance bears all the mental properties and some physical properties that can be attributed to *the self.*

The question about the *causal interaction* between the mental and physical events inevitably arises in debates about any kind dualism. Lowe's account makes the interaction between the mental and the physical possible because the mental properties and some physical properties are attributes of the mental substance. The relationship between the mental decision and the physiological action is due the self and the body being distinct but not separable things. Such actions are characterized by mental and physiological *causes* entirely distinct in their nature. Mental causation is *intentional* because the self intends to produce some action for a certain reason. Physiological causation is "blind" in the sense that physiological processes have no goals.

Within mathematical description of the human temporality dynamics, the representation of the self is determined by the gist of our approach to describing intentional, goal-oriented actions. The logic of mathematical constructions combined with the basic premises of phenomenology enables us to treat the self as a certain bridge between objective reality and the human temporality or, speaking more generally, the mind (cf. Dainton 2008a). Here the metaphor of bridge implies that the self, on the one hand, should share some properties of objective reality, on the other hand, it has common properties with the human temporality. Because the human temporality is not reducible to physical reality in properties, the self as a whole cannot belong to any of them. In this sense, the self is a third independent entity that must be taken into account in describing how the human temporality and human actions implemented in reality are related to each other.

Following the gist of the pattern theory of the self, we single out two constitutive features of the self or two constitutive parts of the self-model:

- the *body-in-the-self* and
- the *mind-in-the-self.*

The relationship between the boy-in-the-self and the mind-in-the-self is accepted to be implemented via unconscious neural processes, which reflects the gist of embodied cognition.

The body-in-the-self is responsible for the movements of the body treated as a certain physical object of environment. The dynamics of the body and its interaction with other physical objects are described by the classical laws of physics. So from the standpoint of external observer, the states of the body and external objects are reflected in the body-in-the-self, but they remain inaccessible to the mind. The mind-in-the-self is responsible for mental phenomena and inherits all the features of the mind including all intentional actions and conscious events of the human temporality. In this sense the body-in-the-self is always anchored in the present of the external world, whereas the min-in-the-self is able to travel along human temporal dimension.

In our further constructions we will accept that the self bears the properties attributed to the mind and some properties attributed to the body. No its own properties characterized by dynamics irreducible to dynamics of the human temporality or external physical objects will be taken into account. Put differently, we will ignore all possible effects that could be caused by the own degree of freedom attributed to the self. Whether the self does not possess its own properties with individual dynamics or these properties are enslaved by the properties of the human temporality and physical objects on temporal scales of complex present requires individual investigation. As an argument for the accepted proposition, we note that time scales characterizing neural processes that cannot directly contributed to cognition are much less than the duration of experiential now (Sects. 2.5 and 2.6). On scales of the experiential now or larger, consciousness and underlying neural mechanics should be closely interconnected.

We want to emphasize that the accepted assumption does not imply the relationship between the body-in-the-self and the mind-in-the-self to be fixed. Just opposite, this relationship should be highly sensitive to the mental states and the state of environment in the present and in the past, reflecting the human capacity for adaption and learning. So it can change in time, however, for this change, the mental states and/or the states of environment must change too.

It should noted that ignoring the self to have its own degree of freedom makes the embodied self and the self of non-Cartesian dualism rather similar in mathematical description. In both the cases, the relationship between the mind and the body is determined by their cumulative contribution. In the case of embodied self, the unconscious relationship between the body-in-the-self and the mind-in-the-self cannot be directly controlled by the mind. In Lowe's account of non-Cartesian dualism, the self is a simple substance (e.g.. Lowe 2018). Under such conditions, in describing mathematically the self the variables of mental states together with the variables of the physical states of the environment are supposed to make up the complete set of required variables.

Finalizing the present section we want to underline once more that the self is a flexible entity. The unity of its characteristic features integrated as a pattern in a situation can change further. Moreover, which and how the features are integrated in constituting the embodied self, gives rise to different properties of the self on short-term and long-term scales as well as under various conditions. For example, the *feeling* of agency and the *judgment* of agency, the *feeling* of ownership and the *judgment* of ownership contribute to the self constitution individually and so their contributions may be different (Synofzik et al. 2008a, b). Besides, the self is a transtemporal entity, so it can behave differently as the temporal scales involved in phenomena under consideration increase.

Below in this chapter we will confine our consideration to the self within the complex present and discuss the corresponding mathematical formalism allowing for the inaccessibility of the body-in-the-self to the mind.

4.2 Complex Present and Dynamical Systems

The remaining part of the present book is aimed at elucidating the characteristic features of mathematical formalism required for describing intentional, goal-oriented actions within the complex present. To accentuate them we will consider a simple situation when pursuing a certain goal the subject governs the motion of a physical object admitting interpretation as a material point. Below we will use the term physical particle or just particle for short to refer to the given physical object.

This situation matches the classical formulation of Newtonian mechanics in describing the laws of particle motion under external forces. The motion of such a physical object is represented by its trajectory $\{x(t)\}$ in the three-dimensional space $\mathbb{R}_3$ specifying the point $x \in \mathbb{R}_3$ where the particle is located at time t. According to Newton's second law, the motion trajectory $\{x(t)\}$ obeys the governing equation

$$m\frac{d^2x}{dt^2} = F\left(x, \frac{dx}{dt} \,\middle|\, t, S_{\text{env}}(t)\right), \tag{4.1}$$

where m is the particle mass quantifying the effect of inertia and $F(\ldots)$ is a force acting on the particle and depending on its current position x and velocity dxd/dt. The force F also depends on the current state $S_{\text{env}}(t)$ of environment comprising other physical objects that interact with the given particle depending on their own position in the space $\mathbb{R}_3$ and motion velocities.

Here we want to emphasize that the force $F(\ldots)$ as a function of its arguments must be *given beforehand* based, e.g., on the basic laws of physical systems. Physical systems governed by equations similar to Eq. (4.1) or its generalizations to the points of the corresponding phase space are often called *dynamical systems*.

The motion of a physical particle affected by human actions formally also obeys Eq. (4.1) but in this case the force F cannot be specified completely turning to the state of environment. So the problems of describing the dynamics of physical objects governed by

- the interaction with surrounding physical objects *only* and
- the interaction with surrounding physical objects *and* human actions

are entirely distinct from each other although the governing Eq. (4.1) seems to take similar forms in the two cases. Below we will demonstrate that the description of physical object motion under human intentional, goal-oriented actions requires a principally new formalism rather than some generalization of Eq. (4.1) to a certain relationship between human response to the current state of object motion taken, maybe, with some time delay.

In this section we present two characteristic examples of goal-oriented actions to be considered further. Their analysis suffices to elucidate the gist of the problem in issue.

4.2.1 Goal-Oriented Actions in the Complex Present: Typical Problems

In developing a theory of the human temporality allowing for its general features it is desirable to clarify:

- typical intentional, goal-oriented actions that admit a closed description on scales of the complex present,
- general features characterizing dynamics of intentional objects that are influential in constructing the mathematical description of the human temporality and its dynamics.

Therefore in this section we discuss two particular examples of human actions within the complex present to illustrate the following features.

First, it is two characteristic types of intentional, goal-oriented actions:

- actions aimed at achieving a final goal and implemented within the complex present from their initiation to completion (Type 1);
- continuous actions aimed at keeping some state of controlled system (Type 2).

It should be noted that the two examples of human actions correspond to two categorical types of human intentions discriminated by Gollwitzer, namely, *goal intentions* and *implementation intentions* (Gollwitzer 1993, 1999, 2014), for a review see also Gollwitzer and Oettingen (2012, 2019). Goal intentions have the structure of "I intend to reach a given goal." Implementation intentions have the structure of "If a given event occurs, then I will react in the corresponding, goal-related way."

Second, it is the order of intentional objects' dynamics, namely, the highest order of time derivatives entering the equation governing the dynamics of a given object described from the standpoint of external observer. In this case, the governing equation is assumed to be written in the form dealing with the object spatial coordinates x only and human actions are considered to be a known beforehand factor, maybe, changing in time. This order of object dynamics, in particular, determines the structure of the minimal phase space required for describing such human actions.

Let us turn to particular examples illustrated in Figs. 4.2 and 4.3 presented here mainly to clarify the noted aspects. It should be pointed out beforehand that these figures only schematically represent the dynamics of systems selected by way of example and the corresponding fragments of the human temporality. The mathematical description of goal-oriented actions generalizing these examples deals with continuous evaluation of perceived information and continuous decision-making process.

4.2.2 Goal-Oriented Actions in Complex Present: Type 1

Figure 4.2 illustrates actions of Type 1 when a desired goal is achieved within the complex present. The left fragment in Fig. 4.2 schematically represents popular psy-

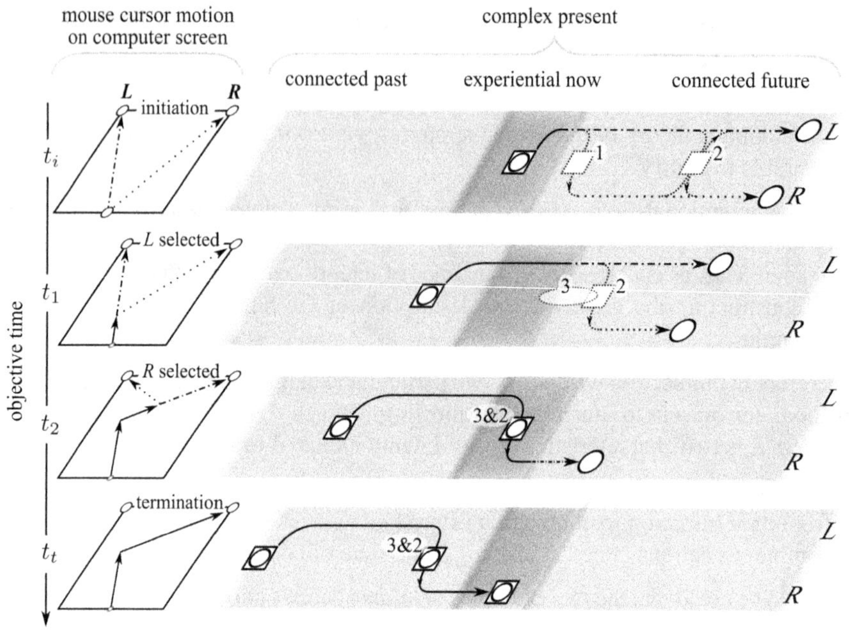

Fig. 4.2 Illustration of Type 1 actions and dynamics of the corresponding intentional network. The left fragment schematically represents *the change of mind* in experiments on binary categorization of an external stimulus (e.g., Resulaj et al. 2009). The mouse cursor movement is recorded and the cursor motion trajectory at different instants of objective time are exhibited by solid lines. The made choice in categorization is visualized via moving the mouse towards the left (L) or right (R) corners of the screen. The expected motion trajectory corresponding to the currently made choice is shown by dot-dashed lines, the alternative actions are depicted with dotted lines. The time instant t_i is the moment when the subject makes choice for the first time and starts to move the mouse. When the mouse courser reaches one of the screen corners (t_t) the experiment is terminated and the current choice is regarded as final. The right fragment illustrates the intentional structure changing as objective time goes on; the L- and R-branches correspond to the selection of the left or right screen corners in visualizing a made choice. Here solid lines depicts the previously made choice or the current choice in the experiential now. Dot-dashed lines illustrate currently selected actions in the connected future, dotted lines represent their alternatives treated as possible in the connected future. Circles, squares, and their unions denote events representing only intentional objects, intentional actions, and both the components, respectively

chological experiments when a subject has to categorize, e.g., a collection of circles moving on the computer screen according to whether they are moving, on the average, to the left or the right. He visualizes the choice by moving the computer mouse cursor to the left or right corner of the screen, which can take a few seconds. The mouse cursor trajectories are recorded and situations shown in Fig. 4.2 are treated as *changes of mind in decision-making* (e.g., Resulaj et al. 2009; Fleming 2016; Van den Berg, Anandalingam, Zylberberg, Kiani, Shadlen and Wolpert 2016).

The right fragment illustrates the corresponding intentional structure changing as objective time goes on. At the moment of initiation the subject selects one of two possible options characterizing his classification of visualized object and starts to move the mouse cursor towards the corresponding screen corner. At this moment the intentional structure, first, contains the event representing initial categorization of visualized object and the beginning of mouse movement which belongs to the experiential now as well as two plausible events of reaching the screen corners. Second, this intentional structure contains possible actions (actions 1 and 2 in Fig. 4.2) of switching between the options within the connected future. It should be noted that:

- these expected actions do not have fixed positions in the human temporal dimension because in the experiments on change of mind, the environment is fixed and the decision on selecting the alternative option is caused by the inner reevaluation of the perceived stimulus,
- there can be only a few expected actions included into the intentional structure because of their similarity and the limit capacity of working memory (Sect. 3.1.5),
- if a possible action has not been implemented (in the experiential now) then it is not kept in memory and ceases to exist as an element of intentional structure.

The latter feature is illustrated by the intentional structure at time t_1 shown in Fig. 4.2. A plausible action 1 belonging initially to the intentional network at time t_i has not been implemented and has ceased to exist at time t_1. The change of mind is demonstrated by the transition from the intentional network at time t_1 to the intentional network at time t_2. Reevaluation of the possible choice options, event 3, is finalized by selecting the alternative, combined event 3&2. Then without a new change of selected option the goal is reached at time t_t; Type 1 actions are completed.

As far as the order of the given object dynamics is concerned, the dynamics of computer mouse cursor—an intentional object admitting interpretation as a material point in virtual reality—is of the first order. It means that the velocity of mouse cursor movement is directly determined by the speed and direction of subject's hand with computer mouse, at least, on time scales under consideration. In physical reality, this situation is met, e.g., in human control over the motion of objects with overdamped dynamics. Leaping ahead we may note that the minimal phase space required for describing the corresponding goal-oriented actions comprises three types of phase variables, the spatial position x of a given object, its velocity v, and the acceleration $a = dx/dt$. In this case, the object velocity v reflects directly subject's actions whereas the acceleration a quantifies the human preference to avoid unreasonably quick, sharp actions. Leaping ahead, we may note that in the given case the acceleration is the control phase variables governed directly by subject's actions.

4.2.3 *Goal-Oriented Actions in Complex Present: Type 2*

Figure 4.3 illustrates intentional network of Type 2 actions related to some long-term process whose dynamics can be based on human actions confined to the complex present. Driver behavior in car-following is a characteristic example of such processes. Intentional structure shown at two moments t_1 and t_2 of objective time illus-

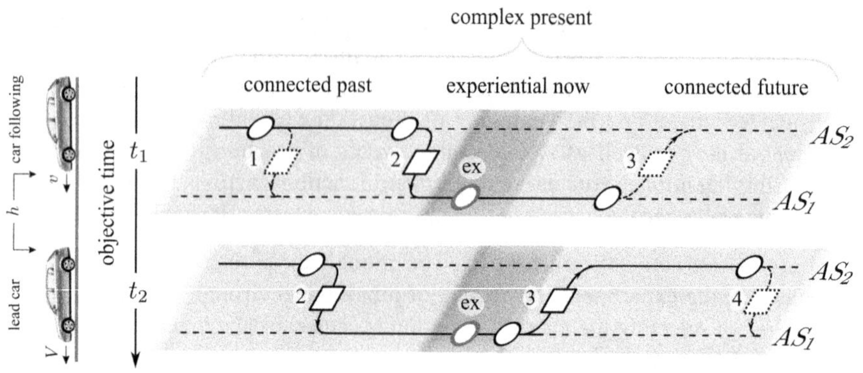

Fig. 4.3 Illustration of Type 2 actions and dynamics of the corresponding intentional network. Car-following is noted as a characteristic example of processes based on this type actions. Two action strategies AS_1 and AS_2 illustrate the driver behavior focused on correcting the current state of car motion or keeping it unchanged. Solid lines depict the currently selected strategy, dashed lines depict its alternative. Circles (1, 2, "ex," 3, 4) denote events representing intentional objects. Circles labeled with "ex" is singled out in red because they correspond to the driver permanent monitoring of the environment state. Squares (1, 2, 3, 4) represent intentional actions. Solid squares represent implemented actions or actions to be implemented with certainty. Dotted squares represent actions either non-implemented or whose implementation is not currently determined

trates its changes as time goes on. Each time this structure contains the event attached to the experiential now (labeled with "ex" in Fig. 4.3) and representing the current state of environment in the mind. It means that subject's attention is permanently focused on the controlled system, e.g., on the motion of driven car and the surrounding cars. Depending on the features of the experiential now, decision on changing or keeping the current action strategy can be made right now or postponed until some event occurs. Figure 4.3 depicts possible transitions between two strategies of driving, AS_1 and AS_2. In the given case, memorizing some intentional actions that were not implemented is reasonable because they can contribute to the learning of driving and the skill acquisition.

The car motion illustrates the object dynamics of the second order caused by the effect inertia. Leaping ahead, we note that under such conditions the minimal phase space required for describing the corresponding goal-oriented actions comprises four types of phase variables:

- the spatial position x of a given object,
- the velocity v,
- the acceleration $a = dx/dt$,
- the jerk $j = da/dt$.

For physical objects with essential inertial the acceleration a reflects directly subject's actions, whereas the jerk j—the control variable—quantifies the human preference to avoid unreasonably quick, sharp actions.

It worthy of noting that the four components' phase space is required for describing human perception of physical object motion within the experiential now in the general case, which reflects the tripartite structure of the experiential now (Sect. 2.4). Besides, three components' components phase met in the previous example may be regarded a particular limit case of the given four components' phase space when the subject just ignores the contribution of jerk to his evaluation of object dynamics.

The matter is that the experiential now is the diachronic units of human temporal dimension and belongs to the basic level of consciousness bearing the properties of E-consciousness and AM-consciousness. So the subject can intentionally make the contribution of jerk to his perception of object dynamics remarkable or delete it from the processing of information provided by the immediate past, the immediate now, and the immediate future individually (Sect. 2.6.6).

Finalizing the present section and turning to the shown dynamics of intentional networks within the complex present (Figs. 4.2 and 4.3), we want to emphasize that they must be categorized as A-type temporal entities. Namely, the experiential now is the reference "point" and all the events are classified using the *before-now-after* terms. Thereby, for describing the human temporality within the complex present at least two-component time structure is required. One time component is objective time which on its own is hidden from the mind.[3] The other is related to intentional network with the fixed reference "point"—the experiential now—and the events appearing, disappearing, and "moving" in the complex present as objective time goes on.[4]

4.3 Action Strategies: The Self and the External Observer

The mind-in-the-self and the body-in-the-self are constituent features of the self being the bridge between the mind and the body. It makes the interaction between the human temporality and object motion in the external world possible (Sect. 4.1.2). The present and following sections are devoted to these issues within the framework of the complex present.

In Sects. 3.1 and 3.4 we discussed in detail intentional, goal-oriented actions within the complex present and drew the conclusion that the intentional network of the human temporality inside the complex present may be conceived of as a set of possible action strategies

[3] According to Gruber et al. (2018) the perceived flow of time can be deconstructed into two levels: a lower level—a perceptual dynamic flow of happening events and an upper level—a cognitive view of past-present-future. Exactly the perceptual completion based on the upper level is responsible for the subjective experience of the lower-level flow of time.

[4] It should be noted that dynamics of intentional network with events appearing and disappearing in working memory implies the corresponding dynamics of memory. Put differently, a cohesive representation of the world requires memories to be altered over time, linked with other memories and eventually integrated into a larger framework. Plausible neural mechanisms governing memory linking, e.g., linking and integration of various memories are discussed, e.g., by Bonasia et al. (2018), Sehgal et al. (2018).

$$\mathfrak{S}_{\mathrm{cp}} = \{\mathcal{AS}\}. \tag{4.2}$$

Each of these action strategies is an ordered sequence of actions aimed at attaining a goal being common for all the actions within one strategy as well as for all the possible strategies in $\mathfrak{S}_{\mathrm{cp}}$. These action strategies have to be implemented within time interval about 10 s—the characteristics duration of complex present (Sect. 3.1).

These strategies are not fully independent of one another because they may have common elements—particular actions. Nevertheless, confining our consideration to goal-oriented actions within the complex present, we have to treat these action strategies as the basic entities of the human temporality. All the properties characterizing human goal-oriented behavior should be attributed to action strategies as whole entities because none of their constituent actions taken separately enables the subject to attain the desired goal.

Speaking about any action strategy $\mathcal{AS}$ we should draw a distinction between its two aspects (Fig. 4.4). One of them is the action strategy $\mathcal{AS}$ from the standpoint of the self—its owner. The other is this action strategy from the standpoint of *external observer*. In both the cases, the issue concerns how the action strategy $\mathcal{AS}$ is described in the corresponding space-time continuum, which enables us to relate the action strategy to the dynamics of intentional object.

The Standpoint of the Self

The self via the mind-in-the-self represents the action strategy $\mathcal{AS}$ as the space-time cloud $\mathbb{C}_{\mathrm{AS}}$ arising via the mental-mental embedding of $\mathcal{AS}$ into the mental images $\mathbb{R}_{[4]} \times t$ of the physical space-time continuum. The mind-in-the-self has a direct access to it and can operate with it. In particular the mind-in-the-self is able to compare the space-time cloud $\mathbb{C}_{\mathrm{AS}}$ of the currently implemented action strategy with the current state of intentional object motion. In Fig. 4.4 the motion state of intentional object at a time t is represented as the object current position and velocity $\{x, v\}_t$. This comparison is due to the *indirect* access of the mind-in-the-self to physical reality via the body-in-the-self.

The Standpoint of the External Observer

The external observer is an abstract notion which does not belong to reality. The external observer cannot be treated as some kind of the self and no intentions can be attributed to it. By definition, the *external observer* is assumed to have direct access to all the physical properties of objects characterizing their position, motion, etc. in the external world

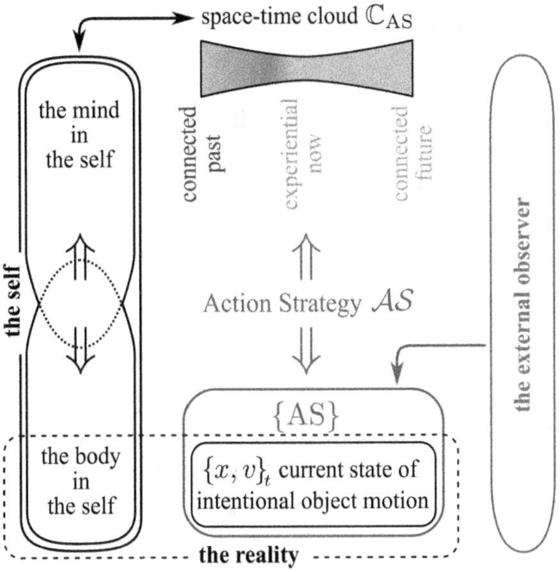

Fig. 4.4 The self and an action strategy $\mathcal{AS}$. The action strategy $\mathcal{AS}$ is represented in the figure center in red. The left fragment schematically depicts the self and its two constituent component—the body-in-the-self and the mind-in-the-self which overlap and, so, interact with each other. The body-in-the-self and the intentional physical object are entities of reality, which is emphasized with dashed rectangular. At each instant t of time the body-in-the-self has a direct access to the current state $\{x, v\}_t$ of intentional object motion. The mind-in-the-self is the owner of the space-time cloud $\mathbb{C}_{AS}$ of the action strategy $\mathcal{AS}$ which is painted in three colors. Blue, orange, and green symbolize the connected past, the experiential now, and the connected future, respectively. The right fragment in magenta illustrates the external observer which, first, has a direct access to the current state of intentional object motion at each instant of time. Second, it accumulates the information about the object motion in the past and the hypothetical future. In this way the external observer is able to record the multitude {AS} of possible implementations of the action strategy $\mathcal{AS}$. Two red arrows around the action strategy $\mathcal{AS}$ directed upward and downward exhibit the relationship between the space-time cloud $\mathbb{C}_{AS}$ owned by the mind and the implementation multitude {AS} accessible via the external observer which does not belong to reality

- at the current instant of time and
- all the instants of time in the past as well as in the hypothetical future.

The external observer and the related multitude of action strategy implementations are actually some abstract entities convenient in mathematical manipulations. In this way we get the possibility of dealing with motion trajectories going through the connected past, the experiential now, and the connected future in mathematical contractions as whole entities determining the properties of human actions.

The external observer represents the action strategy $\mathcal{AS}$ as the multitude {AS} of its possible implementations in reality. Any element AS of this multitude may be regarded as a motion trajectory—a path $\mathrm{MT}_{AS} = \{x(t)\}$ in the physical space along which the intentional object is moved due to the cumulative effect of physical forces and human actions.

We turn to the concept of external observer to deal with the information about the dynamics of intentional object representing its properties in the past, at the current instant of time, and the hypothetical future. In this way not only an individual motion trajectory $\mathrm{MT}_{\mathrm{AS}} = \mathrm{AS}$ as a whole but also their multitude $\{\mathrm{AS}\}$ become available for analysis.

Summary

Any action strategy $\mathcal{AS}$ in the framework of the complex present arises as:

- the space-time cloud $\mathbb{C}_{\mathrm{AS}}$ treated as a whole structureless entity from the standpoint of the mind-in-the-self,
- the multitude $\{\mathrm{AS}\}$ of its possible implementations each of which is treat as a motion trajectory $\mathrm{MT}_{\mathrm{AS}} = \{x(t)\}$ of intentional object within the complex present from the standpoint of the external observer.

4.4 Action Strategies: Un/Conscious Aspects

The purpose of this section is to elucidate the general features characterizing the conscious-unconscious aspects of intentional, goal-oriented actions within the complex present. First, it is essential for describing the human perception of effort required for implementing the desired action strategy on scales of the experiential now and the complex present. Second, in this discussion we intend to clarify the reasons for using optimality criteria in describing human actions on scales of the complex present.

There are several reasons for devoting a whole section to discussing unconscious aspects of action strategy implementation within the complex present. *On the one side*, conscious processes are characterized by the low rate and limited capacity of information processing, whereas unconscious processes are free from these limitations. Therefore the contribution of unconscious mental processes to human actions within the complex present is crucial.

On the other hand, this contribution of unconsciousness poses a doubt as to whether some optimality criteria can be used for describing human actions. The problem is that it is quite natural to suppose that conscious actions can be governed by optimality criteria, whereas it is not clear whether unconscious actions are governed by similar optimality criteria.

Actually the detailed description of this issue is outside the scope of the present book and requires an individual consideration. Nevertheless, our further constructions need this issue to be clarified within the realm of complex present. In this way, we intend to elucidate the possibility of using the notion of optimality for describing goal-oriented actions in the complex present.[5]

In addition, it should be noted that the deliberation—identification of linkages—can be also of unconscious, which is reflected in the notion of *unconscious thought* (Dijksterhuis 2004; Dijksterhuis and Nordgren 2006) and the *deliberation-without-attention* effect (Dijksterhuis et al. 2006). Nevertheless, the use of optimality criteria in developing a common formalism for conscious and unconscious processes cannot be justified by referring to the existence of unconscious deliberation. The matter is that the processes of conscious thought and unconscious thought seem to arise via different mechanisms. For example, for simple problems decision-making produces better results based on conscious thought, whereas decision-making related to complex matters is more efficient in the framework of unconscious thought. For a review of debates about unconscious thought phenomena, a reader may be referred to Calvillo and Penaloza (2009), Strick et al. (2011), Custers (2014), Dijksterhuis and Strick (2016), Abadie and Waroquier (2019), Kong (2019), this issue is also discussed within a wider context by Dijksterhuis and Aarts (2010), Dijksterhuis (2010).

We will justify the possibility of employing common optimality criteria for describing conscious and unconscious processes within the complex present turning to another basis.

4.4.1 Un/Conscious Processes in Complex Present

The notion of the unconscious refers to mental processes including thought processes, memories, interests, motivations, etc. that occur automatically and are not accessible to introspection.[6] A reader may be addressed, e.g., to Westen (1999); Bargh and Morsella (2008) for a general discussion of unconscious actions, to Mills (2014) for philosophical aspects of unconsciousness, and the collection edited by Leuzinger-Bohleber et al. (2017) for conceptual, historical, cognitive, and clinical issues.

Since our analysis is devoted to human goal-oriented behavior on time scales about the complex present duration of order of 10–20 s, we will confine ourselves to discussing human action automaticity and focus on its goal-related features essential for describing the complex present.

At first, let us note the common feature of the conscious and unconscious processes we deal with. Such processes can be categorized as *intentional, goal dependent* (Moors and de Houwer 2007; Moors 2013, 2016). Indeed, citing Moors (2013) we accept the following definition:

---

[5] Recently Lumer (2017, 2019) discussed similar issues concerning the relationship between unconscious motives and automatic actions, on the one side, and agency, intentionality, freedom, and responsibility, on the other side. In particular, Lumer argues that, at least, unconscious elements of goal pursuit can be analyzed turning to the optimal belief theory of intentions and decision-making he developed previously (Lumer 2005).

[6] Below we will use the term *automaticity* and related derivatives mainly with respect to unconscious *actions* and regard the terms *unconscious* and *unconsciousness* as generic names.

> A process is goal dependent when a person has a goal (any goal) and this goal causes the occurrence of the process. The goal may be either the proximal goal to engage in the process or a remote goal that is not about the process. In the first case, the process is goal dependent and intentional; in the second case, the process is just goal dependent. Intentional processes are a subset of goal-dependent processes because intentions are a subset of goals; they are goals to engage in an act or a process. ...Another scenario [justifying intentionality for the remote goal] is that the remote goal causes the proximal goal, and that the latter, in turn, causes the process (p. 166, left column).

In fact this definition reflects the gist of human actions aimed at attaining some proximal goal within the complex present. The given feature makes up a bridge between conscious and unconscious phenomena in issue.

Unconscious processes characterized by intention and goal dependence, especially ones under consideration, belong to a certain class of phenomena that may be described as *adaptive unconscious* complex process (Wilson 2002b). This class represents a bundle of specific processes related to goal pursuit (e.g., Bargh 1994; Gawronski et al. 2006; Hofmann and Wilson 2010) all of which include a definite collection of constituent components, namely, (Hofmann and Wilson 2010, p. 198, itemizing is our):

- the conditions or stimuli that set a process in motion,
- the process itself (i.e., the processing steps and algorithms involved),
- the output of the process, and
- the consequences of the output.

In our case, the complexity of these processes is reflected in that during implementation of an action strategy some of these components can be or become consciously accessible while the others remain unconscious.

4.4.2 Un/Conscious Components of Goal Pursuit: Common and Different Features

Human behavior aimed at attaining some goal—the process of pursuing a goal, or goal pursuit for short—is traditionally assumed to involve reasoning, weighing arguments for and against it as well as its comparison with other goals and, maybe, social norms. Thereby, on the one side, conscious pursuit of a selected goal is effortful because in addition to effort related to physical implementation of this process it includes also mental effort caused by the necessity of

- the permanent control over action execution,
- the permanent monitoring of the environment,
- the correction of action strategies based on the continuous update of information about the continuously changing environment.

On the other side, conscious goal pursuit is endowed with flexibility in its implementation (e.g., Ajzen 1991; Locke and Latham 1990; Latham 2012; Elliot and Fryer

2008). Motives for various actions are also treated traditionally as conscious mental states and motivated behavior is regarded as consciously directed toward some goal (e.g., Fiske 2008; Ryan 2012).

In the 1980s there was a revival of interest in the role of unconscious phenomena in motivation (McClelland et al. 1989), cognition (Kihlstrom 1984, 1987), goal pursuit (Bargh 1990), etc.; for a review a reader may be referred, e.g., to Ferguson et al. (2008), Ferguson and Cone (2013), Gollwitzer et al. (2009), Hassin, Bargh and Zimerman (2009), Hassin, Aarts, Eitam, Custers and Kleiman (2009), Hassin and Sklar (2014), Custers et al. (2012), Moors (2013), Moors (2016), Kihlstrom (2013), Kihlstrom (2019), Huang and Bargh (2014).

Based on evidence for both the conscious and unconscious forms of goal pursuit being governed by the same underlying neural mechanism Huang and Bargh put forward the *consciousness!similarity principle* positing

> a high degree of similarity between conscious and unconscious goals in terms of the various component processes recruited, as well as the outcomes produced. An even stronger hypothesis is that conscious goal operation will share the autonomy of operation that is necessarily observed with unconscious goals (because the latter operate outside of conscious self-control). Thus, even when an individual is consciously pursuing a goal, that person is not necessarily in control of or even aware how engaging in goal pursuit has transformed his or her behavior and experience of the world. (Huang and Bargh 2014, p. 131, left column)

It should be noted that the similarity of conscious and unconscious goal pursuit was recognized earlier, see, e.g., Gollwitzer et al. (2009), Bargh et al. (2010) and reference therein. In particular, in their review Bargh et al. (2010) note that this similarity concerns

- the *outcome* of goal pursuit,
- the *underlying processes*, i.e., unconscious and conscious forms of goal striving in their realization follow the same processing stages,
- the *phenomenal qualities* including persistence in meeting obstacles, resumption of interrupted goal pursuit, and evaluative and motivational consequences of the goal pursuit (see also Ferguson et al. 2008; Bargh and Huang 2009).

Ferguson et al. (2008), Ferguson and Cone (2013) also have argued that unconsciously activated motives possess many properties in common with consciously activated ones.

Applying the similarity principle to human behavior within the complex present, we want to note two characteristic stages of goal-oriented actions that may be attributed to both of the conscious and unconscious implementations (e.g., Dijksterhuis 2010; Bargh et al. 2010; Custers et al. 2012). The two stages possess similar characteristics, however, their essence is different.

The first stage is *setting* a desirable and feasible goal. This stage is characterized by the following:

- *goal content* including the possible positive and negative consequences evaluated by the likelihood of their occurrence and the related emotions;
- *goal structure*, e.g., the goal specificity and the degree of detailing;

- *self-regulation* as a process caused by a discrepancy between the expected future and the current state and enabling efficient goal attainment.[7]

The second stage—*goal striving*—is implementing the goal determined at the previous stage, which includes also the choice of an appropriate means. Whether the desired goal is attained and how efficient is goal striving is determined by several factors; they include:

- *goal content* reflected in creativity, cognitive flexibility, processing of information, and effective coping with failure which is mediated mainly by intrinsic self-organization;
- *goal structure*; here we note only that what is essential within the complex present: (*i*) the feasibility of partitioning the goal into small units and (*ii*) the possibility of learning within which feedback produces valuable cues for improving the current actions;
- *planning of goal implementation*; here we want to note the concept of implementation intentions[8] implying the *if–then* planning of actions which commits the subject to acting in a specific, goal-directed way whenever the crucial cue determined in the *if*-condition is encountered[9];
- *competition* between shielding the goal and monitoring the environment; namely,
  - shielding the goal from other activities,
  - monitoring the environment for information that may necessitate a goal switch (background monitoring).[10]

---

[7] Speaking about the self-regulation of conscious goal setting, i.e., mental strategies advancing the transition from passive behavior to goal commitments, the *theory of fantasy realization* (Oettingen 2000; Oettingen et al. 2009) and the *mindset theory of action phases* (Heckhausen and Gollwitzer 1987; Gollwitzer 1990) should be noted. In particular, the former theory puts forward the concept of *mental contrasting* accentuating the discrepancy between a desired future and obstacles in present reality. The latter theory posits that goal pursuit has multiple stages that people need to successfully navigate to attain a goal.

[8] Implementation intentions may be regarded as an additive component of setting the desired goal. Setting the goal only commits an individual to attain the desired outcome; it does not commit the individual to when, where, and how to act toward the goal.

[9] Implementation intentions also forge a strong association between the specified opportunity and the specified response (Webb and Sheeran 2007, 2008) which automates the initiation of the goal-directed response specified in the *if–then* plan and endows the corresponding actions with immediacy, efficiency, and redundancy of conscious intent. Thereby people no longer have to deliberate about when and how they should act if they have already formed an implementation intention. In other words, forming implementation intentions turns top-down control over goal striving into bottom-up control by the situational cues specified in the *if*-condition (see for a review Bargh et al. 2010).

[10] The *theory of goal systems* by Kruglanski et al. (2002) (see also Kruglanski and Kopetz 2009) develops the concept of goal shielding and contains three major tenets:

- there may be many means to a given goal and there may be many goals linked to a single means;
- given that cognitive resources are limited, investing attention and effort into the currently focal goal is accompanied by removing cognitive resources from competing goals;
- the most salient, vivid, and accessible means are usually selected in goal striving.

Naturally there are basic characteristics in which conscious and unconscious forms of goal pursuit differ from each other, for a review a reader may be referred, e.g., to Gollwitzer et al. (2009), Dijksterhuis (2010), Custers et al. (2012). In particular, they include the following factors:

- the *capacity* of goal pursuit; conscious processes are limited in capacity and not very good at handling large amounts of information (Nørretranders 1998), whereas unconscious processes can handle large amounts of information and are efficient in making complex choices (Dijksterhuis et al. 2006);
- the *reflective/reflexive mechanisms*; according to the *unconscious thought theory* (Dijksterhuis and Nordgren 2006), the implementation of a simple goal is based on conscious top-down processing of information, whereas for a complex goal it is unconscious, bottom-up processing of information[11];
- the *degree of flexibility*; it is held that unconscious goal pursuit is inflexible in contrast to the conscious process able to adapt to changes in environment; however, the concept of *goal representations* to be discussed below, allows for flexibility of unconscious goal pursuit, this flexibility arises at the moment of action initiation (for a review Custers and Aarts 2010; Custers et al. 2012; Aarts and Custers 2012; Mcbride et al. 2012).

Listing the common features of conscious and unconscious modes of goal pursuit and their difference, we have illustrated that both the modes are complex processes and their simple classification in terms of the dichotomy between consciousness and unconsciousness is not feasible. The mental phenomena under consideration can simultaneously exhibit properties some of which may be categorized as conscious and the others as unconscious. For a discussion of the conscious-unconscious categories in the context of goal-oriented behavior, a reader may be referred, e.g., to Kihlstrom (1984, 2013, 2019), Moors (2013, 2016).

As far as the competition between goal shielding and background monitoring is concerned, goal shielding is intensified at the cost of background monitoring when goal striving is at risk of failure and background monitoring is reinstated when goal shielding is no longer needed for goal striving (Goschke and Dreisbach 2008, see also Bargh et al. (2010)).

[11] Nowadays the concept of unconscious thoughts is the subject of on-going debates, for the arguments for and against it a reader may be addressed, e.g., to Lassiter et al. (2009), Strick et al. (2010, 2011), Custers (2014), Newell and Shanks (2014), Vadillo et al. (2015), Zhou et al. (2015), Dijksterhuis and Strick (2016), Hu et al. (2018), Ding et al. (2019). In our description of action strategy implementation within the complex present we also turn to the concept of the two mechanisms governing human behavior based on top-down and bottom-up processing of perceived information. However our interpretation of the unconscious mode of goal pursuit does not require the strong dichotomy between consciousness and unconsciousness, which is clarified in the text.

4.4.3 *Un/Conscious Components of Goal Pursuit: Interrelationship*

In our description, conscious and unconscious processes are closely interwoven and affect each other. In the present section we want to discuss the gist of two rival theories of consciousness being, nowadays, leading accounts of consciousness emergence. In this way we will justify essential propositions of our description of complex present. The two theories are the global workspace theory (e.g., Dehaene and Naccache 2001; Baars 2017; Baars and Alonzio 2018, for a general review) and really a collection of related accounts called the higher-order theories (e.g., Carruthers 2016, 2017b; Gennaro 2018; Brown et al. 2019, for a general review). For comparison of the two approaches as well as the discussion of their possible unification based on metacognition, a reader may be addressed to Cleeremans et al. (2019), Shea and Frith (2019), Graziano et al. (2019).

Emergence of Consciousness via Unconscious Processes

The concept of global workspace takes the central position in our description of goal-oriented actions. We already discussed this concept in Sect. 3.2.5, here referring to Dehaene and Naccache (2001), Hofmann and Wilson (2010), Dehaene (2014) we want to emphasize its following aspects.

- The systems of sensory modalities which are not directly interconnected to each other receive access to each other's content via the global workspace. The global workspace provides a common communication platform where the information of individual sensory systems is broadcast.
- A number of neural subsystems such as working and long-term memory, perceptual systems, evaluation circuits, motor circuits are highly integrated into the global workspace. Thereby they are able to exchange a wide range of potential contents within the global workspace through their mutual interconnections. However, only a tiny fraction of the information from these subsystems is broadcast via the global workspace at a given instant of time.
- Whether the output of a given process is actually recruited into consciousness is determined by several factors reflecting that[12]:
  - information stemming from the corresponding unconscious bottom-up processing has to exceed in amount a certain activation threshold;

[12] In reality, how sensory information becomes accessible to the mind is a multi-factorial complex process exhibiting various properties including adaptation, learning, graded as well as all-or-none behavior; for a detailed discussion a reader may be referred, e.g., to Cleeremans and Jiménez (2002), Perruchet and Vinter (2002), Sandberg et al. (2011), Windey et al. (2014), Windey and Cleeremans (2015). For subcritical bottom-up processing of information, there are unconscious modes of these phenomena (e.g., Hassin, Aarts, Eitam, Custers and Kleiman 2009; Dijksterhuis and Aarts 2010; Custers and Aarts 2010; Custers et al. 2012; Aarts and Custers 2012; Kihlstrom 2013, 2019; Papies and Aarts 2016, for a review).

  - top-down attention is required for this information to enter consciousness via its amplification and maintenance in the global workspace.
- Because the capacity of conscious processes is limited, multiple modular processing outputs permanently compete for the access to the global workspace. As a result the contours of the global workspace are not constant and its states change dynamically, which may be called "stream of consciousness" (Dehaene and Naccache 2001; Hofmann and Wilson 2010).

This concept of consciousness leads to the *three-zone* hierarchical architecture describing the emergence of consciousness (Hofmann and Wilson 2010, a similar representation was proposed also by Cleeremans and Jiménez 2002), we may call it the linear model of unconscious-conscious transitions. Namely,

> [a]t the bottom of the information-processing pyramid is the realm of unconscious processing delegated to modular subsystems. A portion of processing outputs from these modular subsystems may gain the status of phenomenal consciousness. Only a small subset of the rich spectrum of phenomenological experience enters the global workspace through mechanisms of selective attention, thereby gaining the status of access consciousness. Access-conscious information is assumed to be represented in a propositional format and is therefore, in principle, verbally reportable. Behavior can be generated via both unconscious processing and consciously formed action plans (Hofmann and Wilson 2010, p. 203, the caption of Fig. 11.1).

As noted above the transition from the unconscious state of processing sensory information to its conscious state requires attention consciously focused on the given process. Respectively, when attention shifts away from this process it returns to the unconscious state. Thereby, first, the transitions between conscious and unconscious states have to be reversible in the general meaning of this term. In particular, it is illustrated by that in implementing a particular action strategy aimed at a certain goal individuals can shift back and forth seamless between conscious and unconscious processing of sensory information (Dehaene and Naccache 2001; Gollwitzer et al. 2009; Mazzone and Campisi 2013). Second, the possibility of focusing attention on or shifting attention away from a particular processing of information endows the transitions between its conscious and unconscious states with complex properties (e.g., Dijksterhuis and Aarts 2010; Moors 2013, 2016).

The noted properties of unconscious phenomena in goal pursuit argue that the implementation of action strategy within the complex present should be based on the cumulative contribution of conscious and unconscious mechanisms of processing sensory information.

Conscious-Unconscious Transitions and Two Types of Unconscious

Let us elucidate several fundamental issues related to mechanisms via which unconscious goal pursuit is implemented and endowed with flexibility in adapting to a changing environment.

First, it is a bundle of issues reflected in a question as to how a sequence of actions implemented unconsciously can be efficient in attaining the desired goal distant in time. Indeed, the notion of nonconscious processes implies their certain autonomy with respect to the mind. In particular, such a process is not caused by a goal, is unintentional, and does not require attention for implementation (e.g., Moors and De Houwer 2006; Moors 2013). Bargh (1992) defines autonomous processes as processes that once started (irrespective of whether they were started intentionally or unintentionally) do not require conscious guidance or monitoring to run to completion; Logan and Cowan (1984) termed such processes "ballistic." It is quite natural to suppose that, on the one side, the formalism of classical physics may be applied to describing such processes because its paradigmatic aspect is the stimulus-response dynamics termed also the initial-value problem. On the other side, this formalism is inapplicable to describing goal-oriented dynamics.

The proposed mechanism governing unconscious goal pursuit and endowing it with adaptability and flexibility turns to the idea of the mental representation of required actions leading to the desired goal as a whole entity (see e.g., Custers and Aarts 2010, 2014; Custers et al. 2012, 2019, for a detailed discussion and review). Its constituent components represent (*i*) automatic response to detected stimulus unconsciously activating the preparation of actions, (*ii*) detection of positive reward signal indicating to the possibility of attaining the desired goal within the given sequence of actions, and (*iii*) the structure of these actions.[13] The given model of unconscious goal pursuit may be interpreted as the stimulus-response behavior where the initial stimulus indicates the necessity of actions to be activated as well as their efficiency. In the general case, the implementation of this mechanism does not require the contribution of consciousness if the goal pursuit representations are formed and efficient.

The *second* bundle of issues concern how such representations can emerge. There are different accounts of their emergence depending on scales characterizing the corresponding phenomena.

On scales of social phenomena, the emergence of goal pursuit representations seems to be due to implicit learning of implicit motivations governing the unconscious tuning of behavior via human communication (cf., e.g., Ferguson et al. 2008; Ferguson and Cone 2013; Hassin and Sklar 2014; Custers et al. 2019).

On scales of the complex present the contribution of consciousness to learning and skill acquisition should be essential. Moreover, even within regular actions the contribution of phenomena simultaneously bearing the properties of conscious and unconscious processing of sensory information must be crucial. In describing human actions in the complex present, we actually adopt the concept Moors (2009, 2013) calls the *triple-mode view*. Following the gist of the widely accepted approach to describing mental phenomena in terms of the dual-processes theory,[14] the triple-

[13] For a detailed description of implicit and explicit mental representations a reader may be referred to Carlston (2010).

[14] The dual-processes theory of mental phenomena draws a fundamental distinction between two types of psychological processes: one that is unintentional, uncontrollable, unconscious, and effi-

mode view makes a basic distinction between conscious and unconscious (automatic) process and, in addition, distinguishes between bottom-up automatic processes and top-down automatic processes. Actually the triple-mode view inherits the idea of existence of different unconscious (automatic) processes from a number of accounts, in particular, the models proposed by Carver and Scheier (1999, 2002), Cleeremans and Jiménez (2002). Moors (2009, 2013) posits that

> [the] two types of automatic processes come about at different stages in the evolution of a process during practice. In the initial stages of practice, there are bottom-up processes that are triggered by mere stimulus input. These processes are so weak that they remain inaccessible to consciousness and they cannot be controlled (i.e., bottom-up automatic stage). In intermediate stages of practice, processes become so strong that they become consciously accessible. As a result, they can be controlled (i.e., nonautomatic stage). In later stages of practice, processes become so strong and well trained that control is no longer required and they fall out of consciousness again (i.e., top-down automatic stage). Top-down automatic processes are usually unconscious, but they are not inaccessible to consciousness. They can become conscious when attention is directed to them. (Moors 2013, p. 169, left column)

Below we will refer to the unconscious bottom-up processes in terms of *lower-level unconsciousness*, or *L-unconsciousness* for short, and to the unconscious top-down processes in terms of *upper-level unconsciousness*, or *U-consciousness*.

Moors and De Houwer (2006), Moors et al. (2010), Moors (2013) also advocate for the *decompositional view* claiming that mental processes can, in principle, possess any random combination of automatic and nonautomatic features.

The *third* bundle of issues concern the interaction between conscious and unconscious processes in goal pursuit including the transitions between them. The properties of these transitions are crucial for modeling actions on scales exceeding the duration of experiential now because their implementation has to allow for the possibility of required corrections if the anticipated change of the environment deviates from the real situation. Under such conditions, the mental representations of plausible actions include both sensory and motor representations of actions, together with perceptual representations of environmental stimuli for action. As a result, automatic goal-directed processes take the form of a stream of actions. Our perceptions permanently feed automatic processes impinging on motor representations, thereby plans of actions can be activated automatically at each moment and then compete for behavioral expression (Morsella 2009; Mazzone and Campisi 2013).

The *fourth* bundle of issues concern the hierarchy of mental representations of actions and their relationship. The hierarchy base matches unconscious representations followed by representations belonging to E-consciousness (phenomenal consciousness) which, in turn, are followed by higher-order-representations of AM-consciousness (access consciousness). This representation hierarchy underlies all

---

cient, and another that is intentional, controllable, conscious, and inefficient. For a review and detailed discussion of this theory a reader may be addressed to the review by Evans and Stanovich (2013), Gawronski and Creighton (2013), collections of works edited by Sherman et al. (2014), and a resent discussion initiated by Melnikoff and Bargh (2018b) with the following publications by Pennycook et al. (2018), Melnikoff and Bargh (2018a), Osman (2018), Gronchi and Giovannelli (2018).

the higher-order accounts of consciousness. In particular, these higher-order theories posit that phenomenal consciousness can be reductively described in terms of representations that are of higher-order (e.g., Carruthers 2016, see also Rosenthal 2009a, b; Gennaro 2018).

We accept a special version of higher-order theories—the self-representational higher-order theory stating that:

> [a] phenomenally conscious mental state is a state of a certain sort (perhaps with analog/non-conceptual intentional content) which also, at the same time, possesses a higher-order intentional content, thereby in some sense representing *itself* to the person who is the subject of that state. (Carruthers 2016, Sect. 6)

The given proposition is rather natural within our account of the complex present dealing with U-unconsciousness. In fact, the states of U-unconsciousness emerge via AM-consciousness (triple-mode view). Thereby E-conscious states belonging to the "interface" between U-unconsciousness and AM-consciousness have to bear the properties of both the sides. In particular, accepting the self-representational higher-order theory of consciousness we justify the use of the same phase variables for describing human actions within the experiential now as well as the complex present.

Dynamics of Conscious-Unconscious Transitions

As far as mechanisms governing the transitions between conscious and unconscious processing of information as well as transitions between controlled and automatic actions are concerned, they are the matter of ongoing debates. In this context we want to note the following.

First, the transition from unconscious implementation of goal-oriented actions based on their implicit mental representations to consciously implemented actions based on the explicit representations needs a certain moderator (Hofmann and Wilson 2010). Hofmann and Wilson actually proposed that this moderator must be a certain cue simultaneously bearing the properties of unconscious and conscious processes. Put differently, this cue has to belong to the interface between consciousness and unconsciousness, i.e., should be an element of phenomenal consciousness.[15] As a result, conscious attention to the cue can cause the switch from the unconscious state of information processing to its conscious state. Shifting attention away from the given cue should return the information processing to its unconscious state (cf. Fazekas and Nanay 2017; Gatzia and Brogaard 2017). In our further constructing we will regard such phenomenal cues as *effective anchors* of processing the information perceived within the experiential now.

[15] We have already discussed the notion of phenomenal consciousness in Sect. 3.2.5. In particular, within our interpretation phenomenal consciousness is based on unconscious processing of sensory information whose outcome becomes accessible to the mind as just a given result. To emphasize this feature we have introduced the term E-consciousness to denote P-conscious perception of events within the experiential now.

Second, nowadays, there are opposing views on whether the dynamics of transitions between conscious-unconscious states of information processing is a continuous process. The given issue is important for us. Indeed, if the transition between conscious-unconscious states is smooth, it would be possible to speak about the degree of un/consciousness and regard the conscious and unconscious states of information processing as different phases of one common process. If this transition is discontinuous, then, it would be reasonable to assume the conscious and unconscious states are processes to be different in nature and to occur via the all-or-none scenario. In particular, some findings favor a continuous transition from unconscious to conscious processing (e.g., Moutoussis and Zeki 2002; Overgaard et al. 2006; Nieuwenhuis and De Kleijn 2011), whereas others favor a discontinuous one (e.g. Sergent and Dehaene 2004). Based on the analysis of visual experience Windey et al. (2013, 2014), Windey and Cleeremans (2015), Marti and Dehaene (2017) argue for un/consciousness exhibit properties admitting interpretation as graded phenomena as well as properties characterized by the all-or-none emergence. Namely, low-level visual experience is graded, whereas high-level visual experience is all-or-none. To explain this complex behavior of un/conscious perception Overgaard et al. (2006), Windey et al. (2013), Anzulewicz et al. (2015) suppose the transition from unconscious to conscious perception to be influenced by the level of information processing. Bayne et al. (2016), Fazekas and Overgaard (2016), Jonkisz et al. (2017) relate this complexity to the multidimensionality of consciousness including phenomenal, semantic, physiological, and functional dimensions as well as their gradedness.

In our investigations of binary categorization based on the *asymptotics* of psychometric functions, we also have found transitions between the conscious (deliberate) mode and the unconscious (automatic) mode for gray shade categorization, vowel sound categorization, as well as numbers' categorization (Lubashevsky and Watanabe 2016; Lubashevsky et al. 2018, see also Appendix 5.10.1). The obtained shape of psychometric function enables us to suppose that the transitions between the conscious and unconscious modes of categorization may be regarded as phase transitions of second order. In other words, at the phase transition the psychometric function remains continuous but its derivatives exhibit step-wise change. The given type transitions correspond to the supercritical bifurcation of the lower-level consciousness as the classification uncertainty—difficulty of decision-making in categorization—grows (e.g., Arnold 1992; Kelso 1995, 2014; Guastello and Liebovitch 2009). It, in combination with Moors's account of triple-mode view, argues for that:

- as the complexity of information to be processed grows the induced changes in lower-level unconsciousness are gradual, whereas the emergence of conscious looks like step-wise phenomenon admitting the all-or-none interpretation within a rough approximation,
- at the bifurcation point the access of the mind to the processed information should be categorized as E-consciousness (P-consciousness);
- with further increase in the information complexity the degree of consciousness can exhibit considerable increase and in this region the conscious mode is of AM-consciousness type (A-consciousness);

– after bifurcation two modes of information processing have to arise, one of them must be the conscious mode, the other mode may be regarded as unconscious upper-level mode;
– the two modes can coexist with each other and form an arbitrary mixture determined by the degree of attention focused on the features of conscious mode.

The listed items underlie our account of un/conscious mode of information processing within the complex present. For a general account of the relationship between unconscious and conscious phenomena reflecting the modern state of the art in this field, a reader may be referred to Cleeremans (2007), Cleeremans et al. (2019).

4.4.4 Un/Consciousness of Action Strategies: Complex Present

The present section outlines our account of the conscious and unconscious properties of action strategy implementation within the complex present. This account just aggregates the findings discussed in Sects. 4.4.2 and 4.4.3.

First, we suppose that intentional goal-oriented actions on scales of the complex present should be governed by a combination of conscious and upper-level unconscious (U-unconscious) modes of information processing. Indeed, the wide variety of particular goals and the ways of their possible attaining on scales of order of 10 s require consciousness to be involved currently in action implementation or previously in acquiring the corresponding skill.

Second, accepting the triple-mode view and the decompositional view on action automaticity (Sect. 4.4.3) we suppose that any action strategy within the complex present comprises:

- A sequence of particular actions realized individually according to some heuristic plan of implementation, or a *heuristic* for short, determined initially.[16] Heuristic realization can be
  – conscious, i.e., deliberately analyzed during the execution or
  – unconscious, i.e., automatic.

  The U-unconscious mode of action execution is dominant because of the limited capacity of conscious mode and low effort required for executing the unconscious mode.
- Conscious monitoring of action results. The access of the mind to these results is E-conscious, i.e., only the final outcome of the corresponding unconscious processing of sensory information becomes accessible to the mind.

[16] This type heuristic consists of an ordered pattern of elementary actions, i.e., how these actions ordered with respect to one another and stored in memory—the *engram* in Bernstein's terminology (e.g., Latash 2008b, 2012)—and the intensity and time scales of their current execution. A variety of actions in their execution may differ in intensity and time scales whereas have one common engram.

- The E-consciousness experience of system current state and the heuristic implementation of action strategies within the complex present can be described using the same phase variables.

Third, unconscious execution of actions based on their heuristics is of the top-down type and emerged via their previous conscious executions during learning processes (skill acquisition). Therefore the same optimality criteria may be used in describing conscious and unconscious implementations of action strategies.

Fourth, the E-conscious monitoring of action outcomes and the U-unconscious execution of these actions are closely interwoven with each other. The system state as it is currently experienced by the subject can be regarded as the outcome of currently being executed action. Thereby this experience may be regarded as the anchor of U-unconscious action in E-consciousness.

Fifth, when the subject recognizes that the perceived state of system dynamics governed by the currently executed action is no longer appropriate for attaining the desired goal, he selects a new action for implementation.

The five propositions make up our account of *U-unconscious action strategy implementation* within the complex present. The constituent components of this action strategy implementation are:

- the E-conscious monitoring of action outcome combined with the E-conscious representation of the desired goal;
- the variate of actions which can be executed unconsciously (automatically) based on acquired skill; thereby, the implementation of these actions may be regarded as U-unconscious (implicit) goal-oriented behavior.

The given account actually reconciles the reactive (reflexive) approach to describing human actions[17] and the teleological concept[18] of human behavior (goal-oriented actions). In describing intentional, goal-oriented actions within the complex present, the subject

- *explicitly* responds to system state, as he experiences it currently, via activating the execution of a new action or keeping up the action strategy being currently executed (reflexive type behavior),
- *implicitly* (automatically) drives the system to the desired goal which is determined by the current state of system motion and its desired state (teleological process).

This U-unconscious goal pursuit implies the existence of heuristics enabling the automatic implementation of the appropriate actions. The description of how these heuristics emerge during learning processes and skill acquisition is a separate problem; in the present book we will consider that these heuristics have already emerged.

---

[17] In physics this approach to describing system dynamics is often called the initial value problem which reflects the main paradigmatic assumptions accepted in physics.

[18] Teleology (from Greek *telos*, "end," and *logos*, "reason"), explanation by reference to some purpose, end, goal. Traditionally, it was also described as *final causality*, in contrast with explanation solely in terms of *efficient causes*—the origin of a change or an event. The most-celebrated account of teleology was that given by Aristotle when he declared that a full explanation of anything must consider its final cause as well as its efficient, material, and formal causes (Teleology 2015).

Brief digression: As a plausible approach to describing the emergence of action heuristics during learning process, we want to note the *theory of event coding* developed by Hommel (1998), Hommel et al. (2001). This theory allows for perception and action planning to be closely intertwined. Namely, perceptual contents and action plans are coded in a common representational medium by feature codes with distal reference. Both the perceived events (perceptions) and to-be-produced events (actions) are represented within one cognitive structures called event codes. It proposes an explanation of how efficient heuristics can become linked to the perceived stimuli playing the role of external cues in the initiation of these heuristics. The theory of event coding inherits, in particular, the gist of the *ideomotor principle* (for details see Greenwald 1970) as well as the motor theories of cognition broadly discussed a century ago (see, e.g., Scheerer 1984). For a review of basic premises underlying the modern account of ideomotor actions a reader may be addressed to Shin et al. (2010), Hommel (2013b). Schack (2004), Frank (2014, 2016) proposed also an account describing the common contribution of the motor action learning and practice to emergence of cue-heuristic relationship.

The ideomotor type behavior being a constituent component of unconscious goal pursuit becomes flexible due to the implicit or explicit adaptation to changes in environment via learning processes. For details a reader may be referred to, e.g., Hommel (1996, 2007a, b, 2009); the modern state of the art in this field is represented by Rosenbaum (2013), Hommel (2013a, b), Taylor and Ivry (2013), the collection edited by Egner (2017).

The concept related to our understanding of the issue has been put forward by Schlosser (2019) turning to the dual-system theory. How the U-unconscious heuristics and their relationship with E-conscious monitoring of the system dynamics within the experiential now emerge during learning processes is the matter of future research. We suppose that it will require elucidating the dynamics of unconscious and conscious mental representation via E-consciousness. However, this issue is outside the scope of the present book.

4.5 Conclusion

This chapter finalizes our qualitative description of the complex present and two main issues have been the focus of our attention.

First, we have elucidated our approach to describing intentional, goal-oriented actions within the complex present which may be classified as a multi-component time formalism. Our account of human actions contains two constituent components:

- the human temporality with complex temporal structure and
- physical environment: physical objects affected by human actions and other physical objects able to influence the former objects.

On the one hand, the two components possess independent degrees of freedom and, thereby, are characterized by individual dynamics, at least within the phenomenological approach. On the other hand, these components affect each other.

To couple the dynamics of the human temporality with the dynamics of physical environment we introduced the self forming a "bridge" between the inner world of human being and the external world. Within the phenomenological approach, the self with its two constitutive overlapping parts:

- the mind-in-the-self,
- the body-in-the-self

is an entity whose features should be determined based on the available philosophical accounts, psychological and neurological data. Actually within Chaps. 2 and 3, we analyzed the mind-in-the-self exactly from this point of view. The second part of the given Chap. 4 is devoted to the body-in-the-self. Here we also have turned to the concept of external observer to convert the physical description of external objects into entities the self can deal with.

Besides in the given chapter we have presented two characteristic examples illustrating intentional, goal-oriented actions in the complex present. They are:

- type 1 actions caused by goal intentions which met, e.g., in the phenomenon called change of mind in decision making;
- type 2 actions caused by implementation intentions which characterize, e.g., human behavior in car driving.

The two types admit a common description in terms of multi-component time formalism and will be used for illustration of phenomena to be analyzed further.

Second, description of the body-in-the-self is closely interwoven with our understanding of how sensory information is processed via unconscious and conscious channels. Confining our analysis to the complex present we have developed our account of conscious-unconscious components of intentional, goal-oriented actions on scales about 10–20 s. It includes the following propositions.

- Intentional, goal-oriented actions on scales of the complex present are governed by the combination of conscious and upper-level unconscious (U-unconscious) modes of information processing. U-unconsciousness emerges via the conscious mode of information processing and is a result of skill acquisition.
- An action strategy comprises:
  - a sequence of actions executed according to some heuristic implemented either consciously (deliberately) or unconsciously (automatically). The U-unconscious mode of action execution is dominant because of the limited capacity of conscious mode and low effort required for executing the unconscious mode.
  - E-conscious monitoring of action results. The currently experienced state is the outcome of action under execution and, so, may be regarded as the anchor of U-unconscious actions in E-consciousness.

- Due the unconscious information processing being of upper-level
  - the same optimality criteria may be used in describing conscious as well as unconscious implementations of action strategies.
  - the same collection of phase variables may be used in describing the experience of system current state as well as the heuristic implementation of action strategies within the complex present.

The given account actually reconciles the reactive (reflexive) approach to describing human actions and the teleological concept of human goal-oriented behavior. The subject

- *explicitly* responds to system current state and
- *implicitly* (automatically) drives the system to the desired goal.

Here we have finalized the qualitative description of our account of the human temporality within the complex present. Part II will be devoted to developing mathematical formalism based on the discussed properties.

References

Aarts, H., & Custers, R. (2012). Unconscious goal pursuit: Nonconscious goal regulation and motivation. In R. M. Ryan (Ed.), *The Oxford handbook of human motivation* (pp. 232–247). New York, NY: Oxford University Press.

Abadie, M., & Waroquier, L. (2019). Evaluating the benefits of conscious and unconscious thought in complex decision making. *Policy Insights from the Behavioral and Brain Sciences, 6*(1), 72–78.

Ajzen, I. (1991). Attitudes and thought systems. Advances in social cognition. In R. S. Wyer Jr. & T. K. Srull (Eds.), *The content, structure, and operation of thought systems* (Vol. 4, pp. 79–86). Hillsdale, NJ: Lawrence Erlbaum Associates, Publishers.

Anzulewicz, A., Asanowicz, D., Windey, B., Paulewicz, B., Wierzchoń, M., & Cleeremans, A. (2015). Does level of processing affect the transition from unconscious to conscious perception? *Consciousness and Cognition, 36,* 1–11.

Arnold, V. I. (1992). *Catastrophe theory*, third, revised and expanded edition. Berlin: Springer. Translated from the Russian by G. S. Wassermann (Based on a Translation by R. K. Thomas).

Baars, B. J., & Alonzio, A. (2018). The global workspace theory. In R. J. Gennaro (Ed.), *The Routledge handbook of consciousness* (pp. 122–136). New York, NY: Routledge, Taylor & Francis Group.

Baars, B. J. (2017). The global workspace theory of consciousness: Predictions and results. In S. Schneider & M. Velmans (Eds.), *The blackwell companion to consciousness* (2nd ed., pp. 229–242). Chichester, West Sussex, UK: Wiley.

Bargh, J. A. (1994). The four horsemen of automaticity: Awareness, intention, efficiency, and control in social cognition. In R. S. Wyer, Jr., & T. K. Srull (Eds.), *Handbook of social cognition: Volume 1: Basic processes* (2nd edn., pp. 1–40). New York, NY: Psychology Press, Taylor and Francis Group.

Bargh, J. A. (1990). Auto-motives: Preconscious determinants of social interaction. In E. T. Higgins & R. M. Sorrentino (Eds.), *Handbook of motivation and cognition: Foundations of social behavior* (Vol. 2, pp. 93–130). New York, NY: The Guilford Press.

Bargh, J. A. (1992). The ecology of automaticity: Toward establishing the conditions needed to produce automatic processing effects. *The American Journal of Psychology, 105*(2), 181–199.

Bargh, J. A., Gollwitzer, P. M., & Oettingen, G. (2010). Motivation. In S. T. Fiske, D. T. Gilbert, & G. Lindzey (Eds.), *Handbook of social psychology* (5th ed., Vol. 1, pp. 268–316). Hoboken, NJ: Wiley.

Bargh, J. A., & Huang, J. Y. (2009). The selfish goal. In G. B. Moskowitz & H. Grant (Eds.), *The psychology of goals* (pp. 127–150). New York, NY: The Guilford Press.

Bargh, J. A., & Morsella, E. (2008). The unconscious mind. *Perspectives on Psychological Science, 3*(1), 73–79.

Barresi, J., & Martin, R. (2011). History as prologue: Western theories of the self. In S. Gallagher (Ed.), *The Oxford handbook of the self* (pp. 33–56). Oxford, NY: Oxford University Press.

Bayne, T., Hohwy, J., & Owen, A. M. (2016). Are there levels of consciousness? *Trends in Cognitive Sciences, 20*(6), 405–413.

Bonasia, K., Sekeres, M. J., Gilboa, A., Grady, C. L., Winocur, G., & Moscovitch, M. (2018). Prior knowledge modulates the neural substrates of encoding and retrieving naturalistic events at short and long delays. *Neurobiology of Learning and Memory, 153*, Part A, 26–39. MCCS 2018: Time and Memory.

Brown, R., Lau, H., & LeDoux, J. E. (2019). Understanding the higher-order approach to consciousness. *Trends in Cognitive Sciences, 23*(9), 754–768.

Buhrmann, T., & Di Paolo, E. (2017). The sense of agency - a phenomenological consequence of enacting sensorimotor schemes. *Phenomenology and the Cognitive Sciences, 16*(2), 207–236.

Calvillo, D. P., & Penaloza, A. (2009). Are complex decisions better left to the unconscious? further failed replications of the deliberation-without-attention effect. *Judgment and Decision Making, 4*(6), 509–517.

Carlston, D. (2010). Models of implicit and explicit mental representation. In B. Gawronski & B. K. Payne (Eds.), *Handbook of implicit social cognition: Measurement, theory, and applications* (pp. 38–61). New York, NY: The Guilford Press.

Carruthers, P. (2016). Higher-order theories of consciousness. In E. N. Zalta (Ed.), *The stanford encyclopedia of philosophy*, fall (2016th ed.). Metaphysics Research Lab, Stanford University.

Carruthers, P. (2017). Higher-order theories of consciousness. In S. Schneider & M. Velmans (Eds.), *The blackwell companion to consciousness* (2nd ed., pp. 288–297). Chichester, West Sussex, UK: Wiley.

Carruth, A., Gibb, S., & Heil, J. (Eds.). (2018). *Ontology, modality, and mind: Themes from the metaphysics of E. J. Lowe*. Oxford, UK: Oxford University Press.

Carver, C. S., & Scheier, M. F. (1999). Themes and issues in the self-regulation of behavior. In R. S. Wyer Jr. (Ed.), *Perspectives on behavioral self-regulation: Advances in social cognition* (Vol. 12, pp. 1–105). Mahwah, NJ: Lawrence Erlbaum Associates Publishes.

Carver, C. S., & Scheier, M. F. (2002). Control processes and self-organization as complementary principles underlying behavior. *Personality and Social Psychology Review, 6*(4), 304–315.

Cassam, Q. (2011). The embodied self. In S. Gallagher (Ed.), *The Oxford handbook of the self* (pp. 137–156). Oxford, NY: Oxford University Press.

Clark, A. (2015). *Surfing uncertainty: Prediction, action, and the embodied mind*. Oxford, NY: Oxford University Press.

Cleeremans, A., & Jiménez, L. (2002). Implicit learning and consciousness: A graded, dynamic perspective. In R. M. French, & A. Cleeremans (Eds.), *Implicit learning and consciousness: An empirical, philosophical and computational consensus in the making* (pp. 1–40). New York, NY: Psychology Press, the Taylor & Francis Group.

Cleeremans, A., Achoui, D., Beauny, A., Keuninckx, L., Martin, J.-R., Muñoz-Moldes, S., Vuillaume, L., & de Heering, A. (2019). Learning to be conscious. *Trends in Cognitive Sciences*. Online: December 28, 2019.

Cleeremans, A. (2007). Consciousness: The radical plasticity thesis. Progress in brain research. In R. Banerjee & B. K. Chakrabarti (Eds.), *Models of brain and mind: Physical, computational and psychological approaches* (Vol. 168, pp. 19–33). Amsterdam, The Netherlands: Elsevier.

Custers, R., Eitam, B., & Bargh, J. A. (2012). Conscious and unconscious processes in goal pursuit. In H. Aarts, & A. J. Elliot (Eds.), *Goal-directed behavior* (pp. 231–266). New York, NY: Psychology Press, Taylor & Francis Group.

Custers, E. J. F. M. (2014). Unconscious thought and deliberation without attention: A miracle or a mirage? *Perspectives on Medical Education, 3*(3), 155–158.

Custers, R., & Aarts, H. (2010). The unconscious will: How the pursuit of goals operates outside of conscious awareness. *Science, 329*(5987), 47–50.

Custers, R., & Aarts, H. (2014). Conscious and unconscious goal pursuit: Similar functions, different processes? In J. W. Sherman, B. Gawronski, & Y. Trope (Eds.), *Dual-process theories of the social mind* (pp. 386–399). New York, NY: The Guilford Press.

Custers, R., Vermeent, S., & Aarts, H. (2019). Does goal pursuit require conscious awareness? In R. M. Ryan (Ed.), *The Oxford handbook of human motivation* (pp. 269–284). New York, NY: Oxford University Press.

Dainton, B. (2008). *The phenomenal self.* New York: Oxford University Press Inc.

Dehaene, S. (2014). *Consciousness and the brain: Deciphering how the brain codes our thoughts.* New York, NY: VIKING, the Penguin Group.

Dehaene, S., & Naccache, L. (2001). Towards a cognitive neuroscience of consciousness: Basic evidence and a workspace framework. *Cognition, 79*(1–2), 1–37.

Dijksterhuis, A. (2004). Think different: The merits of unconscious thought in preference development and decision making. *Journal of Personality and Social Psychology, 87*(5), 586–598.

Dijksterhuis, A. (2010). Automaticity and the unconscious. In S. T. Fiske, D. T. Gilbert, & G. Lindzey (Eds.), *Handbook of social psychology* (5th ed., Vol. 1, pp. 228–267). Hoboken, NJ: Wiley.

Dijksterhuis, A., & Aarts, H. (2010). Goals, attention, and (un)consciousness. *Annual Review of Psychology, 61*(1), 467–490.

Dijksterhuis, A., Bos, M. W., Nordgren, L. F., & van Baaren, R. B. (2006). On making the right choice: The deliberation-without-attention effect. *Science, 311*(5763), 1005–1007.

Dijksterhuis, A., & Nordgren, L. F. (2006). A theory of unconscious thought. *Perspectives on Psychological Science, 1*(2), 95–109.

Dijksterhuis, A., & Strick, M. (2016). A case for thinking without consciousness. *Perspectives on Psychological Science, 11*(1), 117–132.

Ding, R., Han, Q., Li, R., Li, T., Cui, Y., & Wu, P. (2019). Unconscious versus conscious thought in creative science problem finding: Unconscious thought showed no advantage! *Consciousness and Cognition, 71,* 109–113.

Egner, T. (Ed.). (2017). *The Wiley handbook of cognitive control.* Chichester, West Sussex, UK: Wiley.

Elliot, A. J., & Fryer, J. W. (2008). The goal construct in psychology. In J. Y. Shah & W. L. Gardner (Eds.), *Handbook of motivation science* (pp. 235–250). New York, NY: The Guilford Press.

Evans, J. S. B. T., & Stanovich, K. E. (2013). Dual-process theories of higher cognition: Advancing the debate. *Perspectives on Psychological Science, 8*(3), 223–241.

Fazekas, P., & Nanay, B. (2017). Pre-cueing effects: Attention or mental imagery?. *Frontiers in Psychology, 8,* Article 222 (4 pages).

Fazekas, P., & Overgaard, M. (2016). Multidimensional models of degrees and levels of consciousness. *Trends in Cognitive Sciences, 20*(10), 715–716.

Ferguson, M. J., & Cone, J. (2013). The mind in motivation: A social cognitive perspective on the role of consciousness in goal pursuit. In D. Carlston (Ed.), *The Oxford handbook of social cognition* (pp. 476–496). New York, NY: Oxford University Press.

Ferguson, M. J., Hassin, R., & Bargh, J. A. (2008). Implicit motivation: Past, present, and future. In J. Y. Shah & W. L. Gardner (Eds.), *Handbook of motivation science* (pp. 150–166). New York, NY: The Guilford Press.

Fiske, S. T. (2008). Core social motivations: Views from the couch, consciousness, classroom, computers, and collectives. In J. Y. Shah & W. L. Gardner (Eds.), *Handbook of motivation science* (pp. 3–22). New York, NY: The Guilford Press.

Fleming, S. M. (2016). Decision making: Changing our minds about changes of mind. *eLife, 5*, e14790.

Frank, C. (2014). *Mental representation and learning in complex action: a perceptual-cognitive view on mental and physical practice*. Ph.D. thesis, Bielefeld University, Bielefeld, Germany.

Frank, C. (2016). Learning a motor action "from within": Insights into perceptual-cognitive changes with mental and physical practice. In M. Raab, P. Wylleman, R. Seiler, A.-M. Elbe, & A. Hatzigeorgiadis (Eds.), *Sport and exercise psychology research: From theory to practice* (pp. 91–121). London: Academic Press, Elsevier Inc.

Friston, K. (2010). The free-energy principle: A unified brain theory? *Nature Reviews Neuroscience, 11*(2), 127–138.

Gallagher, S. (2013). A pattern theory of self. *Frontiers in Human Neuroscience*, *7*, Article 443 (7 pages).

Gallagher, S., & Daly, A. (2018). Dynamical relations in the self-pattern. *Frontiers in Psychology, 9*, Article 664 (13 pages).

Gatzia, D. E., & Brogaard, B. (2017). Pre-cueing, perceptual learning and cognitive penetration. *Frontiers in Psychology,8*, Article 739 (4 pages).

Gawronski, B., & Creighton, L. A. (2013). Dual process theories. In D. Carlston (Ed.), *The Oxford handbook of social cognition* (pp. 239–256). New York, NY: Oxford University Press.

Gawronski, B., Hofmann, W., & Wilbur, C. J. (2006). Are "implicit" attitudes unconscious? *Consciousness and Cognition, 15*(3), 485–499.

Gennaro, R. J. (2018). Representational theories of consciousness. In R. J. Gennaro (Ed.), *The Routledge handbook of consciousness* (pp. 107–121). New York, NY: Routledge, Taylor & Francis Group.

Gollwitzer, P. M., Parks-Stamm, E. J., & Oettingen, G. (2009). Living on the edge: Shifting between nonconscious and conscious goal pursuit, Social cognition and social neuroscience. In E. Morsella, J. A. Bargh, & P. M. Gollwitzer (Eds.), *Oxford handbook of human action* (pp. 603–623). New York, NY: Oxford University Press.

Gollwitzer, P. M. (1990). Action phases and mind-sets. In E. T. Higgins & R. M. Sorrentino (Eds.), *Handbook of motivation and cognition: Foundations of social behavior* (Vol. 2, pp. 53–92). New York, NY: The Guilford Press.

Gollwitzer, P. M. (1993). Goal achievement: The role of intentions. *European Review of Social Psychology, 4*(1), 141–185.

Gollwitzer, P. M. (1999). Implementation intentions: Strong effects of simple plans. *American Psychologist, 54*(7), 493–503.

Gollwitzer, P. M. (2014). Weakness of the will: Is a quick fix possible? *Motivation and Emotion, 38*(3), 305–322.

Gollwitzer, P. M., & Oettingen, G. (2012). Goal pursuit. In R. M. Ryan (Ed.), *The Oxford handbook of human motivation* (pp. 208–231). New York, NY: Oxford University Press.

Gollwitzer, P. M., & Oettingen, G. (2019). Goal attainment. In R. M. Ryan (Ed.), *The Oxford handbook of human motivation* (2nd ed., pp. 247–267). New York, NY: Oxford University Press.

Goschke, T., & Dreisbach, G. (2008). Conflict-triggered goal shielding: Response conflicts attenuate background monitoring for prospective memory cues. *Psychological Science, 19*(1), 25–32.

Graziano, M. S. A., Guterstam, A., Bio, B. J.,& Wilterson, A. I. (2019). Toward a standard model of consciousness: Reconciling the attention schema, global workspace, higher-order thought, and illusionist theories. *Cognitive Neuropsychology*, 1–18. Published online: 26 Sep 2019.

Greenwald, A. G. (1970). Sensory feedback mechanisms in performance control: With special reference to the ideo-motor mechanism. *Psychological Review, 77*(2), 73–99.

Gronchi, G., & Giovannelli, F. (2018). Dual process theory of thought and default mode network: A possible neural foundation of fast thinking. *Frontiers in Psychology*, *9*, Article 1237 (4 pages).

Gruber, R. P., Smith, R. P., & Block, R. A. (2018). The illusory flow and passage of time within consciousness: A multidisciplinary analysis. *Timing & Time Perception, 6*(2), 125–153.

Guastello, S. J., & Liebovitch, L. S. (2009). Introduction to nonlinear dynamics and complexity. In S. J. Guastello, M. Koopmans, & M. Koopmans (Eds.), *Chaos and complexity in psychology:*

The theory of nonlinear dynamical systems (pp. 1–40). New York, NY: Cambridge University Press.
Hardcastle, V. G. (2008). *Constructing the self.* Amsterdam: John Benjamins Publishing Company.
Hassin, R. R., Aarts, H., Eitam, B., Custers, R., & Kleiman, T. (2009). Nonconscious goal pursuit and the effortful control of behavior, Social cognition and social neuroscience. In E. Morsella, J. A. Bargh, & P. M. Gollwitzer (Eds.), *Oxford handbook of human action* (pp. 549–566). New York, NY: Oxford University Press.
Hassin, R. R., Bargh, J. A., & Zimerman, S. (2009). Automatic and flexible: The case of nonconscious goal pursuit. *Social Cognition, 27*(1), 20–36.
Hassin, R. R., & Sklar, A. Y. (2014). The human unconscious: A functional perspective. In J. W. Sherman, B. Gawronski, & Y. Trope (Eds.), *Dual-process theories of the social mind* (pp. 299–313). New York, NY: The Guilford Press.
Heckhausen, H., & Gollwitzer, P. M. (1987). Thought contents and cognitive functioning in motivational versus volitional states of mind. *Motivation and Emotion, 11*(2), 101–120.
Hofmann, W., & Wilson, T. D. (2010). Consciousness, introspection, and the adaptive unconscious. In B. Gawronski & B. K. Payne (Eds.), *Handbook of implicit social cognition: Measurement, theory, and applications* (pp. 197–215). New York, NY: The Guilford Press.
Hohwy, J. (2013). *The predictive mind.* Oxford, UK: Oxford University Press.
Hohwy, J. (2018). The predictive processing hypothesis. In A. Newen, L. D. Bruin, & S. Gallagher (Eds.), *The Oxford handbook of 4E cognition* (pp. 129–146). Oxford, UK: Oxford University Press.
Hommel, B. (1998). Perceiving one's own action—and what it leads to, Advances in psychology. In J. S. Jordan (Ed.), *Systems theories and a priori aspects of perception* (Vol. 126, pp. 143–179). Amsterdam: North-Holland, Elsevier Science B.V.
Hommel, B. (2007a). Consciousness and control: Not identical twins. *Journal of Consciousness Studies, 14*(1–2), 155–176.
Hommel, B. (2013a). Dancing in the dark: no role for consciousness in action control. *Frontiers in Psychology, 4,* Article 380 (3 pages).
Hommel, B. (1996). The cognitive representation of action: Automatic integration of perceived action effects. *Psychological Research, 59*(3), 176–186.
Hommel, B. (2007). Feature integration across perception and action: event files affect response choice. *Psychological Research, 71*(1), 42–63.
Hommel, B. (2009). Action control according to TEC (theory of event coding). *Psychological Research Psychologische Forschung, 73*(4), 512–526.
Hommel, B. (2013). Ideomotor action control: On the perceptual grounding of voluntary actions and agents. In W. Prinz, M. Beisert, & A. Herwig (Eds.), *Action science: Foundations of an emerging discipline* (pp. 113–136). Cambridge, MA: The MIT Press.
Hommel, B., Müsseler, J., Aschersleben, G., & Prinz, W. (2001). The theory of event coding (TEC): A framework for perception and action planning. *Behavioral and Brain Sciences, 24*(5), 849–878.
Hu, F., Yu, X., Chu, H., Zhao, L., Jude, U., & Jiang, T. (2018). Similar effects for resting state and unconscious thought: Both solve multi-attribute choices better than conscious thought. *Frontiers in Psychology, 9,* Article 1360 (8 pages).
Huang, J. Y., & Bargh, J. A. (2014). The selfish goal: Autonomously operating motivational structures as the proximate cause of human judgment and behavior. *Behavioral and Brain Sciences, 37*(2), 121–135.
Jonkisz, J., Wierzchoń, M., & Binder, M. (2017). Four-dimensional graded consciousness. *Frontiers in Psychology, 8,* Article 420 (10 pages).
Kelso, J. A. S. (1995). *Dynamic patterns: The self-organization of brain and behavior.* Cambridge, MA: The MIT Press.
Kelso, J. A. S. (2014). The dynamic brain in action: Coordinative structures, criticality, and coordination dynamics. In D. Plenz & E. Niebu (Eds.), *Criticality in neural systems* (pp. 67–104). Weinheim: Wiley-VCH Verlag GmbH & Co. KGaA.

Kihlstrom, J. F. (1984). Conscious, subconscious, unconscious: A cognitive perspective. In K. S. Bowers, & D. Meichenbaum (Eds.), *The unconscious reconsidered* (pp. 149–211). New York: Wiley.

Kihlstrom, J. F. (2019). The motivational unconscious. *Social and Personality Psychology Compass, 13*(5), e12466 (18 pages).

Kihlstrom, J. F. (1987). The cognitive unconscious. *Science, 237*(4821), 1445–1452.

Kihlstrom, J. F. (2013). Unconscious processes. In D. Reisberg (Ed.), *The Oxford handbook of cognitive psychology* (pp. 176–186). New York, NY: Oxford University Press.

Kong, R. (2019). Is deliberation without attention of unconscious thought sufficiently supported with neuroscientific evidence?. *Journal of Psychology & Psychotherapy, 9*(4), Article 364 (3 pages).

Kruglanski, A. W., & Kopetz, C. (2009). The role of goal systems in self-regulation, Social cognition and social neuroscience. In E. Morsella, J. A. Bargh, & P. M. Gollwitzer (Eds.), *Oxford handbook of human action* (pp. 350–367). New York, NY: Oxford University Press.

Kruglanski, A. W., Shah, J. Y., Fishbach, A., Friedman, R., Chun, W. Y., & Sleeth-Keppler, D. (2002). A theory of goal systems. In M. P. Zanna (Ed.), *Advances in experimental social psychology* (Vol. 34, pp. 331–378). San Diego, CA: Academic Press, Elsevier Science.

Lassiter, G. D., Lindberg, M. J., González-Vallejo, C., Bellezza, F. S., & Phillips, N. D. (2009). The deliberation-without-attention effect: Evidence for an artifactual interpretation. *Psychological Science, 20*(6), 671–675.

Latash, M. L. (2008). *Synergy*. Oxford: Oxford University Press.

Latash, M. L. (2012). *Fundamentals of motor control*. London: Academic Press, Elsevier Inc.

Latham, G. P. (2012). *Work motivation: History, theory, research, and practice* (2nd ed.). Thousand Oaks, CA: SAGE Publications Inc.

Leuzinger-Bohleber, M., Arnold, S., & Solms, M. (Eds.). (2017). *The unconscious: A bridge between psychoanalysis and cognitive neuroscience*. London: Routledge, the Taylor & Francis Group.

Locke, E. A., & Latham, G. P. (1990). Work motivation: The high performance cycle. In U. E. Kleinbeck, H.-H. Quast, H. Thierry, & H. Hä (Eds.), *Work motivation (Applied psychology series)* (pp. 3–25). Hillsdale, NJ, US: Lawrence Erlbaum Associates, Publishers.

Logan, G. D., & Cowan, W. B. (1984). On the ability to inhibit thought and action: A theory of an act of control. *Psychological Review, 91*(3), 295–327.

Lowe, E. J. (2014). Why my body is not me: The unity argument for emergentist self-body dualism. In A. Lavazza, & H. Robinson (Eds.), *Contemporary dualism: A defence* (pp. 245–265), chapter 14. New York: Routledge Taylor & Francis Group.

Lowe, E. J. (2018). Non-Cartesian substance dualism. In J. J. Loose, A. J. L. Menuge, & J. P. Moreland (Eds.),*The blackwell companion to substance dualism* (pp. 133–151). Oxford, UK: Wiley.

Lowe, E. J. (2008). *Personal agency: The metaphysics of mind and action*. Oxford: Oxford University Press.

Lowe, E. J. (2010). Substance dualism: A non-cartesian approach. In R. C. Koons & G. Bealer (Eds.), *The waning of materialism* (pp. 439–461). Oxford, NY: Oxford University Press.

Lubashevsky, I., & Watanabe, M. (2016). Statistical properties of gray color categorization: Asymptotics of psychometric function. In *Proceedings of the 47th ISCIE International Symposium on Stochastic Systems Theory and Its Applications Honolulu*, Dec. 5–8, 2015, Honolulu. Institute of Systems, Control and Information Engineers (ISCIE), Kyoto, pp. 41–49. e-print: arXiv:1511.07242.

Lubashevsky, I., Wilson, I., Suzuki, R., & Kato, Y. (2018). Phase transitions in binary categorization: Evidence for dual-system decision making. In F. Müller, L. Ludwigs, & M. Kupper (Eds.), *Proceedings of the 34th Annual Meeting of the International Society for Psychophysics Fechner Day 2018*, Lüneburg, Germany, 20–24 August 2018. Leuphana University, Lüneburg, Germany, pp. 233–239. e-print: arXiv:1910.04511 [physics.med-ph].

Lumer, C. (2019). Unconscious motives and actions – agency, freedom and responsibility. *Frontiers in Psychology, 9*, Article 2777 (16 pages).

Lumer, C. (2005). Intentions are optimality beliefs - but optimizing what? *Erkenntnis, 62*(2), 235–262.
Lumer, C. (2017). Automatic actions: Agency, intentionality, and responsibility. *Philosophical Psychology, 30*(5), 616–644.
Marti, S., & Dehaene, S. (2017). Discrete and continuous mechanisms of temporal selection in rapid visual streams. *Nature Communications*, *8*, Article 1955 (13 pages).
Mazzone, M., & Campisi, E. (2013). Distributed intentionality: A model of intentional behavior in humans. *Philosophical Psychology, 26*(2), 267–290.
Mcbride, J., Boy, F., Husain, M., & Sumner, P. (2012). Automatic motor activation in the executive control of action. *Frontiers in Human Neuroscience*, *6*, Article 82 (14 pages).
McClelland, D. C., Koestner, R., & Weinberger, J. (1989). How do self-attributed and implicit motives differ? *Psychological Review, 96*(4), 690–702.
Melnikoff, D. E., & Bargh, J. A. (2018). The mythical number two. *Trends in Cognitive Sciences, 22*(4), 280–293.
Melnikoff, D. E., & Bargh, J. A. (2018). The insidious number two. *Trends in Cognitive Sciences, 22*(8), 668–669.
Metzinger, T. (2003). *Being no one: The self-model theory of subjectivity*. Cambridge, Massachusetts: The MIT Press.
Metzinger, T. (2009). *The ego tunnel: The science of the mind and the myth of the self.* New York: Basic Books.
Metzinger, T. (2011). The no-self alternative. In S. Gallagher (Ed.), *The Oxford handbook of the self* (pp. 279–296). Oxford, NY: Oxford University Press.
Mills, J. (2014). *Underworlds: Philosophies of the unconscious from psychoanalysis to metaphysics*. London: Routledge, the Taylor & Francis Group.
Moors, A., & De Houwer, J. (2006). Automaticity: A theoretical and conceptual analysis. *Psychological Bulletin, 132*(2), 297–326.
Moors, A., & de Houwer, J. (2007). What is automaticity? an analysis of its component features and their interrelations. In J. A. Bargh (Ed.), *Social psychology and the unconscious: The automaticity of higher mental processes* (pp. 11–50). New York, NY: Psychology Press, the Taylor & Francis Group.
Moors, A. (2009). Automaticity. In T. Bayne, A. Cleeremans, & P. Wilken (Eds.), *The Oxford companion to consciousness* (pp. 91–95). Oxford, UK: Oxford University Press.
Moors, A. (2013). Automaticity. In D. Reisberg (Ed.), *The Oxford handbook of cognitive psychology* (pp. 163–175). New York, NY: Oxford University Press.
Moors, A. (2016). Automaticity: Componential, causal, and mechanistic explanations. *Annual Review of Psychology, 67*(1), 263–287.
Moors, A., Spruyt, A., & De Houwer, J. (2010). In search of a measure that qualifies as implicit: Recommendations based on a decompositional view of automaticity. In B. Gawronski & B. K. Payne (Eds.), *Handbook of implicit social cognition: Measurement, theory, and applications* (pp. 19–37). New York, NY: The Guilford Press.
Morsella, E. (2009). The mechanisms of human action: Introduction and background. Social cognition and social neuroscience. In E. Morsella, J. A. Bargh, & P. M. Gollwitzer (Eds.), *Oxford handbook of human action* (pp. 1–32). New York, NY: Oxford University Press.
Moutoussis, K., & Zeki, S. (2002). The relationship between cortical activation and perception investigated with invisible stimuli. *Proceedings of the National Academy of Sciences, 99*(14), 9527–9532.
Newell, B. R., & Shanks, D. R. (2014). Unconscious influences on decision making: A critical review. *Behavioral and Brain Sciences, 37*(1), 1–19.
Newen, A. (2018). The embodied self, the pattern theory of self, and the predictive mind. *Frontiers in Psychology*, *9*, Article 2270 (14 pages).
Newen, A., Welpinghus, A., & Juckel, G. (2015). Emotion recognition as pattern recognition: The relevance of perception. *Mind & Language, 30*(2), 187–208.

Nieuwenhuis, S., & De Kleijn, R. (2011). Consciousness of targets during the attentional blink: A gradual or all-or-none dimension? *Attention, Perception, & Psychophysics, 73*(2), 364–373.

Nørretranders, T. (1998). *The user illusion: Cutting consciousness down to size*. New York, NY: Viking, the Penguin Group, Translated by Jonathan Sydenham.

Oettingen, G., Mayer, D., Sevincer, A. T., Stephens, E. J., ju Pak, H., & Hagenah, M. (2009). Mental contrasting and goal commitment: The mediating role of energization. *Personality and Social Psychology Bulletin, 35*(5), 608–622.

Oettingen, G. (2000). Expectancy effects on behavior depend on self-regulatory thought. *Social Cognition, 18*(2), 101–129.

Osman, M. (2018). Persistent maladies: The case of two-mind syndrome. *Trends in Cognitive Sciences, 22*(4), 276–277.

Overgaard, M., Rote, J., Mouridsen, K., & Ramsøy, T. Z. (2006). Is conscious perception gradual or dichotomous? a comparison of report methodologies during a visual task. *Consciousness and Cognition, 15*(4), 700–708. Special Issue on Introspection.

Papies, E. K., & Aarts, H. (2016). Automatic self-regulation: From habit to goal pursuit. In K. D. Vohs & R. F. Baumeister (Eds.), *Handbook of self-regulation: Research, theory, and applications* (3rd ed., pp. 203–222). New York, NY: The Guilford Press.

Pennycook, G., Neys, W. D., Evans, J. S. B., Stanovich, K. E., & Thompson, V. A. (2018). The mythical dual-process typology. *Trends in Cognitive Sciences, 22*(8), 667–668.

Perruchet, P., Vinter, A. (2002). The self-organising consciousness: A framework for implicit learning. In R. M. French, & A. Cleeremans (Eds.), *Implicit learning and consciousness: An empirical, philosophical and computational consensus in the making* (pp. 41–67). New York, NY: Psychology Press, the Taylor & Francis Group.

Resulaj, A., Kiani, R., Wolpert, D. M., & Shadlen, M. N. (2009). Changes of mind in decision-making. *Nature, 461,* 263–266.

Rosenbaum, D. A. (2013). Cognitive foundations of action planning and control. In W. Prinz, M. Beisert, & A. Herwig (Eds.), *Action science: Foundations of an emerging discipline* (pp. 89–111). Cambridge, MA: The MIT Press.

Rosenthal, D. M. (2009). Concepts and definitions of consciousness. In W. P. Banks (Ed.), *Encyclopedia of consciousness* (Vol. 1, pp. 157–169). Oxford, UK: Academic Press.

Rosenthal, D. M. (2009). Higher-order theories of consciousness. In A. Beckermann, B. P. McLaughlin, & S. Walter (Eds.), *The Oxford handbook of philosophy of mind* (pp. 239–252). New York: Oxford University Press.

Ryan, R. M. (2012). Motivation and the organization of human behavior: Three reasons for the reemergence of a field. In R. M. Ryan (Ed.), *The Oxford handbook of human motivation* (pp. 3–10). New York, NY: Oxford University Press.

Sandberg, K., Bibby, B. M., Timmermans, B., Cleeremans, A., & Overgaard, M. (2011). Measuring consciousness: Task accuracy and awareness as sigmoid functions of stimulus duration, *Consciousness and cognition, 20*(4), 1659–1675. From Dreams to Psychosis: A European Science Foundation Exploratory Workshop.

Schack, T. (2004). The cognitive architecture of complex movement. *International Journal of Sport and Exercise Psychology, 2*(4), 403–438.

Schechtman, M. (2007). Stories, lives, and basic survival: A refinement and defense of the narrative view. *Royal Institute of Philosophy Supplement, 60,* 155–178.

Schechtman, M. (2011). The narrative self. In S. Gallagher (Ed.), *The Oxford handbook of the self* (pp. 394–416). Oxford, NY: Oxford University Press.

Scheerer, E. (1984). Motor theories of cognitive structure: A historical review. In W. Prinz & A. F. Sanders (Eds.), *Cognition and motor processes* (pp. 77–98). Berlin: Springer.

Schlosser, M. (2019). Dual-system theory and the role of consciousness in intentional action. In B. Feltz, M. Missal, & A. Sims (Eds.), *Free will, causality, and neuroscience* (pp. 35–56). Leiden, The Netherlands: Brill Rodopi.

Sehgal, M., Zhou, M., Lavi, A., Huang, S., Zhou, Y., & Silva, A. J. (2018). Memory allocation mechanisms underlie memory linking across time. *Neurobiology of Learning and Memory, 153*, Part A, 21–25. MCCS 2018: Time and Memory.
Sergent, C., & Dehaene, S. (2004). Is consciousness a gradual phenomenon?: Evidence for an all-or-none bifurcation during the attentional blink. *Psychological Science, 15*(11), 720–728.
Shea, N., & Frith, C. D. (2019). The global workspace needs metacognition. *Trends in Cognitive Sciences, 23*(7), 560–571.
Sherman, J. W., Gawronski, B., & Trope, Y. (Eds.). (2014). *Dual-process theories of the social mind*. New York, NY: The Guilford Press.
Shin, Y. K., Proctor, R. W., & Capaldi, E. J. (2010). A review of contemporary ideomotor theory. *Psychological Bulletin, 136*(6), 943–974.
Strick, M., Dijksterhuis, A., & van Baaren, R. B. (2010). Unconscious-thought effects take place off-line, not on-line. *Psychological Science, 21*(4), 484–488.
Strick, M., Dijksterhuis, A., Bos, M. W., Sjoerdsma, A., van Baaren, R. B., & Nordgren, L. F. (2011). A meta-analysis on unconscious thought effects. *Social Cognition, 29*(6), 738–762.
Synofzik, M., Vosgerau, G., & Newen, A. (2008b). I move, therefore i am: A new theoretical framework to investigate agency and ownership. *Consciousness and Cognition, 17*(2), 411–424. Social Cognition, Emotion, and Self-Consciousness.
Synofzik, M., Vosgerau, G., & Newen, A. (2008). Beyond the comparator model: A multifactorial two-step account of agency. *Consciousness and Cognition, 17*(1), 219–239.
Taylor, J. A., & Ivry, R. B. (2013). Implicit and explicit processes in motor learning. In W. Prinz, M. Beisert, & A. Herwig (Eds.), *Action science: Foundations of an emerging discipline* (pp. 63–87). Cambridge, MA: The MIT Press.
Teleology (2015). Encyclopædia Britannica Ultimate Reference Suite. Encyclopædia Britannica.
Thiel, U. (2011). *The early modern subject: Self-consciousness and personal identity from descartes to hume*. New York: Oxford University Press.
Vadillo, M. A., Kostopoulou, O., & Shanks, D. R. (2015). A critical review and meta-analysis of the unconscious thought effect in medical decision making. *Frontiers in Psychology, 6*, Article 636 (6 pages).
Van den Berg, R., Anandalingam, K., Zylberberg, A., Kiani, R., Shadlen, M. N., & Wolpert, D. M. (2016). A common mechanism underlies changes of mind about decisions and confidence. *eLife, 5*, e12192.
Webb, T. L., & Sheeran, P. (2007). How do implementation intentions promote goal attainment? a test of component processes. *Journal of Experimental Social Psychology, 43*(2), 295–302.
Webb, T. L., & Sheeran, P. (2008). Mechanisms of implementation intention effects: The role of goal intentions, self-efficacy, and accessibility of plan components. *British Journal of Social Psychology, 47*(3), 373–395.
Westen, D. (1999). The scientific status of unconscious processes: Is Freud really dead? *Journal of the American Psychoanalytic Association, 47*(4), 1061–1106.
Wilson, T. D. (2002). *Strangers to ourselves: Discovering the adaptive unconscious*. Cambridge, MA: The Belknap Press of Harvard University Press.
Windey, B., Vermeiren, A., Atas, A., & Cleeremans, A. (2014). The graded and dichotomous nature of visual awareness. *Philosophical Transactions of the Royal Society of London B: Biological Sciences, 369*(1641), 20130282 (11 pages).
Windey, B., & Cleeremans, A. (2015). Consciousness as a graded and an all-or-none phenomenon: A conceptual analysis. *Consciousness and Cognition, 35*, 185–191.
Windey, B., Gevers, W., & Cleeremans, A. (2013). Subjective visibility depends on level of processing. *Cognition, 129*(2), 404–409.
Zhou, J., Zhou, C., Li, J., & Zhang, M. (2015). Cognitive style modulates conscious but not unconscious thought: Comparing the deliberation-without-attention effect in analytics and wholists. *Consciousness and Cognition, 36*, 54–60.

Part II
Temporality-Time Formalism

Chapter 5
Physics of Experiential Now: Effort of Atomic Action

As explained in Part I, the complex present is the minimal fragment of the human temporality admitting a closed description of intentional, goal-oriented actions. Action strategies in this description are the main entities with internal integrity. So all the basic properties characterizing such actions from the standpoint of intentions and goal pursuit should be attributed to action strategies directly.

In our account, action strategies are characterized by their own efficiency, the mental cost of changing the current action strategy for another one, various intrinsic motivations like novelty-seeking, etc. All these properties can be introduced at the level of the complex present. In addition to the noted properties, action strategies are characterized also by the bodily and mental effort required for their implementation. This property—the effort—is special because it can be introduced at the level of the experiential now which is a constituent component of the complex present.

For describing intentional, goal-oriented actions within the complex present we turn to the cost-benefit paradigm. This paradigm understood in the general sense as the effort-reward framework is widely used in describing human intentions for various types of activity including control processes (e.g., Shenhav et al. 2013, 2017; Studer and Knecht 2016; André et al. 2019). The increase in efficiency, the feeling of novelty, etc. play the role of benefit. In this case, the effort of action strategy implementation plays the role of cost. So the effort is a crucial element in developing a mathematical description of the experiential now and, then, the complex present.

For this reason, the present chapter is devoted, first, to the problem of effort within the experiential now. Second, referring to the notion of effort, we intent to propose an account of the experiential now as an individual basic entity of the complex present. As will be seen, the experiential now plays the role of pivot in the human temporality as a whole.

I. Lubashevsky and N. Plavinska, *Physics of the Human Temporality*,
Understanding Complex Systems, https://doi.org/10.1007/978-3-030-82612-3_5

5.1 Atomic Action: Phase Space

A phase space, by definition, is a set of variables that are necessary for describing states of a system under consideration. The first question we meet in our analysis is which variables are needed to describe effort of actions on scales of the experiential now.

The experiential now is the main element of the complex present through which the mind gains E-conscious access to the information about the state of the object affected by the subject. Strictly speaking, the experiential now bears not only E-consciousness but also elements of AM-consciousness. However, we will describe the dynamics of observed objects as it is experienced right now in terms of E-consciousness. This description is based on:

- the primary properties representing E-conscious perception on scales of the immediate now and
- the derivative properties coming into being due to the tripartite structure of the experiential now.

Put differently, in our account the experiential now

- inherits the primary properties from the immediate now with their averaging over time scales of the diachronic unit and
- acquires individual properties derived from the primary ones.

The main difference between the two types is that the mind can affect directly the latter type properties. In particular, the mind can ignore them in evaluating current actions.

In our constructions we will turn to the resource-based account of effort accepting that executing an action requires consuming or mobilizing some vital resources. In this case, the main concern about the cumulative effort required for implementing an action strategy is the amount of remaining resources. No other properties characterizing the effort as a physical quantity are essential for us.

If the resource depletion is not too high, then the perception of current effort should not depend on the amount of used resources at the previous moments. Thereby we may describe the cumulative effort of action strategy implementation as an additive sum of efforts attributed individually to the constituent actions—to be called atomic actions—making up a given action strategy. We will call this proposition the *additivity principle* of action strategy effort.

To describe the effort required for implementing an action strategy we need to introduce the notion of *atomic action* playing the role of basic "brick" in the additivity principle.

The atomic action, on the one hand, should admit the complete description of the current actions as they are experienced by the subject. On the other hand, different instantiations of actomic action must be mutually independent from the standpoint of effort. Therefore, subject's actions within the experiential now and accessible to the mind at the level of E-consciousness may be regarded as atomic action.

The atomic actions are not reduced only to the actions in the framework of the experiential now. Perception of currently executed actions can be also retained in memory and implemented in the connected future.

Speaking about the dynamics of material point governed by human actions, the experience of atomic action includes:

- the perception of the dynamics of material points, i.e., the characteristics of object motion in the real world and
- the conscious perception of the mental state related to the execution of current actions (in particular, attention focused on these actions).

The connected past and the connected future contain many atomic actions; as far as the experiential now is concerned, at each instant of time it contains the only one atomic action. We may say that the atomic action is the eidos of the experiential now and their collection represents the mental traces in the human temporality.

The introduction of the atomic action enables us to regard an action strategy as a sequence of atomic actions. Thereby allowing for the direct independence of atomic actions from one another in their effort, the cumulative effort of action strategy may be represented as the sum over efforts related to the constituent atomic actions.

As noted in Sect. 4.2, we confined our consideration to human goal-oriented actions when the object under control admits the interpretation as a material point moving in the three-dimensional physical space $\mathbb{R}^3 = \{x\}$.[1] In this case, according to ideas discussed in Chap. 3, human perception of external object motion within the experiential now may be treated as some cloud in the four-component phase space $\mathbb{R}_{[4]} = \{x, v, a, j\}$ including:

- the spacial position of observed object $x \in \mathbb{R}^3$,
- the velocity of object motion $v = dx/dt$,
- the acceleration $a = d^2x/dt^2$, and
- the jerk $j = d^3x/dt^3$.

For the observed physical object moving in the space $\mathbb{R}^3$ along a certain trajectory $\mathrm{MT} = \{x(t)\}$ the variables $\{x, v, a, j\}$ take specific values at each instant of time t whose collection $\{x(t), v(t), a(t), j(t)\}_{\mathrm{MT}}$ may be treated as some point of the physical phase space. The exact magnitudes of these values are not accessible to the mind.

In the mental-mental embedding of the observed object into the phase space $\mathbb{R}_{[4]}$ the "point" $\{x(t), v(t), a(t), j(t)\}_{\mathrm{MT}}$ is converted into a fragment $\mathbb{C}_{\mathrm{I}}$ of the space-time cloud $\mathbb{C}_{\mathrm{MT}}$ corresponding to the given instant of time (see Sects. 3.3.3 and 3.3.4).

For the external observer, the subject's actions are completely reflected in the motion state of affected object. Therefore, the collection of variables $\{x, v, a, j\}$ combined with time t may be used as the arguments for describing the subject's

[1] Through the present book the symbol x denotes the three-dimensional spatial point $\mathbf{x} = \{x_1, x_2, x_3\}$. In literature, bold symbols (e.g., $\mathbf{x}$) are usually used to denote such mathematical objects of vector type. We, however, do not follow this tradition here in order not to overload the text where it does not lead to misunderstanding.

actions. For this reason, we will call the collection $\{x, v, a, j\}$ the *external* phase variables of atomic action to emphasize that their specific values characterize the properties of observed object motion in the external world only.

In order to describe the effort of atomic action we also need another type quantities to be called the *internal* phase variables. These quantities characterize the subject's inner state affecting the perception of the object motion. The degree of attention focused on current actions is one of these quantities.

In the present book, we confine our consideration to the case where the degree of attention as the main internal phase variables and accept that, at least, voluntary attention affects action implementation with respect to both accuracy and reaction time (Prinzmetal, McCool and Park 2005; Prinzmetal, Park and Garrett 2005, for a general review see, e.g., Anderson and Druker 2013; Carrasco 2014; Prettyman 2019). Namely, we assume that the characteristic dimensions

$$\{\delta_x, \delta_v, \delta_a, \delta_j\}$$

of the cloud $\mathbb{C}_\mathrm{I}$ quantifying the uncertainties in the perception of object motion state $\{x(t), v(t), a(t), j(t)\}_\mathrm{MT}$ are determined by the degrees $\{A_x, A_v, A_a, A_j\}$ of attention focused on monitoring the corresponding variables $\{x, v, a, j\}$. In other words, these uncertainties $\{\delta_x, \delta_v, \delta_a, \delta_j\}$ are assumed to be a certain function of $\{A_x, A_v, A_a, A_j\}$:

$$\{A_x, A_v, A_a, A_j\} \Rightarrow \{\delta_x(A_x), \delta_v(A_v), \delta_a(A_a), \delta_j(A_j)\}. \tag{5.1}$$

The sum

$$A = A_x + A_v + A_a + A_j \tag{5.2}$$

specifies the degree of attention A focused on monitoring the object motion.

The following premises concerning attention (as a phenomenon as well as its plausible mechanisms) underlie the accepted model for the relationship between perception uncertainty and attention; for introductory review see, e.g., Sternberg and Sternberg (2012, Chap. 4) or Anderson (2015, Chap. 3). Namely, in our constructions:

⁎ The focusing of attention on some cognitive or behavioral phenomenon is understood as the allocation of cognitive resources limited in capacity that are required for processing the corresponding information. The concept of limited cognitive resources justifies the introduction of degree of attention meaning the relative amount of resources involved in processing the corresponding information. In particular, the degree equal to unity implies that all the possible processing resources are used completely.

⁎ The notion of attention is identified with selective attention that is defined as the cognitive process of attending particular phenomena while ignoring or suppressing all other irrelevant ones (e.g., Bater and Jordan 2019, for a brief review).[2]

[2] For a detailed review of the modern state of the art in the theory of selective attention a reader may be referred to Lavie (2010), Khetrapal (2010), Carrasco (2014), Murphy et al. (2016), Murphy

The concept of selective attention allows us to assume the existence of a certain direct relationship between the degree of attention and the variable quantifying the phenomenon or stimulus being the focus of attention. In particular, the gist of selected attention underlies Exp. (5.1).

※ The available amount of limited attentional resources can be divided and allocated according to what the implementation of a particular action requires.[3] We assume that, in this case, neural mechanisms responsible for monitoring different characteristics of action implementation may be regarded as individual ones for each of these characteristics.[4] It enables us, first, to ascribe the individual degrees of attention $\{A_x, A_v, A_a, A_j\}$ to monitoring the characteristics $\{x, v, a, j\}$ of object motion state and to suppose that the corresponding uncertainties $\{\delta_x, \delta_v, \delta_a, \delta_j\}$

---

and Greene (2017) and collection edited by Makovski et al. (2014). In particular, among the recent theories of selective attention we may note the perceptual load theory (Lavie and Tsal 1994; Lavie 1995; Murphy et al. 2016), the dilution theory (Tsal and Benoni 2010), and their hybrid version (Scalf et al. 2013). It is essential that top-down mechanisms contribute substantially to selective attention (e.g., Chen and Cave 2013; Biggs and Gibson 2013; Scalf et al. 2013; Murphy et al. 2016), which justifies the accepted assumption that humans are able consciously to focus attention on desired particular features of controlled processes.

[3] There are two rival models for divided attention (e.g., McDowd 2007; Sternberg and Sternberg 2012). One posits that there is one single pool of attentional resources admitting arbitrary division (Kahneman 1973). The other suggests that there are multiple sources of attention, one for each modality with its own capacity (Navon and Gopher 1979). The real mechanism of allocating attentional resources seems to be more complicated. By way of example, let us note the results obtained within the dual task paradigm when two tasks competing for attentional resources are performed either in a single sensory modality or in two separate modalities. If the interference between two tasks performed in different sensory modalities is weak, then attentional resources are distinct across the sensory modalities. If task interference is strong, then—regardless of whether tasks are performed in different modalities or the same modality—attentional resources are shared across the sensory modalities (for a review see, e.g., Wahn and König 2017; Wahn and Sinnett 2019). In the case we deal with, monitoring the action implementation may be supposed to be performed within a single sensory modality, mainly, the visual modality. It justifies the accepted assumption about the arbitrary division of attentional resources between monitoring different characteristics of action execution, Exp. (5.2), even if neural mechanisms governing this monitoring can be individual for each of them. The introduction of a set of attention degrees admitting individually continuous variations is justified by human ability to distribute attention unequally across multiple tasks (Crowe et al. 2019).

[4] The distinction between the neural mechanisms of monitoring the spacial position and velocity of observed object can be justified turning to the concept of two channels, sustained and transient, determining visual perception which was put forward in the 1970s (e.g., Breitmeyer and Ganz 1976; Legge 1978; Luce 1986; Smith 1995). The transient channel mechanism responds best to rapid temporal changes in perceived stimuli, the sustained channel mechanism responds best to steady or slowly varying stimuli. Nowadays evidence for the existence of these channels within the information processing in high-level visual regions of the brain and their interaction has been found based on fMRI measurements (Stigliani et al. 2017, 2019). The assumption about distinct neural mechanisms governing the monitoring of the acceleration and jerk of object motion is justified by that the perception of the two quantities is due to the tripartite structure of diachronic unit, i.e., the experiential now. Indeed, the perception of acceleration and jerk should be attributed to the different modes of information processing on the temporal scales of diachronic unit. The perception of spatial position and velocity is rooted in the information processing within the synchronic unit of immediate now.

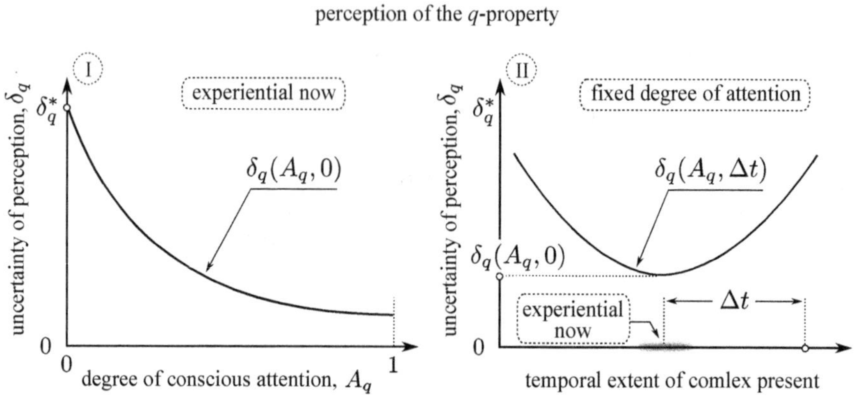

Fig. 5.1 Illustration of the uncertainty δ_q of the q-property perception as a function $\delta_q(A_q, \Delta t)$ of two variables A_q and Δt. Fragment I shows the dependence $\delta_q(A_q, 0)$ on the degree A_q of attention focused on monitoring the q-property within the experiential now ($\Delta t = 0$). Fragment II shows the dependence $\delta_q(A_q, 0)$ on the time lag Δt between a given moment and the experiential now within the complex present for some fixed degree of attention

depend only on the related degrees of attention, Exp. (5.1). Second, the additive sum (5.2) may be treated as the cumulative degree of attention focused on the action implementation as a whole process.

Now let us discuss the plausible relationship between the uncertainty of perception δ_q and the degree A_q of attention focused on monitoring the corresponding property q of the observed object. Here the symbol q stands for one of the phase variables $\{x, v, a, j\}$ specifying the mental-mental embedding of moving object.

For the general reasons, first, the uncertainty of perception δ_q should drop as the degree of attention focused on monitoring the q-property increases. It is illustrated in Fig. 5.1(I), the higher the attention degree A_q, the lower the uncertainty δ_q; it should hold for the experiential now as well as events retained in memory or expected in the connected future. Second, for a fixed degree of conscious attention the uncertainty of perception δ_q has to grow as the time lag Δt between a given moment and the experiential now increases, which is illustrated in Fig. 5.1(II).

In order to propose a plausible ansatz for the $\delta_q(A_q, \Delta t)$-dependence we turn to the scale-free principle of mind functioning (Chater and Brown 1999; Gisiger 2001; Kello et al. 2010). In systems exhibiting scale-free behaviour power-law relationships take the central position (e.g., Uchaikin 2013a, b). By way of example justifying that the experiential now and the complex present should also exhibit power-law regularities, we many note the well-known properties of human reaction time characterized by time scales of the experiential now. Some of these properties are listed in Footnote 4.[4]

[4] The following properties of human reaction time are characterized by power-law regularities. They are:

Following this idea, we accept the $\delta_q(A_q, \Delta t)$-dependence to have a power-law form. Therefore, first, dealing with the experiential now we will make use of the following power-law ansatz for the dependence of the perception uncertainty δ_q on the degree A_q of attention focused on monitoring the q-property

$$\delta_q(A_q, 0) = \delta_q^* \cdot \left[1 + \frac{A_q}{A_q^*}\right]^{-\alpha_q} \qquad \text{for } q = x, v, a, j. \tag{5.3}$$

Here the parameters δ_q^*, A_q^*, and the exponent $\alpha_q > 0$ are the attributes of the corresponding neural mechanism governing the perception of the object motion q-property. In particular, the coefficient δ_q^* represents the perception uncertainty when no attention is *consciously* focused on monitoring the q-property and, thereby, its control is based on the unconscious (automatic) component of the governing mechanism. The parameter $A_q^* \ll 1$ describes the contribution of attentional resources unconsciously allocated to monitoring the q-property and the degree of conscious attention A_q will be considered to change within the interval $0 \leq A_q \leq 1$.

Second, the temporal pattern of the uncertainty $\delta_q(A_q, \Delta t)$ within the complex present will be specified by the ansatz

$$\delta_q(A_q, \Delta t) = \delta_q(A_q, 0) \cdot \left[1 + \epsilon_q \frac{(\Delta t)^2}{\tau_{\text{cp}}^2}\right]^{\beta_q} \qquad \text{for } q = x, v, a, j, \tag{5.4}$$

where Δt is the time lag between a given moment of the complex present and the experiential now,[5] τ_{cp} is the characteristic duration of complex present, the parameter

- the dependence of the mean reaction time on the stimulus intensity above the perception threshold—Piéron's law—(Piéron 1913, 1952); for a detailed discussion of Piéron's law, its various aspects and plausible mechanisms a reader may be referred to Stafford and Gurney (2004), van Maanen et al. (2012), Donkin and Maanen (2014), Medina et al. (2014), Medina and Díaz (2016b), Reina et al. (2018);
- the dependence of the mean value of choice reaction time on the stimulus-goal complexity, which is represented, e.g., by Hick-Hyman's law (Hick 1952; Hyman 1953, see also Reina et al. 2018 for a review);
- the universal linear relationship between the statistical characteristics—trial-to-trial variations and the mean value—of reaction time (Jensen 2006; Wagenmakers and Brown 2007, for a review see, e.g., van Ravenzwaaij et al. 2011; Medina and Díaz 2016a);
- long-term correlations in trial-to-trial variations of individual reaction time which exhibit the $1/f$-noise properties; it should be noted that the $1/f$-noise intrinsic to neural processes (Medina 2009; Medina and Díaz 2012), the self-organized criticality or, more generally, complexity in long-term memory dynamics and cognitive activity (Gilden 1997, 2001; Holden et al. 2009);
- various properties of reaction time reflected power-law decay of working memory and resulting power-law decay of precision (Bays and Husain 2008; Wei et al. 2012; Donkin and Nosofsky 2012; López-Frutos et al. 2014; Reina et al. 2018).

[5] Strictly speaking Δt is specified by the time difference between a given instant of time—which is not accessible to the mind—and the center of the experiential now—which is also not accessible to the mind. Nevertheless due to the complex present being much longer than the experiential now,

$\epsilon_q \sim 1$ and the exponent $\beta_q > 0$ are the attributes of the neural mechanism governing the q-property perception.

Combining Exps. (5.3) and (5.4) together we get the ansatz

$$\delta_q(A_q, \Delta t) = \delta_q^* \cdot \left[1 + \frac{A_q}{A_q^*}\right]^{-\alpha_q} \cdot \left[1 + \epsilon_q \frac{(\Delta t)^2}{\tau_{\mathrm{cp}}^2}\right]^{\beta_q} \quad \text{for } q = x, v, a, j \tag{5.5}$$

which specifies the desired relationship between the dimensions of the space-time cloud $\mathbb{C}_{\mathrm{MT}}$ and the attentional resources consciously allocated for monitoring the dynamics of affected object.

The attentional resources discussed above are involved in monitoring the moving object. This monitoring is supposed to be always conscious. However, the bodily implementation of the corresponding actions can be unconscious[6] as well as conscious. In the latter case a certain additional amount of attentional resources has to be allocated for the performance of these actions.

To take into account this feature we introduce an additional internal phase variable $A_\mathfrak{b}$ of atomic action that specifies the degree of attention focused directly on bodily action execution. Thereby the total amount of attentional resources allocated to the atomic action is specified by the sum

$$A_x + A_v + A_a + A_j + A_\mathfrak{b} \leq 1 \tag{5.6}$$

which must be less than (or equal to) unity. If the actions are implemented automatically (unconsciously) then no attentional resources are required and these actions are experienced as effortless (the details are discussed in the next section).

Finalizing the description of the atomic action phase space we want to group the discussed above features.

The first group of features concerns the description of the space-time cloud $\mathbb{C}_{\mathrm{I}}$ representing the state of object motion as it is perceived by the subject. Namely,

- the external phase variables $\{x, v, a, j\}$ together with the time lap Δt specify the center position of the cloud $\mathbb{C}_{\mathrm{I}}$ in the space-time continuum $\mathbb{R}_{[4]} \times t$,
- the internal phase variables $\{A_x, A_v, A_a, A_j\}$ specify the spatial dimensions $\{\delta_x, \delta_v, \delta_a, \delta_j\}$ of $\mathbb{C}_{\mathrm{I}}$ in $\mathbb{R}_{[4]} \times t$ (Exp. 5.5),
- the temporal dimension of $\mathbb{C}_{\mathrm{I}}$ can be estimated by the characteristic duration of experiential now, i.e., the characteristic duration $\tau_d \sim$ 2–3 s of diachronic unit.[7]

$\tau_{cp} \gg \tau_d$, we may treat the time lag Δt as a value bearing inner uncertainty about the characteristic duration τ_d of experiential now, i.e., the diachronic unit. Within this accuracy the time lag becomes accessible to the mind.

[6] Strictly speaking, the unconsciousness of these actions is of the U-unconscious type, for details, see Sects. 4.4.3 and 4.4.4.

[7] Strictly speaking, the temporal dimension δ_t of the cloud $\mathbb{C}_{\mathrm{I}}$—i.e., the uncertainty in the temporal arrangement of a given atomic action within the complex present—should also depend on the time lag Δt between a given moment of the complex present and the experiential now. It is quite natural to expect a power law similar to Exp. (5.5) to specify this dependence, where $\beta_t \leq 1/2$. The limit

This description concerns the mental-mental embedding which converts the point $\{x, v\}$ of the physical phase space into the cloud $\mathbb{C}_{\text{I}}$ of the mental space. Put differently, it is the direct description of the interaction between the body-in-the-self and the-mind-in-the-self underlying the mental-mental embedding (see Sect. 4.3, Fig. 4.4).

The other group of features concerns the description of attentional resources allocated for implementing the monitoring of moving object and, maybe, the corresponding subject's actions. Namely,

- the internal phase variables $\{A_x, A_v, A_a, A_j\}$ specify the degrees of attention focused on monitoring the corresponding characteristics of object motion,
- the internal phase variable $A_{\mathfrak{b}}$ specifies the degree of attention focused on executing the subject's actions when they are conscious,
- Exp. (5.6) represents the limitation of available attentional resources.

The given group of features concerns the phenomenological description of how the interaction between the body-in-the-self and the-mind-in-the-self is implemented on its own.

5.2 Atomic Action: Effort

The notion of effort generalizes a number of particular ideas reflecting different aspects in human behavior (e.g., Massin 2017). Let us turn to the working definition of effort proposed by Gendolla and Wright (2009) according to which the notion of effort

> refers to two aspects of goal pursuit: persistence and intensity. The former involves resource mobilization extending across time; the latter involves resource mobilization at a point of time. Effort is generated to overcome behavioural impediments, that is, factors that make it more difficult to attain goals. (page 143)

In the present chapter we consider effort understood as the intensity of resource consumption or mobilization within the atomic action.

The aforementioned definition applying to both bodily and mental effort is rooted in the motivational intensity theory (Brehm et al. 1983; Brehm and Self 1989; Wright and Brehm 1989, for a review see Richter et al. 2016; Gendolla et al. 2019; Silvestrini and Gendolla 2019). Its gist is the idea that the amount of effort understood as the investment of resources enabling the execution of desired actions

> is primarily governed by a resource (or energy) conservation principle: given that resources are important for survival, individuals are motivated to avoid wasting them and aim at investing only those that are required for successful task execution.

$\beta_t = 1/2$ corresponds to the logarithmic compression of time structure within working memory advocated, e.g., by Howard (2018). This effect, however, does not contribute substantially to the integral representation of action strategy within the complex present and, therefore, we will ignore it in further constructions.

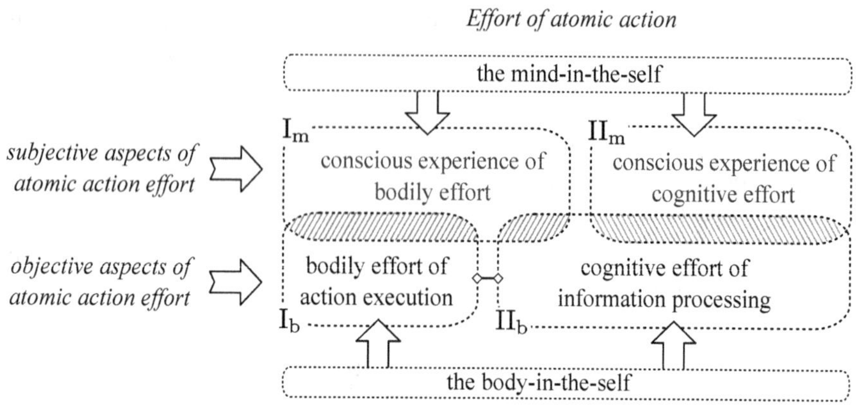

Fig. 5.2 The architecture of atomic action effort consisting of two levels representing the subjective and objective aspects of atomic action and belong to the realms of the mind-in-the-self and the body-in-the-self, respectively. The relationship between the subjective aspects of effort (groups I_m & II_m) and the objective aspects of effort (groups I_b & II_b) are illustrated as the regions where these groups overlap. This architecture reflects the mind-body duality. The shown connection between the groups I_b and II_b denotes the relationship between bodily executed actions and the corresponding information processing by the brain

> That is, people seek to avoid investing more than is required because this would waste resources. (Richter et al. 2016, page 151)

One of the basic premises of the motivational intensity theory is the proposition that effort rises with task difficulty as long as success is possible and the use of necessary amount of effort is justified. As elucidated in Sect. 5.4, we interpret the given proposition as the existence of some fuzzy threshold of action initiation: if the task difficulty exceeds this threshold the probability of initiating the corresponding action tends to zero.

5.2.1 *Dual Structure of Effort*

In order to find out governing parameters in describing the amount and type of mobilized resources, various aspects of human effort should be singled out and considered individually. For example, the intensity of attention may be used as a quantitative measure in the analysis of the psychological aspects of human actions (Kahneman 1973). For a discussion of arguments for and against attention as a measure of effort, a reader may be referred to Bruya and Tang (2018). Speaking about physiological aspects of actions, the energetic resources (Hockey 1997, see also Kruglanski et al. 2012) as well as biochemical resources (e.g., Fibiger and Singer 1989; Newsholme et al. 1992; Noakes 2012) can be related to the amount of effort.

In describing intentional, goal-oriented actions aimed at governing the dynamics of a physical object we have to deal with various aspects of the effort. These aspects may be classified via dividing them into two different levels which, on the one hand, possess their own properties and, on the other hand, are in close interconnection (Fig. 5.2):

- the *objective aspects* characterizing how the effort of atomic action execution manifests itself within the body-in-the-self and
- the *subjective aspects* characterizing how it manifests itself within the mind-in-the-self.

The objective aspects and subjective aspects of atomic action effort will be also referred to as *bodily effort* and *mental effort*, for short, to emphasize their belonging to the realms of the body-in-the-self and the mind-in-the-self.

5.2.2 Characteristic Properties

The given classification reflects a challenging problem in modern psychology concerning the description of human effort from different points of view including (*i*) the *actual* mobilization of physiological resources and (*ii*) the *subjective experience* of bodily and mental efforts. For an introduction to this problem a reader may be referred to Westbrook et al. (2013), Kurzban et al. (2013), Kurzban (2016), Hockey (2013), Shenhav et al. (2017), Silvestrini (2017), Massin (2017), Inzlicht et al. (2018), Bijleveld (2018), André et al. (2019), Silvestrini and Gendolla (2019).

In this subsection we want to discuss in detail a number of issues posed in the publications noted above and forming the background of our description of atomic action effort.

Separability of bodily and mental efforts. The subjective experience (feeling) of effort can be influenced by many other factors in addition to the actual mobilization of resources. So the correspondence between the subjective experience of effort and actual resource mobilization may vary as a function of contextual factors (e.g., Naccache et al. 2005; Marcora 2009; Bijleveld 2018). Turning to Fig. 5.2 we may say that the mental effort of type I_m depends on but is not reduced directly to the bodily effort of type $I_b \bigcup II_b$. As far as the mental effort of type II_m is concerned, it depends on but is not reduced directly to the bodily effort of type II_b. The noted irreducibility is illustrated in Fig. 5.2 as the blocks representing mental effort intersect with but are not contained in the blocks of bodily effort.

Turning to our constructions, this separability argues for the necessity of using individual collections of phase variables for describing bodily effort and mental effort as entities with own properties.

Resource of cognitive effort. Cognitive effort is a special form of effort related *directly* to processing the perceived information under limited computational capacity in contrast to energetic factors related to bodily execution of particular actions (for a review see, e.g., Kurzban et al. 2013; Shenhav et al. 2017). One of the famous accounts of cognitive effort caused by limited computational resources is Kahneman's (1973) capacity model of attention treating attention as the measure of such effort. Kurzban et al. (2013) noted that this type of effort is not literally a resource consumption. It is characterized by the *dynamic* reorganization of processing the perceived information in response to changing task demands but not by *depletion.* To tackle this problem, they put forward the *opportunity cost model* associating the costs and benefits of cognitive effort with the existence of a number of tasks competing for computational resources. Kool and Botvinick (2013, 2014) put forward another account of cognitive effort treated as a *intrinsic* cost attached directly to the exertion of cognitive control.

Naturally there should be biophysical resources underlying the information processing by neural networks in the brain. In particular, the consumption of glucose as the physical energy resource required for mental processes (e.g., Ampel et al. 2018) argues for introducing plausible types of cost for cognitive effort.

Turning to our constructions, the noted features of cognitive effort is the reason for introducing two different categories of effort represented by types $\mathrm{I_b} \bigcup \mathrm{I_m}$ and $\mathrm{II_b} \bigcup \mathrm{II_m}$ in the architecture of atomic action effort (Fig. 5.2). On the one hand, it accentuates the necessity of constructing individual models for effort related to action bodily execution and effort related to the mental control over executed actions. On the other hand, it justifies the use of the same basic notions in both the models.

Attention and mental effort. As indicated in Fig. 5.2 by the link between the blocks $\mathrm{I_b}$ and $\mathrm{II_b}$, the bodily execution of actions and the processing of perceived information related to these actions are closely intertwined. This relationship is flexible and can change in details depending on particular conditions including also the contribution of embodied cognition (perceived as effortless). The noted flexibility must be reflected in the required allocation of limited computational resources in the brain and underlies the introduction of two types $\mathrm{I_m}$ and $\mathrm{II_m}$ of mental effort. Type $\mathrm{I_m}$ corresponds to conscious experience of bodily executed actions and the related effort depending on attention focused on action execution directly. Type $\mathrm{II_m}$ corresponds to conscious experience of the information processing on its own and the related effort depending on attention aimed at the observation of action results for execution control.

In this context, we want to note the ongoing debates about the possibility of attention without consciousness and consciousness in the absence of attention (unconscious attention) (e.g., Koch and Tsuchiya 2007; van Boxtel et al. 2010; Posner 2012; Tsuchiya and Koch 2016; Jennings 2015; Montemayor and Haladjian 2015; Ran et al. 2016; Prasad and Mishra 2019). Nevertheless *selective* attention is closely interwoven with cognitive effort during goal-oriented actions provided all the form of attention, in particular, top-down as well as bottom-up ones are included in consideration (Cohen et al. 2012; Chica and Bartolomeo 2012; Marchetti 2012; Watzl 2017).

Turning to our constructions, the noted features argue for introducing the collection of attention degrees $\{A_x, A_v, A_a, A_j, A_\mathrm{b}\}$ (Sect. 5.1) in the description of metal effort of both the types:

- the type I_m effort controlled by attention of degree A_b aimed at action execution directly,
- the type II_m effort controlled by attention focused on monitoring the results of executed actions divided between monitoring the spacial position of controlled object (with attention degree A_x), its speed (with attention degree A_v), acceleration (with attention degree A_a), and jerk (with attention degree A_x).

Effortless action execution. As has been discussed in Sect. 4.4, human goal-oriented behavior can be governed by automatic mechanisms requiring no attention to particular details in action execution. Such actions can be experienced as effortless.[8] It raises a doubt about the possibility of using the concept of effort for describing goal-oriented actions. To reconcile the automaticity in action execution and the resource conservation principle Silvestrini and Gendolla (2013), Gendolla and Silvestrini (2015) put forward the concept of *bounded effort automaticity*. This concept turns to the idea that the resource conservation principle moderates and limits the effects of action automaticity on effort. Learning and skill acquisition may be regarded as a plausible mechanism underlying the bounded effort automaticity (e.g., Wulf and Lewthwaite 2010). Previously (Sect. 4.4.3) we discussed the *triple-model view* positing that top-down automatic processes come into being via the nonautomatic (conscious) stage (Moors 2009, 2013), which also argues for the concept of bounded effort automaticity. Put differently, automatic actions involved in goal-oriented behavior should be *implicitly* governed by criteria based on effort optimization, at least, approximately.

Turning to our constructions, the presented comment is an argument for employing the same mathematical formalism in the description of goal-oriented actions whose implementation is based on either conscious, deliberate analysis or automatic mechanisms emerging via skill acquisition. Moreover, introducing the degree A_b of attention focused on action execution as a continuous variable changing in within the

[8] There are various mechanisms able to endow action execution with the feeling that a given action does not require any effort for its implementation. In the present book we consider only the action automaticity, i.e., effortless actions executed automatically without attention focused on the control over their implementation. As another mechanism of different nature, we want to note one described by the concept of effortless implementation of actions comprising effortless actions and effortless attention. It was put forward by Csikszentmihalyi (1975, see also Csikszentmihalyi 1990) within the concept of *flow*—a psychological state related to the enjoyment of deep absorption in what one is doing (Nakamura et al. 2019). This state is characterized, in particular, by (*i*) *optimal experience* due to the positive conjunction of cognition, affect, and motivation (see a collection edited by Csikszentmihalyi and Csikszentmihalyi 1988 or a book by Fave et al. 2011) and (*ii*) the *autotelic experience*, i.e., experience having the purpose of its existence within itself (Csikszentmihalyi 1975). For a discussion of this concept a reader may be address also to Bruya (2010a), Csikszentmihalyi and Nakamura (2010), Wulf and Lewthwaite (2010), Schmeichel and Baumeister (2010). Within our concept of E-consciousness we can explain these phenomena turning to the idea that in perceiving the final, integral result of complex neural processes its division into different components becomes problematic. So if positive, pleasurable aspects of executing some action dominate the cost of resource use caused by the action execution, the subject just experiences pleasure and attributes no effort to the given action. There are also a range of phenomena where effort on its own is costly and valued simultaneously (e.g., Inzlicht et al. 2018, for a review). Analysis of these issues and mechanisms via which effortless actions and attention can arise falls outside the scope of the present book.

range $0 \leq A_b \leq 1$, this formalism can describe strictly deliberate actions and purely automatic actions as the limit cases of more general human behavior with partial automatism. It opens a gate toward constructing a theory of phase transition between automatic and deliberate human actions.

Entanglement of conscious-automatic actions. The concept of embodied cognition[9] applied to sport psychology stimulated the appearance of novel ideas about skill acquisition and the interaction between human cognition and action automatic performance. For an introduction to the modern state of the art in this field a reader may be addressed to the collection edited by Cappuccio (2019) as well as reviews by Raab and Araújo (2019), Otte et al. (2019), Pacheco et al. (2019), Montull et al. (2020), Rochat et al. (2020). It turns out that neurophysiological mechanisms[10] of the body functioning can *replace* the cognitive processing of information by the brain in various motor and perceptual tasks (Shapiro and Spaulding 2019). Their mixture gives rise to the entanglement between human actions governed either by cognitive or automatic processes.

In this context we want to note the *Mesh* theory of skill (Christensen et al. 2016, see also Christensen and Sutton 2019; Christensen et al. 2019). This theory proposes a novel account allowing for the basic properties of skill influence on human actions turning to the gist of the dual process approach to cognitive architecture defended, e.g., by Evans (2008, 2011, 2012), Evans and Stanovich (2013).[11] Let us list some of these properties contributing to our concept of mental effort in discriminating the subjective experience of type I_m effort and type II_m effort (Fig. 5.2):

- *Reduced attention and subjective effort*: attention to and subjective experience of action execution can be reduced once the corresponding skill has been acquired.
- *Disruptive attention*: attention to the performance of a highly learned skill can be disruptive.
- *Increased attention and cognitive effort in response to challenge*: challenging situation can give rise to a considerable increase in attention to an action under execution and subjective experience of the related effort.

---

[9] We already discussed the concept of embodied cognition in Sect. 3.2.4 from the general standpoint.

[10] The mirror-neuron system (e.g., Rizzolatti and Craighero 2004; Cattaneo and Rizzolatti 2009) exemplifies such mechanics. It is interesting to note that the contribution of the mirror-neuron system seems to be higher in executing actions than observing these actions (Woodruff and Maaske 2010). In particular, it argues for treating the monitoring of observed object as a conscious process.

[11] For a review of the classic theories of skill acquisition by Paul Morris Fitts (1967), Dreyfus and Dreyfus (1986) treating the skill learning as a progression to an automatic stage a reader may be referred to, e.g., Schmidt and Wrisberg (2008). For a review of other modern rival theories, in particular, the *reinvestment theory* (e.g., Masters and Maxwell 2008), the *execution focus theory* (e.g., Beilock et al. 2004, 2008; Beilock and Gray 2012), and the *attentional focus theory* (e.g., Wulf et al. 2000; Wulf and Prinz 2001; Wulf et al. 2001, also Wulf 2013 for a review) a reader may be referred, e.g., to Cappuccio et al. (2019), Song (2019). As a common aspect of all these theories, they draw a distinction between inner attention to the execution of bodily actions and outer attention focused on monitoring the motion of affected objects. To allow for this effect we have introduced a special variable A_b—the degree of attention focused on action bodily implementation.

These features are reflected in Fig. 5.2 as the intersection of block I_m simultaneously with two blocks I_b and II_b. It symbolizes that for the bodily effort of action execution to be subjectively experienced some amount of attention should be consciously focused on the executed action. Without it, i.e., when an action is executed automatically it can be perceived as effortless.

Turning to our constructions, the noted entanglement of skill and effort experience enables us to posit that the conscious experience of bodily effort can be described as a product of two factors. One of them reflects the physical forces the subject applies—explicitly or implicitly—to the object in governing its dynamics. The other represents the subject's perception of these bodily actions with respect to how difficult they are. We consider the latter factor to be a continuous function $\sigma(A_b)$ of the attention degree A_b such that $\sigma(0) = 0$.

Involuntary attention and bodily effort. Directing attention to an event or stimulus can be either intentional or elicited from outside events as well as some internal conditions. The former type of attention is usually called *voluntary* attention, the latter one is *involuntary* or *automatic* attention. The distinction between voluntary and involuntary attention was recognized in psychology already at the turn of the 19th and 20th centuries (e.g., Eimer et al. 1996). Nowadays it is widely excepted that voluntary and involuntary attention are based on different mechanisms (e.g., Turatto et al. 2000; Hopfinger and Parks 2012). In particular, voluntary attention is of top-down type. As far as involuntary attention is concerned, its early stage is of bottom-up type, whereas the later stage is more subject to top-down control (Sussman et al. 2003; Hopfinger and Parks 2012). The effects caused by voluntary and involuntary attention are also different (Prinzmetal, McCool and Park 2005; Prinzmetal, Park and Garrett 2005; Prinzmetal et al. 2009). Nevertheless, the processing of information controlled by voluntary and involuntary attention can interact (Grubb et al. 2014).

Turning to our constructions, we draw a distinction between voluntary and involuntary attention internally focused on the execution of bodily actions. Here the internal involuntary attention is understood as attention that automatically focused on the current bodily actions and cannot be voluntary shifted by the subject to something different. It does not exclude that the subject can focus additional internal attention to these actions, which meets the aforementioned Mesh theory of skill. Effectively, this involuntary attention will be described by the minimal degree A_b^{min} of attention required for implementing the corresponding bodily actions. The value of A_b^{min} depends, first, on the skill acquired by the subject; the higher the skill, the lower the value of A_b^{min}. Second, the value of A_b^{min} depends on the intensity or difficultly of the bodily effort of action execution. In particular, it reflects the aforementioned skill property about the increase of attention in response to challenge.[12] For the sake of

[12] In this connection it should be noted that an internal focus of attention is not always associated with detrimental effects and, moreover, can even benefit performance. Namely, the attention focus on physical sensations helpful for performance, whereas, a focus on automated processes hampering performance. In particular, it is endorsed by action and motor-centered frameworks (Hanin and Hanina 2009; Carson and Collins 2015). To emphasize this difference in internal focus of attention Schücker et al. (2014), van Ginneken et al. (2017) proposed to differentiate between internal attention

simplicity, in our future constructions we will consider the degree A_b^{min} of involuntary attention constant value. The introduction of the minimal degree of internal attention is reflected in Fig. 5.4.

Multi-channel learning. In line with available theoretical works (Kurzban et al. 2013; Inzlicht et al. 2014) Bijleveld (2018) proposed the *shared cause model* for the relationship between physiological effort and the subjective experience of effort. This model posits that physiological effort and subjective experience of effort are just different and functionally independent facets of effort and their mutual correlations are caused by the same upstream decision-making process (Wegner 2002, 2004; Naccache et al. 2005; Marcora 2009). In other words, physiological effort and the experience of this effort become interrelated due to learning processes dealing with just effort together with its all facets simultaneously. Turning to neurophysiological arguments, André et al. (2019) generalized the shared cause model within the integrated account of effortful control. The main unit involved in learning process, on the one hand, accumulates the information about task constraints, rewards/punishments, time variations in organism's state. On the other hand, this unit governs the dynamics of four parallel "channels:"

- the experience of effort,
- effort signal consciously involved in carrying out current tasks,
- automatic signal playing similar role outside consciousness,
- the feeling of fatigue.

Including the feeling of fatigue opens a gate towards elucidating neural mechanisms enabling the mind to control the use of energetic and/or computational resources when the control is implemented unconsciously. Fatique based on neurophysiological mechanisms becomes accessible to the mind when physiological states get an alarming level, which returns conscious attention to the control process (e.g., Noakes 2012; Hockey 2013).

Turning to our constructions, the presented concept of learning underlies our theory of multi-channel reinforcement learning developed for describing how one cumulative measure of mental effort can be introduced. The problem is that subjective experience of executed actions comprises several channels of information processing. These channels are independent in the sense that the action evaluation within individual channels can be represented in incommensurable terms. The deliberate evaluation of executed actions in terms "efficient-inefficient" compared with the emotion-like evaluation in terms "interesting-boring" illustrates this incommensurability.

Dynamics of effort experience. In evaluating goal-oriented actions sustained for a certain time interval the subjective experience of effort concerns not only the current events but also events retained in memory or anticipated in the forthcoming future. How the subjective feeling of cognitive effort experienced right now—the real-time

---

focus either on physical sensations (the internal monitoring of actions) or on automated processes (the controlling of actions). For a detailed analysis of this issue a reader may be addressed to Vitali et al. (2019), Gottwald et al. (2020).

rating of effort—is related to the anticipated or retrospective ratings of effort is far from deep understanding.[13] Nevertheless, some basic features of this relation playing an essential role in our further construction can be found in publications by Finn (2010), Hoogerheide and Paas (2012), Finn and Miele (2016), Hsu et al. (2018), Dunn et al. (2019), Bambrah et al. (2019).

Let us turn to experiments by Bambrah et al. (2019), where participants were involved, in particular, in performing a sustained attention task of duration in the range of 500–1000 s preceded by practice of duration of 25–100 s. The practice was required for the participants to familiarize themselves with the experimental setup. Just after the practice, the participants were asked to rate the anticipated effort to be required for the following task performance, then they rated the effort during the task performance several times and after finishing the task retrospectively estimated the made effort. As has been found out, the anticipated rating of effort is substantially lower than the real-time ratings whereas the mean real-time rating and the retrospective rating are rather close to each other.

Based on these results, we infer that the *learning* and *adaptation* of the subjective feeling of effort related to keeping attention focused on some process for a certain time interval can be characterized by time scales τ_a estimated as $100 \lesssim \tau_a \lesssim 1000$ s.

Hsu et al. (2018) extending previous works by Finn (2010), Hoogerheide and Paas (2012), Finn and Miele (2016) actually demonstrated substantial nonlinearity in the relationship between the real-time rating of cognitive effort and the retrospective rating of the integral effort required for implementing the corresponding task as a whole. Namely (Hsu et al. 2018),

- the peak and end ratings of cognitive effort are the main factors determining the retrospective judgment about the effort of cognitive task as a whole (the *peak-end effect*),
- the initial and average ratings also partly contribute to the retrospective judgment,
- the retrospective judgment is not the summarized real-time rating (the *duration neglect phenomenon*).

Turning to our constructions, the noted temporal properties of effort experience enable us to posit that the cognitive aggregation of local efforts into the integral amount of effort during task performance is a substantially nonlinear process affected also by memory decay. These features are reflected in that the higher intensity of the locally experienced effort and the closer the corresponding moment of time to the moment of retrospective evaluation of integral effort, the higher its contribution.

---

[13] Because the feeling of effort is typically treated as an unpleasant and aversive emotion (e.g., Kahneman 2011; Kurzban et al. 2013) the properties of time relation between the ratings of effort are assumed to be similar to the corresponding phenomena in the feeling of pain whose understanding is more advanced (e.g., Fredrickson and Kahneman 1993; Kahneman et al. 1993; Redelmeier and Kahneman 1996; Ariely 1998; Kahneman 2000a, b; Miron-Shatz et al. 2009). So the properties of pain feeling often play the role of impetus for particular research concerning subjective experience of cognitive effort.

On scales of the complex present whose integral duration $\tau_d \gtrsim 10$ s, the effort integration is a substantially nonstationary process. The use of retrospective, real-time, and anticipated ratings the same effort enables us:

- to attribute the subjective experience of effort to atomic actions belonging to the connected past, the experiential now, and the connected future;
- to introduce an additional dynamic variable integrating the subjective efforts of atomic actions forming a given action strategy in some nonlinear way.

5.2.3 Mathematical Description: General Features

Let us note the general features of atomic action effort underlying our account of intentional, goal-oriented actions within the complex present. In describing atomic action effort we turn to, first, the effort properties discussed above and, second, the basic approaches to conceptualizing the notion of effort (e.g., Massin 2017, for a review), mainly, the force-based accounts and the resource-based accounts.

Two measures of the same effort. We want to introduce two measures E_b and $\mathbb{E}_\mathrm{m}$ which do not represent a quantitative description of different *components* of human effort, they quantify different *aspects* of the same effort without dividing it into constituent components. Namely, we introduce two different *quantitative measures* of the same effort—*bodily* E_b and *mental* $\mathbb{E}_\mathrm{m}$ measures—for describing the corresponding effects within the scope of the *body*-in-the-self and the *mind*-in-the-self. It should be noted that the bodily measure E_b is an ordinary function whose argument are the body state and the state of object motion. By contrast, the mental measure $\mathbb{E}_\mathrm{m}$ is of the cloud-type rather than a normal function taking some value, which we emphasize using the blackboard typeface. Naturally, these measures are not functionally independent but, nevertheless, are not reducible to each other and, in this sense, are complementary elements of effort description. The introduction of two different measures reflects the dual structure of the self, at least, within phenomenology.

Bodily measure of effort. The bodily measure E_b quantifies the effort of human actions in governing the dynamics of physical object. It is a quantity introduced within the scope of the body-in-the-self and its specific value is not directly accessible to the mind.

The bodily measure E_b is determined by the forces the subject exerts on this object to drive it. The measure E_b quantifies effort of human actions at a given instant of objective time t. For an object treated as the material point, the value of E_b is supposed to be specified by the external phase variables $\{x, v, a\}$—the object spatial position $x(t)$ at time t and its time derivatives, namely, the velocity $v(t) = dx/dt$ and acceleration $a(t) = d^2x/dt^2$—characterizing the physics of object motion.[14]

[14] In Sect. 5.1 we noted that the complete phase space describing the dynamics of material point as it is perceived by subject contains four phase variables, $\{x, v, a, j\}$. However, the bodily measure E_b does not depend explicitly on the jerk $j = d^3x/dt^3$ because Newton's second law operates

Put differently, the bodily measure E_b quantifying effort of human actions as they are implemented in the scope of the body-in-the-self is supposed to be a function

$$E_b = E_b\Big(x, v, a \mid S_{env}(t)\Big). \tag{5.7}$$

Naturally, this function includes also the state $S_{env}(t)$ of environment as an additional argument; the particular form of function (5.7) is determined by a specific system under consideration.

In the next Sect. 5.3 we will elucidate a plausible relationship between the bodily measure E_b and the laws governing the object dynamics within Newtonian mechanics. Leaping ahead, we want to note that the proposed approach to describing bodily aspects of human effort combines the forced-based and resource-based accounts of effort. On the one hand, in constructing the bodily measure E_b, we will turn to the notion of forces applied to a given object and, on the other hand, treat the quantity E_b as a scalar, which is necessary to interpret this type effort as the consumption of some energetic resource. For a comparative analysis of the force-based and resource-based accounts a reader may be addressed to Massin (2017).

Mental measure of effort. The mental measure $\mathbb{E}_m$ of effort is introduced to quantify the experience of effort as a whole. It describes two different types of phenomena represented by blocks I_m and II_m in Fig. 5.2, namely, how the subject experiences:

- the execution of bodily actions in governing object motion, which includes also the processing of *cognitive* information directly related to the bodily implementation of these actions;
- cognitive effort required for processing perceived information reduced, in our case, to the monitoring of object motion.[15]

The individual contribution of these phenomena to $\mathbb{E}_m$ will be quantified by the components $\mathbb{E}^{I}_{m}$, $\mathbb{E}^{II}_{m}$, respectively.

Section 5.4 is devoted to a detailed discussion of the mental measure $\mathbb{E}_m$ so, here, we confine ourselves to general comments elucidating the further constructions. In what follows we will use the terms *bodily effort* and *mental effort*, for short, speaking about how the effort of executing a given action is reflected in the body-in-the-self and the mind-in-the-self, respectively.

First, the properties of mental measure $\mathbb{E}_m$ to be listed below concern mainly the perception of effort within the experiential now. In particular, it is the introduction of two different types of mental effort quasi-additivity contributing to $\mathbb{E}_m$.[16] As far

only with the spatial position x, velocity v, and acceleration a of a material point. The jerk, as an individual phase variable, appear in the mental measure of effort.

[15] In the general case, this type of subjective effort combines all the components of the information conscious processing except for the conscious processing of information concerning bodily implementation of the corresponding actions.

[16] Strictly speaking, the properties of effort described by the components $\mathbb{E}^{I}_{m}$, $\mathbb{E}^{II}_{m}$ are incommensurable, nevertheless, the multi-channel learning enables their effective comparison. This feature allows us to consider the quantities $\mathbb{E}^{I}_{m}$, $\mathbb{E}^{II}_{m}$ additive (in fact, quasi-additive) components of the

as atomic actions not belonging to the experiential now are concerned, only the cumulative measure $\mathbb{E}_\mathrm{m}$ will be attributed to them. The corresponding uncertainty of $\mathbb{E}_\mathrm{m}$ depends, in general, on the time leg between the time instant in issue and the experiential now.

Second, because human perception is characterized by intrinsic uncertainty, the mental measure $\mathbb{E}_\mathrm{m}$ together with its constituent components $\mathbb{E}^\mathrm{I}_\mathrm{m}$, $\mathbb{E}^\mathrm{II}_\mathrm{m}$ cannot be quantities taking some specific numerical values for a given action. For this reason we will interpret them as one-dimensional clouds—regions of $\mathbb{R}$ with blurred boundaries. These cloud-type measures have to be related, in some way, to clouds characterizing the subject's perception of moving object and the mental state of the subject himself. Put differently, the cloud-type measure $\mathbb{E}_\mathrm{m}$ should admit a certain representation that can be written in the general form as

$$\mathbb{E}_\mathrm{m} = \widehat{\mathcal{E}}\left\{\mathbb{C}_\mathrm{I} \times \mathbb{C}_\mathrm{A}\right\}, \tag{5.8}$$

where $\widehat{\mathcal{E}}$ is some, maybe, nonlinear operator and the symbol $\times$ denotes the Cartesian product of

- the space-time cloud $\mathbb{C}_\mathrm{I}$ describing the mental-mental embedding of observed object and
- the attention-degree cloud $\mathbb{C}_\mathrm{A}$ in the space $\{A_x, A_v, A_a, A_j, A_\mathrm{b}\}$ describing human perception of computational resource allocation—mental workload.

The center of the cloud $\mathbb{E}_\mathrm{m}$—the mean value of mental measure—quantifies numerically how the subject evaluates the effort on the average, the cloud blurriness quantifies the uncertainty of this evaluation.

Third, since the bodily execution of actions and the monitoring of object motion are different components of goal-oriented behavior, we assume that the relationship between the cloud-type measure $\mathbb{E}_\mathrm{m}$ and its constitutive components $\mathbb{E}^\mathrm{I}_\mathrm{m}$ and $\mathbb{E}^\mathrm{II}_\mathrm{m}$ can be written as

$$\mathbb{E}_\mathrm{m} = \mathbb{E}^\mathrm{I}_\mathrm{m} \oplus \mathbb{E}^\mathrm{II}_\mathrm{m}\,. \tag{5.9}$$

Here the symbol $\oplus$ denotes that equality (5.9) may be understood as the arithmetic sum[16] with respect to the mean values of the cloud-type measures entering this expression. As far as the blur of the cloud $\mathbb{E}_\mathrm{m}$ is concerned, it is determined by the basic properties of stimulus perception relating its uncertainty to the mean stimulus intensity. In other words, the blur of $\mathbb{E}_\mathrm{m}$ is not the direct result of arithmetical operations with the blur of the components, which, however, does not exclude that it can be the case under some special conditions.

Fourth, when some skill in executing actions of a given type is acquired, the conscious and automatic mechanisms together govern their implementation. Because the automatic implementation of actions is usually perceived as effortless, we will assume that the mental measure of effort related to bodily actions is determined by:

---

cumulative mental measure, which is elucidated in Sect. 5.5 partly devoted to nonlinear fusion of the two components of mental effort.

- the bodily measure $E_\mathfrak{b}$ of effort integrated in some way over the space-time cloud $\mathbb{C}_\mathrm{I}$ and
- the degree $A_\mathfrak{b}$ of attention *consciously* focused on these actions.

It enables us to represent this relationship in the general form as

$$\mathbb{E}^\mathrm{I}_\mathrm{m} = \widehat{\mathcal{E}}_\mathfrak{b}(E_\mathfrak{b}, A_\mathfrak{b})\,\{\mathbb{C}_\mathrm{I}\}\,, \tag{5.10}$$

where $\widehat{\mathcal{E}}_\mathfrak{b}(E_\mathfrak{b}, A_\mathfrak{b})$ is a certain operator which quantifies the contribution of the "point" $\{x, v, a, j\} \in \mathbb{C}_\mathrm{I}$ based on the corresponding bodily measure of effort, $E_\mathfrak{b}(x, v, a, j)$, and obeys the condition

$$\widehat{\mathcal{E}}_\mathfrak{b}(E_\mathfrak{b}, A_\mathfrak{b})\Big|_{A_\mathfrak{b}=0} \equiv 0\,. \tag{5.11}$$

Fifth, phase transitions between effortful conscious and effortless automatic performance of bodily actions can occur when the current state of object motion—as it is experienced by the subject—enables (or seems to enable) the subject to govern it within

- an *action strategies* executed automatically (for transitions from conscious to automatic performance),
- a more efficient *action strategy* executed consciously (for transitions from automatic to conscious performance).

These phase transitions underlie the intermittency of goal-oriented actions—a basic property of human behavior—to be discussed further.

In the next sections we, at first, will construct the bodily measure of human effort, then the mental measure incorporating the bodily measure and playing the leading role in developing the theory of human goal-oriented behavior.

5.3 Bodily Measure of Effort and Newtonian Mechanics

The body-in-the-self, on the one hand, is responsible for the direct or indirect interaction with external physical objects in implementing goal-oriented behavior. This interaction is based on particular movements of the body or its elements, muscle tension, etc. On the other hand, the body-in-the-self is responsible for automatic actions as well as embodied cognition relating motor behavior with E-conscious perception of current actions.

Therefore, the bodily measure of human effort required for executing a given atomic action has to describe

- the physical force the subject needs to apply—directly or indirectly—to the controlled object to drive it along the desired trajectory,
- the intensity of the corresponding neural signals endowing the mind with E-conscious perception of these actions.

The former aspect reflects the basic physical regularities, in particular, the laws of Newtonian mechanics underlying the dynamics of a controlled physical object. So, the bodily measure $E_{\mathfrak{b}}$ of effort has to include the basic mathematical elements of Newtonian mechanics, e.g., the forces caused by the interaction between the object controlled by human actions and the other physical objects of environment.

The latter aspect reflects the basic features of how humans experience effort required for implementing atomic actions. Within the resource-based account accepted in our constructions this effort should be of the scalar form.

The aforementioned general properties of action effort within the scope of the body-in-the-self prompt us to turn to the following particular constructions of the bodily measure $E_{\mathfrak{b}}$ of atomic action effort.

5.3.1 Human Control over Dynamics of Material Point

As noted in the beginning of Sect. 4.2, we have confined our consideration mainly to human actions when a controlled object can be regarded as a material point moving in the space $\mathbb{R}^3$. The dynamics of this material point is supposed to be governed by the laws of Newtonian mechanics.

Under these conditions in describing the object motion along the trajectory $\{x(t)\}$ ($x \in \mathbb{R}^3$) we may turn to Newton's second law, i.e., Eq. (4.1), where the force F acting on the object consists of two components. One of them is the force F_e determined by the interaction between the given object and other physical objects of environment. This component is assumed to be specified as a known beforehand function similar to one entering the right-hand side of Eq. (4.1). The other component F_b is due to subject's intentional actions and can be used as a quantitative measure of bodily effort within the force-based account of human effort. From the subject's side, the force F_b is required to implement the object motion trajectory $\{x(t)\}$ at a given instant of objective time t.

In these terms, the equation governing the dynamics of controlled object may be written in the form

$$m\frac{d^2x}{dt^2} = F_e\left(x, \frac{dx}{dt}, t \,\middle|\, S_{\text{env}}(t)\right) + F_b\,, \tag{5.12a}$$

and, then, rewritten as

$$F_b\left(x, \frac{dx}{dt}, \frac{d^2x}{dt^2} \,\middle|\, t, S_{\text{env}}(t)\right) \stackrel{\text{def}}{=} m\frac{d^2x}{dt^2} - F_e\left(x, \frac{dx}{dt}, t \,\middle|\, S_{\text{env}}(t)\right), \tag{5.12b}$$

which actually specifies the force $F_b(\ldots)$ caused by subject's *intentional* actions at time t needed for moving the object along the path $\{x(t)\}$. Expression (5.12b) may be treated as the definition of the functional dependence of the force $F_b(\ldots)$ on the properties of object motion, namely, on the current position $x(t)$ of moving object, its instant speed dx/dt, and acceleration d^2x/dt^2. We want to emphasize that two factors contribute to the given expression. One is the physics of interaction between the objects of the given system reflected in the dependence of $F_e(\ldots)$ on x and dx/dt which is assumed to be known beforehand. The other factor is subject's intention to drive the object along the desired path $\{x(t)\}$ which determines the current value of acceleration d^2x/dt^2.

From the standpoint of the external observer having direct access to all the physical characteristics of object motion in the space $\mathbb{R}^3$, the force $F_b(\ldots)$—a vector in the space $\mathbb{R}^3$—is a natural quantity specifying the intensity and the direction of subject's actions.[17]

From the standpoint of the subject, in evaluating the corresponding bodily effort as a certain resource depletion, the bodily measure $E_\mathfrak{b}$ must be a scalar related in some way to the force F_b. It is quite natural to accept that the stronger the force $F_b(\ldots)$, the higher the required effort for driving the object along the path $\{x(t)\}$.[18] Moreover, when the motion of the object is governed solely by its physical interaction with surrounding objects, i.e., when $F_b = 0$ at the given instant of time t, no subject's intentional actions are *locally* required. Strictly speaking, even in this case some subject's effort could be necessary, e.g., for monitoring the object motion to respond in the right way to events anticipated in the future. However, such effort has to be attributed to the mind-in-the-self. So within the realm of the body-in-the-self the no-effort condition implies the equality $F_b = 0$.

To be specific, we turn to the following ansatz for the relationship between the force-based and resource-based accounts of human effort

$$E_\mathfrak{b} = \langle F_b | \, \mathbf{B} \, | F_b \rangle \,, \tag{5.13}$$

where the positive-definite scalar $E_\mathfrak{b}$ quantifying the mean magnitude of the vector F_b in averaging over its components is treated as the desired *bodily measure* of atomic action effort in driving the point-like object. In Exp. (5.13) $\mathbf{B}$ is a certain positive-definite quadratic form allowing for a possible difference in contribution of the F_b-components as well as their mutual correlations. The form matrix $\mathbf{B}$ on its own may depend on the characteristics of object motion, i.e., in the general case the matrix

$$\mathbf{B} = \mathbf{B}\left(x, \frac{dx}{dt}, \frac{d^2x}{dt^2} \,\middle|\, t\right). \tag{5.14}$$

---

[17] It should be noted that the muscle strength F_{muscle} required for the force F_b to arise may be much less; muscle strength required for turning the steering wheel in driving a heavy truck illustrates this situation. However, it does not contradict the use of the force F_b for quantifying human intentional actions. The relationship between F_{muscle} and F_b actually determines the force units that should be used in categorizing the force F_b in terms "easy-difficult."

[18] The given statement does not contradict the nonmonotonous dependence of effort on task difficulty postulated within the motivational intensity theory (Brehm et al. 1983; Brehm and Self 1989; Wright and Brehm 1989, for a review see Richter et al. 2016; Gendolla et al. 2019; Silvestrini and Gendolla 2019). This non-monotonicity effectively allows for the fact that humans avoid executing tasks whose difficulty exceeds a certain threshold.

Here the direct dependence of the matrix **B** on time t allows for physiological and emotional factors, e.g., fatigue or boredom.

Expressions (5.12b) and (5.13) specify the desired functional dependence of the bodily measure of atomic action effort on the parameters of the object motion trajectory $\{x(t)\}$, namely,

$$E_{\mathrm{b}} = E_{\mathrm{b}}\left(x, \frac{dx}{dt}, \frac{d^2x}{dt^2} \middle| t, S_{\mathrm{env}}(t)\right). \tag{5.15}$$

Expression (5.15), on the one hand, bears the information about the physics of controlled object, on the other hand, it reflects subject's intention in driving the object along the desired path $\{x(t)\}$.

It should be noted that the constructed bodily measure of effort does not depend on the object jerk $j = d^3t/dt^3$—the phase variable of the extended phase space $\mathbb{R}_{[4]}$ matching the highest order time derivative of x. Put differently, the list of arguments of function (5.15) merely does not contain the jerk. This result does not contradict the proposition we justified in Part I that the four variable phase space $\mathbb{R}_{[4]} = \{x, v, a, j\}$ is generally required for describing goal-oriented human actions. As will be clarified below, the jerk j as a phase variable appears in the description of human preference for avoiding abrupt time variations in the control parameter via which the subject affects directly the controlled object. For example, without a real necessity drivers prefer not to abruptly hit the brakes. However, the corresponding effort should be attributed to the mind-in-the-self rather than the body-in-the-self. It is due to this preference reflects directly human intention rather than stems from the physics of controlled object.

Brief digression: Plausible Ansätze for the $E_{\mathrm{b}}(F_b)$-Relationship. Strictly speaking, the variety of plausible expressions for the relationship between the force-based and resource-based accounts of effort within the scope of the body-in-the-self is wider than the relationship specified by Exp. (5.13). For example, dealing with the coordinate system in the space $\mathbb{R}^3$ oriented such that the positive-definite matrix **B** take the diagonal form $\mathbf{B} = \{B_1, B_2, B_3\}$ the plausible ansatz $E_{\mathrm{b}}(F_b)$ can be written as

$$E_{\mathrm{b}} = \left[B_1\left|F_{b,1}\right|^p + B_2\left|F_{b,2}\right|^p + B_3\left|F_{b,3}\right|^p\right]^q, \tag{5.16}$$

where $\{F_{b,1}, F_{b,2}, F_{b,3}\}$ are the spacial components of the vector F_b, and the exponents $p > 0$, $q > 0$ are the model parameters. However, for $p \neq 2$ Exp. (5.16) loses its simplicity and takes a rather complex form in the generally oriented coordinate system. So without clear reasons we see no necessity of over-complicating our constructions, which justified the given choice of $p = 2$ in writing Exp. (5.13). The case of $q \neq 1$ can be reduced to that of $q = 1$ just by modifying the corresponding relationship between the bodily measure of effort and mental measure of effort. Put differently, we may treat the bodily measure E_{b} as a physical stimulus whereas the mental measure $\mathbb{E}_{\mathrm{m}}$ as the derived magnitude of this stimulus experience related to each other by the exponential Stevens' law (Sect. 1.3.1). For this reason we confined ourselves to ansatz (5.13) in further constructions.

Brief digression: Objects with Overdamped Dynamics. The overdamped dynamics of physical objects implies that the effect of object inertia is ignorable, which is formally described by setting $m = 0$ in the governing equation (5.12a). As a result in expression (5.12b) for the force F_b of subject's intentional actions the term containing the object acceleration $a = d^2x/dt^2$ disappears. We note this limit because in virtual reality the motion of objects controlled by human actions, e.g., via mouse movements is simulated rather often within inertialess models.

In this case, first, the bodily measure $E_\mathfrak{b} = E_\mathfrak{b}(x, dx/dt)$ is functionally specified by the two phase variables x and $v = dx/dt$. Second, the acceleration $a = dx/dt$ becomes the phase variable allowing for human preference for avoiding abrupt variations in the main control parameter. Therefore the structure of human effort in the case of control over object motion characterized by overdamped dynamics is similar to that of the general case. The main difference is only the reduction of extended phase space $\mathbb{R}_{[4]} \to \mathbb{R}_{[3]} = \{x, v, a\}$.

Brief digression: Effort and Work. It should be emphasized that the notion of human effort we deal with and the notion of work used on physics seem to be rather similar. Both of them are related to the depletion of some energetic resources. However, they are entirely distinct from each other in the basic properties.

Indeed, in physics, work is related to the product of a force F_e acting, e.g., on an object and the corresponding displacement of this object. No displacement implies no work and, so, no change in energy. For human actions, just applying a force on an object even without its displacement is experienced as an effort by the subject. This effort can be explained as a certain depletion of energetic resources caused by the corresponding metabolic processes.

5.3.2 *Human Control over Complex Object Dynamics*

In the present book to elucidate the gist of mathematical formalism required for describing the human temporality, we have confined our consideration to the dynamics of point-like physical objects—the material points—affected by human actions. The constructed bodily measure of atomic action effort, Exp. (5.13), describes this simplest situation when the subject exerts the force F_s directly upon a point-like physical object and the analysis of its dynamics is reduced to the investigation of the object motion trajectory $\{x(t)\}$ in the space $\mathbb{R}^3$.

However, the situation can be more complex, e.g., balancing a tray with glasses. Dealing with a complex object O consisting of various elements the subject can exert a force upon one element while focusing attention on another element. It raises a doubt about the *principal* applicability of the obtained expression (5.13) or similar expressions to describing bodily effort for complex objects. The goal of this section playing the role of some digression from the mainstream is to dispel this doubt. A sophisticated construction of the bodily measure of human effort for complex objects is a problem deserving an individual consideration.

Appendix 5.8 is devoted to constructing the bodily measure of human effort for complex objects within a rather general framework. Here we present only its main cornerstones.

The dynamics of a complex object O is described in terms of its degrees of freedom and the corresponding generalized coordinates $q = \{q_k\}_{k=1}^{k=M}$ forming the M-dimensional space $\mathbb{R}^M$. The generalized coordinates are divided into two classes:

- the class $q^+ = \{q_k\}_{k=1}^{k=M^*}$ of generalized coordinates matching the *monitored* degrees of freedom being in the focus of the subject's attention, where $1 \leq M^* < M$ (usually $M^* \ll M$),
- the class $q^- = \{q_k\}_{k=1+M^*}^{k=M}$ of generalized coordinates called *freely variative* or just *variative* that match the *neglected* degrees of freedom with respect to which the subject's intention remain indifferent.

The introduction of variative variables reflects the considerable redundancy (or abundance) in human actions which is one of crucial problems in human motor control (e.g., Latash 2008b, 2012).

Following a way similar to one given rise to Exp. (5.13), we introduce the bodily measure of human action $E_\mathfrak{b}$ required for driving the object O with the desired acceleration $\ddot{q}^+$ in the space of monitored degrees of freedom as a positive-definite quadratic form

$$E_\mathfrak{b} = \langle f \,|\, \mathbf{B} \,|\, f \rangle$$

defined in the space of all the forces $\{f\}$ the subject exerts upon the object, Exp. (5.104). The configuration O_F of the points where the forces are applied to the object O are assumed to be fixed. Then, the bodily measure of human effort is represented as the additive sum of two terms, Exp. (5.107)

$$E_\mathfrak{b} = \left\langle \Delta_{\ddot{q}}^{\mathcal{P}} \,\middle|\, \mathbf{B}_\mathcal{P} \,\middle|\, \Delta_{\ddot{q}}^{\mathcal{P}} \right\rangle + \langle f_0 | \, \mathbf{B}_0 \, | f_0 \rangle \,. \tag{5.17}$$

Here the vector $\Delta_{\ddot{q}}^{\mathcal{P}}$ belonging to the space of monitored degrees of freedom specifies the difference between the desired acceleration

$$\ddot{q} = \{\ddot{q}_k\}_{k=1}^{k=M^*} = \frac{d^2 Q_\mathcal{P}(t)}{dt^2}$$

of the object motion along the path $\mathcal{P} = Q_\mathcal{P}(t) \in q^+$ within the space q^+ and the corresponding acceleration the external physical forces

$$F_e\big(q, \dot{q} \mid S_\text{env}(t)\big)$$

would cause if human actions were absent. Namely, see Exp. (5.109),

$$\Delta_{\ddot{q}}^{\mathcal{P}} = \widehat{J}(q)\frac{d^2 Q_{\mathcal{P}}}{dt^2} - P^{+} F_e\big(q, \dot{q} \mid S_{\text{env}}(t)\big)\,, \tag{5.18}$$

where $\widehat{J}(q)$ is the diagonal matrix with the corresponding components describing the effect of object inertia and P^{+} is the operator of projection in the space q onto the subspace q^{+}. The form matrix $\mathbf{B}_{\mathcal{P}}(q, \dot{q}, O_F)$ defined within the subspace q^{+}, as previously, allows for the individuality of different degrees of freedom in human perception of required effort and their possible correlations.

The second term in Exp. (5.17) takes into account the human effort related to the redundancy in human actions. Namely, when a force f_0 applied by the subject to the intentional object belongs to so-called null-space (e.g., Latash et al. 2007; Latash 2008b; Ranganathan et al. 2014) it has no effect on the object motion in the space of monitored degrees of freedom, on the one hand, but affects the subject's perception. The matrix $\mathbf{B}_0$ specifies a positive-definite quadratic form defined within the null-space.

Therefore, the first term in Exp. (5.17) may be regarded as the minimal bodily effort required for the subject to implement the object desired motion.

In *conclusion*, comparing Exps. (5.13) and (5.17) with each other we may state the following. First, the bodily measure of human

- effort related to direct intentional actions aimed at driving an object with simple structure, e.g., the material point along the desired path,
- minimal effort related to indirect intentional actions aimed at driving a complex (compound) object with monitored and neglected degrees of freedom

are related to the desired object acceleration and the forces caused by the interaction between the given object and the other physical objects of its environment via expressions having a similar form.

Second, the redundancy of intentional actions with complex objects gives rise to additional human effort not related directly to implementing the desired motion of intentional object. Strictly speaking there is no analogy in the case of direct intentional actions. Nevertheless this effect can be taken into account, at least, effectively within the notion of action strategy cloud. Indeed, the effect of null-space forces on the object desired motion may be described in terms of the multitude of its equivalent implementations underlying the notion of action strategy cloud (Sect. 3.3.4).

Therefore, the description of intentional, goal-oriented actions with point-like material objects is also able to capture the basic features of indirect intentional actions with objects having complex spatial structures.

5.4 Mental Measure of Effort: Individual Components

The realm of the mind-in-the-self is the ground on which the subject evaluates intentional actions in making a decision about keeping or changing the current action strategy. Therefore in our description of atomic actions, the mental measure $\mathbb{E}_\mathrm{m}$ of effort is the basic element.

As noted in Sect. 5.2 (Sect. 5.2.3), speaking about mental effort of a given atomic action we need to draw a distinction between action within the experiential now and atomic actions belonging to the connected past or the connected future.

Mental effort within the experiential now is rather rich in property. In particular, there can be singled out two types of its component already noted in Sect. 5.2 and also illustrated in Fig. 5.2. First, it is the mental measure $\mathbb{E}^\mathrm{I}_\mathrm{m}$ related to E-conscious perception of bodily effort. Put differently, the mental measure $\mathbb{E}^\mathrm{I}_\mathrm{m}$ quantifies the effect of the body-in-the-self on the mind-in-the-self as a bottom-up causal process converting neural signals generated by sense organs into entities of E-consciousness. Second, it is the mental measure $\mathbb{E}^\mathrm{II}_\mathrm{m}$ quantifying conscious effort related to monitoring the current dynamics of controlled object. Although this monitoring is conscious, the perception of quantities such as the spacial position x, velocity v, acceleration a, and jerk j of controlled object is E-conscious too. So, all the variables used in describing the mental measure of effort as a whole or its constituent parts have to be treated as quantities accessible to the mind via E-consciousness. In this sense we may treat the conversion of bodily effort into mental effort as a phenomenon admitting a description within the realm of psychophysics. Moreover, a description of mental effort in terms of mental images may be categorized as a certain formalism of inner psychophysics (Sect. 1.3).

The main purpose of the present section is to construct rather general ansätze for the two mental measure components $\mathbb{E}^\mathrm{I}_\mathrm{m}$, $\mathbb{E}^\mathrm{II}_\mathrm{m}$. However, before passing directly to the constituent components $\mathbb{E}^\mathrm{I}_\mathrm{m}$ and $\mathbb{E}^\mathrm{II}_\mathrm{m}$ of the mental measure $\mathbb{E}_\mathrm{m}$ (within the experiential now), we want to discuss the general features of $\mathbb{E}_\mathrm{m}$ partly noted in Sect. 5.2.

5.4.1 *Cloud-Type Structure of Mental Measure: General Features*

Because of the intrinsic uncertainty of mental images of observed objects, the mental measure $\mathbb{E}_\mathrm{m}$ (as well as $\mathbb{E}^\mathrm{I}_\mathrm{m}$ and $\mathbb{E}^\mathrm{II}_\mathrm{m}$) cannot be a function taking particular values. This mental measure may be regarded as a certain cloud—a fuzzy region in the space $\mathbb{R}^E = \{e \geq 0\}$, where e is a non-negative[19] real number.

[19] The execution of some action on its own can be pleasurable. In this case, the corresponding measure of effort may be treated as a cloud able to penetrate also in the region of negative values in the space $\mathbb{R}^E = \{e\}$, where now e is a just real number, which however will not be considered in the present book.

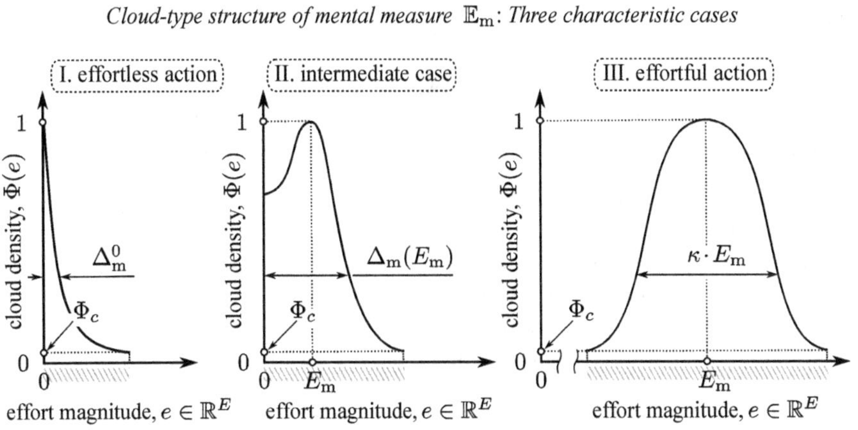

Fig. 5.3 Cloud-type mental measure $\mathbb{E}_m$ represented as the distribution of the cloud density $\Phi(e)$ (y-axis) in the space of effort magnitudes $e \in \mathbb{R}^E$ (x-axis). The cloud density is normalized such that its maximum be equal to unity, $\Phi_{max} = 1$, in this case, the minimal value of cloud density is denoted as $\Phi_c \ll 1$. Hatched stripes depict the regions of the $\mathbb{E}_m$-confinement in the space $\mathbb{R}^E$. The depicted three characteristic forms of the cloud $\mathbb{E}_m$ illustrate the transformation of the $\mathbb{E}_m$-structure including its characteristic width $\Delta_m(E_m)$ as the magnitude E_m meeting the maximum of the cloud density increases. The limit cases I and III can be categorized as the effortless actions and the effortful actions. Case II illustrates the intermediate situation. In case I the value of E_m is located at the origin, $E_m = 0$, so it is not shown in the figure. The limit values of the $\mathbb{E}_m$-cloud width are denoted as $\Delta^0_m = \Delta_m(E_m)|_{E_m=0}$ and $\kappa E_m = \Delta_m(E_m)|_{E_m \gg \Delta^0_m/\kappa}$. In case III the x-axis is shown with a break to compress the large distance between the origin and the region of cloud localization. Dotted lines serve as guide for the eye

In the present section we discuss the general properties the cloud-type mental measure $\mathbb{E}_m$ exhibits. Leaping ahead, we note that, first, the sense of agency and the sense of ownership[20] are accepted in our constructions of the mental measure as ones of the basic premises. Second, the universality of inner psychophysics (Sect. 1.3.2) implying the scale-free principle of mind functioning is another premise underlying our constructions.

Figure 5.3 illustrates the characteristic properties we attribute to the mental measure $\mathbb{E}_m$ of atomic action effort which take into account also the noted premises. The cloud-type measure $\mathbb{E}_m$ is treated as the distribution of its density $\Phi(s)$ in the space $\mathbb{R}^E$:

$$\mathbb{E}_m(e) = \frac{1}{Z}\,\Phi(e)\,, \tag{5.19}$$

where the cloud density $\Phi(s)$ is assumed to be normalized such that its maximum be equal to unity, i.e.,

$$\max_{e \in \mathbb{R}^E} \Phi(e) = \Phi(E_m) = 1 \tag{5.20}$$

[20] We discussed the sense of agency and the sense of ownership in the context of their contribution to the properties of mental images in Sect. 3.2.3.

and attained at the only one effort magnitude E_m for a given situation. The quantity E_m will be also referred to as the *effort mean magnitude*, for short.[21] How this normalization is related to the self's ownership and what conditions specify the normalization coefficient Z will be clarified below. In these terms the properties of the mental measure $\mathbb{E}_m$ can be elucidated turning to the following aspects:

- the universality of the cloud density structure;
- the mental measure of atomic action effort in the complex present;
- the levels of consciousness in action effort evaluation;
- the confinement of mental measure;
- the mental measures as the composition of constituent components;
- the comparison of different actions in effort of their implementation.

Let us discuss them in more details.

Universality of the cloud density structure. The cloud $\mathbb{E}_m$, i.e., the cloud density $\Phi(e)$ is characterized by a universal form depending on its proximity to the origin $e = 0$ representing formally the no-effort state (effortless action) in the mind-in-the-self. Three characteristic forms of the cloud density $\Phi(e)$ are shown in Fig. 5.3, which illustrates the transformation of $\Phi(e)$ as the mean magnitude E_m of effort increases.

- Panel I depicts the limit form when the subject, even perceiving some effort near the perception threshold (fuzzy threshold) Δ^0_m, categorizes a given action as effortless. The maximum of this cloud density is located at $e = 0$.
- Panel III depicts the limit form when the subject clearly recognizes that the implementation of a given action requires some finite effort. It is the case of effortful actions. Turning to the gist of inner psychophysics, we relate the width Δ_m of the cloud $\mathbb{E}_m(e)$ to the mean magnitude E_m via some coefficient κ, i.e., we set $\Delta_m = \kappa E_m$.[22]
- Panel II illustrates the intermediate case, when at some moments a given action is characterized as an action requiring a certain finite effort for its execution, whereas at other moments its execution effort is ignored. In this case, the E_m-dependence of the cloud width Δ_m is some crossover between its limits values. Accepting, e.g.,

---

[21] Strictly speaking, the quantity E_m can differ from the real mean value $\overline{E}$ obtained via averaging the effort magnitude e over the cloud $\mathbb{E}_m(e)$ in some way. However, the quantity $\overline{E} = \overline{E}(E_m)$ is specified directly by the value of E_m and closely approximated by E_m except for a small neighborhood of the origin. So we prefer not to discriminate terminologically between E_m and $\overline{E}$ in order to avoid unnecessary text overloading. Where it can lead to a possible inconsistency we will avoid using the term *mean magnitude* of effort.

[22] For the subjective evaluation of bodily effort E_b playing the role of external stimulus it is reasonable to equate the coefficient κ with the Ekman fraction κ_e (see Sect. 1.3.1 and Sect. 1.3.2) For the mental effort related to the monitoring of object dynamics the situation is more intricate because the origin of this effort is purely internal. It is the continuous use of limited computational resource in tracking the dynamics of observed object and the corresponding coefficient κ can be much larger than κ_e. We return to this issue in Sect. 5.4.6.

ansatz (1.1b) we write[23]

$$\Delta_{\mathrm{m}}(E_{\mathrm{m}}) = \Delta_{\mathrm{m}}^{0} + \kappa E_{\mathrm{m}} \,. \tag{5.21}$$

Expression (5.21) actually specifies the $\Delta_{\mathrm{m}}(E_{\mathrm{m}})$-dependence for mental effort experienced right now, i.e., mental effort of action currently executed when subject's attention is mainly focused on its execution.

Reduction of attention to a currently executed action can make the cloud $\mathbb{E}_{\mathrm{m}}$ wider. A similar widening should be the case when the effort atomic action in the connected past or the connected future is evaluated. So Exp. (5.21) specifies the limit value of uncertainty in evaluating the effort magnitude. For this reason we suppose that the universality of the cloud-type mental measure $\mathbb{E}_{\mathrm{m}}$ admits interpretation as the cloud density having a universal form $\Phi(e \mid E_{\mathrm{m}}, \Delta_{\mathrm{m}})$ containing two parameters—the mean magnitude E_{m} and the uncertainty Δ_{m} in effort evaluation.

Mental measure of atomic action effort in the complex present. Uncertainty in estimating mental effort of atomic actions based on memory or anticipation must be higher than this uncertainty at the moments when these actions are under execution.[24]

In order to describe this effect, similar to how it was done in Sect. 5.1, we introduce an additional argument δt—the time led between an instant of time under consideration and the experiential now. Then the dependence $\Delta_{\mathrm{m}}(\delta t)$ is described by the ansatz for the uncertainty in subject's evaluation of mental effort within the complex present:

$$\Delta_{\mathrm{m}}(\delta t) = \Delta_{\mathrm{m}}\Big|_{\delta t=0} \cdot \left[1 + \epsilon_{\mathrm{m}} \frac{(\delta t)^2}{\tau_{\mathrm{cp}}^2}\right]^{\beta_{\mathrm{m}}} . \tag{5.22}$$

Here, as previously, τ_{cp} is the characteristic duration of complex present, the parameter $\epsilon_{\mathrm{m}} \sim 1$ and the exponent $\beta_{\mathrm{m}} > 0$ are the attributes of the neural mechanism governing the mental effort perception. The uncertainty $\Delta_{\mathrm{m}}|_{\delta t=0}$ of subject's evaluation of effort at the moment when the corresponding action was executed and just after that in retaining in short-term memory is specified by Exp. (5.21) as the limit case.

Levels of consciousness in action effort evaluation. Within the framework of inner psychophysics accepted in our constructions the mental measure $\mathbb{E}_{\mathrm{m}}$ of effort describes the image of atomic action with respect to the use of the corresponding resource for its implementation. This image is an entity belonging to the realm of the

---

[23] Using this ansatz we tacitly accept Stevens' law for the relationship between the intensity of physical stimulus and the magnitude of its perception (Sect. 1.3). It also implies that the range of analyzed effort is characterized by the mental commensurability of the involved values (Lubashevsky 2019).

[24] As an argument for this statement we may note that the interstimulus time interval separating presentation of standard and comparison stimuli affects essentially the just noticeable difference (JND) which is reflected in the value of the Weber fraction (e.g., Baird 1997, p. 114). In particular, according to experiments by Graf et al. (1974) on brightness discrimination the interstimulus interval of 10 s can cause 20%–120% increase of the Weber fraction. Time perception characterized by the Weber fraction growing with the duration of standard signal (Grondin 2012) seems to be a phenomenon of similar nature.

mind-in-the-self and, as such, possesses individual properties attributed to it directly. Turning to the concept of mental space (Lubashevsky 2019) we suppose that:

- The quantity E_{m} is determined by the action as it is represented in its embodied implementation combined with the neural processing of the related information. The value E_{m} is accessible to the mind as a quantity of the E-consciousness level.
- The uncertainty in evaluating the atomic action effort—the characteristic width Δ_{m} of the cloud $\mathbb{E}_{\mathrm{m}}$—results from conscious analysis of perceived effort including the conscious comparison of similar actions. So the value Δ_{m} has to be categorized as a quantity of the AM-consciousness level.

Put differently, the mean magnitude E_{m} of experienced effort originates from the corresponding sensory modality generating neural signals bearing the information about the given action. Then this information is broadcast by attention focused on action execution via the global workspace (Sect. 3.2.5).

The uncertainty in effort evaluation—the width Δ_{m} of the cloud $\mathbb{E}_{\mathrm{m}}$—is mainly determined by the global workspace functioning except, maybe, a close proximity to its limit capacity corresponding to the cloud width Δ_{m} specified by Exp. (5.21). Therefore, the uncertainty Δ_{m} may be supposed to be determined by the properties of the global mental state.

Confinement of mental measure. As elucidated in Appendix 5.10 based on our experimental investigations, the cloud $\mathbb{E}_{\mathrm{m}}$ must be confined to a certain region in the space $\mathbb{R}^E$ with sharp boundary $\partial_c\mathbb{E}_{\mathrm{m}}$ (Fig. 5.3) determined by the transition between:

- AM-conscious evaluation of the corresponding effort with respect to its magnitude; the characteristic magnitude of this uncertainty is quantified by the width Δ_{m} of the cloud $\mathbb{E}_{\mathrm{m}}$,
- the automatic recognition that a given action cannot require efforts e such that $|e - E_{\mathrm{m}}| \gg \Delta_{\mathrm{m}}$; the nonconscious neural processes underlying this recognition are responsible for the emergence of this sharp boundary.

The given boundary $\partial_c\mathbb{E}_{\mathrm{m}}$ is characterized by the cloud density $\Phi_c \ll 1$. The latter inequality reconciles the fuzziness of the mental measure $\mathbb{E}_{\mathrm{m}}$ within the realm of the mind-in-the-self and the existence of the sharp boundary $\partial_c\mathbb{E}_{\mathrm{m}}$ caused by nonconscious neural processes.

Mental measures as the composition of constituent components. As discussed in Sect. 5.2, the cumulative mental measure $\mathbb{E}_{\mathrm{m}}$[25] of effort comprises two constituent

[25] As it will be discussed in detail in Sect. 5.5.1, effort related to action execution and effort related to monitoring the corresponding results are experienced simultaneously. However, the possibility of their quantitative evaluation in the same units is not a trivial consequence of this fact. Indeed their individual experience can be governed by different and partly independent neural processes. So when a subject just experiences the action implementation including its control the mean effort magnitudes $E_{\mathrm{m}}^{\mathrm{I}}$ and $E_{\mathrm{m}}^{\mathrm{II}}$ are incommensurable. Only when this experience is involved into decision-making on keeping the corresponding action strategy or changing it for another one, the two processes effectively merge together. Exactly this fusion underlies the possibility of introducing the cumulative mental measure of $\mathbb{E}_{\mathrm{m}}$ and using the same units for quantifying the individual effort magnitudes $E_{\mathrm{m}}^{\mathrm{I}}$, $E_{\mathrm{m}}^{\mathrm{II}}$ of the two processes.

components: (*i*) the mental measure $\mathbb{E}^{\mathrm{I}}_{\mathrm{m}}$ quantifying effort related to bodily execution of a given action and (*ii*) the mental measure $\mathbb{E}^{\mathrm{II}}_{\mathrm{m}}$ quantifying effort of monitoring the results of this action. Postponing the justification of the properties characterizing the composition $\mathbb{E}_{\mathrm{m}} = \mathbb{E}^{\mathrm{I}}_{\mathrm{m}} \oplus \mathbb{E}^{\mathrm{II}}_{\mathrm{m}}$ to Sect. 5.5.1, let us here just summarize these properties as follows.

- The effort mean magnitude E_{m}, as well as the corresponding mean magnitudes $E^{\mathrm{I}}_{\mathrm{m}}$, $E^{\mathrm{II}}_{\mathrm{m}}$ are just given to the mind as quantities of E-consciousness.
- The uncertainty Δ_{m} in evaluating the perceived effort is a property of a certain *mental state* characterized by attention focused on recognizing the key features of some process or phenomenon. In the case under consideration, it is the cumulative mental effort required for executing a given action. Therefore the uncertainty in evaluating effort required separately for the bodily execution of action and the monitoring of system dynamics is determined by the uncertainty in evaluating their cumulative result.

The noted features of effort perception enable us to interpret the integration of the components $\mathbb{E}^{\mathrm{I}}_{\mathrm{m}}$ and $\mathbb{E}^{\mathrm{II}}_{\mathrm{m}}$ into $\mathbb{E}_{\mathrm{m}}$,

$$\mathbb{E}^{\mathrm{I}}_{\mathrm{m}} \oplus \mathbb{E}^{\mathrm{II}}_{\mathrm{m}} \to \mathbb{E}_{\mathrm{m}}, \quad \text{(in Sect. 5.2 it is described by Exp. (5.9))}$$

- as the arithmetic sum of the corresponding effort mean magnitudes[26]

$$E^{\mathrm{I}}_{\mathrm{m}} + E^{\mathrm{II}}_{\mathrm{m}} = E_{\mathrm{m}}, \tag{5.23}$$

- whereas the widths $\Delta^{\mathrm{I}}_{\mathrm{m}}$, $\Delta^{\mathrm{II}}_{\mathrm{m}}$ of the corresponding clouds are determined by the width Δ_{m} of the cloud $\mathbb{E}_{\mathrm{m}}$, i.e.,

$$\Delta^{\mathrm{I}}_{\mathrm{m}} = \Delta^{\mathrm{II}}_{\mathrm{m}} = \Delta_{\mathrm{m}}. \tag{5.24}$$

In other words, the cloud-type measure E_{m} differs from its constituent components $E^{\mathrm{I}}_{\mathrm{m}}$ and $E^{\mathrm{I}}_{\mathrm{m}}$ mainly in the mean magnitudes of required effort, the width of these clouds is practically the same.

Comparison of different actions in effort of their implementation. The extension of the cloud-type measure $\mathbb{E}_{\mathrm{m}}$ along the space $\mathbb{R}^{E}$ makes it necessary to develop a special formalism for comparing two different actions with respect to effort of their execution. This formalism should enable us to describe which action requires "smaller" or "greater" effort for its implementation. When this distinction cannot be made, these events have to be categorized as "equivalent" in effort.

[26] In fact, this arithmetic sum should be regarded as the symbolical representation for the nonlinear fusion of the two components of mental effort. This fusion units the individual quantitative characteristics of effort perception via different channels into one quantity, Sect. 5.5.

The gist of our approach is to turn to the cloud overlapping as the ground of such comparison.[27] The proposed probabilistic formalism includes the following items.

Common base: For two different actions to be consciously compared concerning execution effort their evaluation should be implemented within the same state of the mind-in-the-self. As a result, two cloud-type measures $\mathbb{E}_{m,1}$ and $\mathbb{E}_{m,2}$ can be discriminated mainly based on the difference in their mean magnitudes $E_{m,1}$ and $E_{m,2}$. The widths of these clouds are identical, $\Delta_{m,1} = \Delta_{m,2}$.[28] Within the present book we confine our analysis to the case when this width is determined by the magnitude specified by Exp. (5.21) for the action of the largest magnitude $E_m = \max\{E_{m,1}, E_{m,2}\}$.

Cloud normalization: There are various problems requiring the comparison of two clouds via their overlapping under different conditions. Therefore, the complete description of clouds, e.g., similar to ones shown in Fig. 5.3, including their

- "spatial" structure (the cloud density $\Phi(e)$ for mental effort, the structure of space-time clouds for the mental-mental embedding of observed objects, etc.) and
- the scaling cofactor, i.e., the normalization factor

seems to be impossible in the general case. The "spatial" structure reflects the basic properties attributed to the corresponding entity and, so, bears a meaningful information. By contrast, the scaling cofactor is merely a mathematical element introduced to describe formally how the subject operates with the density of the analyzed clouds, which may change from one case to another.

The ownership of mental entities by the self (Sect. 3.2.3) opens a gate towards tackling this problem. Namely,

- "spatial" structure of clouds bears the factual information about their properties;
- the scaling (normalization) factor may be introduced in different ways depending on particular situation to reconcile the related mathematical conditions and subject's particular operations with the clouds.

For these reasons, we may normalize the cloud-type mental measure $\mathbb{E}_m$ shown in Fig. 5.3 such that the maximum of its density be equal to unity, Exp. (5.20).

Cloud comparison: The cloud normalization based on the ownership of mental entities by the self enables us to introduce the ordering of the $\mathbb{E}_m$-clouds in the following way (Appendix 5.9.1):

[27] It should be noted that there are also other approaches to ordering fuzzy entities, in particular, the formalism of fuzzy numbers dealing with linear relations (e.g., Bede 2013; Prokopowicz and Ślęzak 2017a, b). However, from our point of view, a formalism dealing with cloud overlapping is more relevant to the problem under consideration where there are no point-like entities in the description of human perception. In this case, the cloud overlapping may be introduced as some formal binary operation.

[28] The comparison of two clouds $\mathbb{E}_{m,1}$ and $\mathbb{E}_{m,2}$ can be implemented also under weakened conditions when (*i*) their widths $\Delta_{m,1} = \Delta_{m,2}$ are identical for $E_{m,1} = E_{m,2}$ and (*ii*) their difference $\Delta_{m,1} - \Delta_{m,2}$ does not grow too fast with the difference $|E_{m,1} - E_{m,2}|$. This issue is analyzed in Appendix 5.9.1.

- The subject's comparison of two clouds $\mathbb{E}_{m,1}(e)$ and $\mathbb{E}_{m,2}(e)$—the mental measures of two actions—is a mental operation denoted by the symbol $\langle \mathbb{E}_{m,1} | \mathbb{E}_{m,2} \rangle$ as well as $\langle 1 | 2 \rangle$ for short. In terms of cloud density, Exp. (5.19), it is specified as the cloud overlapping

$$\langle \mathbb{E}_{m,1} | \mathbb{E}_{m,2} \rangle = \frac{1}{Z_1 Z_2} \int_0^\infty de\, \Phi_1(e)\, \Phi_2(e)\,, \tag{5.25a}$$

where for $i = 1, 2$

$$Z_i = \left[\int_0^\infty de \left[\Phi_i(e) \right]^2 \right]^{1/2}. \tag{5.25b}$$

For different clouds their overlapping $\langle \mathbb{E}_{m,1} | \mathbb{E}_{m,2} \rangle < 1$. The comparison of a cloud $\mathbb{E}_m(e)$ with itself corresponds to $\langle \mathbb{E}_m | \mathbb{E}_m \rangle = 1$, which meets the statement that any cloud overlaps with itself for 100%.
- The probability P_{12} that the subject will treat two actions, action 1 and action 2, as equivalent in effort is determined by the overlapping of the clouds $\mathbb{E}_{m,1}(e)$ and $\mathbb{E}_{m,2}(e)$ and can be written as

$$P_{12} = G\big(\langle \mathbb{E}_{m,1} | \mathbb{E}_{m,2} \rangle \big), \tag{5.26}$$

where $G(x)$ is a certain monotonically increasing function of its argument x such that $G(0) = 0$ and $G(1) = 1$. The expression

$$G(x) = x^{\alpha_m}, \tag{5.27}$$

where the exponent $\alpha_m > 0$ is a positive number regarded as a model parameter, can be used as a plausible ansatz allowing for hierarchical scale-free organization of mind functioning. Naturally, a more complex ansätze are also possible.
- The probabilistic ordering of the clouds $\{\mathbb{E}_m\}$ or, what is the same, the ordering of actions according to their effort is based on the relations "$\prec$", "$\simeq$", and "$\succ$" introduced in the following way. In the comparison of two actions:
    - action 1 and action 2 are considered equivalent in effort of their implementation, i.e.,

$$\mathbb{E}_{m,1} \simeq \mathbb{E}_{m,2} \qquad \text{with a certain probability } P_{12}, \tag{5.28a}$$

    - action 1 is accepted to require less effort than action 2 for their implementation, i.e.,

$$\mathbb{E}_{m,1} \prec \mathbb{E}_{m,2} \qquad \text{with the probability } 1 - P_{12} \text{ iff } E_{m,1} < E_{m,2}, \tag{5.28b}$$

- action 1 is accepted to require more effort than action 2 for their implementation, i.e.,

$$\mathbb{E}_{\mathrm{m},1} \succ \mathbb{E}_{\mathrm{m},2} \qquad \text{with the probability } 1 - P_{12} \text{ iff } E_{\mathrm{m},1} > E_{\mathrm{m},2}. \tag{5.28c}$$

The probabilistic comparison (5.28) of actions tacitly implies that if the subject rejects the categorization of two actions as equivalent, he is able to recognize with certainty which one of the relations $E_{\mathrm{m},1} < E_{\mathrm{m},2}$ and $E_{\mathrm{m},1} > E_{\mathrm{m},2}$ is the case. Put differently, if the difference in the effort mean magnitudes $E_{\mathrm{m},1}$, $E_{\mathrm{m},2}$ is near the subject's perception threshold (fuzzy threshold) and the subject cannot classify them into categories, e.g., $\mathbb{E}_{\mathrm{m},1} \prec \mathbb{E}_{\mathrm{m},2}$ or $\mathbb{E}_{\mathrm{m},1} \simeq \mathbb{E}_{\mathrm{m},2}$ with certainty, in no case he will consider the category $\mathbb{E}_{\mathrm{m},1} \succ \mathbb{E}_{\mathrm{m},2}$ to be a plausible candidate for classification. The categorization of two actions as equivalent in effort is actually a result of rejecting their categorization in terms of less or more effort is required for implementing a given action. The question about the real identity of two actions in their mean magnitudes, $E_{\mathrm{m},1} = E_{\mathrm{m},2}$, cannot be posed within the realm of the mind-in-the-self. This identity is just not feasible for continuous variables describing properties of different mental entities.

Besides, within this understanding of relations "$\prec$", "$\simeq$", and "$\succ$," the relation "$\precsim$" meaning "$\prec$" or "$\simeq$" (or the relation "$\succsim$" meaning "$\succ$" or "$\simeq$") possesses the properties usually attributed to the order relations, namely, it is reflexive, antisemitic, and transitive.

5.4.2 Mental Measure $\mathbb{E}^I_m$: Bodily Effort Experience

In Sect. 5.1 we introduced the degrees $\{A_{\mathrm{b}}, A_x, A_v, A_a, A_j\}$ of attention as a limited computational resource allocated for monitoring

- the details of bodily actions,
- the perceived dynamics of controlled object, i.e., time variations and accuracy of the phase variables $\{x, v, a, j\}$.

In this and the following subsections we will consider situations when focusing attention on some action, physical or mental one, opens a gate to sensing the effort related to its execution, i.e., makes the given action effortful.[29] The mechanisms enabling effortless attention are related to a special class of phenomena (see Footnote 8) and deserve individual consideration.

In the case of inner attention focused on bodily execution of an analyzed action, the subject just starts to feel the effort related to the action implementation. Naturally

[29] Close association between attention and effort is the case in many situations. In his book *Attention and Effort* Daniel Kahneman (1973) examined a large number of studies on a variety of aspects of attention, including divided attention, task interference, and the role of perception. He argued for the possibility of quantifying attention and, even, put forward the notion of effort as the *subjective* measure of attention intensity.

even in the given case attention on its own should also have a cost. However such own effort of attention can be integrated into the feeling of action effort as a whole without dividing it into constituent components.[30] In this way the consideration of attention-related effort is reduced to the description presented below in the present section.

In the case of external attention, here it is attention related to monitoring the dynamics of controlled object, the subject feels effort required for keeping the focus on the object continuously during the control process. This effort may be regarded as the own effort of attention. Sections 5.4.3–5.4.6 are devoted to exactly this situation in the context of monitoring the object dynamics.

In the present subsection we consider how the bodily effort belonging to the realm of the body-in-the-self becomes accessible to the mind-in-the-self, which is described by the cloud-like measure $\mathbb{E}^{\mathrm{I}}_{\mathrm{m}}$. The neural processes governing motor behavior are mainly hidden from the mind. Only the integral results related to action execution can rise to the level of the mind-in-the self. In particular, it concerns how easy or difficult these actions are and the degree to which the corresponding biophysical resources are depleted. These results are just given to the mind like a certain physical stimulus via sensory modalities.

So, turning to the gist of psychophysics (see Sect. 1.3) we describe the perception of bodily effort as the conversion of a certain stimulus characterized by the "intensity" $E_{\mathfrak{b}}$—the bodily measure of effort—into the corresponding mental image characterized by the cloud $\mathbb{E}^{\mathrm{I}}_{\mathrm{m}}$. This conversion is determined by the inner state of the self described in our analysis by the degree $A_{\mathfrak{b}}$ of attention consciously focused on the corresponding details of subject's motor behavior. Figure 5.4 illustrates the given psychophysical type concept.

As explained in Sect. 5.4.1, the mental measure $\mathbb{E}^{\mathrm{I}}_{\mathrm{m}}$ is characterized by the mean magnitude $E^{\mathrm{I}}_{\mathrm{m}}$ and the width $\Delta^{\mathrm{I}}_{\mathrm{m}}$. The properties of these characteristics, however, are different. The quantity $E^{\mathrm{I}}_{\mathrm{m}}$ belongs to the level of E-consciousness. As noted above the subject can intentionally affect the perception of bodily effort via focusing on or shifting the focus away from a given bodily action. This effect is reducible to changing the capacity of the corresponding channel of information processing. By contrast, the uncertainty in evaluating the bodily effort, i.e., the width of the cloud $\mathbb{E}^{\mathrm{I}}_{\mathrm{m}}$ is an entity of AM-consciousness. Indeed this uncertainty arises via conscious comparison of two (or more) actions similar in properties. Its limit value specified by Exp. (5.21) seems to be determined by the unconscious neural processing of perceived information when attention is completely focused on the action comparison. Nevertheless, this uncertainty stems from the functioning of a given inner state of the self and if other mental tasks are also under implementation the resulting uncertainty

[30] The fusion of bodily effort and own effort of attention becomes most essential and plays the main role in evaluating the corresponding actions for involuntary event-driven attention. This fusion has to include also neural effort quantifying the information processing required for implementing muscle actions as well as their control in some unconscious way. For discussion of this issue, in particular, with respect to strategy of arm movement and the related neural effort for joint coordination a reader may be referred to Shimansky and Rand (2013), Dounskaia and Shimansky (2016), Shimansky and Dounskaia (2018).

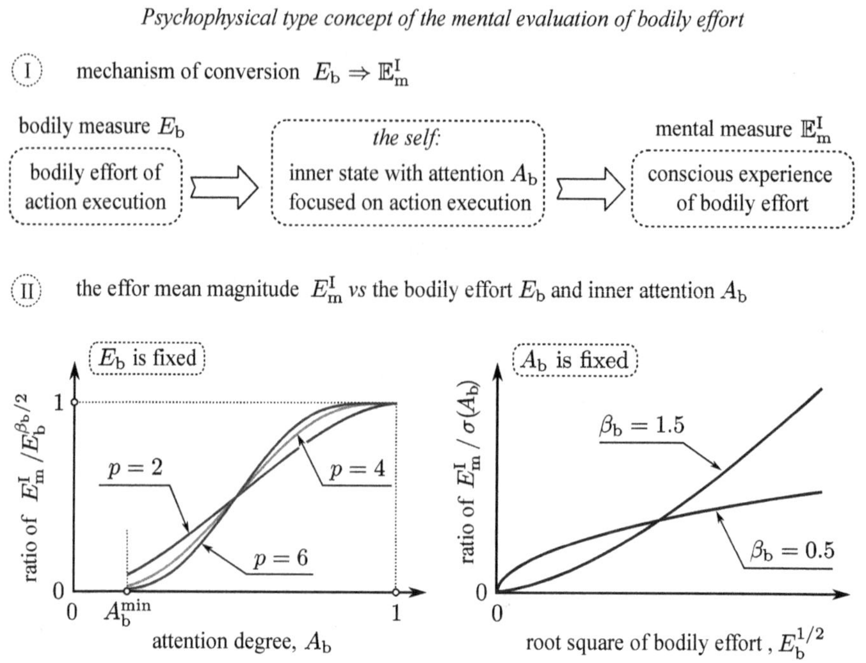

Fig. 5.4 Illustration of the mechanism implementing the perception and mental evaluation of bodily effort. *Fragment I*: The constituent components of the conversion $E_{\rm b} \Rightarrow \mathbb{E}^{\rm I}_{\rm m}$. *Fragment II*: The characteristic dependence of the effort mean magnitude $E^{\rm I}_{\rm m}$ on the bodily effort $E_{\rm b}$ and the degree $A_{\rm b}$ of inner attention to bodily actions (including voluntary attention and involuntary attention treated as a lower boundary $A^{\rm min}_{\rm b}$). This dependence is specified by ansatz (5.29) together with Exp. (5.30) and the cofactor $e^{\rm I}_{\rm m} = 1$. The left panel shows the effort mean magnitude $E^{\rm I}_{\rm b}$ (as the ratio $E^{\rm I}_{\rm m}/E^{\beta_{\rm b}/2}_{\rm b}$) vs. the attention degree $A_{\rm b}$ for a fixed value of bodily effort $E_{\rm b}$; the right panel shows the effort mean magnitude $E^{\rm I}_{\rm b}$ (as the ratio $E^{\rm I}_{\rm m}/\sigma(A_{\rm b})$) vs. the square root of bodily effort $E_{\rm b}$ for a fixed attention degree $A_{\rm b}$. The shown curves correspond to different values of the exponents p and $\beta_{\rm b}$ entering Exps. (5.29) and (5.30)

can be higher. In other words, the cloud width $\Delta^{\rm I}_{\rm m}$ must be a function of the inner state of the self. For this reason we have accepted (Sect. 5.4.1) that the cloud $\mathbb{E}_{\rm m}$ and its constituent components $\mathbb{E}^{\rm I}_{\rm m}$, $\mathbb{E}^{\rm II}_{\rm m}$ are characterized by the same width. Put differently, the uncertainty of the effort mental evaluation should be attributed to the mind-in-the-self as a whole entity; the consequences of this feature will be analyzed individually in the special Sect. 5.5.1. In the present section we confine our analysis to the relationship between:

- the bodily effort of action in the experiential now, i.e., the value $E_{\rm b}$—the bodily measure of current actions—and
- the mean magnitude $E^{\rm I}_{\rm m}$ of the $\mathbb{E}^{\rm I}_{\rm m}$-cloud.

Exactly the $E^{\rm I}_{\rm m}(E_{\rm b})$-relationship will be described below turning to the gist of inner psychophysics.

As illustrated in Fig. 5.4(I), the experience of bodily effort is determined by the bodily effort raised up to the level of the mind-in-the-self where the inner state of the self plays the role of converting unit. The *inner* attention, i.e., attention A_b focused intentionally or automatically (involuntary) on the motion and the state of the body itself controls directly how the subject perceives and evaluates the corresponding effort. We have already discussed this issue in Sect. 5.2. Here we accept that for bodily effort related to executing an action to be consciously recognized, i.e., to be accessible to the mind, some inner attention is required. Put differently, no attention, no experienced effort, which does not exclude subject's awareness of the corresponding actions.[31]

For this reason we employ the following ansatz for the relationship between the quantities E^I_m and E_b:

$$E^\mathrm{I}_\mathrm{m} = e^\mathrm{I}_\mathrm{m} \cdot \sigma(A_\mathrm{b}) \cdot E_\mathrm{b}^{\beta_\mathrm{b}/2}. \tag{5.29}$$

In constructing this ansatz we, first, have turned to the gist of Stevens' law with the exponent β_b relating

- the intensity of a physical stimulus—here the characteristic force the subject exerts on the governed object which, by virtue of (5.13), scales with of the bodily measure E_b of effort as $E_\mathrm{b}^{1/2}$ and
- the magnitude of its perception—here the effort mean magnitude E^I_m.

Second, the capacity of the converting unit—the inner state of the self—is described as a cofactor $\sigma(A_\mathrm{b})$ depending on the degree A_b of inner attention. The proportionality coefficient e^I_m may be selected arbitrary, in particular, it may be set equal to unity, $e^\mathrm{I}_\mathrm{m} = 1$. Indeed, the role of the coefficient e^I_m is reduced to the conversion of effort units between the realms of the body-in-the-self and the mind-in-the-self, whereas the units of subjective effort are ill-defined and only relative changes in subject effort matter.

It should be noted that the bodily effort E_b is a quantity introduced at the level of the body-in-the-self, i.e., in each particular case it takes some value. By contrast, the mean magnitude E^I_m is a parameter of the cloud-type measure $\mathbb{E}^\mathrm{I}_\mathrm{m}$; this parameter is not accessible to the mind directly. Therefore for small values of E_b the subject is not able to reliably recognize the difference E_b and its zero value. This feature enable us to suppose that the exponent β_b is larger than unity, $\beta_\mathrm{b} > 1$.[32] Thereby, the dependence E^I_m on the square root of bodily effort $E_\mathrm{b}^{1/2}$ takes the desired convex shape. Figure 5.4(II, right panel) illustrates the $E^\mathrm{I}_\mathrm{m}(E_\mathrm{b})$-dependence $\beta_\mathrm{b} > 1$ and

[31] Even some actions are fully automated and so perceived as effortful it does not exclude that after a certain time of their repetition fatigue accumulated nonconsciously becomes conscious and gives rise to conscious attention to these actions (e.g., Burnley and Jones 2016).

[32] For example, Stevens' law for stimuli related to bodily actions such as pressure on palm, muscle force, heaviness, vocal effort is characterized by the exponent β_s taking the values $\beta_s =$ 1.1, 1.7, 1.45, 1.1, respectively (e.g., Gescheider 1997). The difference of the Stevens exponent β_s for, e.g., loudness ($\beta_s = 0.67$) and brightness ($\beta_s = 0.5$) from the noted values seems to admit an explanation that under the normal conditions such stimuli related to bodily actions are characterized by low intensities.

$\beta_\mathrm{b} < 1$ to emphasize their difference for small values of E_b and to justify the choice of $\beta_\mathrm{b} > 1$.

The degree A of any kind attention, *as it is perceived by the subject*, is also an entity of the mind-in-the-self. Within this understanding the degree A_b of inner attention must be of cloud-type and admit description in terms like 'little' or 'full' but cannot ba a particular number or percent. So in the presented description the value A_b must be regarded as either[33]:

- the mean degree of the cloud-type attention—the entity of the mind-in-the-self or
- the real number quantifying the computational resource allocated to a given task at the level of the body-in-the-self; in this case, the degree A_b of inner attention is not accessible to the mind.

Below in the text we will not draw a distinction between the two interpretations of the attention degree keeping in mind mainly the latter interpretation. The matter is that the attention degree A_b (and degree of attention of other types) arises mainly in expressions similar to Exp. (5.29) relating bodily effort and effort mean magnitude, i.e., quantities accessible to the mind only averaging over the corresponding clouds.

For this reason, as in the case of the $E^\mathrm{I}_\mathrm{m}(E_\mathrm{b})$-relationship, the capacity $\sigma(A_\mathrm{b})$ cannot exhibit remarkable variations in the relative small neighborhoods of the limit values $A_\mathrm{b} = A^\mathrm{min}_\mathrm{b} \ll 1$ and $A_\mathrm{b} = 1$.[34] The following functional ansatz allows for this feature and is actually illustrated in Fig. 5.4(II, left panel)

$$\sigma \cdot (1 - \sigma) = 2^{2(p-1)} \left[A_\mathrm{b} \cdot (1 - A_\mathrm{b})\right]^p, \tag{5.30}$$

where the exponent $p > 1$ is a model parameter. For $A_\mathrm{b} \ll 1$ and $1 - A_\mathrm{b} \ll 1$ the capacity is scaled as $\sigma \propto A^p_\mathrm{b}$ or $1 - \sigma \propto (1 - A_\mathrm{b})^p$, respectively. As seen in Fig. 5.4(II, left panel) for $p \sim 6$ this ansatz looks like the $\sigma(A_\mathrm{b})$ dependence having some fuzzy threshold near the limit values $A_\mathrm{b} = 0, 1$.

5.4.3 *Mental Measure $\mathbb{E}^{II}_m$: Speed-Accuracy Tradeoff*

Now we turn to the second component of experienced effort—mental effort with the measure $\mathbb{E}^\mathrm{II}_\mathrm{m}$—which is directly caused by the *conscious* processing of perceived information. The present subsection and the next ones are devoted to constructing a specific expression for the component $\mathbb{E}^\mathrm{II}_\mathrm{m}$ of mental measure within several steps.

In our constructions of the mental measure $\mathbb{E}^\mathrm{II}_\mathrm{m}$ of effort related to monitoring the object motion, we turn to the multifactorial psychological phenomenon called the *speed-accuracy tradeoff*. The speed-accuracy tradeoff is generally understood as

[33] Naturally, the relationship between attention and action execution is much more complex, in particular, there are various dimensions of this relationship (e.g., Wulf 2013; Deubel 2014; Lohse 2015; Pageaux 2016, for a review).

[34] The inequality $A^\mathrm{min}_\mathrm{b} \ll 1$ implies that the subject's skill is high.

> [t]he complex relationship between an individual's willingness to respond slowly and make relatively fewer errors compared to their willingness to respond quickly and make relatively more errors. (Zimmerman 2018, p. 3250, right collomn)

There are many particular instantiations of the speed-accuracy tradeoff, for details a reader may be referred, e.g., to Johnson and Proctor (2004), Agam et al. (2013), Heitz (2014), Standage et al. (2015). Evolutionary and neurological background of speed-accuracy tradeoff and similar tradeoffs is discussed, e.g., by Duckworth et al. (2018). Here we consider a few aspects of this tradeoff to elucidate our constructions.

First of all, it is Fitts' law (Fitts 1954)—a general law of human sensory-motor behavior that relates the movement time τ to the distance of movement D with the precision $\pm\delta D$; the farther or more accurately the movement is performed, the longer time is required for this movement. In mathematical terms Fitts' law is usually written as:

$$\tau = k_1 + k_2 \cdot I_d\,, \tag{5.31}$$

where k_1 and $k_2 > 0$ stand for adjustable constants and the index of difficulty I_d is some increasing function of the ratio $D/\delta D$

$$I_d = f\left(\frac{D}{\delta D}\right). \tag{5.32}$$

A number of different models for dependence (5.32) have been proposed in literature (e.g., Plamondon and Alimi 1997; Guiard and Olafsdottir 2011; Goldberg et al. 2015, for a review) comprising different classes, in particular,

- the original logarithmic model $f(x) = \log_2(x)$ (Fitts 1954) and its derivatives like Welford' (1960, 1968) model $f(x) = \log_2(x+1)$ (with $k_1 = -k_2$) and
- power-law ansätze like $f(x) = x^g$ ($g > 0$ is a constant) (Meyer et al. 1988, 1990) or even Kvålseth's (1980) power-law model for τ, i.e., model (5.31) with $k_1 = 0$.

Which particular form of Fitts' law should be selected is the matter of ongoing debates because the practical range of the ration $D/\delta D$ that can be actually investigated in the laboratory is rather narrow (Guiard and Olafsdottir 2011; Rioul and Guiard 2012) and the experimental data-sets are characterized by strong scattering (e.g., Gori et al. 2018).

In our constructions we turn to the ansatz proposed by Guiard et al. (2011), Guiard and Rioul (2015) to approximate relationship (5.31) using the weighted homographic function. Namely, based on experimental data-sets collected within the time-minimization paradigm (Fitts 1954), the spread-minimization paradigm (Schmidt et al. 1979), and the dual-minimization paradigm (Guiard et al. 2011) the following function is used to fit the data-points on the plane $\{\tau, \delta D/D\}$[35]

[35] In fact, ansatz (5.33) was proposed by Guiard et al. (2011), whereas Guiard and Rioul (2015) employed its modified version, namely, the expression that can be written as

$$\tau^{\beta_F} \left(\frac{\delta D}{D} \right)^{1-\beta_F} = g_F \,, \tag{5.33}$$

where the exponent $0 < \beta_F < 1$ and the constant g_F are the model parameter which can be found via fitting the experimental data. This relationship formally exhibits two singularities, first, $\delta D/D \to \infty$ for $\tau \to 0$ and, second, $\delta D/D \to 0$ for $\tau \to \infty$. It implies that the range of possible values of $\delta D/D$ and τ must be bounded from below and above by certain finite quantities.

Turning to the data presented by Guiard and Rioul (2015) we see that the analyzed values of τ may be regarded as the time scales of the experiential now bounded from below by the characteristic duration τ_s of the immediate now, $\tau \gtrsim \tau_s$, and from above by the characteristic duration τ_d of experiential now, $\tau \lesssim \tau_d$.[36] Thereby, the analyzed range of τ is specified by the inequality

$$\tau_s \lesssim \tau \lesssim \tau_d \,. \tag{5.34a}$$

The values limiting the variations of the ratio $\delta D/D$

$$\kappa \lesssim \frac{\delta D}{D} \lesssim K \tag{5.34b}$$

must be related to the corresponding values limiting the values of τ via the Fitts' law approximated by ansatz (5.33), namely,

---

$$(\tau - \tau_s)^{\beta_F} \left(\frac{\delta D}{D} - \kappa \right)^{1-\beta_F} = g_F \,,$$

where, as above, the time scale τ_s and the quantity κ specify the limit capacity in subject's reaction to external stimuli and the reproduction of aimed movement of mean amplitude D, respectively. Formally speaking, the given representation of Fitts' law does not require restrictions imposed on the upper boundary limiting possible values of $\delta D/D$ and τ. However the difference between the given ansatz and ansatz (5.33) is not too essential because in close proximity to the asymptotics $\delta D/D = \kappa$ and $\tau = \tau^*$ the uncertainty caused by the probabilistic properties of human actions is dominating and the particular form of a used ansatz does not matter. Moreover, in the noted experiments the analyzed time scales do not exceed 2 s (see, e.g., Guiard and Rioul (2015)).

Our analysis is devoted to the experiential now, so, here we explicitly deal with the fixed range of possible time scales bounded from above by $\tau_d \sim 2$–3 s. Therefore in our constructions we confined ourselves to ansatz (5.33) to avoid possible misconceptions implicitly caused by the assumed behavior of the $\delta D/D$-τ-dependence near these asymptotics.

[36] As summarized in Sect. 2.6, the time scales of the immediate now—one of the synchronic units composing the diachronic unit—is characterized by time scales ranging

- from the perception threshold $\tau_0 \sim 20$–60 ms below which all the temporal information is completely lost in human perception and
- up to the mean duration of the immediate now $\tau_s \lesssim 0.5$ s.

The characteristic duration of experiential now—the duration of diachronic unit—is about $\tau_d \sim 2$ s. According to Guiard and Rioul (2015) the time intervals characterizing the duration of aimed movements studied experimentally within different paradigms belong to the interval from 100 ms up to 2 s, which justifies the made statement.

$$\tau_s^{\beta_F} K^{1-\beta_F} = g_F \qquad \text{and} \qquad \tau_d^{\beta_F} \kappa^{1-\beta_F} = g_F\,. \tag{5.35}$$

The quantity κ admits interpretation as the limit capacity of the subject's reproduction of aimed movement when (*i*) attention is focused completely on these actions and (*ii*) there is no strict requirements imposed on the rate of action implementation like "as quickly as possible." In this case, it is quite natural to relate the accuracy of movement reproduction to the just noticeable difference in discriminating the "property" D exhibited by two instances of this movement (Sect. 1.3.1). Thereby we may suppose that the quantity κ is the Weber fraction, $\kappa \sim \kappa_w$, of the corresponding stimulus—aimed movement over the distance D.

In this case, the latter equality in Exp. (5.35) gives us the estimate of the constant g_F, demonstrating that Fitts' law approximated by Exp. (5.33) possesses only one own parameter—the exponent β_F. The former equality gives us the estimate

$$K = \kappa \left(\frac{\tau_d}{\tau_s}\right)^{\beta_F/(1-\beta_F)}. \tag{5.36}$$

The value K estimates the lower accuracy of movement amplitude at the boundary of perception. In this case, the movements are so quick that the processing of perceived information and the movement control based on it can be implemented only unconsciously.

Ansatz (5.33) is justified within the resource-allocation theory of Fitts' law developed by Guiard et al. (2011), Guiard and Rioul (2015) based on a number premises underlying also our constructions. In particular, they can be represented as follows.

- The speed-accuracy tradeoff stems from the competition for the same limited resource—attention—between two concurrent processes aimed at minimizing the reaction time τ and maximizing the accuracy of movement reproduction.
- When 100% of the resource is invested in the corresponding aimed-movement task—in our notions, when the degree of attention focused on the corresponding task is equal to unity—equality (5.33) holds. In this sense the curve specified by Exp. (5.33) on the plane $\{\tau, \delta D/D\}$ represents the front of performance, the region located above the given curve should contain scattered experimental points because "it is also possible to do worse" (Guiard and Rioul 2015).
- In such actions resource-allocation strategy is rather flexible. Humans are able to modulate the proportion in which they allocate their resource to the mutually incompatible speed and accuracy configurations. In particular, the layer on the plane $\{\tau, \delta D/D\}$ containing curves fitting experimental points obtained under different experimental paradigms is relatively narrow (Guiard and Rioul 2015, Fig. 11). The aforementioned paradigms are distinct from one another in that the subject has to

  - o minimize the duration of his movements provided the required accuracy of aimed movements is fixed (the time-minimization paradigm, (Fitts 1954));
  - o reproduce his actions as accurate as possible within a given time interval (the spread-minimization paradigm, (Schmidt et al. 1979));

- minimize both the duration of action and maximize its accuracy with various degrees of imbalance, which is enhanced via special instructions (the dual-minimization paradigm, (Guiard et al. 2011)).

It should be noted that these premises fall into the general scope of tradeoffs in dual-task performance. So relationship (5.33) may be regarded as a particular form of the performance operating characteristic (Norman and Bobrow 1975b, 1976) called also an attention operating characteristic (Alvarez et al. 2005) (for a detailed discussion see also Proctor and Zandt 2017).

This kind flexibility of resource-allocation strategy enables us to suppose that the subject *perceives* all the actions represented by different points of relationship (5.33) for a fixed value of g_F as equivalent in effort. In other words, the locus on the plane $\{\tau, \delta D/D\}$ specified by function (5.33) corresponds to one mental state characterized by a fixed degree of attention focused on a given task. This mental state is degenerated with respect to dividing attention between minimizing the reaction time and maximizing the movement reproducibility. This degeneracy becomes an intrinsic property of such mental states if the attention division is implemented automatically. It implies that the attention division is a result of unconscious response to some external stimulus or internal details of unconscious processing of perceived information. Moreover, we suppose that efficiency of controlling the movement reproducibility is determined by two factors. First, it is the attention focused on the movement control, which explicitly determines the ratio $\delta D/D$. The second factor is the attention focused on the control over the speed of aimed movements. It affects the control over the movement accuracy implicitly because the efficiency of the $\delta D/D$-control depends not only on attention focused on it directly but also on the rate at which this control should be implemented. Broadly speaking, in this interpretation Fitts' law describes the $D/\delta D$-control efficiency proportional to the degree of attention focused on the movement accuracy directly, whereas the rate of movement implementation specifies the coefficient of this proportionality.[37] Exactly this interpretation of the speed-accuracy tradeoff to be accepted bellow can explain why Fitts' law admits representation (5.33) as the product of two terms.

In this case, the relationship between

- the reaction time τ_k, i.e., the "speed" of subject's actions,
- the accuracy of movement reproduction, i.e., $\delta D/D$, and
- the degree A of attention focused on implementing a given task

can be written in the form of equality (5.33) where the right-hand side is assumed to be a certain function of A, i.e., $g_F \Rightarrow g(A)$. Fitts's law is included into this relationship as the limit $g(A) \to g_F$ for $A \to 1$. For a general discussion of the relationship

[37] Speaking strictly, this interpretation of Fitts' law corresponds directly to the time-minimization paradigm of the corresponding experiments where the control over the movement speed is dominant. However, this interpretation does not exclude the spread-minimization paradigm, the subject just tries to adopt the movement speed such that the control over aimed movements with the required accuracy δD be feasible.

between speed-accuracy tradeoff and attention a reader may be referred, e.g., to Liu et al. (2009), Salinas et al. (2014).

5.4.4 Mental Measure $\mathbb{E}_m^{II}$: Power-Law of Working Memory Load

In order to propose a plausible approximation for the function $g(A)$, we turn to the dependence between the number of similar objects retained in the visual short-term memory and the precision with which each object is storied in memory. In this case, attention which affects memorizing and keeping the objects in memory is divided between them, so the degree A_m of attention devoted to each object individually can be estimated as $A_m \sim 1/m$, where m is the number of these objects. For the general reasons, it could be expected that the larger the number m of objects, the lower the degree A_m of individual attention, and, so the lower the precision P_m with which the objects are retained in the visual short-term memory.[38] As demonstrated by Bays and Husain (2008), Donkin and Nosofsky (2012), Sewell et al. (2014), Donkin et al. (2016), Smith et al. (2016, 2018), Lilburn et al. (2019)

$$P_m \sim \frac{P_1}{m^{\beta_M}}\,, \tag{5.37}$$

other conditions being equal. Typically the exponent β_M changes in the range $\beta_M \in [0.5, 0.75]$ (for a review see Smith et al. 2018).

As an addition argument for treating the noted scaling law (5.37) in terms of divided attention, we note increase in the uncertainty of time duration judgment caused by the difficulty of the parallel nontemporal task found within the experimental paradigm of dual-task prospective timing (Brown 1997, for a review see, also, Block et al. 2010; Grondin 2010; Block and Gruber 2014; Matthews and Meck 2016). Because there is no special modality of time perception the noted effect should be regarded as a regularity of inner psychophysics.

Expression (5.37) can be also understood as the power-law dependence of the uncertainty δD in perceiving and memorizing the property D of such objects on the degree A_m of individual attention responsible for mental operations with one of these objects:

$$\frac{D}{\delta D} \propto A_m^{\beta_M}\,. \tag{5.38}$$

Expressions (5.33) and (5.38) enable us to write the generalized Fitts' law relating the speed-accuracy tradeoff, i.e., the division of attention between optimizing the

[38] Following the cited literature here the term "precision" P is used in the sense of either the inverse variance δD of the errors in discriminating a property D of retained objects or the inverse thickness δD of the region where the corresponding psychometric function exhibits strong variation, i.e., $P \sim D/\delta D$ (e.g., Bays and Husain 2008).

speed and accuracy of human actions to the degree A of total attention focused on these actions as

$$\frac{\delta D}{\kappa D} = \left(\frac{\tau_d}{\tau}\right)^{\beta_F/(1-\beta_F)} A^{-\beta_M}, \tag{5.39a}$$

$$A = \left(\frac{\tau_d}{\tau}\right)^{\beta_F/[(1-\beta_F)\beta_M]} \left(\frac{\kappa D}{\delta D}\right)^{1/\beta_M}, \tag{5.39b}$$

or in the form which can be regarded as the mathematical description of the *speed-accuracy tradeoff under partial attention*

$$\left(\frac{\tau_d}{\tau}\right)^{\beta_F} \left(\frac{\kappa D}{\delta D}\right)^{(1-\beta_F)} = A^{(1-\beta_F)\beta_M}. \tag{5.39c}$$

The first expression is just the direct combination of Exps. (5.33) and (5.38), whereas the second one being equivalent to it is more suitable for describing the uncertainty of space-time clouds in the phase space $\mathbb{R}_{[4]} \times t$.

It should be emphasized that setting $\tau = \tau_d$ in Exp. (5.39a) we may rewrite it in the form

$$\delta D = \kappa D \cdot A^{-\beta_M}, \tag{5.40}$$

which admits interpretation as *Weber's law under partial attention.*

5.4.5 Mental Measure $\mathbb{E}^{II}_m$: Nonstationary Speed-Accuracy Tradeoff

Expressions (5.39) were obtained for situations when an aimed movement possesses an initial point—where the movement starts—and a terminal point—where the movement finishes. Below we will consider similar relations for situations when the terminal point does not exist. Namely, for describing mental effort caused by the monitoring of moving object we need to know the relation between:

- the uncertainties in the perception of motion characteristics—position, speed, acceleration, etc.,
- the uncertainty of time localization—i.e., the time interval within which the required information is accumulated by the corresponding sensory organ, and
- the degree of attention focused on the monitoring—the amount of computational resources allocated for processing this information.

The neural mechanism governing the speed-accuracy tradeoff analyzed in Sects. 5.4.3 and 5.4.4 and the mechanism governing the desired relationship must be very similar to each other if not are just identical in nature. Indeed, these situations seem to be equivalent except for the fact that may be illustrated turning to the following example.

A material point moving in the physical space cannot stay at one point as time goes on. Thereby, the uncertainty in the mental-mental embedding of moving material point reflects purely mental aspects as well as the physical uncertainty in spatial localization of moving object on finite time scales. This non-stationarity is reflected in the section title.

The purpose of the present section is to generalize the speed-accuracy tradeoff under partial attention, Exp. (5.39c), to allow for the effect of physical uncertainty caused by the permanent, regular change of the phase variable under consideration. In the next section we will apply the results to be obtained here to describing the relation between dimensions of material point eidos and attention focused on it. In the present section, confining our analysis to a characteristic example, we elucidate the basic features of this generalized tradeoff.

Let us consider the perception of the subject monitoring the motion of a material point. The used system coordinates is attached to a certain reference point playing the role of origin. Let D and v denote the current position and the speed of moving object whose motion the subject analyzes with respect to the reference point.

When the object speed is ignorable, subject's perception is assumed to be characterized by the uncertainty δD of object position which via Exp. (5.39) is related to:

- the current distance D between the object and the reference point,
- the degree A of attention focused on monitoring the object motion, and
- the characteristic duration τ of time interval within which the perceived information is integrated before the result becomes accessible to the mind.

As previously, the integration time τ is supposed to meet the inequality $\tau_s \lesssim \tau \lesssim \tau_d$ characterizing time scales of the experiential now. In turn the Weber (Ekman) fraction κ and the related coefficient K determining the interval (5.34b) of possible values of the uncertainty δD within the speed-accuracy tradeoff with $A = 1$ are assumed to meet inequality (5.36).

The uncertainty in the spatial position of moving object caused by its motion becomes noticeable when the object speed v gets values such that

$$\rho \tau v \sim \delta D \,,$$

where the uncertainty δD is specified by Exp. (5.39a) and the cofactor $\rho \lesssim 1$.

Dealing with the direct observations of moving object this condition may be interpreted as the criterion comparing

- the uncertainty δD arising in the mental-mental embedding of the material point, i.e., the x-width of the material point eidos caused by purely mental processes and
- the motion smear $\rho v \tau$ of the mental image representing the moving object as it is perceived by the subject via visual modality and caused by unconscious processing of visual information.

The cofactor ρ has been introduced actually to allow for the contribution of various deblurring mechanisms the human visual modality possesses to reduce the motion smear substantially, which is taken into account by the cofactor $\rho \ll 1$.[39]

If the subject, seeing the moving object without remarkable blur, for some other reasons has to average the visual information on time scales τ then in the mental-mental embedding of the material point moving with the speed v the uncertainty of its localization cannot be much less than τv.

Therefore, in our constructions to quantify the effect of object motion on the mental-mental embedding we, first, estimate the effective motion smear as $\rho\tau v$, where the coefficient $\rho \lesssim 1$. Second, we posit that the uncertainty δD in the spatial position of moving object as it is perceived by the subject within the experiential now cannot be less than the effective motion smear, i.e., the condition

$$\delta D \geq \rho\tau v \tag{5.41}$$

must be fulfilled.

Figure 5.5 represents a semi-quantitative approach to describing the object motion effect on the speed-accuracy tradeoff. Within this approach the uncertainty δD is assumed to be specified by relationship (5.39a) provided condition (5.41) holds, otherwise we set $\delta D = \rho\tau v$. The layer $\tau_s \leq \tau \leq \tau_d$ represents the experiential now

---

[39] The integration of visual information by the visual modality together with the brain is usually implemented on time scales $\tau \sim 100$ ms or 200 ms if the saccadec eyes' movements play an essential role (e.g., Purves et al. 2004). However, the motion smear—the blur of point-like object moving with speed v—as it is *perceived* by humans does not coincide with the motion smear τv in a photograph taken with the same integration period $\tau \sim 100$ ms (shutter speed) (Burr 1980). The matter is that the perceived motion smear is a result of the multi-step processing of visual information based on a variety of bottom-up and top-down mechanisms including attention, expectation, memory, prior experience (for a review see Burr and Thompson 2011; Nishida 2011; Spering and Montagnini 2011; Watson and Ahumada 2011; Carrasco 2014; Öğmen and Herzog 2016; Rucci et al. 2018; Perez et al. 2020). Eye and head movements also contribute to the perception of motion smear (e.g., Bedell et al. 2010, for a review).

An explanation of how the motion smear can be depressed in the human visual system was put forward by Hammett (1997), Hammett et al. (1998, 2003) turning to the idea of non-linear contrast-response processing of visual information that effectively 're-sharpens' the image. There are also other concepts of deblurring, for example, the motion smear suppressed from awareness (Burr et al. 1986; Morgan and Benton 1989; Burr and Morgan 1997; Marinovic and Arnold 2013). For a review of plausible underlying mechanisms a reader may be referred to Breitmeyer and Öğmen (2006), Watson and Ahumada (2011), Spering and Montagnini (2011) In these mechanisms the space-time pattern of visual images takes the central position in the deblurring. In particular, it explains why for a single small-size object (a single dot) its smear is highly pronounced in comparison with visual images of their dense systems (Breitmeyer and Öğmen 2006, for a review see Chen et al. 1995; Purushothaman et al. 1998, Chap. 6). Moreover, focusing on the object spatial position it is rather natural to fix a certain point of environment near the moving object and to stare at it. In this case, the perceived motion smear must be much more pronounced in comparison with the motion smear when eyes track the moving object.

In the present analysis we confine our consideration to the mental effort of monitoring the dynamics of material point and, thereby, may estimate the perceived motion smear as $\rho\tau v$ with the cofactor $\rho \lesssim 1$.

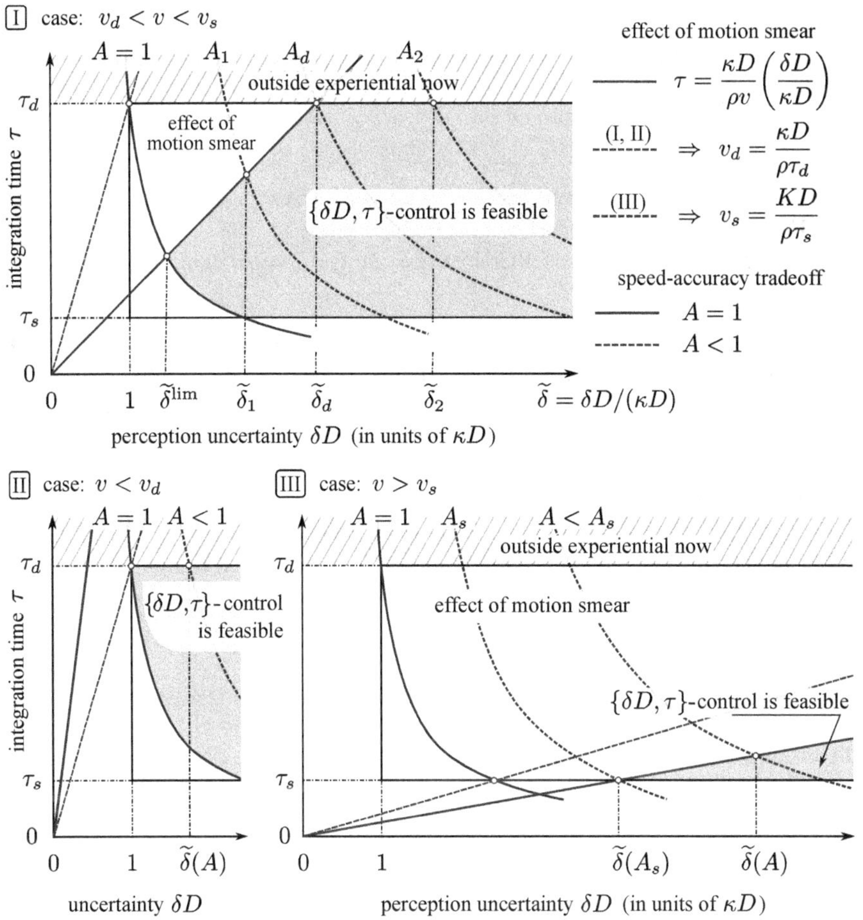

Fig. 5.5 Illustration of the speed-accuracy tradeoff under partial attention when the effective motion smear contributes substantially to the uncertainty δD in quantifying the spatial position of material point in its mental-mental embedding (Panel I). Panels II and III illustrate the limit cases when the object motion effect is ignorable because of small values of object speed or the object speed is so high that the subject's control over the object motion is not feasible. Here the symbols D and v denote the position and velocity of moving object as they are mentally evaluated by the subject within the experiential now, A is the degree of attention focused on monitoring the moving object. The y-axis shows the integration time τ of sensory information and the layer $\tau_s \leq \tau \leq \tau_d$ represents time scales of the experiential now bounded from below by the characteristic duration τ_s of the immediate now synchronic unit and from above by the characteristic duration τ_d of the diachronic unit. The x-axis shows the uncertainty δD normalized to its limit value κD. Red lines depict the boundary of the region where the motion smear determines the uncertainty δD (for different values of the object velocity v). Blue lines are the locus of speed-accuracy tradeoff for different values of the attention degree A. Shaded regions represent regions where the control over the object spatial position with accuracy δD is feasible. The coefficients κ and K characterize the speed-accuracy tradeoff, Exps. (5.36) and (5.50)

in its temporal properties. On the $\{\delta D, \tau\}$-plane the points representing the pairs $\{\delta D, \tau\}$ of particular values of uncertainty δD and integration time τ for which there is a degree A of attention required for their simultaneous realization are specified as follows. The point $\{\delta D, \tau\}$ should lie

- inside the layer of the experiential now;
- above the locus of the speed-accuracy tradeoff for $A = 1$ given by Exp. (5.39a), in Fig. 5.5 this locus is shown with solid blue lines;
- to the right from the line specified by the equality $\delta D = \rho\tau v$, in Fig. 5.5 it is solid red lines.

In Fig. 5.5 these points make up shaded regions. When the degree A of attention that can be focused on the object monitoring is less than 1, the corresponding locus of speed-accuracy tradeoff is shown with dashed blue lines.

The dashed read lines specify two velocity thresholds v_d and v_s. In the case of $v < v_d$ (Panel II) the object motion does not affect the uncertainty δD and so the object motion effect is ignorable. In the case of $v > v_s$ (Panel III) the object speed is so high that the control over the object position is of minor importance. Indeed, under such conditions the minimum of uncertainty δD_{mim} that can be attained is

$$\delta D_{\text{min}} = \rho\tau_s v > K D\,.$$

Expression (5.36) gives us the estimate $K \sim 0.5$–1 for $\kappa = 0.05$ (the characteristic value of Weber or Ekman fraction), $\tau_d/\tau_s = 10$–20, and $\beta_F = 0.5$. So, in this case, the uncertainty δD of mental-mental embedding is about the characteristic distance D itself or even exceed it. Therefore the feasibility of controlling the object position becomes doubtful and the main effect of object motion on the uncertainty δD is represented by the case of $v_s < v < v_d$ (Panel I).

As follows from the given analysis, in studying the effect of object motion on the uncertainty δD in the mental-mental embedding of moving material point we may

- confine ourselves to the case of $v < v_s$,
- ignore the lower boundary of the experiential now lay ($\tau = \tau_s$) or, put differently, formally use Exp. (5.39) subject to the condition $\tau \leq \tau_d$ only.

The presented approach describes the motion effect on the uncertainty δD turning to the step-wise dependence illustrated in Fig. 5.5 (Panel I). Namely, the value of δD is given by Exp. (5.39a) as a function $\delta D = \delta D(\tau)$ decreasing with increase in the integration time τ until the value $\delta D = \rho\tau v$ is gotten provided $\tau \leq \tau_d$. After that the value of δD does not change. This behavior is rather artificial, so, we want to propose a model free from such behavior and dealing with the given limits cases as asymptotics.

To do this we want to emphasize that Exp. (5.39a) specifies the limit capacity of localizing the mental-mental embedding in the space provided the attention degree A and the integration time τ are fixed. When in addition a regular drift of observed object is present the difference

$$|\delta D - \rho\tau v|$$

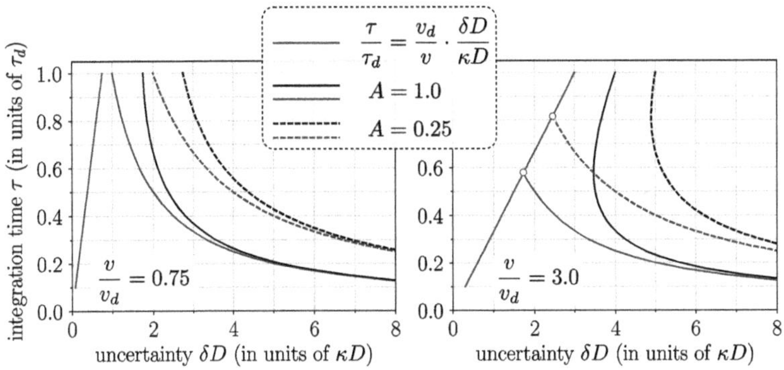

Fig. 5.6 Comparison of two approaches to describing the nonstationary speed-accuracy tradeoff allowing for the effective motion smear. Blue and black lines are the locus on the $\{\delta D, \tau\}$-plane representing the relationship between (*i*) the uncertainty δD in the spatial position of moving object as it is perceived by subject and (*ii*) the integration time τ of the neural processing of visual information. In plotting the object speed v and the attention degree A were fixed. Blue lines correspond to the step-wise approximation (illustrated in Fig. 5.5), black lines represent model (5.44a). Solid lines (blue and black) match the case when attention is focused completely ($A = 1$) on monitoring the object motion, dashed lines represent the case of partial attention ($A < 1$). The red lines depict the left boundary of the region of possible values $\{\delta D, \tau\}$. The left panel illustrates the case of object slow motion, $v < v_d$ (which correspond to Panel II in Fig. 5.5), the right panel illustrates the case of object fast motion, $v > v_d$ (Panel I in Fig. 5.5). The velocity threshold v_d is specified by Exp. (5.43). In plotting the exponents $\beta_F = 0.5$ and $\beta_M = 0.5$ were used

may be assumed to be specified by this limit capacity, which forms a plausible way of introducing the uncertainty δD in the spatial position for moving material point. Then turning to Exp. (5.39a) the desired relationship between the uncertainty δD in the spatial position of material point moving with speed v, the mean value of this position D, the degree A of attention focused on integrating sensory information within time τ can be written as

$$\left(\frac{\tau_d}{\tau}\right)^{\beta_F} \left(\frac{\kappa D}{|\delta D - \rho\tau v|}\right)^{(1-\beta_F)} = A^{(1-\beta_F)\beta_M}. \tag{5.42}$$

Expression (5.42) may be regarded as the mathematical description of the *nonstationary speed-accuracy tradeoff under partial attention*.

Formally expression (5.42) possesses the singularity when $\delta D \to \rho\tau v$, which, however, is of minor importance because it does not give rise to artificial results. Indeed, first, the analyzed time scales are bounded from above by the duration τ_d of experiential now and the ratio $\tau_d/\tau > 1$ for all the values of integration times τ in question. Second, the degree A of attention cannot exceed unity, $A \leq 1$. Therefore in all the case the difference $\delta D - \rho\tau v$ is bounded from below by a finite quantity, namely $\delta D - \rho\tau v > \kappa D$.

Using the velocity threshold v_d defined as[40]

$$v_d = \frac{(\kappa D)}{\rho \tau_d} \tag{5.43}$$

expression (5.42) can be rewritten in the form

$$\frac{\delta D}{\kappa D} = \frac{v}{v_d}\frac{\tau}{\tau_d} + \left(\frac{\tau_d}{\tau}\right)^{\beta_F/(1-\beta_F)} A^{-\beta_M}, \tag{5.44a}$$

or in the form to be used in the following constuctions

$$A = \left(\frac{\tau_d}{\tau}\right)^{\beta_F/[(1-\beta_F)\beta_M]} \left(\frac{\kappa D}{|\delta D - \rho\tau v|}\right)^{1/\beta_M}. \tag{5.44b}$$

Expressions (5.44a) and (5.44b) can be regarded the generalization of expressions (5.39a) and (5.39b) allowing for the effective motion smear.

Figure 5.6, first, illustrates the effect of object motion on the speed-accuracy tradeoff represented as the locus specified by expression (5.44a) on the $\{\delta D, \tau\}$-plane. Second, it compares two approaches, the step-wise approximation represented in Fig. 5.5 and model (5.42) allowing for a smooth crossover between the limits cases when the effective motion smear is ignorable or dominant. In particular, as seen in Fig. 5.6 (right panel) when the object speed v exceeds the threshold v_d ($v > v_d$) there can emerge two branches of the dependence $\delta D(\tau)$, increasing and decreasing ones. In the case of slow motion this dependence is monotonic.

5.4.6 *Mental Measure $\mathbb{E}^{II}_{m}$: Effort of Monitoring and Attention*

The relationship between the mean magnitude $E^{\mathrm{II}}_{\mathrm{m}}$ of this effort and attention A consciously focused on the corresponding activity have already been discussed in Sect. 5.2. The special situation considered here is that the monitoring of object dynamics is analyzed from the standpoint of speed-accuracy tradeoff and the corresponding effort experienced by the subject.

Accepting the scale-free organization of the functioning of the mind, we have to suppose that for small values of attention degree A the relationship $E^{\mathrm{II}}_{\mathrm{m}}(A)$ should be approximated by the power-law $E^{\mathrm{II}}_{\mathrm{m}}(A) \propto A^{\beta_{E0}}$ with some exponent $\beta_{E0} > 0$.[41]

[40] The velocity threshold v_d has been also specified in Fig. 5.5.

[41] Turning to the same arguments used in Sect. 5.4.2 to justify the dependence of mental effort on bodily effort and inner attention, we have consider that the exponent $\beta_{E0} \geq 1$. We also may consider that there is some critical value of the attention degree $A_c \ll 1$ such that for $A < A_c$ the corresponding information processing is implemented automatically (unconsciously). Then the transition from the automatic processing to the conscious processing should be of the second

When the current actions require complete attention (i.e., the attention degree $A = 1$) the possibility of monitoring actions is limited by the bounded capacity of human cognition. In this case, further increase in the accuracy of the subject's actions becomes impossible. We describe this effect as extremely strong growth in the experienced effort; in terms of mean characteristics it implies that $E^{\rm II}_{\rm m}(A) \to \infty$ when $A \to 1$. The following ansatz will be used to describe the crossover between the two limit cases

$$E^{\rm II}_{\rm m}(A) = e^{\rm II}_{\rm m} \frac{A^{\beta_{E0}}}{(1-A)^{\beta_{E1}}}, \tag{5.45}$$

where the constant $e^{\rm II}_{\rm m}$ and the exponent $\beta_{E1} > 0$ are model parameters. As in the case of subjective effort of action bodily execution, the cofactor $e^{\rm II}_{\rm m}$ may be treated as a certain unit of action monitoring effort and cannot be specified directly in dealing with this type effort separately from other types of mental effort.

It is worthy of noting that the given assumption about the singularity of experienced effort when $A \to 1$ meets, at least qualitatively, the premises of Brehm's motivational intensity theory (see Sect. 5.2). This theory posits the existence of some threshold in action difficulty such that humans reject the execution of actions whose difficulty exceeds this threshold.

Self-Consistency of Effort-Accuracy Tradeoff Theory

As far as the uncertainty in the effort of monitoring is concerned, the situation is rather complex. The matter is that attention to the process of monitoring, i.e., computational resources allocated for monitoring object dynamics determines simultaneously:

- the accuracy of this process, i.e., the accuracy δD in monitoring the object property D,
- the corresponding subjective effort, i.e., $E^{\rm II}_{\rm m}$,
- the accuracy of effort experience, i.e., the width $\Delta^{\rm II}_{\rm m}$ of the cloud $\mathbb{E}^{\rm II}_{\rm m}$.

By way of example, let us accept that the monitoring is aimed mainly at control over the property D; in Sect. 5.4.3 it was the amplitude of aimed movements. So, the accuracy δD of this control and the degree A of required attention are related by expression (5.39) where we may set $\tau = \tau_d$. For $A \ll 1$, i.e., when the control over the property D is rather far for its limit capacity,

$$E^{\rm II}_{\rm m} \propto \left(\frac{\kappa D}{\delta D}\right)^{\beta_{E0}/\beta_M} \tag{5.46}$$

order (Sect. 5.10). In this case, for $A > A_c$ the $E^{\rm II}_{\rm m}(A)$-dependence must be of the form $E^{\rm II}_{\rm m}(A) \propto (A - A_c)^{\beta_{E0}}$, where the exponent $\beta < 1$. Which case holds requires individual analysis, so in the present book we confine our consideration to the ansatz $E^{\rm II}_{\rm m}(A) \propto A^{\beta_{E0}}$ with the exponent $\beta_{E0} > 0$ as an approximation allowing for the two particular cases.

by virtue of Exp. (5.45). Expression (5.46) describes the mental effort required for monitoring the time variations in the quantity D within the accuracy δD.

The aforesaid advocates for that the perception threshold of changes in the relative accuracy of monitoring should match the threshold in recognizing the difference in the mental effort of monitoring. The given statement may be regarded as the requirement of *self-consistency* of the theory describing the relationship the experienced effort of monitoring and the accuracy of monitoring.

By virtue of the obtained expression (5.46), the self-consistency condition immediately leads to the estimate $\beta_{E0} \sim \beta_M$. Therefor, in order to avoid unnecessary increase in the number of model parameters we accept that

$$\beta_{E0} = \beta_M. \tag{5.47}$$

Taking into account the available estimates of the exponent β_M (Sect. 5.4.4), we see that the exponent β_{E0} may be estimated as $\beta_{E0} < 1$, which argues for the existence of second-order phase transitions in the effort perception (see Footnote 41).

It is also worthy of noting that the obtained relationship (5.46) enables us to estimate the uncertainty $\Delta^{\mathrm{II}}_{\mathrm{m}}$ in evaluating the effort of monitoring as

$$\Delta^{\mathrm{II}}_{\mathrm{m}} = \kappa_{2m} E^{\mathrm{II}}_{\mathrm{m}}. \tag{5.48}$$

Here the ratio κ_{2m} describes the uncertainty in effort evaluation which also may be treated as the second-type uncertainty of mental-mental embedding of point-like physical object. Namely, it is the just noticeable difference in the change in the width of the corresponding eidos (Appendix 5.9.2).

5.4.7 Mental Measure $\mathbb{E}^{II}_{m}$: Multi-channel Speed-Accuracy Tradeoff

Now we can formulate our model for the mean magnitude $E^{\mathrm{II}}_{\mathrm{m}}$ of mental effort; in the case under consideration, it is the magnitude where the density of the cloud $\mathbb{E}^{\mathrm{II}}_{\mathrm{m}}$ attains its maximum. This effort is determined directly by the dimensions of the space-time cloud $\mathbb{C}_{\mathrm{I}}$ representing the mental-mental embedding of moving material point into the space $\mathbb{R}_{[4]} \times t$ (see Fig. 3.4). Dealing with the effort of monitoring on time scales of the experiential now, we may confine our consideration to the projection of the space-time cloud $\mathbb{C}_{\mathrm{I}}$ onto the phase space $\mathbb{R}_{[4]} = \{x, v, a, j\}$ characterizing the dynamics of material point as it is perceived by the subject (Sect. 3.3.1). This projection denoted by the symbol $\mathbb{C}_{\mathbb{R}_{[4]}}$ and to be called also the space cloud $\mathbb{C}_{\mathbb{R}_{[4]}}$ is illustrated in Fig. 5.7.

The real characteristics of object motion—the current position x of this object, its speed v, acceleration a, and jerk j as well as the current instant of time t specify the center of the cloud $\mathbb{C}_{\mathbb{R}_{[4]}}$ (pointed out by white dots in Fig. 5.7) are not accessible

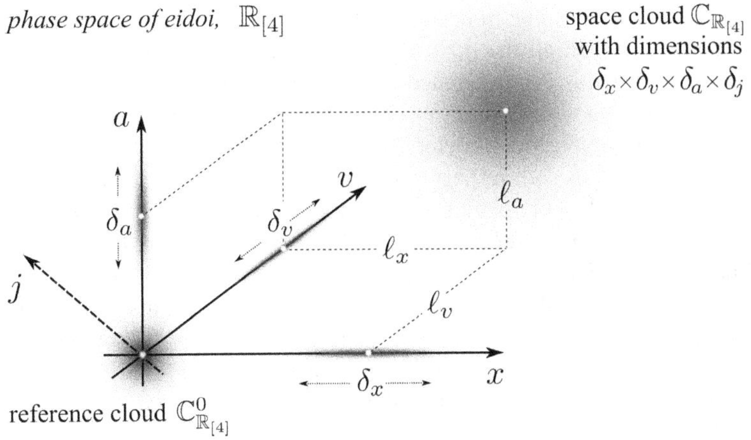

Fig. 5.7 Illustration of the space cloud $\mathbb{C}_{\mathbb{R}_{[4]}}$ with its dimensions $\{\delta_x, \delta_v, \delta_a, \delta_j\}$ and arrangement in the phase space $\mathbb{R}_{[4]} = \{x, v, a, j\}$ of eidoi. Here only three axes—the axes of spatial position x, velocity v, and acceleration a—are depicted in detail, the axis of jerk j is shown schematically with the dashed, directed line. The scales characterizing the position of the space cloud $\mathbb{C}_{\mathbb{R}_{[4]}}$ in the phase space $\mathbb{R}_{[4]}$ relative to some reference space cloud $\mathbb{C}^0_{\mathbb{R}_{[4]}}$ are denoted by the corresponding symbols $\{\ell_x, \ell_v, \ell_a, \ell_j\}$. White dots point out the centers of the corresponding fuzzy regions—the space cloud $\mathbb{C}_{\mathbb{R}_{[4]}}$ and its projections onto the axes x, v, a, j as well as the reference cloud $\mathbb{C}^0_{\mathbb{R}_{[4]}}$. These points are shown to guide the reader's eye only, their positions are not accessible to the mind

to the mind. The mind can operate only with the cloud $\mathbb{C}_{\mathbb{R}_{[4]}}$ as a whole entity. To elucidate our construction of the mean mental effort $E^{\mathrm{II}}_{\mathrm{m}}$ we need to emphasis the following items starting from a brief comment.

Brief comment: Mathematical eidoi of mental images In Sect. 3.3.3 we introduced two types of mathematical eidoi. First, it is the material point as an eidos of Newtonian mechanics. The second type is the space-time cloud as an eidos of physics of the human mind. Below we will speak about the eidoi of cloud type only. We will deal with space-time clouds paying special attention to their space-time structures and draw the sharp distinction between the notions of mental image and mathematical eidos of cloud type. It should be noted that there can be singled out two constituent components of the space-time cloud: the space cloud and the time cloud—different projections of the space-time cloud.

Phase Space $\mathbb{R}_{[4]}$ of Eidoi: Scale-Free Properties

Analyzing the properties of the cloud $\mathbb{C}_{\mathrm{I}} \subset \mathbb{R}_{[4]} \times t$ and its projection $\mathbb{C}_{\mathbb{R}_{[4]}}$ onto the phase space $\mathbb{R}_{[4]}$ of eidoi, it is necessary to take into account that there are no temporal and spatial units (scales) given beforehand. Put differently, there are no temporal and spatial units that can be initially selected for describing mental phenomena in subject's perception in all the possible cases even if we confine ourselves to the experiential now.[42] It is a consequence of the hierarchical, scale-free organization of the mind functioning.

The choice of suitable units is determined by how the space cloud $\mathbb{C}_{\mathbb{R}_{[4]}}$ is arranged in the phase space $\mathbb{R}_{[4]}$ with respect to the eidoi of other objects playing the role of reference points in the environment. For simplicity, here let us consider a situation when there is only one reference eidos $\mathbb{C}^0_{\mathbb{R}_{[4]}}$ shown in Fig. 5.7 at the center of coordinates. The subject based on acquired skill can mentally evaluate or just perceive as E-conscious quantity the distance between the clouds $\mathbb{C}_{\mathbb{R}_{[4]}}$ and $\mathbb{C}^0_{\mathbb{R}_{[4]}}$ in the phase space $\mathbb{R}_{[4]}$. In this case, the characteristic distances $\{\ell_x, \ell_v, \ell_a, \ell_j\}$ between the space cloud $\mathbb{C}_{\mathbb{R}_{[4]}}$ and the reference cloud $\mathbb{C}^0_{\mathbb{R}_{[4]}}$ can play the role of units for quantifying the dimensions of the cloud $\mathbb{C}_{\mathbb{R}_{[4]}}$, i.e., the accuracy of monitoring the dynamics of observed object (Fig. 5.7).

In Sect. 5.4.3 we have discussed the properties of the speed-accuracy tradeoff which argue for that only the ratio $D/\delta D$ between the mean amplitude D of aimed movement and the accuracy δD of movement reproducibility affect the human perception of performed actions. Turning to this evidence, we suppose that in describing the mental effort in monitoring, i.e., in constructing the mental measure $\mathbb{E}^{\text{II}}_{\text{m}}$ the dimensions of the space cloud $\mathbb{C}\mathbb{R}_{[4]}$ must be taken in the units of $\{\ell_x, \ell_v, \ell_a, \ell_j\}$. Put differently, dimensionless quantities such as

$$\left\{\frac{\delta_x}{\ell_x}, \frac{\delta_v}{\ell_v}, \frac{\delta_a}{\ell_a}, \frac{\delta_j}{\ell_j}\right\} \text{ and cross-ralted quanitities simila to } \frac{\delta_x}{\tau_d \ell_v} \tag{5.49}$$

characterizing the current state of the monitoring should be regarded as the basic variables in the description of the corresponding mental effort.

Phase Space $\mathbb{R}_{[4]}$ of Eidoi: Limit Capacity of Monitoring

Particular values of quantities (5.49) are determined by two factors—the unconscious processing of information and attention voluntarily focused on their perception. When attention is entirely focused, e.g., on perceiving the spatial position of moving object, the recognized magnitude of δ_x/ℓ_x should get its minimum. If, in addition, this object does not move too fast this minimum is determined by the unconscious processing of information. This limit value to be denoted as κ_x admits

[42] To elucidate this statement we may note the difference in quantifying physical stimuli and their mental images described within inner psychophysics. In quantifying the intensity of physical stimuli arbitrary relevant units can be used. The choice of these units is the issue of convenience only. For example, in the children's cartoon "38 parrots" the length of the same boa-constrictor is measured by parrots, monkies, and elephant calves. In quantifying the mental images of perceived stimuli the choice of units of measurement, i.e., the choice of some mental entity for comparison with other mental images in magnitude is crucial. The matter is that even the mental images of physical stimuli of the same nature can become incommensurable in magnitude (Lubashevsky 2019).

interpretation as a physiological characteristics which can be found experimentally. In a similar way we introduce the quantities κ_v, κ_v, κ_a, and κ_j, namely,

$$\begin{aligned}
\kappa_x &= \lim_{A_x \to 1} \frac{\delta_x}{\ell_x}\,, \text{ provided } \ell_v \ll \ell_v^c\,, \\
\kappa_v &= \lim_{A_v \to 1} \frac{\delta_v}{\ell_v}\,, \text{ provided changes in } \ell_x \text{ are ignored and } \ell_a \ll \ell_a^c\,, \\
\kappa_a &= \lim_{A_a \to 1} \frac{\delta_a}{\ell_a}\,, \text{ provided changes in } \ell_x, \ell_v \text{ are ignored and } \ell_j \ll \ell_j^c\,, \\
\kappa_j &= \lim_{A_j \to 1} \frac{\delta_j}{\ell_j}\,, \text{ provided changes in } \ell_x, \ell_v, \text{ and } \ell_a \text{ are ignored}\,.
\end{aligned} \tag{5.50}$$

Here the quantities $\{\ell_v^c, \ell_a^c, \ell_j^c\}$ denote the corresponding thresholds when the object position, velocity, and acceleration do not change too fast and, so, their variations within the experiential now do not contribute to the perception of these characteristics of moving object. The jerk j seems to be the highest time derivative of spatial position that humans are able to perceive within the experiential now. Therefore the last line in Exp. (5.50) contains no restrictions on the higher variable dj/dt. Quantities (5.50) are actually some analogy to the Ekman fraction describing inner properties of perceived stimuli (Sect. 1.3.1).

Phase Space $\mathbb{R}_{[4]}$ of Eidoi: Multi-channel Processing of Visual Information

When the control over object dynamics requires the monitoring of several phase variables $\{x, v, a, j\}$, the limit values of cloud uncertainty given by Exp. (5.50) cannot be attained for two reasons. First, it is the competition for computational resource—attention—between processes responsible for extracting the corresponding information from neural signals generated within visual modality. Second, these phase variables are not independent quantities in the physical world. For example, the velocity of moving object determines the rate of temporal changes in its spatial position. The same concerns the object velocity and its acceleration. The perceived information is integrated over some finite interval within its neurophysiological processing before the mind gains access to the results representing the observed objects. Therefore, e.g., the velocity exceeding some threshold should affect the perceived spatial position of the given object.

In Sects. 5.4.3–5.4.5 we considered a similar situation where the relationship between the accuracy of aimed movements and the speed of information processing is caused by the interaction as well as competition of the two processes sharing the same pool of computational resources. Finally, we proposed the account of nonstationary speed-accuracy tradeoff under partial attention that deals with the rate of information integration, the perception accuracy, attention focused on monitoring object motion, and the regular change of tracked quantity in time. In the case under consideration,

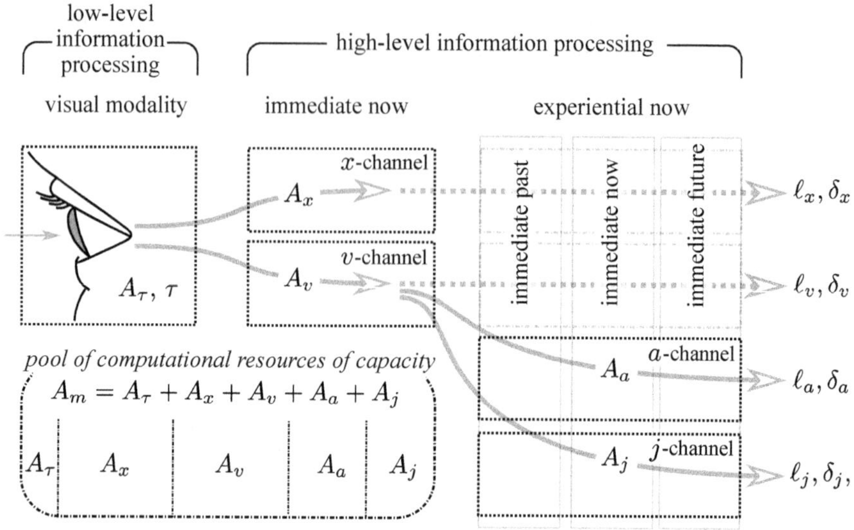

Fig. 5.8 Schematic illustration of the three-stage model for processing visual information that (*i*) specifies the multi-channel speed-accuracy tradeoff and (*ii*) relates the dimensions of eidos $\mathbb{C}_\mathrm{I}$ in the space $\mathbb{R}_{[4]}$ to attention required for monitoring material point motion. The symbols A_q $(q = r, x, v, a, j)$ denote the degree of attention, i.e., the amount of computational resources allocated for the corresponding channels and belonging to a pool of computational resources of capacity A_m. The quantities $\{\ell_q, \delta_q\}$ $(q = x, v, a, j)$ are the final output of information processing. Solid gray lines illustrate the "active" processing of information, whereas dashed lines denote some post-processing of already processed information like an inheritance of already extracted features, maybe, with some averaging

we actually meet four particular instantiations of this type speed-accuracy tradeoff which are united by the low-level integration of perceived information.

The gist of our approach to describing the speed-accuracy aspects in monitoring object motion on scales of the experiential now may be reduced to the following premises illustrated in Fig. 5.8. Namely, within the experiential now the processing of visual information comprises three basic stages responsible for extracting different properties characterizing the dynamics of observed material point. This division of the information processing is not strict because its steps can overlap. The given three-stage representation is used to single out the characteristic features of information processing which underlie our theoretical constructions and enable us to formulate the basic premises in a clear manner.

Speaking about this complex processing of visual information as a whole we may note the following:

- The four phase variables $\{x, v, a, j\}$—describing the eidos of moving material point are finally attributed to the level of the experiential now and may be regarded as quantities of E-consciousness. In particular, the characteristics of this eidos as the cloud $\mathbb{C}_{\mathbb{R}_{[4]}}$ in the space $\mathbb{R}_{[4]}$ (Fig. 5.7)—the mean values $\{\ell_x, \ell_v, \ell_a, \ell_j\}$

of these variables and the uncertainties $\{\delta_x, \delta_v, \delta_a, \delta_j\}$ in the mental evaluation—determine the subject's perception of the eidos $\mathbb{C}_{\mathbb{R}_{[4]}}$ as whole. In Fig. 5.8 it is illustrated as the quantities $\{\ell_q, \delta_q\}$ $(q = x, v, a, j)$ related to the final output of complex process.

- The given information processing is arranged in a hierarchical way. On one hand, it enables the sensory system (sensory modality combined with the brain) to extract *automatically* sophisticated details concerning the dynamics of observed object—up to the third derivative of spatial position of moving object. Thereby, human perception of moving objects initially, i.e., at the level of E-consciousness becomes rather rich in properties. On the other hand, a single source used by human sensory system endows the extracted properties with the mutual self-consistency.

The details of the given information processing are reflected in its three-stage structure (Fig. 5.8) represented by:

- **The low-level conversion of retinal images into the corresponding neural signals**. Already at this stage there can be singled out two channels of information processing aimed at extracting the details of slow and rapid changes in observed properties. We relate them to the perception of object spatial position and the speed of motion, respectively, and will refer to them as the x-*channel* and v-*channel*.

  Strictly speaking, the pace $1/\tau$ of information "updating" is determined by the following stages of unconscious integration of visual information within the immediate now and the immediate past. The duration τ of this cumulative integration can be intentionally influenced by the mind via top-down processes (Sect. 2.6.3). We allow for this feature via introducing the degree A_r of attention focused on intensifying the information processing as a whole. In this sense the pace of information "updating" and the related attention degree A_τ may be attributed to both these low-level and high-level stages. However, to emphasize that the given characteristics (τ and A_τ) concern mainly the processing of raw information we attribute them to the low-level stage. In Fig. 5.8 we have illustrated this aspect via placing the attention degree A_τ and the time τ of information "updating" inside the visual modality block. In this way we accentuate that all the channels of information processing are related with one another via their common part.
- **The high-level unconscious integration of visual information on scales of the immediate now**—one of three basic synchronic units forming the diachronic unit, i.e., the experiential now (see Sect. 2.6, in particular, Fig. 2.6). Within this stage, the extraction of information concerning the position of the observed object and its speed—the "active" phase of information processing—is mainly completed and its output contains the quantities $\{\ell_x, \ell_v, \delta_x, \delta_v\}$ averaged on scales, e.g., the characteristic duration τ_s of immediate now. Here we used the term "active" processing of information because it implies close interaction between the sensory modality and the brain (Sect. 2.5).

  When the desired duration τ of cumulative integration exceeds the duration τ_s of immediate now, this output is passed to the immediate past where it is accumulated on scales τ. Finally, the result of averaging is raised up to the level of the experiential now and then becomes accessible to the mind as E-conscious quantities. In

Fig. 5.8 this type transformation is illustrated by dashed gray lines. So the result $\{\ell_x, \ell_v, \delta_x, \delta_v\}$ is attributed to the level of the experiential now. However, the corresponding processing of information can be conceived of the process implemented within the immediate now based on the interaction between the sensory modality and the brain during the time τ. For this reason, we attribute the degrees A_x and A_v of required attention to the level of immediate now (Fig. 5.8) and consider the x-channel and v-channel to be mechanisms of information processing within the immediate now.

- **The high-level integration of visual information within the experiential now** which is based on processing the information within the immediate past, the immediate now, and the immediate future. The output of the previous stage play the role of the input for the given stage. In addition to the relatively "passive" conversion of the position- and velocity-data within the x-channels and v-channels, there are two new channels of "active" information processing (Fig. 5.8). They are due to the tripartite structure of the diachronic unit, i.e., the experiential now. We call them the a-channel and j-channel because they imply detecting the difference of the first and second order in the speed of moving object as it is perceived at different parts of integration interval.

  The output of this information processing is the mean acceleration ℓ_a and the uncertainty δ_a of its perception (a-channel) as well as the mean jerk ℓ_j and its uncertainty δ_j (j-channel) that can be attributed to the level of the experiential now only. The degrees A_a and A_j of attention focused on processing visual information through the a-channel and j-channels should be also attributed directly to the stage of experiential now.

The given channels of information processing, on the one hand, are related to one another via the common source of raw information and the duration of integration. On the other hand, the mechanisms extracting the information about the object position, velocity, acceleration, and jerk must be different. Indeed, speaking in mathematical terms, they must be based on different operations with the perceived time pattern of object motion. We understand the mutual independence of the x-channel, v-channel, a-channel, and j-channel exactly in this sense. For this reason, we posit that these channels require individual computational resources to be allocated to implement their functioning.

Therefore, the total amount of attention related to monitoring a moving object and distribution over the these channels meet the equality

$$A_m = A_r + A_x + A_v + A_a + A_j = \sum_q A_q \quad (q = r, x, v, a, j), \tag{5.51}$$

where the cumulative degree A_m of attention related to object monitoring plays the role of the capacity of the computational resource pool allocated for the monitoring as a whole (Sect. 2.5). As a special assumption, we accept that the amount of computational resources A_r required for just intensifying the information processing up to the pace $1/\tau$ cannot be as large as the amount of resources $\sum_{q\neq r} A_q$ related to processing the perceived information. Indeed, otherwise the monitoring of object motion is problematic and the possibility of efficient control over object dynamics is doubtful. In this case, turning to the results of Sect. 5.4.5 we may accept also the following premises:

- The integration time τ not exceeding the duration τ_d of experiential now, $\tau \lesssim \tau_d$, admits interpretation as independent phase variable of information processing. Therefore the degree A_q of attention focused on the information processing via a q-channel ($q = x, v, a, j$) specifies the relationship between the integration time τ and the output of the given channel independently of the other channels. We call it *partial speed-accuracy tradeoff* within the complex processing of visual information in monitoring the object motion.
- The competition of the channels for computational resources is described by the condition
$$A_x + A_v + A_a + A_j \le A_m\,, \tag{5.52}$$
where the attention degree A_m is the capacity of resource pool under a given conditions.

These collection of premises form the gist of the *multi-channel speed-accuracy tradeoff*.

As possible neurophysiological arguments in favor of the accepted premises, we want to note the following. Each time humans move the eyes to track a target, retinal image motion and physical motion become decoupled. Therefore during evolution, the brain has acquired the ability to recover, e.g., the speed of moving objects and their spatial position discounting the eye movements. The repertoire of eyes' movements (together with head movements) is rather wide, which enables humans to cope with complex visual information and to extract from it various properties of observed motion. In particular, this repertoire includes the "fixation" and "tracking" modes related to staring at a fixed point in the environment or tracking moving object, saccades—rapid, ballistic movements of the eyes that abruptly change the point of fixation—and smooth pursuit movements (e.g., Purves et al. 2004, Chap. 19). Mixing these modes of eyes' movements within the experiential now makes it possible to accentuate different aspects of physical motion of observed object such as the spatial position of object and its speed, to get some compromise between the accuracy of position and speed perception, etc.

The idea to describe the complexity of visual system functioning in terms of channels, we accepted in our model, is not new. The concept of two channels, sustained and transient, determining visual perception was put forward in the 1970s (e.g., Breitmeyer and Ganz 1976; Legge 1978; Luce 1986; Smith 1995). The transient channel mechanisms respond best to rapid temporal changes in perceived stimuli, the sustained channel mechanisms respond best to steady or slowly varying stimuli. Nowadays evidence for the existence of these channels within the information processing in high-level visual regions of the brain and their interaction has been found based on fMRI measurements (Stigliani et al. 2017, 2019). Moreover, these fMRI investigations captured differences in the channel properties including their nonlinearity on scales from 33 ms to 20 s.

These channels are not independent (e.g., von Grünau 1978). Moreover, during the last decades there has been accumulated firm evidence in favor of that motion perception and eyes' actions in the visual system are organized into the same pathway (e.g., Franz et al. 2000; Spering and Montagnini 2011; Schütz et al. 2011, for a review). Put differently, visual perception of motion and different eyes's actions are two facets of the same origin. It is the gist of the model proposed by Stone and Krauzlis (2003) for the relationship between pursuit eye movements and speed perception. A strong similarity of tracking and perceiving object motion at their early stage were found in many neurophysiological studies. However, at the later state dissociation between them was also found (see also Hughes 2018, for recent studies). It means that pursuit and perception are different processes strongly interacting with each other rather than reducible to one of them (e.g., Hafed and Krauzlis 2010, for a detailed discussion).

A relationship between the perceived position of moving object and its velocity can be analyzed, at least implicitly, via studying the interception of tracking moving targets and the speed judgments governed by eyes' movements. As demonstrated by Goettker et al. (2019), Goettker, Braun and Gegenfurtner (2019), oculomotor behavior and interception performance are related appropriately on the average, whereas trial-to-trial data are scattered uniformly within some region. It argues for that perceived speed and perceived position are determined by processing the visual information via two channels interacting with each other rather than being mutually reducible.

As far as the channel of acceleration perception is concerned, both psychophysical (Gottsdanker 1956; Gottsdanker et al. 1961) and neurophysiological studies (Price et al. 2005, 2006) have revealed that the visual system has no mechanism of the acceleration direct perception (for a review see also Traschütz et al. 2012). So visual system has to infer this information by integrating and comparing speed over time, which also reflected in our model.

The noted facts underlie our model of hierarchical system of channels possessing their own properties and functioning which are based on one common source of visual information.

The stated above premises concerning the hierarchical processing of visual information enable us write directly the relationship between the parameters of the cloud $\mathbb{C}_{\mathbb{R}_{[4]}}$, the amount of computation resources—attention degree A_m—cumulatively allocated for monitoring the object motion, and the time τ of integration of raw visual information. However, before writing the desired expression we need to note the following:.

- As also illustrated in Fig. 5.8, the first pair of phase variables—the object position x and velocity v—are not equivalent to the second pair of phase variables—the object acceleration a and jerk j. The position x and velocity v become accessible to the mind as a result of unconscious processing of sensory information within the immediate now implying strong interaction of the sensory modality and the brain in the information processing. The object acceleration a and jerk j are the results of processing time variations in the output of the integration of visual information within the immediate now. Therefore, accuracy in the perception of the acceleration a and jerk j has to be much lower in comparison with the perception of the position x and velocity v. For this reason, the feasibility of recognizing a temporal change in the acceleration within the experiential now is doubtful.

Therefore in describing the partial speed-accuracy tradeoff within the x- and v-channels we turn to the concept of nonstationary speed-accuracy tradeoff under partial attention (Sect. 5.4.5), whereas dealing with the a- and j-channels we confine the description to the speed-accuracy tradeoff under partial attention (Sect. 5.4.4).

- In Sects. 5.4.3–5.4.5 we considered the situations when for the analyzed quantity D the limit capacity of its perception is given by Weber's law (1.1a) (Ekman's law, Exp. (1.4)). It is justified until the mean value of D becomes rather small and the uncertainty in the perception of this quantity takes some finite value Δ independent of D. Therefore, in order to allow for such small values of analyzed quantity within the given concepts of speed-accuracy tradeoff we turn to the modified Weber's law (1.1b) and replace
$$\kappa D \Rightarrow \kappa D + \Delta^0 \tag{5.53}$$
in the corresponding equations.
- Different channels of information processing form the hierarchical processing of the same raw information with multicomponent output, which is illustrated in Fig. 5.8. In some sense, this hierarchical information processing is one complex process and its division into different channels reflects the corresponding aspects of data analysis rather than the main steps of processing stream. In particular:

  - The integration time τ of the raw visual information actually determines the completeness of the process as a whole, which is quantified by the ratio $(\tau/\tau_d)^{\beta_F}$. For $\tau = \tau_d$ the information processing should be completed and the perception uncertainties have to attain their limit values characterized by Exps. (5.50) (within modification 5.53), provided all the computational resources would be allocated for the corresponding channel individually.
  - When only some part of computational resources is allocated for a q-channel ($q = x, v, a, j$), i.e., the attention degree $A_q < 1$, only some part of working memory is accessible to the q-channel. The memory operations with blocks containing information about the object position, velocity, etc. should be rather similar in properties. Therefore the effect of memory limitation seems to be reduced to just slowing the information processing via the q-channel.
  - The dynamics of mental image of perceived object with respect to the uncertainties in its properties like perceived spatial position, velocity, et. should be governed by the universal laws of inner psychophysics (Sect. 1.3.2).

Therefore for the sake of simplicity we may accept that all the four channels are characterized by the same collection of exponents for their partial speed-accuracy tradeoffs.[43]

Using formula (5.44b) for the partial speed-accuracy tradeoff within the x- and v-channels and formula (5.39b) for the a- and j-channels, we can write the desired relationship as

$$A_m = \left(\frac{T_d}{T}\right)^{\alpha_T}\left[\left(\frac{\kappa_x \ell_x + \Delta_x^0}{\delta_x - \rho_x T \ell_v}\right)^{\alpha_m} + \left(\frac{\kappa_v \ell_v + \Delta_v^0}{\delta_v - \rho_v T \ell_a}\right)^{\alpha_m} + \left(\frac{\kappa_a \ell_a + \Delta_a^0}{\delta_a}\right)^{\alpha_m} + \left(\frac{\kappa_j \ell_j + \Delta_j^0}{\delta_j}\right)^{\alpha_m}\right], \tag{5.54}$$

where we have introduced the exponents

$$\alpha_T = \frac{\beta_F}{(1-\beta_F)\beta_M}, \qquad \alpha_m = \frac{1}{\beta_M}, \tag{5.55}$$

and the parameters $\rho_x \sim 1$ and $\rho_v \sim 1$ can be different.

The given expression for the multi-channel speed-accuracy tradeoff implies that the total attention devoted to monitoring the dynamics of controlled object is *automatically* divided between monitoring individually the spatial position x of object, its velocity v, acceleration a, and jerk j. Only the total attention with the degree A_m is *consciously* controlled by the mind.

It also should be noted that, first, under some conditions the object acceleration can be perceived by the subject directly. For example, in car-driving the subject perceives the car acceleration or braking via the inertial forces. Second, the motion jerk can be determined by subject's intentional actions and, in this case, it play the role of the main control parameter. Pressing the gas or brake pedals in car driving exemplifies this situation (Lubashevsky 2017; Lubashevsky and Morimura 2019). In this case, the jerk is also accessible to the mind directly. Expression (5.54) can take these effects into account explicitly via renormalization of the fractions κ_a and κ_j as well as limit values Δ_a^0 and Δ_j^0. When the acceleration or jerk are perceivable directly the accuracy of their perception must be much higher, which is reflected in the substantially decrease of these quantities in magnitude.

[43] The similarity of the partial speed-accuracy tradeoffs in their exponents does not exclude a complex relationship between the mean values (e.g., ℓ_x and ℓ_v) of the perceived properties of moving object. The a speed-curvature power law relating, in particular, the amplitude of human hand movements and their speed (Lacquaniti et al. 1983, for a review see, e.g., Viviani and Flash 1995; Gribble and Ostry 1996; Huh and Sejnowski 2015; James et al. 2020) exemplifies such complex relations.

5.4.8 Mental Measure $\mathbb{E}^{II}_{m}$: Quasi-entropy of Atomic Eidos

In physics of statistical systems—many-particle ensembles—the notion of entropy takes a central position. This entropy is the energetic measure (in units of temperature) of the uncertainty in the realization of a given macroscopic state of statistical system. In statistical systems any macroscopic state e represents a multitude of N_e different, equiprobable microscopic states which are identical, i.e., indistinguishable in properties at macro-level. Then for a given macroscopic state e Boltzmann's entropy $S_{\mathrm{B}}(e)$ is defined as

$$S_{\mathrm{B}}(e) = k_{\mathrm{B}} \ln N_e$$

where k_{B} is the Boltzmann constant.

Speaking about the mean magnitude $E^{\mathrm{II}}_{\mathrm{m}}$ of mental measure $\mathbb{E}^{\mathrm{II}}_{\mathrm{m}}$ related to monitoring object dynamics, we face up to a situation looking similar. The dimensions of the cloud $\mathbb{C}_{\mathbb{R}_{[4]}}$ represent the uncertainty in the mental-mental embedding of the moving material point. Expression (5.54) relates these dimensions to attention A_m required for monitoring the object motion within this accuracy and Exp. (5.45), in its turn, relates these dimensions to the mental effort required for implementing this monitoring. Combing the noted expressions together and taking into account equality (5.47) and definition (5.55) we can write

$$E^{\mathrm{II}}_{\mathrm{m}} = e^{\mathrm{II}}_{\mathrm{m}} \frac{A_m^{1/\alpha_m}}{(1 - A_m)^{\beta_{E1}}}, \qquad \text{where}$$
$$A_m = \left(\frac{\tau_d}{\tau}\right)^{\alpha_\tau} \sum_{q=x,v,a,j} \left(\frac{\kappa_q \ell_q + \Delta^0_q}{\delta_q - \rho_q \tau \ell_{q^*}}\right)^{\alpha_m} < 1, \tag{5.56}$$

$\rho_a = 0$ and $\rho_j = 0$, and for $q = x,\ v$ the index $q^* = v,\ a$. The proportionality coefficient $e^{\mathrm{II}}_{\mathrm{m}}$, i.e., its meaning and particular magnitude will be elucidated in the next section. Expression (5.56) directly relates the mean magnitude $E^{\mathrm{II}}_{\mathrm{m}}$ of mental measure $\mathbb{E}^{\mathrm{II}}_{\mathrm{m}}$ to the uncertainty in monitoring the material point motion represented by the eidos $\mathbb{C}_{\mathrm{I}}$, i.e., the cloud $\mathbb{C}_{\mathbb{R}_{[4]}} \times \tau$ representing the mental-mental embedding of moving material point into the space $\mathbb{R}_{[4]} \times t$. We call this eidos the *atomic eidos* to accentuate its relation to subject's atomic action at governing the material point dynamics including the monitoring of object motion with the integration time τ of perceived information. So the quantity $E^{\mathrm{II}}_{\mathrm{m}}$ represents space-time aspects of subject's perception.

It should be noted that space-time uncertainties can take only such values that the attention degree A_m be less than unity, $A_m < 1$. Otherwise, these uncertainties are not achievable within human perception. The locus

$$A_m\big(\{\ell_x, \ell_v, \ell_a, \ell_j, \tau\}\big) = 1$$

in the space of phase variables $\mathbb{R}_{\mathbb{C}_\mathrm{I}} = \big\{\{\ell_x, \ell_v, \ell_a, \ell_j, \tau\}\big\}$ corresponds to the limit capacity of human perception and may be regarded as the multi-channel speed-accuracy tradeoff dealing with space-time aspects of human perception. The possibility of reaching this locus seems to be doubtful because some computational resources are inevitably allocated for unconscious (or automatic) processing of perceived information. In our model it is taken into account in the terms of singularity $E^{\mathrm{II}}_{\mathrm{m}} \to \infty$ when $A_m \to 1$.

We introduce the quantity S_m defined as

$$S_\mathrm{m}\big(\mathbb{C}_\mathrm{I}\big) = -E^{\mathrm{II}}_{\mathrm{m}}\big(\mathbb{C}_{\mathbb{R}_{[4]}} \times \tau\big) \tag{5.57}$$

to be called the *quasi-entropy of atomic eidos*. The term "entropy" has been used to emphasis, first, that the quantity S_m is an effort type measure of the uncertainty in subject's perception of the motion state of observed object. Second, the role of this measure in quantifying current state of human actions is similar to the role of Boltzmann's entropy in physics of statistical systems. The prefix "quasi-" has been added to underline that the notions of quasi-entropy and Boltzmann's entropy describe entirely distinct phenomena.

Below we will elucidate the similarity and difference of the quasi-entropy and Boltzmann's entropy.

Similarity

- The quasi-entropy S_m as well as Boltzmann's entropy S_B are certain measures of uncertainty in the relationship

atomic eidos $\mathbb{C}_\mathrm{I}$	$\xleftarrow{\text{mental-mental embedding}}$	moving material point
$\Updownarrow$		$\Updownarrow$
system macro-state	$\xleftarrow{\text{statistical averaging}}$	system micro-state

  Within the mental-mental embedding a multitude of motion states of observed material point are mapped on the eidos $\mathbb{C}_\mathrm{I}$ which are regarded by the subject as identical. In statistical physics the multitude of system micro-states forming one macro-state are identified at the macro-level.

  Like Boltzmann's entropy, the quasi-entropy S_m increases as the corresponding uncertainty grows. Namely, the wider the cloud $\mathbb{C}_\mathrm{I}$, the higher the uncertainty in evaluating the object motion i.e., the large the quantities $\{\delta_x, \delta_v, \delta_a, \delta_j\}$, the lower the effort of motion monitoring, the large the value of S_m.
- As elucidated in the next section, the mean magnitude E_m of mental effort combing
  - the subject's perception of bodily effort related to executing a given action and
  - the mental effort caused by monitoring the motion of controlled object

contributes to the preference of selecting the corresponding action strategy. The probability P_{str} of choosing this action strategy takes the form similar to the Gibbs distribution function in thermodynamic systems, namely, $\ln P_{\text{str}} \propto -(E^{\text{I}}_{\text{m}} + E^{\text{II}}_{\text{m}}) = -(E^{\text{I}}_{\text{m}} - S_{\text{m}}(\mathbb{C}_{\text{I}}))$. In this case, the proportionality coefficient e^{II}_{m}, see Exp. (5.56), quantifies the balance between the perception of bodily effort and the effort of monitoring.

- As follows from the previous item the preference in choosing a given action strategy is higher as the corresponding value of quasi-entropy is larger, other conditions being equal. It qualitatively matches the second law of thermodynamics about the increase of Boltzmann's entropy of isolated system.

Difference

- In order to compare the mathematical forms of the quasi-entropy $S_{\text{m}}(\mathbb{C}_{\text{I}})$ and Boltzmann's entropy $S_{\text{B}}(e)$ we can *formally* treat the volume $\ell_x \times \ell_v \times \ell_a \times \ell_j$ of the space cloud $\mathbb{C}_{\mathbb{R}[4]}$ as a quantity measuring the number of states of material point motion in the physical space. Formally accepting this interpretation of the $\mathbb{C}_{\mathbb{R}[4]}$-cloud volume we just obtain a common base for comparing the quasi-entropy $S_{\text{m}}(\mathbb{C}_{\text{I}})$ and Boltzmann's entropy $S_{\text{B}}(e)$ as functions of micro-level states. Then turning to Exp. (5.57) actually relating the quasi-entropy $S_{\text{m}}(\mathbb{C}_{\text{I}})$ to the dimensions of the space cloud $\mathbb{C}_{\mathbb{R}[4]}$ we see that the quasi-entropy S_{m} as a function of $\ell_x \times \ell_v \times \ell_a \times \ell_j$ does not have a logarithmic form characterizing Boltzmann's entropy.[44]
- The introduced concept of quasi-entropy and Boltzmann's entropy describe phenomena entirely distinct from each other in nature. For a statistical system its macro-level properties are determined by the cumulative contribution of many microscopic states equiprobable in realization. So Boltzmann's entropy as well as entropy of other types is one of the basic quantities characterizing the result of averaging some property over possible micro-level realizations of a given statistical system.

  As far as the quasi-entropy $S_{\text{m}}(\mathbb{C}_{\text{I}})$ is concerned, we deal with the eidos $\mathbb{C}_{\text{I}}$—the entity belonging to the realm of the mind-in-the-self and existing on its own. The eidos $\mathbb{C}_{\text{I}}$—the space-time cloud—is the basic entity the mind directly operates with, it is characterized by internal integrity and is not reduced to any ensemble of motion states of physical object. The uncertainty of the space-time cloud $\mathbb{C}_{\text{I}}$ describes the properties of the body-in-the-self when the subject selects or evaluate some motion state of physical object. All the motion states of physical object that can be related to the space-time cloud $\mathbb{C}_{\text{I}}$ are indistinguishable at the realm of the mind-in-the-self. The properties of the eidos $S_{\text{m}}(\mathbb{C}_{\text{I}})$ are the cumulative result of

[44] This discrepancy, however, does not preclude the use of the term "entropy." Indeed, there is a rich variety of non-equilibrium physical systems whose probabilistic properties are described within the concept of nonextensive statistics. In particular, theories of this type widely turn to the notions of non-additive Tsallis's entropy (e.g., Tsallis 2001, for a review), its generalization (Tsekouras and Tsallis 2005) as well as the family of quantum entropies (e.g., Chehade and Vershynina 2019, for introductory review) whose properties are quite distinct from that of Boltzmann's entropy.

many trial-and-error events when the cloud $S_{\mathrm{m}}(\mathbb{C}_{\mathrm{I}})$ is coupled with some state of object motion. However, because the subject is not able to differentiate between them the experienced properties are treated as random, which endows the mind-in-the-self with a special mechanism of randomness rooted in the nature of space-time clouds.

Finalizing this discussion of the introduced quasi-entropy we want to emphasize that it is the quantity belonging to the realm of the human temporality. Indeed, the space-time cloud $\mathbb{C}_{\mathrm{I}}$—the argument of the function (5.56) specifying the quasi-entropy S_{m}—is an entity of the human temporality whose spatial and temporal extensions are essential and cannot be ignored.

5.5 Mental Measure of Effort: Fusion of Efforts

The present section is devoted to constructing a cumulative measure $\mathbb{E}_{\mathrm{m}} = \mathbb{E}^{\mathrm{I}}_{\mathrm{m}} \oplus \mathbb{E}^{\mathrm{II}}_{\mathrm{m}}$ quantifying the effort fusion. In order to do this we need to go outside the framework of the experiential now and raise our analysis to the level of the complex present, then, return to experiential now again.

In the previous sections we considered the mental measure $\mathbb{E}^{\mathrm{I}}_{\mathrm{m}}$ of bodily effort and the mental measure $\mathbb{E}^{\mathrm{II}}_{\mathrm{m}}$ of monitoring in the framework of the experiential now. Bodily effort as it is experienced by the subject and effort of monitoring represent the feelings of actions and may be conceived of as outputs of two independent channels (Channel I and Channel II) via which the subject experiences the same action. The mutual independence of the two channels implies that their governing neural mechanisms are independent in implementation, i.e., they are based on different neural processes in the brain combined with the corresponding sensory modalities. We have constructed the expressions specifying the mean magnitudes $E^{\mathrm{I}}_{\mathrm{m}}$ and $E^{\mathrm{II}}_{\mathrm{m}}$ of subjective effort related to subject's action on a physical object, Exp. (5.29), and the accuracy in the monitoring of its motion, Exp. (5.56). In these constructions, however, the units of quantifying effort of bodily actions and effort of monitoring remain mutually independent. It is reflected in the presence of the coefficients $e^{\mathrm{I}}_{\mathrm{m}}$ and $e^{\mathrm{II}}_{\mathrm{m}}$ in Exp. (5.29) and Exp. (5.56) which may *independently* take arbitrary positive values. Therefore, the fusion of the two mental measures $\mathbb{E}^{\mathrm{I}}_{\mathrm{m}}$ and $\mathbb{E}^{\mathrm{II}}_{\mathrm{m}}$ cannot be described within the analysis confined to the experiential now only.

Previously, Sect. 5.2.3, we supposed that the resulting mental measure $\mathbb{E}_{\mathrm{m}}$ of the cumulative effort can be approximated by the algebraic sum for the mean magnitudes, i.e., $E_{\mathrm{m}} = E^{\mathrm{I}}_{\mathrm{m}} + E^{\mathrm{II}}_{\mathrm{m}}$ when the coefficients $e^{\mathrm{I}}_{\mathrm{m}}$ and $e^{\mathrm{II}}_{\mathrm{m}}$ are interrelated in a certain way. Below we will substantiate this feature, clarify the underlying mechanisms, and specify the relationship between the coefficients $e^{\mathrm{I}}_{\mathrm{m}}$ and $e^{\mathrm{II}}_{\mathrm{m}}$. This relationship actually determines the common units of measuring the subjective effort of bodily actions and the mental effort of monitoring.

5.5.1 Effort Fusion and Top-Down Causal Relations

In Sect. 3.2.5 we introduced the notion of E-consciousness which is characterized by that the mind recognizes observed phenomena but cannot analyze them. Put differently, the subject just experiences the observed phenomena without their deliberate analysis. We posits that the effort fusion occurs via a complex E-conscious mechanism. The mathematical description of this mechanism with a certain generalization will be given in Sect. 5.6.

Within the experiential now the subject perceives effort related to the bodily actions and effort caused by the monitoring separately. In the connected past these efforts are fused into one cumulative effort because, first, working memory is limited in capacity. Second, in selecting an action strategy for implementation the subject deals with the effort integrated over all its constituent actions and effort division into different types is just redundant. In the connected future efforts attributed to anticipated actions are based on information accumulated in memory on long-term scales and so these efforts are of the cumulative form.

Before passing directly to the discussion of fusion mechanism, we want to note that there are two aspects of this process. One of them is the fusion itself which implies that when two cloud-type entities fuse, their result—also a cloud-type entity—may be of another kind than the constituent components causing its emergence. The other aspect concerns the channels understood as the collections of neural mechanisms underlying the effort fusion. These channels are integrated, meaning that the corresponding neural mechanisms are combined into one larger group.

In our account of the effort fusion we accept the following. The initial incomparability of the two types of cognitive effort implies that for describing their cooperative effects we need 2D space of effort magnitudes

$$\mathbb{R}_{\{e_1 e_2\}} = \mathbb{R}_{e_1} \times \mathbb{R}_{e_2} \,,$$

where $\mathbb{R}_{e_1}$ and $\mathbb{R}_{e_2}$ are 1D spaces of magnitudes representing the experienced efforts separately. In the realm of the body-in-the-self these efforts are represented individually as distributions $\mathbb{E}^{\mathrm{I}}_{\mathrm{m}}(e_1)$ and $\mathbb{E}^{\mathrm{II}}_{\mathrm{m}}(e_2)$ in the corresponding 1D spaces (Sect. 5.4.1). It should be reminded that these distributions are supposed to be of a certain fixed form. This form is the characteristic feature of the two types of effort simultaneously possessing aspects belonging to the body-in-the-self and the mind-in-the-self. These aspects must be closely intertwined via the self.

Without fusion, effects implying the simultaneous contribution of subjective efforts require, for their description, the product

$$\mathbb{E}^{\mathrm{I,II}}(e_1, e_2) = \mathbb{E}^{\mathrm{I}}_{\mathrm{m}}(e_1) \cdot \mathbb{E}^{\mathrm{II}}_{\mathrm{m}}(e_2)$$

representing the 2D image of the two-types of effort in the realm of the body-in-the-self. Within the same scope the effort fusion may be written as a certain function

$$e_{12} = F_\omega(e_1, e_2) \tag{5.58}$$

mapping the plane $\mathbb{R}_{\{e_1e_2\}}$ on 1D space $\mathbb{R}_{e_{12}}$. Here ω stands for parameters of function (5.58) and the function $F_\omega(e_1, e_2)$ will be referred to as the *kernel of fusion*. The 1D function $\Psi(e_{12})$ defined in $\mathbb{R}_{e_{12}}$ as

$$\Psi(e_{12}) = \iint\limits_{\mathbb{R}_{\{e_1e_2\}}} de_1 de_2 \delta\big[e_{12} - F_\omega(e_1, e_2)\big]\mathbb{E}^{\mathrm{I,II}}(e_1, e_2) \tag{5.59}$$

(here $\delta[\ldots]$ is the Dirac δ-function) plays the role of the source in the body-in-the-self for the fusion-induced effort $\mathbb{E}_\mathrm{m}$. The distribution $\Psi(e_{12})$ is not necessarily of the required form $\mathbb{E}_\mathrm{m}(e_{12})$, so, at the final stage of effort fusion the function $\Psi(e_{12})$ is converted into $\mathbb{E}_\mathrm{m}(e_{12})$ (for details see, Sect. 5.6.1).

It should be accentuated that the kernel of fusion F_ω is not given initially and results from the skill the subject acquires on long-term scales dealing with a given system, we will return to this issue below.

When the experienced magnitudes of efforts are retained in working memory, they are immediately converted into a single cloud $\mathbb{E}_\mathrm{m}(e_{12})$ of magnitudes as it has been described above, which represents the cumulative effort of action implementation. Section 5.6.2 will present the mathematical description of this conversion in more details. Here for short we will use the symbolical form of effort fusion $\mathbb{E}_\mathrm{m} = \mathbb{E}^\mathrm{I}_\mathrm{m} \oplus \mathbb{E}^\mathrm{II}_\mathrm{m}$, where the cloud-type measures $\mathbb{E}^\mathrm{I}_\mathrm{m}$ and $\mathbb{E}^\mathrm{II}_\mathrm{m}$ quantify the output of Channels I and II.

Speaking about the human temporality of the complex present, the effort fusion may be conceived of as a continuous process taking place at the "boundary" between the experiential now and the connected past. In a simplified representation, a new instant of fusion-induced effort $\mathbb{E}_\mathrm{m}$ is

- generated via the fusion of currently *experienced* efforts $\mathbb{E}^\mathrm{I}_\mathrm{m}$ and $\mathbb{E}^\mathrm{II}_\mathrm{m}$ within the experiential now,
- retained in working memory as a certain fixed form distribution $\mathbb{E}_\mathrm{m}(e_{12})$ in the space of effort magnitudes $\mathbb{R}_{e_{12}}$ with some mean value E_m and some width Δ_m;
- given to the mind-in-the-self as an indivisible entity with internal integrity and, then,
- attributed by the mind-in-the-self to the just implemented action belonging to the connected past and adjacent to the experiential now.

In terms of mean magnitudes the effort fusion may be represented as a certain mapping

$$\{E^\mathrm{I}_\mathrm{m}, E^\mathrm{II}_\mathrm{m}\} \mapsto E_\mathrm{m} \tag{5.60a}$$

or, in the functional form,

$$E_\mathrm{m} = \mathfrak{F}\big(E^\mathrm{I}_\mathrm{m}, E^\mathrm{II}_\mathrm{m}\big)\,. \tag{5.60b}$$

According to the accepted concept of cloud-type measure of cognitive effort (Sect. 5.4.1), the width $\Delta_{\mathrm{m}}(E_{\mathrm{m}})$ of the cloud $\mathbb{E}_{\mathrm{m}}(e_{12})$ in 1D e_{12}-space is determined by the mean magnitude E_{m} of the cumulative effort rather the widths $\Delta^{\mathrm{I}}_{\mathrm{m}}$, $\Delta^{\mathrm{II}}_{\mathrm{m}}$ of its constituent components.

Therefore relationship (5.60) is at the heart of the mathematical description of effort fusion. If the function $\mathfrak{F}(\ldots)$ is known, then, together with the dependence $\Delta_{\mathrm{m}}(E_{\mathrm{m}})$ it enables us to reconstruct the fusion $\mathbb{E}_{\mathrm{m}} = \mathbb{E}^{\mathrm{I}}_{\mathrm{m}} \oplus \mathbb{E}^{\mathrm{II}}_{\mathrm{m}}$.

Conversion (5.60) should be implemented unconsciously or, speaking more strictly, automatic. The matter is that conversion (5.60) does not emerge via local mechanisms confined to the experiential now. Although it is *implemented* within the experiential now; the mechanism *responsible* for its *emergence* belongs, at least, to the level of the complex present. Indeed, the particular details of how two quantities $\{E^{\mathrm{I}}_{\mathrm{m}}, E^{\mathrm{II}}_{\mathrm{m}}\}$ are fused into one E_{m} can be justified only based on the skill of implementing action strategies in controlling the object dynamics. This skill is acquired during numerous trial-by-trial actions and their retrospective evaluation on scales larger than the duration of complex present.

So, even if at the stage of learning and skill acquisition the comparison of efforts related to bodily actions and the monitoring was conscious, then their fusion becomes automatic (unconscious). In other words, turning to the triple-mode view[45] on conscious-unconscious relations proposed by Moors (2009, 2013) we have to categorize conversion (5.60) as a top-down unconscious (automatic) relationship. Whence it follows that conversion (5.60) reflects:

- *the general features of the given type actions* as a whole ensemble. Put differently, the dependence $E_{\mathrm{m}} = \mathfrak{F}(E^{\mathrm{I}}_{\mathrm{m}}, E^{\mathrm{II}}_{\mathrm{m}})$ in its emergence and particular form is the generic property of this type actions including some optimality of human actions. The given optimality can be recognized and understood only at the level of the complex present. Conversion (5.60) exemplifies top-down causal relations where the properties of higher order fragment of the human temporality—the complex present in the case under consideration—directly determine some properties of lower-level fragment—the experiential now.
- *loss of universality*, i.e., dependence on particular details of given type human actions.[46] Naturally, it concerns also actions with different objects and in different environment.

We suppose that this type top-down automatic relations underlie the concepts of *unconscious goal pursuit* (e.g., Bargh 1989; Bargh et al. 2001; Eitam et al. 2008;

[45] We already discussed Moors' triple-mode account of conscious-unconscious transitions in Sect. 4.4.3.

[46] In physics such relations are often called *constitutive equations* or *constitutive relations*. They deal with meso-level approximations of relationships between two physical quantities that are specific to a given material or substance. In other words, constituent equations are based on particular micro-level physical processes which remain hidden within a meso-level description, nevertheless, give rise to relations between meso-level quantities. In this sense conversion (5.60) is a result of long-term evolution (on scales of the complex present) involving multitude of conscious-unconscious transitions caused by learning process before the relationship $E_{\mathrm{m}} = E_{\mathrm{m}}(E^{\mathrm{I}}_{\mathrm{m}}, E^{\mathrm{II}}_{\mathrm{m}})$ gets its stable form.

Hassin, Bargh and Zimerman 2009; Hassin, Aarts, Eitam, Custers and Kleiman 2009; Gollwitzer et al. 2009; Custers and Aarts 2010; Aarts and Custers 2012; Custers et al. 2012).

Summarizing the present discussion, we come to the conclusion that the fusion $\mathbb{E}_{\rm m} = \mathbb{E}^{\rm I}_{\rm m} \oplus \mathbb{E}^{\rm II}_{\rm m}$ of

- *experienced* effort of bodily executed action quantified by the cloud-type measure $\mathbb{E}^{\rm I}_{\rm m}$ and
- *experienced* effort of monitoring the action results quantified by cloud-type measure $\mathbb{E}^{\rm II}_{\rm m}$

is characterized by the following features.

- The effort fusion is implemented at the boundary between the experiential now and the connected past.
- The effort fusion should be regarded as a E-conscious phenomenon. In particular, the details of how conversion (5.60) is implemented and the width $\Delta_{\rm m}(E_{\rm m})$ of the cloud-type effort $\mathbb{E}_{\rm m}$ is related to its mean magnitude $E_{\rm m}$ are hidden from the mind, only the final result $\mathbb{E}_{\rm m}$ becomes accessible to the mind as a whole entity.
- The fusion-induced effort quantified by the cloud-type measure $\mathbb{E}_{\rm m}$ is of another type than the constituent components; its type can be categorized in term of *effort-as-evaluated*. The fusion-induced effort is related to how

  (i) the information about effort related to implementing a given type of atomic actions is aggregated in the realm of the body-in-the-self,
  (ii) the mind attributes the result to a given atomic action in the realm of the mind-in-the-self,
  (iii) the mind orders atomic actions according to the effort of their implementation such that for any two pairs of efforts $\{E^{\rm I}_{\rm m,1}, E^{\rm II}_{\rm m,1}\}$ and $\{E^{\rm I}_{\rm m,2}, E^{\rm II}_{\rm m,2}\}$ one of the three probabilistic relations (Sect. 5.4.1)

$$\mathbb{E}_{\rm m,1} \prec \mathbb{E}_{\rm m,2}\,, \qquad \mathbb{E}_{\rm m,1} \simeq \mathbb{E}_{\rm m,2}\,, \qquad \mathbb{E}_{\rm m,1} \succ \mathbb{E}_{\rm m,2}\,,$$

  holds; here the cloud-type measures $\mathbb{E}_{\rm m,1}$ and $\mathbb{E}_{\rm m,2}$ represent subject's evaluation of the cumulative effort of the two atomic actions.

- The effort fusion is a complex process of achieving the tradeoff in feeling efforts within the experiential now and evaluating these efforts on long-term scales. In the present book we suppose that this tradeoff is achieved and the fusion kernel (5.58) is self-consistent in relating experience to evaluation. For this reason we will treat the relationship $E_{\rm m} = \mathfrak{F}\big(E^{\rm I}_{\rm m}, E^{\rm II}_{\rm m}\big)$ (Exp. 5.60b) as a given phenomenological regularity, which can be reconstructed based on experimental statistical data using, e.g., the aforementioned relations.

Leaping ahead, we want to note that conversion (5.60) underlies the possibility of introducing a common effort-type unit for measuring subject's experience of currently implemented action. It makes two types effort comparable, first, at the level of the complex present and, then, at the level of the experiential now via top-down relations in the realm of the mind-in-the-self.

5.5.2 *Fatigue and Fusion-Induced Effort*

In order to clarify the physiological and psychological roots of the proposed concept of effort fusion let us turn to psychological phenomena where accumulation of effort plays a crucial role. As a characteristic example, we may consider fatigue induced by physical and mental exertions. Indeed, on one hand, such fatigue is the result of effort accumulation during a certain time interval, thereby, it characterizes the averaged properties of the corresponding kind effort. On the other hand, fatigue is related to a critical state of the body-in-the-self and, in this case, the effort properties have to become pronounced. Put differently, we intend to discuss the details of fatigue relevant to subject's actions within the experiential now whose properties are averaged within the complex present or, even, outside it. For a general analysis of the relationship between effort and fatigue a reader may be referred to Hockey (2011, 2013), Kanfer (2011).

Fatigue is a rather general notion and integrates in itself many aspects. How these aspects are aggregated into classification categories—*fatigue dimensions*—depends on particular phenomena under consideration or goals underlying these classifications (e.g., Dittner et al. 2004; Abd-Elfattah et al. 2015; Phillips 2015). The classification diversity is partly caused by the fact that there is no single mechanism responsible for fatigue being a complex multi-factorial phenomenon; a reader may be referred to Gandevia (2001), Enoka and Duchateau (2008), Enoka et al. (2011), Boyas and Guével (2011) for a review of mechanisms governing muscle fatigue and to Boksem and Tops (2008), Dongen et al. (2011), Van der Linden (2011), Lorist and Faber (2011), Shenhav et al. (2017), Qi et al. (2019) for that of mental fatigue (fatigue induced by mental exertion).

Within our constructions, we consider a special case of human actions where fatigue manifests itself as:

- performance decrements induced solely by accumulation of effort caused by actions executed over relatively long time interval;[47]
- the impairment of endurance performance—the ability to perform physical or mental work over an extended period of time;
- a cooperative phenomenon combining in itself the results of simultaneous physical and mental exertions;
- complex phenomena with internal integrity.

[47] Fatigue phenomena induced by some external, depressive circumstances, aversive emotions, or medical causes are not included into consideration.

Actually the last item in the given list reflects the essence of our interest in discussing fatigue phenomena in the present book. The preceding items specify the characteristic features of the fatigue phenomena involved into consideration.

Below we will summarize psychological and neurophysiological data to elucidate the possibility of whether simultaneous physical and mental exertions can fuse together and induce fatigue with internal integrity. This integrity implies that this fatigue is experienced by the subject without its conscious division into the constituent components. Going in this way we intend to clarify two interrelated issues:

1. whether the know properties of physical fatigue and mental fatigue as separated phenomena are similar and, thereby, admit their union within one concept,
2. whether there are known psychological phenomena whose properties may be treated as arguments in favor of the fusion of different kinds of effort into one entity with internal integrity.

As far as **Issue 1** is concerned, neural processes associated with fatigue induced by mental exertion (Wang et al. 2016) as well as physical exertion (Taylor et al. 2000) take place in brain areas upstream of the primary motor cortex. It argues for a plausible neural relationship between the two kinds of fatigue (e.g., Newsholme et al. 1992; Di Giulio et al. 2006).

We confine ourselves to discussing the impairment of endurance performance which may be regarded as the main fatigue dimension in the case of human goal-oriented behavior within an extended period of time.

Concerning the effects of muscle fatigue on bodily executed actions, the impairment of endurance performance is well pronounced. For details a reader may be referred, e.g., to Coyle (1995) and Morris et al. (2008).

Mental fatigue confined to the realm of mental phenomena is also a well known phenomenon (e.g., Boksem and Tops 2008, for a review) reflecting in the impairment of a verity of mental processes like sustaining attention and action monitoring (e.g., Van der Linden et al. 2003; Lorist et al. 2005; Boksem et al. 2005, 2006).

Comparing mental and physical fatigue with respect to their effects on bodily executed actions, we note the following. Numerous experimental investigations (e.g., Marcora et al. 2009; Pageaux et al. 2013, 2015; Martin et al. 2014; Rozand et al. 2014, see al so Pageaux and Lepers 2016, 2018; Cutsem et al. 2017; Martin et al. 2018; McMorris et al. 2018 for a review) demonstrated that mental and physical exertions resemble each other in impairing subsequent endurance performance. Besides, the impairment of endurance performance induced by mental exertion is implemented via (*i*) a substantial increase in human sensitivity to experienced effort and (*ii*) impairment of motivation or the willingness to exert effort. In turn, physical activities can speed up mental fatigue (Tanaka et al. 2016; Xu et al. 2018), e.g., muscle fatigue concurrently degrades attentional resources (Lorist et al. 2002; Stephenson et al. 2019).

A decrease in endurance performance caused by mental exertion occurs via accumulation of a special biochemical compound—cerebral adenosine in the brain (e.g., Rozand et al. 2014; Martin et al. 2018, for a review).

Therefore, we may suppose that physical and mental exertion gradually change the physiological state of the body. In particular, this change is reflected in the subject's sensitivity to the consumption of physiological resources required for action implementation. This sensitivity is a function of subject's physiological state and grows in time provided the exertion is sustained. These changes in the body state, on the one hand, can be perceived by the subject, on the other hand, the underlying neurological processes are nonconscious. For this reason, the two kinds of fatigue and the induced decrease in endurance performance should be categorized as E-conscious type phenomena.

Within **Issue 2** we want to discuss two examples arguing for the emergence of new effort properties exhibited by human actions combining in their implementation physical and mental constituent components.

First, it is cognitive-motor dual-tasks involving physical and mental actions concurrently implemented. It turns out that the subjective effort related to such tasks and the resulting fatigue effects are not reduced to just "algebraic sums" of the individual components (e.g., Chang et al. 2012; Mehta and Parasuraman 2013; Mehta and Agnew 2014; Cullen and Agnew 2016; Guzmán-González et al. 2020, for details). Such emergent properties should depend not only on their own properties of physical and mental exertions but also on their interrelation (e.g., Diekfuss et al. 2016). According to experiments by Hachard et al. (2020) studying the influence of cognitive exertion on balance control, because of mental fatigue, postural regulatory mechanisms can cease to be automatic. The latter affects human perception of bodily actions essentially and can give rise to effortless-effortful transitions. These and similar cognitive aspects self-regulation are also discussed, e.g., by Brick et al. (2016).

The second example is a phenomenon called *ego-depletion*[48] which is parallel but overlapping with the mental fatigue discussed above. Ego-depletion is a special type of mental fatigue (see, e.g., Inzlicht and Berkman 2015; Clarkson et al. 2016; Giboin and Wolff 2019, for justifying this statement). Its characteristic feature is that the capacity to exert mental effort depends on a *global* resource whose depletion leads to impairing the performance in subsequent tasks requiring self-control (Hagger et al. 2010; Giboin and Wolff 2019, and Sosa and Howell 2020 for an introduction to the problem).

Because ego-depletion as a cognitive phenomenon was put forward fairly recently and concerns subtle effects in human behavior, up to now its properties are the matter

[48] The concept of ego-depletion as an aspect of the self-control was put forward by Baumeister et al. (1998), Muraven et al. (1998).

of on-going debates.[49] Our interest in ego-depletion within the present book is related to its novel account elaborated by Martela et al. (2016). This account unites[50]

- the *strength model* of self-control (Baumeister et al. 2000, 2007, see also Baumeister et al. 2018, for a review) taking central position in the concept of ego-depletion and
- the *self-determination theory* (Deci and Ryan 1985, 2000; Ryan and Deci 2000, see also Lavrusheva 2020; Legault 2020b, for a review) which operates with subjective vitality.[51]

The self-determination theory discriminates between (e.g., Legault 2020b)

- the *self-control* based on controlled (non-self-determined) motivations,
- the *autonomous self-regulation* based on autonomous (self-determined) motivations.

Autonomous self-regulation is less depleting with respect to subjective vitality (e.g., Muraven et al. 2007; Muraven 2008; Muraven et al. 2008; Legault and Inzlicht 2013) which reflects the fact that autonomy itself is a motivation (Deci and Ryan 2000; Ryan and Deci 2000, see also Legault 2020a for introductory). Moreover, experiencing autonomy in actions is associate with pleasant emotions (e.g., Ryan and Deci 2008). Naturally, subject's actions under particular conditions can be characterized by some degree of autonomy, which endows various instantiations of ego-depletion with complexity in property.

We relate the autonomous self-regulation to the automaticity in subject's actions. Indeed, such actions in their reproduction should be characterized by stable patterns

[49] These debates reflect several problems related to the concept of ego-depletion. The diversity of neurophysiological mechanisms underlining fatigue appearance together with related self-regulation processes doubts the possibility of introducing a global, i.e., universal source of self-control (Evans et al. 2016). To cope with this diversity an alternative account called the *process model* (Inzlicht and Schmeichel 2012) focusing on changes in motivation and attention has been proposed (Inzlicht and Berkman 2015; Inzlicht et al. 2014, for a review) as well as extended concept of resources is discussed, e.g., by Evans et al. (2016), Loschelder and Friese (2016), Baumeister and Vohs (2016), Baumeister et al. (2018). The latter is partly caused by the failure in finding a decrease in blood glucose levels after exerting self-control (e.g., Finley et al. 2019; Zahn et al. 2019).

The concept of ego-depletion also has met a number of challenges caused by its "fragility," i.e., reproducibility or irreproducibility depending on variety of factors including subject's emotions, cognitive biases like believing that willpower relies on a limited resource, etc. (e.g., Carter et al. 2015; Dang et al. 2017; Job and Walton 2017; Friese et al. 2018; Wenzel et al. 2019). It indicates a certain crisis in the theory of ego-depletion and raises a question about new research approaches to understanding the concept of ego-depletion and describing self-control (Lurquin and Miyake 2017; Dang and Hagger 2019; Inzlicht and Friese 2019; Englert 2019).

For the current state of the art in the theory of ego-depletion and discussion of the arguments for and against this notion a reader may be addressed to Muraven et al. (2019).

[50] The possibility of this integration is due to the two approaches actually dealing with the same aspect of human behavior (Ryan and Deci 2008).

[51] Subjective vitality is understood as the experience of having positive energy available to subject's intentional actions (Ryan and Frederick 1997). It plays the role of energetic-type resource. For elucidating the nature of subjective vitality and mechanics of its emergence a reader may be addressed to Martela et al. (2016).

which, maybe, formed previously under the direct control of the mind but now are fixed in the inner collection of action scenarios. So the subject responding to a given situation can just reproduce one of the pre-prepared scenarios without its detailed analysis by the mind.

In Sect. 3.2.5 we introduced the notion of E-consciousness—the level of consciousness where the mind just receives the final information about perceived stimuli from nonconscious processes but does have access to the details in information processing. In this case, autonomy of human actions and their automaticity are two facets of one phenomenon and may be regarded as entities belonging to the interface between the mind-in-the-self and the body-in-the-self (cf. Schütz et al. 2020, concerning the concept of the self and the ego-depletion phenomenon).

In the case of human actions governed by autonomous (automatic) process of self-regulation, all the particular factors must be merged into one entity with internal integrity from the standpoint of the mind. If, in this case, the cumulative effect gives rise to pleasant feelings or emotions, the subject perceives these actions as effortless and not inducing the depletion of subjective vitality. Oppositely, in the case of self-control, when the subject recognizes the individual contributions of the involved factors separately, the corresponding actions are clearly perceived as associated with subjective effort, and their implementation can induce mental fatigue perceived as ego-depletion—the depletion of subjective vitality.

Whether two successive actions are governed by self-control or autonomous self-regulation depends on a number of particular conditions. First, it explains that specific cases where the effect of ego-depletion was not detected cannot be used as arguments against the existence of ego-depletion. Second, these cases are arguments for E-consciousness of autonomous (automatic) actions governed by self-regulation when all the factors—e.g., the physical and mental constituent components—are experienced by the subject as a single entity with internal integrity.

The integral accounts of self-regulation and ego-depletion by Ampel et al. (2016), Berkman et al. (2016), Francis and Inzlicht (2016) also single our similar aspects as the pivot points for generalization.

Summarizing this discussion, we may note that

- cognitive effort of a bodily executed action with a physical object quantified individually by the cloud-type measure $\mathbb{E}^{\mathrm{I}}_{\mathrm{m}}$ and
- mental effort associated with monitoring the action results quantified individually by the cloud-type measure $\mathbb{E}^{\mathrm{II}}_{\mathrm{m}}$

as the constituent components of subject's experience

- can be related to fatigue induced by mental and physical exertions with cross-influence,
- merge together such that the subject perceives their combination as a single entity with internal integrity belonging to the level of E-consciousness.

In the case under consideration, this cross-influence of exertions means that the induced fatigue exhibits properties of both muscle fatigue and cognitive fatigue simultaneously, even if under some conditions only bodily execution of actions or monitoring their results separately determines subject's tiredness. The fusion of the two types of subjective effort can emerge after a certain time required for the subject to learn via trial-and-error practice the given repertoire of actions and to acquire the skill.

5.5.3 *Time-to-Fatigue as a Measure of Fusion-Induced Effort*

This subsection is devoted to constructing the common measurement units to be used in quantitative comparison of the subjective effort of bodily executed actions, the effort of monitoring the action results, and the corresponding fusion-induced effort. Fatigue will take the central position in our account.

Let us turn to the concept of *pacing* used in sport psychology. Pacing[52] characterizes non-reflex exercises taking a relatively long time where effort of required intensity is distributed over time more or less uniformly to facilitate task accomplishment (e.g., Edwards and Polman 2012, 2013; Smits et al. 2014; Brick et al. 2016, for a review). Actually we regard the cost-benefit type properties attributed to atomic actions as well as action strategies as a result of their paced performance on relative long time scales. For this reason we confine our discussion to fatigue and perceived effort related to paced performance.

Nowadays, debates about understanding perceived effort and the underlying neurophysiological mechanisms are still ongoing (Razon et al. 2012; Halperin and Emanuel 2019; Balagué et al. 2020, for a review). It may be explained by the multidimensional structure of effort perception (Tenenbaum et al. 1999, also Razon et al. 2012 for a review) including:

- the *sensory-discriminative* dimension,
- the *cognitive-evaluative* dimension,
- the *motivational-affective* dimension,

which can be perceived distinctly during exercise and, thus, cannot be aggregated into one dimension in the general case (Hutchinson and Tenenbaum 2006).[53] In our

[52] Along with other meanings, the verb "pace" means to walk first in one direction then in another many times, especially because you are nervous (e.g., Longman Dictionary of Contemporary English, 5th edition). Like in sport psychology, we use the term *pacing* with respect to human goal-oriented actions within the complex present to emphasize that the subject has to repeat actions of a given repertoire many times for learning and acquiring the skill in their implementation.

[53] Moreover, attention (e.g., Hutchinson and Tenenbaum 2007; Stanley et al. 2007; Balagué et al. 2012), other cognitive factors (Tenenbaum 2001; Tenenbaum and Hutchinson 2007), as well as transitions between conscious-unconscious stages of paced exercises (e.g., Ekkekakis 2009a, b; Edwards and Polman 2013; Micklewright et al. 2016) affect the contribution of these dimensions to effort perception. This diversity of factors in perceived effort can be explained turning to that the sensation of fatigue must be the conscious output of central governor control mechanisms that regulate the exercise behavior to ensure the protection of whole body homeostasis (Noakes et al. 2005; Noakes 2012; Venhorst et al. 2018c).

consideration of subjective effort, we actually deal with the sensory-discriminative and cognitive-evaluative dimensions.

Following the approaches of Abbiss et al. (2015), Pageaux (2016), Venhorst et al. (2018b) to understanding the notions of effort and exertion, we draw a distinction between them. Exertion is regarded as the accumulation of effort which leads to fatigue. Moreover, perceived exertion responsible for exercise-induced fatigue admits interpretation in terms of a real dynamic processes exhibiting short- and long-term fluctuations on scales from 1 s to 100 s (or even longer) (Aragonés et al. 2013; Balagué et al. 2020).

In order to quantify the two types of subjective effort as they are evaluated on long-term scales, we turn to a fundamental property of exercise performance—the hyperbolic relationship between work rate and *time-to-exhaustion* also called *time-to-fatigue*. For the first time this relationship was reported by Monod and Scherrer (1965), nowadays, there is a vast literature about it including possible generalizations and the underlying neural mechanisms. For details a reader may be addressed to Fukuba et al. (2003); de Lucas et al. (2013); Burnley and Jones (2016); Poole et al. (2016); Nicolò et al. (2019).

Naturally, this hyperbolic relationship must become more complicated when the time-to-fatigue is related to perceived effort rather than work rate measured in physical units. Nevertheless, e.g., the phenomenon called *fatigue-induced spontaneous termination point* as an non-equilibrium phase transitions (Hristovski and Balagué 2010) can be a suitable tool for quantifying subjective effort. Indeed, "termination emerges as a consequence of the spontaneous loss of intentional or volitional task stability. That is, as a result of the abrupt dissolution of volition to continue the task itself that was maintained through the coupling between psychobiological components" (Balagué et al. 2020, p. 961). So, measuring the time to the fatigue-induced terminal point, it becomes possible to quantify subjective effort in temporal units (cf. Balagué et al. 2015; Vázquez et al. 2016; Venhorst et al. 2018a). In this connection, we want to note the dual-task experiments by Holgado et al. (2019) challenged the idea that self-paced exercise is regulated by top-down processing. Below we will refer to this result in constructing the measure of fusion-induced effort.

We also want to note that the concept of cognitive cost in evaluating actions in terms of, e.g., ego-deletion can be understood in a wider scope including behavioural economics (Kool and Botvinick 2013, 2014; Shenhav et al. 2017; Kool and Botvinick 2018).

The model to be accepted for constructing the description of effort fusion comprises the following propositions.

F1: The effort fusion is an E-conscious process emerged via perceiving numerous actions related to a given type goal-oriented behavior and evaluating their results. It implies:

- the aggregation of information on long-term scales (exceeding the duration of complex present) which is required for learning and skill acquisition,
- the membership of actions to a certain repertoire $\mathfrak{A} = \{a\}$ characterizing the given goal-oriented behavior.

So the effort fusion as the emerged regular process reflects the properties of the action repertoire $\mathfrak{A}$ rather the properties of atomic actions or action strategies individually.

F2: The bodily execution of actions and the monitoring of their results are implemented via different neural mechanisms—Channel I and Channel II—and require individual energy-type vital resources. On the one side, subject's actions gradually deplete these resources. On the other side, the resources are also restored in some way, such that in the realm of the body-in-the-self their amount is kept, on average, at the levels $\mathcal{E}^{\mathrm{I}}$ and $\mathcal{E}^{\mathrm{II}}$, respectively.

- *On long-term scales*, due to the adaptation ability, the resources can be reallocated between the two channels such that the levels $\mathcal{E}^{\mathrm{I}}$ and $\mathcal{E}^{\mathrm{II}}$ meet some conditions determined by the action repertoire $\mathfrak{A}$.
- *On short-term scales* the capacity of resource reallocation, i.e., their integration is limited and depends on the efficiency of currently executed actions.[54]

F3: The resource allocation $\{\mathcal{E}^{\mathrm{I}}, \mathcal{E}^{\mathrm{II}}\}$ is *balanced* such that without resource restoration the paced execution of actions from the repertoire $\mathfrak{A}_a$ depletes both the components simultaneously. This condition implies a certain self-consistency of subject's actions with respect to the compromise between the resource use and the efficiency of achieving the goal.

F4: Let us single out a certain family $\mathfrak{K} = \{a\}$ of actions in the set $\mathfrak{A}$ that is specified as follows. If in the realm of the body-in-the-self the execution of a given action a is characterized by the resource depletion with rates r^{I}_a and r^{II}_a, then

$$\mathfrak{K} = \left\{a : \; \frac{r^{\mathrm{I}}_a}{r^{\mathrm{II}}_a} = \frac{\mathcal{E}^{\mathrm{I}}}{\mathcal{E}^{\mathrm{II}}}\right\}. \tag{5.61}$$

The action family $\mathfrak{K}$ will be called the *kernel* of the repertoire $\mathfrak{A}$. The implementation of actions from the kernel $\mathfrak{K}$ matches the balanced use of resources.

F5: The accepted self-consistency of the relationship between efforts-as-experienced and efforts-as-evaluated causes us to suppose that an action

- meeting the balanced use of resources within the body-in-the-self
- must be also experienced-and-evaluated as an action meeting the balanced use of resources within the mind-in-the-self.

In other words, definition (5.61) of the kernel $\mathfrak{K}$ in terms of depletion rates $\{r^{\mathrm{I}}_a, r^{\mathrm{I}}_a\}$ should admit a similar interpretation in terms of the mean magnitudes $\{E^{\mathrm{I}}_{\mathrm{m}}(a), E^{\mathrm{II}}_{\mathrm{m}}(a)\}$ of the corresponding subjective efforts. Namely, definition (5.61) and the condition

$$\frac{E^{\mathrm{I}}_{\mathrm{m}}(a)}{E^{\mathrm{II}}_{\mathrm{m}}(a)} = \Lambda \quad \text{for any } a \in \mathfrak{K}, \tag{5.62}$$

where Λ is some constant, have to be equivalent representations of the kernel $\mathfrak{K}$.

[54] Below we will note two phenomena where similar integration is the case.

Due to the independence of Channels I and II in neural mechanisms governing the subject's perception of the two types of effort, expressions (5.61) and (5.62) are equivalent if the mean magnitudes of subjective efforts and the rates of resource depletion are related as

$$E^{\mathrm{I}}_{\mathrm{m}} = \varkappa^{\mathrm{I}} \cdot \left(r^{\mathrm{I}}\right)^{\beta_r} \qquad \text{and} \qquad E^{\mathrm{II}}_{\mathrm{m}} = \varkappa^{\mathrm{II}} \cdot \left(r^{\mathrm{II}}\right)^{\beta_r}, \tag{5.63}$$

where the coefficients $\varkappa^{\mathrm{I}}$, $\varkappa^{\mathrm{II}}$, and the exponent β_r are the parameters of effort fusion.[55] Expressions (5.63) have the form of the classic Stevens' law of psychophysics, Exp. (1.3). Their identity in the exponent value must be due the corresponding stimuli as they are implemented in the body-in-the-self belonging to same class of perceived stimuli.

F6: Let us select a certain action a_U in the kernel $\mathfrak{K}$, i.e., $a_U \in \mathfrak{K}$ to be used as reference entity in the repertoire $\mathfrak{A}$ for comparing other actions in subjective efforts. The implementation of the action a_U as it is experienced via Channels I and II is associated with subjective efforts of the mean magnitudes $U^{\mathrm{I}}_{\mathrm{m}}$ and $U^{\mathrm{II}}_{\mathrm{m}}$ as well as the resource depletion rates r^{I}_U and r^{II}_U meeting condition (5.61).[56] Then, by virtue of (5.63), for any action a the corresponding mean magnitudes $E^{\mathrm{I}}_{\mathrm{m}}$ and $E^{\mathrm{II}}_{\mathrm{m}}$ of the two types of subjective effort and the resource depletion rates r^{I} and r^{II} are related as

$$\frac{r^{\mathrm{I}}}{r^{\mathrm{I}}_u} = \left(\frac{E^{\mathrm{I}}_{\mathrm{m}}}{U^{\mathrm{I}}_{\mathrm{m}}}\right)^{1/\beta_F} \qquad \text{and} \qquad \frac{r^{\mathrm{II}}}{r^{\mathrm{II}}_u} = \left(\frac{E^{\mathrm{II}}_{\mathrm{m}}}{U^{\mathrm{II}}_{\mathrm{m}}}\right)^{1/\beta_F}. \tag{5.64}$$

F7: Fatigue is understood as the *state* of the body-in-the-self when one of the vital resources (or both of them) is depleted and the sensitivity to additional exertions—physical or mental ones—becomes so high that the subject is ready to cease his activity.

F8: *Effort comparability*: The perception-evaluation self-consistency, which emerges via learning and skill acquisition, relates the local experience of a given action a to its evaluation as a typical element of paced performance considered from

---

[55] It should be noted that at this level of the perception-evaluation self-consistency the two channels of action perception remain mutually incomparable yet. Indeed the coefficients $\varkappa^{\mathrm{I}}$ and $\varkappa^{\mathrm{II}}$ (containing also the coefficients $e^{\mathrm{I}}_{\mathrm{m}}$ and $e^{\mathrm{II}}_{\mathrm{m}}$ entering expressions (5.29) and (5.56) for the mean magnitudes $E^{\mathrm{I}}_{\mathrm{m}}$ and $E^{\mathrm{II}}_{\mathrm{m}}$ of subjective efforts) may independently take arbitrary values.

[56] It should be noted that the degree $A_{\mathfrak{b}}$ of attention focused on bodily actions enters the relationship between the bodily effort $E_{\mathfrak{b}}$ and its subjective experience $E^{\mathrm{I}}_{\mathrm{m}}$ via the cofactor $\sigma(A_{\mathfrak{b}})$ dependent on $A_{\mathfrak{b}}$, see Exp. (5.29). Due to the skill acquisition the subject prefers to implement bodily actions under conscious control characterized by a certain fixed attention degree $A_{\mathfrak{b},\mathfrak{A}}$ for the given repertoire $\mathfrak{A}$ of actions. So in describing actions meeting the balanced use of resources we will implicitly assume the equality $A_{\mathfrak{b}} = A_{\mathfrak{b},\mathfrak{A}}$ to hold. When the subject evaluates his actions with different amount of attention, he can underestimate or overestimate the situation. For example, it is so when no attention is focused on bodily action, $A_{\mathfrak{b}} = 0$ (and $A^{\min}_{\mathfrak{b}} = 0$ too), and these automatic actions seem to be effortless.

the standpoint of final result. In particular, this relation admits the following representation.

- The effort of a given action $a \in \mathfrak{C}$ as it is *evaluated* by the subject is quantified in terms of the *inverse time-to-fatigue* $1/\mathcal{T}(a)$. In this case, the subjective effort becomes a quantity of inverse time dimension.
- The effort of action a implementation as it is *perceived* via Channel I (Channel II) can be mathematically measured in terms of the *inverse time-to-fatigue* $1/\mathcal{T}^{\mathrm{I,II}}(a)$ *when the depletion of the corresponding vital resource is dominant*. Indeed, $\mathcal{T}^{\mathrm{I,II}} = \mathcal{E}^{\mathrm{I,II}}/r^{\mathrm{I,II}}$ and by virtue of (5.61) and (5.64)

$$\frac{\mathcal{T}_u}{\mathcal{T}^{\mathrm{I}}} = \left(\frac{E^{\mathrm{I}}_{\mathrm{m}}}{U^{\mathrm{I}}_{\mathrm{m}}}\right)^{1/\beta_F} \quad \text{and} \quad \frac{\mathcal{T}_u}{\mathcal{T}^{\mathrm{II}}} = \left(\frac{E^{\mathrm{II}}_{\mathrm{m}}}{U^{\mathrm{II}}_{\mathrm{m}}}\right)^{1/\beta_F}, \tag{5.65}$$

  where $\mathcal{T}_u = \mathcal{T}^{\mathrm{I}}_u = \mathcal{T}^{\mathrm{II}}_u$.
- During skill acquisition, first, the coefficients $e^{\mathrm{I}}_{\mathrm{m}}$ (see Exp. 5.29) and $e^{\mathrm{II}}_{\mathrm{m}}$ (see Exp. 5.56) are coupled with each other, i.e., their ratio $e^{\mathrm{I}}_{\mathrm{m}}/e^{\mathrm{II}}_{\mathrm{m}}$ takes a special value such that the equality $U^{\mathrm{I}}_{\mathrm{m}} = U^{\mathrm{II}}_{\mathrm{m}} = U_{\mathrm{m}}$ hold. Put differently

$$\frac{U^{\mathrm{I}}_{\mathrm{m}}}{U^{\mathrm{II}}_{\mathrm{m}}} = 1 \quad \Rightarrow \quad \frac{e^{\mathrm{I}}_{\mathrm{m}}}{e^{\mathrm{II}}_{\mathrm{m}}}. \tag{5.66}$$

  Condition (5.66) determines the common units for measuring the two types of subjective effort, i.e., makes Channels I and II mutually comparable.

  Second, the graduate adaptation enables the modification of the relationships (5.29) and (5.56) such that the exponent β_F become equal to unity, $\beta_F = 1$. Below, for simplicity, we will assume that expressions (5.29) and (5.56), specifying the functional dependence of the effort magnitudes $E^{\mathrm{I}}_{\mathrm{m}}$ and $E^{\mathrm{II}}_{\mathrm{m}}$ on the motion characteristics of controlled object and attention distribution, were initially constructed meeting this requirement. Otherwise, it can be done mainly via renormalization of the corresponding exponents.

  Third, after the previous step, expressions (5.65) may be regarded as a simple linear scaling of the plane $\{E^{\mathrm{I}}_{\mathrm{m}}, E^{\mathrm{II}}_{\mathrm{m}}\}$. Because the coefficients $e^{\mathrm{I}}_{\mathrm{m}}$ and $e^{\mathrm{II}}_{\mathrm{m}}$ even after coupling (5.66) are determined up to a common arbitrary cofactor, we may accept now that the quantities $E^{\mathrm{I}}_{\mathrm{m}}$ and $E^{\mathrm{II}}_{\mathrm{m}}$ are just the inverse time-to-fatigue determined by the corresponding vital resources. In this case, for an action a^{I} (a^{II}), experienced only via Channel I (Channel II), the value $E^{\mathrm{I}}_{\mathrm{m}}$ ($E^{\mathrm{II}}_{\mathrm{m}}$) describes its effort understood as inverse time-to-fatigue within division into the particular vital resources. For the action a^{I} (a^{II}) one of these resources is not involved in use at all.

Summarizing the given model fragments, the perception-evaluation self-consistency enables

- quantifying the subjective effort as the inverse time-to-fatigue and
- measuring the two types of subjective effort associated with action implementation experienced via Channels I and II also in units of inverse time-to-fatigue.

Below we will assume all the components of action effort to be measured in units of inverse time-to-fatigue.

F9: *Effort fusion model*: The relationship between

- the effort E_m attributed to a given action $a \in \mathfrak{A}$ as a result of its evaluation (i.e., *effort-as-evaluated*) and
- the magnitudes E^I_m, E^II_m of effort associated with the action a implementation experienced via Channel I and Channel II, respectively, (i.e., *effort-as-experienced*)

will be called the *effort fusion*. By this term we want to emphasize that this relationship stems from long-term learning and skill acquisition, i.e., it is a result of complex emergent process. In this context, the quantity E_m will be also called *fusion-induced effort* or, speaking more strictly, the *mean magnitude of fusion-induced effort*.

The particular form of the relationship $E_\mathrm{m} = \mathfrak{F}(E^\mathrm{I}_\mathrm{m}, E^\mathrm{II}_\mathrm{m})$ depends on the possibility of short-term resource reallocation when one of the vital resources comes close to being exhausted. We will refer to this resource allocation as to the *channel integration*. When the resource use is balanced, i.e., $E^\mathrm{I}_\mathrm{m} = E^\mathrm{II}_\mathrm{m} = E^\mathrm{I,II}_\mathrm{m}$ and the two resources is exhausted simultaneously, their interaction has no effect and $E_\mathrm{m} = E^\mathrm{I,II}_\mathrm{m}$. How the relationship $E_\mathrm{m} = \mathfrak{F}(E^\mathrm{I}_\mathrm{m}, E^\mathrm{II}_\mathrm{m})$ changes as the degrees of channel integration increases is demonstrated in Fig. 5.9 (left panel) with curves 1–4—the locus of $\mathfrak{F}(E^\mathrm{I}_\mathrm{m}, E^\mathrm{II}_\mathrm{m}) = E^\mathrm{I,II}_\mathrm{m}$ on the plane $\{E^\mathrm{I}_\mathrm{m}, E^\mathrm{II}_\mathrm{m}\}$ for a fixed value of $E^\mathrm{I,II}_\mathrm{m}$.

- *No channel integration*: In this case, fatigue occurs when the vital resource depleted with higher rate is exhausted, so

$$E_\mathrm{m} = \max\left(E^\mathrm{I}_\mathrm{m}, E^\mathrm{II}_\mathrm{m}\right), \tag{5.67a}$$

  which is illustrated in Fig. 5.9 (left panel) by curve 1.
- *Full channel integration*: In this case, the two resources form actual one pool, so

$$E_\mathrm{m} = E^\mathrm{I}_\mathrm{m} + E^\mathrm{II}_\mathrm{m}, \tag{5.67b}$$

  which is illustrated in Fig. 5.9 (left panel) by curve 4.
- *Partial channel integration*: If, at lease, partial resource redistribution (short-term reallocation) is feasible, the depletion of the resource being near its exhaustion can be retarded through the other resource. It is quiet natural to expect that the larger amount of the latter resource is available after its own depletion, the stronger the former resource is retarded.

  As illustrated in Fig. 5.9 (left panel), this effect is reflected in a deviation ΔE of curves 2 and 3 from the straight line $E^\mathrm{II}_\mathrm{m} = E^\mathrm{I,II}_\mathrm{m}$ which increases as the variable E^I_m decreases. In order to describe the effort fusion for actions admitting partial integration of channels we will accept the ansatz[57]

[57] The use of ansatz (5.67c) can be justified as follows. Perceiving the subjective effort of bodily actions and the mental effort of monitoring require computational resource in the brain. So at first

$$E_{\rm m} = \left[\left(E^{\rm I}_{\rm m}\right)^{1/\gamma_E} + \left(E^{\rm II}_{\rm m}\right)^{1/\gamma_E}\right]^{\gamma_E}, \tag{5.67c}$$

where the exponent $0 < \gamma_E < 1$ is treated as the *degree of channel integration*. The cases of no-channel-integration and full-channel-integration are included within ansatz (5.67c) as its limits $\gamma_E \to 0$ and $\gamma_E \to 1$, respectively.

Ansatz (5.67c) is the final result describing the effort fusion in terms of effort mean magnitudes. In the case of resource balanced use the channel integration has no effect and all the curves 1–4 have to meet at one point, which is illustrated in Fig. 5.9 (left panel) by a small whole circle.

F10: *Cloud-type magnitude of fusion-induced effort*: Although the recognition of fusion-induced effort emerges via evaluation and adaptation of multiple actions experienced via Channels I and II, finally fusion-induced effort becomes a E-conscious entity that the mind-in-the-self also can experience. It is due to the acquired skill converting AM-conscious interrelations into E-conscious phenomena. For this reason, the fusion-induced effort must be also experienced as a cloud of magnitudes—the cloud-type magnitude $\mathbb{E}_{\rm m}$ of fusion-induced effort. This cloud-type magnitude is accessible to the mind only as a whole entity. In the realm of the body-in-the-self, it is characterized by the mean value $E_{\rm m} = \mathfrak{F}\left(E^{\rm I}_{\rm m}, E^{\rm II}_{\rm m}\right)$ and the width $\Delta_{\rm m}(E_{\rm m})$ determined by $E_{\rm m}$.

Figure 5.9 (right panel) illustrates the cloud-type magnitude $\mathbb{E}_{\rm m}$ on the plane $\{E^{\rm I}_{\rm m}, E^{\rm II}_{\rm m}\}$ as well as in the domain $\mathbb{C}_{\mathfrak{A}}$ representing the main part of action repertoire $\mathfrak{A}$—atomic actions executed relatively often. Actually the characteristic properties of these actions determine the basic features of effort fusion. In the domain $\mathbb{C}_{\mathfrak{A}}$ the dependence $E_{\rm m} = \mathfrak{F}\left(E^{\rm I}_{\rm m}, E^{\rm II}_{\rm m}\right)$ can be approximated as

---

step some amount $A_{\rm b}$ and $A_{\rm m}$ of this resource is allocated to them individually with the possibility of reallocation. The possibility of this reallocation is argued by Verschooren et al. (2019). However, the neural mechanisms governing the two type perception phenomena—Channels I and II—are different. So at the next step these resources are rearranged individually in a special way such that the capacity of information processing be maximized. Here the capacity is understood as a certain amount of operations to be implemented before the resource inner organization is destroyed. This capacity $\mathcal{E}$ should depend on the available amount of computational resource sub-linearly, i.e., the capacity of two channels is specified as $\mathcal{E}_{\rm b} = \kappa_{\rm b} A^p_{\rm b}$ and $\mathcal{E}_{\rm m} = \kappa_{\rm m} A^p_{\rm m}$, where the exponent $p < 1$ and $\kappa_{\rm b,m}$ are some proportionality coefficients. Exactly the quantities $\mathcal{E}_{\rm b}$, $\mathcal{E}_{\rm m}$ were used as the vital resources allocated for Channels I and II. Reallocation of computational resource can be implemented rather quickly provided $A_{\rm b} + A_{\rm m} = A$, where A is constant amount of computational resource allocated initially. Therefore the learning of operating a paced implementation of a given action a gives rise to skill to exhaust the Channels I and II simultaneously. In this case, the action a implementation is described by the equality

$$\left[\left(\frac{E_{\rm b}}{\kappa_{\rm b}}\right)^{1/p} + \left(\frac{E_{\rm m}}{\kappa_{\rm m}}\right)^{1/p}\right] t^{1/p} = A\,,$$

where t is the time-to-exhaustion and $E_{\rm b,r}$ are the rates of information processing via Channels I and II playing the role of subjective efforts. The obtained expression after normalization leads directly to Exp. (5.67c).

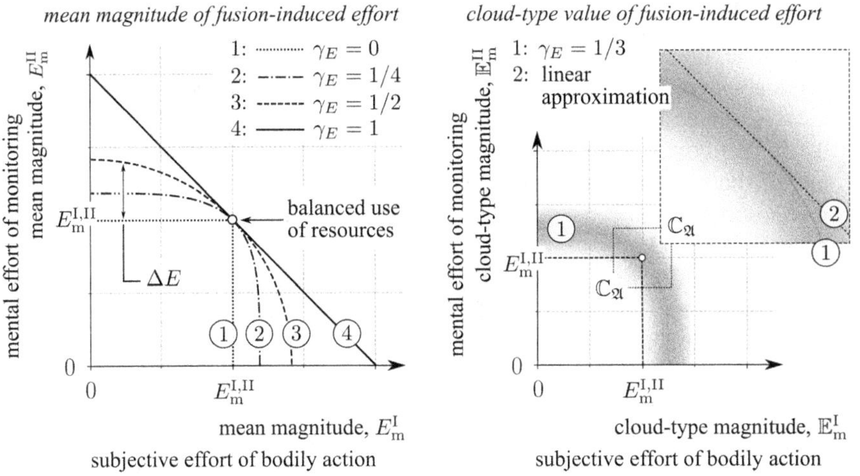

Fig. 5.9 Illustration of the effort fusion properties. Curves 1–4 in the left panel are the loci specified by the fixed mean magnitude of fusion-induced effort $E_m^{I,II} = \mathfrak{F}(E_m^I, E_m^{II})$ on the plane $\{E_m^I, E_m^{II}\}$. Ansatz (5.67c) was used in plotting curves 1–4, the corresponding parameters are shown. The right panel illustrates the cloud-type properties of fusion-induced effort shown as a similar locus (curve 1 corresponding to $\gamma_E = 1/3$) with blurred points on the plane $\{E_m^I, E_m^{II}\}$. The inset zooms in on the neighborhood $\mathbb{C}_{\mathfrak{A}}$ of the point representing the balanced use of resources, dashed line 2 in the inset is the linear approximation of the dependence $E_m^{I,II} = \mathfrak{F}(E_m^I, E_m^{II})$ with fixed $E_m^{I,II}$

$$E_m = g_1\left(E_m^I + E_m^{II}\right) + g_2\frac{1}{E_m^{I,II}}\left(E_m^I - E_m^{II}\right)^2, \tag{5.68}$$

where the coefficients

$$g_1 = 2^{\gamma_E - 1} \qquad \text{and} \qquad g_2 = 2^{\gamma_E - 3} \cdot \frac{(1-\gamma_E)}{\gamma_E}. \tag{5.69}$$

As should be, for $\gamma_E = 1$—the case of channel full integration—the second nonlinear term vanishes.

The second term in Exp. (5.68) is small while $|E_m^I - E_m^{II}| \ll E_m^{I,II}$, i.e., a given action in its implementation does not deviate too far from the resource balanced use. The fusion-induced effort as subjective effort of any type is characterized by the cloud-type magnitude, which also weakens the effect of nonlinear term as illustrated in the inset in Fig. 5.9 (right panel).

Ansatz (5.67c) and its approximation (5.68) finalize our construction of effort fusion. In summary we want to emphasize the following.

- The emergence of effort fusion is a complex multifactorial process uniting various fragments of the human temporality different in time scales. On the one side, the effort fusion is based on two types of subjective effort as they are experienced via two channels—each of them is a collection of intertwined neural mechanisms. The corresponding fragment of the human temporality is the experiential now. On the other side, the effort fusion is based on long-term learning and skill acquisition dealing with subjective effort as it is evaluated from the standpoint of fatigue. Here the role of fatigue is to signal to the mind that the body is at a critical state making the further continuation of current activity problematic. The complex present is intermediate fragment of the human temporality where perception and evaluation meet each other and the perception-evaluation self-consistency is attained via entanglement of bottom-up and top-down causal processes.

- The effort fusion is a nonlinear phenomenon reflecting complexity of channel integration. The properties of fusion-induced effort characterize the collection of actions involved in the current type goal-oriented behavior—the action repertoire—as a whole rather than these actions individually. The main part of repertoire is related to actions implemented relatively often and meeting the resource balanced use.
- Only in the case of channel full integration, the fusion-induced effort is the algebraic sum of its constituent components—the subjective effort of bodily executed actions and the mental effort of monitoring their results. However, for the main part of repertoire the effort fusion can be approximately described by the algebraic sum of constituent components. The remaining nonlinear component describes the channel entanglement. The nonlinear effect of partial channel integration on effort fusion can be weakened by cloud-type nature of fusion-induced effort.
- As far as psychological data arguing for our account of channel integration are concerned, we may note a wide class of phenomena united under the name multisensory (multi-modal) perception.[58] Besides, experiments by Longman et al. (2017) seem to directly demonstrate the incompleteness of channel integrity.

[58] There are a wide variety of psychological phenomena combined within the notions of *multisensory (multimodal) integration* or *crossmodal interaction* with the *binding problem* being in the heart of these phenomena. Broadly speaking, multisensory integration concerns the range of issues of how various information about the outer world received via different sensory modalities is integrated into a single experience. Phenomena and effects of multisensory integration have been investigated during the last century and nowadays there is a vast literature on these issue. For the modern state of the art in this scope, a reader may be referred to collections edited by Stein (2012), Murray and Wallace (2012), Shi and Mueller (2015), Sathian and Ramachandran (2020) as well as reviews by Bayne and Spence (2013), Spence (2018), Wallace et al. (2020). For a discussion focused on the binding problem on its own a reader may be addressed to Singer (2006), Herzog (2009), Schmidt (2009), Burwick (2014). The neural mechanisms governing multisensory integration are reviewed, e.g., by Stein and Stanford (2008), Stein et al. (2014), Cuppini et al. (2018), Stein and Rowland (2020).

It is worthy of noting that nowadays investigations of multisensory integration effects went beyond purely academic research and have many commercial applications. For example, the color-fragrance correspondences (Edwards 2010) caused by crossmodal interaction are essential factors in perfume production (e.g., Kim 2011). The odor-color-shape crossmodal interactions affect the wine preferences of humans (e.g., Heatherly et al. 2019).

By synthesizing information from two or more modality-specific neural inputs the brain of humans (and higher animals too) is able to create something entirely new, which is the gist of multisensory integration. This integration greatly enhances human perception and behavior by combining information across the senses into one experience. Stein and Meredith (1993, see also Rowland et al. 2015 for a review) put forward three basic principles of multisensory integration.

5.5.4 *Fusion-Induced Effort as Analogy to Free Energy in the Mind-in-the-Self*

In this subsection we present arguments for treating the fusion-induced effort as a certain analogy to free-energy in thermodynamics. As will be elucidated, in the realm of the mind-in-the-self, first, the fusion-induced effort reflects a certain tradeoff between sensation of bodily executed actions and perception uncertainty. Second, the minimum $E_{\mathrm{m}}^{\min}$ of the fusion-induced effort describes in some sense optimal actions. Therefore, in selecting a new action for implementation instead of the currently executed action associated with the cumulative effort E_{m}, the subject can gain an effort decrease not exceeding the difference $(E_{\mathrm{m}} - E_{\mathrm{m}}^{\min})$. The latter feature is the reason for calling the fusion-induced effort also the *accessible effort*.

Before passing directly to the fusion-induced effort, let us remind the basic features of free-energy which in physics of statistical systems near thermodynamic equilibrium is one of the basic notions. The global or local minima of free energy describe equilibrium or metastable states. If there are several metastable states characterized by the corresponding values $\{\mathcal{F}_i\}$ of free energy, then the probability of finding the system at a state i is

$$P_i \propto \exp -\mathcal{F}_i/T\,,$$

where T is temperature (in energetic units). The free energy consists of two part $F = U - TS$ which reflects the tradeoff between

- ordering the system arrangement caused by regular dissipative processes trying to minimize the system energy U and
- disordering the system arrangement caused by stochastic transitions of the system between possible micro-level states composing the given macro-level state, which is described by the entropy S.

In this case, the temperature T plays the role of common measurement unit for comparing the two factors. The term free energy reflects the fact that in getting its equilibrium, the system can integrally perform maximum work equal to the decrease in its free energy.

On scales of the experiential now the subject evaluates atomic actions based on two components—the subjective effort of their bodily execution $\mathbb{E}_{\mathrm{m}}^{\mathrm{I}}$ and the mental effort

For a discussion of neural mechanisms underlying these principles a reader may be addressed to Stein et al. (2009), Stein and Rowland (2011).

A reader may be referred to Angelaki et al. (2009), Seilheimer et al. (2014) for a discussion of Bayesian formalism in describing multisensory integration, to Beck et al. (2011), Sabes (2011), Ursino et al. (2014), Chandrasekaran (2017), Cuppini et al. (2018), Miller and Rowland (2018) for the modeling this phenomena using the technique of neural networks. Besides, general issues relating the problem of multisensory perception to bottom-up and top-down phenomena, attention, memory, etc. are discussed, in particular, by Choi et al. (2018) and in collection edited by Hartcher-O'Brien et al. (2017). Crossmodal integration for perception and action is analyzed by Lalanne and Lorenceau (2004).

of their monitoring $\mathbb{E}^{\rm II}_{\rm m}$.[59] Beyond the experiential now, the two types of subjective effort are fused into a cumulative effort $\mathbb{E}_{\rm m} = \mathbb{E}^{\rm I}_{\rm m} \oplus \mathbb{E}^{\rm II}_{\rm m}$ due to the skill acquired by the subject previously.

Within linear approximation[60] of the mean magnitude $E_{\rm m}$ of this fusion-induced effort, see Exp. (5.68), we may write

$$E_{\rm m} = g_0 \left(E^{\rm I}_{\rm m} + E^{\rm II}_{\rm m}\right), \tag{5.70}$$

where $g_0 > 0$ is a certain cofactor and the mean magnitudes $E^{\rm I}_{\rm m}$, $E^{\rm II}_{\rm m}$ of efforts experienced via Channels I and II are specified by Exps. (5.29) and (5.56). As noted in the previous Sect. 5.5.3, in the framework of effort fusion, the coefficients $e^{\rm I}_{\rm m}$ and $e^{\rm II}_{\rm m}$ entering the expressions specifying the functional dependence of the mean magnitudes $E^{\rm I}_{\rm m}$, $E^{\rm II}_{\rm m}$ become determined within a common multiplier. So the coefficient g_0 in Exp. (5.70) may be supposed to represent this multiplier. Up to the given point of our constructions, this coefficient is not determined, all the other quantities entering the expressions for $E^{\rm I}_{\rm m}$, $E^{\rm II}_{\rm m}$ are determined in the sense that they have clear physical or psychological meaning and can be figured out, at in principle, from experimental data.

The term $E^{\rm I}_{\rm m}$ describes the subject's perception of the current motion state of the controlled object. So, the arguments of this term functional form are the motion characteristics $\{x, v, a, j\}$[61] of controlled object and the degree $A_{\rm b}$ of attention focused on the bodily executed action. In other words, $E^{\rm I}_{\rm m}$ is just a function of system state uniting the motion state of controlled external object and the physiological and mental states of the self. The term $E^{\rm II}_{\rm m}$ is actually the effort-type measure of uncertainty in human perception, in Sect. 5.4.8 we have categorized it as the quasi-entropy $S_{\rm m}$ of atomic action (atomic eidos), namely, $S_{\rm m} = -E^{\rm II}_{\rm m}$. In simplified form, the arguments of the quasi-entropy may be treated as a collection $\{A_x, A_v, A_a, A_j\}$ of degrees of attention focused on monitoring the variables $\{x, v, a, j\}$ of object motion within a required accuracy $\{\delta_x, \delta_v, \delta_a, \delta_j\}$.

In these terms Exp. (5.70) may be interpreted as

$$E_{\rm m} = g_0 \cdot \Bigg[\underbrace{E^{\rm I}_{\rm m}(x, v, a, j, |A_{\rm b})}_{\substack{\text{effot-type measure of}\\ \text{subject's perception of}\\ \text{object dynamics}}} - \underbrace{S_{\rm m}(A_x, A_v, A_a, A_j)}_{\substack{\text{effot-type measure of}\\ \text{subject's state required}\\ \text{for monitoring object motion}\\ \text{with accuracy } \{\delta_x, \delta_v, \delta_a, \delta_j\}}} \Bigg]. \tag{5.71}$$

[59] As previously, we will refer to the subjective effort of action bodily execution and the mental effort of monitoring the action results as to subjective efforts of action implementation experienced via Channel I and Channel II, respectively.

[60] The nonlinear interaction between the constituent components of the fusion-induced effort (e.g., described by the second term on the right hand side in Exp. 5.68) can contribute to the properties of the fusion-induced effort on scale of complex present, which is outside the scope of current discussion.

[61] In the present book we consider the dynamics of external object treated as the material point whose motion is described by the spacial position x, the motion velocity v, acceleration a, and jerk $j = da/dt$.

As elucidated in Sect. 5.5.3, the effort fusion is a complex process of optimizing the use of available vital resources. It enables to state the following.

- The fusion-induced effort reflects the tradeoff between
  - implementing a particular action in governing the motion of the controlled object within acceptable accuracy and
  - uncertainty in controlling the results of this action.

  The goal of this trade is to balance the use of computational resources required for implementing these actions under the control of the mind-in-the-self.

So, the fusion-induced effort, first, has a form similar to that of free energy and, second, reflects the tradeoff of two competing factors. In the case of the fusion-induced effort, it is the accuracy-uncertainty tradeoff, for free energy it is the order-disorder tradeoff.

Within the experiential now the subject's choice of actions for implementation is specified by their individual preference $\mathcal{Q}$ based on minimizing their efforts.[62] The probability $\mathcal{P}_a$ of selecting a given action a for implementation is related to its preference $\mathcal{Q}_a = -E_\mathrm{m}$[63] as

$$\mathcal{P}_a \propto \exp\left\{\mathcal{Q}_a\right\} = \exp\Big\{-g_0\big(E^\mathrm{I}_\mathrm{m} - S_\mathrm{m}\big)\Big\}. \tag{5.72}$$

Therefore, comparing with free energy, we see that the fusion-induced effort does have similar properties within the experiential now. Namely:

- The minimum $E^\mathrm{min}_\mathrm{m}$ of the mean magnitude of the fusion-induced effort describes the action characterized by the optimal implementation from the standpoint of saving vital resource.
- How often different actions are selected for implementation is described by probability (5.72) have the same exponential form.
- Changing the currently executed action associated with the effort E_m for another one, the subject is able to gain decrease in effort not exceeding the difference $(E_\mathrm{m} - E^\mathrm{min}_\mathrm{m})$.

The choice of actions for their implementation is characterized by some uncertainty including two factors:

- uncertainty characterizing the limit capacity in discriminating two actions in their efforts; it is the width $\Delta_\mathrm{m}(E_\mathrm{m})$ of the cloud-type magnitude $\mathbb{E}_\mathrm{m}$ in the body-in-the-self which is determined by the mean magnitude E_m;

---

[62] In fact the subject operates with action strategies rather than individual actions. The next chapter is devoted to the description of subject's choice of action strategies where the mathematical constructions presented here will be included as particular elements.

[63] In fact we should write $\mathcal{Q}_a \propto -E_\mathrm{m}$, which, however, is reduced to the equality via renormalizing the coefficient g_0.

- the effort magnitude Δ_E whose value is ignorable within the given action goal-oriented actions; the value Δ_E should be regarded as a given parameter within the experiential now.

We take both the factors setting

$$g_0 = \frac{1}{\Delta_E + \Delta_{\mathrm{m}}(E_{\mathrm{m}})}. \tag{5.73}$$

Thereby, the description of effort fusion no longer contains undetermined parameters. In the context of free energy the coefficient g_0 admits interpretation as inverse temperature.

As a particular example illustrating applications of the fusion-induced effort, let us note the performance of Indian dance "The Golden God" by a famous Chechen dancer Makhmud Esambayev (1924–2000). In this performance the dancer, squatting initially on the floor, imitated sunrise and finally sunset getting up and down during one and a half minutes such that his movements be imperceptibly for the eyes.

In this case, first, the body slow movements are controlled by counteracting muscle forces and for a small velocity v we may approximate the former term on the right hand side in Exp. (5.70) as[64]

$$g_0 E^{\mathrm{I}}_{\mathrm{m}} = E_0 + \frac{v}{\overline{V}},$$

where E_0 is a constant and $\overline{V}$ is a characteristic speed of this type motions reflecting subject's skill. Second, we way suppose that the monitoring of body movements on time scales of the experiential now, $\tau \sim \tau_d$, is reduced only to control over the body velocity v and the required accuracy ϵv is substantially lower than one specified by the limit capacity of velocity perception, i.e., $\epsilon \gg \kappa_v$ (see Exp. 5.56). Besides we suppose that the subject's capacity of the control over the body movement is not exhausted, which implies that the related degree of attention $A_{\mathrm{m}} \ll 1$. Under these conditions the latter term on the right hand side in Exp. (5.70) admits the approximation (see Exp. 5.56)

$$g_0 E^{\mathrm{II}}_{\mathrm{m}} = g_{\mathrm{m}} \frac{\Delta^0_v}{\epsilon v},$$

where coefficient g_{m} is determined by the acquired skill in controlling the body motion. Combining the two approximation, we obtain the desired expression for the fusion-induced effort

$$E_{\mathrm{m}} = E_0 + E_v\,, \qquad \text{where} \qquad E_v(v) = \frac{v}{\overline{V}} + g_{\mathrm{m}} \frac{\Delta^0_v}{\epsilon v}. \tag{5.74}$$

[64] This expression stems for Exps. (5.13) and (5.29) within the assumption that body movement is also accompanied by viscous friction.

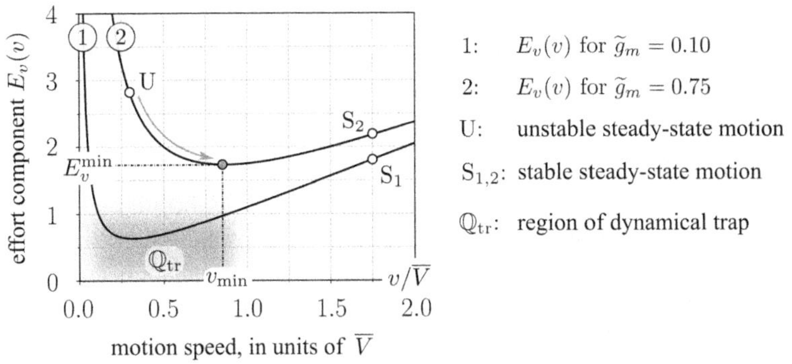

Fig. 5.10 Illustration of the properties of steady-state motion governed by the subject's perception of the fusion-induced effort within model (5.74). The present figure focuses on the relationship between the velocity dependence of the fusion-induced effort $E_v(v)$ and the characteristic modes of steady state motion. The shown parameter $\widetilde{g}_m = g_m \Delta_v^0 / \epsilon \overline{V}$

In the presented form a change in the fusion-induced effort that the subjects takes into account in evaluating action implementation is about unity. In other words, the fusion-induced effort is measure in units of $1/g_0$, Exp. (5.73).

Figure 5.10 illustrates possible situations plotting the component $E_v(v)$ of the fusion-induced effort. As seen, there are three typical situations characterizing the desired steady-state motion with a speed v:

- The current steady-state motion is stable, which is characterized by that (*i*) the corresponding point, S_1 or S_2, belongs to the increasing branch of the curve $E_v(v)$ ($v > v_{\min}$) and (*ii*) the related magnitude $E = E_v(v) \gg 1$ ($g_0 E \gg 1$ in dimensional form).
- The current steady-state motion is unstable, which is characterized by that (*i*) the corresponding point, U, belongs to the decreasing branch of the curve $E_v(v)$ ($v < v_{\min}$) and (*ii*) the related magnitude $E = E_v(v) \gg 1$ ($g_0 E \gg 1$ in dimensional form). In this case, instead of steady state motion the motion should consist of an irregular sequence of alternate fragments of two types:

  - the fragments of motion with speed $v_{\min}$ meeting the minimum of the fusion-induced effort $E_m^{\min} = E_0 + E_v^{\min}$ and
  - the fragments of no motion and no monitoring also corresponding to another minimum of the fusion-induced effort.

  This type of motion obeys the human intermittent control paradigm which is considered to be effective and physiological (e.g., Loram et al. 2011).

- The current state of uniform motion belongs the region of dynamical trap, $\mathbb{Q}_{tr}$—the neighborhood of the minimum point $v = v_{\min}$—characterized by the inequality $E_v(v) - E_v^{\min} \lesssim 1$ ($g_0(E_v(v) - E_v^{\min}) \lesssim 1$ in dimensional form). In this region all the state of motion are treated by the subject as equivalent which can give rise to phase transitions of novel type not met in physical systems (e.g., Lubashevsky 2016, 2017; Lubashevsky and Morimura 2019). It should be emphasized that this type dynamics admitting the description in terms of long-lived states is unique to human behavior where the contribution of the mind is crucial.

Concluding the discussion presented in this section, we want to emphasis that in physics of statistical systems near thermodynamic equilibrium the formalism of free energy is efficient in describing complex processes and phase transitions. As demonstrated by the given rather simple example, the formalism of fusion-induced effort also turns out to be efficient in describing emergent phenomena in the realm of the mind-in-the-self. In particular, applying this formalism to the analysis of slow body movements within the scope of the experiential now such phenomena as (*i*) the intermittent human behavior in controlling unstable systems and (*ii*) the emergence of multitude of long-lived states can be related to the extremal properties of the fusion-induced effort even within this simple model.

5.6 Emergence of Fusion-Induced Cloud: Beyond Experiential Now

In this section we intend to represent a new concept of endless cloud cycle in the mind. Through the "window" of the experiential now the mind-in-the-self gains access to reality via sensory modalities belonging to the realm of the body-in-the-self. The mind-in-the-self deals with perceived stimuli as elementary structureless clouds. Rich diversity of these cloud causes the mind to fuse them into clouds that aggregate information and can be retained in memory. It should be noted that these fusion-induced clouds may be of another kind than their sources—the elementary clouds of the experiential now. The fusion-induced clouds spread to the past and to the anticipated future. Simultaneously newly emerged fusion-induced clouds are continuously compared by the mind with the clouds related to the past and anticipated experience. This process goes beyond the experiential now and, in particular, endows the self with the ability to correct the cloud fusion.

5.6.1 *Stimulus-Induced Clouds: Experiential Now*

Within the experiential now, the subject perceives various physical stimuli and inner stimuli coming from the lower level of neural system, we will denote such stimuli using the symbol $\mathcal{S}$. The quantitative and qualitative properties[65] of a stimulus $\mathcal{S}$ will be denoted by the symbols s and σ, respectively. These stimuli via hierarchical neural

[65] We use the term *quantitative property* to refer to properties of stimuli and their mental images that can vary in wide ranges without affecting the categorical classification of a given object. The stimulus intensity and the magnitude of its perception exemplify quantitative properties. The term

processes are converted into the forms determining directly their mental images the mind can operate with.

The mental images of these stimuli possess dualistic nature. Namely, let us consider the mental images of a stimulus $\mathcal{S}$ from the standpoint of the mind and the body.

- In the realm of the *mind-in-the-self* we meet this image as a structureless entity. Mathematical eidoi, introduced in Sect. 3.3.3, are some analogy to such entities. Therefore we will call these entities *clouds* too and denote them by the symbol $\mathbb{C}$ or $\mathbb{C}\rangle$. It should be noted that the term "cloud" belongs to a vocabulary appropriate for speaking about the overlapping of such entities, the degree of their equality, their blur, etc. In no case a cloud $\mathbb{C}$ can be conceived of as a point or consisting of points.

  Dealing with stimuli, the mind can operates only with clouds and the result of any mental operation with clouds is a cloud-type entity again. Speaking about *direct* operations with clouds, i.e., without creating intermediate cloud-like entities, the mind only initiates such operations which are implemented in the realm of the body-in-the-self. For example, a binary operation with two clouds $\mathbb{C}_1$, $\mathbb{C}_2$ admits only symbolic representations at the level of the mind-in-the-self, i.e.,

$$\widehat{\mathfrak{F}}\{\mathbb{C}_1, \mathbb{C}_2\} \overset{\text{the mind}}{\longmapsto} \mathbb{C}_{\mathfrak{F}}\,, \tag{5.75}$$

  where $\widehat{\mathfrak{F}}$ is a (nonlinear) operator and the result $\mathbb{C}_{\mathfrak{F}}$ is a new cloud-type entity which may possess individual properties irreducible to the properties of the clouds $\mathbb{C}_1$, $\mathbb{C}_2$. Naturally, at the level of the body-in-the-self map (5.75) has a particular integral form.
- In the realm of the *body-in-the-self* the image of the stimulus $\mathcal{S}$ reflected in the mind-in-the-self as the cloud $\mathbb{C}$ can be conceived of as some distribution $\Phi(m, \mu)$ in the space $\{m, \mu\}$ representing the subject's perception of quantitative and qualitative properties $\{s, \sigma\}$ of the stimulus $\mathcal{S}$. The perception magnitude m of the stimulus intensity s and the perceived color pattern μ of light spectrum illustrate the set of quantities $\{m, \mu\}$.

  The distribution $\Phi(m, \mu)$ emerges via processing the information about the perceived stimulus $\mathcal{S}$ occurring via a certain neural mechanism to be called the *Channel* associated with the stimulus $\mathcal{S}$, which is written as

$$\text{Channel}_{\mathcal{S}}(\chi): \quad \mathcal{S}(s, \sigma) \overset{\text{the body}}{\longmapsto} \Phi(m, \mu \,|\, \chi)\,, \tag{5.76}$$

---

qualitative properties implies properties whose change can cause a change in the classification of perceived stimuli and their mental images. For example, changes in light frequency are are reflected in the associated colors. In Stevens's terminology (e.g., Stevens 1957) it is *prothetic* and *metathetic* properties, respectively. Stimuli different only in quantitative properties are mutually comparable. By contrast, when a difference in qualitative properties gives rise to different categorization of perceived stimuli, their mutual comparison can become problematic.

where the symbol χ stands for individual parameters of the Channel$_{\mathcal{S}}$. We have included the channel parameter χ explicitly in the list of arguments of the function Φ to accentuate its dependence on the channel properties which can be affected by the mind via, e.g., attention reallocation.

The distribution $\Phi(m, \mu)$ is located inside a certain domain of the space $\mathbb{R}_{\{m,\mu\}}$ which is centered at a certain point $\{\overline{m}, \overline{\mu}\}$ and has dimensions $\{\Delta_m, \Delta_\mu\}$. The quantities $\{\overline{m}, \overline{\mu}\}$ and $\{\Delta_m, \Delta_\mu\}$ characterize the mean magnitudes of perceived properties and subject's uncertainty in their perception. It should be noted that although the subject "feels" these quantities, directly they are not accessible to the mind.

Within our phenomenological description of the mind-body problem, the relationship between a cloud $\mathbb{C}$ as an entity of the mind-in-the-self and the corresponding distribution $\Phi(m, \mu)$ as an entity of the body-in-the-self is due to the two parts of the self being combined into the self with internal integrity. In particular, because the emergence of the distribution $\Phi(m, \mu)$ must be a nonlinear processes aimed at efficient implementation of various mental operations, we suppose, first, that the function $\Phi(m, \mu \,|\, \overline{m}, \overline{\mu}, \Delta_m, \Delta_\mu)$ for a given Channel$_{\mathcal{S}}$ is of a certain universal form symbolically written as

$$\Phi(m, \mu \,|\, \overline{m}, \overline{\mu}, \Delta_m, \Delta_\mu) = \Phi_{\mathcal{S}}\left(\frac{m - \overline{m}}{\Delta_m}, \frac{\mu - \overline{\mu}}{\Delta_\mu}\right). \tag{5.77}$$

In this case, the dependence of the distribution $\Phi(m, \mu)$ on the channel properties χ is hidden in a plausible dependence of parameters $\overline{m}(\chi)$, $\overline{\mu}(\chi)$, $\Delta_m(\chi)$, and $\Delta_\mu(\chi)$ on the channel properties.

Second, when a certain mental operation, e.g., operation (5.75) is implemented in the realm of the body-in-the-self and the finally generated distribution

$$\Psi_{\mathfrak{F}} = \widehat{\mathfrak{F}}_{\mathfrak{b}}\{\Phi_1, \Phi_2\} \qquad \text{(here } \widehat{\mathfrak{F}}_{\mathfrak{b}} \text{ is a particular integral operator)}$$

does not have the required universal form $\Phi_{\mathfrak{F}}$, the self initiates the transformation to be called the *body-mind raising*:

$$\mathfrak{S}: \quad \Psi_{\mathfrak{F}} \overset{\text{the self}}{\mapsto} \Phi_{\mathfrak{F}}\,. \tag{5.78}$$

Below we will use the symbol $\mathfrak{S}$ to denote this transformation.

The two properties of the $\mathbb{C}$-Φ-relationship, in particular, enable us to represent mental operation in the diagram form

$$\begin{array}{ccccccc}
 & & \mathbb{C}_1;\mathbb{C}_2 & \xrightarrow{\text{the mind}} & \{\mathbb{C}_1,\mathbb{C}_2\} & \xrightarrow{\text{the mind: }\widehat{\mathfrak{F}}} & \mathbb{C}_{\mathfrak{F}} \\
 & & \uparrow\;\uparrow & & \downarrow & & \text{the self }(\mathfrak{S})\text{:}\uparrow \Phi_{\mathfrak{F}} \\
\mathcal{S}_1;\mathcal{S}_2 & \overset{\text{Channel 1}}{\underset{\text{Channel 2}}{\Longrightarrow}} & \Phi_1;\Phi_2 & & \{\Phi_1,\Phi_2\} & \xrightarrow{\text{the body: }\widehat{\mathfrak{F}}_b} & \Psi_{\mathfrak{F}}
\end{array} \tag{5.79}$$

which illustrates the gist of the mind-body interaction in mental operations with clouds.

The noted properties of the stimulus-induced clouds enable us to categorize them as entities of E-consciousness. Indeed, the mind has access to them as whole entities only. The mind on its own can initiate operations similar (5.75) but the details are implemented in the realm of the body-in-the-self.

5.6.2 *Fusion-Induced Clouds: Beyond Experiential Now*

The arguments given above for the E-consciousness of stimulus-induced clouds hold also for, e.g., the cloud $\mathbb{C}_{\mathfrak{F}}$ generated by the mental operation (5.75), as it is illustrate in diagram (5.79). However, there is a fundamental difference between stimulus-induced clouds and clouds similar to $\mathbb{C}_{\mathfrak{F}}$. The former are rooted in the actions of outer stimuli or inner stimuli originating from the lower level of neural processes. In this sense their emergence is due to bottom-up casual processes governed by neural mechanisms.

The latter arise due to mental actions and, thus, have to be categorized as consequences of top-down causal processes. The present subsection focuses on these clouds emerging via mental operations by the mind.

Let us consider two stimulus-induced clouds $\mathbb{C}_1$ and $\mathbb{C}_2$ emerging via *different* Channels 1 and 2. Without saving these clouds in memory they exist while their sources—a pair of stimuli $\mathcal{S}_1$, $\mathcal{S}_2$—are active, so the existence of $\mathbb{C}_1$ and $\mathbb{C}_2$ is confined to the experiential now. However, mental operations with these clouds by the mind can initiate the emergence of a new cloud $\mathbb{C}_{\mathfrak{F}}$ arising via upper level neural mechanisms as illustrated in diagram (5.79). It is quite natural to suppose that a cloud generated in this way will be retained automatically in memory and, thus, its existence is not confined to the experiential now only. Via the mind-in-the-self this cloud can be attributed to any instant of the connected past, the connected future, and, naturally, the experiential now is not excluded.

Let us define a special mental operation with the two stimulus-induced clouds $\mathbb{C}_1$ and $\mathbb{C}_2$ to be called a *cloud fusion* and denoted as

$$\mathbb{C}_{\mathfrak{F}} = \widehat{\mathfrak{F}}\{\mathbb{C}_1, \mathbb{C}_2\} \quad \text{or} \quad \mathbb{C}_{\mathfrak{F}}\big\rangle = \mathbb{C}_1\rangle \times \mathbb{C}_2\rangle\,.$$

The results of cloud fusion will be categorized as *fusion-induced clouds.*

The fusion-induced cloud $\mathbb{C}_{\mathfrak{F}}$ may have a meaning and/or properties irreducible to that of the constituent components $\mathbb{C}_1$ and $\mathbb{C}_2$. In this case, the mind based on an initially existing general image of the corresponding entity creates or modifies a token-type image represented by cloud $\mathbb{C}_{\mathfrak{F}}$ in the mind-in-the-self. The creation of the fusion-induced cloud $\mathbb{C}_{\mathfrak{F}}$ based on the stimulus-induced clouds $\mathbb{C}_1$, $\mathbb{C}_2$ enables the subject to experience some feelings initially caused by the stimuli $\mathcal{S}_1$ and $\mathcal{S}_2$ but now explicitly associated with this entity.

Since the clouds $\mathbb{C}_1$ and $\mathbb{C}_2$ are caused by different channels, they have to be mutually incomparable on their own. In other words, there are no common measurement units initially available for the pair-wise comparison of the magnitudes $\{m_1, \mu_1\}$ and $\{m_2, \mu_2\}$ quantifying the properties of perceived stimuli.

In the realm of the body-in-the-self the clouds $\mathbb{C}_1$ and $\mathbb{C}_2$ are represented as the distributions $\Phi_1(m_1, \mu_1)$ and $\Phi_2(m_2, \mu_2)$. The mutual incomparability of $\mathbb{C}_1$ and $\mathbb{C}_2$ prompts us to consider the quantities $\{m_1, \mu_1\}$, $\{m_2, \mu_2\}$ independent variables and turn to the Cartesian product

$$\mathbb{R}_{12} = \{m_1, \mu_1\} \times \{m_2, \mu_2\}$$

as the initial phase space of cloud fusion. The product $\Phi_1(m_1, \mu_1) \cdot \Phi_2(m_2, \mu_2)$ can be used to determine the fuzzy domain (domain with blurring boundaries) matching the fusion-induced cloud $\mathbb{C}_{\mathfrak{F}}$

$$\mathbb{D}_{\mathfrak{F},12} = \{m_1, \mu_1, m_2, \mu_2\} : \\ \Phi_1(m_1, \mu_1) \cdot \Phi_2(m_2, \mu_2) \lesssim \max\left[\Phi_1(m_1, \mu_1) \cdot \Phi_2(m_2, \mu_2)\right].$$

There should be some reasons for fusing the clouds $\mathbb{C}_1$ and $\mathbb{C}_2$ into the cloud $\mathbb{C}_{\mathfrak{F}}$ being an entity of another type. For example, the fusion can be based on considering the stimuli $\mathcal{S}_1$, $\mathcal{S}_2$, and the entity related to the cloud $\mathbb{C}_{\mathfrak{F}}$ from the same aspect such as efficiency, pleasure, etc. Let this entity at the level of the body-in-the-self is described by a distribution $\Phi_{\mathfrak{F}}(p, \pi)$ characterized by its own quantitative properties p and qualitative properties π. Then the underline reason—the governing mechanism of fusion—may be represented as a certain function relating the quantities $\{p, \pi\}$ to $\{m_1, \mu_1, m_2, \mu_2\}$

$$\begin{pmatrix} p \\ \pi \end{pmatrix} = F_\omega \begin{pmatrix} \{m_1, m_2\} \\ \{\mu_1, \mu_2\} \end{pmatrix} = \begin{pmatrix} F_{\omega,p}(m_1, m_2) \\ F_{\omega,\pi}(\mu_1, \mu_2) \end{pmatrix}. \tag{5.80}$$

Here ω stands for a collection of intrinsic parameters characterizing the function F which enables the fusion tuning based on the subject's skill acquired with time. The introduced quantity ω also allows for the effect of how attention is divided between the Channels I and II. The function F_ω will be called the *fusion kernel*.

As illustration of the fusion kernel, let the stimuli $\mathcal{S}_1$ and $\mathcal{S}_2$ be some different actions executed simultaneously and the quantities m_1 and m_2 describe the subjective effort related to these actions, respectively. The stimuli $\mathcal{S}_1$ and $\mathcal{S}_2$ are represented by the clouds $\mathbb{C}_1$ and $\mathbb{C}_2$ in the mind-in-the-self and by one-dimensional distributions $\Phi_1(m_1)$ and $\Phi_2(m_2)$ in the body-in-the-self. Then let the fusion of $\mathbb{C}_1$ and $\mathbb{C}_2$ describes the cumulative effort p evaluated from the standpoint of integral efficiency of these actions on long-term scales. In this case, function (5.80) could be written as one-dimensional relationship

$$p = F_\omega(m_1, m_2) = \omega_1 m_1 + \omega_2 m_2$$

where the coefficients ω_1, ω_2 play the role of trial units of measurements enabling the subject to compare the stimuli $\mathcal{S}_1$, $\mathcal{S}_2$ in their perception. Accumulating the experience and tuning the fusion kernel the subject can make (*i*) the *perception* of the experiential now and (*ii*) its mental *evaluation* based on long-term experience consistent with each other. In this case, the coefficients ω_1, ω_2 should take some stable values which may be treated as the common units of measurements and the stimulus $\mathcal{S}_1$, $\mathcal{S}_2$ become comparable in the realm of the mind-in-the-self. We will return to this issue in the next subsection.

The fusion kernel (5.80) enables us to represent the result of fusion within the body-in-the-self as the distribution

$$\Psi_{\mathfrak{F}}(p, \pi) = \int\limits_{\mathbb{R}_{12}} dm_1\, d\mu_1\, dm_2\, d\mu_2\, \Phi_1(m_1, \mu_1) \cdot \Phi_2(m_2, \mu_2) \\ \times \delta\big[p - F_{\omega,p}(m_1, m_2)\big] \cdot \delta\big[\pi - F_{\omega,\pi}(\mu_1, \mu_2)\big], \qquad (5.81)$$

where $\delta[\ldots]$ is the Dirac δ-function. In the general case the function $\Psi_{\mathfrak{F}}(p, \pi)$ does not have the required form

$$\Phi_{\mathfrak{F}}(p, \pi) = \Phi_{\mathfrak{F}}\left(\frac{p - \overline{p}}{\Delta_p}, \frac{\pi - \overline{\pi}}{\Delta_\pi}\right). \qquad (5.82)$$

Therefore, at the last step of fusing the clouds $\mathbb{C}_1$, $\mathbb{C}_2$ the self has to initiate the body-mind raising of the function $\Psi(p, \pi)$, which can be written as the minimization problem

$$\min_{\lambda, \overline{p}, \overline{\pi}, \Delta_p, \Delta_\pi} \iint\limits_{p,\pi} dp\, d\pi \left[\Psi_{\mathfrak{F}}(p, \pi) - \lambda \Phi_{\mathfrak{F}}\left(\frac{p - \overline{p}}{\Delta_p}, \frac{\pi - \overline{\pi}}{\Delta_\pi}\right)\right]^2. \qquad (5.83)$$

Minimization (5.83) specifies the particular values of quantities $\overline{p}$, $\overline{\pi}$, Δ_p, Δ_π characterizing the properties of the cloud $\mathbb{C}_{\mathfrak{F}}$.

Summarizing the aforesaid about the cloud fusion we want to emphasized the following.

- Fusion-induced clouds stem from top-down causal processes between the mind-in-the-self and the body-in-the-self. So they do not need the presence of external physical stimuli or internal stimuli coming from the lower level of neural system.

Thereby fusion-induced clouds in their existence are not confined to the experiential now. Moreover the mind may attribute fusion-induced clouds to any instant of time inside complex present—including the connected past, the experiential now, and the connected future—as well as beyond it.

- The cloud fusion being a process with dualistic nature is governed as a whole by the self. In particular, as illustrated in diagram (5.79),
  - the mind-in-the-self (*i*) initiates the fusion and (*ii*) determines its essence reflected in the fusion kernel;
  - the body-in-the-self implements the fusion described by transformation (5.81), the realization of this transformation in the brain and information processing by artificial neural networks seem to have much in common;
  - the final body-mind raising, Exp. (5.83), may be conceived of as the cooperative action of the body-in-the-self and the mind-in-the-self under the direct control of the self.
- The self plays a crucial role in describing the cloud fusion. First, referring to the self—the entity with internal integrity—which units the mental and physiological phenomena, we may suppose that the mind turning to a cloud $\mathbb{C}$ can initiate some process with its representation $\Phi(m, \mu)$ in the body. Second, the body-mind raising implies the complete control over the distribution $\Phi(m, \mu)$ at all its points $\{m, \mu\}$. So the self as the owner of all mental entities and their physiological "shadows" (Sect. 3.2.3) is one of the central concepts in our phenomenological description of goal-oriented actions.

5.6.3 *Dynamics of Fusion-Induced Clouds Within and Beyond Experiential Now*

The fusion-induced clouds are initially created within the connected past in close proximity to the experiential now. As time goes on, the mind based on memory and the anticipation ability distributes them over complex present and beyond it. As a result, the formed family of fusion-induced clouds $\{\mathbb{C}_{\mathfrak{F}}(t)\}$ bears the information about the state of environment—here the collection of stimuli $\{S_i(t)\}$—in the past, the current instant, and a plausible future anticipated within some time horizon. It enables the mind

- to integrate the information about the properties of the environment in some aggregated form on time scales exceeding also the duration of complex present;
- to anticipate the state of acting stimuli and their variations in the near future;
- to operate with two types of entities closely intertwined with each other, namely, the current stimulus-induced clouds $\{\mathbb{C}_i\}$ and their fusion-induced cloud $\mathbb{C}^*_{\mathfrak{F}}$ formed previously and placed by the mind into the experiential now.

As a result, the dynamics of stimulus-induced clouds and fusion-induced clouds within and beyond the experiential now possesses complex properties.

The mind permanently deals with clouds of two types separately and together simultaneously. The properties of stimulus-induced clouds reflect the particular neural mechanism of stimulus conversion into these clouds. Mathematically it is taken into account via the dependence of the function $\Phi(m, \mu \mid \chi)$ (Exp. 5.76) on the channels properties χ. In other words, the properties of this type clouds represent stimuli as they are *experienced.*

The properties of fusion-induced clouds reflect (*i*) the basic features of the underlying conception and (*ii*) the particular parameters of accepted model of how the experienced stimuli are aggregated within the cloud fusion. Mathematically it is reflected in that the fusion kernel F_ω (Exp. 5.80) and the function $\Phi_{\mathfrak{F}}(p, \pi)$ (Exp. 5.82) contain certain parameters. Because the family of fusion-induced clouds in emergence is related to relatively long-term scales the properties of these clouds represent stimuli as they are *evaluated* beyond not only the experiential now but also complex present.

If in comparing the experiencing and evaluation of stimuli the mind finds a discrepancy then several scenarios can be implemented:

- This discrepancy is caused by a change in the external world and the subject has to respond to this changes in some way.
- This discrepancy is caused by a permanent inconsistency between the evaluation of stimuli based on their aggregation within currently fused stimuli and the long-term evaluation of their effects. In this case, the model of cloud fusion should be modified to eliminate the inconsistency, which can need a number of iterations. In the case of success:
  - The mechanism of cloud fusion becomes self-consistent and the stimulus aggregation within the cloud fusion adequately represents the subject's evaluation of his actions.
  - The stimuli initially incomparable in the mind become comparable due to the emergence of effective units of measurements specified by the cloud fusion regularities.
  - Because the skill acquisition usually needs time much longer than the duration of complex present, the cloud fusion parameters may be treated as some constants characterizing the collection of actions in the situation under consideration as a whole.

  It should be noted that the parameters of self-consistent cloud fusion, in particular, the common units of stimulus measurement are not universal, they depend substantially on particular situation and the type of actions. In this sense such regularities of cloud fusion should be categorized as the constitutive relations (see Footnote 46).
- This discrepancy is caused by a not relevant perception of stimuli. This situation is rather similar to the previous one and may be analyzed from the same standpoint. In particular, the subject's response should give rise to the emergence of self-consistent cloud fusion including also the adaptable features of neural mechanisms governing the individual processing of perceived stimuli.

The multiscale dynamic interaction between stimulus-induced clouds and fusion-induced clouds may be treated as a certain tradeoff between the stimulus sensation within the experiential now and the stimulus evaluation beyond it. This tradeoff is a particular implementation of the general mechanism called predictive coding paradigm (Sect. 2.6.3) governing the consistency between the human inner world and reality. The presented discussion of this tradeoff has led us to the concept of self-consistent cloud fusion emerging via skill acquisition. It has been accepted as a basic premise in the description of the cumulative subjective effort in the given chapter. In particular, the self-consistent cloud fusion allows us to introduce the common units of measurements for different types of subjective effort, which enables their mental comparison.

The cloud fusion is actually a hierarchical process and in the given section here we have considered its lower level—the emergence of fusion-induced clouds from stimulus-induced clouds. The construction of higher level of hierarchy can be actually implemented in the same way illustrated by diagram (5.79). Higher level clouds emerge via the fusion of lower level clouds when the mind-in-the-self initiate the fusion process, the body-in-the-self implements it and, finally, the self initiates the body-mind raising of the result. In this case, the self-consistent cloud fusion endows the mind with the ability to compare mental entities of different nature when they simultaneously affect human actions of a certain type. We will turn to this concept in the next Chapter for comparing the implementation effort and the efficiency of action strategies.

Generalizing the presented discussion we want put forward the principle of cloud-fusion self-consistency.

Principle of cloud-fusion self-consistency:

The mental evaluation of human goal-directed actions is a complex processes including the simultaneous contribution of various factors different in nature and, thus, being initially incomparable. The latter means that there are no initially given units of measurements for comparing these factors with one another. When these factors are combined by *the self* into one cloud fusion hierarchy, the subject's skill acquisition on temporal scales exceeding substantially the duration of complex present endows the cloud fusion hierarchy with self-consistency. As a result:

- the common system of measurement units emerges and these factors become mutually comparable in the mind,
- these units and other parameters of the cloud fusion characterize all the possible actions as well as action strategies as a whole collection of entities rather than their individual features;
- the cloud occupying the top level of cloud fusion hierarchy integrates in itself all the factors; as far as human actions are concerned, this cloud represents their main stimulus which, in its existence, should be attributed to the mind-in-the-self;
- the implementation of the top level cloud in the body-in-the-self can be used as the general functional for the mathematical description of goal-oriented actions.

Because the dynamics of cloud fusion implies a permanent analysis of the sensation-evaluation tradeoff by the mind, fusion-induced clouds should be generally categorized as entities of AM-consciousness.[66] However, when the cloud fusion attains the self-consistency state, the cloud fusion can be implemented automatically. In this case, fusion-induced clouds become entities of E-consciousness. In other words, the hierarchical system of fusion-induced clouds belongs to E-consciousness except for transient processes endowing it temporally with AM-conscious features.

The mathematical analysis of the emergence and evolution of hierarchical cloud fusion requires consideration of human goal-oriented behavior beyond complex present and is outside the scope of the present book.

5.7 Conclusion

In this chapter we have proposed an account of the experiential now as an individual entity of the complex present that plays the role of vehicle in the human temporality. We have considered effort—one of the basic aspects of human actions—whose analysis can be confined to the experiential now. In our theory, effort determines the implementation cost of action strategies. For this reason we elucidated the effort properties and have developed the multi-component theory of subjective effort.

This theory is based on:

- The model for bodily executed actions described in the realm of the body-in-the-self. This model analyses human actions aimed at governing some external physical object (material point). It deals with the law of Newtonian mechanics and converts vector-type forces into scalar-type effort. This effort characterizes the intensity of neurophysiological processes related to the corresponding actions and playing the role of input to the mind-in-the-self.
- The model for effort of actions as they experienced by the subject. Here we have drawn a distinction between two types of subjective effort:
  - the effort related to the subject's perception of bodily executed actions,
  - the effort related to monitoring the object dynamics caused by these actions; we call this type effort *mental effort* to emphasize that it originates from the human inner world.

Both the models describe effort in terms of cloud-type magnitudes.

[66] The notion of AM-consciousness as well as E-consciousness was introduced in Sect. 3.2.5. AM-consciousness denotes the level of consciousness admitting mental operations related to synthesis and analysis of some phenomenon. E-consciousness implies that the mind has access only to basic properties of this phenomenon but its analysis is impossible. In the case of cloud fusion, a fusion-induced cloud is of AM-consciousness when the mind analysis and compares intermediate instantiations of this cloud, when the cloud fusion is implemented automatically, the fusion-induced cloud becomes an entity of E-consciousness.

- A number of known psycho-physiological phenomena such as the speed-accuracy tradeoff and the power-law of working memory load underlying the construction of mental effort model.
- Description of mental effort admitting interpretation as quasi-entropy allowing for uncertainty in human perception.
- The two types of subjective effort united within the concept of effort fusion. The effort fusion mechanism includes, in particular, the concepts of vital resource depletion and fatigue as a state of the self.
- The fusion-induced effort as some analogy to free energy in thermodynamic systems. The properties of fusion-induced effort extrema determine the probabilistic feature of human actions similar to that of free energy in statistical systems.
- The subjective effort of bodily executed action and the mental effort of monitoring the action results categorized individually as the type *effort-as-experienced*. The fusion-induced effort belonging to the type *effort-as-evaluated*.
- The fact that due to learning and skill acquisition the two types of subjective effort and the fusion-induced effort achieve the mutual self-consistency. In particular, the self-consistency gives rise to:
  - the two types of subjective effort become mutually comparable,
  - the fusion-induced effort admits interpretation not only as *effort-as-evaluated* but also *effort-as-experienced*.

Based on the constructed notion of effort fusion, we have proposed a new concept of endless cloud cycle in the mind. Through the "window" of the experiential now, the mind-in-the-self gains access to reality as a rich collection of elementary structureless clouds which are merged into fusion-induced clouds and spread to the past and to the anticipated future. Simultaneously newly emerged fusion-induced clouds are continuously compared by the mind with the clouds related to the past and anticipated experience. This process goes beyond the experiential now and, in particular, endows the self with the ability to correct the cloud fusion.

Appendix A

Appendices to Chapter 5: Properties of Cognitive Effort

5.8 Bodily Measure of Human Effort: Complex Objects with Holonomic Constraints

In Sect. 5.3 we constructed the bodily measure of human effort assuming the subject to act directly on an intentional object treated as a material point. The situation, however, may be more complicated when an intentional object consists of several parts interconnected via rigid bonds. In this case, the subject can exert a force F_s on one part while focusing attention on the dynamics of another part. Moreover the integral force $\mathcal{F}_s$ exerted by the subject on such a compound object may have a more complex structure than a 3D-force $F_s \in \mathbb{R}^3$ applied to a material point. In particular, the integral force $\mathcal{F}_s = \{F_s^\alpha\}$ may comprise several 3D-forces applied to the object at different points. Turning the steering wheel by both the hands in car-driving illustrates such cases.

Elements of the human body, e.g., the arms may be also included into this description as components of a compound object. In this case the forces $\{F_s^\alpha\}$ should be related to the muscles activated during intentional actions aimed at driving the corresponding physical object in the desired way.

The present Appendix is devoted to constructing the bodily measure of human effort required for governing the motion of a complex (compound) object when its dynamics admits the description in terms of systems with holonomic constraints (e.g., Goldstein et al. 2014). A pendulum with the pivot point attached to a movable cart (a paradigmatic setup in pendulum balancing tasks) and a rigid body able to move in space and revolve around itself exemplify such systems.

5.8.1 Holonomic Constraints and Generalized Coordinates

Let us consider a complex object O as a certain system consisting of material points, $O = \{Q_i\}_{i=1}^{i=N}$, with holonomic constraints imposed on their positions $\{x_i\}$ in the space $\mathbb{R}^3$. Here we understood holonomic constraints as a collection of geometric relations

$$\left\{f_\beta(x_1, x_2, \dots, x_N) = 0\right\}. \tag{5.84}$$

These constraints are due to strong internal interaction between the system elements that enables relations (5.84) but cannot be specified directly in the framework of such an approach. Therefore the description of system dynamics dealing directly with the constituent material points becomes impossible. To cope with this problem the description of complex object motion is confined to possible degrees of freedom determined by conditions (5.84). In the subspace of the complete system space $\mathbb{R}^{3N}$ that is based on the degrees of freedom, the motion of this complex object is governed entirely by its interaction with *external* objects, including human actions.

In order to specify the response of the object O to forces the subject exerts upon it, the following aspects should be discussed in more detail.

First, we need to introduce a collection of generalized coordinates $q = \{q_k\}_{k=1}^{k=M}$ representing the object degrees of freedom. The number M of these generalized coordinates is bounded from below by the dimension of physical space $\mathbb{R}^3$ and cannot exceed the dimension of the complete system space $\mathbb{R}^{3N}$, i.e., $3 \leq M \leq 3N$. In particular, the inequality $M \ll N$ holds for rigid bodies. The generalized coordinates $\{q_k\}$

– completely specify the spatial positions of all the object elements

$$x_i = x_i(q_1, q_2, \ldots, q_M) \quad \text{for } i = 1, 2, \ldots, N\,,$$

– are mutually independent variables whose time variations are caused by *external* forces and the effect of inertia rather than constraints (5.84),
– enable us to represent the equation governing the dynamics of this complex object in the form

$$J_k(q)\,\ddot{q}_k = F_{e:k}\big(q, \dot{q} \mid S_{\text{env}}(t)\big) + \sum_{k'=1}^{M} \Lambda_{kk'}(q)\mathcal{F}_{s:k'}\,. \tag{5.85}$$

In Eq. (5.85) $J_k(q)$ is the measure of object inertia with respect to the k-degree of freedom, $F_{e:k}(\ldots)$ is the k-component of the generalized force caused by the interaction of the given object with external physical objects making up the environment, the symbol $S_{\text{env}}(t)$ stands for the current state of this environment. The matrix $\widehat{\Lambda}(q) = \|\Lambda_{kk'}(q)\|$ describes the effect of k'-component $\mathcal{F}_{s:k'}$ of the integral force $\mathcal{F}_s$ on the object dynamics with respect to the k-degree of freedom.

When the integral force $\mathcal{F}_s$ comprises several 3D-forces $\{F_s^\alpha \in \mathbb{R}^3\}$ the subject applies to different points of object O, its k-component $\mathcal{F}_{s:k}$ can be represented as a linear form of the 3D-elements

$$\mathcal{F}_{s:k} = \sum_{\alpha} \sum_{i=1}^{3} \eta_k^{\alpha,i}(q) F_{s:i}^{\alpha}\,, \tag{5.86}$$

where the coefficients $\{\eta_k^{\alpha,i}(q)\}$, in the general case, are some functions of the object current state. In particular, if the k-degree of freedom is related to the motion of object O as a whole in the physical space $\mathbb{R}^3$ along one of its axes, Exp. (5.86) is just the sum of the corresponding components of the constituent 3D-forces $\{F_s^\alpha\}$. When the k-degree of freedom describes the object rotation around some axis, Exp. (5.86) represents the relationship between the corresponding moment of integral force and the 3D-forces $\{F_s^\alpha\}$. In this case the coefficients $\{\eta_k^{\alpha,i}(q)\}$ bear the information about the arrangement of the 3D-forces with respect to the axis of rotation.

Second, turning to human intentional actions there can be singled out two types of generalized coordinates and the related degrees of freedom, we will call them *monitored* q^+ and *neglected* q^- coordinates/degree of freedom. The two types are distinct from each other in that subject's *intentional* actions are aimed at controlling the object motion in the dimensions corresponding the monitored degrees of freedom whereas remain indifferent to the neglected ones. For example, in balancing an inverted pendulum the subject has to take actions to keep the pendulum near its upright position but may remain indifferent to the position of the pendulum as a whole in space.

For the sake of convenience we order the generalized coordinates such that some number M^* divides their sequence into the two classes:

$$q^+ = \{q_k^+\} \stackrel{\text{def}}{=} \{q_k\}_{k=1}^{k=M^*} \quad \text{and} \quad q^- = \{q_k^-\} \stackrel{\text{def}}{=} \{q_k\}_{k=M^*+1}^{k=M}.$$

In the general case $1 \leq M^* \leq M$; when $M^* = M$ all the generalized coordinates are of the monitored type.

Third, the analysis of human effort from the standpoint of external observer may be confined to the current instant of time. Besides, the object motion in the past of physical reality cannot be affected by human actions the present. Therefore in constructing the bodily measure of human effort describing subject's intention in driving the object along the desired path

$$\mathcal{P} \stackrel{\text{def}}{=} \{q_k^+(t) = Q_{\mathcal{P}:k}(t)\} \tag{5.87}$$

by exerting the force $\mathcal{F}_s$ on it, we may accept that:

- The path $\mathcal{P}$ determines the time dependence of only the monitored coordinates, the time dependence of neglected coordinates may be arbitrary and obey any requirement selected for some other reason.
- In evaluating the subject's actions in terms of the applied force $\mathcal{F}_s$, only a fragment of the path $\mathcal{P}$ corresponding to a small neighborhood of moments near the current instant of time t may be taken into account.
- At the current instant of time t the state $S_O(t) = \{q_k, \dot{q}_k\}_t$ of object motion belongs to the path $\mathcal{P}$, namely, for $k = 1, 2, \dots, M^*$

$$q_k|_t = Q_{\mathcal{P}:k}(t) \quad \text{and} \quad \dot{q}_k|_t = \frac{dQ_{\mathcal{P}:k}(t)}{dt}.$$

  This assumption is actually based on the standard proposition that if the current state of a given physical system is known the system history is of minor importance. Thereby there is no sense to consider a path $\mathcal{P}'$ of object motion that does not go through the current state of this object. If it happens the path has to be corrected.

As a result the second order time derivatives of the generalized coordinates $\{\ddot{q}_k\}$ are the main quantities that should be used for constructing the bodily measure of human effort required for driving the given complex object along the path $\mathcal{P}$ at the current instant of time. In particular, from the standpoint of external observer the object motion may be regarded as the implementation of the desired path $\mathcal{P}$ when the condition

$$\ddot{q}_k|_t = \frac{d^2 Q_{\mathcal{P}:k}(t)}{dt^2} \tag{5.88}$$

holds for $k = 1, 2, \dots, M^*$.

5.8.2 *Redundancy of Human Actions*

Before passing directly to constructing the desired expression for the bodily measure of human effort generalizing Exp. (5.13), we need to emphasize that there is an essential difference in describing human actions on

- a material point (Sect. 5.3), and
- a complex object with holonomic constraints.

This difference is reflected in the following details.

First, for a complex object there may be degrees of freedom related to the neglected type coordinates q^-. On the one hand, subject's intentions are indifferent with respect to time variations in the neglected coordinates. On the other hand, these time variations can contribute to the human perception of related effort.

The introduction of holonomic constraints implies that any dimension of object motion not confined by these constraints—a degree of freedom—has to be accessible to external impacts independently of the other degrees of freedom. Therefore the matrix $\widehat{\Lambda}(q)$ entering the governing equation (5.85) must be nondegenerate, which enables us to rewrite Eq. (5.85) as

$$\mathcal{F}_{s:k} = \sum_{k'=1}^{M} \Lambda^{-1}_{kk'}(q) \left[J_{k'}(q)\, \ddot{q}_{k'} - F_{e:k'}\big(q, \dot{q} \mid S_{\text{env}}(t)\big)\right] \tag{5.89a}$$

or in the vector form

$$\mathcal{F}_s = \widehat{\Lambda}^{-1}(q)\Delta_{\ddot{q}}\,, \tag{5.89b}$$

where we have introduced the vector $\Delta_{\ddot{q}} = \{\Delta_{\ddot{q}:k}\}_{k=1}^{k=M}$ with components

$$\Delta_{\ddot{q}:k} = J_k(q)\, \ddot{q}_k - F_{e:k}\big(q, \dot{q} \mid S_{\text{env}}(t)\big)\,. \tag{5.90}$$

and the symbol $\widehat{\Lambda}^{-1}(q) = \|\Lambda^{-1}_{kk'}(q)\|$ stands for the inverse of the matrix $\widehat{\Lambda}(q)$.

The implementation of the desired path $\mathcal{P}$ requires the integral force $\mathcal{F}_s$ determined by the condition

$$\mathcal{F}_s = \mathcal{F}_s^+ + \mathcal{F}_s^-\,. \tag{5.91}$$

The first term $\mathcal{F}_s^+$ in this expansion is specified by the q^+-component of the vector $\Delta_{\ddot{q}}$ meeting condition (5.88), i.e., in the component-wise form

$$\mathcal{F}^+_{s:k} = \sum_{k'=1}^{M^*} \Lambda^{-1}_{kk'}(q) \left[J_{k'}(q)\, \frac{d^2 Q_{\mathcal{P}:k'}(t)}{dt^2} - F_{e:k'}\big(q, \dot{q} \mid S_{\text{env}}(t)\big)\right] \tag{5.92a}$$

or in the vector form

$$\mathcal{F}_s^{+} = \widehat{\Lambda}^{-1}(q)\Delta_{\ddot{q}}^{\mathcal{P}}\,, \tag{5.92b}$$

where, by definition, the components of the vector $\Delta_{\ddot{q}}^{\mathcal{P}} = \{\Delta_{\ddot{q}:k}^{\mathcal{P}}\}$ are given by the expression

$$\Delta_{\ddot{q}:k}^{\mathcal{P}} = \begin{cases} J_k(q)\,\dfrac{d^2 Q_{\mathcal{P}:k'}(t)}{dt^2} - F_{e:k}\big(q, \dot{q} \mid S_{\text{env}}(t)\big), & k \in [1, M^*], \\ 0, & k \in (M^*, M]. \end{cases} \tag{5.93}$$

The second term $\mathcal{F}_s^{-}$ related to the q^{-}-component of the vector $\Delta_{\ddot{q}}$, i.e.,

$$\mathcal{F}_{s:k}^{-} = \sum_{k'=M^*+1}^{M} \Lambda_{kk'}^{-1}(q)\left[J_{k'}(q)\,\ddot{q}_{k'} - F_{e:k'}\big(q, \dot{q} \mid S_{\text{env}}(t)\big)\right] \tag{5.94}$$

is not determined by the $\mathcal{P}$-path implementation. Below we will specify it turning to the criterion of effort minimality.

Second, as noted above, in governing the dynamics of complex object the subject can exert several 3D-forces $\{F_s^{\alpha}\}$ upon the object that contribute to the integral force $\mathcal{F}_s$ with different weights. Moreover, these weights may be individual for the components $\{\mathcal{F}_{s:k}\}$ of $\mathcal{F}_s$ and also depend on the spatial arrangement of 3D-forces with respect to the given object. The number N_{α} of these 3D-forces is determined by the bounded capacity of human simultaneous control over the object dynamics via various channels and can be less or greater than the number M of the object degrees of freedom. For example, $N_{\alpha} < M$ in balancing an inverted pendulum via moving a cart bearing the pendulum pivot point; in governing the car motion via the steering wheel kept by two hands, $N_{\alpha} > M$. Assuming the configuration O_F of the points where the forces $\{F_s^{\alpha}\}$ are applied to the object O to be fixed we can rewrite relationship (5.86) in the coordinate-wise form

$$\mathcal{F}_{s:k} = \sum_{\gamma=1}^{N_s} G_{k\gamma}(q, O_F) f_{\gamma} \tag{5.95a}$$

or in the vector form

$$\mathcal{F}_s = \widehat{G}(q, O_F) f\,. \tag{5.95b}$$

Here, for simplicity, we have combined all the 3D-forces and their spatial components within one set $\{f_{\gamma}\}$ of quantities labeled from start-to-end with the index γ running from 1 to $N_s = 3N_{\alpha}$. The matrix $\widehat{G}(q, O_F) = \|G_{k\gamma}(q, O_F)\|$ of $M \times N_s$-size represents relation (5.86) in the introduced terms and can depend on the system current state q and the point configuration O_F. In the present analysis the matrix $\widehat{G}(q, O_F)$ is assumed to be given beforehand.

Third, the desired path $\mathcal{P}$ (5.87) is supposed to be physically implementable within the given configuration O_F of force application points. It means that there are such values of $\{f_\gamma^{\mathcal{P}}\}$ that the corresponding integral force $\mathcal{F}_s(f^{\mathcal{P}})$ specified by Exp. (5.95b) meets condition (5.91). Put differently, we suppose that there is a vector $f^{\mathcal{P}} = \{f_\gamma^{\mathcal{P}}\}$ such that for

$$\Delta_{\ddot{q}} = \widehat{\mathcal{G}}(q, O_F) f^{\mathcal{P}} \tag{5.96}$$

the equality

$$\Delta_{\ddot{q}}^{\mathcal{P}} = \widehat{P}^{+} \Delta_{\ddot{q}} = \widehat{\mathcal{G}}_{\mathcal{P}}(q, O_F) f^{\mathcal{P}} \tag{5.97}$$

holds. Here we have introduced the operators

$$\widehat{\mathcal{G}}(q, O_F) = \widehat{\Lambda}(q) \cdot \widehat{G}(q, O_F) \quad \text{and} \quad \widehat{\mathcal{G}}_{\mathcal{P}}(q, O_F) = \widehat{P}^{+} \cdot \widehat{G}(q, O_F), \tag{5.98}$$

the projection operator $\widehat{P}^{+} = \| P_{kk'}^{+} \|$ mapping the vector $\Delta_{\ddot{q}}$ onto its q^{+}-component, i.e.,

$$P_{kk'}^{+} = \begin{cases} 1, & \text{if } k = k' \text{ and } 1 \leq k \leq M^*, \\ 0, & \text{othewise,} \end{cases}$$

and have taken into account (5.92b).

In fact we accept a stronger assumption that the given path $\mathcal{P}$ is not only implementable but its implementation is also characterized by substantial degeneracy. It means that in the linear space of all the possible forces $f = \{f_\gamma\}$ the subject can exert on the object O, i.e.,

$$\mathfrak{S} = \{f \,:\, f_\gamma \in (-\infty, +\infty)\,, \text{ for } \gamma = 1, 2, \ldots, N_s\} \tag{5.99}$$

there is a null-subspace $\mathfrak{S}_0$ of the operator $\widehat{\mathcal{G}}_{\mathcal{P}}(q, O_F)$, i.e.,

$$\mathfrak{S}_0 = \mathfrak{S}_0(q, O_F) = \{f \,:\, \widehat{\mathcal{G}}_{\mathcal{P}}(q, O_F) f = 0\}. \tag{5.100}$$

The dimension M_0 of the null-subspace $\mathfrak{S}_0$—the number of its degrees of freedom—is determined by (*i*) the division of the generalized coordinates q into the classes q^{+} and q^{-} as well as (*ii*) the configuration O_F of force application points. In particular, when the first factor is main, $M_0 = N_s - M^*$ by virtue of (5.97). Calculating the dimension M_0 in the general case requires an individual analysis and below we will consider M_0 to be just a given quantity.

As a characteristic example arguing for this degeneracy we note the redundancy of human motor degrees of freedom with respect to the dimension of typical motor tasks. It endows us with high adaptability and flexibility in various actions and response to changing environment; for discussion of motor redundancy/abundance and related issues of human control a reader may be referred, e.g., to Latash (2008b, 2012) as well as a collection edited by d'Avella et al. (2016). As an additional argument for the constructive role of redundancy we note the hypothesis posed by Singh et al. (2016) that redundancy aids in motor learning and that the redundant component of motor variability is not noise, see also Dhawale et al. (2017).

The introduced null-subspace $\mathfrak{S}_0 \subset \mathfrak{S}$ enables us to construct its complement $\mathfrak{S}_0^{\perp}$ in the space $\mathfrak{S}$, i.e.,

$$\mathfrak{S}_0 \oplus \mathfrak{S}_0^{\perp} = \mathfrak{S} \quad \text{and} \quad \mathfrak{S}_0 \cap \mathfrak{S}_0^{\perp} = \emptyset\,.$$

Then any vector $f \in \mathfrak{S}$ admits the unique expansion

$$f = f_0 + f_0^{\perp}\,, \tag{5.101}$$

where its components obey the relations

$$\begin{aligned} f_0 &\in \mathfrak{S}_0\,, \ \text{so}\ \widehat{\mathcal{G}}_{\mathcal{P}}(q, O_F)\, f_0 = 0\,, \\ f_0^{\perp} &\in \mathfrak{S}_0^{\perp}\,, \ \text{so}\ \widehat{\mathcal{G}}_{\mathcal{P}}(q, O_F)\, f_0^{\perp} \neq 0\,. \end{aligned}$$

It should be emphasized that the construction of the complementary subspace $\mathfrak{S}_0^{\perp}$ is not unique and any set of linearly independent vectors not belonging to the null-subspace $\mathfrak{S}_0$ may be used.

In its turn the vector $f^{\mathcal{P}} = \{f_{\gamma}^{\mathcal{P}}\}$ causing the object O to move along the desired path $\mathcal{P}$ is not unique too, any force $f = f^{\mathcal{P}} + f_0$ with component $f_0 \in \mathfrak{S}_0$ produces the same effect. However, for a given null-subspace $\mathfrak{S}_0$ the vector $f^{\mathcal{P}}$ admits a special representation being unique. Namely, let us select such a vector f_0 that the sum $f^{\mathcal{P}\perp} = f^{\mathcal{P}} + f_0 \in \mathfrak{S}_0^{\perp}$ contain no component of the null-subspace. Under these conditions the vector $f^{\mathcal{P}\perp}$ is uniquely specified by Exp. (5.97), which allows us write

$$f^{\mathcal{P}\perp} = \widehat{\mathcal{G}}_{\mathcal{P}}^{-1\perp}(q, O_F)\, \Delta_{\ddot{q}}^{\mathcal{P}}\,, \tag{5.102}$$

where $\widehat{\mathcal{G}}_{\mathcal{P}}^{-1\perp}(q, O_F)$ is the inverse of the operator $\widehat{\mathcal{G}}_{\mathcal{P}}(q, O_F)$ *within* the subspace $\mathfrak{S}_0^{\perp}$. It should be noted that the choice of the force $f^{\mathcal{P}\perp}$ specifies also the second order time derivatives of the neglected type coordinates q^{-} via Exp. (5.96). Thereby any subject's actions implementing the desired motion of object O are represented by the forces $f^{\mathcal{P}} = \{f_{\gamma}^{\mathcal{P}}\}$ given by the expression

$$f^{\mathcal{P}} = f^{\mathcal{P}\perp} + f_0\,, \tag{5.103}$$

where $f_0 \in \mathfrak{S}_0$ is any vector belonging to the null-subspace $\mathfrak{S}_0$.

5.8.3 Bodily Measure of Human Effort

The redundancy in the implementation of object motion along the desired path can be analyzed and partly reduced turning to the notion of human effort. The construction of the corresponding bodily measure is the purpose of this final subsection.

There are several criteria proposed for specifying additional constraints imposed on human actions that reduce the redundancy stemming from the physical constraints of analyzed system, they are often called *optimality principles* (e.g.., Butz et al. 2007). Among these criteria we may note movement smoothness (Flash and Hogan 1985) or end-point accuracy and comfort (Harris and Wolpert 1998; Weigelt et al. 2006), for reviews on optimality principles, see Engelbrecht (2001), Todorov and Jordan (2002), Todorov (2004). However, speaking about the criteria focused on the current state of motion—the standpoint of external observer—we turn to a local-in-time cost of motion dealing with the object acceleration in the space of monitored coordinates, cf., for example, Todorov and Jordan (2002, including Supplementary Notes) and also Liu and Todorov (2007). It should be emphasized that criteria of selecting an optimal strategy of actions may be irreducible to the minimization of bodily effort because on global scales purely mental phenomena can play a significant role (e.g., Todorov and Jordan 2002).

Following the spirit of Sect. 5.3 we specify the bodily measure of human effort E_s currently required for driving the object O by exerting the forces $f = \{f_\gamma\}$ upon it as a positive-definite quadratic form

$$E_s \overset{\text{def}}{=} \langle f \,|\, \mathbf{B} \,|\, f \rangle \,. \tag{5.104}$$

The symmetric positive-definite matrix $\mathbf{B} = \| B_{\gamma\gamma'} \|$ allows for a possible difference in the contribution of the $\{f_\gamma\}$-components to human perception of effort as well as their mutual correlations in this contribution. For example speaking about arm movement control, its cost may include not only different forms of muscle effort but also neural effort quantifying the information processing required for joint coordination (for a review see Dounskaia and Shimansky 2016; Shimansky and Dounskaia 2018). Naturally the matrix $\mathbf{B}(q, \dot{q}, O_F)$ on itself can depend on the current state of object motion and how the subject applies his forces to the object O.

The positive-definite quadratic form (5.104) may be used as a metric, which converts the linear space $\mathfrak{S}$ into a metric space. It enables us to construct the complement $\mathfrak{S}_0^{\perp}$ of the null-subspace $\mathfrak{S}_0$ such that the two subspaces be orthogonal to each other; in the following constructions it is supposed to be done. In this case the matrix $\mathbf{B}$ takes the block-diagonal form

$$\mathbf{B} = \left\| \begin{matrix} \mathbf{B}_0^{\perp} & \vdots & \mathbf{0} \\ & \cdots & \\ \mathbf{0} & \vdots & \mathbf{B}_0 \end{matrix} \right\| \tag{5.105}$$

where the matrices $\mathbf{B}_0^{\perp}$ and $\mathbf{B}_0$ determine the positive-definite quadratic forms within the subspaces $\mathfrak{S}_0^{\perp}$ and $\mathfrak{S}_0$, respectively. Within this partition of the space $\mathfrak{S}$ the bodily measure of human effort (5.104) required for driving the object O along the path $\mathcal{P}$ becomes the additive sum of two terms

$$E_s = \left\langle f^{\mathcal{P}\perp} \right| \mathbf{B}_0^{\perp} \left| f^{\mathcal{P}\perp} \right\rangle + \langle f_0 | \, \mathbf{B}_0 \, | f_0 \rangle \,, \tag{5.106}$$

by virtue of (5.103) and (5.105). Then Exp. (5.102) enables us to rewrite the obtained expression as

$$E_s = \left\langle \Delta_{\ddot{q}}^{\mathcal{P}} \right| \mathbf{B}_{\mathcal{P}} \left| \Delta_{\ddot{q}}^{\mathcal{P}} \right\rangle + \langle f_0 | \, \mathbf{B}_0 \, | f_0 \rangle \,, \tag{5.107}$$

where the positive-definite symmetric matrix

$$\mathbf{B}_{\mathcal{P}}(q, \dot{q}, O_F) = [\widehat{\mathcal{G}}_{\mathcal{P}}^{-1\perp}(q, O_F)]^{\mathrm{T}} \cdot \mathbf{B}_0^{\perp}(q, \dot{q}, O_F) \cdot \widehat{\mathcal{G}}_{\mathcal{P}}^{-1\perp}(q, O_F) \,. \tag{5.108}$$

Here the symbol A^{T} stands for the transpose of the matrix/operator A.

Expression (5.107) is the main result of Section 5.8 specifying the bodily measure of human effort E_s required for driving the object O along the desired path $\mathcal{P} = \{q^+(t) = Q_{\mathcal{P}}(t)\}$ (see Exp. (5.87)).

The first term in (5.107) represents the minimal effort required to move the object with acceleration $\ddot{q}^+$ when, in addition, the external forces F_e caused by the object interaction with other physical objects in its environment act on the object. The two factors are taken into account within the vector $\Delta_{\ddot{q}}^{\mathcal{P}}$ which confined to the space of monitored coordinates q^+ can be written in the form

$$\Delta_{\ddot{q}}^{\mathcal{P}} = \widehat{J}(q) \frac{d^2 Q_{\mathcal{P}}}{dt^2} - P^+ F_e\big(q, \dot{q} \mid S_{\text{env}}(t)\big) \,, \tag{5.109}$$

where $\widehat{J}(q)$ is the diagonal matrix with the corresponding components describing the effect of object inertia.

The second term in (5.107) represents the null-type subject's effort caused by exerting forces upon the object O that gives rise to no change in the time variations of the monitored degrees of freedom. The growth of this type "redundant" effort with the subject's forces belonging the null-subspace $\mathfrak{S}_0$, on the one hand, may be reduced via learning process. On the other hand, it enables a smooth correction of the currently used action strategy.

It should be noted that the notion of null-space is met in the theory of motor task learning and understood as the collection of control variables whose changes do not affect performance (Ranganathan et al. 2014). Beside the theory of uncontrolled manifold also turns to the notion of null-space (e.g., Latash et al. 2007; Latash 2008b). The structural learning hypothesis (Braun et al. 2009, 2010) assumes the brain to learn gradually the structure of a task by (*i*) focusing on the subset of available motor control variables that influence task performance and (*ii*) avoiding control variable of null-space. However, when the goal is to maintain stability selectively exploring the null-space may become important since the goal is to avoid any changes in the movement outcome (Ranganathan et al. 2014).

The particular forms of the matrices/operators $\mathbf{B}_{\mathcal{P}}$ and $\mathbf{B}_0$ are determined by the "physics" of intentional object as well as human locomotion. As a result they must depend substantially on the particular details of a system under consideration. Within the present book they are regarded as mathematical objects given beforehand.

5.9 Formalism of Cloud Comparison

5.9.1 1D-Clouds: Ordering in Magnitude

The given Appendix is focused on 1D-clouds representing mental entities appearing, e.g., (*i*) in the conscious evaluation of effort required for implementing bodily actions (Sect. 5.4.1) or (*ii*) the cognitive evaluation of physical stimuli with respect to the magnitude quantifying their perception (Sect. 1.3). The purpose of this Appendix is to develop the formalism describing the comparison of such clouds in ordering them according to their magnitude. Here magnitude is understood as a generic term describing the amount of effort, the perceived intensity of physical stimuli, etc.

Let us consider 1D-clouds $\{\mathbb{C}\}$ characterized by the following properties:

- Each cloud $\mathbb{C}$ is a region of the space $\mathbb{R}^M = \{m : -\infty < m < \infty\}$ with fuzzy (blurred) boundaries.[67]
- A cloud $\mathbb{C}$ is represented by its density function $\phi(m)$ such that

$$\phi(m) = \frac{1}{Z}\,\Phi(m)\,, \qquad \text{and} \qquad Z^2 = \langle\Phi|\,\Phi\rangle \stackrel{\text{def}}{=} \int_{-\infty}^{\infty} dm\big[\Phi(m)\big]^2\,. \tag{5.110}$$

Here the function $\Phi(m)$ specifies the spatial structure (m-structure) of the cloud $\mathbb{C}$ and meets the condition

$$\max_{m\in\mathbb{R}^M} \Phi(m) = 1\,, \tag{5.111}$$

the normalizing constant Z enables the equality $\langle\phi|\,\phi\rangle$ 1. Below the function $\Phi(m)$ will be also called the cloud density and where it could lead to some misunderstanding the function $\phi(m)$ will be referred to as the normalized cloud density.

[67] In Sect. 5.4.1 we dealt with the cloud-type mental measure $\mathbb{E}_{\mathrm{m}}(e)$ of effort, where the argument e—effort value—is a non-negative real number $e \geq 0$. However confining our analysis to effortful actions we may consider $e \in \mathbb{R}^M$, which justifies the application of the formalism to be developed here to describing mental effort. The problem of transitions between effortful and effortless actions does not belong to scope of this Appendix.

- The main parameters M and Δ of a cloud $\mathbb{C}$ or, speaking more strictly, the cloud density $\Phi(m)$ determine the position of its maximum on the m-axis and the characteristic width of the given cloud, respectively, which is illustrated in Fig. 5.11.
- The functional form $\Phi(m) = \Phi(m|M, \Delta)$ of the cloud density is assumed to be the same for all the clouds $\{\mathbb{C}\}$.[68]
- The parameters M and Δ are interrelated with each other such that

$$\Delta = \Delta(M) \tag{5.112}$$

  is a certain increasing function considered to be given beforehand.

One of the plausible ansätze for dependence (5.112) is the linear relationship

$$\Delta(M) = \kappa M \quad \text{for } M > 0 \tag{5.113}$$

representing regularities of inner psychophiscis such as Ekman's law (see Sects. 1.3.1 and 1.3.2, Exp. (1.4)) or the mental images of perceived stimuli (for details see Lubashevsky 2019). Turning to the discussion of the mental effort properties in Sect. 5.4.1 we may consider another ansatz

$$\Delta(M) = \Delta_0 \,. \tag{5.114}$$

where Δ_0 is some constant.

Due to dependence (5.112) the maximum density magnitude M is the only one free parameter of the cloud parametrization. So, at first, let us describe a model for cloud ordering by the subject that operates with the maximum density magnitude M. In this way we elucidate the gist of ordering such clouds as objects with blurred boundaries. However, the precise value of M is not accessible to the mind. Therefore, at the next step, we rewrite this model turning to quantities admitting the direct access to the mind.

Cloud Ordering Based on the Density Maximum Magnitude

We accept the following conception of how the subject compares two clouds $\mathbb{C}_1$ and $\mathbb{C}_2$ with respect to which one should be categorized as possessing lower (higher) magnitudes. The subject is *formally* supposed to be able to compare clouds $\mathbb{C}_1$ and $\mathbb{C}_2$ in this way when he can recognize the difference $\delta M = M_1 - M_2$ in the position of the density maxima (Fig. 5.11).

---

[68] This proposition reflects the universality of inner psychophysics and the scale-free structure of the mind functioning (Sect. 1.3.2).

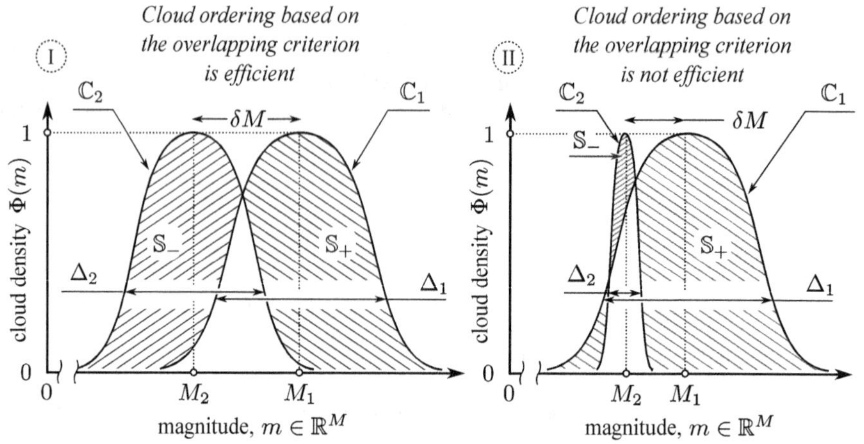

Fig. 5.11 The illustration of, first, the characteristic form of the clouds $\{\mathbb{C}\}$ and its parametrization with two quantities M and Δ. Here M is the magnitude at which the cloud density $\Phi(m)$ attains its maximum set equal to unity, Δ is the characteristic width of a given cloud $\mathbb{C}$. Second, it is the illustration of the cloud ordering along the m-axis based on the position of the cloud density maximum M. The regions hatched with blue and red lines, $\mathbb{S}_-$ and $\mathbb{S}_+$, represent the parts of two clouds $\mathbb{C}_1$ and $\mathbb{C}_2$ where the cloud $\mathbb{C}_1$ or the cloud $\mathbb{C}_2$ dominates in density. The arrangement of the regions $\mathbb{S}_-$ and $\mathbb{S}_+$ in the space $\mathbb{R}^M$ determines the m-ordering of the clouds $\mathbb{C}_1$ and $\mathbb{C}_2$

From the standpoint of external observer[69] the ordering of two *different* clouds $\mathbb{C}_1$ and $\mathbb{C}_2$ is actually a binary relation[70] such that

$$M_1 < M_2 \quad \text{or} \quad M_1 > M_2\,. \tag{5.115}$$

For the subject with the bounded capacity of cognition, the situation is more complicated.

When the difference δM is rather small the subject cannot order the magnitudes M_1 and M_2 and has to categorize the clouds as equivalent, $\mathbb{C}_1 \simeq \mathbb{C}_2$. Near the perception threshold the subject cannot state *with certainty* that, e.g., $M_1 > M_2$ for the situation when, in fact, $M_1 > M_2$ but he clearly recognizes that the opposite inequality $M_1 < M_2$ is not the case. Therefore, in the realms of the mind-in-the-self the binary relation (5.115) has to convert into the three-category classification of subject's perception:

$\mathfrak{C}_<$ $(\mathbb{C}_1 \prec \mathbb{C}_2)$: the relation $M_1 < M_2$ can be accepted with some degree of certainty,
$\mathfrak{C}_>$ $(\mathbb{C}_1 \succ \mathbb{C}_2)$: the relation $M_1 > M_2$ can be accepted with some degree of certainty,

---

[69] The external observer, by definition, is assumed to have complete access to all the details of the clouds $\{\mathbb{C}\}$.

[70] In the general case the strict equality $M_1 = M_2$ for *different* clouds $\mathbb{C}_1$ and $\mathbb{C}_2$ is of infinitely low probability.

$\mathfrak{C}_=$ ($\mathbb{C}_1 \simeq \mathbb{C}_2$): both the classification categories $\mathfrak{C}_<$ and $\mathfrak{C}_>$ are rejected.[71]

Dealing with an individual pair $\{\mathbb{C}_1, \mathbb{C}_2\}$ of clouds analyzed at a given trial, the three classes of categorization do not intersect with one another. It means that the subject selects one and only one of them each time he wants to categorize the given pair. However, at the level of ensembles of such pairs, i.e. for the collection $\{\mathbb{C}\} \times \{\mathbb{C}\}$ of all the possible pairs these classes can intersect. In particular, within the sequence of trials, in categorizing a given pair $\{\mathbb{C}_1, \mathbb{C}_2\}$ different classes may be selected with certain probabilities (choice frequencies)

$$P_<(M_1, M_2)\,, \quad P_>(M_1, M_2)\,, \quad P_=(M_1, M_2)$$

depending on the cloud parameters M_1, M_2.[72]

In our constructions we suppose that the bounded capacity of human cognition gives rise mainly to the uncertainty in the cloud classification between the categories $\mathfrak{C}_<$ and $\mathfrak{C}_=$ or between the categories $\mathfrak{C}_>$ and $\mathfrak{C}_=$ but the categories $\mathfrak{C}_<$ and $\mathfrak{C}_>$ are never mixed. Put differently, the probabilistic relations introduced at the level of cloud ensembles:

$\mathfrak{C}_\leq$ ($\mathbb{C}_1 \precsim \mathbb{C}_2$): $\mathfrak{C}_< \bigcup \mathfrak{C}_=$ defined for $M_1 < M_2$ as

$$\mathbb{C}_1 \precsim \mathbb{C}_2 = \begin{cases} \mathbb{C}_1 \prec \mathbb{C}_2\,, & \text{with the probability } P_<(M_1, M_2) \\ \mathbb{C}_1 \simeq \mathbb{C}_2\,, & \text{with the probability } P_=(M_1, M_2) \end{cases} \tag{5.116a}$$

$\mathfrak{C}_\geq$ ($\mathbb{C}_1 \succsim \mathbb{C}_2$): $\mathfrak{C}_> \bigcup \mathfrak{C}_=$ defined for $M_1 > M_2$ as

$$\mathbb{C}_1 \succsim \mathbb{C}_2 = \begin{cases} \mathbb{C}_1 \succ \mathbb{C}_2\,, & \text{with the probability } P_>(M_1, M_2) \\ \mathbb{C}_1 \simeq \mathbb{C}_2\,, & \text{with the probability } P_=(M_1, M_2) \end{cases} \tag{5.116b}$$

are mutually exclusive. In other words, there is no one pair $\{\mathbb{C}_1, \mathbb{C}_2\}$ of different clouds belonging with some finite probabilities to both the classes $\mathfrak{C}_\leq$ and $\mathfrak{C}_\geq$. In particular, for this reason these relations can be used for ordering the set $\{\mathbb{C}\}$ of clouds.

In the probabilistic terms the mutual exclusivity of the classification categories $\mathfrak{C}_\leq$ and $\mathfrak{C}_\geq$ is reduced to the conditions

$$\begin{aligned} P_<(M_1, M_2) + P_=(M_1, M_2) = 1 \quad P_>(M_1, M_2) = 0 \quad \text{for } M_1 < M_2\,, \\ P_>(M_1, M_2) + P_=(M_1, M_2) = 1 \quad P_<(M_1, M_2) = 0 \quad \text{for } M_1 > M_2\,. \end{aligned} \tag{5.117}$$

---

[71] In the realm of this mental categorization, the class $\mathfrak{C}_=$ is not the image of the case $M_1 = M_2$. Its emergence is cased by the bounded capacity of mental classification in terms of "less than" or "greater than."

[72] In the general case, these probabilities must depend on the cloud widths Δ_1 and Δ_2 too as individual arguments. However in the case under consideration, the width Δ of a cloud $\mathbb{C}$ is assumed to be a certain function of the density maximum magnitude M, Exp. (5.112). So only the quantities M_1 and M_2 are independent variables.

Assuming the continuity of these functions we immediately get the equalities

$$P_{=}(M_1, M_1) = 1\,, \qquad P_{<}(M_1, M_1) = 0\,, \qquad P_{>} = (M_1, M_1) = 0\,, \tag{5.118}$$

which must hold for the constructed relations to have meaning.

Cloud Ordering Based on the Cloud Overlapping

The use of the density maximum magnitudes M_1 and M_2 for ordering the corresponding clouds $\mathbb{C}_1$ and $\mathbb{C}_2$ along the m-axis is actually the mathematical background of this operation. The quantities M_1 and M_2 on their own are some specific numbers inaccessible to the mind. So to describe the mental comparison of such clouds implemented at the levels of the mind-in-the-self and the body-in-the-self we need another formalism. It should be emphasized that in the following constructions the assumption (5.112) about the relationship between the cloud width Δ and the density maximum magnitude M is crucial. In particular, if both of the cloud parameters Δ and M were independent, these theoretical constructions would be different.[73]

Let us turn to a rather general idea that the subject compares two clouds $\mathbb{C}_1$ and $\mathbb{C}_2$ dealing with them as whole entities. In this case the cloud overlapping seems to be the only one quantity allowing for such mental operations and being accessible to the mind directly. Indeed when the region of the cloud location is actually the same for both the clouds $\mathbb{C}_1$ and $\mathbb{C}_2$, i.e., their overlapping is about 100%, the subject just cannot recognize the difference in their positions on the m-axis. When the difference $\delta M = M_1 - M_2$ becomes essential whereas the cloud widths Δ_1 and Δ_2 do not differ drastically, the subject is able to identify two similar regions $\delta_1\mathbb{C}$ and $\delta_2\mathbb{C}$ where the cloud $\mathbb{C}_1$ or the cloud $\mathbb{C}_2$ dominates in density. Figure 5.11 I represents these regions by the fragments $\mathbb{S}_-$ and $\mathbb{S}_+$ of the plot where the difference $\Phi_1(m) - \Phi_2(m)$ of cloud densities takes negative or positive values, $\Phi_1(m) - \Phi_2(m) \lessgtr 0$, respectively. When the subject recognizes the appearance of the two regions, he is able to order the clouds $\mathbb{C}_1$ and $\mathbb{C}_2$ along the m-axis, which is the essence of the mechanism governing the cloud ordering. As demonstrated below, the appearance of the regions $\delta_1\mathbb{C}$ and $\delta_2\mathbb{C}$ and their growth can be quantified by the degree of cloud overlapping instead of the difference δM in the density maximum magnitudes.

Figure 5.11 illustrates that, first, in the case when the cloud overlapping as the criterion of cloud proximity to each other is efficient, the fragments $\mathbb{S}_-$ and $\mathbb{S}_+$ representing the regions $\delta_1\mathbb{C}$ and $\delta_2\mathbb{C}$ must be similar in form and area. Indeed, as shown in Fig. 5.11 II, the low degree of cloud overlapping can be caused by the difference of the clouds in form rather than the position of the density maximum on the m-axis. In particular, the cloud width Δ must not change substantially when the

[73] Moreover, in the case when $M_1 \approx M_2$ whereas $\Delta_1 \ll \Delta_2$, the comparison of the clouds $\mathbb{C}_1$ and $\mathbb{C}_2$ as *whole entities* with respect to which one is characterized by lower (higher) magnitudes is an ill-defined problem.

density maximum magnitude M varies on scales about Δ. Ansatz (5.113) meets the letter requirement for $\kappa \ll 1$.

Second, their area related directly to the mean value of the difference $\Phi_1(m) - \Phi_2(m)$ inside them can quantify the cloud proximity; the larger their area, the larger the distance between the clouds $\mathbb{C}_1$ and $\mathbb{C}_2$.

In mathematical terms the cloud overlapping is specified by the expression

$$\langle \mathbb{C}_1 | \, \mathbb{C}_2 \rangle \stackrel{\text{def}}{=} \frac{1}{Z_1 Z_2} \int_{-\infty}^{\infty} dm \, \Phi_1(m) \Phi_2(m) = \int_{-\infty}^{\infty} dm \, \phi_1(m) \phi_2(m) \,. \tag{5.119}$$

Using the same construction let us quantify the areas of the fragments $\mathbb{S}_-$ and $\mathbb{S}_+$ in terms of quantities

$$S_- = \frac{1}{2} \int_{-\infty}^{\infty} dm \big[\phi_1(m) - \phi_2(m)\big]^2 \Theta_-\big(\phi_1(m) - \phi_2(m)\big) \,, \tag{5.120a}$$

$$S_+ = \frac{1}{2} \int_{-\infty}^{\infty} dm \big[\phi_1(m) - \phi_2(m)\big]^2 \Theta_+\big(\phi_1(m) - \phi_2(m)\big) \,, \tag{5.120b}$$

where the step-wise functions $\Theta_-(\ldots)$ and $\Theta_+(\ldots)$ are defined as

$$\Theta_-(x) = \begin{cases} 0, & x \geq 0 \\ 1, & x < 0 \end{cases} \qquad \text{and} \qquad \Theta_+(x) = \begin{cases} 1, & x \geq 0 \\ 0, & x < 0 \end{cases} .$$

It should be noted that, strictly speaking, the quantities S_- and S_+ are related to the difference between the normalized cloud densities $\phi_1(m) - \phi_2(m)$ rather than the cloud densities $\Phi_1(m) - \Phi_2(m)$. However, we have decided to use the quantities S_- and S_+ introduced via Exps. (5.120) to avoid the over-complication of our constructions without a particular reason. The matter is that the replacement of ϕ_i for Φ_i in Exps. (5.120) leads to the difference insignificant in comparison with the uncertainty of subject's effort perception. Another reason is discussed in Sect. 5.9.2.

Due to the equality $\Theta_-(x) + \Theta_+(x) = 1$, the quantities S_-, S_+, and cloud overlapping degree $\langle \mathbb{C}_1 | \, \mathbb{C}_2 \rangle$ are related as

$$S_- + S_+ = 1 - \langle \mathbb{C}_1 | \, \mathbb{C}_2 \rangle \tag{5.121}$$

and, thus,

$$S_- \approx S_+ \approx \frac{1}{2}\Big(1 - \langle \mathbb{C}_1 | \, \mathbb{C}_2 \rangle\Big) \,. \tag{5.122}$$

Figure 5.12 illustrates the dependence of the quantities S_1, S_2, and the cloud overlapping $\langle 1 | \, 2 \rangle$ on the difference δM obtained numerically for the clouds of the Gaussian

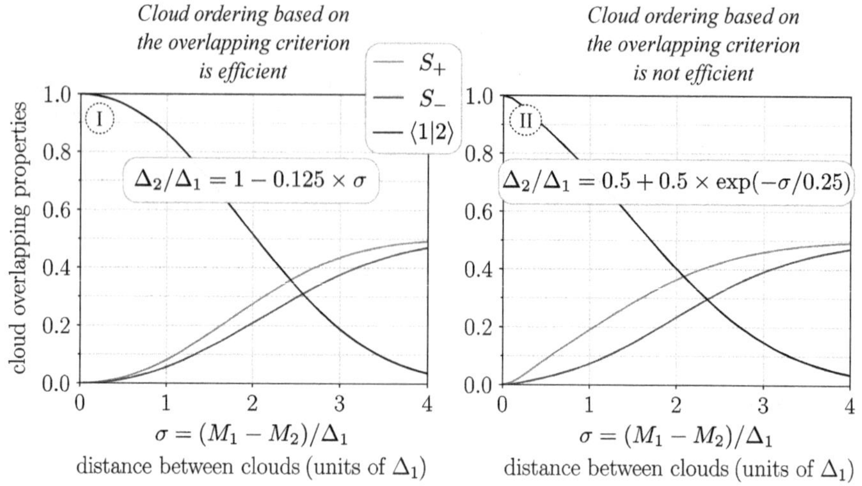

Fig. 5.12 The dependence of S_1, S_2, and the cloud overlapping $\langle 1|\ 2\rangle$ on the difference δM obtained numerically for the clouds of the Gaussian form

form

$$\phi(m) = \frac{1}{(2\pi\Delta^2)^{1/4}} \exp\left\{-\frac{(m-M)^2}{4\Delta^2}\right\} \tag{5.123}$$

meeting the condition $\langle\phi|\,\phi\rangle$ 1. Panel I exhibits the obtained results for ansatz (5.113) with the coefficient $\kappa = 0.125$. For comparison, Panel II exhibits these dependencies when the cloud width $\Delta(M)$ varies with the density maximum magnitude M strongly and, as a result, the cloud overlapping criterion becomes inefficient in ordering the clouds (see Fig. 5.11 II).

In summary, in the given Appendix we have demonstrated that the 1D-clouds $\{\mathbb{C}\}$ whose width $\Delta(M)$:[74]

- on the one hand, is directly determined by the density maximum magnitude M of the corresponding cloud,
- on the other hand, depends on M weakly or, speaking more strictly, changes weakly for variations δM in the density maximum magnitude about the cloud width, $\delta M \sim \Delta$

can be ordered as whole entities along the m-axis. This ordering implies the possibility of cloud pairwise classification with respect to which one of two clouds is lower (higher) in magnitude or their equivalence.

[74] The $\Delta(M)$-dependence specified by ansatz (5.113) with the coefficient κ about the Ekman fraction, $\kappa \sim \kappa_e \approx 0.03$ and, naturally, by ansatz (5.114) meets these requirements. It should be noted that ansatz (5.113) reflects the basic regularities of physical stimulus perception (Sect. 1.3) and ansatz (5.114) is used in the description of mental effort (Sect. 5.4.1).

Within the mind-in-the-self this probabilistic classification of a given pair of clouds $\{\mathbb{C}_1, \mathbb{C}_2\}$ is reduced to the choice between the following categories:

- $\mathbb{C}_1 \simeq \mathbb{C}_2$ with the probability $P_=$ and
- one of two categories with the probability $1 - P_=$:
  - $\mathbb{C}_1 \prec \mathbb{C}_2$ or
  - $\mathbb{C}_1 \succ \mathbb{C}_2$

  depending on the relation between the density maximum magnitudes of these clouds, namely, $M_1 < M_2$ or $M_1 > M_2$, respectively, whose particular values are inaccessible to the mind.

The probability $P_=$ is determined by the cloud overlapping whose degree $\langle\mathbb{C}_1|\,\mathbb{C}_2\rangle$ is accessible to the mind and given by Exp. (5.119). The general properties of the cloud overlapping enable us to set that the probability $P_=$ is a certain function of $\langle\mathbb{C}_1|\,\mathbb{C}_2\rangle$, namely,

$$P_= = G\big(\langle\mathbb{C}_1|\,\mathbb{C}_2\rangle\big), \tag{5.124}$$

where $G(x)$ is a monotonically increasing function of its argument x such that $G(0) = 0$, $G(1) = 1$, e.g.,

$$G(x) = x^{\alpha_m}, \tag{5.125}$$

where the exponent $\alpha_m > 0$ is a model parameter. Naturally, more complex ansätze for the $G(x)$-dependence are possible.

5.9.2 1D-Clouds: Comparison by Uncertainty

Let us discuss the comparison of two clouds $\mathbb{C}_1$ and $\mathbb{C}_2$ having the same mean magnitude M—the magnitude of cloud density maximum—but being different in their widths $\Delta_1 \neq \Delta_2$ (Fig. 5.13(I)). These clouds can represent the mental images of two physical objects or mental images rooted in purely mental phenomena like experienced effort of monitoring object dynamics. The given clouds are described in the same way as in Sect. 5.9.1 except for that relationship (5.112) no longer holds. Namely, a cloud $\mathbb{C}_i$ $(i = 1, 2)$ is represented by the density $\Phi_i(m)$, Exp. (5.110), distributed in the space of magnitudes $\mathbb{R}^M$.

As in Sect. 5.9.1 their comparison by the subject is described via the difference between the normalized cloud densities $\phi_1(m) - \phi_2(m)$, namely,

$$S = \frac{1}{2}\int\limits_{-\infty}^{\infty} dm\big[\phi_1(m) - \phi_2(m)\big]^2 = 1 - \langle\mathbb{C}_1|\mathbb{C}_1\rangle\,, \tag{5.126}$$

where the overlapping $\langle\mathbb{C}_1|\mathbb{C}_1\rangle$ of the $\mathbb{C}_1$ and $\mathbb{C}_2$ may be regarded as the measure of their proximity. So the probability of identifying the two clouds, $\mathbb{C}_1 \simeq \mathbb{C}_2$, is again

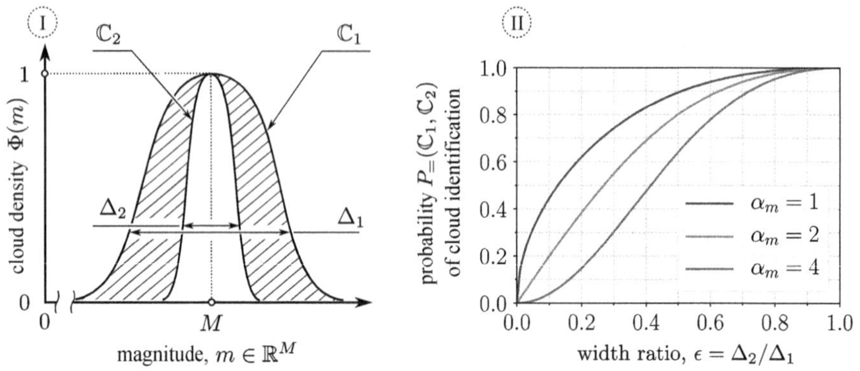

Fig. 5.13 Illustration of a plausible mechanism describing subject's comparison of the clouds $\mathbb{C}_1$ and $\mathbb{C}_2$ having the same mean magnitude M and different width Δ_1 and Δ_2 in the space $\mathbb{R}^M$ of magnitudes. Panel I visualizes the cloud difference as the hatched region between the cloud densities. Panel illustrates the probability of cloud identification $P_=(\mathbb{C}_1, \mathbb{C}_2)$ specified by Exp. (5.128) *vs.* the width ratio $\epsilon = \Delta_2/\Delta_1$. The case of the exponent $\alpha_m = 2$ is considered in the text, whereas here also the cases $\alpha_m = 1$ and $\alpha_m = 4$ are shown for comparison

have form (5.124)

$$P_=(\mathbb{C}_1, \mathbb{C}_2) = \big(\langle \mathbb{C}_1 | \, \mathbb{C}_2 \rangle\big)^{\alpha_m} \tag{5.127}$$

and for ansatz (5.123)

$$P_=(\mathbb{C}_1, \mathbb{C}_2) = \left(\frac{2\epsilon}{1+\epsilon^2}\right)^{\alpha_m/2} . \tag{5.128}$$

To determine the particular value of the exponent α_m a more sophisticated analysis of mental comparison is required, which does not belong to the scope of the present book. Here, turning to some analogy with quantum mechanics, we illustrate the dependence of the identification probability $P_=$ on the ratio $\epsilon = \Delta_2/\Delta_1$ set the exponent $\alpha_m = 2$; the result is shown in Fig. 5.13 (II).

It should be noted that in order to quantify the difference between the clouds $\mathbb{C}_1$ and $\mathbb{C}_2$, in principle, the density $\Phi_i(m)$ can be used instead of the normalized density $\phi_i(m)$ (within the relevant normalization of difference (5.126)). The choice of the normalized density $\phi_i(m)$ reflects the assumption that the subject focuses on the region where the cloud with the minimal width is located. Indeed, as it follows from Exp. (5.110), at the magnitude M where all the cloud densities take their maxima the smaller the width, the higher the value of the normalized density. Constructing a more sophisticated description of the cloud comparison by the subject is outside the scope of this book.

The conducted analysis enables us to pose a question about *the second-type uncertainty* in human cognition. By way of example, let us consider the perception of some physical stimulus under influence of a parallel task whose performance also needs attention. In the given case the uncertainty Δ of stimulus perception depends on the task difficulty D, i.e., $\Delta = \Delta(D)$. The posed question is when this effect becomes recognizable as the task difficulty grows gradually. In other words, when the subject starts to recognize that the accuracy of stimulus perception is remarkable decreased. Put differently, the question concerns the just noticeable difference in the change of just noticeable difference of perceived stimulus intensity, which is a reason for calling it second-type uncertainty of stimulus perception.

Treating the mental image of perceived stimulus as a certain cloud $\mathbb{C}$ of mental space (Lubashevsky 2019), we may use the formalism of the given Appendix to quantify this second-type uncertainty $\widetilde{\Delta}$. Indeed, dealing with two clouds $\mathbb{C}_1$ and $\mathbb{C}_2$ with the same mean magnitude M the difference in their widths, $\widetilde{\Delta} = |\Delta_1 - \Delta_2|$, becomes recognizable when the probability of identifying the two clouds is about 0.5, i.e., $P_=(\mathbb{C}_1, \mathbb{C}_2) \sim 0.5$. Turning to Fig. 5.13 (II) we see that the desired threshold (fuzzy threshold) admits the estimate

$$\widetilde{\Delta} = \kappa_{2m}\, \Delta \tag{5.129}$$

where the fraction $\kappa_{2m} \sim 0.5$.

When the difficulty of the parallel task is quantified as the effect it has on the signal perception, the perceived variance δD in the task difficulty D must correspond to the recognizable difference in the accuracy of stimulus perception. When it follows that the estimates

$$\frac{\widetilde{\Delta}}{\Delta} \sim \frac{d \ln \Delta}{d \ln D} \cdot \frac{\delta D}{D} \sim \kappa_{2m}. \tag{5.130}$$

must hold. Thereby the relationship $\Delta(D)$ should be of the power-law form

$$\Delta(D) \propto D^g \quad \text{with the exponent } g \sim 1. \tag{5.131}$$

In particular, the resulting estimate $\delta D \sim \kappa_{2m} D$ enables us to call the coefficient κ_{2m} the Ekman fraction of second-type perception threshold.

In the main text of the book we turn to these results within the analysis of the mental effort of monitoring, including the corresponding uncertainty of its experience.

5.10 1D-Clouds of Mental Images: Confinement

The notion of 1D-clouds $\{\mathbb{C}\}$ introduced above describes the mental images of perceived physical stimuli or similar images as entities of the mind-in-the-self which bears an intrinsic uncertainty in the attributed magnitudes. This uncertainty reflects the bounded capacity of human cognition and is responsible for a certain fuzziness of such mental images. Put differently, each cloud $\mathbb{C}$ may be conceived of as a certain region of the space $\mathbb{R}^M$ with blurred boundaries, which justifies the introduction of the cloud density function continuously changing in the space $\mathbb{R}^M$.

The clouds $\{\mathbb{C}\}$ represent conscious evaluation of perceived objects and processes. So, for the general reason, it is clear that a given cloud $\mathbb{C}$ cannot contain points rather distant from its center. Therefore, on the one hand, a cloud is fuzzy object, on the other hand, there should be some boundary separating magnitudes that can be attributed to the corresponding image from those related to the image in no way.

The present Appendix is devoted to reconciling the blur (fuzziness) of 1D-clouds with the existence of their infinitely thin boundaries.

5.10.1 Phase Transitions in Binary Categorization

Below we present the results of our experiments (summarized by Lubashevsky et al. 2018) on binary categorization of stimuli from two different sensory modalities, auditory and visual ones. The goal of this presentation is to demonstrate that such categorization is governed by unconscious neural mechanisms together with mental processes of higher level. Since the binary categorization and the evaluation of perceived stimuli are closely related to each other in governing mechanisms, we may expand the found properties of binary categorization upon the phenomena under consideration.

Leaping ahead we note that the found transitions between deliberate and automatic stages of categorization propose a solution to the seeming contradiction between the fuzziness of these clouds and their possession of strict boundaries. The pivot point of this solution is related to the dual-processing account of human behavior widely used in cognitive psychology. It holds that there are two distinct processing systems available for cognitive tasks. System 1 is fast, automatic and non-conscious, System 2 is slow, controlled and conscious. For a detailed discussion a reader may be addressed, e.g., to Rustichini (2008), Evans (2008, 2011), Evans and Stanovich (2013), Barrouillet (2011), Kahneman (2011), Gómez-Chacón et al. (2014), Handley and Trippas (2015). Within the realm of the mind-in-the-self the analyzed mental images are characterized by fuzziness. The transitions between System 1 and System 2 endow these images with strict boundaries.

Now let us discuss the particular properties found in our experiments and their setups.

Gray color categorization: Each trial of color categorization was implemented as follows. A random integer $I \in [0, 255]$ is generated and some area on PC monitor is filled with the gray color $G(I) := \text{RGB}(I, I, I)$. Then a subject has to classify the visualized gray shade $G(I)$ according to his/her perception into two possible categories, "light gray" and "dark gray." A made choice is recorded via pressing one of two joystick buttons. Then a mosaic pattern of various shades of gray is visualized for 500 ms to depress a possible interference between color perception in successive trials that can be caused by human iconic memory. After that a new number I is generated and the next trial starts. The moment when a subject presses the button are also recorded, which gives the decision time in the current trial.

The experiments we presented here were set up as follows (Lubashevsky and Watanabe 2016). Four subjects, two female and two male students of age 21–22 were involved. The experiments spanned 5 successive days, for each subject the total number of data records was 2,000 data-points per day and finally 10,000 for 5 days. One day set comprised four blocks of 15 min experiments separated by 3 min rest. For each subject the total data-set aggregates all the records collected during 5 days. No special instructions were given to the subjects about the necessity to make decision in selecting categories as quickly as possible.

Figure 5.14 illustrates the obtained results which allow us to draw the following conclusions.

1. In the log-normal scales the asymptotics of psychometric functions for both the categories can be approximated by a linear dependence on the shade number I.
2. As the analyzed shade of gray penetrates deeper in the region of rare events (rare choice of inappropriate category) this linear asymptotics is replaced by the probability of choosing the inappropriate category that does not decrease or even can increase with the further change in the shade number I. The transition between these modes of the choice probability behavior is well pronounced, which is noted by arrows in Fig. 5.14.
3. The increase of the mean decision time in the region of choice uncertainty typically exceeds 1 s, which is substantially longer than the upper boundary of human response delay controlled by purely physiological processes. It argues for a significant contribution of mental processes to the categorization in the region of its essential uncertainty.

Vowel sound categorization: Using the "Create Sound from Vowel Editor..." menu item in Praat acoustic analysis software (Boersma and Weenink 2018), a total of 68 sound files were synthesized. Each sound was a 100-millisecond vowel with a different second-formant (F2) frequency, which ranged from 810 Hz ([o]) to 2,150 Hz ([e]) in 20 Hz steps. The first-formant (F1) was kept constant at 400 Hz for all sounds. So, the first sound's (F1, F2) values in Hz were (400, 810); the second sound's values were (400, 830), the third sound's values were (400, 850), and so on up to the 68th sound, which had values (400, 2150). Thus, the 68 sounds varied along a single dimension—that of F2. The pitch of all sounds was kept constant at 140 Hz—a pitch within the normal speaking range of a human voice. The 68 unique audio files were randomly ordered to make a "cycle," and this was done 12 times to make 12 different cycles. A randomly ordered group of 12 cycles was considered 1 trial and contained $68 \times 12 = 816$ vowel sounds. Each subject completed 10 trials (each with a random order of cycles) with a short break between each one, for a total of 8,160 vowel sounds.

Two healthy subjects (female and male) with no reported hearing problems participated in this research. They were both Japanese 4th-year undergraduate students who were in their early twenties. E-prime 2.0 software (*Psychology Software Tools*) and

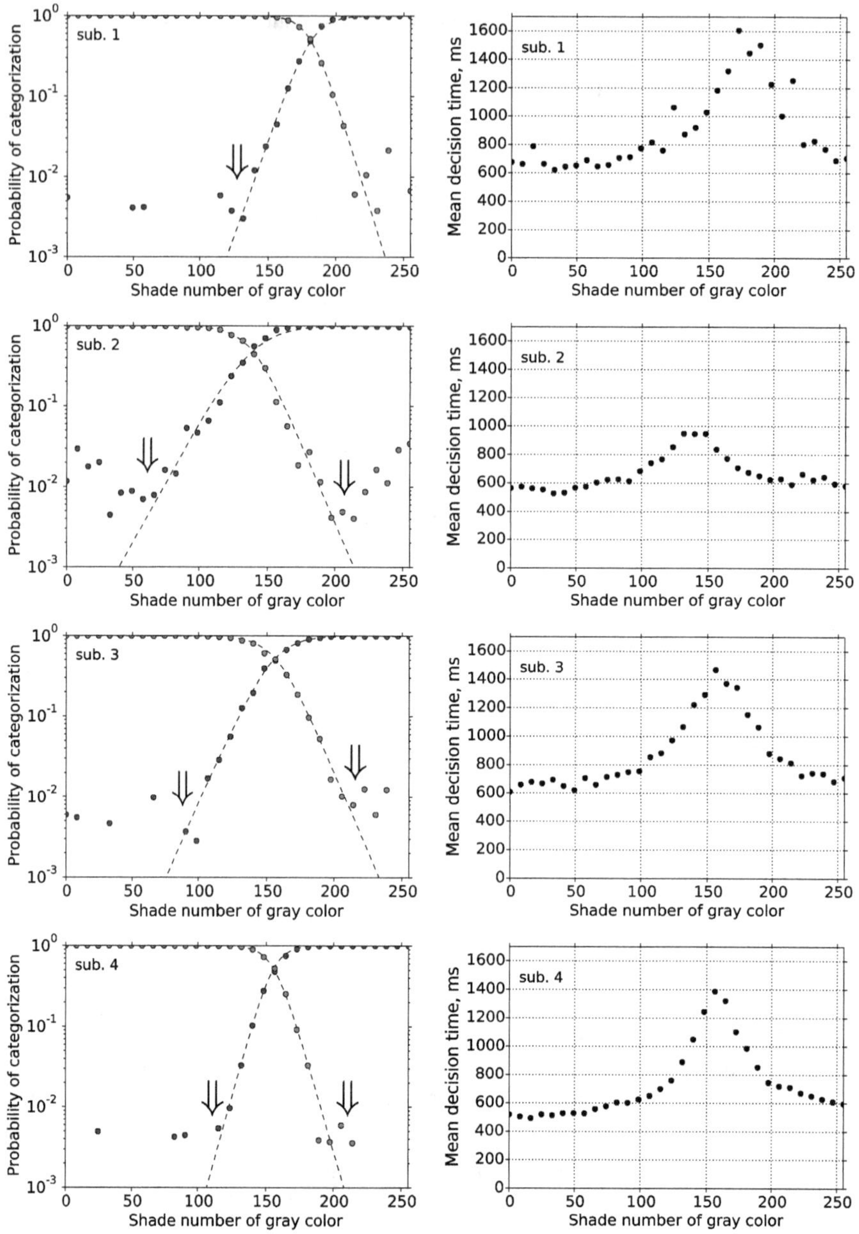

Fig. 5.14 *Left column*: Psychometric functions of categorization of shades of gray, data points corresponding to the "light-gray" category and the "dark-gray" category are shown in red and blue, respectively. Dashed lines represent fitting functions given by the logistic approximation. Arrows point to cusps in the psychometric function behavior. *Right column*: Mean decision time in gray color categorization depending on the shade number of gray. After Lubashevsky and Watanabe (2016)

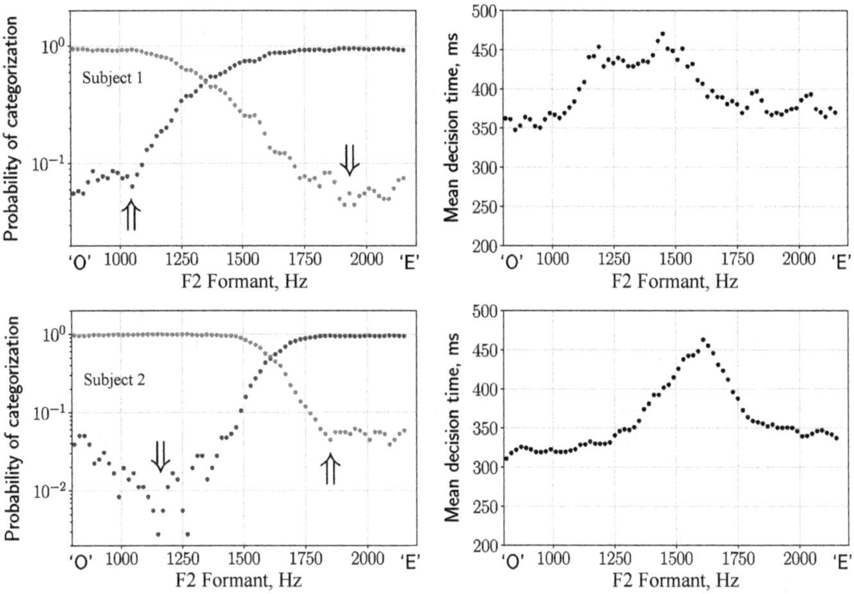

Fig. 5.15 *Left column*: Psychometric functions of vowel sound categorization, data points corresponding to [o] and [e] categories are shown in red and blue, respectively. Arrows point to cusps in the psychometric function behavior. *Right column*: Mean decision time in vowel sound categorization depending on the F2 Formant. After Lubashevsky et al. (2018)

a 5-button "Chronos" multi-functional response and stimulus device were used for recording subject actions. The Chronos device has millisecond accuracy and consistent sound output latencies across machines. Subjects listened to the stimuli through JVC headphones and were instructed to react as quickly as possible to identify the stimuli that they heard.

Figure 5.15 exhibits the obtained results which also argue for Conclusions 1 and 2 stated in the previous subsection for the color categorization. The main difference is that the mean decision time in its maximum attains the upper boundary of the response delay time usually attributed to the information processing by neural networks. However, the similarity of the asymptotic behavior of psychometric functions in both the cases allows us to hypothesize that mental processes play a significant role also in the sound categorization in the region of its substantial uncertainty. The reaction of subjects instructed to respond as quickly as possible just corresponds to the minimal delay of the conscious response coinciding with the maximum of physiological delay such that no gap in the human response delay time appears.

Conclusion: As a plausible explanation of the presented results we may accept the following (Lubashevsky et al. 2018).

– The choice between the auditory and visual categories is governed by a universal central mechanism (cf. Sect. 1.3.2). It is characterized by the linear asymptotics of psychometric functions in log-normal scales, which corresponds to significant uncertainty in categorization.

- There is a sharp transition between the linear asymptotic behavior of psychometric functions and their behavior when the choice of appropriate category is obvious. We relate the linear asymptotic behavior to conscious choice governed by slow System 2; it is characterized by the regular growth of decision time as the choice uncertainty increases. The choice of appropriate category when it is obvious seems to be governed by automatic, fast System 1. It explains the relative increase in the probability of choosing the inappropriate category by human errors in motor behavior (pressing a joystick button) when the conscious control is depressed.
- When the choice uncertainty is high, mental processes seem to affect substantially the decision time and its maximum can be used to quantify the relative contribution of conscious and unconscious processes as a whole. In particular, the shorter the maximal decision time, the higher the contribution of unconscious component, the higher the probability of motor errors in the case of System 1 control.

5.10.2 Boundary of 1D-Clouds

As far as the concept of 1D-clouds in the description of physical stimulus perception and mental evaluation of bodily effort is concerted, we introduced such mental entities for describing conscious (mainly AM-conscious) estimation of observed situation. The presented experimental results illustrate the properties of binary categorization which in its essence is closely related to physical stimulus perception.

The found transitions between the asymptotics of psychometric functions demonstrate that we can single out two modes of the mental evaluation of perceived stimuli which are entirely distinct in properties. They are the automatic categorization (System 1) and the deliberate evaluation (System 2). The transitions between them occur when the uncertainty in decision making becomes rather low and gets some threshold understood in mathematical sense—a quantity corresponding to step-wise transitions in some property. The existence of such thresholds is a typical feature for physical systems rather than mental operations. Turing to the case under consideration, we suppose the emergence of the found threshold to be rooted in the neural mechanisms governing the processing of perceived information. The detailed functioning of these mechanisms is not accessible to the mind and only the results of information processing become accessible to the mind at the level of E-consciousness.

The analyzed 1D-clouds represent the conscious evaluation of perceived stimulus intensity or, speaking more strictly, the magnitude of the corresponding mental images. The cloud width Δ characterizes the uncertainty in evaluating this magnitude. So for a given cloud $\mathbb{C}$ with the density maximum magnitude M, values m of magnitude such that $|m - M| \gg \Delta$ cannot be related to the cloud $\mathbb{C}$ based on AM-conscious operations. Put differently, the cloud $\mathbb{C}$ does not contain the points with magnitudes $|m - M| \gg \Delta$ for certainty. The cloud density function $\Phi(m)$ characterizes the degree $1 - \Phi(m)$ of certainty that a given point with the magnitude m

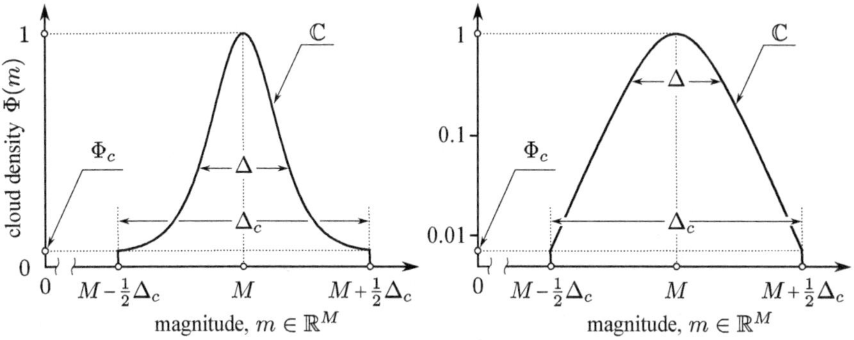

Fig. 5.16 The confinement of 1D-cloud $\mathbb{C}$ in the space $\mathbb{R}^M$. The figure depicts the cloud $\mathbb{C}$ as the cloud density pattern $\Phi(m)$ along the m-axis whose maximum is located at M and characteristic width is Δ. The left and right panels differ from each other in that the Φ-axis is presented in the normal and logarithmic scales, respectively. The value $\Phi_c \ll 1$ bounds from below all the possible values of cloud density, which immediately leads to the confinement of the cloud $\mathbb{C}$ within the region $[M - \Delta_c/2, M + \Delta_c/2]$ of thickness Δ_c

does not belong to the cloud $\mathbb{C}$. Therefore, the presented results about binary categorization enable us to accept that there is some critical value $\Phi_c \ll 1$ such that the density of any cloud $\mathbb{C}$ cannot be less than $\Phi_c \ll 1$. In this case such a cloud $\mathbb{C}$ takes on a boundary $\partial\mathbb{C}$ specified by the condition

$$\mathbb{C} \Rightarrow \partial\mathbb{C} = \{m : \Phi_{\mathbb{C}}(m) = \Phi_c\}. \tag{5.132}$$

and becomes confined in the space $\mathbb{R}^M$ (Fig. 5.16). Turing to the experimental results presented in Figs. 5.14 and 5.15 the value of Φ_c can be estimated as $\Phi_c \sim 0.01$.

References

Aarts, H., & Custers, R. (2012). Unconscious goal pursuit: Nonconscious goal regulation and motivation. In R. M. Ryan (Ed.), *The Oxford handbook of human motivation* (pp. 232–247). New York, NY: Oxford University Press.

Abbiss, C. R., Peiffer, J. J., Meeusen, R., & Skorski, S. (2015). Role of ratings of perceived exertion during self-paced exercise: What are we actually measuring? *Sports Medicine, 45*(9), 1235–1243.

Abd-Elfattah, H. M., Abdelazeim, F. H., & Elshennawy, S. (2015). Physical and cognitive consequences of fatigue: A review. *Journal of Advanced Research, 6*(3), 351–358. Editors and International Board Member collection.

Agam, Y., Carey, C., Barton, J. J. S., Dyckman, K. A., Lee, A. K. C., Vangel, M., & Manoach, D. S. (2013). Network dynamics underlying speed-accuracy trade-offs in response to errors. *PLOS ONE, 8*(9), e73692 (10 pages).

Alvarez, G. A., Horowitz, T. S., Arsenio, H. C., DiMase, J. S., & Wolfe, J. M. (2005). Do multielement visual tracking and visual search draw continuously on the same visual attention resources? *Journal of Experimental Psychology: Human Perception and Performance, 31*(4), 643–667.

Ampel, B. C., Muraven, M., & McNay, E. C. (2018). Mental work requires physical energy: Self-control is neither exception nor exceptional. *Frontiers in Psychology, 9*, Article 1005 (11 pages).

Ampel, B. C., O'Malley, E. E., & Muraven, M. (2016). Self-control and motivation: Integration and application. In E. R. Hirt, J. J. Clarkson & L. Jia (Eds.), *Self-regulation and ego control* (pp. 125–141). Cambridge, MA: Elsevier Inc.

Anderson, J. R. (2015). *Cognitive Psychology and Its Implications* (8th ed.). New York, NY: Worth Publishers, A Macmillan Education Company.

Anderson, B., & Druker, M. (2013). Attention improves perceptual quality. *Psychonomic Bulletin & Review, 20*(1), 120–127.

André, N., Audiffren, M., & Baumeister, R. F. (2019). An integrative model of effortful control. *Frontiers in Systems Neuroscience, 13*, Article 79 (22 pages).

Angelaki, D. E., Gu, Y., & DeAngelis, G. C. (2009). Multisensory integration: Psychophysics, neurophysiology, and computation. *Current Opinion in Neurobiology, 19*(4), 452–458.

Aragonés, D., Balagué, N., Hristovski, R., Pol, R., & Tenenbaum, G. (2013). Fluctuating dynamics of perceived exertion in constant-power exercise. *Psychology of Sport and Exercise, 14*(6), 796–803.

Ariely, D. (1998). Combining experiences over time: The effects of duration, intensity changes and on-line measurements on retrospective pain evaluations. *Journal of Behavioral Decision Making, 11*(1), 19–45.

Baird, J. C. (1997). *Sensation and judgment: Complementarity theory of psychophysics*. Mahwah, NJ: Erlbaum.

Balagué, N., Hristovski, R., & García-Retortillo, S. (2020). Perceived exertion: Dynamic psychobiological model of exercise-induced fatigue. In G. Tenenbaum & R. C. Eklund (Eds.), *Handbook of sport psychology, volume II: Exercise, methodologies, & special topics* (4th ed., pp. 950–965). Hoboken, NJ: Wiley.

Balagué, N., Hristovski, R., Aragonés, D., & Tenenbaum, G. (2012). Nonlinear model of attention focus during accumulated effort. *Psychology of Sport and Exercise, 13*(5), 591–597.

Balagué, N., Hristovski, R., García, S., Aguirre, C., Vázquez, P., Razon, S., & Tenenbaum, G. (2015). Dynamics of perceived exertion in constant-power cycling: Time- and workload-dependent thresholds. *Research Quarterly for Exercise and Sport, 86*(4), 371–378.

Bambrah, V., Hsu, C.-F., Toplak, M. E., & Eastwood, J. D. (2019). Anticipated, experienced, and remembered subjective effort and discomfort on sustained attention versus working memory tasks. *Consciousness and Cognition, 75*, 102812 (16 pages).

Bargh, J. A. (1989). Conditional automaticity: Varieties of automatic influence in social perception and cognition. In J. S. Uleman & J. A. Bargh (Eds.), *Unintended thought* (pp. 3–51). New York, NY: The Guilford Press.

Bargh, J. A., Gollwitzer, P. M., Lee-Chai, A., Barndollar, K., & Trötschel, R. (2001). The automated will: Nonconscious activation and pursuit of behavioral goals. *Journal of Personality and Social Psychology, 81*(6), 1014–1027.

Barrouillet, P. (2011). Dual-process theories and cognitive development: Advances and challenges. *Developmental Review, 31*(2–3), 79–85. Special Issue: Dual-Process Theories of Cognitive Development.

Bater, L. R., & Jordan, S. S. (2019). Selective attention. In V. Zeigler-Hill & T. K. Shackelford (Eds.), *Encyclopedia of personality and individual differences* (pp. 1–4). Cham: Springer.

Baumeister, R. F., & Vohs, K. D. (2016). Strength model of self-regulation as limited resource. In M. P. Z. James M. Olson (Ed.), *Advances in experimental social psychology* (Vol. 54, pp. 67–127). Cambridge, MA: Elsevier Inc.

Baumeister, R. F., Bratslavsky, E., Muraven, M., & Tice, D. M. (1998). Ego depletion: Is the active self a limited resource? *Journal of Personality and Social Psychology, 74*(5), 1252–1265.

Baumeister, R. F., Muraven, M., & Tice, D. M. (2000). Ego depletion: A resource model of volition, self-regulation, and controlled processing. *Social Cognition, 18*(2), 130–150.

Baumeister, R. F., Tice, D. M., & Vohs, K. D. (2018). The strength model of self-regulation: Conclusions from the second decade of willpower research. *Perspectives on Psychological Science, 13*(2), 141–145.

Baumeister, R. F., Vohs, K. D., & Tice, D. M. (2007). The strength model of self-control. *Current Directions in Psychological Science, 16*(6), 351–355.

Bayne, T., & Spence, C. (2013). Multisensory perception. In M. Matthen (Ed.), *The Oxford handbook of philosophy of perception* (pp. 603–620). Oxford, UK: Oxford University Press.

Bays, P. M., & Husain, M. (2008). Dynamic shifts of limited working memory resources in human vision. *Science, 321*(5890), 851–854.

Beck, J. M., Latham, P. E., & Pouget, A. (2011). Marginalization in neural circuits with divisive normalization. *Journal of Neuroscience, 31*(43), 15310–15319.

Bede, B. (2013). *Mathematics of fuzzy sets and fuzzy logic*. Berlin: Springer.

Bedell, H. E., Tong, J., & Aydin, M. (2010). The perception of motion smear during eye and head movements. *Vision Research, 50*(24), 2692–2701.

Beilock, S. L., Bertenthal, B. I., Hoerger, M., & Carr, T. H. (2008). When does haste make waste? speed-accuracy tradeoff, skill level, and the tools of the trade. *Journal of Experimental Psychology: Applied, 14*(4), 340–352.

Beilock, S. L., Bertenthal, B. I., Mccoy, A. M., & Carr, T. H. (2004). Haste does not always make waste: Expertise, direction of attention, and speed versus accuracy in performing sensorimotor skills. *Psychonomic Bulletin & Review, 11*(2), 373–379.

Beilock, S. L., & Gray, R. (2012). From attentional control to attentional spillover: A skill-level investigation of attention, movement, and performance outcomes. *Human Movement Science, 31*(6), 1473–1499.

Berkman, E. T., Kahn, L. E., & Livingston, J. L. (2016). Valuation as a mechanism of self-control and ego depletion. In E. R. Hirt, J. J. Clarkson & L. Jia (Eds.), *Self-regulation and ego control* (pp. 255–279). Cambridge, MA: Elsevier Inc.

Biggs, A. T., & Gibson, B. S. (2013). Learning to ignore salient color distractors during serial search: Evidence for experience-dependent attention allocation strategies. *Frontiers in Psychology, 4*, Article 326 (13 pages).

Bijleveld, E. (2018). The feeling of effort during mental activity. *Consciousness and Cognition, 63*, 218–227.

Block, R. A., & Gruber, R. P. (2014). Time perception, attention, and memory: A selective review. *Acta Psychologica, 149*(Supplement C), 129–133. Including Special section articles of Temporal Processing Within and Across Senses - Part-2.

Block, R. A., Hancock, P. A., & Zakay, D. (2010). How cognitive load affects duration judgments: A meta-analytic review. *Acta Psychologica, 134*(3), 330–343.

Boersma, P., & Weenink, D. (2018). Praat: Doing phonetics by computer [computer program], retrieved 11 May 2018 from http://www.praat.org. Version 6.0.40.

Boksem, M. A. S., Meijman, T. F., & Lorist, M. M. (2005). Effects of mental fatigue on attention: An ERP study. *Cognitive Brain Research, 25*(1), 107–116.

Boksem, M. A. S., Meijman, T. F., & Lorist, M. M. (2006). Mental fatigue, motivation and action monitoring. *Biological Psychology, 72*(2), 123–132.

Boksem, M. A. S., & Tops, M. (2008). Mental fatigue: Costs and benefits. *Brain Research Reviews, 59*(1), 125–139.

Boyas, S., & Guével, A. (2011). Neuromuscular fatigue in healthy muscle: Underlying factors and adaptation mechanisms. *Annals of Physical and Rehabilitation Medicine, 54*(2), 88–108.

Braun, D. A., Aertsen, A., Wolpert, D. M., & Mehring, C. (2009). Motor task variation induces structural learning. *Current Biology, 19*(4), 352–357.

Braun, D. A., Mehring, C., & Wolpert, D. M. (2010). Structure learning in action. *Behavioural Brain Research, 206*(2), 157–165.

Brehm, J. W., & Self, E. A. (1989). The intensity of motivation. *Annual Review of Psychology, 40*(1), 109–131.

Brehm, J. W., Wright, R. A., Solomon, S., Silka, L., & Greenberg, J. (1983). Perceived difficulty, energization, and the magnitude of goal valence. *Journal of Experimental Social Psychology, 19*(1), 21–48.

Breitmeyer, B. G., & Ganz, L. (1976). Implications of sustained and transient channels for theories of visual pattern masking, saccadic suppression, and information processing. *Psychological Review, 83*(1), 1-36.

Breitmeyer, B. G., & Öğmen, H. (2006). *Visual masking: Time slices through conscious and unconscious vision* (2nd ed.). Oxford, New York: Oxford University Press.

Brick, N. E., MacIntyre, T. E., & Campbell, M. J. (2016). Thinking and action: A cognitive perspective on self-regulation during endurance performance. *Frontiers in Physiology, 7*, Article 159 (7 pages).

Brown, S. W. (1997). Attentional resources in timing: Interference effects in concurrent temporal and nontemporal working memory tasks. *Perception & Psychophysics, 59*(7), 1118–1140.

Bruya, B., & Tang, Y.-Y. (2018). Is attention really effort? revisiting Daniel Kahneman's influential 1973 book attention and effort. *Frontiers in Psychology, 9*, Article 1133 (10 pages).

Bruya, B. (2010). Introduction: Toward a theory of attention that includes effortless attention and action. In B. Bruya (Ed.), *Effortless attention: A new perspective in the cognitive science of attention and action* (pp. 1–28). Cambridge, MA: The MIT Press.

Burnley, M., & Jones, A. M. (2016). Power-duration relationship: Physiology, fatigue, and the limits of human performance. *European Journal of Sport Science, 18*(1), 1–12.

Burr, D. C., & Morgan, M. J. (1997). Motion deblurring in human vision. *Proceedings of the Royal Society of London. Series B: Biological Sciences, 264*(1380), 431–436.

Burr, D. (1980). Motion smear. *Nature, 284*(5752), 164–165.

Burr, D., Ross, J., & Morrone, M. (1986). Seeing objects in motion. *Proceedings of the Royal Society of London - Biological Sciences, 227*(1247), 249–265.

Burr, D., & Thompson, P. (2011). Motion psychophysics: 1985–2010. *Vision Research, 51*(13), 1431–1456.

Burwick, T. (2014). The binding problem. *Wiley Interdisciplinary Reviews: Cognitive Science, 5*(3), 305–315.

Butz, M. V., Herbort, O., & Hoffmann, J. (2007). Exploiting redundancy for flexible behavior: Unsupervised learning in a modular sensorimotor control architecture. *Psychological Review, 114*(4), 1015–1046.

Cappuccio, M. L. (Ed.). (2019). *Handbook of embodied cognition and sport psychology*. Cambridge, MA: The MIT Press.

Cappuccio, M. L., Gray, R., Hill, D. M., Mesagno, C., & Carr, T. H. (2019). The many threats of selfconsciousness: Embodied approaches to choking under pressure in sensorimotor skills. In M. L. Cappuccio (Ed.), *Handbook of embodied cognition and sport psychology* (pp. 101–155). Cambridge, MA: The MIT Press.

Carrasco, M. (2014). Spatial covert attention: Perceptual modulation. In A. C. K. Nobre & S. Kastner (Eds.), *The Oxford handbook of attention* (pp. 183–230). Oxford, UK: Oxford University Press.

Carson, H. J., & Collins, D. (2015). The fourth dimension: A motoric perspective on the anxiety-performance relationship. *International Review of Sport and Exercise Psychology, 9*(1), 1–21.

Carter, E. C., Kofler, L. M., Forster, D. E., & McCullough, M. E. (2015). A series of meta-analytic tests of the depletion effect: Self-control does not seem to rely on a limited resource. *Journal of Experimental Psychology: General, 144*(4), 796–815.

Cattaneo, L., & Rizzolatti, G. (2009). The mirror neuron system. *Archives of Neurology, 66*(5), 557–560.

Chandrasekaran, C. (2017). Computational principles and models of multisensory integration. *Current Opinion in Neurobiology, 43*, 25–34.

Chang, Y., Labban, J., Gapin, J., & Etnier, J. (2012). The effects of acute exercise on cognitive performance: A meta-analysis. *Brain Research, 1453*, 87–101.

Chater, N., & Brown, G. D. (1999). Scale-invariance as a unifying psychological principle. *Cognition, 69*(3), B17–B24.

Chehade, S. S., & Vershynina, A. (2019). Quantum entropies. *Scholarpedia, 14*(2), 53131. revision #190401.
Chen, Z., & Cave, K. (2013). Perceptual load vs. dilution: The roles of attentional focus, stimulus category, and target predictability. *Frontiers in Psychology, 4*, Article 327 (14 pages).
Chen, S., Bedell, H. E., & Öğmen, H. (1995). A target in real motion appears blurred in the absence of other proximal moving targets. *Vision Research, 35*(16), 2315–2328.
Chica, A. B., & Bartolomeo, P. (2012). Attentional routes to conscious perception. *Frontiers in Psychology, 3*, Article 1 (12 pages).
Choi, I., Lee, J.-Y., & Lee, S.-H. (2018). Bottom-up and top-down modulation of multisensory integration. *Current Opinion in Neurobiology, 52*, 115–122.
Christensen, W., & Sutton, J. (2019). Mesh: Cognition, body, and environment in skilled action: A new introduction to "cognition in skilled action". In M. L. Cappuccio (Ed.), *Handbook of embodied cognition and sport psychology* (pp. 157–164). Cambridge, MA: The MIT Press.
Christensen, W., Sutton, J., & McIlwain, D. J. F. (2016). Cognition in skilled action: Meshed control and the varieties of skill experience. *Mind & Language, 31*(1), 37–66.
Christensen, W., Sutton, J., & McIlwain, D. (2019). Cognition in skilled action: Meshed control and the varieties of skill experience. In M. L. Cappuccio (Ed.), *Handbook of embodied cognition and sport psychology* (pp. 165–198). Cambridge, MA: The MIT Press.
Clarkson, J. J., Otto, A. S., Hassey, R., & Hirt, E. R. (2016). Perceived mental fatigue and self-control. In E. R. Hirt, J. J. Clarkson, & L. Jia (Eds.), *Self-regulation and ego control* (pp. 185–202). London: Elsevier, Academic Press.
Cohen, M. A., Cavanagh, P., Chun, M. M., & Nakayama, K. (2012). The attentional requirements of consciousness. *Trends in Cognitive Sciences, 16*(8), 411–417.
Coyle, E. F. (1995). Integration of the physiological factors determining endurance performance ability. *Exercise and Sport Sciences Reviews, 23*(1), 25–64.
Crowe, E. M., Howard, C. J., Attwood, A. S., & Kent, C. (2019). Goal-directed unequal attention allocation during multiple object tracking. *Attention, Perception, & Psychophysics, 81*(5), 1312–1326.
Csikszentmihalyi, M. (1975). *Beyond boredom and anxiety: The experience of play in work and games*. The Jossey-Bass behavioral science series. San Francisco, CA: Jossey-Bass Publishers.
Csikszentmihalyi, M. (1990). *Flow: The psychology of optimal experience*. New York, NY: Harper & Row Publishes.
Csikszentmihalyi, M., & Csikszentmihalyi, I. S. (Eds.). (1988). *Optimal experience: Psychological studies of flow in consciousness*. Cambridge, UK: Cambridge University Press.
Csikszentmihalyi, M., & Nakamura, J. (2010). Effortless attention in everyday life: A systematic phenomenology. In B. Bruya (Ed.), *Effortless attention: A new perspective in the cognitive science of attention and action* (pp. 179–190). Cambridge, MA: The MIT Press.
Cullen, R. H., & Agnew, M. J. (2016). Comparing different measures of overall workload in a multimodal postural/auditory dual-task environment. *IIE Transactions on Occupational Ergonomics and Human Factors, 4*(2–3), 115–127.
Cuppini, C., Stein, B. E., & Rowland, B. A. (2018). Development of the mechanisms governing midbrain multisensory integration. *The Journal of Neuroscience, 38*(14), 3453–3465.
Custers, R., Eitam, B., & Bargh, J. A. (2012). Conscious and unconscious processes in goal pursuit. In H. Aarts & A. J. Elliot (Eds.), *Goal-directed behavior* (pp. 231–266). New York, NY: Psychology Press, Taylor & Francis Group.
Custers, R., & Aarts, H. (2010). The unconscious will: How the pursuit of goals operates outside of conscious awareness. *Science, 329*(5987), 47–50.
Cutsem, J. V., Marcora, S., Pauw, K. D., Bailey, S., Meeusen, R., & Roelands, B. (2017). The effects of mental fatigue on physical performance: A systematic review. *Sports Medicine, 47*(8), 1569–1588.
Dang, J., & Hagger, M. S. (2019). Time to set a new research agenda for ego depletion and self-control. *Social Psychology, 50*(5–6), 277–281.

Dang, J., Liu, Y., Liu, X., & Mao, L. (2017). The ego could be depleted, providing initial exertion is depleting. *Social Psychology, 48*(4), 242–245.

d'Avella, A., Giese, M., Ivanenko, Y. P., Schack, T., & Flash, T. (Eds.). (2016). *Modularity in motor control: From muscle synergies to cognitive action representation*. Frontiers in computational neuroscience. Lausanne: Frontiers Media SA.

de Lucas, R. D., de Souza, K. M., Costa, V. P., Grossl, T., & Guglielmo, L. G. A. (2013). Time to exhaustion at and above critical power in trained cyclists: The relationship between heavy and severe intensity domains. *Science & Sports, 28*(1), e9–e14.

Deci, E. L., & Ryan, R. M. (1985). *Intrinsic motivation and self-determination in human behavior*. New York, NY: Springer Science+Business Media.

Deci, E. L., & Ryan, R. M. (2000). The "what" and "why" of goal pursuits: Human needs and the self-determination of behavior. *Psychological Inquiry, 11*(4), 227–268.

Deubel, H. (2014). Attention and action. In A. C. K. Nobre & S. Kastner (Eds.), *The Oxford handbook of attention* (pp. 865–). Oxford, UK: Oxford University Press.

Dhawale, A. K., Smith, M. A., & Ölveczky, B. P. (2017). The role of variability in motor learning. *Annual Review of Neuroscience, 40*(1), 479–498.

Di Giulio, C., Daniele, F., & Tipton, C. M. (2006). Angelo Mosso and muscular fatigue: 116 years after the first congress of physiologists: IUPS commemoration. *Advances in Physiology Education, 30*(2), 51–57.

Diekfuss, J. A., Ward, P., & Raisbeck, L. D. (2016). Attention, workload, and performance: A dual-task simulated shooting study. *International Journal of Sport and Exercise Psychology, 15*(4), 423–437.

Dittner, A. J., Wessely, S. C., & Brown, R. G. (2004). The assessment of fatigue: A practical guide for clinicians and researchers. *Journal of Psychosomatic Research, 56*(2), 157–170.

Dongen, H. P. A. V., Belenky, G., & Krueger, J. M. (2011). Investigating the temporal dynamics and underlying mechanisms of cognitive fatigue. In P. Ackerman (Ed.), *Cognitive fatigue: Multidisciplinary perspectives on current research and future applications* (pp. 127–147). Washington, DC: American Psychological Association.

Donkin, C., Kary, A., Tahir, F., & Taylor, R. (2016). Resources masquerading as slots: Flexible allocation of visual working memory. *Cognitive Psychology, 85*, 30–42.

Donkin, C., & Maanen, L. V. (2014). Piéron's law is not just an artifact of the response mechanism. *Journal of Mathematical Psychology, 62–63*, 22–32.

Donkin, C., & Nosofsky, R. M. (2012). A power-law model of psychological memory strength in short- and long-term recognition. *Psychological Science, 23*(6), 625–634.

Dounskaia, N., & Shimansky, Y. (2016). Strategy of arm movement control is determined by minimization of neural effort for joint coordination. *Experimental Brain Research, 234*(6), 1335–1350.

Dreyfus, H. L., & Dreyfus, S. E. (1986). *Mind over machine: The power of human intuition and expertise in the era of the computer*. New York, NY: The Free Press.

Duckworth, R. A., Potticary, A. L., & Badyaev, A. V. (2018). On the origins of adaptive behavioral complexity: Developmental channeling of structural trade-offs. *Advances in the study of behavior* (Vol. 50, pp. 1–36). Elsevier.

Dunn, T. L., Inzlicht, M., & Risko, E. F. (2019). Anticipating cognitive effort: Roles of perceived error-likelihood and time demands. *Psychological Research, 83*(5), 1033–1056.

Edwards, M. (2010). *Fragrances of the World 2010: Parfums du Monde* (26th (Anniversary Edition). Sydney, Australia: Michael Edwards.

Edwards, A., & Polman, R. (2012). *Pacing in sport and exercise: A psychophysiological perspective*. New York: Nova Science Publishers Inc.

Edwards, A. M., & Polman, R. C. J. (2013). Pacing and awareness: Brain regulation of physical activity. *Sports Medicine, 43*(11), 1057–1064.

Eimer, M., Nattkemper, D., Schröger, E., & Prinz, W. (1996). Involuntary attention. Attention. In O. Neumann & A. F. Sanders (Eds.), *Handbook of perception and action* (Vol. 3, pp. 155–184). London: Academic Press.

Eitam, B., Hassin, R. R., & Schul, Y. (2008). Nonconscious goal pursuit in novel environments: The case of implicit learning. *Psychological Science, 19*(3), 261–267.

Ekkekakis, P. (2009a). The dual-mode theory of affective responses to exercise in metatheoretical context: I. initial impetus, basic postulates, and philosophical framework. *International Review of Sport and Exercise Psychology, 2*(1), 73–94.

Ekkekakis, P. (2009b). The dual-mode theory of affective responses to exercise in metatheoretical context: II. bodiless heads, ethereal cognitive schemata, and other improbable dualistic creatures, exercising. *International Review of Sport and Exercise Psychology, 2*(2), 139–160.

Engelbrecht, S. E. (2001). Minimum principles in motor control. *Journal of Mathematical Psychology, 45*(3), 497–542.

Englert, C. (2019). The self-regulation of human performance: A critical discussion and future directions for self-control research. *Performance Enhancement & Health, 6*(3–4), 156–157.

Enoka, R. M., Baudry, S., Rudroff, T., Farina, D., Klass, M., & Duchateau, J. (2011). Unraveling the neurophysiology of muscle fatigue. *Journal of Electromyography and Kinesiology, 21*(2), 208–219.

Enoka, R. M., & Duchateau, J. (2008). Muscle fatigue: What, why and how it influences muscle function. *The Journal of Physiology, 586*(1), 11–23.

Evans, J. S. B. T. (2011). Dual-process theories of reasoning: Contemporary issues and developmental applications. *Developmental Review, 31*(2–3), 86–102. Special Issue: Dual-Process Theories of Cognitive Development.

Evans, J. S. B. T. (2008). Dual-processing accounts of reasoning, judgment, and social cognition. *Annual Review of Psychology, 59,* 255–278.

Evans, J. (2012). Questions and challenges for the new psychology of reasoning. *Thinking & Reasoning, 18*(1), 5–31.

Evans, D. R., Boggero, I. A., & Segerstrom, S. C. (2016). The nature of self-regulatory fatigue and "ego depletion". *Personality and Social Psychology Review, 20*(4), 291–310.

Evans, J. S. B. T., & Stanovich, K. E. (2013). Dual-process theories of higher cognition: Advancing the debate. *Perspectives on Psychological Science, 8*(3), 223–241.

Fave, A. D., Massimini, F., & Bassi, M. (2011). *Psychological selection and optimal experience across cultures*. Netherlands, Dordrecht: Springer.

Fibiger, W., & Singer, G. (1989). Biochemical assessment and differentiation of mental and physical effort. *Work & Stress, 3*(3), 237–247.

Finley, A. J., Tang, D., & Schmeichel, B. J. (2019). Sweet nothings: No effects of self-control exertion on blood glucose levels. *Social Psychology, 50*(5–6), 322–331.

Finn, B. (2010). Ending on a high note: Adding a better end to effortful study. *Journal of Experimental Psychology: Learning, Memory, and Cognition, 36*(6), 1548–1553.

Finn, B., & Miele, D. B. (2016). Hitting a high note on math tests: Remembered success influences test preferences. *Journal of Experimental Psychology: Learning, Memory, and Cognition, 42*(1), 17–38.

Fitts, P. M. (1954). The information capacity of the human motor system in controlling the amplitude of movement. *Journal of Experimental Psychology, 47*(6), 381–391.

Flash, T., & Hogan, N. (1985). The coordination of arm movements: An experimentally confirmed mathematical model. *The Journal of Neuroscience, 5*(7), 1688–1703.

Francis, Z. L., & Inzlicht, M. (2016). Proximate and ultimate causes of ego depletion. In E. R. Hirt, J. J. Clarkson & L. Jia (Eds.), *Self-regulation and ego control* (pp. 373–398). Cambridge, MA: Elsevier Inc.

Franz, V. H., Gegenfurtner, K. R., Bülthoff, H. H., & Fahle, M. (2000). Grasping visual illusions: No evidence for a dissociation between perception and action. *Psychological Science, 11*(1), 20–25.

Fredrickson, B. L., & Kahneman, D. (1993). Duration neglect in retrospective evaluations of affective episodes. *Journal of Personality and Social Psychology, 65*(1), 45–55.

Friese, M., Loschelder, D. D., Gieseler, K., Frankenbach, J., & Inzlicht, M. (2018). Is ego depletion real? an analysis of arguments. *Personality and Social Psychology Review, 23*(2), 107–131.

Fukuba, Y., Miura, A., Endo, M., Kan, A., Yanagawa, K., & Whipp, B. J. (2003). The curvature constant parameter of the power-duration curve for varied-power exercise. *Medicine & Science in Sports & Exercise, 35*(8), 1413–1418.

Gandevia, S. C. (2001). Spinal and supraspinal factors in human muscle fatigue. *Physiological Reviews, 81*(4), 1725–1789.

Gendolla, G. H. E., & Silvestrini, N. (2015). Bounded effort automaticity: A drama in four parts. In G. H. Gendolla, S. L. Koole & M. Tops (Eds.), *Handbook of biobehavioral approaches to self-regulation* (pp. 271–286). New York: Springer Science+Business Media.

Gendolla, G. H. E., & Wright, R. A. (2009). Effort. In D. Sander & K. Scherer (Eds.), *The Oxford companion to emotion and the affective science* (pp. 134–135). New York, NY: Oxford University Press.

Gendolla, G. H. E., Wright, R. A., & Richter, M. (2019). Advancing issues in motivation intensity research: Updated insights from the cardiovascular system. In R. M. Ryan (Ed.), *The Oxford handbook of human motivation* (2nd ed., pp. 373–392). New York, NY: Oxford University Press.

Gescheider, G. A. (1997). *Psychophysics: The fundamentals* (3rd ed.). Mahwah, NJ: Lawrence Erlbaum Associates.

Giboin, L.-S., & Wolff, W. (2019). The effect of ego depletion or mental fatigue on subsequent physical endurance performance: A meta-analysis. *Performance Enhancement & Health, 7*(1-2), 100150 (11 pages).

Gilden, D. L. (2001). Cognitive emissions of 1/f noise. *Psychological Review, 108*(1), 33–56.

Gilden, D. L. (1997). Fluctuations in the time required for elementary decisions. *Psychological Science, 8*(4), 296–301.

Gisiger, T. (2001). Scale invariance in biology: Coincidence or footprint of a universal mechanism? *Biological Reviews of the Cambridge Philosophical Society, 76*(2), 161–209.

Goettker, A., Braun, D. I., & Gegenfurtner, K. R. (2019). Dynamic combination of position and motion information when tracking moving targets. *Journal of Vision, 19*(7), Article 2 (22 pages).

Goettker, A., Brenner, E., Gegenfurtner, K. R., & de la Malla, C. (2019). Corrective saccades influence velocity judgments and interception. *Scientific Reports, 9*, 5395 (12 pages).

Goldberg, K., Faridani, S., & Alterovitz, R. (2015). Two large open-access datasets for Fitts' law of human motion and a succinct derivation of the square-root variant. *IEEE Transactions on Human-Machine Systems, 45*(1), 62–73.

Goldstein, H., Poole, C. P., & Safko, J. L. (2014). *Classical mechanics* (3rd ed.). Harlow: Pearson Education Limited.

Gollwitzer, P. M., Parks-Stamm, E. J., & Oettingen, G. (2009). Living on the edge: Shifting between nonconscious and conscious goal pursuit. In E. Morsella, J. A. Bargh, & P. M. Gollwitzer (Eds.), *Oxford handbook of human action*, Social cognition and social neuroscience (pp. 603–623). New York, NY: Oxford University Press.

Gómez-Chacón, I. M., García-Madruga, J. A., Vila, J. Ó., Elosúa, M. R., & Rodríguez, R. (2014). The dual processes hypothesis in mathematics performance: Beliefs, cognitive reflection, working memory and reasoning. *Learning and Individual Differences, 29*, 67–73.

Gori, J., Rioul, O., & Guiard, Y. (2018). Speed-accuracy tradeoff: A formal information-theoretictransmission scheme (FITTS). *ACM Transactions on Computer-Human Interaction, 25*(5), 1–33.

Gottsdanker, R. M. (1956). The ability of human operators to detect accerelation of target motion. *Psychological Bulletin, 53*(6), 477–487.

Gottsdanker, R., Frick, J. W., & Lockard, R. B. (1961). Identifying the acceleration of visual targets. *British Journal of Psychology, 52*(1), 31–42.

Gottwald, V. M., Owen, R., Lawrence, G. P., & McNevin, N. (2020). An internal focus of attention is optimal when congruent with afferent proprioceptive task information. *Psychology of Sport and Exercise, 47*, 101634 (17 pages).

Graf, V., Baird, J. C., & Glesman, G. (1974). An empirical test of two psychophysical models. *Acta Psychologica, 38*(1), 59–72.

Gribble, P. L., & Ostry, D. J. (1996). Origins of the power law relation between movement velocity and curvature: Modeling; the effects of muscle mechanicsand limb dynamics. *Journal of Neurophysiology, 76*(5), 2853–2860.

Grondin, S. (2012). Violation of the scalar property for time perception between 1 and 2 seconds: Evidence from interval discrimination, reproduction, and categorization. *Journal of Experimental Psychology: Human Perception and Performance, 38*(4), 880–89.

Grondin, S. (2010). Timing and time perception: A review of recent behavioral and neuroscience findings and theoretical directions. *Attention, Perception, & Psychophysics, 72*(3), 561–582.

Grubb, M. A., White, A. L., Heeger, D. J., & Carrasco, M. (2014). Interactions between voluntary and involuntary attention modulate the quality and temporal dynamics of visual processing. *Psychonomic Bulletin & Review, 22*(2), 437–444.

Guiard, Y., & Olafsdottir, H. B. (2011). On the measurement of movement difficulty in the standard approach to Fitts' law. *PLoS ONE, 6*(10), e24389 (15 pages).

Guiard, Y., & Rioul, O. (2015). A mathematical description of the speed/accuracytrade-off of aimed movement. In *Proceedings of the 2015 British HCI Conference on - British HCI '15* (pp. 91–100). ACM Press.

Guiard, Y., Olafsdottir, H. B., & Perrault, S. T. (2011). Fitt's law as an explicit time/error trade-off. In *Proceedings of the 2011 Annual Conference on Human Factors in Computing Systems - CHI'11* (pp. 1619–1628). ACM Press.

Guzmán-González, B., Bustos-Briones, C., Calatayud, J., Tapia, C., Torres-Elgueta, J., García-Massó, X., & Cruz-Montecinos, C. (2020). Effects of dual-task demands on the complexity and task performance of submaximal isometric handgrip force control. *European Journal of Applied Physiology, 120*(6), 1251–1261.

Hachard, B., Noé, F., Ceyte, H., Trajin, B., & Paillard, T. (2020). Balance control is impaired by mental fatigue due to the fulfilment of a continuous cognitive task or by the watching of a documentary. *Experimental Brain Research, 238*(4), 861–868.

Hafed, Z. M., & Krauzlis, R. J. (2010). Interactions between perception and smooth pursuit eye movements. In U. J. Ilg & G. S. Masson (Eds.), *Dynamics of visual motion processing: Neuronal, behavioral, and computational approaches* (pp. 189–211). New York: Springer Science+Business Media, LLC.

Hagger, M. S., Wood, C., Stiff, C., & Chatzisarantis, N. L. D. (2010). Ego depletion and the strength model of self-control: A meta-analysis. *Psychological Bulletin, 136*(4), 495–525.

Halperin, I., & Emanuel, A. (2019). Rating of perceived effort: Methodological concerns and future directions. *Sports Medicine, 50*(4), 679–687.

Hammett, S. T. (1997). Motion blur and motion sharpening in the human visual system. *Vision Research, 37*(18), 2505–2510.

Hammett, S. T., Georgeson, M. A., Bedingham, S., & Barbieri-Hesse, G. S. (2003). Motion sharpening and contrast: Gain control precedes compressive non-linearity? *Vision Research, 43*(10), 1187–1199.

Hammett, S. T., Georgeson, M. A., & Gorea, A. (1998). Motion blur and motion sharpening: Temporal smear and local contrast non-linearity. *Vision Research, 38*(14), 2099–2108.

Handley, S. J., & Trippas, D. (2015). Dual processes and the interplay between knowledge and structure: A new parallel processing model. In B. H. Ross (Ed.), *Psychology of Learning and Motivation* (pp. 33–58). Oxford, UK: Elsevier.

Hanin, Y., & Hanina, M. (2009). Optimization of performance in top-level athletes: An action-focused coping approach. *International Journal of Sports Science & Coaching, 4*(1), 47–91.

Harris, C. M., & Wolpert, D. M. (1998). Signal-dependent noise determines motor planning. *Nature, 394*(6695), 780–784.

Hartcher-O'Brien, J., Soto-Faraco, S., & Adam, R. (2017). Editorial: A matter of bottom-up or top-down processes: The role of attention in multisensory integration. *Frontiers in Integrative Neuroscience, 11*, Article 5 (3 pages).

Hassin, R. R., Aarts, H., Eitam, B., Custers, R., & Kleiman, T. (2009). Nonconscious goal pursuit and the effortful control of behavior. In E. Morsella, J. A. Bargh, & P. M. Gollwitzer (Eds.),

Oxford handbook of human action, Social cognition and social neuroscience (pp. 549–566). New York, NY: Oxford University Press.

Hassin, R. R., Bargh, J. A., & Zimerman, S. (2009). Automatic and flexible: The case of nonconscious goal pursuit. *Social Cognition, 27*(1), 20–36.

Heatherly, M., Dein, M., Munafo, J. P., & Luckett, C. R. (2019). Crossmodal correspondence between color, shapes, and wine odors. *Food Quality and Preference, 71*, 395–405.

Heitz, R. P. (2014). The speed-accuracy tradeoff: History, physiology, methodology, and behavior. *Frontiers in Neuroscience, 8*, Article 150 (19 pages).

Herzog, M. (2009). Binding problem. In M. D. Binder, N. Hirorawa, & U. Windhorst (Eds.), *Encyclopedia of neuroscience* (pp. 388–391). Berlin: Springer-Verlag GmbH.

Hick, W. E. (1952). On the rate of gain of information. *Quarterly Journal of Experimental Psychology, 4*(1), 11–26.

Hockey, G. R. J. (2011). A motivational control theory of cognitive fatigue. In P. Ackerman (Ed.), *Cognitive fatigue: Multidisciplinary perspectives on current research and future applications* (pp. 167–187). Washington, DC: American Psychological Association.

Hockey, R. (2013). *The psychology of fatigue: Work, effort and control*. Cambridge, UK: Cambridge University Press.

Hockey, G. R. J. (1997). Compensatory control in the regulation of human performance under stress and high workload: A cognitive-energetical framework. *Biological Psychology, 45*(1–3), 73–93.

Holden, J. G., Van Orden, G. C., & Turvey, M. T. (2009). Dispersion of response times reveals cognitive dynamics. *Psychological Review, 116*(2), 318–342.

Holgado, D., Zabala, M., & Sanabria, D. (2019). No evidence of the effect of cognitive load on self-paced cycling performance. *PLOS ONE, 14*(5), e0217825 (11 pages).

Hoogerheide, V., & Paas, F. (2012). Remembered utility of unpleasant and pleasant learning experiences: Is all well that ends well? *Applied Cognitive Psychology, 26*(6), 887–894.

Hopfinger, J. B., & Parks, E. L. (2012). Involuntary attention. In G. R. Mangun (Ed.), *The neuroscience of attention: Attentional control and selection* (pp. 30–53). New York, NY: Oxford University Press.

Howard, M. W. (2018). Memory as perception of the past: Compressed time in mind and brain. *Trends in Cognitive Sciences, 22*(2), 124–136.

Hristovski, R., & Balagué, N. (2010). Fatigue-induced spontaneous termination point – nonequilibrium phase transitions and critical behavior in quasi-isometric exertion. *Human Movement Science, 29*(4), 483–493.

Hsu, C.-F., Propp, L., Panetta, L., Martin, S., Dentakos, S., Toplak, M. E., & Eastwood, J. D. (2018). Mental effort and discomfort: Testing the peak-end effect during a cognitively demanding task. *PLOS ONE, 13*(2), e0191479 (17 pages).

Hughes, A. E. (2018). Dissociation between perception and smooth pursuit eye movements in speed judgments of moving Gabor targets. *Journal of Vision, 18*(4), 1–19.

Huh, D., & Sejnowski, T. J. (2015). Spectrum of power laws for curved hand movements. *Proceedings of the National Academy of Sciences, 112*(29), E3950–E3958.

Hutchinson, J. C., & Tenenbaum, G. (2006). Perceived effort–can it be considered gestalt? *Psychology of Sport and Exercise, 7*(5), 463–476.

Hutchinson, J. C., & Tenenbaum, G. (2007). Attention focus during physical effort: The mediating role of task intensity. *Psychology of Sport and Exercise, 8*(2), 233–245.

Hyman, R. (1953). Stimulus information as a determinant of reaction time. *Journal of Experimental Psychology, 45*(3), 188–196.

Inzlicht, M., & Berkman, E. (2015). Six questions for the resource model of control (and some answers). *Social and Personality Psychology Compass, 9*(10), 511–524.

Inzlicht, M., & Friese, M. (2019). The past, present, and future of ego depletion. *Social Psychology, 50*(5–6), 370–378.

Inzlicht, M., & Schmeichel, B. J. (2012). What is ego depletion? toward a mechanistic revision of the resource model of self-control. *Perspectives on Psychological Science, 7*(5), 450–463.

Inzlicht, M., Schmeichel, B. J., & Macrae, C. N. (2014). Why self-control seems (but may not be) limited. *Trends in Cognitive Sciences, 18*(3), 127–133.

Inzlicht, M., Shenhav, A., & Olivola, C. Y. (2018). The effort paradox: Effort is both costly and valued. *Trends in Cognitive Sciences, 22*(4), 337–349.

James, L., Davies, T. G. E., Lim, K. S., & Reynolds, A. (2020). Do bumblebees have signatures? demonstrating the existence of a speed-curvature power law in Bombus terrestris locomotion patterns. *PLOS ONE, 15*(1), e0226393.

Jennings, C. D. (2015). Consciousness without attention. *Journal of the American Philosophical Association, 1*(2), 276–295.

Jensen, A. R. (2006). *Clocking the mind: Mental chronometry and individual differences*. Amsterdam, The Netherlands: Elsevier.

Job, V., & Walton, G. M. (2017). Lay theories of self-control. In C. M. Zedelius, B. C. N. Müller, & J. W. Schooler (Eds.), *The science of lay theories: How beliefs shape our cognition, behavior, and health* (pp. 47–69). Cham, Switzerland: Springer International Publishing AG.

Johnson, A., & Proctor, R. W. (2004). *Attention: Theory and practice*. London: SAGE Publications Inc.

Kahneman, D. (2011). *Thinking, fast and slow*. New York: Farrar, Straus and Giroux.

Kahneman, D. (1973). *Attention and Effort*. Englewood Cliffs, NJ: Prentice-Hall.

Kahneman, D. (2000). Evaluation by moments: Past and future. In D. Kahneman & A. Tversky (Eds.), *Choices, Values, and Frames* (pp. 693–708). Cambridge, UK: Cambridge University Press.

Kahneman, D. (2000). Experienced utility and objective happiness: A moment-based approach. In D. Kahneman & A. Tversky (Eds.), *Choices, Values, and Frames* (pp. 673–692). Cambridge, UK: Cambridge University Press.

Kahneman, D., Fredrickson, B. L., Schreiber, C. A., & Redelmeier, D. A. (1993). When more pain is preferred to less: Adding a better end. *Psychological Science, 4*(6), 401–405.

Kanfer, R. (2011). Determinants and consequences of subjective cognitive fatigue. In P. Ackerman (Ed.), *Cognitive fatigue: Multidisciplinary perspectives on current research and future applications* (pp. 189–207). Washington, DC: American Psychological Association.

Kello, C. T., Brown, G. D. A., Ferrer-i-Cancho, R., Holden, J. G., Linkenkaer-Hansen, K., Rhodes, T., & Orden, G. C. V. (2010). Scaling laws in cognitive sciences. *Trends in Cognitive Sciences, 14*(5), 223–232.

Khetrapal, N. (2010). Load theory of selective attention and the role of perceptual load: Is it time for revision? *European Journal of Cognitive Psychology, 22*(1), 149–156.

Kim, Y.-J. (2011). Can eyes smell? cross-modal correspondences between color hue-tone and fragrance family. *Color Research & Application, 38*(2), 139–156.

Koch, C., & Tsuchiya, N. (2007). Attention and consciousness: Two distinct brain processes. *Trends in Cognitive Sciences, 11*(1), 16–22.

Kool, W., & Botvinick, M. (2013). The intrinsic cost of cognitive control. *Behavioral and Brain Sciences, 36*(6), 697–698.

Kool, W., & Botvinick, M. (2014). A labor/leisure tradeoff in cognitive control. *Journal of Experimental Psychology: General, 143*(1), 131–141.

Kool, W., & Botvinick, M. (2018). Mental labour. *Nature Human Behaviour, 2*(12), 899–908.

Kruglanski, A. W., Bélanger, J. J., Chen, X., Köpetz, C., Pierro, A., & Mannetti, L. (2012). The energetics of motivated cognition: A force-field analysis. *Psychological Review, 119*(1), 1–20.

Kurzban, R. (2016). The sense of effort. *Current Opinion in Psychology, 7*, 67–70.

Kurzban, R., Duckworth, A., Kable, J. W., & Myers, J. (2013). An opportunity cost model of subjective effort and task performance. *Behavioral and Brain Sciences, 36*(6), 661–679.

Kvålseth, T. O. (1980). An alternative to Fitts' law. *Bulletin of the Psychonomic Society, 16*(5), 371–373.

Lacquaniti, F., Terzuolo, C., & Viviani, P. (1983). The law relating the kinematic and figural aspects of drawing movements. *Acta Psychologica, 54*(1–3), 115–130.

Lalanne, C., & Lorenceau, J. (2004). Crossmodal integration for perception and action. *Journal of Physiology-Paris, 98*(1–3), 265–279.

Latash, M. L. (2008). *Synergy*. Oxford: Oxford University Press.

Latash, M. L. (2012). *Fundamentals of motor control*. London: Academic Press, Elsevier Inc.

Latash, M. L., Scholz, J. P., & Schöner, G. (2007). Toward a new theory of motor synergies. *Motor Control, 11*(3), 276–308.

Lavie, N. (1995). Perceptual load as a necessary condition for selective attention. *Journal of Experimental Psychology: Human Perception and Performance, 21*(3), 451–468.

Lavie, N. (2010). Attention, distraction, and cognitive control under load. *Current Directions in Psychological Science, 19*(3), 143–148.

Lavie, N., & Tsal, Y. (1994). Perceptual load as a major determinant of the locus of selection in visual attention. *Perception & Psychophysics, 56*(2), 183–197.

Lavrusheva, O. (2020). The concept of vitality. Review of the vitality-related research domain. *New Ideas in Psychology, 56*, 100752 (14 pages).

Legault, L. (2020a). Need for autonomy. In V. Zeigler-Hill & T. K. Shackelford (Eds.), *Encyclopedia of personality and individual differences* (pp. 3118–3120). Cham, Switzerland: Springer Nature Switzerland AG.

Legault, L. (2020b). Self-determination theory. In V. Zeigler-Hill & T. K. Shackelford (Eds.), *Encyclopedia of personality and individual differences* (pp. 4694–4702). Cham, Switzerland: Springer Nature Switzerland AG.

Legault, L., & Inzlicht, M. (2013). Self-determination, self-regulation, and the brain: Autonomy improves performance by enhancing neuroaffective responsiveness to self-regulation failure. *Journal of Personality and Social Psychology, 105*(1), 123–138.

Legge, G. E. (1978). Sustained and transient mechanisms in human vision: Temporal and spatial properties. *Vision Research, 18*(1), 69–81.

Lilburn, S. D., Smith, P. L., & Sewell, D. K. (2019). The separable effects of feature precision and item load in visual short-term memory. *Journal of Vision, 19*(1), 2, 1–18.

Liu, D., & Todorov, E. (2007). Evidence for the flexible sensorimotor strategies predicted by optimal feedback control. *Journal of Neuroscience, 27*(35), 9354–9368.

Liu, C. C., Wolfgang, B. J., & Smith, P. L. (2009). Attentional mechanisms in simple visual detection: A speed-accuracy trade-off analysis. *Journal of Experimental Psychology: Human Perception and Performance, 35*(5), 1329–1345.

Lohse, K. R. (2015). On attentional control: A dimensional framework for attention in expert performance. In J. Baker & D. Farrow (Eds.), *Routledge handbook of sport expertise* (pp. 38–49). London: Routledge, The Taylor & Francis Group.

Longman, D., Stock, J. T., & Wells, J. C. K. (2017). A trade-off between cognitive and physical performance, with relative preservation of brain function. *Scientific Reports, 7*, Article 13709 (6 pages).

López-Frutos, J. M., Poch, C., García-Morales, I., Ruiz-Vargas, J. M., & Campo, P. (2014). Working memory retrieval differences between medial temporal lobe epilepsy patients and controls: A three memory layer approach. *Brain and Cognition, 84*(1), 90–96.

Loram, I. D., Gollee, H., Lakie, M., & Gawthrop, P. J. (2011). Human control of an inverted pendulum: is continuous control necessary? is intermittent control effective? is intermittent control physiological? *The Journal of Physiology, 589*(2), 307–324.

Lorist, M. M., Boksem, M. A. S., & Ridderinkhof, K. R. (2005). Impaired cognitive control and reduced cingulate activity during mental fatigue. *Cognitive Brain Research, 24*(2), 199–205.

Lorist, M. M., & Faber, L. G. (2011). Consideration of the influence of mental fatigue on controlled and automatic cognitive processes and related neuromodulatory effects. In P. Ackerman (Ed.), *Cognitive fatigue: Multidisciplinary perspectives on current research and future applications* (pp. 105–126). Washington, DC: American Psychological Association.

Lorist, M. M., Kernell, D., Meijman, T. F., & Zijdewind, I. (2002). Motor fatigue and cognitive task performance in humans. *The Journal of Physiology, 545*(1), 313–319.

Loschelder, D. D., & Friese, M. (2016). Moderators of the ego depletion effect. In E. R. Hirt, J. J. Clarkson & L. Jia (Eds.), *Self-regulation and ego control* (pp. 21–42). Cambridge, MA: Elsevier Inc.

Lubashevsky, I. (2016). Human fuzzy rationality as a novel mechanism of emergent phenomena. In C. H. Skiadas & C. Skiadas (Eds.), *Handbook of applications of chaos theory* (pp. 827–878). London: CRC Press, Taylor & Francis Group.

Lubashevsky, I. (2019). Psychophysical laws as reflection of mental space properties. *Physics of Life Reviews, 31*, 276–303. Special issue: Physics of Mind.

Lubashevsky, I., & Watanabe, M. (2016). Statistical properties of gray color categorization: Asymptotics of psychometric function. In *Proceedings of the 47th ISCIE International Symposium on Stochastic Systems Theory and Its Applications Honolulu, Dec. 5-8, 2015, Honolulu*, Institute of Systems, Control and Information Engineers (ISCIE), Kyoto (pp. 41–49). e-print: arXiv:1511.07242.

Lubashevsky, I., Wilson, I., Suzuki, R., & Kato, Y. (2018). Phase transitions in binary categorization: Evidence for dual-system decision making. In F. Müller, L. Ludwigs & M. Kupper (Eds.), *Proceedings of the 34th Annual Meeting of the International Society for Psychophysics Fechner Day 2018, Lüneburg, Germany, 20–24 August 2018*, Leuphana University, Lüneburg, Germany (pp. 233–239). e-print: arXiv:1910.04511 [physics.med-ph].

Lubashevsky, I. (2017). *Physics of the human mind*. Cham: Springer International Publishing AG.

Lubashevsky, I., & Morimura, K. (2019). Physics of mind and car-following problem. In B. S. Kerner (Ed.), *Complex dynamics of traffic management, encyclopedia of complexity and systems science series* (pp. 559–592). New York, NY: Springer Science+Business Media, LLC.

Luce, R. D. (1986). *Response times: Their role in inferring elementary mental organization*. Oxford, NY: Oxford University Press.

Lurquin, J. H., & Miyake, A. (2017). Challenges to ego-depletion research go beyond the replication crisis: A need for tackling the conceptual crisis. *Frontiers in Psychology, 8*, Article 568 (5 pages).

Makovski, T., Humphreys, G., & Hommel, B. (Eds.). (2014). *Early and late selection: Effects of load, dilution and salience*, Frontiers Media SA, Frontiers in Psychology.

Marchetti, G. (2012). Against the view that consciousness and attention are fully dissociable. *Frontiers in Psychology, 3*, Article 36 (14 pages).

Marcora, S. (2009). Perception of effort during exercise is independent of afferent feedback from skeletal muscles, heart, and lungs. *Journal of Applied Physiology, 106*(6), 2060–2062.

Marcora, S. M., Staiano, W., & Manning, V. (2009). Mental fatigue impairs physical performance in humans. *Journal of Applied Physiology, 106*(3), 857–864.

Marinovic, W., & Arnold, D. H. (2013). An illusory distortion of moving form driven by motion deblurring. *Vision Research, 88*, 47–54.

Martela, F., DeHaan, C., & Ryan, R. (2016). On enhancing and diminishing energy through psychological means: Research on vitality and depletion from self-determination theory. In E. R. Hirt, J. J. Clarkson, & L. Jia (Eds.), *Self-regulation and ego control* (pp. 67–85). London: Elsevier, Academic Press.

Martin, K., Meeusen, R., Thompson, K. G., Keegan, R., & Rattray, B. (2018). Mental fatigue impairs endurance performance: A physiological explanation. *Sports Medicine, 48*(9), 2041–2051.

Martin, K., Thompson, K. G., Keegan, R., Ball, N., & Rattray, B. (2014). Mental fatigue does not affect maximal anaerobic exercise performance. *European Journal of Applied Physiology, 115*(4), 715–725.

Massin, O. (2017). Towards a definition of efforts. *Motivation Science, 3*(3), 230–259.

Masters, R., & Maxwell, J. (2008). The theory of reinvestment. *International Review of Sport and Exercise Psychology, 1*(2), 160–183.

Matthews, W. J., & Meck, W. H. (2016). Temporal cognition: Connecting subjective time to perception, attention, and memory. *Psychological Bulletin, 142*(8), 865–907.

McDowd, J. M. (2007). An overview of attention: Behavior and brain. *Journal of Neurologic Physical Therapy, 31*(3), 98–103.

McMorris, T., Barwood, M., Hale, B. J., Dicks, M., & Corbett, J. (2018). Cognitive fatigue effects on physical performance: A systematic review and meta-analysis. *Physiology & Behavior, 188*, 103–107.

Medina, J. M. (2009). 1^{α} noise in reaction times: A proposed model based on piéron's law and information processing. *Physical Review E, 79*(1), 011902 (8 pages).

Medina, J. M., & Díaz, J. A. (2016a). Fluctuation scaling in the visual cortex at threshold. *Physical Review E, 93*(5), 052403 (11 pages).

Medina, J. M., & Díaz, J. A. (2016b). Noise-induced transition in human reaction times. *Journal of Statistical Mechanics: Theory and Experiment, 2016*(9), 093502.

Medina, J. M., Díaz, J. A., & Norwich, K. H. (2014). A theory of power laws in human reaction times: Insights from an information-processing approach. *Frontiers in Human Neuroscience, 8*, Article 621 (4 pages).

Medina, J. M., & Díaz, J. A. (2012). 1/f noise in human color vision: The role of s-cone signals. *Journal of the Optical Society of America A, 29*(2), A82–A95.

Mehta, R. K., & Agnew, M. J. (2014). Subjective evaluation of physical and mental workload interactions across different muscle groups. *Journal of Occupational and Environmental Hygiene, 12*(1), 62–68.

Mehta, R. K., & Parasuraman, R. (2013). Effects of mental fatigue on the development of physical fatigue: A neuroergonomic approach. *Human Factors: The Journal of the Human Factors and Ergonomics Society, 56*(4), 645–656.

Meyer, D. E., Abrams, R. A., Kornblum, S., Wright, C. E., & Smith, J. E. K. (1988). Optimality in human motor performance: Ideal control of rapid aimed movements. *Psychological Review, 95*(3), 340–370.

Meyer, D. E., Keith-Smith, J. E., Kornblum, S., Abrams, R. A., & Wright, C. E. (1990). Speed-accuracy trade-offs in aimed movements: Toward a theory of rapid voluntary action. In M. Jeannerod (Ed.), *Attention and performance XIII* (pp. 173–226). NJ: Lawrence Erlbaum Associates Inc., Hillsdale.

Micklewright, D., Kegerreis, S., Raglin, J., & Hettinga, F. (2016). Will the conscious-subconscious pacing quagmire help elucidate the mechanisms of self-paced exercise? new opportunities in dual process theory and process tracing methods. *Sports Medicine, 47*(7), 1231–1239.

Miller, R. L., & Rowland, B. A. (2018). Multisensory integration how the brain combines information across the senses. In D. A. A. Moustafa (Ed.), *Computational models of brain and behavior* (pp. 215–228). Hoboken, NJ: Wiley.

Miron-Shatz, T., Stone, A., & Kahneman, D. (2009). Memories of yesterday's emotions: Does the valence of experience affect the memory-experience gap? *Emotion, 9*(6), 885–891.

Monod, H., & Scherrer, J. (1965). The work capacity of a synergic muscular group. *Ergonomics, 8*(3), 329–338.

Montemayor, C., & Haladjian, H. H. (2015). *Consciousness, attention, and conscious attention*. Cambridge, MA: The MIT Press.

Montull, L., Vázquez, P., Rocas, L., Hristovski, R., & Balagué, N. (2020). Flow as an embodied state. informed awareness of slackline walking. *Frontiers in Psychology, 10*, Article 2993 (11 pages).

Moors, A. (2009). Automaticity. In T. Bayne, A. Cleeremans, & P. Wilken (Eds.), *The Oxford companion to consciousness* (pp. 91–95). Oxford, UK: Oxford University Press.

Moors, A. (2013). Automaticity. In D. Reisberg (Ed.), *The Oxford handbook of cognitive psychology* (pp. 163–175). New York, NY: Oxford University Press.

Morgan, M. J., & Benton, S. (1989). Motion-deblurring in human vision. *Nature, 340*(6232), 385–386.

Morris, M. G., Dawes, H., Howells, K., Scott, O. M., & Cramp, M. (2008). Relationships between muscle fatigue characteristics and markers of endurance performance. *Journal of Sports Science & Medicine, 7*(4), 431–436.

Muraven, M., Tice, D. M., & Baumeister, R. F. (1998). Self-control as a limited resource: Regulatory depletion patterns. *Journal of Personality and Social Psychology, 74*(3), 774–789.

Muraven, M. (2008). Autonomous self-control is less depleting. *Journal of Research in Personality, 42*(3), 763–770.

Muraven, M., Buczny, J., & Law, K. F. (2019). Ego depletion: Theory and evidence. In R. Ryan (Ed.), *The Oxford handbook of human motivation* (2nd ed., pp. 113–134). New York, NY: Oxford University Press.

Muraven, M., Gagné, M., & Rosman, H. (2008). Helpful self-control: Autonomy support, vitality, and depletion. *Journal of Experimental Social Psychology, 44*(3), 573–585.

Muraven, M., Rosman, H., & Gagné, M. (2007). Lack of autonomy and self-control: Performance contingent rewards lead to greater depletion. *Motivation and Emotion, 31*(4), 322–330.

Murphy, G., & Greene, C. M. (2017). Load theory behind the wheel: Perceptual and cognitive load effects. *Canadian Journal of Experimental Psychology/Revue canadienne de psychologie expérimentale, 71*(3), 191–202.

Murphy, G., Groeger, J. A., & Greene, C. M. (2016). Twenty years of load theory–where are we now, and where should we go next? *Psychonomic Bulletin & Review, 23*(5), 1316–1340.

Murray, M. M., & Wallace, M. T. (Eds.). (2012). *The neural bases of multisensory processes*. Boca Raton, FL: CRC Press, Taylor & Francis Group.

Naccache, L., Dehaene, S., Cohen, L., Habert, M.-O., Guichart-Gomez, E., Galanaud, D., & Willer, J.-C. (2005). Effortless control: Executive attention and conscious feeling of mental effort are dissociable. *Neuropsychologia, 43*(9), 1318–1328.

Nakamura, J., Tse, D. C., & Shankland, S. (2019). Flow: The experience of intrinsic motivation. In R. M. Ryan (Ed.), *The Oxford handbook of human motivation* (2nd ed., pp. 169–185). New York, NY: Oxford University Press.

Navon, D., & Gopher, D. (1979). On the economy of the human-processing system. *Psychological Review, 86*(3), 214–255.

Newsholme, E. A., Blomstrand, E., & Ekblom, B. (1992). Physical and mental fatigue: Metabolic mechanisms and importance of plasma amino acids. *British Medical Bulletin, 48*(3), 477–495.

Nicolò, A., Sacchetti, M., Girardi, M., McCormick, A., Angius, L., Bazzucchi, I., & Marcora, S. M. (2019). A comparison of different methods to analyse data collected during time-to-exhaustion tests. *Sport Sciences for Health, 15*(3), 667–679.

Nishida, S. (2011). Advancement of motion psychophysics: Review 2001–2010. *Journal of Vision, 11*(5), 11 (1–53).

Noakes, T. D. (2012). Fatigue is a brain-derived emotion that regulates the exercise behavior to ensure the protection of whole body homeostasis. *Frontiers in Physiology, 3*, Article 82 (13 pages).

Noakes, T. D., Gibson, A. S. C., & Lambert, E. V. (2005). From catastrophe to complexity: A novel model of integrative central neural regulation of effort and fatigue during exercise in humans: Summary and conclusions. *British Journal of Sports Medicine, 39*(2), 120–124.

Norman, D. A., & Bobrow, D. G. (1975). On data-limited and resource-limited processes. *Cognitive Psychology, 7*(1), 44–64.

Norman, D. A., & Bobrow, D. G. (1976). On the analysis of performance operating characteristics. *Psychological Review, 83*(6), 508–510.

Öğmen, H., & Herzog, M. H. (2016). A new conceptualization of human visual sensory-memory. *Frontiers in Psychology, 7*, Article 830 (15 pages).

Otte, F. W., Millar, S.-K., & Klatt, S. (2019). Skill training periodization in "specialist" sports coaching–an introduction of the "PoST" framework for skill development. *Frontiers in Sports and Active Living, 1*, Article 61 (17 pages).

Pacheco, M. M., Lafe, C. W., & Newell, K. M. (2019). Search strategies in the perceptual-motor workspace and the acquisition of coordination, control, and skill. *Frontiers in Psychology, 10*, Article 1874 (24 pages).

Pageaux, B., & Lepers, R. (2016). Fatigue induced by physical and mental exertion increases perception of effort and impairs subsequent endurance performance. *Frontiers in Physiology, 7*, Article 587 (9 pages).

Pageaux, B., Marcora, S. M., Rozand, V., & Lepers, R. (2015). Mental fatigue induced by prolonged self-regulation does not exacerbate central fatigue during subsequent whole-body endurance exercise. *Frontiers in Human Neuroscience, 9*, Article 67 (12 pages).

Pageaux, B. (2016). Perception of effort in exercise science: Definition, measurement and perspectives. *European Journal of Sport Science, 16*(8), 885–894.

Pageaux, B., & Lepers, R. (2018). The effects of mental fatigue on sport-related performance. Progress in brain research. In S. Marcora & M. Sarkar (Eds.), *Sport and the brain: The science of preparing, enduring and winning, Part C* (Vol. 240, pp. 291–315). Cambridge, MA: Elsevier, Academic Press.

Pageaux, B., Marcora, S. M., & Lepers, R. (2013). Prolonged mental exertion does not alter neuromuscular function of the knee extensors. *Medicine & Science in Sports & Exercise, 45*(12), 2254–2264.

Paul Morris Fitts, M. I. P. (1967). *Human performance*. Belmont, Calif: Brooks/Cole Pub. Co.

Perez, D. C., Cook, S. M., & Peterson, M. A. (2020). Prior experience alters the appearance of blurry object borders. *Scientific Reports, 10*, 5821 (13 pages).

Phillips, R. O. (2015). A review of definitions of fatigue - and a step towards a whole definition. *Transportation Research Part F: Traffic Psychology and Behaviour, 29*, 48–56.

Piéron, H. (1913). Ii. recherches sur les lois de variation des temps de latence sensorielle en fonction des intensités excitatrices. *L'année psychologique, 20*(1), 17–96.

Piéron, H. (1952). *The sensations: Their functions, processes, and mechanisms*. New Haven, CT, US: Yale University Press.

Plamondon, R., & Alimi, A. M. (1997). Speed/accuracy trade-offs in target-directed movements. *Behavioral and Brain Sciences, 20*(2), 279–303.

Poole, D. C., Burnley, M., Vanhatalo, A., Rossiter, H. B., & Jones, A. M. (2016). Critical power: An important fatigue threshold in exercise physiology. *Medicine & Science in Sports & Exercise, 48*(11), 2320–2334.

Posner, M. I. (2012). Attentional networks and consciousness. *Frontiers in Psychology, 3*.

Prasad, S., & Mishra, R. K. (2019). The nature of unconscious attention to subliminal cues. *Vision, 3*(3), 38 (18 pages).

Prettyman, A. (2019). Perceptual precision. *Philosophical Psychology, 32*(6), 923–944.

Price, N. S. C., Crowder, N. A., Hietanen, M. A., & Ibbotson, M. R. (2006). Neurons in v1, v2, and PMLS of cat cortex are speed tuned but not acceleration tuned: The influence of motion adaptation. *Journal of Neurophysiology, 95*(2), 660–673.

Price, N. S. C., Ono, S., Mustari, M. J., & Ibbotson, M. R. (2005). Comparing acceleration and speed tuning in macaque MT: Physiology and modeling. *Journal of Neurophysiology, 94*(5), 3451–3464.

Prinzmetal, W., McCool, C., & Park, S. (2005). Attention: Reaction time and accuracy reveal different mechanisms. *Journal of Experimental Psychology: General, 134*(1), 73–92.

Prinzmetal, W., Park, S., & Garrett, R. (2005). Involuntary attention and identification accuracy. *Perception & Psychophysics, 67*(8), 1344–1353.

Prinzmetal, W., Zvinyatskovskiy, A., Gutierrez, P., & Dilem, L. (2009). Voluntary and involuntary attention have different consequences: The effect of perceptual difficulty. *Quarterly Journal of Experimental Psychology, 62*(2), 352–369.

Proctor, R. W., & Zandt, T. V. (2017). *Human factors in simple and complex systems* (3rd ed.). Boca Raton, FL: CRC Press Taylor & Francis Group.

Prokopowicz, P., & Ślęzak, D. (2017a). Ordered fuzzy numbers: Definitions and operations. In P. Prokopowicz, J. Czerniak, D. Mikołajewski, Ł. Apiecionek & D. Ślęzak (Eds.), *Theory and applications of ordered fuzzy numbers: A tribute to professor Witold Kosiński* (pp. 57–79). Springer Open.

Prokopowicz, P., & Ślęzak, D. (2017b). Ordered fuzzy numbers: Sources and intuitions. In P. Prokopowicz, J. Czerniak, D. Mikołajewski, Ł. Apiecionek & D. Ślęzak (Eds.), *Theory and applications of ordered fuzzy numbers: A tribute to professor Witold Kosiński* (pp. 47–55). Springer Open.

Purushothaman, G., Öğmen, H., Chen, S., & Bedell, H. E. (1998). Motion deblurring in a neural network model of retino-cortical dynamics. *Vision Research, 38*(12), 1827–1842.

Purves, D., Augustine, G. J., Fitzpatrick, D., Hall, W. C., Lamantia, A.-S., McNamara, J. O., & Williams, S. M. (Eds.). (2004). *Neuroscience* (3rd ed.). Sunderland, MA: Sinauer Associates Inc., Publishers.

Qi, P., Ru, H., Gao, L., Zhang, X., Zhou, T., Tian, Y., et al. (2019). Neural mechanisms of mental fatigue revisited: New insights from the brain connectome. *Engineering, 5*(2), 276–286.

Raab, M., & Araújo, D. (2019). Embodied cognition with and without mental representations: The case of embodied choices in sports. *Frontiers in Psychology, 10*, Article 1825 (12 pages).

Ran, G., Chen, X., Cao, X., & Zhang, Q. (2016). Prediction and unconscious attention operate synergistically to facilitate stimulus processing: An fMRI study. *Consciousness and Cognition, 44*, 41–50.

Ranganathan, R., Wieser, J., Mosier, K. M., Mussa-Ivaldi, F. A., & Scheidt, R. A. (2014). Learning redundant motor tasks with and without overlapping dimensions: Facilitation and interference effects. *Journal of Neuroscience, 34*(24), 8289–8299.

Razon, S., Hutchinson, J., & Tenenbaum, G. (2012). Effort perception. In G. Tenenbaum, R. C. Eklund, & A. Kamata (Eds.), *Measurement in sport and exercise psychology* (pp. 265–278). Champaign, IL: Human Kinetics.

Redelmeier, D. A., & Kahneman, D. (1996). Patients' memories of painful medical treatments: Real-time and retrospective evaluations of two minimally invasive procedures. *Pain, 66*(1), 3–8.

Reina, A., Bose, T., Trianni, V. and Marshall, J. A. R. (2018). Psychophysical laws and the super-organism. *Scientific Reports, 8*(1), 4387 (8 pages).

Richter, M., Gendolla, G. H. E., & Wright, R. A. (2016). Three decades of research on motivational intensity theory. In A. J. Elliot (Ed.), *Advances in motivation science* (Vol. 3, pp. 149–186). Oxford, UK: Elsevier Inc.

Rioul, O., & Guiard, Y. (2012). Power vs. logarithmic model of Fitts' law: A mathematical analysis. *Mathématiques et sciences humaines, 199*, 85–96.

Rizzolatti, G., & Craighero, L. (2004). The mirror-neuron system. *Annual Review of Neuroscience, 27*(1), 169–192.

Rochat, N., Hacques, G., Ganière, C., Seifert, L., Hauw, D., Iodice, P., & Adé, D. (2020). Dynamics of experience in a learning protocol: A case study in climbing. *Frontiers in Psychology, 11*, Article 249 (20 pages).

Rowland, B. A., Stein, B. E., & Stanford, T. R. (2015). Multisensory integration, principles of. In J. D. Wright (Ed.), *International encyclopedia of the social & behavioral sciences* (2nd ed., Vol. 16, pp. 94–102). Oxford, UK: Elsevier.

Rozand, V., Pageaux, B., Marcora, S. M., Papaxanthis, C., & Lepers, R. (2014). Does mental exertion alter maximal muscle activation?, *Frontiers in Human Neuroscience, 8*, Article 755 (10 pages).

Rucci, M., Ahissar, E., & Burr, D. (2018). Temporal coding of visual space. *Trends in Cognitive Sciences, 22*(10), 883–895.

Rustichini, A. (2008). Dual or unitary system? two alternative models of decision making. *Cognitive, Affective, & Behavioral Neuroscience, 8*(4), 355–362.

Ryan, R. M., & Deci, E. L. (2000). Self-determination theory and the facilitation of intrinsic motivation, social development, and well-being. *American Psychologist, 55*(1), 68–78.

Ryan, R. M., & Deci, E. L. (2008). From ego depletion to vitality: Theory and findings concerning the facilitation of energy available to the self. *Social and Personality Psychology Compass, 2*(2), 702–717.

Ryan, R. M., & Frederick, C. (1997). On energy, personality, and health: Subjective vitality as a dynamic reflection of well-being. *Journal of Personality, 65*(3), 529–565.

Sabes, P. N. (2011). Sensory integration for reaching: Models of optimality in the context of behavior and the underlying neural circuits. In A. M. Green, C. E. Chapman, J. F. Kalaska, & F. Lepore (Eds.), *Enhancing performance for action and perception: Multisensory integration, neuroplasticity and neuroprosthetics, Part I*. Progress in Brain Research (Vol. 191, pp. 195–209). Amsterdam, The Netherlands: Elsevier.

Salinas, E., Scerra, V. E., Hauser, C. K., Costello, M. G., & Stanford, T. R. (2014). Decoupling speed and accuracy in an urgent decision-making task reveals multiple contributions to their trade-off. *Frontiers in Neuroscience, 8*, Article 85 (17 pages).

Sathian, K., & Ramachandran, V. S. (Eds.). (2020). *Multisensory perception: From laboratory to clinic*. London: Elsevier, Academic Press.

Scalf, P., Torralbo, A., Tapia, E., & Beck, D. (2013). Competition explains limited attention and perceptual resources: Implications for perceptual load and dilution theories. *Frontiers in Psychology, 4*, Article 243 (9 pages).

Schmeichel, B. J., & Baumeister, R. F. (2010). Effortful attention control. In B. Bruya (Ed.), *Effortless attention: A new perspective in the cognitive science of attention and action* (pp. 29–50). Cambridge, MA: The MIT Press.

Schmidt, R. A., Zelaznik, H., Hawkins, B., Frank, J. S., & Quinn (Jr.), J. T. (1979). Motor-output variability: A theory for the accuracy of rapid motor acts. *Psychological Review, 86*(5), 415–451.

Schmidt, T. (2009). Perception: The binding problem and the coherence of perception. In W. P. Banks (Ed.), *Encyclopedia of consciousness* (Vol. 2, pp. 147–158). Oxford, UK: Elsevier, Academic Press.

Schmidt, R. A., & Wrisberg, C. A. (2008). *Motor learning and performance: A situation-based learning approach* (4th ed.). Champaign, IL: Human Kinetics Inc.

Schücker, L., Knopf, C., Strauss, B., & Hagemann, N. (2014). An internal focus of attention is not always as bad as its reputation: How specific aspects of internally focused attention do not hinder running efficiency. *Journal of Sport and Exercise Psychology, 36*(3), 233–243.

Schütz, A. C., Braun, D. I., & Gegenfurtner, K. R. (2011). Eye movements and perception: A selective review. *Journal of Vision, 11*(5), 9–9.

Schütz, A., Fehn, T., & Baumeister, R. F. (2020). Self. In V. Zeigler-Hill & T. K. Shackelford (Eds.), *Encyclopedia of Personality and Individual Differences* (pp. 4628–4637). Cham, Switzerland: Springer Nature Switzerland AG.

Seilheimer, R. L., Rosenberg, A., & Angelaki, D. E. (2014). Models and processes of multisensory cue combination. *Current Opinion in Neurobiology, 25*, 38–46.

Sewell, D. K., Lilburn, S. D., & Smith, P. L. (2014). An information capacity limitation of visual short-term memory. *Journal of Experimental Psychology: Human Perception and Performance, 40*(6), 2214–2242.

Shapiro, L., & Spaulding, S. (2019). Embodied cognition and sport. In M. L. Cappuccio (Ed.), *Handbook of embodied cognition and sport psychology* (pp. 3–21). Cambridge, MA: The MIT Press.

Shenhav, A., Botvinick, M., & Cohen, J. (2013). The expected value of control: An integrative theory of anterior cingulate cortex function. *Neuron, 79*(2), 217–240.

Shenhav, A., Musslick, S., Lieder, F., Kool, W., Griffiths, T. L., Cohen, J. D., & Botvinick, M. M. (2017). Toward a rational and mechanistic account of mental effort. *Annual Review of Neuroscience, 40*(1), 99–124.

Shi, Z., & Mueller, H. J. (Eds.). (2015). *Multisensory perception and action: Psychophysics, neural mechanisms, and applications*. Frontiers in Integrative Neuroscience, Frontiers SA Media.

Shimansky, Y. P., & Dounskaia, N. (2018). Inclusion of neural effort in cost function can explain perceptual decision suboptimality. *Behavioral and Brain Sciences, 41*, e242.

Shimansky, Y. P., & Rand, M. K. (2013). Two-phase strategy of controlling motor coordination determined by task performance optimality. *Biological Cybernetics, 107*(1), 107–129.

Silvestrini, N. (2017). Psychological and neural mechanisms associated with effort-related cardiovascular reactivity and cognitive control: An integrative approach. *International Journal of Psychophysiology, 119*, 11–18. Special Issue: The Psychophysiology of Motivation: Body and Brain in Action.

Silvestrini, N., & Gendolla, G. H. E. (2013). Automatic effort mobilization and the principle of resource conservation: One can only prime the possible and justified. *Journal of Personality and Social Psychology, 104*(5), 803–816.

Silvestrini, N., & Gendolla, G. H. (2019). Affect and cognitive control: Insights from research on effort mobilization. *International Journal of Psychophysiology, 143,* 116–125.
Singer, W. (2006). Consciousness and the binding problem. *Annals of the New York Academy of Sciences, 929*(1), 123–146.
Singh, P., Jana, S., Ghosal, A., & Murthy, A. (2016). Exploration of joint redundancy but not task space variability facilitates supervised motor learning. *Proceedings of the National Academy of Sciences, 113*(50), 14414–14419.
Smith, P. L., Corbett, E. A., Lilburn, S. D., & Kyllingsbæk, S. (2018). The power law of visual working memory characterizes attention engagement. *Psychological Review, 125*(3), 435–451.
Smith, P. L. (1995). Psychophysically principled models of visual simple reaction time. *Psychological Review, 102*(3), 567–593.
Smith, P. L., Lilburn, S. D., Corbett, E. A., Sewell, D. K., & Kyllingsbæk, S. (2016). The attention-weighted sample-size model of visual short-term memory: Attention capture predicts resource allocation and memory load. *Cognitive Psychology, 89,* 71–105.
Smits, B. L. M., Pepping, G.-J., & Hettinga, F. J. (2014). Pacing and decision making in sport and exercise: The roles of perception and action in the regulation of exercise intensity. *Sports Medicine, 44*(6), 763–775.
Song, J.-H. (2019). The role of attention in motor control and learning. *Current Opinion in Psychology, 29,* 261–265.
Sosa, N. E., & Howell, J. L. (2020). Ego depletion. In V. Zeigler-Hill & T. K. Shackelford (Eds.), *Encyclopedia of personality and individual differences* (pp. 1252–1254). Cham, Switzerland: Springer Nature Switzerland AG.
Spence, C. (2018). Multisensory perception. In J. T. Wixted & J. T. Serences (Eds.), *Stevens' handbook of experimental psychology and cognitive neuroscience, volume 2: sensation, perception, & attention* (4th ed., pp. 625–680). Hoboken, NJ.: Wiley.
Spering, M., & Montagnini, A. (2011). Do we track what we see? common versus independent processing for motion perception and smooth pursuit eye movements: A review. *Vision Research, 51*(8), 836–852.
Stafford, T., & Gurney, K. N. (2004). The role of response mechanisms in determining reaction time performance: Piéron's law revisited. *Psychonomic Bulletin & Review, 11*(6), 975–987.
Standage, D., Wang, D.-H., Heitz, R. P., & Simen, P. (2015). Toward a unified view of the speed-accuracy trade-off. *Frontiers in Neuroscience, 9,* Article 139 (3 pages).
Stanley, C. T., Pargman, D., & Tenenbaum, G. (2007). The effect of attentional coping strategies on perceived exertion in a cycling task. *Journal of Applied Sport Psychology, 19*(3), 352–363.
Stein, B. E., & Rowland, B. A. (2011). Organization and plasticity in multisensory integration: Early and late experience affects its governing principles. In A. M. Green, C. E. Chapman, J. F. Kalaska, & F. Lepore (Eds.), *Enhancing performance for action and perception: Multisensory integration, neuroplasticity and neuroprosthetics, Part I.* Progress in brain research (Vol. 191, pp. 145–163). Amsterdam, The Netherlands: Elsevier.
Stein, B. E., Rowland, B., Laurienti, P., & Stanford, T. R. (2009). Multisensory convergence and integration. In L. R. Squire (Ed.), *Encyclopedia of neuroscience* (pp. 1119–1124). Elsevier, Academic Press.
Stein, B. E. (Ed.). (2012). *The new handbook of multisensory processing.* Cambridge, MA: The MIT Press.
Stein, B. E., & Meredith, M. A. (1993). *The merging of the senses.* Cambridge, MA: The MIT Press.
Stein, B. E., & Rowland, B. A. (2020). Neural development of multisensory integration. In K. Sathian & V. S. Ramachandran (Eds.), *Multisensory perception: From laboratory to clinic* (pp. 57–87). London: Elsevier, Academic Press.
Stein, B. E., & Stanford, T. R. (2008). Multisensory integration: Current issues from the perspective of the single neuron. *Nature Reviews Neuroscience, 9*(4), 255–266.
Stein, B. E., Stanford, T. R., & Rowland, B. A. (2014). Development of multisensory integration from the perspective of the individual neuron. *Nature Reviews Neuroscience, 15*(8), 520–535.

Stephenson, M. L., Ostrander, A. G., Norasi, H., & Dorneich, M. C. (2019). Shoulder muscular fatigue from static posture concurrently reduces cognitive attentional resources. *Human Factors: The Journal of the Human Factors and Ergonomics Society, 62*(4), 589–602.

Sternberg, R. J., & Sternberg, K. (2012). *Cognitive psychology* (6th ed.). Belmont, CA: Wadsworth, Cengage Learning.

Stevens, S. S. (1957). On the psychophysical law. *Psychological Review, 64*(3), 153–181.

Stigliani, A., Jeska, B., & Grill-Spector, K. (2019). Differential sustained and transient temporal processing across visual streams. *PLOS Computational Biology, 15*(5), e1007011 (26 pages).

Stigliani, A., Jeska, B., & Grill-Spector, K. (2017). Encoding model of temporal processing in human visual cortex. *Proceedings of the National Academy of Sciences, 114*(51), E11047–E11056.

Stone, L. S., & Krauzlis, R. J. (2003). Shared motion signals for human perceptual decisions and oculomotor actions. *Journal of Vision, 3*(11), 725–736.

Studer, B., & Knecht, S. (2016). A benefit-cost framework of motivation for a specific activity. In B. Studer & S. Knecht (Eds.), *Motivation: Theory, neurobiology and applications*, Progress in brain research (Vol. 229, pp. 25–47). Amsterdam, Netherlands: Elsevier B.V.

Sussman, E., Winkler, I., & Schröger, E. (2003). Top-down control over involuntary attention switching in the auditory modality. *Psychonomic Bulletin & Review, 10*(3), 630–637.

Tanaka, M., Ishii, A., & Watanabe, Y. (2016). Neural effect of physical fatigue on mental fatigue: A magnetoencephalography study. *Fatigue: Biomedicine Health & Behavior, 4*(2), 104–114.

Taylor, J. L., Allen, G. M., Butler, J. E., & Gandevia, S. C. (2000). Supraspinal fatigue during intermittent maximal voluntary contractions of the human elbow flexors. *Journal of Applied Physiology, 89*(1), 305–313.

Tenenbaum, G. (2001). A social-cognitive perspective of perceived exertion and exertion tolerance. In R. N. Singer, H. A. Hausenblas, & C. Janelle (Eds.), *Handbook of sport psychology* (2nd ed., Vol. 2, pp. 810–820). New York, NY: Wiley.

Tenenbaum, G., Fogarty, G., Stewart, E., Calcagnini, N., Kirker, B., Thorne, G., & Christensen, S. (1999). Perceived discomfort in running: Scale development and theoretical considerations. *Journal of Sports Sciences, 17*(3), 183–196.

Tenenbaum, G., & Hutchinson, J. C. (2007). A social-cognitive perspective of perceived and sustained effort. In G. Tenenbaum & R. C. Eklund (Eds.), *Handbook of sport psychology* (3rd ed., pp. 560–577). Hoboken, NJ: Wiley.

Todorov, E. (2004). Optimality principles in sensorimotor control. *Nature Neuroscience, 7*(9), 907–915.

Todorov, E., & Jordan, M. I. (2002). Optimal feedback control as a theory of motor coordination. *Nature Neuroscience, 5*(11), 1226–1235.

Traschütz, A., Zinke, W., & Wegener, D. (2012). Speed change detection in foveal and peripheral vision. *Vision Research, 72,* 1–13.

Tsal, Y., & Benoni, H. (2010). Diluting the burden of load: Perceptual load effects are simply dilution effects. *Journal of Experimental Psychology: Human Perception and Performance, 36*(6), 1645–1656.

Tsallis, C. (2001). Nonextensive statistical mechanics and thermodynamics: Historical background and present status. In S. Abe & Y. Okamoto (Eds.), *Nonextensive statistical mechanics and its applications* (pp. 3–98). Berlin: Springer.

Tsekouras, G. A., & Tsallis, C. (2005). Generalized entropy arising from a distribution of q indices. *Physical Review E, 71*(4), 046144 (8 pages).

Tsuchiya, N., & Koch, C. (2016). The relationship between consciousness and top-down attention. In O. Gosseries, S. Laureys, & G. Tononi (Eds.), *The neurology of consciousness: Cognitive neuroscience and neuropathology* (2nd ed., pp. 71–91). Elsevier, San Diego, CA: Academic Press.

Turatto, M., Benso, F., Facoetti, A., Galfano, G., Mascetti, G. G., & Umiltà, C. (2000). Automatic and voluntary focusing of attention. *Perception & Psychophysics, 62*(5), 935–952.

Uchaikin, V. V. (2013a). *Fractional derivatives for physicists and engineers: Volume I. Background and theory*. Beijing: Higher Education Press; Berlin: Springer.

Uchaikin, V. V. (2013b). *Fractional derivatives for physicists and engineers: Volume II. Applications*. Beijing: Higher Education Press; Berlin: Springer.

Ursino, M., Cuppini, C., & Magosso, E. (2014). Neurocomputational approaches to modelling multisensory integration in the brain: A review. *Neural Networks, 60,* 141–165.

van Boxtel, J. J. A., Tsuchiya, N., & Koch, C. (2010). Consciousness and attention: On sufficiency and necessity. *Frontiers in Psychology, 1*, Article 217 (13 pages).

Van der Linden, D. (2011). The urge to stop: The cognitive and biological nature of acute mental fatigue. In P. Ackerman (Ed.), *Cognitive fatigue: Multidisciplinary perspectives on current research and future applications* (pp. 149–164). Washington, DC: American Psychological Association.

Van der Linden, D., Frese, M., & Meijman, T. F. (2003). Mental fatigue and the control of cognitive processes: Effects on perseveration and planning. *Acta Psychologica, 113*(1), 45–65.

van Ginneken, W. F., Poolton, J. M., Masters, R. S., Capio, C. M., Kal, E. C., & van der Kamp, J. (2017). Comparing the effects of conscious monitoring and conscious control on motor performance. *Psychology of Sport and Exercise, 30,* 145–152.

van Maanen, L., Grasman, R. P. P. P., Forstmann, B. U., & Wagenmakers, E.-J. (2012). Piéron's law and optimal behavior in perceptual decision-making. *Frontiers in Neuroscience, 5*, Article 143 (15 pages).

van Ravenzwaaij, D., Brown, S., & Wagenmakers, E.-J. (2011). An integrated perspective on the relation between response speed and intelligence. *Cognition, 119*(3), 381–393.

Vázquez, P., Hristovski, R., & Balagué, N. (2016). The path to exhaustion: Time-variability properties of coordinative variables during continuous exercise. *Frontiers in Physiology, 7*, Article 37 (8 pages).

Venhorst, A., Micklewright, D., & Noakes, T. D. (2018a). Perceived fatigability: Utility of a three-dimensional dynamical systems framework to better understand the psychophysiological regulation of goal-directed exercise behaviour. *Sports Medicine, 48*(11), 2479–2495.

Venhorst, A., Micklewright, D., & Noakes, T. D. (2018b). Towards a three-dimensional framework of centrally regulated and goal-directed exercise behaviour: A narrative review. *British Journal of Sports Medicine, 52*(15), 957–966.

Venhorst, A., Micklewright, D. P., & Noakes, T. D. (2018c). The psychophysiological determinants of pacing behaviour and performance during prolonged endurance exercise: A performance level and competition outcome comparison. *Sports Medicine, 48*(10), 2387–2400.

Verschooren, S., Schindler, S., Raedt, R. D., & Pourtois, G. (2019). Switching attention from internal to external information processing: A review of the literature and empirical support of the resource sharing account. *Psychonomic Bulletin & Review, 26*(2), 468–490.

Vitali, F., Tarperi, C., Cristini, J., Rinaldi, A., Zelli, A., Lucidi, F., Schena, F., Bortoli, L., & Robazza, C. (2019). Action monitoring through external or internal focus of attention does not impair endurance performance. *Frontiers in Psychology, 10*, Article 535 (10 pages).

Viviani, P., & Flash, T. (1995). Minimum-jerk, two-thirds power law, and isochrony: Converging approaches to movement planning. *Journal of Experimental Psychology: Human Perception and Performance, 21*(1), 32–53.

von Grünau, M. W. (1978). Interaction between sustained and transient channels: Form inhibits motion in the human visual system. *Vision Research, 18*(2), 197–201.

Wagenmakers, E.-J., & Brown, S. (2007). On the linear relation between the mean and the standard deviation of a response time distribution. *Psychological Review, 114*(3), 830–841.

Wahn, B., & König, P. (2017). Is attentional resource allocation across sensory modalities task-dependent? *Advances in Cognitive Psychology, 13*(1), 83–96.

Wahn, B., & Sinnett, S. (2019). Shared or distinct attentional resources? confounds in dual task designs, countermeasures, and guidelines. *Multisensory Research, 32*(2), 145–163.

Wallace, M. T., Woynaroski, T. G., & Stevenson, R. A. (2020). Multisensory integration as a window into orderly and disrupted cognition and communication. *Annual Review of Psychology, 71*(1), 193–219.

Wang, C., Trongnetrpunya, A., Samuel, I. B. H., Ding, M., & Kluger, B. M. (2016). Compensatory neural activity in response to cognitive fatigue. *The Journal of Neuroscience, 36*(14), 3919–3924.

Watson, A. B., & Ahumada, A. J. (2011). Blur clarified: A review and synthesis of blur discrimination. *Journal of Vision, 11*(5), 10 (21 pages).

Watzl, S. (2017). *Structuring mind: The nature of attention and how it shapes consciousness*. Oxford, UK: Oxford University Press.

Wegner, D. M. (2002). *The illusion of conscious will*, Bradford Books. Cambridge, MA: The MIT Press.

Wegner, D. M. (2004). Précis of the illusion of conscious will. *Behavioral and Brain Sciences, 27*(5), 649–659.

Weigelt, M., Kunde, W., & Prinz, W. (2006). End-state comfort in bimanual object manipulation. *Experimental Psychology, 53*(2), 143–148.

Wei, Z., Wang, X.-J., & Wang, D.-H. (2012). From distributed resources to limited slots in multiple-item working memory: A spiking network model with normalization. *Journal of Neuroscience, 32*(33), 11228–11240.

Welford, A. T. (1960). The measurement of sensory-motor performance: Survey and reappraisal of twelve years' progress. *Ergonomics, 3*(3), 189–230.

Welford, A. T. (1968). *Fundamentals of skill*. London: Methuen & Co Ltd.

Wenzel, M., Lind, M., Rowland, Z., Zahn, D., & Kubiak, T. (2019). The limits of ego depletion: A crossover study on the robustness of performance deterioration in consecutive tasks. *Social Psychology, 50*(5–6), 292–304.

Westbrook, A., Kester, D., & Braver, T. S. (2013). What is the subjective cost of cognitive effort? load, trait, and aging effects revealed by economic preference. *PLoS ONE, 8*(7), e68210 (8 pages).

Woodruff, C. C., & Maaske, S. (2010). Action execution engages human mirror neuron system more than action observation. *NeuroReport, 21*(6), 432–435.

Wright, R. A., & Brehm, J. W. (1989). Energization and goal attractiveness. In L. A. Pervin (Ed.), *Goal concepts in personality and social psychology* (pp. 169–210). Hillsdale, NJ: Lawrence Erlbaum Associates Inc.

Wulf, G. (2013). Attentional focus and motor learning: A review of 15 years. *International Review of Sport and Exercise Psychology, 6*(1), 77–104.

Wulf, G., & Lewthwaite, R. (2010). Effortless motor learning??: An external focus of attention enhances movement effectiveness and efficiency. In B. Bruya (Ed.), *Effortless attention: A new perspective in the cognitive science of attention and action* (pp. 75–101). Cambridge, MA: The MIT Press.

Wulf, G., McNevin, N. H., Fuchs, T., Ritter, F., & Toole, T. (2000). Attentional focus in complex skill learning. *Research Quarterly for Exercise and Sport, 71*(3), 229–239.

Wulf, G., & Prinz, W. (2001). Directing attention to movement effects enhances learning: A review. *Psychonomic Bulletin & Review, 8*(4), 648–660.

Wulf, G., Shea, C., & Park, J.-H. (2001). Attention and motor performance: Preferences for and advantages of an external focus. *Research Quarterly for Exercise and Sport, 72*(4), 335–344.

Xu, R., Zhang, C., He, F., Zhao, X., Qi, H., Zhou, P., Zhang, L., & Ming, D. (2018). How physical activities affect mental fatigue based on EEG energy, connectivity, and complexity. *Frontiers in Neurology, 9*, Article 915 (13 pages).

Zahn, D., Gomille, L. K., Grammes, J., Gottschling, P., Fottner, C., Weber, M. M., et al. (2019). The effects of self-control on glucose utilization in a hyperinsulinemic euglycemic glucose clamp. *European Journal of Health Psychology, 26*(4), 111–119.

Zimmerman, M. E. (2018). Speed-accuracy tradeoff. In J. S. Kreutzer, J. DeLuca, & B. Caplan (Eds.), *Encyclopedia of clinical neuropsychology* (2nd ed., pp. 3250–3251). Cham, Switzerland: Springer International Publishing AG.

Chapter 6
Physics of Complex Present: Properties of Action Strategy Cloud

Speaking about the temporality of human actions, we emphasize their temporal nonlocality in the mind. Human goal-oriented behavior implies that

- *the past events* that are retained in memory (including also bodily memory as the acquired skill) and contribute to the evaluation of the current situation,
- *the experienced system state* that is experienced right now (i.e., belonging to the experiential now and playing the role of the current state),
- *the goal of further actions* that the subject intends to implement and whose time course can be anticipated based on the acquired skill

are equipollent and mutually independent elements of any mathematical formalism that can be constructed for describing human *goal-oriented* behavior. In this formalism, all the events contributing to the subject's perception of the current state and the desired goal are described by their mental-mental embedding into the phase space of mental images. As a result, they are represented by some space-time clouds rather than spatial points and point-like instants of time.

The development of such a formalism for the complex present is the main purpose this chapter. As noted before, the complex present is the minimal fragment of the human temporality within which human goal-oriented behavior admits a closed description. This closure is reflected in that the description of action strategies to be developed in the given chapter operates with the properties attributed to these action strategies within the complex present.

We want to emphasize that we will consider the space-time clouds of action strategies from the perspectives of the mind-in-the-self, the body-in-the-self, and physical reality. The realms of the mind-in-the-self and the body-in-the-self are the main focus of our account, whereas physical reality serves as the origin of the entities the mind deals with. Therefore, the space-time clouds of action strategies will be considered from the standpoints of:

- mental objects belonging to the realm of the mind-in-the-self,

I. Lubashevsky and N. Plavinska, *Physics of the Human Temporality*, Understanding Complex Systems, https://doi.org/10.1007/978-3-030-82612-3_6

– spatial and temporal distributions over the human temporality which are treated as entities belonging to the realm of the body-in-the-self,
– motion trajectories of the controlled object in physical space which represent the implementations of action strategies in reality but are inaccessible within both the realms of the self.

6.1 Multitude of Optimal Action Strategies: New Interpretation of Ostrogradsky's Formalism

In the previous chapter we proposed the conception of how space-time clouds emerge within the experiential now. Following the gist of Sect. 5.6 we treat the space-time cloud of action strategy as a certain integral entity belonging to the realm of the self. One aspect of this entity represents the corresponding action strategy in the realm of the mind-in-the-self, the other does this in the realm of the body-in-the-self. Within the body-in-the-self the space-time cloud of action strategy becomes a entity with internal structure.

In the present book we confine our consideration to the external object described as a material point. In this case, the space-time cloud of action strategy can be treated as a certain fuzzy region in the mental image $\mathbb{R}_{[4]} \times t$ of the real space-time continuum. Formally this fuzzy region is described by several *mathematical* characteristics which are useful in specifying the action strategy clouds in the realm of the body-in-the self but in no way accessible to the mind. One of these characteristics is the strictly optimal (perfect) action strategy to be denoted as $\mathrm{AS}_{\mathrm{opt}}$. It may be conceived of as a certain center-line joining the current motion state of the observed object (material point) to the center position of some fuzzy region representing the desired goal of subject's actions. We want to note that this goal has to be treated as a certain space-time cloud because of the intrinsic uncertainty of human perception.

In the present section we will develop the mathematical description of possible optimal action strategies as a certain multitude and analyze their properties. The concept of effort fusion, the uncertainty in evaluating the current state of object motion, and the principle of the cloud-fusion self-consistency formulated in the previous chapter underlie our constructions.

6.1.1 Goal of Subject's Actions and Its Space-Time Cloud in Complex Present

Before passing directly to the quantitative description of action strategies we need (*i*) to specify the goal of subject's actions in more detail and (*ii*) to clarify the quantitative criterion of attaining this goal. Keeping in mind the examples of goal-oriented actions noted in Sect. 4.2, we will consider below subject's actions aimed at reaching a certain spatial point x_g. In the case illustrating the change of mind in decision-making (Sect. 4.2.2) the point x_g is just an upper corner of monitor screen. In the case of the car-following (Sect. 4.2.3) it is the optimal headway distance—a certain stationary point in the system of coordinates attached to the lead car.

Goal of subject's actions: In analyzing subject's actions, however, we do not deal with the spatial point x_g directly. The desired goal, as it is reflected in subject's actions, is the mental image of the point x_g described by its mental-mental embedding into the space $\mathbb{R}_{[4]}$. In this way the point x_g converts into the space cloud $\mathbb{C}^g_{[4]}$—a certain neighborhood of the point $r_g = \{x_g,\ v_g = 0,\ a_g = 0,\ j_g = 0\}$. In describing subject's actions we will assume that the control of the object motion state is characterized by the accuracy δ_x, δ_v, δ_a, and δ_j in monitoring the object position x, velocity v, acceleration a, and jerk j. So the space cloud $\mathbb{C}^g_{[4]}$ of action goal may be regarded as a fuzzy region

$$\mathbb{C}^g_{[4]} = \begin{Bmatrix} |x - x_g| \lesssim \delta_x, \\ |\, v\,| \lesssim \delta_v, \\ |\, a\,| \lesssim \delta_a, \\ |\, j\,| \lesssim \delta_j \end{Bmatrix} \subset \mathbb{R}_{[4]}. \tag{6.1}$$

Criterion of attaining the goal: In the introduced terms, attaining the goal of subject's actions can be treated as driving the object into the space cloud $\mathbb{C}^g_{[4]}$. It should be noted that getting the cloud $\mathbb{C}^g_{[4]}$ means not only reaching the neighborhood $|x - x_g| \lesssim \delta_x$ but also reducing the velocity, acceleration, and jerk to the perception thresholds δ_v, δ_a, δ_j. It enables the object to reside in the region $\mathbb{C}^g_{[4]}$ for a relatively long time.

As far as the time moment when the object enters the region $\mathbb{C}^g_{[4]}$ is concerned,[1] the requirement imposed on this event is much weaker. The object must be driven into the region $\mathbb{C}^g_{[4]}$ within the complex present, the particular moment when it happens does not matter. The given requirement is actually the condition enabling us to confine the consideration of subject's goal-oriented actions to the complex present.

The stated criteria of attaining the desired goal can be represented as reaching the *goal space-time cloud* $\mathbb{C}^g$ defined as a fuzzy region in the space $\mathbb{R}_{[4]} \times t$ that has the form of fuzzy "cylinder" with the base $\mathbb{C}^g_{[4]}$ and extended over the time t through the complex present,

$$\mathbb{C}^g = \begin{Bmatrix} |x - x_g| \lesssim \delta_x, \\ |\, v\,| \lesssim \delta_v, \\ |\, a\,| \lesssim \delta_a, \\ |\, j\,| \lesssim \delta_j \end{Bmatrix} \times \{|t - t_0| \lesssim \tau_{cp}\} \subset \mathbb{R}_{[4]} \times t. \tag{6.2}$$

[1] Because the cloud $\mathbb{C}^g_{[4]}$ is the fuzzy region, i.e., it has blurred boundaries, the moment when the object is regarded as entering the region $\mathbb{C}^g_{[4]}$ is a random variable. The probability of this event is high when inequalities (6.1) hold. So if the object initial position is rather distant from the space cloud $\mathbb{C}^g_{[4]}$, the moment of entering this region is an well-defined event. Otherwise, a more detailed probabilistic description is necessary.

Here t_0 is the current moment of objective time which is not accessible to the mind and perceived as some center of the experiential now, $\tau_{cp} \sim 10$ s is the characteristic duration of complex present (in fact, it is $2\tau_{cp}$).[2]

It should be noted that the goal space-time cloud $\mathbb{C}^g$ may be confined to the connected future only. However, because the experiential now and the connected future (as well as the connected past) are fuzzy regions, this confinement leads to unnecessary over-complication of the analysis. Therefore we have stretched the goal space-time cloud across the complex present as a whole.

As elucidated in the following subsection, this interpretation enables us, in describing the subject's intentions to reach the goal, to deal with some functional distributed over the complex present and attributed to the action strategy as a whole.

6.1.2 Efficiency of Action Strategies and Its Components

We ascribe to each action strategy $\mathcal{AS}$, understood as a space-time cloud $\mathbb{C}_{\text{AS}}$ in the space $\mathbb{R}_{[4]} \times t$, a certain quantity $\mathcal{E}_{\text{AS}}$ to be called the *efficiency* of the action strategy $\mathcal{AS}$. Although the efficiency $\mathcal{E}_{\text{AS}}$ is a mathematical characteristics of $\mathcal{AS}$ which is not accessible to the mind directly, it is reflected in ordering the possible action strategies in their selection for implementation. The higher the efficiency, the higher the probability of selecting the corresponding action strategy. Mathematically in this ordering of action strategies, their efficiency values underlie the probabilistic relations similar to ones described in Sect. 5.4.1.

The present subsection is devoted to constructing a rather general expression for action strategy efficiency employing the gist of cost-benefit tradeoffs. Namely, the subject is assumed to *try* to select an action strategy for implementation that enables him to attain the desired goal within minimizing the efforts required for it.

In principle, there should be two components of the action strategy efficiency which initially are not comparable. One of them is the *cost* of action strategy implementation $\mathcal{E}^e_{\text{AS}}$ integrating the effort of atomic actions forming a given action strategy $\mathcal{AS}$ (the superscript *e* at the symbol $\mathcal{E}^e_{\text{AS}}$ stands for the component of action strategy efficiency related to **e**ffort). The second component $\mathcal{E}^g_{\text{AS}}$ quantifies the *necessity of attaining the goal* of actions within the complex present (the superscript *g* at the symbol stands for the component of action strategy efficient related to **g**oal). As elucidated in the previous subsection, attaining the goal of subject's actions can be interpreted as reaching the goal space-time cloud $\mathbb{C}^g$.

If, within the selected action strategy, the object (material point) has been driven into the region $\mathbb{C}^g$, the subject's goal-oriented actions are assumed to be successfully accomplished. As a result, subject's active behavior is halted and only monitoring the object spatial position remains relevant. In this case, the subject's actions aimed at governing the object motion can be initiated again when the object leaves the region $\mathbb{C}^g$ for some reason. If it turns out that within the complex present the selected action

[2] For details of the time-structure of the complex present see Sect. 3.1.

strategy does not enable the subject to drive the object into the region $\mathbb{C}^g$, a new action strategy should be selected.

In quantifying a given action strategy $\mathcal{AS}$ using a single measure of its efficiency, we face up to a situation similar to one met in constructing the fusion-induced effort (Sect. 5.5). In the framework of an action strategy considered individually, the two components $\mathcal{E}^e_{\text{AS}}$ and $\mathcal{E}^g_{\text{AS}}$ are not comparable. Moreover, at the current stage of theory development, the particular form of the second terms is not clear. It is possible to state beforehand only that this term should quantitatively discriminate the situations when the object is located inside the region $\mathbb{C}^g_{[4]}$ or outside it.

As in the case of fusion-induced effort, in order to understand, even qualitatively, how the two components become comparable we should shift our view beyond the complex present and take into account learning and skill acquisition. Generally, we may state that the component $\mathcal{E}^g_{\text{AS}}$ has to reflect the properties of the component $\mathcal{E}^e_{\text{AS}}$. Indeed, the former quantifies the subject's intention to attain the goal, i.e., quantifies in some sense the benefit of subject's actions. The latter quantifies the cost of driving the object into the region $\mathbb{C}^g$. The cost of actions is determined by the vital resource depletion. So turning to the gist of cost-benefit tradeoff, we come to conclusion that this kind of benefit should integrate in some way the cost of actions. The possibility of using the two terms for describing action strategies within the complex present stems from the principle of cloud-fusion self-consistency (Sect. 5.6.3) generalized to the level of the space-time clouds of action strategies.

Let us demonstrate that the two terms can be aggregated in one expression stemming from (*i*) the cognitive measure of bodily effort (in fact, its mean magnitude) (Sect. 5.3 and Sect. 5.4.2), (*ii*) the formula for quasi-entropy (Sect. 5.4.8), and (*iii*) the general ansatz for fusion-induced effort (Sect. 5.5.3). We will do this via accepting a number of propositions.

Proposition 1 *The introduced goal space-time cloud $\mathbb{C}^g$ enables us to quantify the progress in attaining the desired goal within a given action strategy $\mathcal{AS}$ as the minimization of some functional*

$$\mathcal{E}^g_{AS} \sim \int\limits_{t \in CP} G\left(\frac{|x - x_g|}{\delta_x}, \frac{|v|}{\delta_v}, \frac{|a|}{\delta_a}, \frac{|j|}{\delta_j}\right) dt\,, \tag{6.3}$$

where the function $G(\ldots) > 0$ takes the minimum inside the region $\mathbb{C}^g_{[4]}$. Here the time dependence of variables $x(t)$, $v(t)$, $a(t)$, and $j(t)$ reflects the time course of implementing the action strategy $\mathcal{AS}$ in the physical world. If the component-wise accuracy δ_x, δ_v, δ_a, and δ_j of object monitoring is fixed, the function $G(\ldots)$ may be approximated by the ansatz

$$G\left(\frac{|x - x_0|}{\delta_x}, \frac{|v|}{\delta_v}, \frac{|a|}{\delta_a}, \frac{|j|}{\delta_j}\right) = g_b\left[\epsilon_x\left(\frac{|x - x_g|}{\delta_x}\right)^{\gamma_b} + \epsilon_v\left(\frac{|v|}{\delta_v}\right)^{\gamma_b} + \epsilon_a\left(\frac{|a|}{\delta_a}\right)^{\gamma_b} + \epsilon_j\left(\frac{|j|}{\delta_j}\right)^{\gamma_b} + 1\right], \tag{6.4}$$

where the coefficients $\epsilon_q > 0$ $(q = x, v, a, j)$ *meeting the equality*

$$\epsilon_x + \epsilon_v + \epsilon_a + \epsilon_j = 1$$

are the weights describing the relative contribution of different channels in processing the information about the object motion proximity to the region $\mathbb{C}^g_{[4]}$*; the cofactor* $g_b > 0$ *as well as the exponent* $\gamma_b > 0$ *will be determined below. In constructing Exp. (6.4) we have turned to Stevens' law (1.3) for quantifying the subject's perception of the object deviation from the region* $\mathbb{C}^g_{[4]}$*, which underlies the exponential form of Exp. (6.4). Because the object spatial position* x*, velocity* v*, acceleration* a*, and jerk* j *are just different components of one stimulus—the object deviation from the desired goal—we describe their perception by the same Stevens exponent* γ_b*. It should be noted that the nonconscious processing of information in extracting the estimates of the object position, velocity, acceleration, and jerk should be different, which is the reason to accept the additive form (6.4) of the function* $G(\dots)$*. These estimates become accessible to the mind at the level of E-consciousness. A generalization of Exp. (6.4) to the case when the accuracy in monitoring the object motion depends on the distance between the object and the region* $\mathbb{C}^g_{[4]}$ *will be constructed below.*

The region of integration in Exp. (6.4) denoted symbolically as $t \in CP$ *implies the integration over the complex present (CP). The subject deals with action strategies as whole entities stretched across the complex present from the connected past to the connected future. Therefore integral (6.3) contains all the time moments* $|t - t_0| \lesssim \tau_d$ *whose contribution should gradually decrease as the time difference* $|t - t_0|$ *increases. In the next subsections we will elucidate the mathematical meaning of the given description of integration over the complex present.*[3]

Proposition 2 *In their essence, monitoring*

- *the motion state of a controlled object within the desired accuracy and*
- *monitoring the current position of this object relative to the goal space-time cloud* $\mathbb{C}^g$

are different facets of the same mental process—the analysis of perceived information about the object position in the space $\mathbb{R}_4 \times t$*. So the component* $\mathcal{E}^g_{AS}$ *(Exp. 6.3) has to be reducible in some way to the quasi-entropy (5.57) or, what is the same, the mental effort* E^{II}_m *(Exp. 5.56) of monitoring the object dynamics. In particular, it implies that the additive contribution of*

- *bodily actions in governing the object dynamics and*

[3] In principle, we could definite integral (6.3) directly as

$$\int\limits_{t\in \mathrm{CP}} G(\dots)dt \overset{\text{def}}{=} \int\limits_{-\infty}^{\infty} \exp\left\{-\left(\frac{|t-t_0|}{\tau_d}\right)^p\right\} G(\dots)dt$$

with the exponent $p > 0$. However, this type definition leads to unnecessary over-complications in our constructions.

– *this type mental processes of controlling the object position*

to the efficiency of action strategy implementation should be a properties of the self-consistent fusion of the components $\mathcal{E}^e_{AS}$ and $\mathcal{E}^g_{AS}$. Indeed, this additivity enables the subject to avoid the undesirable interference of the two factors.

Proposition 3 *As discussed in Sect. 5.2.2 the accumulation of experienced effort within the complex present (see Sect. 5.2.2 and next one) is characterized by the peak-end effect reflected in the domination of higher values of effort with respect to lower ones. It implies that the integral cost of action strategy implementation as it is experienced within a particular case*

$$\mathcal{E}^e_{AS} \sim \int_{t \in CP} \Phi(E_m) dt \,. \tag{6.5}$$

has to contain the integrand $\Phi(E_m)$ being a certain function of the fusion-induced effort E_m that grows with E_m faster than a linear one. Appealing again to Stevens' law we accept the exponential ansatz

$$\Phi(E_m) \propto E_m^{\gamma_e} \tag{6.6}$$

with the exponent $\gamma_e > 1$. Turning to the general form of effort fusion, formula (5.67c), we see that the bodily executed actions and the monitoring of object position contribute to the efficiency component $\mathcal{E}^e_{AS}$ in an additive manner when $\gamma_e \gamma_E = 1$. Below in our constructions we will accept this equality to hold and denote the corresponding value of the exponent γ_e as[4]

$$\gamma_e = \gamma \overset{def}{=} \frac{1}{\gamma_E} \,. \tag{6.7}$$

In this case, the efficiency component $\mathcal{E}^e_{AS}$ of action strategy implementation caused by the cumulative fusion-induced effort takes the form

$$\mathcal{E}^e_{AS} \sim \int_{t \in CP} \left[\left(E^I_m\right)^\gamma + \left(E^{II}_m\right)^\gamma \right] dt \,. \tag{6.8}$$

Proposition 4 *When the object is located near the region $\mathbb{C}^g_{[4]}$, i.e.,*

$$|x - x_g| \sim \delta_x \,, \ |v| \sim \delta_v \,, \ |a| \sim \delta_a \,, \ \text{and} \ |j| \sim \delta_j \,, \tag{6.9}$$

the monitoring of its dynamics and the monitoring of its proximity to the region $\mathbb{C}^g_{[4]}$ should be the same. Under this condition, (i) the function $G(\ldots)$ given by Exp. (6.4) and (ii) the value $\left(E^{II}_m\right)^\gamma$ specified by Exp. (5.56) for the mental effort of monitoring

[4] In particular, in this way we have established the relationship between the peak-end effect in nonlinear accumulation of experienced effort and the partial channel integration in the fusion of different types of mental effort.

the object dynamics within the accuracy $\{\delta_x, \delta_v, \delta_a, \delta_j\}$, *take similar forms when, first,*[5]

$$\gamma = \alpha_m = \frac{1}{\beta_M} \in \left[\tfrac{4}{3}, 2\right], \tag{6.10}$$

where $\beta_M \in [0.5, 0.75]$ *is the exponent of the power-law describing memory load (Sect. 5.4.4). In this case, by virtue of (5.56) and (5.44b), we can write*

$$\left(E_m^{II}\right)^\gamma \propto A_m = A_x + A_v + A_a + A_j\,, \tag{6.11a}$$

where

$$\begin{aligned} A_x &= \left(\frac{\kappa_x|x - x_g|}{\delta_x}\right)^\gamma, \quad & A_v &= \left(\frac{\kappa_v|v|}{\delta_v}\right)^\gamma, \\ A_a &= \left(\frac{\kappa_a|a|}{\delta_a}\right)^\gamma, \quad & A_j &= \left(\frac{\kappa_j|j|}{\delta_j}\right)^\gamma. \end{aligned} \tag{6.11b}$$

In writing Exp. (6.11) as a particular case of formula (5.56) we have

- *set* $\tau = \tau_d$ *and ignored the terms* $\rho_q \tau_s \ell_{q*}$ *because considerable changes in the phase variables of object dynamics occur on the time scales of the complex present rather than the scales of the experiential now,*
- *accepted the inequality* $A_m \ll 1$ *to hold because in the case under consideration attention should be divided among many aspects of subject's actions,*
- *assumed that the accuracy* δ_q *in monitoring the phase variable* $q = x, v, a, j$ *exceeds substantially its limit accuracy of perception when attention is completed focused on it, i.e.,* $\delta_q \gg \Delta_q^0/\kappa_q$.

Second, in the analyzed case:

- the interpretation of the quantities $|x - x_g|$, $|v|$, $|a|$, $|j|$ as the uncertainties in evaluating the phase variables of governed object, Exp. (6.9), and
- the estimation of the uncertainties δ_x, δ_v, δ_a, and δ_j specified by Weber's law (1.1a) within partial attention, Exp. (6.11b),

are consistent when

$$A_x = \kappa_x^\gamma\,, \quad A_v = \kappa_v^\gamma\,, \quad A_a = \kappa_a^\gamma\,, \quad A_j = \kappa_j^\gamma\,. \tag{6.12}$$

Equalities (6.12) may be regarded as the minimal degrees of attention required for monitoring the object dynamics inside the region $\mathbb{C}^g_{[4]}$ within the component-wise accuracy $\{\delta_x, \delta_v, \delta_a, \delta_j\}$.

When the object is located outside the region $\mathbb{C}^g_4$ and

[5] Supposing equality (6.10) to hold, we actually accept that the peak-end effect in nonlinear accumulation of experienced effort, the partial channel integration in the fusion of different types of mental effort, and the division of working memory in operating with several similar elements are closely related in neural mechanisms.

$$|x - x_g| \gg \delta_x\,,\ |v| \gg \delta_v\,,\ |a| \gg \delta_a\,,\ |j| \gg \delta_j\,,$$

the degrees of attention focused on monitoring the object dynamics must be increased to keep the required accuracy, which is described by Exp. (6.11b). In this case,

- the resulting dependence of the mental effort of monitoring on the object phase variables, which is specified by Exp. (6.11a), and
- the function $G(\ldots)$ quantifying the subject's intentions to drive the object into the region $\mathbb{C}^g_{[4]}$, Exp. (6.4),

take actually the same form, which justifies our initial statement. Put differently, the mathematical expression for the component $\mathcal{E}^e_{\text{AS}}$ can incorporate the mathematical expression for the component $\mathcal{E}^g_{\text{AS}}$ within some renormalization of coefficients of proportionality.

Generalization: When the object deviates rather far from the region $\mathbb{C}^g_{[4]}$, the monitoring of its dynamics within the accuracy $\{\delta_x, \delta_v, \delta_a, \delta_j\}$ can be unnecessary because it leads to excessive mental effort. It prompt us to generalize the obtained results to the case when the accuracy $\{\widetilde{\delta}_x, \widetilde{\delta}_v, \widetilde{\delta}_a, \widetilde{\delta}_j\}$ of monitoring the dynamics of governed object depends on its state of motion and obeys the condition

$$\delta_q \ll \widetilde{\delta}_q \ll \ell_q$$

for $q = x, v, a, j$ and $\ell_q = |x - x_g|, |v|, |a|, |j|$, respectively. Let us employ the following ansatz

$$\widetilde{\delta}_q = \delta_q^{1-\beta_q} \cdot \ell_q^{\beta_q}\,, \tag{6.13a}$$

accompanied with the replacement

$$\ell_q \to \sqrt{\delta_q^2 + \ell_q^2} \tag{6.13b}$$

allowing for a plausible crossover when the objects comes close to the region $\mathbb{C}^g_{[4]}$. Here the exponents $\beta_q \in (0, 1)$ are model parameters which may be different for different phase variables. It should be noted that this difference between the pair $\{x, v\}$ of phase variable and the pair $\{a, j\}$ of phase variables may be pronounced. The matter is that the perception of the former pair of variables is based on the information processing within the immediate now, i.e., the central synchronic unit of the diachronic unit—the experiential now. The perception of the latter pair of variables is due to the cumulative contribution of the three constituent synchronic units of the diachronic unit—the immediate past, the immediate now, and the immediate future (Sect. 2.6).

The substitution of expressions (6.11b) into expression (6.11a) after the replacement $\delta_q \to \widetilde{\delta}_q$ yields the formula

$$\left(E_{\rm m}^{\rm II}\right)^{\gamma} = g_{\rm m}\left[\epsilon_x\left(\frac{\sqrt{(x-x_g)^2+\delta_x^2}}{\delta_x}\right)^{(1-\beta_x)\gamma} + \epsilon_v\left(\frac{\sqrt{v^2+\delta_v^2}}{\delta_v}\right)^{(1-\beta_v)\gamma}\right.$$
$$\left. + \epsilon_a\left(\frac{\sqrt{a^2+\delta_a^2}}{\delta_a}\right)^{(1-\beta_a)\gamma} + \epsilon_j\left(\frac{\sqrt{j^2+\delta_j^2}}{\delta_j}\right)^{(1-\beta_j)\gamma}\right] \quad (6.14)$$

with the coefficients, by virtue of (5.56),

$$g_{\rm m} = \left(e_{\rm m}^{\rm II}\right)^{\gamma}\left(\kappa_x^{\gamma}+\kappa_v^{\gamma}+\kappa_a^{\gamma}+\kappa_j^{\gamma}\right), \qquad \epsilon_q = \frac{\kappa_q^{\gamma}}{\kappa_x^{\gamma}+\kappa_v^{\gamma}+\kappa_a^{\gamma}+\kappa_j^{\gamma}}.$$

Expression (6.14) is the desired generalization. It simultaneously describes

- the contribution of mental effort of monitoring the object dynamics to the efficiency component $\mathcal{E}_{\rm AS}^{e}$ and,
- the subject's intention in attaining the desired goal, the component $\mathcal{E}_{\rm AS}^{g}$.

To find a plausible estimate of the exponent $\beta_{...}$ the stability analysis of subject's actions in driving the object into the region $\mathbb{C}_{[4]}^{g}$ is required.[6]

Summarizing the obtained results, we come to the following **conclusions**:

- The component $\mathcal{E}_{\rm AS}^{g}$ of action strategy efficiency allowing for the subject's intention to attend the desired goal within the complex present can be incorporated into the efficiency component $\mathcal{E}_{\rm AS}^{e}$ caused by the integral effort related to action strategy implementation.

In our further constructions this incorporation will be assumed to be already done. In particular, the coefficient $g_{\rm m}$ in Exp. (6.14) is assumed to be renormalized such that allow for also the component $\mathcal{E}_{\rm AS}^{g}$, i.e., Exp. (6.3).

- For a given action strategy $\mathcal{AS}$ whose *particular* implementation in the real space-time continuum is *mathematically* described by a trajectory

$$P_{\rm AS} \stackrel{\rm def}{=} \left\{x(t),\ v(t) = \frac{dx}{dt},\ a(t) = \frac{d^2x}{dt^2},\ j(t) = \frac{d^3x}{dt^3}\right\}_{t\in{\rm CP}}, \quad (6.15)$$

the efficiency can be calculated (cf. Exp. 6.8) as the integral

$$\mathcal{E}_{\rm AS}\{P_{\rm AS}\} = \int\limits_{t\in{\rm CP}} \left[\left(E_{\rm m}^{\rm I}\right)^{\gamma} + \left(E_{\rm m}^{\rm II}\right)^{\gamma}\right] dt > 0\,. \quad (6.16)$$

The possibility of introducing a cofactor of proportionality into equality (6.16) and its meaning will be analyzed in the following sections.

[6] The analysis of object motion governed by subject's action to be presented in Sect. 6.1.8 exemplifies this stability analysis.

Here we meet an internal contradiction. In the analysis of choice, it is usually assumed that the higher the efficiency of an option, the more probable its choice. In our case, however, the situation is opposite. The choice of an action strategy is more probable when the value of integral (6.16) used for evaluating its efficiency is smaller. To avoid this cognitive dissonance, we will call such quantities *AS-functionals*.[7]

- The trajectory P_{AS} of action strategy $\mathcal{AS}$, Exp. (6.15), being a mathematical element in our constructions does not belong to the realm of the mind-in-the-self as well as the realm of the body-in-the-self. However, the value of $\mathcal{E}_{\mathrm{AS}}\{P_{\mathrm{AS}}\}$ (Exp. 6.16) in a blurred form caused by perception uncertainty is accessible to the mind via E-consciousness.

  In the realm of the self any action strategy $\mathcal{AS}$ has a stable representation in the form of its space-time cloud $\mathbb{C}_{\mathrm{AS}} \in \mathbb{R}_4 \times t$. So the perceived value $\mathcal{E}_{\mathrm{AS}}\{P_{\mathrm{AS}}\}$ can be attributed by the mind to the cloud $\mathbb{C}_{\mathrm{AS}}$ only. After integration of perceived values $\big\{\mathcal{E}_{\mathrm{AS}}\{P_{\mathrm{AS}}\}\big\}$ over a number of action strategy implementations the mean value of efficiency $\mathcal{E}_{\mathrm{AS}}$, becoming now the property of action strategy $\mathcal{AS}$ explicitly, can be represented as

$$\mathcal{E}_{\mathrm{AS}} = \Big\langle \mathcal{E}_{\mathrm{AS}}\{P_{\mathrm{AS}}\} \Big\rangle_{P_{\mathrm{AS}}}, \tag{6.17}$$

  where the symbol $\langle \ldots \rangle$ stands for the averaging over all the possible implementations of action strategy $\mathcal{AS}$ in the real space-time continuum. The particular mathematical description of this averaging will be elucidated below. It should be noted beforehand that the value $\mathcal{E}_{\mathrm{AS}}$ given by Exp. (6.17) is the mean magnitude of the corresponding cloud-type measure of the action strategy efficiency.

6.1.3 Action-Strategy Lagrangian

In the previous subsection we constructed AS-functional (6.16) which, after its averaging (6.17), can be used for quantifying subject's preference in selecting action strategies for implementation. Namely, the smaller the value $\mathcal{E}_{\mathrm{AS}}$, the higher the probability of the action strategy choice.

The argument of the AS-functional is the trajectory P_{AS} (6.15) treated as a particular implementation of action strategy $\mathcal{AS}$ in the real space-time continuum within the complex present. In the present subsection—leaving aside all the issues of embedding the trajectory P_{AS} into the connected past, the experiential now, and the connected future—we analyze the extremal properties of AS-functional (6.16) as a *mathematical* problem.

At the first step, we will formally regard the integration over the complex present as the integration over some time interval $[t_1, t_2]$ within the possibility of passing to

[7] In principle we may use the negative values of AS-functionals to quantify the action strategy efficiency. However, it is not also free from a cognitive dissonance because the efficiency is usually assumed to be a positive quantity.

limit $t_2 \to \infty$:

$$\int\limits_{t \in \mathrm{CP}} (\dots)\, dt \Longrightarrow \int\limits_{t_1}^{t_2} (\dots)\, dt\,. \tag{6.18}$$

In the framework of the given replacement, the minimization problem

$$\min_{P_{\mathrm{AS}}} : \quad \mathcal{E}_{\mathrm{AS}}\{P_{\mathrm{AS}}\} = \int\limits_{t_1}^{t_2} \left[\left(E_{\mathrm{m}}^{\mathrm{I}}\right)^{\gamma} + \left(E_{\mathrm{m}}^{\mathrm{II}}\right)^{\gamma} \right] dt \tag{6.19}$$

becomes similar to the Lagrangian formulation of Newtonian mechanics turning to the principle of least action. For this reason we will call the integrand

$$L\left(x, v = \frac{dx}{dt}, a = \frac{d^2x}{dt^2}, j = \frac{d^3x}{dt^3}\right) \stackrel{\mathrm{def}}{=} \left(E_{\mathrm{m}}^{\mathrm{I}}\right)^{\gamma} + \left(E_{\mathrm{m}}^{\mathrm{II}}\right)^{\gamma} \tag{6.20}$$

the *Lagrangian of action strategy* $\mathcal{AS}$ or *AS-Lagrangian*.

However, in contrast to the Lagrangian of Newtonian mechanics, the AS-Lagrangian in the list of its arguments contains the acceleration a and jerk j—higher order time derivatives of spatial position x—as *independent* phase variables. As elucidated below, this difference impacts upon problem (6.19) drastically changing its crux in comparison with Newtonian mechanics. In particular, the minimization of AS-functional (6.19) is no longer the initial value problem.

Brief digression: Lagrangian mechanics

Joseph-Louis Lagrange (born Giuseppe Luigi Lagrangia or Giuseppe Ludovico De la Grange Tournier, 1736–1813), French mathematician and astronomer, developed the original and effective approach to Newtonian mechanics. For an introduction to the Lagrangian mechanics a reader may be referred to books by Landau and Lifshitz (1976), Goldstein et al. (2014). Within the Lagrangian formalism the basic equations of Newtonian mechanics stem from the minimization of the functional

$$\min_{\{x(t)\}} : \quad \int\limits_{t_1}^{t_2} L\,(x, v)\; dt$$

called the action. The integrand $L(x, v)$ depending on the spatial coordinate x and the velocity $v = dx/dt$ is called the Lagrangian. Its particular dependence on the variables x and v reflects the properties of a system under consideration.

It is essential that the Lagrangian $L(x, v)$ depends only on the two variables, x and v. In this case, the principle of least action gives rise to the following equation for action extremals—the *Euler-Lagrange* equation:

$$\frac{d}{dt}\left(\frac{\partial L}{\partial v}\right) - \frac{\partial L}{\partial x} = 0\,,$$

which leads to the Newton's second law describing the acceleration $a = d^2x/dt^2$ of object motion as a some function of the spatial position and velocity, $a = F(x, v)$. The Euler-Lagrange equation resolved with respect to the acceleration a, gives rise to the second order differential equation—Newton's second law of motion:

$$\frac{d^2x}{dt^2} = F\left(x, \frac{dx}{dt}\right).$$

This result illustrates the unique feature of Lagrangian approach to Newtonian mechanics. Although the principle of least action deals with virtual trajectories of object motion as entities belonging to the finite time interval (t_1, t_2), constructing the action extremals—the real trajectory of object motion—is the *initial value problem.* It means that if the state of motion $\{x_1, v_1\}$ is know at the initial instant t_1 of time, the trajectory of object further motion can be reconstructed directly solving the corresponding equation of motion via a sequence of time steps.

Below the results to be obtained will be based on the general form of the AS-Lagrangian $L(x, v, a, j)$. However, to elucidate some details in a clear way let us specify a particular form of the AS-Lagrangian to be used in this case. It includes several steps.

First, in describing the bodily effort (e.g., muscle effort) we suppose that the force $F_e(x, v)$ caused by the interaction between the governed object (material point) and its environment (Sect. 5.3.1) can be approximated by a linear function

$$F_e(x, v) = -k_v v + k_x(x - x_g)\,. \tag{6.21}$$

Here the coefficients $k_v \geq 0$ and $k_x \geq 0$ may be regarded as parameters characterizing the effects of viscous friction and the external forces causing the object deviation from the unstable equilibrium $x = x_g$ without subject's control. It is the case when subject's actions are aimed at keeping the object near the point x_g of unstable equilibrium. The car-following problem can be also considered within this approach setting $k_v = 0$ and $k_x = 0$, which means that the driver's intentional actions control completely the car motion.

Second, the quadratic form $\mathbf{B}$ in Exp. (5.13) converting the force F_b, which the subject exerts on the object, into the bodily effort is suppose to be diagonal with the same cofactor B.[8] Under this condition the bodily measure of action effort is specified by the expression

$$E_\text{b} = B \cdot \left[ma - F_e(x, v)\right]^2 = B \cdot \left[ma + k_v v - k_x(x - x_g)\right]^2 \tag{6.22}$$

[8] Here we want to remind that the object position x, velocity v, acceleration a, and j are generally supposed to be vectors of the three dimensional physical space $\mathbb{R}^3$.

by virtue of (5.12b) and (6.21).

Third, dealing with the relationship between the bodily effort E_b and its cognitive evaluation, or strictly speaking, its mean magnitude E^I_m of subject's perception, we (*i*) employ Exp. (5.29) with $\beta_\text{b} = 1$[9], (*ii*) assume that internal attention focused on bodily actions is completely involuntary and, thus, (*iii*) set the attention degree A_b equal to some fixed value, $A_\text{b} = A^\text{min}_\text{b}$.

Under these conditions, by virtue of (5.29) and (6.22), the relationship between the cognitive evaluation of bodily executed action E^I_b and the corresponding phase variables $\{x, v, a, j\}$, i.e. the characteristics of the corresponding fragment of the trajectory P_AS can be written as

$$E^\text{I}_\text{m} = e^\text{I}_\text{m}\sigma\big(A^\text{min}_\text{b}\big)\sqrt{B}\cdot\big|ma + k_v v - k_x(x - x_g)\big|\,. \tag{6.23}$$

Forth, in Sect. 6.1.2 we have obtained estimate (6.10) of the exponent γ relating the perception magnitude E^I_b of bodily action effort to the corresponding component of the AS-Lagrangian. To be specific let us use the upper value of $\gamma = 2$. In this case, the first component of AS-Lagrangian related to the cognitive effort of bodily executed actions is

$$L^\text{I}(x, v, a, j) = \big[e^\text{I}_\text{m}\sigma\big(A^\text{min}_\text{b}\big)\big]^2 B\cdot\big[ma + k_v v - k_x(x - x_g)\big]^2\,. \tag{6.24}$$

Fifth, in describing the mental effort of the object monitoring we will discriminate between the object spatial position x and velocity v, on the one side, and the object acceleration a and jerk j, on the other side. As also noted in the previous subsection, the first pair $\{x, v\}$ of phase variables becomes accessible to the mind via processing perceived information within the immediate now—the central synchronic unit of the diachronic unit, i.e. the experiential now. The perception of the second pair $\{a, j\}$ is due to the information processing based on the cumulative contribution of all the three synchronic units—the immediate past, the immediate now, and the immediate future (see Sect. 2.6).

Therefore, the capacity of monitoring and adaptation to change caused by objection motion should be much higher for the former pair of phase variables. For this reason, in Exp. (6.14) we set $\beta_a = \beta_j = 0$ and keep some finite values of the exponents $\beta_x = \beta_v \in [0, 1/2)$. As a result, the second component of AS-Lagrangian related to the mental effort of object monitoring given by Exp. (6.14) can be rewritten as

$$L^\text{II}(x, v, a, j) = g_\text{m}\Bigg(\epsilon_x\left[\frac{(x - x_g)^2}{\delta_x^2} + 1\right]^{\gamma_s/2} + \epsilon_v\left[\frac{v^2}{\delta_v^2} + 1\right]^{\gamma_s/2} + \epsilon_a\left[\frac{a^2}{\delta_a^2} + 1\right] + \epsilon_j\left[\frac{j^2}{\delta_j^2} + 1\right]\Bigg), \tag{6.25}$$

[9] The exponent $\beta_\text{b} = 1$ implies that the subject's perception of bodily effort and the force which the subject exerts on the object are related by the linear form of Stevens' law, see also discussion around expression (5.29).

where we have introduced the exponent $\gamma_s = (1 - \beta_{x,v})\,\gamma \in (1, 2]$.

Six, at the current stage of theory development, a proportionality cofactor for the AS-Lagrangian is not well-defined, so, we may set it for convenience. It enables us to write the desired expression for the AS-Lagrangian as

$$L(x, v, a, j) = \frac{g}{2}\left[ma + k_v v - k_x(x - x_g)\right]^2 + \frac{\epsilon_x}{2}\left[\frac{(x - x_g)^2}{\delta_x^2} + 1\right]^{\gamma_s/2} + \frac{\epsilon_v}{2}\left[\frac{v^2}{\delta_v^2} + 1\right]^{\gamma_s/2} + \frac{\epsilon_a}{2}\left[\frac{a^2}{\delta_a^2} + 1\right] + \frac{\epsilon_j}{2}\left[\frac{j^2}{\delta_j^2} + 1\right]. \quad (6.26)$$

Here the coefficient g has been formally introduced via the equality

$$g = \left[e^{\mathrm{I}}_{\mathrm{m}}\sigma\left(A^{\min}_{\mathrm{b}}\right)\right]^2 g_{\mathrm{m}}^{-1}, \quad (6.27)$$

which reflects the common unity of measuring the cognitive effort of bodily executed actions and the mental effort of monitoring which emerges via effort fusion. However, below we will get some estimate of the coefficient g in studying the dynamics of action strategy implementation which also reflect the effort fusion properties.

Finalizing the present section, we want to add the following comments.

The first term in the AS-Lagrangian (6.26) is similar in structure to the functional criterion in comparing the constrained and unconstrained motions within Gauss' principle of least constraint (e.g., Lanczos 1986; Vujanovic and Jones 1989, for historical review see Ramm 2011).

The AS-Lagrangian (6.26) is written for the subject's actions affecting the motion of an object governed by the laws of Newtonian mechanics. However, when the viscous friction is rather strong, the effect of inertia may be ignored. In this case, the external force including the force the subject exerts on the object controls directly the object velocity. A similar situation can be the case when the subject deals with some virtual object. Under such conditions, subject's actions can be also described by the given AS-Lagrangian after (*i*) setting the object mass $m = 0$, (*ii*) eliminating the last terms in Exp. (6.26) via setting $\epsilon_j = 0$, and (*iii*) reducing the four-variable space $\mathbb{R}_{[4]}$ to the tree-variable space $\mathbb{R}_{[3]} = \{\{x, v, a\}\}$. Below such objects will be referred to as *objects with overdamped dynamics*.

6.1.4 Equations of Extremals: Temporal Boundary Value Problem

In this subsection we will continue our analysis of extremal properties of AS-functional (6.16) within the framework of the *mathematical* problem (6.19). The obtained expression (6.26) for the AS-Lagrangian $L(x, v, a, j)$ enables us to rewrite the minimization problem (6.19) in the form

$$\min_{P_{AS}}: \quad \mathcal{E}_{AS}\{P_{AS}\} = \int_{t_1}^{t_2} L\left(x, \frac{dx}{dt}, \frac{d^2x}{dt^2}, \frac{d^3x}{dt^3}\right) dt \,. \tag{6.28}$$

In comparing different realizations of the motion trajectory P_{AS} (Exp. 6.15) via the corresponding values of functional (6.28), the following should be taken into account.

Two types of phase variables: The phase variables $\{x, v, a, j\}$ characterizing the object dynamics are not equivalent in the context of subject's actions aimed at controlling the object motion. The subject can change the current state of object motion only via some kind of motor behavior. The latter takes time required for recognition and realization of corresponding bodily actions which can be estimated from below as the duration of the immediate now. So the subject intentionally and directly can change only the strength of the force he exerts on the object. In the case under consideration, it implies that the subject intentionally and directly can affect only the jerk j.[10] Changes in the other phase variables $\{x, v, a\}$ are derivatives of changes in the jerk j. For this reason we divide the phase variables into two classes different in property:

Inner phase variables: It is the group of variables comprising the object position x, velocity v, and acceleration a. The subject is not able to affect these phase variables directly, so their time variations are smooth and cannot exhibit step-wise changes on scales of the experiential now. The dynamics of inner phase variables is governed by the cumulative contribution of physical regularities and subject's intentional actions.

Control phase variable: It is the jerk j of object motion. The subject is able to affect directly the jerk according to his intentions. So the jerk can exhibit step-wise changes on scales of the experiential now and continuous time variations in the jerk reflect subject's intentions. In other words, the jerk dynamics is controlled directly by the subject's intentional actions.

In the case of objects with overdamped dynamics, the acceleration a becomes the control phase variable, whereas the class of inner phase variables comprises the object position x and velocity v only.

The initial state of action strategy implementation: The employed replacement (6.18) implies that the lower boundary t_1 of integral (6.28) represents the moment when the action strategy $\mathcal{AS}$ has been selected for implementation. In this

---

[10] In reality, there are situations where the relationship between the jerk j of controlled object and subject's intentional bodily actions is also implicit. For example, in driving a car the driver directly affects the car motion via pressing the gas or brake pedal or turning the steering wheel. So in modeling, e.g., the car-following problem the jerk should be replaced by another variable—the time derivative of pedal position (Lubashevsky 2017; Lubashevsky and Morimura 2019). In the present book, however, we confine our constructions to the direct relationship between the jerk and subject's intentions, which is sufficient to elucidate the basic feature of the problem under consideration.

sense, a particular realization $P_{\mathrm{AS}}(t)$ of action strategy $\mathcal{AS}$ in the space $\mathbb{R}_{[4]} \times t$ is already given to the subject and he cannot change its initial properties. The moment of action strategy implementation belongs to the connected past or the immediate past. As a result:

- The corresponding values of the inner phase variables, i.e., the object spatial position x, velocity v, and acceleration j

$$\{x_1, v_1, a_1\} = \{x, v, a\}_{t=t_1} \tag{6.29}$$

  have to be treated as fixed quantities—the *initial conditions* for the trajectory $P_{\mathrm{AS}}(t)$ within the given mathematical description.
- The subject is able to change the jerk j in a step-wise manner within experiential now. Therefore a smooth trajectory $P_{\mathrm{AS}}(t)$ actually represents the fragment of action strategy the subject intends to realize smoothly. So the value j_1 of the jerk within *continuous approximation*

$$j(t) \to j_1 \quad \text{for} \quad t \to t_1 + 0$$

  is not specified directly by the initial state of action strategy implementation and should not treated as a fixed initial value given beforehand. Generally speaking, the quantity j_1 is determined by the action strategy implementation as a whole entity within the complex present.

The terminal state of action strategy implementation: The upper boundary t_2 of integral (6.28) introduced via replacement (6.18) represents the temporal horizon within which the subject can predict the results of his actions. This horizon plays the role of the upper blurred boundary of the complex present. Beyond this horizon, the subject can only roughly approximate the expected dynamics of the controlled object. Therefore, the related events are not taken into account by the subject in evaluating the efficiency and other properties of a given action strategy. In this evaluation only the instant of time within the complex present are essential.

In the analyzed case, it means that if a given action strategy enables the subject to drive the controlled object into the region $\mathbb{C}^g_{[4]}$ within the complex present, then no correction is required. In terms of trajectories $\{P_{\mathrm{AS}}\}$ this requirement can be reformulated as the condition that the desired trajectory should merely enter the region $\mathbb{C}^g_{[4]}$ on time scales about τ_{cp}, after that the trajectory is just assumed to reach the goal in future.

Within this approach, the minimization problem (6.28) admits the modification including three consistent steps:

- We pass to the limit $t_2 \to \infty$ with the conditions

$$\lim_{t\to\infty} x = x_g\,, \quad \lim_{t\to\infty} v = 0\,, \quad \lim_{t\to\infty} a = 0\,, \tag{6.30}$$

imposed on the inner phase variables. It should be noted that the last two equalities in (6.30) are the direct consequences of the first equality for the given goal of subject's actions.

- We have not included here a similar condition imposed on the jerk j, i.e., $\lim_{t\to\infty} j = 0$ because it is a result of subject's intentional actions rather than a consequence of mechanical laws of object motion.
- We specify the coefficient g of AS-Lagrangian (6.26) such that the trajectory $P_{\text{opt}}(t)$ giving the solution to the minimization problem (6.28) with $t_2 \to \infty$ be characterized by temporal increments Δ_t of order of $1/\tau_{cp}$.[11]

Below we will formally refer to equalities (6.30) as the boundary conditions imposed on the inner phase variables at the upper boundary of the complex present.

E-conscious closure of inner phase variables: As elucidated in Sects. 2.6 and 3.3 four phase variable—three inner variables $\{x, v, a\}$ and one control variable j—are accessible to the mind via E-consciousness. It means that these quantities are just given to the mind within the experiential now, the details of their extraction from perceived information remain hidden from the mind. All the other characteristics of object dynamics are mentally constructed based on the perception of these variables. To emphasize this feature we (i) introduced the *phase space of mental images* (Sect. 3.3.1) $\mathbb{R}_{[4]} = \{x, v, a, j\}$ and (ii) accepted the principle of *physical closure of mental-mental embedding* (Sect. 3.3.2), which underlie our analysis of trajectories $\{P_{\text{AS}}\}$ within four variables' phase space.

In the context of action strategy efficiency, these concepts enable us to posit that:

- All the characteristics of action strategy efficiency, that are attributed to the subject's perception within the experiential now, can be quantified using the AS-Lagrangian $L(x, v, a, j)$ being a function of the phase variables.[12] Naturally, internal variables quantifying the subject's mental state, e.g., the pair-wise accuracy $\{\delta_x, \delta_v, \delta_a, \delta_j\}$ in monitoring the variables $\{x, v, a, j\}$ of object dynamics also enter the AS-Lagrangian as additional arguments.
- Within the experiential now, the object spatial position x, velocity v, acceleration a, and jerk j are perceived individually as mutually independent characteristics of object dynamics. Actually, it is a reason for treating the AS-Lagrangian as an explicit function of the arguments $\{x, v, a, j\}$.
- All the other quantities characterizing the subject's perception of object motion are mentally reconstructed based on these phase variables of object dynamics and subject's mental state. In other words, these quantities belong to AM-consciousness. In particular, all the higher order time derivatives of object spatial position x:

---

[11] At this point in our mathematical description of action strategy implementation within the complex present, we actually finalize the construction of the AS-Lagrangian, specify the parameters of effort fusion, and elucidate the meaning of integration over the region $t \in$ CP.

[12] It does not exclude that this quantification will require complex mathematical constructions dealing with the AS-Lagrangian, including integration, averaging, etc., or using other mathematical objects incorporating, maybe, implicitly, the AS-Lagrangian.

$$\frac{d^4x}{dt^4}, \quad \frac{d^5x}{dt^5}, \quad \frac{d^6x}{dt^6}, \quad \dots$$

must be some functions of $\{x, v, a, j\}$ and $\{\delta_x, \delta_v, \delta_a, \delta_j\}$ and, so, cannot be treated as additional phase variables. Below the pair-wised uncertainties $\{\delta_x, \delta_v, \delta_a, \delta_j\}$ specified by the accuracy in monitoring the object dynamics near the desired goal position are assumed to be fixed quantities given beforehand.

These statements will be referred to as the *E-conscious closure* of the phase space $\mathbb{R}_{[4]}$ of mental images.

Returning to the minimization problem (6.28), the E-conscious closure of the phase space $\mathbb{R}_{[4]}$ implies that we have no reason to ascribe some individual values to the higher order time derivatives d^nx/dt^n for $n = 4, 5, 6, \dots$ independent of the values ascribed to the variables x, v, a, and j.

Higher-order Euler-Lagrange equation of extremals and temporal boundary value problem: Taking into account the aforementioned statements about the initial and terminal conditions of the trajectories $\{P_{\text{AS}}\}$, the minimization problem (6.28) is reduced to finding the trajectory P_{opt} for which the variational derivative

$$\frac{\delta \mathcal{E}_{\text{AS}}}{\delta P_{\text{AS}}} = 0\,, \tag{6.31}$$

where the variations $\{\delta P_{\text{AS}}(t)\}$ are localized individually within some finite region inside (t_1, ∞).

As demonstrated in Appendix 6.5.2 in the general case, Eq. (6.31), in its turn, is reduced the following differential equation usually called the higher-order Euler-Lagrange equation

$$\frac{\partial L}{\partial x} - \frac{d}{dt}\frac{\partial L}{\partial v} + \left[\frac{d}{dt}\right]^2 \frac{\partial L}{\partial a} - \left[\frac{d}{dt}\right]^3 \frac{\partial L}{\partial j} = 0\,. \tag{6.32}$$

The solution $P_{\text{opt}}(t)$ to this equation is the strictly optimal implementation AS_{opt} of action strategy $\mathcal{AS}$ in the space $\mathbb{R}_{[4]} \times t$.

Equation (6.32) is a of the sixth order in time derivative. So if we tried to solve Eq. (6.32) following the standard technique used in physics in describing dynamical processes, we would face up to a challenge. This technique known in mathematics as the *initial value problem* deals with the phase space of the corresponding physical system whose points represent possible states of this system with definite properties. The equations governing the dynamics of this system enable one to find the system displacement in the phase space in infinitesimal time step. The repetition of this procedure creates the motion trajectory of this system in its phase. Thereby, the initial state occupied by the system—the initial conditions—and the known physical regularities of the given system determine its further dynamics.

In our case, if we tried to solve Eq. (6.32) as an initial value problem, we would need to operate with the six-dimensional phase space

$$x\,,\quad v = \frac{dx}{dt}\,,\quad a = \frac{d^2x}{dt^2}\,,\quad j = \frac{d^3x}{dt^3}\,,\quad \frac{d^4x}{dt^4}\,,\quad \frac{d^5x}{dt^5}\,.$$

However, according to the E-conscious closure of the phase space $\mathbb{R}_{[4]}$, we have no properties of the subject's perception of object motion *within the experiential now* that can be referred to in ascribing individual values to

$$\frac{d^4x}{dt^4}\,,\quad \frac{d^5x}{dt^5}\,.$$

independently of the variables $\{x, v, a, j\}$.

This challenge is overcome turning to the concept of the *temporal boundary value problem*. Beforehand, we want to note that the possibility of turning to the conception of **temporal** boundary value problem is due to the basic feature of the mind—the ability to anticipate future events and respond to them in an appropriate way leading to the desired goal. In other words, the conception of the goal-oriented behavior opens a gate toward dealing with not only the current states but also the anticipated states related to goals to be attained in future.

In the case under consideration, the control variable—the jerk $j = d^3x/dt^3$ being the highest order time derivative among the phase variables $\{x, v, a, j\}$—admits step-wise variations reflecting the subject's intentions. Therefore, first, the initial values (6.29) of the inner phase variables $\{x, v, a\}$ are the quantities given to the subject at the initial time $t = t_1$. Second, their final values (6.30) are the quantities that specify the object final state that should be reached at the upper boundary of the complex present. So, in mathematical terms, the Euler-Lagrange equation (6.32) should be treated as the temporal boundary problem containing six variables

$$\begin{cases} x_{t=t_1} &= x_1 \\ v_{t=t_1} &= v_1 \\ a_{t=t_1} &= a_1 \end{cases} \bigcup \begin{cases} x_{t\to\infty} &= x_g \\ v_{t\to\infty} &= 0 \\ a_{t\to\infty} &= 0 \end{cases} \Longrightarrow \quad j_{\text{opt}}(t) \text{ for } t \in (t_1, \infty)\,, \tag{6.33}$$

where the corresponding time dependence of the jerk $j_{\text{opt}}(t)$ specifies the strictly optimal implementation AS_{opt} of the action strategy $\mathcal{AS}$. When the time dependence $j_{\text{opt}}(t)$ has been found, the corresponding trajectory $P_{\text{opt}}(t)$ is reconstructed directly via the explicit integration of the relations

$$\frac{d^3x_{\text{opt}}}{dt^3} = j_{\text{opt}}(t)\,,\quad \frac{d^2v_{\text{opt}}}{dt^2} = j_{\text{opt}}(t)\,,\quad \frac{da_{\text{opt}}}{dt} = j_{\text{opt}}(t)\,.$$

For objects with overdamped dynamics, these statements also hold within the replacements $\{x, v, a\} \to \{x, v\}$ and $j \to a$. Besides, the last term in the Euler-Lagrange equation (6.32) should be also omitted.[13]

We want to emphasize that the constructed temporal boundary value problem with the underlying categorization of the phase variables is not confined to the analyzed case only. As demonstrated in Appendix 6.5.3, this approach holds in the general case when the phase space of a system in issue contains time derivatives higher than the first order derivative. For example, if the highest order time derivative is $x^{(n-1)}$ with $n = 3, 4, \ldots$, then, the collection $\{x, \dot{x}, \ddot{x}, \ldots, x^{(n-2)}\}$ has to be treated as the space of inner phase variables, whereas the highest order time derivative $x^{(n-1)}$ plays the role of control phase variable. The corresponding higher-order Euler-Lagrange equation (6.239) should be subject to the initial and terminal conditions imposed on the inner phase variables of object motion.

6.1.5 Manifold of Optimal Action Strategies

As constituent mathematical elements, the temporal boundary value problem (6.33) contains the initial $\{x_1, v_1, a_1\}$ and terminal $\{x_g, 0, 0\}$ states of object motion treated as points belonging to the space of inner phase variables, $\mathbb{R}_L = \{x, v, a\}$.[14] The initial state of object motion may be treated as an arbitrary point of the space $\mathbb{R}_L$. The terminal motion state is a fixed parameter of the analyzed system. It enables us to introduce the multitude $\Xi_{\text{opt}} = \{\text{AS}_{\text{opt}}\}$ of optimal action strategies, where each action strategy AS_{opt} (*i*) matches a point $\{x, v, a\} \in \mathbb{R}_L$ playing the role of the initial state of object motion, (*ii*) leads to one common terminal point, and (*iii*) obeys the Euler-Lagrange equation (6.32).

We suppose that there is only one solution to the temporal boundary value problem (6.33) when the initial point $\{x, v, a\}$ is given. So, in order to parameterize the multitude $\Xi_{\text{opt}} = \Xi_{\text{opt}}(x, v, a)$, the inner phase variables $\{x, v, a\}$ can be used. In this case, the multitude Ξ_{opt} of optimal action strategies specified by the temporal boundary value problem (6.33) for the Euler-Lagrange equation (6.32) is characterized by the following temporal self-consistency.[15]

Temporal self-consistency of optimal action strategies: For any initial state of object motion $\{x_i, v_i, a_i\}$, the trajectory $P_{\text{opt}}(t \mid x_i, v_i, a_i, t_i)$ representing the optimal action strategy AS_{opt} has the property that for any time moment $t' > t_i$, the fragment $P^{>}_{\text{opt}}(t) = P_{\text{opt}}(t > t')$ represents the optimal action strategy $\text{AS}^{>}_{\text{opt}}$ originating from the object motion state $\{x', v', a'\} = \{x(t), v(t), a(t)\}_{t=t'}$ at time t'. Put differently,

---

[13] This type of the temporal boundary value problem was considered for the car-following problem by Lubashevsky et al. (2003).

[14] Below we will call the space of the inner phase variables the Lagrange space $\mathbb{R}_L$.

[15] The temporal self-consistency of optimal action strategies is actually a direct analogy to Bellman's principle of optimality (Bellman 1957), for discussion of this principle see, e.g., Bertsekas (2017).

first, any trajectory $P_{\text{opt}}(t)$ obeying the Euler-Lagrange equation (6.32) and leading to the terminal point $\{x_g, 0, 0\}$ as $t \to \infty$, represents a certain family of optimal action strategies $\mathfrak{F}_P = \{\text{AS}_{\text{opt}}\}_P$ whose initial points lie on the trajectory $P_{\text{opt}}(t)$. Second, no two trajectories of this type intersect each other.

As a result, all the optimal action strategies leading to one terminal point—reaching which is the goal of subject's actions—can be divided into the classes of equivalence according to the following criterion. Two optimal actions strategies AS^1_{opt} and AS^2_{opt} are considered equivalent, $\text{AS}^1_{\text{opt}} \sim \text{AS}^2_{\text{opt}}$, if they belong to the same family $\mathfrak{F}_P$. Otherwise, they have no common points except for the terminal one. Exactly the family $\mathfrak{F}_P = \{\text{AS}_{\text{opt}}\}_P$ may be regarded as the representation of the corresponding action strategy AS^P_{opt}, where the components $\{\text{AS}_{\text{opt}}\}_P$ are just its particular implementations depending on the initial conditions.

In this case, the multitude Ξ_{opt} admits interpretation as a hyper-surface

$$\Xi_{\text{opt}}(x, v, a) = \{x, v, a, j(x, v, a)\} \subset \mathbb{R}_{[4]} = \mathbb{R}_L \times j\,, \tag{6.34}$$

where the inner phase variables $\{x, v, a\}$ are its local coordinates. Therefore, we will call the multitude Ξ_{opt} the *manifold of optimal action strategies.*

Finalizing this subsection, we want to emphases that the ability of the subject to implement an action strategy in some optimal manner, at least approximately, is a result of acquiring skill for a long-term learning process. As a result, the subject's skill is not confined to effective actions aimed at attaining the desired goal from a *certain fixed* state of object motion. The acquired skill concerns *all the typical states* of object motion. Put differently, the subject responds to the current situation but now he responds is a result of long-term experience accumulated beyond the scope of the complex present. In particular, it is the gist of the principle of cloud-fusion self-consistency (Sect. 5.6.3) relating the experiential now to other fragments of the human temporality including the complex present.

Thereby, the properties of subject's skill characterizes the optimal action strategy manifold $\Xi_{\text{opt}}(x, v, a) = \{\text{AS}_{\text{opt}}\}$ as a whole rather than individual features of a single action strategy AS_{opt}.

6.1.6 Action-Strategy Hamiltonian: Ostrogradsky's Formalism

The Russian mathematician Mikhail V. Ostrogradsky (1801–1862) seems to be the first who generalized the formalism of Hamiltonian description to systems with higher-order time-derivatives (Ostrogradsky 1850). As clear from the presented description of action strategies, Ostrogradsky's formalism plays a crucial role in our constructions within, however, a new interpretation.

Usually Ostrogradsky's formalism is employed for studying the dynamics of physical systems described in the framework of the Lagrangian mechanics with higher-order time-derivatives, which has a long history. Speaking about researches during the last decades, we may note, by way of example, the investigations of such *physical* systems within the context of relativistic quantum mechanics (Pons 1989), canonical formalism of Hamiltonian description (Nakamura and Hamamoto 1996; Andrzejewski 2014; Masterov 2016), field theory (Andrzejewski et al. 2010; Cruz et al. 2016; Dai 2020), a number of issues related to Ostrogradsky instability (Stephen 2008; Motohashi and Suyama 2015; Woodard 2015b; Swanson 2019), systems with infinite order of time derivatives (Morozov 2008), physics of gravitation (Hamamoto and Nakamura 2000; Woodard 2009; Mukherjee 2010). The challenge, these investigations face up to, is the problem typically called the Ostrogradsky's ghost—the appearance of auxiliary variables related to higher-order time-derivatives not presented in the initial description of a system in issue. For a review of Ostrogradsky ghost and approaches to copying with it, a reader may be referred, e.g., to Chen et al. (2013), Motohashi et al. (2016, 2018a, b), Aoki and Motohashi (2020), Ganz and Noui (2020).

In our case, we meet another situation in dealing with a different interpretation of the phase variables entering the AS-functional, on the one hand, and the phase variables emerging in the description of its extremals and the corresponding Hamiltonian theory, on the other hand.

The description of goal-oriented actions should be categorized as a teleological problem. Here, we regard teleology as a doctrine that the goal to be attained in future possesses a certain causal power on the current actions (for a discussion of this problem see, e.g., Di Paolo 2005; Droege 2009; Allen and Neal 2020, as well as Footnote 18). Actually the presented analysis demonstrates that the Ostrogradsky's formalism is highly appropriate to describing human goal-oriented behavior. It is reflected in that the number of variables emerging in the corresponding Hamiltonian description fits exactly

- the number of inner phase variables appearing in the initial and terminal conditions and
- the number of control phase variables governing the properties of action strategies as whole entities.

In describing action strategies with respect to their efficiency in attaining the goal, it is desirable to have some property that can quantify the proximity a given action strategy to the optimal one, even if the latter is not accessible to the mind directly. The *Hamiltonian* of action strategy $\mathcal{AS}$ or just the *AS-Hamiltonian* of its optimal implementation $\mathrm{AS}_{\mathrm{opt}}$ can be employed for constructing the desired property. Moreover, this property turns our to be intensive one, i.e., independent of the duration of action strategy as well as its completeness.

The present subsection is devoted to constructing the AS-Hamiltonian. The construction of the Hamiltonian based on the Lagrangian containing higher-order time-derivatives was elaborated by Ostrogradsky (1850). Nowadays, his construction is reviewed in many publications (e.g., Pons 1989; Stephen 2008; Motohashi and

Suyama 2015; Woodard 2015b). However, the Hamiltonian description of systems with higher-order time-derivatives is not met widely in textbooks on theoretical physics. Therefore, we represented Ostrogradsky's formalism in Appendices 6.5 and 6.6, so, in this subsection, we will present the final results only.

The AS-Hamiltonian $H(p_x, p_v, p_a, x, v, a)$ is a function of six arguments comprising the object spatial position x, velocity v, acceleration a, and three new phase variables called the *generalized momenta* p_x, p_v, p_a. The generalized momenta combined with the corresponding inner phase variables form the pairs of canonical variables

$$\{x, p_x\}, \quad \{v, p_v\}, \quad \{a, p_a\} \tag{6.35}$$

of the AS-Hamiltonian $H(p_x, p_v, p_a, x, v, a)$.

By definition, the AS-Hamiltonian $H(p_x, p_v, p_a, x, v, a)$ and the AS-Lagrangian $L(x, v, a, j)$ are related by the expression

$$H(p_x, p_v, p_a, x, v, a) = p_x v + p_v a + p_a j - \big[L(x, v, a, j) - L_0\big]. \tag{6.36}$$

Here, for the sake of convenience, we have subtracted the term L_0 from the AS-Lagrangian which gives the value of the AS-Lagrangian when the goal is reached,

$$L_0 \stackrel{\text{def}}{=} L(x = x_g, v = 0, a = 0, j = 0)\,.$$

In the framework of the description dealing with the object spatial position x and its time-derivatives, the generalized momenta are some functions of $\{x, v, a, j, d^4x/dt, d^5x/dt\}$ with additional condition imposed on the relationship between the control phase variables j specified as follows:

$$p_a = \frac{\partial L(x, v, a, j)}{\partial j} \quad \Longrightarrow \quad j = j(x, v, a, p_a)\,, \tag{6.37a}$$

$$p_v = \frac{\partial L(x, v, a, j)}{\partial a} - \frac{d}{dt}\frac{\partial L(x, v, a, j)}{\partial j}\,, \tag{6.37b}$$

$$p_x = \frac{\partial L(x, v, a, j)}{\partial v} - \frac{d}{dt}\frac{\partial L(x, v, a, j)}{\partial a} + \left[\frac{d}{dt}\right]^2 \frac{\partial L(x, v, a, j)}{\partial j}\,. \tag{6.37c}$$

Expression (6.37a) takes a special position in our constructions. It specifies the direct relationship between the variables x, v, a, j, and p_a. Being resolved with respect to the jerk j[16] it determines the mapping of the Hamiltonian variables onto the Lagrangian variables:

[16] It should be emphasized that the control phase variable—here it is the jerk j of object dynamics—does not belong to the canonical variables of the AS-Hamiltonian.

$$\begin{aligned}
x &\longmapsto x \\
v &\longmapsto v \\
a &\longmapsto a \\
p_x & \\
p_v & \\
p_a &\longmapsto j = j(x, v, a, p_a)
\end{aligned} \tag{6.38}$$

The time course of the optimal action strategy AS_{opt}, i.e., the time pattern of the trajectory $P_{opt}(t)$ is described by the classical Hamilton's equations:

$$\begin{aligned}
\frac{dx}{dt} &= \frac{\partial H}{\partial p_x} = v\,, & \frac{dp_x}{dt} &= -\frac{\partial H}{\partial x}\,, \\
\frac{dv}{dt} &= \frac{\partial H}{\partial p_v} = a\,, & \frac{dp_v}{dt} &= -\frac{\partial H}{\partial v}\,, \\
\frac{da}{dt} &= \frac{\partial H}{\partial p_a} = j\,, & \frac{dp_a}{dt} &= -\frac{\partial H}{\partial a}\,.
\end{aligned} \tag{6.39}$$

It is worthy of noting that the identity

$$\left.\frac{\partial H}{\partial x}\right|_{v,a,p_x,p_v,p_a} = -\left.\frac{\partial L}{\partial x}\right|_{v,a,j}$$

together with the upper right equation of (6.39) and relationship (6.37c) lead, as it must, to the higher-order Euler-Lagrange equation (6.32). Naturally, the AS-Hamiltonian is the first integral of "motion" along the trajectory $P_{opt}(t)$, i.e., the equality

$$\frac{d}{dt}\,H\Big[p_x(t), p_v(t), p_a(t), x(t), v(t), a(t)\Big]_{P_{opt}} = 0 \tag{6.40}$$

holds.

This type Hamilton's equations are usually applied to describing various problems in physics of systems with higher-order time-derivatives. We want to emphasize that these equations also describe the properties of optimal action strategy AS_{opt}, which is an essentially different problem. In our case, solving the given Hamilton's equations is the temporal boundary value problem (6.33) rather than an initial value problem. We have no initial conditions to be imposed on the generalized momenta, their values must be such that the generated time course of the inner phase variables $x(t)$, $v(t)$, and $a(t)$ lead to the terminal conditions (6.30).

For the AS-Lagrangian $L(x, v, a, j)$ specified by Exp. (6.26) the relationship between the jerk j and the momentum p_a takes the special form

$$j = \frac{\delta_j^2}{\epsilon_j}\,p_a\,, \tag{6.41}$$

by virtue of (6.37a). Substituting (6.41) into (6.36) and, then, rearranging the terms, we can write the AS-Hamiltonian as

$$H(p_x, p_v, p_a, x, v, a) = p_x v - \frac{\epsilon_v}{2}\left[\left(\frac{v^2}{\delta_v^2} + 1\right)^{\gamma_s/2} - 1\right] + a\left[p_v - \frac{\epsilon_a}{2}\frac{a}{\delta_a^2}\right] + \frac{\delta_j^2}{2\epsilon_j} p_a^2$$
$$- \frac{\epsilon_x}{2}\left[\left(\frac{(x - x_g)^2}{\delta_x^2} + 1\right)^{\gamma_s/2} - 1\right] - \frac{g}{2}\left[ma + k_v v - k_x(x - x_g)\right]^2. \qquad (6.42)$$

For objects with overdamped dynamics, in the developed Hamiltonian description the list of canonical variables has to be reduced, namely,

$$\{x, v, a, p_x, p_v, p_a\} \Rightarrow \{x, v, p_x, p_v\},$$

Exp. (6.37c) and the bottom line in (6.39) should be omitted, the AS-Lagrangian becomes a function of the argument $\{x, v, a\}$, in Exps. (6.37a) and (6.37b) the replacements $j \to a$, $a \to v$, and $p_a \to p_v$, $p_v \to p_x$ have to be fulfilled. As far as the particular form of the AS-Hamiltonian $H(x, v, p_x, p_v)$ corresponding to the AS-Lagrangian stemming from Exp. (6.26) is concerned, relationship (6.41) must be rewritten as

$$a = \frac{\delta_a^2}{\epsilon_a} p_v \qquad (6.43)$$

and the desired Hamiltonian is

$$H(p_x, p_v, x, v) = p_x v - \frac{\epsilon_v}{2}\left[\left(\frac{v^2}{\delta_v^2} + 1\right)^{\gamma_s/2} - 1\right] + \frac{\delta_a^2}{2\epsilon_a} p_v^2$$
$$- \frac{\epsilon_x}{2}\left[\left(\frac{(x - x_g)^2}{\delta_x^2} + 1\right)^{\gamma_s/2} - 1\right] - \frac{g}{2}\left[k_v v - k_x(x - x_g)\right]^2. \qquad (6.44)$$

The obtained expressions will be used in the next subsection in the analysis of particular properties of AS_{opt}.

Finalizing the present subsection, we want to accentuate that Ostrogradsky's formalism provides a natural mathematical language for describing human goal-oriented behavior as a teleological problem. It has enabled us to construct the Hamiltonian description applicable to systems with higher-order time-derivatives, which opens a gate toward the quantitative description of action strategies as whole entities. It is important that the corresponding number of variables arising in this Hamiltonian description is exactly the number of variables required for specifying

- the initial and terminal conditions of this temporal boundary value problems and
- the direct subject's control over the action strategy implementation.

6.1.7 Hamiltonian Description of Optimal Action Strategies

In the present subsection, we reformulate the main elements in the developed mathematical description of optimal action strategies in terms of Hamiltonian dynamics to represent them in the form to be used in further constructions.

Beforehand, we want to accentuate that the results and expressions obtained previously are tacitly assumed to deal with the three-dimensional position x, velocity v, acceleration a, and jerk j—the vectors belonging to the 3D physical space $\mathbb{R}^3$, i.e.,

$$x,\ v,\ a,\ j \in \mathbb{R}^3.$$

Even if the written expressions may look like one-dimensional models, their generalization to the three-dimensional space $\mathbb{R}^3$ is rather evident. Below, to elucidate the crucial aspects of our constructions without a considerable over-complication of the text, we will confine our analysis to systems admitting an one-dimensional description of object motion in physical space. Put differently, in what follows, the object spatial position x, velocity v, acceleration a, and jerk j will be treated as one-dimensional variables,

$$x,\ v,\ a,\ j \in \mathbb{R}.$$

Hamiltonian representation of temporal boundary value problem: The temporal boundary value problem (6.33) is formulated in terms of the variables $\{x, v, a, j\}$ of the space $\mathbb{R}_{[4]}$—the phase space of mental images. To deal with subject's goal-oriented behavior, we need to reformulate this problem in terms of the canonical variables $\{x, v, a, p_x, p_v, p_a\}$ of the action strategy Hamiltonian description. The initial conditions within representation (6.33) directly specify the inner phase variables $\{x, v, a\}$, i.e., the generalized coordinates of the canonical variables, at the moment of action strategy initiation. At the terminal state attained at $t \to \infty$ and corresponding to the unstable stationary point of the object dynamics, the terminal conditions are reduced to zero-values of the generalized momenta $\{p_x, p_v, p_a\}$, by virtue of expressions (6.37) and the form of AS-Lagrangian $L(x, v, a, j)$ (Exp. 6.26).

Therefore, within the Hamiltonian representation, the temporal boundary value problem (6.33) is reduced to the conditions

$$\begin{Bmatrix} x|_{t=t_1} &= x_1 \\ v|_{t=t_1} &= v_1 \\ a|_{t=t_1} &= a_1 \end{Bmatrix} \bigcup \begin{Bmatrix} p_x|_{t\to\infty} &= 0 \\ p_v|_{t\to\infty} &= 0 \\ p_a|_{t\to\infty} &= 0 \end{Bmatrix} \Longrightarrow \begin{Bmatrix} x(t),\ p_x(t) \\ v(t),\ p_v(t) \\ a(t),\ p_a(t) \end{Bmatrix} \text{ for } t \in (t_1, \infty). \tag{6.45}$$

accompanied by Hamilton's equations (6.39). Whence it follows, in particular, that for any optimal action strategy $\mathrm{AS}_{\mathrm{opt}}$ at the moment of its initiation, the current motion state $\{x, v, a\}$ of the controlled object directly specifies the corresponding values of the generalized momenta $\{p_x, p_v, p_a\}$, i.e.,

$$\begin{aligned} p_x &= \pi_x^{\text{opt}}(x, v, a)\,, \\ p_v &= \pi_v^{\text{opt}}(x, v, a)\,, \\ p_a &= \pi_a^{\text{opt}}(x, v, a)\,, \end{aligned} \tag{6.46}$$

where $\pi_q^{\text{opt}}(x, v, a)$ $(q = x, v, a)$ are some functions determined by the particular form of the Hamiltonian $H(x, v, a, p_x, p_v, p_a)$.

Hamiltonian representation of the optimal action strategy manifold: Due to the temporal self-consistency of optimal action strategies (stated in Sect. 6.1.5), relationship (6.46) holds for any optimal action strategy AS_{opt} at any time moment of its implementation. Thereby, expressions (6.46) actually specify the three-dimensional manifold of optimal action strategies $\Xi_{\text{opt}}^H(x, v, a)$ in the six-dimensional Hamilton space $\mathbb{R}^H = \{x, v, a, p_x, p_v, p_a\}$. The Hamiltonian representation $\Xi_{\text{opt}}^H(x, v, a)$ of the optimal action strategy manifold and its representation (6.34) in the phase space $\mathbb{R}_{[4]}$ of mental images are related via the general expression (6.37a) or its particular form (6.41). This relationship can be represented in the symbolic form

$$\mathbb{R}_H : \Xi_{\text{opt}}^H = \begin{cases} p_x = \pi_x^{\text{opt}}(x, v, a) \\ p_v = \pi_v^{\text{opt}}(x, v, a) \\ p_a = \pi_a^{\text{opt}}(x, v, a) \end{cases} \Rightarrow \mathbb{R}_{[4]} : \Xi_{\text{opt}} = \big\{ j = j_{\text{opt}}(x, v, a) \big\}, \tag{6.47}$$

where, by definition,

$$j_{\text{opt}}(x, v, a) = j\big[x, v, a, p_a = \pi_a^{\text{opt}}(x, v, a)\big]. \tag{6.48}$$

The latter expression combines (6.37a) and (6.46). Expression (6.47) also admits interpretation as the embedding of the optimal action strategy manifold Ξ^H, constructed in the Hamilton space $\mathbb{R}_H$, into the phase space $\mathbb{R}_{[4]}$ of mental images.

The embedding (6.47) reduces the temporal boundary value problem (6.45) to the initial value problem

$$\frac{dx}{dt} = v\,, \quad \frac{dv}{dt} = a\,, \quad \frac{da}{dt} = j_{\text{opt}}(x, v, a) \tag{6.49}$$

to be treated as the *inner dynamics* of the optimal action strategy manifold Ξ_{opt}^H. The term "inner dynamics" has been introduced to emphasize that if the system state as a point of the Hamilton space $\mathbb{R}_H$ lies on the manifold Ξ_{opt}^H at a given instant of time, it will remain on the manifold Ξ_{opt}^H as time goes on. For this reason, the system dynamics on the the manifold Ξ_{opt}^H can be completely analyzed within the projection $\mathbb{R}_H \to \mathbb{R}_{[4]}$.

6.1.8 Zero-Hamiltonian Manifold: Stable and Unstable Branches

The value of Hamiltonian H is conserved through the motion along any trajectory described by Eqs. (6.39). At the goal point it is equal to zero according to the construction of this Hamiltonian, Exp. (6.36). Therefore, the zero-value of Hamiltonian, $H = 0$, can be attributed to the optimal action strategy manifold Ξ^H_{opt} as a whole.

The five-dimensional hyper-surface in the Hamilton space $\mathbb{R}_H$ specified by the condition $H(x, v, a, p_x, p_v, p_a) = 0$ includes in itself the manifold Ξ^H_{opt}. As it must, if the system state belongs to this hyper-surface, then it remains on the hyper-surface further. However, in the general case, the system dynamics on the hyper-surface cannot be described within the projection $\mathbb{R}_H \to \mathbb{R}_{[4]}$. Nevertheless, there are two three-dimensional manifolds $\Xi_s(x, v, a)$ and $\Xi_u(x, v, a)$ belonging to this hyper-surface that admit the description of the system dynamics within the projection $\mathbb{R}_H \to \mathbb{R}_{[4]}$. In other words, they admit the introduction of inner-dynamics. The union $\Xi = \Xi_s \cup \Xi_u$ will be referred to as the *zero-Hamiltonian manifold* or *zero-manifold*. The first component Ξ_s of zero-Hamiltonian manifold is the optimal action strategy manifold $\Xi^H_{\text{opt}}(x, v, a)$ characterized, according to its construction, by the system dynamics leading to the goal point $\{x = x_g, v = 0, a = 0\}$. For this reason, below, the manifold $\Xi_s \equiv \Xi^H_{\text{opt}}$ will be referred to as the *stable branch* of zero-Hamiltonian manifold (or the *stable manifold*) and the optimal action strategy manifold interchangeably, depending on context. The other component Ξ_u is characterized by that on it any trajectory of object motion governed by Hamilton's equations (6.39) goes away from the goal point. For this reason, component Ξ_u will be referred to as the *unstable branch* of the zero-Hamiltonian manifold or the *unstable* manifold.

The present subjection is aimed at a detailed analysis of the properties of the stable and unstable manifolds Ξ_s and Ξ_s underlying our further constructions.

Spectral properties of object dynamics near the goal point: First of all, let us study the eigenvalues of Hamilton's equations (6.39) at the goal point

$$x = x_g\,, \qquad v = 0\,, \qquad a = 0\,, \qquad p_x = 0\,, \qquad p_v = 0\,, \qquad p_a = 0$$

for weak perturbations Δ_q $(q = x, v, a, p_x, p_v, p_a)$

$$x(t) = x_g + \exp\left\{\frac{\lambda}{\tau_{cp}}t\right\}\Delta_x\,, \qquad p_x(t) = \exp\left\{\frac{\lambda}{\tau_{cp}}t\right\}\Delta_{p_x}\,,$$

$$v(t) = \exp\left\{\frac{\lambda}{\tau_{cp}}t\right\}\Delta_v\,, \qquad p_v(t) = \exp\left\{\frac{\lambda}{\tau_{cp}}t\right\}\Delta_{p_v}\,,$$

$$a(t) = \exp\left\{\frac{\lambda}{\tau_{cp}}t\right\}\Delta_a\,, \qquad p_a(t) = \exp\left\{\frac{\lambda}{\tau_{cp}}t\right\}\Delta_{p_a}\,,$$

where the dimensionless increment λ is the eigenvalue we are seeking for. Using the standard routine, we get the following eigenvalue equation

$$\frac{\gamma_s}{2}\left(\epsilon_x - \epsilon_v\lambda^2\right) + \lambda^4\left(\epsilon_a - \epsilon_j\lambda^2\right) + \widetilde{g}\left[\left(f_a\lambda^2 - f_x\right)^2 - f_v^2\lambda^2\right] = 0\,. \quad (6.50)$$

In obtaining Exp. (6.50), we have introduced the dimensionless parameters

$$\widetilde{g} = \frac{g\widetilde{k}^2\delta_x^2}{\tau_{cp}^2}\,, \quad f_a = \frac{m}{\widetilde{k}\tau_{cp}}\,, \quad f_x = \frac{k_x\tau_{cp}}{\widetilde{k}}\,, \quad f_v = \frac{k_v}{\widetilde{k}} \quad (6.51)$$

where the introduced coefficient $\widetilde{k}$ is selected such that it meet, e.g., the condition

$$\frac{m}{\tau_{cp}} + k_x\tau_{cp} + k_v \sim \widetilde{k}\,, \quad (6.52)$$

as well as accepted, for simplicity, the following equalities

$$\delta_x = \delta_v\tau_{cp} = \delta_a\tau_{cp}^2 = \delta_j\tau_{cp}^3\,, \quad (6.53)$$

to hold. Equalities (6.53) mean a certain self-consistency of component-wise accuracy in monitoring the motion characteristics of the controlled object within the complex present.

For objects with overdamped dynamics, the spectral properties of the corresponding Hamilton's system are also described by the eigenvalue equation (6.50) where $\epsilon_j = 0$ and $f_a = 0$.

As it must be according to the general theory of Hamiltonian systems (e.g., Meyer and Offin 2017), the eigenvalue equation (6.50) contains only even degrees of the increment λ. Thereby, if λ is the eigenvalue of the analyzed system, i.e., λ is a root of Eq. (6.50), so are $-\lambda$ as well as $\overline{\lambda}$ and $-\overline{\lambda}$, if, addition, λ is a complex number.[17]

Effort fusion, terminal condition, and spectral properties: Now we can complete the description of two issued which actually were postponed up to this moment of our constructions.

Problem 1, it is the finding of the common units in quantifying the cognitive effort of bodily executed action and mental effort of monitoring the resulting object motion. We have supposed that the effort fusion endows the two types of effort with comparability implying the emergence of common units, which underlies the introduction of the coefficient g, Exp. (6.27). The particular value of the coefficient g remained undetermined.

Problem 2, it is the quantitative description of the terminal state of action strategy implementation. In Sect. 6.1.4, we introduced the concept of terminal state of action strategy implementation, which led to the temporal boundary value problem (6.33) or

[17] Here and below the bar above some number, e.g., $\overline{\lambda}$, denotes its complex conjugate.

its Hamiltonian representation (6.45). Among the stated three conditions (Sect. 6.1.4), the last one is qualitative; the introduced quantity—the time increment Δ_t specifying the confinement of subject's actions to the complex present—was not related directly to the system parameters at that stage of our constructions.

The constructed eigenvalue equation (6.50) enables us to relate these two quantities, g and Δ_t, to particular values of system parameters. Starting from Problem 2, we impose the following two conditions on the roots of Eq. (6.50).

- All the dimensionless eigenvalues λ (the roots) of the eigenvalue equation (6.50) in magnitude should be about unity, $|\lambda| \sim 1$. In the dimensional form this proposition converts into the requirement that the optimal implementation of the action strategy $\mathcal{AS}$ be confined to time scales τ_{cp}—the characteristic duration of the complex present.[18]
- The eigenvalue equation (6.50) and, consequently, the linear space of Hamilton's system (6.39) attached to the goal point is not degenerated. In our description, it is reduced to the requirement that the spectrum of Eq. (6.50) contain six eigenvalues, a pair $\{\pm\lambda_r\}$ real roots and a group $\{\pm\lambda_c, \pm\overline{\lambda}_c\}$ of four complex roots.[19] In the case of objects with overdamped dynamics, the spectrum contains only the group $\{\pm\lambda_c, \pm\overline{\lambda}_c\}$ of four complex roots.

The two conditions are fulfilled when

$$\widetilde{g} = \frac{g\widetilde{k}^2\delta_x^2}{\tau_{cp}^2} \sim 1\,. \tag{6.54}$$

On the one hand, estimate (6.54) resolves Problem 2, for these system parameters the time increment $\Delta_t \sim \lambda/\tau_{cp}$ takes the required values, $\Delta_t \sim 1/\tau_{cp}$. On the other hand, estimate (6.54) specifies the relationship between the cognitive effort of bodily executed actions and the effort of monitoring, which gives the desired solution to Problem 1. Namely, the coefficient g is estimated as

$$g \sim \frac{\tau_{cp}^2}{\widetilde{k}^2\delta_x^2}\,.$$

Thereby, the description of optimal action strategy, see Exp. (6.28), becomes an well-definite problem whose parameters can be estimated turning to well-understood psychophysical properties of human perception.

Manifold of zero-value Hamiltonian:

So let us analyze the system dynamics in detail for Hamiltonian (6.42) with the exponent $\gamma_s = 2$.[20] In this case, Hamilton's equations (6.39) become linear. As a

[18] The exponent $\gamma_s < 2$ has to make this localization within the complex present more pronounced.

[19] As follows directly from the form of the eigenvalue equation (6.50), it cannot possess pure imaginary roots.

[20] The exponent $\gamma_s = 2$ corresponds to the situation when the accuracy in monitoring the dynamics of controlled object does not depend on the proximity of its motion state to the desired goal. However,

result, the space $\mathbb{R}_H$ may be regarded as a six-dimensional linear Hamilton space with vector dynamics governed by the equation

$$\frac{dz}{dt} = \mathbf{J}\frac{\partial H}{\partial z} = \mathbf{J}\mathbf{H}z\,, \tag{6.55}$$

where the real vector z and the matrix (operator) $\mathbf{J}$ have the forms

$$z = \begin{bmatrix} x \\ v \\ a \\ p_x \\ p_v \\ p_a \end{bmatrix}, \qquad \mathbf{J} = \left[\begin{array}{ccc|ccc} 0 & 0 & 0 & 1 & 0 & 0 \\ 0 & 0 & 0 & 0 & 1 & 0 \\ 0 & 0 & 0 & 0 & 0 & 1 \\ \hline -1 & 0 & 0 & 0 & 0 & 0 \\ 0 & -1 & 0 & 0 & 0 & 0 \\ 0 & 0 & -1 & 0 & 0 & 0 \end{array}\right]. \tag{6.56}$$

The Hamiltonian $H(z)$ given by Exp. (6.42) with the exponent $\gamma_s = 2$ determines the symmetric matrix (operator) $\mathbf{H} = \left[H_{ij}\right]_{i,j\in[1,6]}$ via the equality

$$H(z) = \langle z|\mathbf{H}|z\rangle \stackrel{\text{def}}{=} \frac{1}{2}z^{\mathrm{T}}\mathbf{H}z \;\Rightarrow\; H_{ij} = \frac{\partial^2 H(z)}{\partial z_i \partial z_j}\,, \tag{6.57}$$

where the symbol z^{T} stands for the transpose of the vector z. The corresponding symplectic bilinear form $\omega(z_1, z_2)$ called also the *symplectic inner product* is defined by the expression

$$\omega(z_1, z_2) = \langle z_1|\,\mathbf{J}\,|z_2\rangle = \overline{z}_1^{\mathrm{T}}\mathbf{J}z_2\,. \tag{6.58}$$

It should be noted that in the theory of linear Hamiltonian systems (e.g., Meyer and Offin 2017) the symplectic form $\omega(z_1, z_2)$ is usually introduced without complex conjugation. Here we prefer definition (6.58) to use Dirac's bra-ket notation in its standard meaning.

The analyzed above spectral properties of Hamilton's equations (6.39) near the goal point actually specify the eigenvalues $\{\lambda\}$ (in units of τ_{cp}^{-1}) and the corresponding eigenvectors $\{z_\lambda\}$ of the matrix $\mathbf{JS}$. According to the accepted model of the Hamilton's spectral properties, the matrix $\mathbf{JS}$ possesses six eigenvalues, a pair $\{-\lambda_0, \lambda_0\}$ of real eigenvalues, $\lambda_0 \sim 1$, and a group of four complex eigenvalues $\{-\lambda_c, -\overline{\lambda}_c, \lambda_c, \overline{\lambda}_c\}$, where $\lambda_c = \lambda_r + i\lambda_i$ and $\lambda_r \sim 1$ and $\lambda_i \lesssim 1$. These eigenvalues are the roots of the eigenvalue equation (6.50) and in our further analysis are assumed to be known.

Two eigenvectors z_{λ_1} and z_{λ_2} whose eigenvalues $\overline{\lambda}_1 + \lambda_2 \neq 0$ meet the condition (e.g., Meyer and Offin 2017)

---

even in the nonlinear case, $\gamma_s < 2$, Hamiltonian (6.42) and the Hamilton's equations (6.39) admit linear approximation. So the zero-manifold Ξ to be constructed below is just a linear approximation of its general form which may be regarded as a certain continuation into a region of nonlinear dynamics. Constructing the general form of Ξ is an individual problem beyond the scope of the present book.

$$\omega(z_{\lambda_1}, z_{\lambda_2}) = 0\,. \tag{6.59}$$

This property prompts us to introduce two three-dimensional linear subspaces, i.e., two three-dimensional manifolds in the six-dimensional Hamilton space $\mathbb{R}_H$:[21]

- the linear span Ξ_s of the eigenvectors $z_{s0} = z_{-\lambda_0}$, $z_{sc} = z_{-\overline{\lambda}_c}$, $\overline{z}_{sc} = z_{-\lambda_c}$:

$$\Xi_s(\zeta_0, \zeta_c, \overline{\zeta}_c) = \left\{\zeta_0 z_{s0} + \zeta_c z_{sc} + \overline{\zeta}_c \overline{z}_{sc}\right\} \text{ for } \zeta_0 \in \mathbb{R},\ \zeta_c \in \mathbb{C}\,, \tag{6.60a}$$

- the linear span Ξ_u of the eigenvectors $z_{u0} = z_{\lambda_0}$, $z_{uc} = z_{\lambda_c}$, $\overline{z}_{uc} = z_{\overline{\lambda}_c}$:

$$\Xi_u(\zeta_0, \zeta_c, \overline{\zeta}_c) = \left\{\zeta_0 z_{u0} + \zeta_c z_{uc} + \overline{\zeta}_c \overline{z}_{uc}\right\} \text{ for } \zeta_0 \in \mathbb{R},\ \zeta_c \in \mathbb{C}\,. \tag{6.60b}$$

Here we have taken into account that for the given real-value Hamiltonian the eigenvectors of the eigenvalues λ and $\overline{\lambda}$ are related as

$$z_{\overline{\lambda}} = \overline{z}_\lambda$$

and in decompositions (6.60), the coefficients at the corresponding complex eigenvectors should be related via complex conjugation. In the presented form the variables $\{\zeta_0, \zeta_c, \overline{\zeta}_c\}$ play the role of parametrization arguments of these manifolds.

The constructed linear manifolds are characterized by the following properties:

- Their dimension is half the dimension of the Hamilton space $\mathbb{R}_H$ and for any two vectors $z_1,\ z_2 \in \Xi_s$ or Ξ_u

$$\omega(z_1, z_2) = 0 \tag{6.61}$$

by virtue of (6.59). For this reason they are classified as Lagrange subspaces of the space $\mathbb{R}_H$ (e.g., Meyer and Offin 2017).
- Any vector of the space $\mathbb{R}_H$ can be expanded into its components belonging to the manifolds Ξ_s, Ξ_u, so their union reconstructs the Hamilton space $\mathbb{R}_H$, i.e.,

$$\Xi_s \bigoplus \Xi_u = \mathbb{R}_H\,.$$

This feature is usually referred to as the Lagrange splitting of a Hamilton space (e.g., Meyer and Offin 2017).
- Any point on the manifold Ξ_s whose motion is governed by Hamilton's equations (6.39) goes along Ξ_s towards the goal point—the stationary point of this equations. So Hamiltonian H (6.42) (see also Exp. 6.57) takes zero-value at any point of the manifold Ξ_s. It is due to the basic eigenvectors of the linear manifold Ξ_s possess the eigenvalues with negative real parts.
- Any point on the manifold Ξ_u whose motion is governed by Hamilton's equations (6.39) goes along Ξ_u away from the goal point. Because this point can start

[21] A detailed description of this Hamiltonian system is presented in Appendix 6.7.

its motion in any close proximity to the goal point, Hamiltonian H (6.42) has also to take zero-value at any point of the manifold Ξ_u. This motion away from the goal point is due to the basic eigenvectors of the linear manifold Ξ_u possess the eigenvalues with positive real parts.

Taking into account these properties, we see that the Lagrange subspaces Ξ_s and Ξ_u of the Hamilton space $\mathbb{R}_H$ are the *stable* and *unstable branches* of the zero-manifold introduced previously.[22]

Below we want to discuss several aspects of the stable branch Ξ_s of zero-manifold clarifying its role in the description of human goal-oriented actions.

The stable-manifold-based model of human control: Finalizing the present subsection, we want to note that the classical paradigm of human control assumes human actions to be *consciously* aimed at driving a controlled object to the desired state. The concept of stable manifolds, which can be introduced in various ways, posits that subject's *conscious* actions are aimed at keeping the motion of the controlled object near the corresponding stable manifold. In this case, subject's automatic actions or just physical regularities, on their own, drive the object to the desired state without conscious control from the subject.

For the first time, the paradigm of human control assuming the keeping of the object motion near a stable manifold seems to be put forward by Scholz and Schöner (1999), for a review a reader may be referred to Latash (2008b, 2012), Huys et al. (2014). To cope with the redundancy of motor actions, Scholz and Schöner (1999) introduced the concept of *uncontrolled manifold* to emphasis that perturbations in task performance can be divided into two types. Some perturbations cause unexpected displacements of the object along the uncontrolled manifold, the others cause the object deviation from the manifold. In the former case, no correction is necessary because such perturbations do not affect the task achievement, which is exactly the characteristic feature of the stable manifold Ξ_s. Bottaro et al. (2008), Asai et al. (2009, 2013), Suzuki et al. (2012) also turned to a similar concept of stable manifold in describing human intermittent control. In their account, the stable and unstable manifolds represent the motion of object with inertia along the separatrices of its unstable equilibrium, which is described by Newtonian mechanics, for a review see Suzuki et al. (2016), Yoshikawa et al. (2016).

The account of the stable and unstable manifolds, Ξ_s and Ξ_u, and their role in human control, we have presented in this subsection, may be regarded as a generalization of the previous models. One of the most crucial features of this generalization is the mechanism governing the emergence of the stable manifold. As demonstrated, the emergence of the stable and unstable branches Ξ_s and Ξ_u of zero-manifold is

[22] It should be noted that, according to the presented analysis, the stability or instability of the manifolds Ξ_s and Ξ_u is understood as the direction of inner dynamics toward or outwards the goal point rather than decrease or increase of some deviation in object motion from Ξ_s and Ξ_u. Even the strict motion along, e.g., the stable branch Ξ_s of zero-manifold is a result of subject's intentional actions, so is the correction of current action when the object motion deviates from Ξ_s. Therefore, the notion of stability of intentional human actions requires special consideration, in the following sections we will return to this issue.

a basic feature of human goal-oriented actions. Their emergence is due to the fact that all the stationary points representing the desired goals of human actions in their mathematical description are saddles independently of whether the effect of inertia is essential or the object dynamics is overdamped.[23] Put differently, speaking about human control we may posit that the stable manifold characterizing such actions describes the cumulative contribution of

- inner mechanisms governing the human perception of bodily executed actions including the monitoring of their results,
- physical regularities governing the dynamics of controlled object in the external world,

rather than the two factors taken separately.

The constructed concept of the stable manifold Ξ_s enables us to recognize in more detail the common and different aspects of intentional and automatic actions as well as their role in human control and goal-oriented behavior.

- Automatic actions reflecting (*i*) the general details of subject's actions on the controlled object and (*ii*) the physical regularities of object motion can be conceived of as the object motion along the stable branch $\Xi_s(x, v, a)$ of zero-manifold or, at least, in its proximity.

  If the object motion does not deviate essentially from the stable manifold $\Xi_s(x, v, a)$, no correction is necessary. In this case, an action strategy implementation, initiated at a certain instant of time in response to the current situation, is finished after the desired goal having been attained. It means that between the instants of initiation and completion of action strategy implementation, the subject only monitors the object motion without active response to the changing situation, which is usually called the open-loop control (e.g., Franklin et al. 2018).
- When the object motion deviates substantially from the stable manifold $\Xi_s(x, v, a)$, for the subject to correct the current situation it is enough to return the object motion to *any point* on the manifold $\Xi_s(x, v, a)$. This correction is an conscious, intentional action. In other words, subject's conscious, intentional actions in implementing a give action strategy are focused on keeping the object motion near the stable manifold $\Xi_s(x, v, a)$, which automatically drives the object to the desired goal. It is a reason for calling this conception the *stable-manifold-based model* of human control.

It should be emphasized that the developed concept of the stable manifold describes the strictly optimal implementations of action strategies. So, although the stable manifold is a useful element of the mathematical description, the mind is not able to manipulate with it directly. The mind operates with action strategy clouds and to elucidate their properties we need to understand the characteristic features of action strategy implementations located near the stable manifold but not belonging to it, which is the purpose of the next section.

---

[23] In this context we want to note also the concept of metastability of the brain functioning assuming the presence of a certain multitude of saddle points describing the dynamics of neural processes (e.g., Kelso 2009a, b, 2012).

6.2 Space-Time Structure of Action Strategy Cloud: Ostrogradsky Instability

In the previous section we have analyzed the properties of optimal implementation of action strategies which plays an essential role in describing subject's actions but is inaccessible to the mind. In the present section we discuss real implementations of action strategies which reflect the basic properties of human perception and motor behavior.

First of all, let us clarify the key aspects of the problem under consideration. We will use the term *hypothetical ideal subject* concerning the optimal implementations of action strategies. To develop an appropriate mathematical formalism for describing real implementations of actions strategies, we need to confine our consideration to situations that:

- are met rather often in reality and form some interesting class of phenomena characterizing human behavior,
- reflect the basic features of human perception and motor actions,
- are characterized by certain regularities, maybe, at the level of statistical description.[24]

Put differently, these situations, on the one hand, should admit a relatively simple description to elucidate the gist of the posed problems in a tractable, representative manner. On the other hand, we must not "throw the baby out with the bathwater."

Below we will deal with real, i.e., imperfect implementations of action strategies turning to the concept of the *skilled subject*. In this way we turn to the gist of the principle of cloud-fusion self-consistency stated in Sect. 5.6.3. It implies that fragments of the human temporality different in temporal scales are characterized by a certain self-consistency resulting from long-term learning and skill acquisition. The actions of the skilled subject on scales of the experiential now, the complex present, and beyond are assume to be related to each other in some reasonable way. These relations are not perfect because of the intrinsic uncertainty in human perception and motor behavior, the memory limited capacity, etc. So the actions of the skilled subject may be regarded as approximately optimal.

Put differently, in what follows, we will analyze *the implementations of action strategies located near the optimal implementations within certain proximity determined by the bounded capacity of human cognition*. All the other action strategies and their implementations that evidently are not reasonable will not be included in consideration.

At the beginning, we want to describe imperfect implementations of action strategies as they may be recorded by the external observer. For a given action strategy $\mathcal{AS}$, its implementations convert into the cloud $\mathbb{C}_{\mathrm{AS}}$ in the self. In the body-in-the-self the cloud $\mathbb{C}_{\mathrm{AS}}$ is represented as a certain space-time structure in $\mathbb{R}_{[4]} \times t$. In the mind-in-the-self, the cloud $\mathcal{C}_{\mathrm{AS}}$ is a certain indivisible entity with internal integrity.

[24] These regularities may be categorized as *ceteris paribus* laws; their discussion within the given context can be found, e.g., in Lubashevsky (2017).

The final goal of this section is the construction of the governing equation for the space-time structure of the cloud $\mathbb{C}_{\mathrm{AS}}$.

6.2.1 *Imperfection of Action-Strategy Implementation: Two Sources*

As noted above, the subject is not able to drive the controlled object such that its motion state be on the stable manifold Ξ_s during the action strategy implementation. It is caused by the limited capacity of monitoring the object motion and reproducing motor actions. The intrinsic uncertainty in

- perceiving the characteristics of object motion—the spatial position x, velocity v, acceleration a, and jerk j,
- reproducing a sequence of actions retained in memory and planned for realization in the connected future

makes actually any implementation of a given action strategy $\mathcal{AS}$ imperfect. In other words, any action strategy implementation based on an initially fixed sequence of actions at the beginning leads toward the desired goal but, then, inevitably goes away from it. The purpose of the present subsection is to consider in detail these sources of imperfectness and their mathematical description.

Broadly speaking, there are two factors contributing to the implementation flaw of action strategies. They can be reduced to the subject's capacity in controlling the time course of (*i*) the initiation stage of action strategy implementation and (*ii*) its main stage. Let us consider these stages together with the corresponding factors individually.

Action strategy initiation: When the subject makes the decision to implement a new action strategy $\mathcal{AS}$, at first, he needs to initiate it. Namely, the subject has to transfer the current state of object motion into the state corresponding to the beginning stage of the action strategy $\mathcal{AS}$. This transformation should be completed within the experiential now when the subject has recognized that the action strategy being under implementation must be corrected.

In the scope of the body-in-the-self, the motion state of controlled object is specified by its spatial position x, velocity v, acceleration a, and jerk j. As elucidated in Sect. 6.1.4, the first three quantities $\{x, v, a\}$ are the inner phase variables which cannot be changed step-wise by subject's actions. Moreover, the subject tries not to change the inner phase variables substantially within the experiential now via very strong changes in the jerk not to loose control over the situation. The jerk j is the control phase variable and the subject changes it sharply within the experiential now to adapt the motion state of controlled object, e.g., to a newly selected action strategy. Thereby, the long-term time pattern of the jerk $j(t)$ is partitioned into the sequence of alternate fragments containing:

– sharp and short variations on scales of the experiential now bounded from above by the duration τ_d of experiential now[25] and
– smooth and relatively long variations whose duration is comparable with the duration τ_{cp} of complex present.

The motion state of controlled object that belongs to the strictly optimal implementation $\mathrm{AS}_{\mathrm{opt}}$ of action strategy $\mathcal{AS}$—the trajectory $P_{\mathrm{opt}}(t)$ in the space $\mathbb{R}_{[4]} \times t$—is actually specified by the last equation of (6.45) describing the attachment of this state to the stable manifold $\Xi_s(x, a, j)$. It enables us to represent this state as the condition

$$\{x, v, a, j\}_{\mathrm{AS}_{\mathrm{opt}}} \in \Xi_s \quad \Leftrightarrow \quad j = j_{\mathrm{opt}}(x, v, a)\,, \tag{6.62}$$

where, by definition,

$$j_{\mathrm{opt}}(x, v, a) \stackrel{\mathrm{def}}{=} \delta_j \left[\omega^s_{ax} \frac{(x - x_g)}{\delta_x} + \omega^s_{av} \frac{v}{\delta_v} + \omega^s_{aa} \frac{a}{\delta_a} \right]; \tag{6.63}$$

the coefficients $\{\omega^s_{\dots}\}$ are given by Exp. (6.289) (Appendix 6.7.3).

The initiation of action strategy can be described in terms of

$$\frac{dj}{dt} = \frac{1}{\tau_d} V_{\mathrm{ini}}(t)\,, \tag{6.64}$$

where the function $V_{\mathrm{ini}}(t)$ to be called the *initiation force* is localized on scales of the experiential now. For the optimal implementation $\mathrm{AS}_{\mathrm{opt}}$ of action strategy to be initiated, the condition

$$\frac{1}{\tau_d} \int\limits_{t \in \mathrm{EN}} V_{\mathrm{ini}}(t) dt = j_{\mathrm{opt}}(x, v, a) - j_0 \tag{6.65}$$

must hold. Here j_0 is the current value of the jerk j determined, e.g., by the implementation of the preceding action strategy and the symbol EN stands for the region of integration related to the experiential now with the characteristic duration $\tau_d \sim 2$–3 s (Sect. 2.6).

In reality, equality (6.65) never can be attained because of uncertainty in human perception. We suppose that after acquiring the skill, the subject becomes able to initiate the action strategy $\mathcal{AS}$ within the accuracy δ_j, i.e.,

[25] Following the concept of "action points" proposed by Todosiev (1963), Todosiev and Barbosa (1963/64) for modeling the car-following, we will call these short fragments *action points* representing the events when the subject intentionally changes one action strategy for another one. The other fragments may be conceived of as fragments of subject's automatic actions. We will return to this issue in Sect. 6.3 when describing the probabilistic dynamics of action strategies.

$$\frac{1}{\tau_d} \int_{t \in \text{EN}} V_{\text{ini}}(t) dt = j_{\text{opt}}(x, v, a) + \delta_j \xi_j \,, \tag{6.66}$$

where ξ_j is a random (uncontrollable) variable characterized by the amplitude about unity; we will refer to such variables as the *initiation noise*. The subject cannot recognize the difference between the desired jerk j_{opt} and the result $j = j_{\text{opt}} + \delta_j \xi_j$ of his actions. So, even if he were able to reproduce the sequence of actions related to the strictly optimal implementation AS_{opt} of action strategy $\mathcal{AS}$ starting from the point $\{x, v, a, j_{\text{opt}}\}$, the subject would never drive the object into the desired goal point. It is caused by the wrong starting position of the expected trajectory P_{opt}, i.e., $j|_{t=t_{\text{ini}}} \neq j_{\text{opt}}(x, v, a)|_{t=t_{\text{ini}}}$.

It should be noted that in the present analysis of initiation noise, we tacitly assumed that the subject has access to the precise values of the variables $\{x, v, a\}$. In reality, the subject perceives this values within the accuracy $\{\delta_x, \delta_v, \delta_a\}$, which also contributes to the uncontrolled difference between the current value j of jerk and the value of j_{opt}. Expression (6.63) enables us to aggregate this effect into the initiation noise.

Summarizing the aforesaid, we introduce the notion of the *initiation uncertainty of action strategy implementations* as attaching the current state of object motion to a certain point σ_s on the stable manifold $\Xi_s(x, v, a,)$,

$$\sigma_s \stackrel{\text{def}}{=} \left\{x_s, v_s, a_s, j_s = j_{\text{opt}}(x_s, v_s, a_s)\right\} \in \Xi_s \,,$$

within the component-wise accuracy $\delta_x, \delta_v, \delta_a, \delta_j$ meeting the self-consistency condition (6.53). Here we deal with the manifold Ξ_s embedded into the space $\mathbb{R}_{[4]}$. Below we will call the fuzzy region

$$\mathbb{C}_{\text{ini}[4]} = \frac{(x - x_s)^2}{\delta_x^2} + \frac{(v - v_s)^2}{\delta_v^2} + \frac{(a - a_s)^2}{\delta_a^2} + \frac{(j - j_s)^2}{\delta_j^2} \lesssim 1 \tag{6.67}$$

in the space $\mathbb{R}_{[4]}$ the *cloud of action strategy initiation*.

The initiation uncertainty is the **first factor** responsible for the implementation flaw of action strategies. As far as its mathematical description within Ostrogradsky's formalism is concerned, the process of action strategy initiation, strictly speaking, does not belong to the scope of this formalism. Nevertheless, formally we can include the action strategy initiation into the Hamiltonian description based on Eqs. (6.39) and, in particular, on Eq. (6.55) after a certain generalization of the AS-Hamiltonian $H(x, v, a, p_x, p_v, p_a)$, Exp. (6.42) and, respectively, Exp. (6.57). Namely, to allow for the initiation process, the AS-Hamiltonian $H(x, v, a, p_x, p_v, p_a)$ is modified as

$$H^{(1)}(x, v, a, p_x, p_v, p_a, t) = H_0(x, v, a, p_x, p_v, p_a) - \frac{\epsilon_j}{\tau_{cp}\delta_j^2} a V_{\text{ini}}(t) \,, \tag{6.68}$$

where $H_0(\ldots)$ is the Hamiltonian given by Exp. (6.42) and relationship (6.41) between the jerk j and the momentum p_a has been taken into account. The replacement $H_0(\ldots) \Rightarrow H^{(1)}(\ldots)$ leads to the only one modification of Hamilton's equations (6.39) (see also 6.55)

$$\frac{dp_a}{dt} = -\frac{\partial H_0}{\partial a} + \frac{\epsilon_j}{\tau_d \delta_j^2} V_{\text{ini}}(t)\,, \tag{6.69}$$

which contains directly the initiation term (6.64). In particular, by virtue of (6.66), the initiation noise in the control variable j gives rise to the uncertainty in the initiated value of the momentum p_a about

$$\delta_{p_a} = \frac{1}{\delta_j} = \frac{\tau_{cp}^3}{\delta_x}\,. \tag{6.70}$$

The last estimate follows from the accepted relations (6.53) caused by the self-consistency of component-wise accuracy in monitoring object motion. As it must be, the product of the uncertainty δ_{p_a} of the momentum p_a and the uncertainty δ_j of the jerk j is a value about unity, which reflects the structure of Hamiltonian (6.36) containing the term $p_a j$ as a constituent element.

It should be emphasized that the action strategy initiation is a process governed by the monitoring of object motion state rather than the bodily execution of the corresponding motor actions. Indeed, when the subject recognizes that the currently executed action strategy must be corrected, he selects a new action strategy and transforms the system state as it is required to start its implementation. So although the initiation process may include automatic fragments, its neural mechanisms should be different from ones underlying subject's motor behavior in action strategy implementation which is described by Ostrogradsky's formalism.

This feature is taken into account in the Hamiltonian modification (6.68) with the initiation force $V_{\text{ini}}(t)$ depending explicitly only on physical time t. In principle, the initiation force $V_{\text{ini}}(t)$ can depend on the current motion state of controlled object as it is perceived by the subject, in addition to the anticipated motion in the connected future and the information accumulated within the connected past. So, even the initiation force $V_{\text{ini}}(t)$ reflects in some way the current state of object motion, this dependence comes from another process being external with respect to the action execution described by Ostrogradsky's formalism.

Automatic implementation of action strategies: On the one hand, (*i*) the temporal boundary value problem (6.33) for the Euler-Lagrange equation (6.32) and, then, (*ii*) the equivalent temporal boundary value problem (6.45) based on Hamilton's equations (6.39) make up the mathematical description of the optimal implementation AS_{opt} of a given action strategy $\mathcal{AS}$. In particular, the constructed stable manifold $\Xi_s(x, v, a)$ (Sect. 6.1.8) describes this optimal implementation AS_{opt} in terms of the initial value problem dealing with only the inner phase variables of object dynamics, Eq. (6.49).

On the other hand, the skilled subject implements action strategies mainly automatically. It implies that the subject is able to estimate the current situation, select an appropriate action strategy, and arrange the required actions in a specific way within the connected future.

To reconcile the two approaches to describing the optimal implementation $\mathrm{AS}_{\mathrm{opt}}$ we suppose that the hypothetical ideal subject

- detects precisely the current spatial position x, velocity v, and acceleration a of the controlled object;
- initiates the implementation $\mathrm{AS}_{\mathrm{opt}}$ of the action strategy $\mathcal{AS}$ such that the jerk take the value $j = j_{\mathrm{opt}}(x, v, a)$;
- creates the plan of his further actions reduced to implementing the time patter $j(t)$ meeting the inner dynamics (6.49) of the stable manifold $\Xi_s(x, v, a)$ and follows it strictly.

The corresponding time variations of the momenta $p_x(t)$ and $p_v(t)$ are determined by the time variations of the components

$$\zeta_0 e^{-\lambda_0 t}\, z_{s0}\,, \quad \zeta_c e^{-\lambda_c t}\, z_{sc}\,, \quad \overline{\zeta}_c e^{-\overline{\lambda}_c t}\, \overline{z}_{sc}$$

belonging to the manifold Ξ_s, i.e., the linear span of the eigenvectors z_{s0}, z_{sc}, $\overline{z}_{sc}$. In particular, by virtue of (6.39) for Hamiltonian (6.42) with the exponent $\gamma_s = 2$, the initial values of the momenta p_x and p_v corresponding to $\mathrm{AS}_{\mathrm{opt}}$ can be related to the initial values of the inner phase variables of motion as their linear combinations represented in Appendix 6.7. Their time pattern may be regarded as the plan of subject's intentional actions within the connected future giving rise to $\mathrm{AS}_{\mathrm{opt}}$.

The real subject is not able to realize all these items. However, we accept that the skilled subject is able to realize these actions approximately such that a considerable deviation of the object motion from the desired state can be reduced to a substantial degree. It implies that the acquired skill enables the subject to

- initiate an appropriate action strategy $\mathcal{AS}$ such that its initial point $\{x, v, a, j\}$ be not too far from the stable manifold Ξ_s ,
- recognize the sequence of further actions required for driving the object towards the desired goal and estimate their intensity and duration.

Here our attention is focused on the latter issue.

The subject's plan of the further actions and its particular realization is always influenced by the intrinsic uncertainty of human perception and motor behavior. Therefore, even if after the initiation the initial point $\{x, v, a, j\}$ were located on the stable manifold Ξ_s, the produced trajectory of object motion would never get the goal. This is the **second factor** responsible for the implementation flaw of action strategy.

As will be discussed more thoroughly in the next subsection, we may regard the generalized momenta p_x and p_v as also *individual* control variables in addition to p_a just proportional to the jerk j. The main difference between the control variables p_x and p_v, on the one side, and the control variable p_a, on the other side, is that the former pair concerns the future actions, whereas via the latter variable the subject affects the current state of objection motion.

Within the optimal implementations of action strategies the object motion is reduced to the inner dynamics on the stable manifold Ξ_s, which implies a liner relationship between the momenta p_x, p_v, and the motion variables $\{x, v, a\}$. The coefficients of this relationship are about unity if the time is measured in units of τ_{cp} (Appendix 6.7). In dimensional units, by virtue of the structure of Hamiltonian (6.36), the corresponding uncertainties in specifying the values of p_x and p_v can be estimates as

$$\delta_{p_x}\Big|_{\Xi_s} \sim \frac{1}{\delta_v} = \frac{\tau_{cp}}{\delta_x}, \qquad \delta_{p_v}\Big|_{\Xi_s} \sim \frac{1}{\delta_a} = \frac{\tau_{cp}^2}{\delta_x}.$$

For real action strategy implementations, the momenta p_x and p_v describe the subject's automatic actions planned beforehand based on the acquired skill. So, the accuracy within which the momenta p_x and p_v are determined depends on several factors. Uncertainty in perceiving the object motion is only one of them. Uncertainty in reproducing motor actions and anticipating the object dynamics are other factors. Therefore, the given values of the uncertainties δ_{p_x} and δ_{p_v} are actually their estimates from below. To allow for the other factors, we introduce a certain coefficient $\varkappa > 1$ in the estimates

$$\delta_{p_x} = \frac{\varkappa}{\delta_v} = \frac{\varkappa\tau_{cp}}{\delta_x}, \qquad \delta_{p_v} = \frac{\varkappa}{\delta_a} = \frac{\varkappa\tau_{cp}^2}{\delta_x}. \tag{6.71}$$

to be accepted in our further constructions. The coefficient $\varkappa$ is regarded as a model parameter to be specified in the next subsection.

As far as the mathematical description of imperfect implementation of action strategies is concerned, we want to note the following two aspects illustrating the gist of further constructions.

First, at the moment of action strategy initiation, the momenta p_x and p_v may be assumed to take some values from the region

$$\frac{(p_x - \pi_x^s)^2}{\delta_{p_x}^2} + \frac{(p_v - \pi_v^s)^2}{\delta_{p_v}^2} \lesssim 1, \tag{6.72}$$

where $\pi_x^s = \pi_x^s(x, v, a)$ and $\pi_v^s = \pi_x^v(x, v, a)$ are the momenta p_x and p_v for the inner dynamics of the stable manifold Ξ_s.[26]

Second, during the action strategy implementation the time variations in the momenta p_x and p_v are also determined within the accuracy δ_{p_x} and δ_{p_v} on time scales about the duration τ_d of the experiential now. To allow for this effect we, following a way similar to one led to the Hamiltonian generalization (6.68), can modify the Hamiltonian $H(x, v, a, p_x, p_v, p_a, t)$ as

[26] The particular forms of the functions $\pi_x^s(x, v, a)$ and $\pi_x^v(x, v, a)$ are specified in the next subsection, Exp. (6.75) based on Appendix 6.7.3.

$$H^{(2)}(x, v, a, p_x, p_v, p_a, t) = H_0(x, v, a, p_x, p_v, p_a) - \frac{\delta_{p_x}}{\tau_d} x W_x(t) - \frac{\delta_{p_v}}{\tau_d} v W_v(t) . \quad (6.73)$$

Here, as before, $H_0(\ldots)$ is the Hamiltonian given by Exp. (6.42) and $W_x(t)$, $W_p(t)$ are random functions of physical time t with amplitude about unity and temporal correlations on scales about τ_d. Besides, at the moment of action strategy initiation these functions can exhibit strong variations on scales of τ_d responsible for the initial value of the momenta p_x and p_v meeting condition (6.72).

The next subsection takes into account both the factors responsible for the implementation flaw of action strategies in constructing the Hamiltonian description of subject's actions.

6.2.2 *Hamiltonian Description of Action Strategy Implementations*

In the previous subsection we discussed the factors making the *optimal implementation* of action strategies impossible and noted how these factors can be incorporated into our account. The purpose of the present subsection is to elaborate a formalism for describing real, i.e., imperfect subject's actions turning to the ideas and elements of the theory of Hamiltonian systems we used in Sect. 6.1. Beforehand we want to note the following:

- In this subsection we continue our analysis of action strategy implementations treated as object motion trajectories $\{P_{\text{AS}}(t)\}$, e.g.,

$$P_{\text{AS}}(t) = \left\{x(t),\ v(t) = \frac{dx}{dt},\ a(t) = \frac{d^2x}{dt^2},\ j(t) = \frac{d^3x}{dt^3}\right\}$$

  representing a given action strategy $\mathcal{AS}$ in the space $\mathbb{R}_{[4]} \times t$. So this description belongs to the realm of external observer—such trajectories are not elements that can be attributed to the realm of the self including the mind-in-the-self and the body-in-the-self.
- We call the formalism to be developed also the Hamiltonian description of action strategy implementations by the skilled subject. Although this formalism looks pretty similar to the Hamiltonian description constructed in Sect. 6.1.6, it does not deal with strictly optimal trajectories $\{P_{\text{opt}}(t)\}$ and is not reducible to Ostrogradsky's approach in mathematical aspects as well as the underlying premises.

The following premises underlie our account of subject's real actions.

Action strategy initiation: The subject initiates an action strategy implementation $P_{\text{AS}}(t)$ with a state $z_{\text{ini}} \in \mathbb{R}_H$

$$P_{\text{AS}}\big|_{\text{ini}} = z_{\text{ini}}^{\text{T}} = \{x, v, a, p_x, p_v, p_a\}_{\text{ini}}$$

belonging to the fuzzy region

$$\mathbb{C}_H^{\text{ini}}\{z\} = \frac{(x-x_s)^2}{\delta_x^2} + \frac{(v-v_s)^2}{\delta_v^2} + \frac{(a-a_s)^2}{\delta_a^2} + \frac{(p_x-\pi_x^s)^2}{\delta_{p_x}^2} + \frac{(p_v-\pi_v^s)^2}{\delta_{p_v}^2} + \frac{(p_a-\pi_a^s)^2}{\delta_{p_a}^2} \lesssim 1 \tag{6.74}$$

to be called also the *cloud of action strategy initiation* in the Hamilton space $\mathbb{R}_H$. Here the uncertainties $\delta_{p_x}, \delta_{p_v}, \delta_{p_a}$ of the momenta p_x, p_v p_a given by estimates (6.71), (6.69) are treated as model parameters. The coordinates $\{x_s, v_s, a_s\}$ represent the current state of object real motion which is accessible only to the external observer. The quantities π_x^s, π_v^s, π_a^s are the value the corresponding momenta take at the point $\sigma_s(x_s, v_s, a_s)$ lying on the stable manifold Ξ_s and specified by the coordinate $\{x_s, v_s, a_s\}$. Therefore these quantities are determined by the functions describing the embedding of the stable manifold Ξ_s into the Hamilton space $\mathbb{R}_H$. As demonstrated in Appendix 6.7.3 the values $\{\pi_x^s, \pi_v^s, \pi_a^s\}$ are related to the inner phase variables of object motion via the following expressions,

$$\pi_x^s = \frac{\epsilon_j}{\varkappa}\delta_{p_x}\left[\omega_{xx}^s\frac{(x_s-x_g)}{\delta_x} + \omega_{xv}^s\frac{v_s}{\delta_v} + \omega_{xa}^s\frac{a_s}{\delta_a}\right], \tag{6.75a}$$

$$\pi_v^s = \frac{\epsilon_j}{\varkappa}\delta_{p_v}\left[\omega_{vx}^s\frac{(x_s-x_g)}{\delta_x} + \omega_{vv}^s\frac{v_s}{\delta_v} + \omega_{va}^s\frac{a_s}{\delta_a}\right], \tag{6.75b}$$

$$\pi_a^s = \epsilon_j\delta_{p_a}\left[\omega_{ax}^s\frac{(x_s-x_g)}{\delta_x} + \omega_{av}^s\frac{v_s}{\delta_v} + \omega_{aa}^s\frac{a_s}{\delta_a}\right], \tag{6.75c}$$

where the values of the coefficients $\{\omega_{\dots}^s\}$ are summarized by Exp. (6.291) with coefficients specified by Exp. (6.292).

Control variables of action strategy implementations: As discussed in the previous subsection, within Ostrogradsky's formalism the momenta p_x and p_v have no apparent physical meaning, which is often emphasized in the terms of Ostrogradsky's ghost. These variables are required to describe the temporal boundary value problem (6.45) within the strictly optimal approximation employing the corresponding Hamilton's equations (6.39). It implies that the hypothetical ideal subject is able to create the perfect plan of his future actions—the time patterns $p_x(t)$ and $p_v(t)$ as well as $p_a(t)$—and strictly following it to get the desired goal. Although the variable p_a has a clear physical meaning—the jerk of object motion—it is a characteristics of subject's actions rather than a physical quantity determined directly by the physical regularities of object dynamics. Put differently, the ideal subject is able to find the appropriate initial values $\{p_x, p_v, p_a\}_{\text{ini}}$ of the momenta that enable him to drive the object to the desired goal from the current state $\{x, v, a\}_{\text{ini}}$ of object motion. These subject's actions are described by the right-hand side equations of system (6.39); the left-hand equations of system (6.39) are merely the kinematic relations.

At the next step of developing our account of skilled subject behavior, we want to reveal the plausible meaning of the momenta p_x, p_v, p_a. In this way we face up to a challenge caused by the feature of Hamiltonian systems known under the name *canonical transformations* (see, e.g., Landau and Lifshitz 1976; Goldstein et al. 2014). In Newtonian mechanics the equations of motion can be treated as the extremals of the action functional dealing with a certain Lagrangian $L(x, v, t)$ (we discussed this issue previously, see Sect. 6.1.3). However, these extremals are invariant with respect, in particular, to the replacement of the Lagrangian $L(x, v, t)$ via adding the time derivative of any function $G(x, t)$

$$L(x, v, t) \Rightarrow L(x, v, t) + \frac{dG(x, t)}{dt}.$$

In fact, via integration over time, the contribution of such terms is reduced to the modification of boundary conditions, whereas the extremals are determined by the inner properties of action functional. As a result, there is a wide variety of canonical variables $\{q, p\}$ and the corresponding Hamiltonians $H(q, p)$ describing the same physical system.

In describing the strictly optimal action implementations, we also meet this situation. In the AS-Lagrangian $L(x, v, a, j)$ specified by Exp. (6.26) the replacement

$$\frac{g}{2}\Big[ma + k_v v - k_x(x - x_g)\Big]^2 \Rightarrow \frac{g}{2}\Big[m^2a^2 + (k_v^2 + 2mk_x)v^2 + k_x^2(x - x_g)^2\Big] \tag{6.76}$$

does not affect the optimal implementation of action strategies in their basic features. In particular, the eigenvalue equation (6.50) and the inner dynamics of the stable (and unstable) manifold governed by Eq. (6.49) are invariants. This replacement gives rise only to the change of the momenta p_x and p_v, i.e., is reduced to an *equivalent* transformation

$$\{x, v, a, p_x, p_v, p_a\} \Rightarrow \{x, v, a, P_x, P_v, p_a\}$$

where P_x, P_v are new momenta. Naturally, the form of the Hamiltonian $H(\dots)$ also has to change respectively,

$$H(x, v, a, p_x, p_v, p_a) \Rightarrow H(x, v, a, P_x, P_v, p_a).$$

In other words, in describing the mathematical properties of strictly optimal implementation of action strategies, particular meanings of the momenta p_x and p_v are not essential. So a specific Hamiltonian representation may be selected for convenience reasons.

The situation essentially changes when we want to modify Hamilton's equations to allow for imperfect actions of the skilled subject who can follow the optimal actions *approximately*. This modification concerns mainly the generalized momenta p_x, p_v, and p_a. Indeed, the inner phase variable x, v, a, and also the jerk j are the quantities characterizing the subject's perception of object motion within the experiential now and belong to the level of E-consciousness. So the difference between the hypothetical ideal subject and the real subject in describing the object motion is that the real subject cannot precisely record these quantities. By contrast, the momenta p_x, p_v, and p_a describe the subject's intentions in current and future actions based on information retained in memory and accumulated in his skill.

Therefore the desired generalization requires the understanding of the "meaning" of these momenta, i.e., the particular reasons behind the subject's actions reflected in the dynamics of the momenta. There are two aspects arguing for the feasibility of this generalization and prompting us on how to construct the appropriate formalism.

First, as will be elucidated below, in mathematical terms the subject's imperfect actions can be interpreted as transitions—partly random, partly regular—between the stable and unstable manifolds Ξ_s, Ξ_u. The inner dynamics of these manifolds is invariant with respect to canonical transformations, affecting only the embedding of the manifolds into the Hamilton space. In turn, the description of such transitions should not be too sensitive to the details in plausible generalization.

Second, a specific form of the AS-Lagrangian $L(x, v, a, j)$ can be selected turning to a clear physical meaning of its terms which specifies the form of the resulting Hamiltonian. The AS-Lagrangian given by Exp. (6.26) meets this requirement. Its first term *directly* quantifies the force which the subject has to exert on the object to drive it in the desired way; the other terms also *directly* quantify the effort of monitoring.

So in constructing our account of real implementations of action strategies, we may start from the developed Hamiltonian description of the subject's optimal behavior. In this context, all the momenta $\{p_x, p_v, p_a\}$ should be regarded as control variables which, however, are not equivalent in their role. The proposed interpretation of these momenta is as follows.

- The momentum p_a—being actually the jerk $j = da/dt$ within the given Hamiltonian description (Exp. (6.41))—is the control variable describing subject's actions right now, i.e., within the experiential now in affecting the object dynamics. Put differently, the momentum p_a describes the subject's direct response to the current situation via changing the acceleration. The subject recognizes the necessity of changing the momentum p_a via comparing the effort of monitoring and the effort of action bodily execution and trying to get their balance. It is directly reflected in Hamilton's equation
$$\frac{dp_a}{dt} = -\frac{\partial H}{\partial a},$$
where the Hamiltonian H given by Exp. (6.42) is used. The presence of the momentum p_v in this equation is related to the anticipation of how the current actions will affect the dynamics of the velocity in the nearest future.
- The other two variables—the momenta p_x and p_v—describe subject's intentions to correct (modify) his actions in the nearest future. In terms of constituent compo-

nents of the experiential now, we may suppose that (*i*) the momentum p_a represents the subject's actions within the immediate now, whereas (*ii*) the role of the momenta p_x, p_v should be attributed to the immediate future and, maybe, the immediate past as a source of accumulated information (Sect. 2.6). Mathematically it is described by Hamilton's equations for the Hamiltonian (6.42)

$$\frac{dp_v}{dt} = -\frac{\partial H}{\partial v}, \qquad \frac{dp_x}{dt} = -\frac{\partial H}{\partial x}.$$

These equations are considered to describe the subject's current actions in planning and correcting his future behavior based on anticipating the variations in the balance between the two types of effort related to changes in the velocity and position of controlled object.

The skilled subject is assumed to be able to reproduce the optimal behavior *only approximately*. This ability is a consequence of long-term learning and skill acquisition. Its emergence may be regarded as another facet of the principle of cloud-fusion self-consistency (Sect. 5.6.3) assuming the integrity of the human temporality reflected in the self-consistency of its various fragments. Such actions are characterized by

- subject's response mainly to the current situation, whereas how the subject responds is determined by the acquired skill,
- a certain similarity to the actions governed by the right-hand side equations of system (6.39) because, in particular, these equations reflect some physical regularities of object dynamics,
- smooth changes in subject's actions on the temporal scales of the experiential now,
- uncontrollable and unrecognized deviations from the time variations governed by the Hamilton's equations
- in principle, some regular bias from the strictly optimal behavior especially if this behavior is complex in structure.

We will explicitly consider the first three of the noted features in describing the subject's imperfect actions. In the case under consideration, the subject's behavior is not too complicated in structure, which is justified by the properties of the Hamiltonian system eigenvalues (Sect. 6.1.8). They correspond to smooth regular motion towards the goal within the complex present. Therefore, for simplicity, we will ignore this bias.

Summarizing this discussion of control variables, we want to note the following. First, the dynamics of the control variables $\{p_x, p_v, p_a\}$—which will be called, as previously, the momenta characterizing the dynamics of the inner phase variables $\{x, v, a\}$—is treated as an initial value problem admitting also the presence of some noise.

Second, taking into account the results obtained in Sect. 6.2.1, in particular, Exp. (6.68) and Exp. (6.73), we turn to the Hamilton's equations (6.39) within the modification

$$H(x, v, a, p_x, p_v, p_a, t) = H_0(x, v, a, p_x, p_v, p_a) \\ - \frac{\delta_{p_x}}{\tau_d} x W_x(t) - \frac{\delta_{p_v}}{\tau_d} v W_v(t) - \frac{\delta_{p_a}}{\tau_d} a W_a(t), \quad (6.77)$$

where the Hamiltonian $H_0(\ldots)$ is specified by Exp. (6.42) with $\gamma_s = 2$, the random functions of time $W_x(t)$, $W_v(t)$, $W_a(t)$ with the mean equal to zero, the amplitude about unity and correlations on time scales about τ_d effectively allow for uncertainty in subject's implementation of action strategy.

Hamiltonian Description of Action Strategy Implementation by the Skilled Subject

Now we can formulate the proposed account of action strategy implementations from the *standpoint of external observer*. Using term "external observer" we want to emphasis that the subject's actions based on his intentions can be recorded, at least in principle, as a dynamical system with noise in the Hamilton space $\mathbb{R}_H$. Namely, the external observer is supposed to have direct access to (*i*) the physical characteristics of the object motion $\{x, v, a\}$ and (*ii*) the subject's motor behavior described by the momenta $\{p_x, p_v, p_a\}$. The latter type access may be elucidated in the following way.

- The momentum p_a is just the jerk j within some proportionality. So via a regularity $R_a(x, v, a)$ in subject's behavior reflected in the Hamilton's equation

$$\frac{dp_a}{dt} = -\frac{\partial H}{\partial a} = R_a(x, v, a) - p_v + \text{noise}_a$$

  when the object motion is at the state $\{x, v, a\}$, the external observer is able to record the momentum p_v.
- Then, via another regularity $R_v(x, v, a)$ related to the Hamilton's equation

$$\frac{dp_v}{dt} = -\frac{\partial H}{\partial v} = R_v(x, v, a) - p_x + \text{noise}_v$$

  the external observer records the moment p_x.
- After that using the statistical data analysis to get rid of the noise (uncertainty) effect as well as assuming a certain regularity $R_x(x, v, a)$ to be hidden in the Hamilton's equation

$$\frac{dp_x}{dt} = -\frac{\partial H}{\partial x} = R_v(x, v, a) + \text{noise}_x\,,$$

  the external observer can, at least in principle, reconstruct the dynamical variables p_x, p_v, and p_a as well as the regularities $R_q(x, v, a)$ $(q = x, v, a)$.

The proposed account of action strategy implementations comprises the following elements.

- **Initial conditions:** After selecting an action strategy $\mathcal{AS}$ the subject initiates its implementation such that the system initial state represented by the point

$$z_{\text{ini}}^{\text{T}} = \{x, v, a, p_x, p_v, p_a\}_{\text{ini}}$$

of the Hamilton space $\mathbb{R}_H$ be in the region $\mathbb{C}_H^{\text{ini}}$ specified by Exp. (6.74). In other words, the initial condition imposed on the action strategy implementation at time t_{ini} is of the form

$$\{x, v, a, p_x, p_v, p_a\}_{t=t_{\text{ini}}} \in \mathbb{C}_H^{\text{ini}} . \tag{6.78a}$$

Beside, because any Hamiltonian is determined up to an additive constant, we select this constant such that

$$H\big(x, v, a, p_x, p_v, p_a, t\big)_{t=t_{\text{ini}}} = H_0\big(x, v, a, p_x, p_v, p_a\big)_{t=t_{\text{ini}}} \stackrel{\text{def}}{=} H_0 . \tag{6.78b}$$

- **Time-dependent Hamiltonian:** In order to meet the initial condition (6.78b) for the Hamiltonian $H(x, v, a, p_x, p_v, p_a, t)$ determined by Exp. (6.77) we can transform it as

$$\begin{aligned} H(x, v, a, p_x, p_v, p_a, t) &= H_0(x, v, a, p_x, p_v, p_a) \\ &\quad - \sum_{q=x,v,a} \frac{\delta_{p_q}}{\tau_d} \Big\{ q W_q(t) - \big[q W_q(t)\big]_{t=t_{\text{ini}}} \Big\} . \end{aligned} \tag{6.79}$$

- **Governing equations:** The system dynamics is described by Hamilton's equations (6.39). In the shortened form they can be rewritten as

$$\frac{dq}{dt} = \frac{\partial H}{\partial p_q}, \qquad \frac{dp_q}{dt} = -\frac{\partial H}{\partial q} \qquad \text{for } q = x, v, a. \tag{6.80}$$

We want to emphasis once more that the given description of action strategy implementations by the skilled subject is not equivalent to the temporal boundary value problem (6.33) and the corresponding Ostrogradsky's formalism. In particular, the generated trajectory

$$P_{\text{AS}}(t) = \{x(t), v(t), a(t), p_x(t), p_v(t), p_a(t)\}$$

in the Hamilton space $\mathbb{R}_H$ does not necessarily lead to the desired goal,

$$z_g = \{x = x_g,\ v = 0,\ g = 0,\ p_x = 0,\ p_v = 0,\ p_a = 0\}$$

as time goes on. So when the trajectory $P_{\text{AS}}(t)$ deviates substantially from the goal point z_g, the subject has to terminate the current implementation of action strategy $\mathcal{AS}$

and select another action strategy. As far as Ostrogradsky's formalism is concerned, the relations (6.37) between the inner variables of object dynamics $\{x, v, a\}$ and the corresponding momenta $\{p_x, p_v, p_a\}$ are also broken because of the noise presence.

In accordance with the general properties of Hamiltonian systems, the time variations of the Hamiltonian $H(x, v, a, p_x, p_v, p_a, t)$ along the trajectory $P_{\mathrm{AS}}(t)$ obey the equation

$$\frac{dH}{dt} = \frac{\partial H}{\partial t} = -\sum_{q=x,v,a} \frac{\delta_{p_q}}{\tau_d} q \frac{dW_q(t)}{dt}, \tag{6.81a}$$

by virtue of (6.79). Integrating this equation we get

$$H(t) = H_0 - \sum_{q=x,v,a} \frac{\delta_{p_q}}{\tau_d} \int_{t_{\mathrm{ini}}}^{t} dt'\, q(t') \frac{dW_q(t')}{dt'} . \tag{6.81b}$$

The corresponding variations of the momenta $\{p_q\}$ can be calculated integrating the second Hamilton's equation of system (6.80) which gives us the expression

$$p_q(t) = p_q(t)_{t=t_{\mathrm{ini}}} - \int_{t_{\mathrm{ini}}}^{t} dt' \left.\frac{\partial H_0}{\partial q}\right|_{t'} + \frac{\delta_{p_q}}{\tau_d} \int_{t_{\mathrm{ini}}}^{t} dt'\, W_q(t') . \tag{6.82}$$

The inner variables $\{x, v, a\}$ of object motion are not affected directly by the implementation noise.

The concept of the skilled subject implies that:

- The implementation noise, $\{W_q(t)\}$, is not too strong such that the motion state of controlled object deviates substantially from the desired goal in the second stage of action strategy implementation. At least, it is the case when the initial state of object motion is rather far from the goal.
- In the second stage of action strategy implementation, when the deviation from the goal is evident, the subject does not try to gradually correct the current state of motion. Instead, he prepares himself for selecting a new action strategy for implementation starting from the current state of motion.
- During the first stage of action strategy implementation, when the characteristics $\{x, v, a\}$ of object motion are rather close to their initial values, the subject is not able to recognize the deviation of the current state of motion control from the optimal one. Thereby he treats the current state of motion control as the optimal one and sees no reasons for deviating from the standard regularities described by the Hamiltonian H_0.
- In estimating the proximity of current actions to the optimal implementation of action strategies, the subject deals with integral characteristics averaged in some way over all the phase variables rather than their individual values.

Therefore, as a certain approximation, we may consider that (*i*) the main changes in the Hamiltonian H and (*ii*) uncontrolled changes in the momenta p_x, p_v, p_a caused by subject's uncertainty in reproducing motor behavior are located within the first stage. During the second stag the changes in the Hamiltonian H are relatively weak and the time-variations of the momenta are mainly due to the regularities described by the Hamiltonian H_0.

Initiation noise approximation: The noted features can be taken into account within the *effective* approximation assuming that:

- The implementation flaw of action strategies is mainly caused by erroneous initiation of actions strategies such that the initial state z_{int} is deviated from the stable manifold Ξ_s in the Hamilton space $\mathbb{R}_H$. The region of initiation $\mathbb{C}_H^{\text{ini}}$ as previously is described by Exp. (6.74). However, the uncertainties δ_{p_x}, δ_{p_v}, and δ_{p_a} of the momenta p_x, p_v, and p_a should be increased to allow for uncontrolled changes in these momenta (see Exp. 6.82) during the first stage of action strategy implementations. We does this via the increased values of cofactor $\varkappa_{\text{eff}} \gg 1$ entering the expressions

$$\delta_{p_x} = \frac{\varkappa_{\text{eff}}\tau_{cp}}{\delta_x}\,, \quad \delta_{p_v} = \frac{\varkappa_{\text{eff}}\tau_{cp}^2}{\delta_x}\,, \quad \delta_{p_a} = \frac{\varkappa_{\text{eff}}\tau_{cp}^3}{\delta_x}\,. \tag{6.83}$$

  Here, for simplicity, we have introduced the same cofactor $\varkappa_{\text{eff}}$ for all the momenta; this cofactor aggregates in itself also the cofactor $\varkappa$ introduced previously (see Exp. 6.71). Because the cofactor $\varkappa_{\text{eff}}$ takes into account the accumulation of the control noise $\{W_q(t)\}$ $(q = x, v, a,)$ on scales about τ_{cp} (see Exp. 6.82) the cofactor $\varkappa_{\text{eff}}$ can be estimated as

$$\varkappa_{\text{eff}} \sim \sqrt{\frac{\tau_{cp}}{\tau_d}}\,\varkappa\,. \tag{6.84}$$

  In the case $\varkappa_{\text{eff}} \gg 1$ the uncertainty of the momenta $\{p_a, p_v, p_a\}$ is the main factor responsible for the implementation flaw of action strategies and the perception uncertainty of the quantities $\{x, v, a\}$ is ignorable.

  To accentuate this feature we call the case $\varkappa_{\text{eff}} \gg 1$ the limit of **dual-manifold initiation** implying that at the moment of initiation the inner phase variables of object motion are located at the center of the region $\mathbb{C}_{\text{ini}}^H\{z\}$, i.e.,

$$\{x, v, a\} = \{x_s, v_s, a_s\}\,.$$

  Therefore the component of the initial vector z_{ini} belonging to the unstable manifold Ξ_u is caused solely by the difference in the momenta $(p_x - \pi_x^s)$, $(p_v - \pi_v^s)$, and $(p_a - \pi_a^s)$.
- After initialization, the subject actions are governed by the Hamilton's equations with the Hamiltonian H_0 specified by Exp. (6.42). Although these subject's actions are described by Hamiltonian H_0 used previously for constructing the strictly optimal behavior of the ideal subject, the generated trajectory $P(t)$ does not lead to the goal because of the flawed initiation of action strategy.

Below we will continue constructing our account of action strategy implementation within the initiation noise approximation accepting the limit of dual-manifold initiation for simplicity. This approximation is sufficient for elucidating the characteristic features of human goal-oriented actions within the complex present. As far as the approach presented in this subsection to describing the skilled subject's behavior is concerned on its own, we want to point out the possibility of going beyond the initiation noise approximation.

Beyond initiation noise approximation: Dynamics of Hamiltonian systems with noise has been studied in physics for a long time. In context of our investigations, we want to note a special type of dynamical systems—the port-Hamiltonian systems (van der Schaft and Jeltsema 2014, for a review)—describing the internal dynamics of physical systems and their interaction with the surroundings which are strongly influenced by conservation laws. The introduction of noise in such systems was considered, e.g., by Vladimirov and Petersen (2018). For mathematical aspects of introducing noise in Hamiltonian systems preserving their symplectic structure a reader may be addressed to, e.g., Milstein et al. (2002a, b), Cresson and Darses (2007), Chetrite and Gawędzki (2008), Lázaro-Camí and Ortega (2008), Wang et al. (2009), Deng et al. (2014), Birrell and Wehr (2018).

In our account of action strategy implementation we deal with an atypical situation due to the used Hamiltonian being based on Ostrogradsky's formalism for systems with higher-order time-derivatives. It enabled us to turn to the concept of zero-manifold Ξ which assumes the subject's *intentional* actions to be aimed mainly at keeping the system dynamics near its stable branch Ξ_s. In these terms the Hamilton's equations (6.80) with the noisy Hamiltonian (6.79) describes

- the inner dynamics of the stable (Ξ_s) and unstable (Ξ_u) branches of the manifold Ξ governed by regular and random "forces" and
- the continuous random transitions $\Xi_s \rightleftarrows \Xi_u$ between the two branches of the manifold Ξ.

If the implementation noise $\{W_x(t), W_v(t), W_a(t)\}$ is not strong, the contribution of random "forces" to the inner dynamics of the manifolds Ξ_s and Ξ_u is not essential. It is due to their regular dynamics is characterized by eigenvalues with essential real parts. However the transitions $\Xi_s \rightleftarrows \Xi_u$ should affect the system dynamics substantially. In the main steam of our constructions we confined ourselves to the initiation noise approximation sufficient for the present goal of our research.

In this brief digression we focus attention on the possibility of generalizing the developed description of the skilled subject's actions. This generalization which may be referred to as a *dual-manifold approach* could comprise the following components.

- Partitioning the system into two components belonging to the stable and unstable manifolds Ξ_s, Ξ_u individually.
- The inner dynamics of the stable and unstable manifolds Ξ_s, Ξ_u governed by the noiseless Hamiltonian H_0.
- Stochastic transitions $\Xi_s \rightleftarrows \Xi_u$ governed by individual regularities including subject's intentional actions such that:

- the transitions $\Xi_s \rightleftarrows \Xi_u$ are depressed when the system comes close to the origin and
- the transitions $\Xi_s \leftarrow \Xi_u$ to the stable manifold Ξ_s from to the unstable manifold Ξ_u are intensified when the system deviates essentially from the origin.

These transitions affect the system partitioning between the stable and unstable manifolds Ξ_s, Ξ_u, endowing the system with complex dynamics.

The dual-manifold approach, first, could form a new problem in the theory of stochastic Hamiltonian systems that

- singles out characteristics features related to the present of higher-order time-derivatives as phase variables,
- bears different types of noise reflecting the properties of human automatic and voluntary actions.

Second, the dual-manifold approach can open a gate toward the description of action strategy implementations as a certain process homogeneous in its structural elements.

We have only noted this prospective generalization, its detailed development is outside the scope of the present book.

6.2.3 Temporal Criterion of Proximity to Action Strategy Optimal Implementations and Ostrogradsky Instability

As elucidated in Sect. 6.2.1, almost all implementations of action strategies by the skilled subject are unstable in the standard understanding of instability. Practically any trajectory representing the subject's automatic actions finally goes away from the desired goal. This automaticity implies:

i: the initiation of an acceptable action strategy and, then,
ii: its implementation according to an available plan of actions, i.e., as a sequence of actions determined beforehand.

It seems that on the scale of the complex present, actually on the scales of the connected future less than 10 s, the deliberate realization of a sequence of actions is just not feasible. Only changing action strategies can be a deliberate conscious process. Therefore we need a special criterion and the corresponding quantitative measure to evaluate whether a given implementation of action strategy may be acceptable or should be rejected by the skilled subject.

Proximity to the optimal behavior: To elucidate criteria for evaluating possible actions of the skilled subject, we need to remind the premises underlying the behavior of ideal subject. The ideal subject has been assumed to minimize the integral cumulative effort required for these actions, which is specified by AS-functional (6.28). Here we faced up to the fact that for this functional its extremals reduced to the temporal boundary value problem (6.32) and (6.33) belong to a special class of mathematical models. Such models imply that the current actions of the ideal subject are affected directly by the terminal goal belonging to the future. For human behavior this effect of the future on the present is not something anomalous. Indeed, this effect is the gist of goal-oriented actions and reflects the basic human ability to anticipate future events. So this model for the ideal subject's behavior does capture the essence of human goal-oriented actions and may be used as the zeroth-approximation. The Ostrogradsky's formalism leading to the Hamiltonian description of ideal subject behavior reduces the temporal boundary value problem (6.32), (6.33) to the *effective* initial value problem. The resulting initial value problem contains new independent variables having no evident meaning; they serve to connect the current actions to the goal remote in time. However, the given approach poses a challenging question about how the real subject with bounded capacity of cognition can realize such actions.

The proposed concept of skilled subject allows for the real features of human behavior caused by the bounded capacity of cognition and implies that the subject can realize the optimal actions only approximately. The model constructed in Sect. 6.2.2 for automatic subject's actions deals with the noisy Hamilton's equations. It also captures the essence of human goal-oriented behavior; the subject responds to the current situation but how he does this is determined by his skill acquired within long-term learning.

The developed Hamiltonian description is no longer reducible to the original temporal boundary value problem. It is illustrated by the fact that almost all the trajectories of object motion described by this Hamiltonian model formally go away from the desired goal point. On the one hand, it is a typical feature of automatic human actions reflecting the characteristic features of open-loop control (e.g., Franklin et al. 2018). On the other hand, this trajectory divergence does not enable us to quantify the proximity between

- a trial implementation $P_{\mathrm{AS}}(t)$ of an action strategy $\mathcal{AS}$ by the skilled subject and
- its optimal version $P_{\mathrm{opt}}(t\,|\,\mathcal{AS})$ employing AS-functional (6.28).

The two implementations mathematically are not close to each other.

The noted features have demonstrated that human goal-oriented behavior, at least, considered within the concept of action strategies in the complex present is characterized by the inevitable divergence from the optimal behavior. Exactly this divergence causes the action strategy implementations to take the form of a sequence of alternate fragments of automatic actions and their corrections. This instability is the fundamental consequence of Ostrogradsky's ghost and the general properties of Hamiltonian systems, we call it the *Ostrogradsky instability*.[27] The Ostrogradsky instability is the reason why we pose a question about the necessity of a new criterion for the proximity of action strategy implementations to the optimal ones rather than the criterion dealing with the AS-functional (6.28).

[27] The notion of the Ostrogradsky instability we discussed here, reflects the aspects of human goal-oriented behavior. Other aspects of this instability met in physical systems with higher-order time-derivatives are discussed, e.g., by Woodard (2009, 2007, 2015b), Stephen (2008), Chen et al. (2013), Motohashi and Suyama (2015), Swanson (2019), Aoki and Motohashi (2020), Ganz and Noui (2020).

The confinement to the complex present implicitly imposed on the original temporal boundary value problem (for details see Sect. 6.1.4) opens a gate to responding to this challenge. This confinement implies that the subject ignores all possible trends in the object dynamics on temporal scales exceeding the duration τ_{cp} of complex present. Therefore, if an action strategy, as it is currently implemented, leads toward the goal on scales about τ_{cp}, the subject cannot recognize the details of its further deviation from the goal. As a result, he has to consider this action strategy efficient and the current state of object motion to be stable. This is the gist of the *temporal criterion of the proximity to the optimal behavior* to be used further in our constructions.

As far as the corresponding measure for quantifying the proximity of a given action strategy implementation $P_{\text{AS}}(t)$ to the corresponding optimal implementation $P_{\text{opt}}(t|\mathcal{AS})$ is concerned, we impose on it the following requirements. Namely, we posit that the desired measure:

- should quantify the proximity of $P_{\text{AS}}(t)$ to $P_{\text{opt}}(t|\mathcal{AS})$ in terms of a time interval during which the given implementation $P_{\text{AS}}(t)$ leads to the desired goal within the perceived accuracy;
- being of temporal type, has to be an intensive property of $P_{\text{AS}}(t)$ rather than extensively one[28]; in other words, this measure should characterize the action strategy implementation $P_{\text{AS}}(t)$ as a whole entity independently of the duration of its temporal parts;
- should characterize directly the subject's *planned* actions, i.e., the quality of actions' automaticity; it implies that this measure quantifies the efficiency of a given action strategy at the moment of its initiation and does not change in the course of its implementation;
- must aggregate in some way the values of all the control phase variables—the momenta $\{p_x, p_v, p_a\}$—within one quantity because the subject is not able to evaluate their individual dynamics separately.

In the next subsections we will demonstrate that within the initiation noise approximation (Sect. 6.2.2) the value of the Hamiltonian H_0 given by Exp. (6.42) (or, respectively, by Exp. (6.44) for objects with overdamped dynamics) can form the bases of this measure. In the present subsection we want to elaborate in more detail the proposed temporal criterion of the proximity to the optimal behavior. For this purpose we need to introduce two clouds—fuzzy regions in the Hamilton space $\mathbb{R}_H$ possessing special properties.

Goal cloud in the Hamilton space: It is a transformation of the goal space-time cloud $\mathbb{C}^g$ in the space $\mathbb{R}_{[4]} \times t$ (Sect. 6.1.1) into the corresponding fuzzy region $\mathbb{C}^g_H$ to be called also the *goal cloud in the Hamilton space* $\mathbb{R}_H$. In structure, the goal cloud $\mathbb{C}^g_H$ is similar to the cloud of action strategy initiation $\mathbb{C}^{\text{ini}}_H$ (Exp. 6.74). Actually it is the region $\mathbb{C}^{\text{ini}}_H$ moved to the origin of the space $\mathbb{R}_H$ with shift to $x = x_g$,

[28] By definition, a quantity (property) whose magnitude is independent of the extent of the system is called *intensive*; a quantity (property) whose magnitude is additive for subsystems is called *extensive* (IUPAC 1993, p. 7).

$$\mathbb{C}_H^g = \frac{(x - x_g)^2}{\delta_x^2} + \frac{v^2}{\delta_v^2} + \frac{a^2}{\delta_a^2} + \frac{p_x^2}{\delta_{p_x}^2} + \frac{p_v^2}{\delta_{p_v}^2} + \frac{p_a^2}{\delta_{p_a}^2} \lesssim 1 \,. \tag{6.85}$$

The subject considers the object to have gotten the desired goal, when its current state of motion belongs to the region $\mathbb{C}_H^g$ with respect to the inner phase variables $\{x, v, a\}$. In this case, no control over the object dynamics is required, so with respect to the control phase variables $\{p_x, p_v, p_a\}$ the goal cloud $\mathbb{C}_H^g$ is also centered at $\{p_x, p_v, p_a\} = \{0, 0, 0\}$.

It should be noted that the goal cloud $\mathbb{C}_H^g$ can contain a special part

$$\mathfrak{C}_H^g = \left[\frac{(x - x_g)^2}{\delta_x^2} + \frac{v^2}{\delta_v^2} + \frac{a^2}{\delta_a^2} \lesssim 1\right] \bigcup \left[p_x = 0,\, p_v = 0,\, p_a = 0\right]$$

representing the situation when the subject halts the control over the object dynamic governed, after that, by the physical regularities only. In the present book not to overload our constructions we do not consider this feature.[29]

We want to emphasize that driving the object directly to the goal (the goal cloud $\mathbb{C}_H^g$) meets the classical paradigm of describing human control. The subject's *intentional* actions are described in terms of driving the object toward the goal point. Because of uncertainty in subject's perception, when the object comes into the region $\mathbb{C}_H^g$ the subject cannot recognize the appropriate actions, so should complete the current action strategy implementation and wait until the object leaves the region $\mathbb{C}_H^g$ again.

Stable manifold cloud in the Hamilton space: The concept of the stable manifold Ξ_s in the Hamilton space $\mathbb{R}_H$ reflects the modern paradigm of human control operating with action strategies rather than the states of object motion. We already discussed this paradigm in Sect. 6.1.8 within the stable-manifold-based model (Sect. 6.1.8). In this paradigm the subject's optimal actions are treated as driving the object strictly along the stable manifold Ξ_s. So if some perturbation preserves the system position on the manifold Ξ_s, no correction is required. When perturbations cause the system deviation from the manifold Ξ_s, the subject's *conscious* actions are aimed at returning the system to the manifold Ξ_s. After that, the subject's automatic (unconscious) actions reflected in the inner dynamics of the manifold Ξ_s drive the object to the desired goal.

From the standpoint of the skilled subject, the current state of system motion—specified by the inner phase variables $\{x, v, a\}$ and the control phase variables $\{p_x, p_v, p_a\}$—has to be treated as belonging to the stable manifold Ξ_s if there is a point $\sigma_s(x_s, v_s, a_s)$ on the manifold Ξ_s for which condition (6.74) holds. This criterion can be represented in terms of the fuzzy neighborhood of the manifold Ξ_s in the Hamilton space $\mathbb{R}_H$

[29] This situation described by the concept of dynamical traps was analyzed, e.g., in Lubashevsky (2016, 2017), Lubashevsky and Morimura (2019).

$$\mathbb{C}^s_\Xi = \min_{\sigma_s}\left[\frac{(x-x_s)^2}{\delta_x^2} + \frac{(v-v_s)^2}{\delta_v^2} + \frac{(a-a_s)^2}{\delta_a^2} + \frac{(p_x-\pi^s_x)^2}{\delta_{p_x}^2} + \frac{(p_v-\pi^s_v)^2}{\delta_{p_v}^2} + \frac{(p_a-\pi^s_a)^2}{\delta_{p_a}^2}\right] \lesssim 1 \qquad (6.86)$$

to be called the *cloud of the stable manifold* Ξ_s in the Hamilton space $\mathbb{R}_H$ or just the *stable manifold cloud*. When an action strategy $\mathcal{AS}$ is initiated at some point z_{ini} of the Hamilton space $\mathbb{R}_H$ then, according to condition (6.74), its implementation is initially located inside the stable manifold cloud $\mathbb{C}^s_\Xi$. If during the course of action strategy implementation the state of system motion remains in the region $\mathbb{C}^s_\Xi$, the subject cannot discriminate between the current implementation AS of the action strategy $\mathcal{AS}$ and its strictly optimal one AS_{opt}. As a result, the subject has no reason to interrupt the action strategy implementation until the current state of system motion goes out of the region $\mathbb{C}^s_\Xi$. In other words, the stable manifold cloud $\mathbb{C}^s_\Xi$ contains all the possible trajectories

$$P_{\text{AS}}(t) = \{x(t), v(t), a(t), p_x(t), p_v(t), p_a(t)\} \in \mathbb{C}^s_\Xi \times t$$

treated individually as a single implementation of an action strategy.

When the system state goes out of the cloud $\mathbb{C}^s_\Xi$ rather far from the goal cloud $\mathbb{C}^g_H$, the subject has to interrupt the current implementation P_{AS} of the action strategy $\mathcal{AS}$ and select a new action strategy $\mathcal{AS}'$ for implementation. To allow for such events in our account and, then, to describe them mathematically let us introduce the notion of the *action point boundary* of the $\mathbb{C}^s_\Xi$-cloud to be denoted by the symbol $\partial\mathbb{C}^s_\Xi$. Here the term *action point* is understood as the event of the subject changing one action strategy for another. The action point boundary $\partial\mathbb{C}^s_\Xi$ is a certain periphery of the cloud $\mathbb{C}^s_\Xi$ whose points are from the stable manifold Ξ_s at distances exceeding the characteristic thickness of the cloud $\mathbb{C}^s_\Xi$ by a factor $\theta_\Xi \gtrsim 1$. We will call the coefficient θ_Ξ the *fuzzy threshold of action points.*[30] In terms of Exp. (6.86) the fuzzy region $\partial\mathbb{C}^s_\Xi$ can be specified as

$$\partial\mathbb{C}^s_\Xi = \min_{\sigma_s}\left[\frac{(x-x_s)^2}{\delta_x^2} + \frac{(v-v_s)^2}{\delta_v^2} + \frac{(a-a_s)^2}{\delta_a^2} + \frac{(p_x-\pi^s_x)^2}{\delta_{p_x}^2} + \frac{(p_v-\pi^s_v)^2}{\delta_{p_v}^2} + \frac{(p_a-\pi^s_a)^2}{\delta_{p_a}^2}\right] \sim \theta_\Xi\,. \qquad (6.87)$$

In principle, the fuzzy threshold can be a large quantity, $\theta_\Xi \gg 1$. In this case, the action point cloud $\partial\mathbb{C}^s_\Xi$ should be regarded as a satellite of the stable manifold cloud $\mathbb{C}^s_\Xi$ rather then its periphery. To allow for this situation we will call $\partial\mathbb{C}^s_\Xi$ also

[30] We use the term "fuzzy threshold" instead of just "threshold" because usually the notion *threshold* is understood as a position of step-wise boundary separating two regions where some property or function exhibits distinct behavior. Below, where it does not lead to a possible misunderstanding, we will use also the term "threshold" in the meaning of fuzzy threshold.

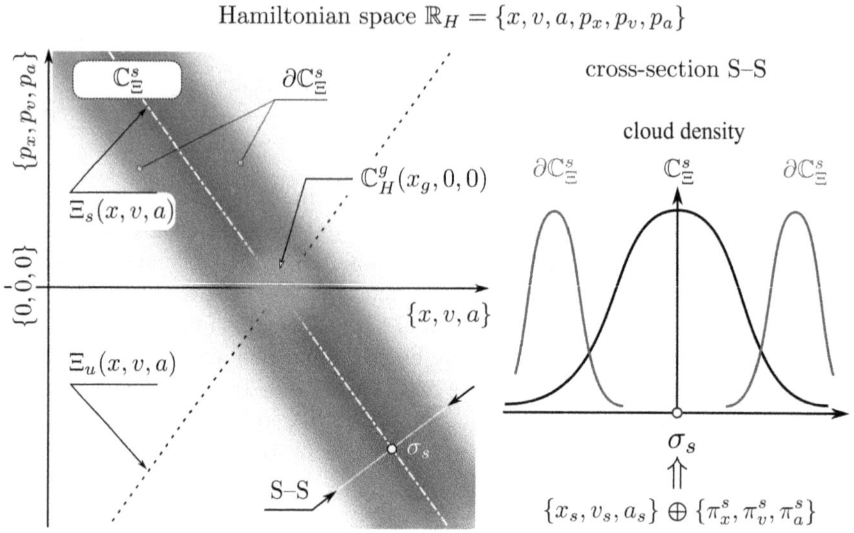

Fig. 6.1 Schematic illustration of the arrangement of the stable manifold cloud $\mathbb{C}^s_\Xi$ (blurred gray stripe) with the action point boundary $\partial\mathbb{C}^s_\Xi$ (two blurred red stripes) as its peripheral part, and the goal cloud $\mathbb{C}^g$ (green blurred sphere) in the Hamilton space $\mathbb{R}_H$. The Hamilton space is composed of (*i*) the space $\{x, v, a\}$ of the inner phase variables specifying the object motion state (the x-axis in figure) and (*ii*) the space $\{p_x, p_v, p_a\}$ of the control phase variables (the y-axis in figure). The goal cloud $\mathbb{C}^g_H$ is centered at the point $\{x_g, 0, 0\}$ of the $\{x, v, a\}$-space. Dashed white and black lines represent the stable (Ξ_s) and unstable (Ξ_u) manifolds. The right insert illustrates the structure of the stable manifold cloud in the cross-section S–S centered at the point $\sigma_s \in \Xi_s$ with the coordinates $\{x_s, v_s, a_s, \pi^s_x, \pi^s_v, \pi^s_a\}$

the *action point layer*. Dealing with a situation characterized by the limit $\theta_\Xi \gg 1$, we need to draw a distinction between the uncertainty in perception of the stable manifold position in the Hamilton space $\mathbb{R}_H$ and the acceptable deviations from the stable manifold Ξ_s.

Figure 6.1 illustrates the mutual arrangement of the three clouds and its elements. Using the introduced notions we may say that action strategies are:

- initiated inside the stable manifold cloud $\mathbb{C}^s_\Xi$,
- interrupted (terminated) inside the action point boundary $\partial\mathbb{C}^s_\Xi$,
- completed inside the goal cloud $\mathbb{C}^g_H$.

In this sense the stable manifold cloud $\mathbb{C}^s_\Xi$ may be regarded as *the cloud of all possible implementations* $\{P_{\mathrm{AS}}(t)\}$ *of action strategies in the Hamilton space* $\mathbb{R}_H$.

It is worthy of noting that the stable manifold cloud $\mathbb{C}^s_\Xi$ and the goal cloud $\mathbb{C}^g_H$ accompanied by a certain external process affecting the action strategy choice form the appropriate mathematical vocabulary for coping with such problems as the change of mind noted in Sect. 4.2.2. The stable manifold cloud $\mathbb{C}^s_\Xi$ and the action point boundary $\partial\mathbb{C}^s_\Xi$ may be regarded as the basic elements in describing such problems as the car-following (Sect. 4.2.3). Besides the combination of the two clouds admits a description of such problems beyond the complex present.

6.2.4 *Hamiltonian Description of Dual-Manifold Initiation of Action Strategies*

In the initiation noise approximation (Sect. 6.2.2), the imperfectness of subject's actions is described as the simultaneous initiation of two components of the system state. One of them belongs to the stable manifold Ξ_s and its dynamics leads to the desired goal point. The other component belongs to the unstable manifold Ξ_u and represents the deviation of the system state from the optimal one. This component increases in magnitude as time goes on and finally the induced deviation becomes so essential that the subject has to intentionally terminate the current action strategy implementation. Exactly the time increase of this component is taken into account within the temporal criterion of the proximity to the optimal action strategy implementation.

The present subsection is devoted to the mathematical description of this two-manifold representation of system dynamics within the framework of the Hamiltonian account of the skilled subject (Sect. 6.2.2).

Let us consider an implementation AS of a given action strategy $\mathcal{AS}$ treated as a trajectory $P_{\text{AS}}(t)$ in the Hamilton space $\mathbb{R}_H$:

$$P_{\text{AS}}(t) = \big\{x(t),\ v(t),\ a(t),\ p_x(t),\ p_v(t),\ p_a(t)\big\}, \quad t \in \text{CP},$$

where the symbol CP stands for the complex present as a fragment of the human temporality. Accepting, for simplicity, the limit of dual-manifold initiation (Sect. 6.2.2), we may suppose that at some instant of time before the initiation of $P_{\text{AS}}(t)$ is completed, the system state

$$z_s = \{x, v, a,\ p_x = \pi_x^s,\ p_v = \pi_v^s,\ p_a = \pi_a^s\} \in \Xi_s$$

as a point of the Hamilton space $\mathbb{R}_H$ belongs to the stable branch Ξ_s of the zero-Hamiltonian manifold. Here the corresponding values $\{\pi_x, \pi_v, \pi_a\}$ are the functions of $\{x, v, a\}$ given by Exp. (6.75). At the point z_s the value of the Hamiltonian $H(\ldots)$ is equal to zero, $H(z_s) = 0$.

At the final stage of initiation, the initiation noise composed of three mutually independent components

$$\delta_{p_x} W_x^{\text{ini}}, \quad \delta_{p_v} W_p^{\text{ini}}, \quad \delta_{p_a} W_a^{\text{ini}}$$

disturbs mainly the dynamics of the momenta $\{p_x, p_v, p_a\}$ and causes their magnitudes to deviate from $\{\pi_x, \pi_v, \pi_a\}$ for distances about $\{\delta_{p_x}, \delta_{p_v}, \delta_{p_a}\}$ within a narrow region $\mathrm{EN}_{\mathrm{ini}}$—a neighborhood of the initiation moment t_{ini}. The corresponding variations in the inner phase variables $\{x, v, a\}$ of object motion are ignorable. Thereby the initiation noise also causes the Hamiltonian $H(\ldots)$ to take a value H_{ini} related to the induced changes in the momenta as

$$H_{\mathrm{ini}} = H_0(x, v, a, p_x, p_v, p_a) = (p_x - \pi_x^s)v + (p_v - \pi_v^s)a + \frac{\delta_j^2}{2\epsilon_j}(p_a^2 - \pi_a^{s2}) \tag{6.88}$$

to be called the *initiation-induced Hamiltonian*. In obtaining this expression we have used Hamiltonian (6.42) supposing the inner phase variables $\{x, v, a\}$ to exhibit no sharp jumps within the region $\mathrm{EN}_{\mathrm{ini}}$. It should be noted that the obtained expression is rather general, only the particular form of its term containing the momentum p_a is inherited by Exp. (6.88).[31]

Below we will treat the initiation-induced Hamiltonian H_{ini} as a random quantity related to the initiation noise comprising three mutually independent components $\{W_x^{\mathrm{ini}}, W_v^{\mathrm{ini}}, W_a^{\mathrm{ini}}\}$ of unit amplitude via

$$\begin{aligned} p_x &= \pi_x(x, v, a) + \delta_{p_x} W_x^{\mathrm{ini}}, \\ p_v &= \pi_v(x, v, a) + \delta_{p_v} W_v^{\mathrm{ini}}, \\ p_a &= \pi_a(x, v, a) + \delta_{p_a} W_a^{\mathrm{ini}}. \end{aligned} \tag{6.89}$$

Because at the stable and unstable manifolds Ξ_s and Ξ_u the Hamiltonian $H(\ldots)$ takes zero value, the value of H_{ini} not equal to zero

- does not change in system motion along the trajectory $P_{\mathrm{AS}}(t)$, which is a particular instantiation of the general conservation law of Hamiltonian systems,
- indicates that just after the initiation the initial point of the trajectory $P_{\mathrm{AS}}(t)$ contains two non-zero components z_s and z_u belonging to the stable and unstable manifolds,
- quantifies their product via some bilinear form (to be specified below) at the action strategy initiation as well as during the further system motion,
- thereby, quantifies the relationship between
  - decrease in the component $z_s(t)$ with time growth caused the inner dynamic of the stable manifold Ξ_s and
  - increase in the component $z_u(t)$ with time growth caused the inner dynamic of the unstable manifold Ξ_u.

Comparing these properties with the requirements imposed on plausible measures of the proximity to optimal behavior (Sect. 6.2.3) we infer that the initiation-induced Hamiltonian H_{ini} can be employed as the basic element in constructing the desired measure. It itself cannot be used as such a measure because (*i*) the Hamiltonian H_{ini} can take positive as well as negative values depending on the initiation noise and (*ii*) its particular value is not accessible to the mind.

[31] In AS-Lagrangian (6.26) it is the term containing the jerk j.

To represent the further constructions concisely, let us introduce the following notations. Any vector z of the Hamilton space $\mathbb{R}_H$ can be expanded relative to two Lagrange subspaces $\varXi_s$ and $\varXi_u$ (Sect. 6.1.8); in the present book we usually refer to them as the stable and unstable manifolds. Both the subspaces (manifolds) admit the parametrization using the inner phase variables $\{x, v, a\}$ of the object motion state and their embedding into the Hamilton space

$$\mathbb{R}_H = \mathbb{R}_L \times \mathbb{R}_L \quad \Longrightarrow \quad z = \begin{bmatrix} x \\ v \\ a \\ p_x \\ p_v \\ p_a \end{bmatrix} = \begin{bmatrix} Q \\ P \end{bmatrix},$$

can be represented as

$$z_s(x, v, a) = \begin{bmatrix} x \\ v \\ a \\ \pi_x^s \\ \pi_v^s \\ \pi_a^s \end{bmatrix} = \begin{bmatrix} Q \\ \varPi_s = \widehat{\varOmega}_s Q \end{bmatrix} \quad \text{and} \quad z_u(x, v, a) = \begin{bmatrix} x \\ v \\ a \\ \pi_x^u \\ \pi_v^u \\ \pi_a^u \end{bmatrix} = \begin{bmatrix} Q \\ \varPi_u = \widehat{\varOmega}_u Q \end{bmatrix} \tag{6.90}$$

Here $Q^{\mathrm{T}} = \{x, v, a\}$ and $P^{\mathrm{T}} = \{p_x, p_v, p_a\}$ are treated as vectors (their transposes) of the 3D Euclidean space to be called the Lagrange space and denoted as $\mathbb{R}_L$, the symbols $\widehat{\varOmega}_{s,u}$ stand for two real-value 3×3-matrices

$$\widehat{\varOmega}_w = \begin{bmatrix} \omega_{xx}^w & \omega_{xv}^w & \omega_{xa}^w \\ \omega_{vx}^w & \omega_{vv}^w & \omega_{va}^w \\ \omega_{ax}^w & \omega_{av}^w & \omega_{aa}^w \end{bmatrix}, \quad w = s, u \tag{6.91}$$

treated as linear operators acting in the Lagrange space $\mathbb{R}_L$. Referring to property (6.61) of Lagrange subspaces $\varXi_s$ and $\varXi_u$, we may assert that for any two vectors z_1 and z_2 belonging together to one of the two subspaces, their symplectic inner product (6.58) equal to zero,

$$\overline{z}_1^{\mathrm{T}} \mathbf{J} z_2 = \langle z_1 | \mathbf{J} | z_2 \rangle = \big\langle Q_1 | \widehat{\varOmega}_w Q_2 \big\rangle - \big\langle \widehat{\varOmega}_w Q_1 | Q_2 \big\rangle = 0$$

for $w = s, u$. Whence it immediately follows that the operators $\widehat{\varOmega}_{s,u}$ are Hermitian (i.e., the real-value matrices $\widehat{\varOmega}_{s,u}$ are symmetric)

$$\widehat{\varOmega}_s^{\dagger} = \widehat{\varOmega}_s, \quad \text{and} \quad \widehat{\varOmega}_u^{\dagger} = \widehat{\varOmega}_u. \tag{6.92}$$

Here and below we use the conjugate transposition $(\ldots)^{\dagger}$ instead of just the transposition $(\ldots)^{\mathrm{T}}$ because the vectors of $\mathbb{R}_H$ and $\mathbb{R}_L$ can contain complex-value eigenvectors and for real-value elements the two operators are equivalent. Expressions (6.291), (6.292) derived in Appendix 6.7.3 specify the matrices $\widehat{\Omega}_{s,u}$ for Hamiltonian (6.42). Expressions (6.75) illustrate these matrices written in the dimensionless form. Below we will refer to the operators $\widehat{\Omega}_{s,u}$ as to the *generators* of the stable and unstable manifolds Ξ_s and Ξ_u, they determine the embedding of the given manifolds into the Hamilton space $\mathbb{R}_H$.

The Hamiltonian $H(z)$ given by Exp. (6.57) can be rewritten in the form containing its components treated as linear operators acting on the vectors of the Lagrange space $\mathbb{R}_L$

$$H(z) = \begin{bmatrix} Q \\ P \end{bmatrix}^{\dagger} \begin{bmatrix} \widehat{H}_{qq} & \widehat{H}_{qp} \\ \widehat{H}_{pq} & \widehat{H}_{pp} \end{bmatrix} \begin{bmatrix} Q \\ P \end{bmatrix}. \tag{6.93}$$

The Hamiltonian components $\widehat{H}_{(\ldots)}$ are operators in $\mathbb{R}_L$ with properties

$$\widehat{H}^{\dagger}_{qq} = \widehat{H}_{qq}\,, \qquad \widehat{H}^{\dagger}_{pp} = \widehat{H}_{pp}\,, \qquad \widehat{H}^{\dagger}_{qp} = \widehat{H}_{pq}\,.$$

For the Hamiltonian (6.42)

$$\widehat{H}_{pp} = \begin{bmatrix} 0 & 0 & 0 \\ 0 & 0 & 0 \\ 0 & 0 & h_a \end{bmatrix}, \qquad \widehat{H}_{qp} = \widehat{H}^{\dagger}_{pq} = \begin{bmatrix} 0 & 0 & 0 \\ 1 & 0 & 0 \\ 0 & 1 & 0 \end{bmatrix},$$

where we have introduced the notion $h_a = H_{p_a p_a}$. As it must, the inner dynamics of the manifolds Ξ_s and Ξ_u reflecting the basic properties of the Hamilton's equations, e.g., represented in form (6.55), is described by the equations ($w = s, u$)

$$\Xi_w: \quad \frac{dQ}{dt} = \widehat{D}_w Q\,, \tag{6.94}$$

where we have introduced the operators $\widehat{D}_{s,u}$ governing the inner dynamics of the manifolds $\Xi_{s,w}$,

$$\widehat{D}_w = \widehat{H}_{pq} + \widehat{H}_{pp}\widehat{\Omega}_w = \begin{bmatrix} 0 & 1 & 0 \\ 0 & 0 & 1 \\ h_a\omega^w_{ax} & h_a\omega^w_{av} & h_a\omega^w_{aa} \end{bmatrix}. \tag{6.95}$$

Equation (6.94) is a direct consequence of the Hamilton's equations for the variables $Q^{\dagger} = \{x, v, a\}$. The Hamilton's equations for the generalized momenta lead to the condition imposed on the operators $\widehat{\Omega}_w$ and $\widehat{D}_w$,

$$\widehat{\Omega}_w\widehat{D}_w + \widehat{H}_{qq} + \widehat{H}_{qp}\widehat{\Omega}_w = 0\,. \tag{6.96}$$

For the system state $z = z_s + z_u$ containing both the components of the manifolds Ξ_s and Ξ_u, the Hamiltonian $H(z)$ given by Exp. (6.57) within representation (6.93) specifies the initiation-induced Hamiltonian H_{ini} as

$$H_{\text{ini}} = \frac{1}{2} z^\dagger \mathbf{H} z = Q_u^\dagger \widehat{H}_{qq} Q_s + Q_u^\dagger \widehat{H}_{qp} P_s + P_u^\dagger \widehat{H}_{pq} Q_s + P_u^\dagger \widehat{H}_{pp} P_s \,.$$

Here we have taken into account that the Hamiltonian $H(z)$ is equal to zero at the manifolds Ξ_s and Ξ_u.[32] Within the Lagrange space $\mathbb{R}_L$ the initiation-induced Hamiltonian can be represented as the operator $\widehat{H}_{us}$ uniting the components z_s and z_u into one entity z,

$$H_{\text{ini}}(z_u, z_a) = \left\langle Q_u \left| \widehat{H}_{us} \right| Q_s \right\rangle \,, \tag{6.97}$$

where the operator

$$\widehat{H}_{us} = \widehat{H}_{qq} + \widehat{H}_{qp} \widehat{\Omega}_s + \widehat{\Omega}_u \widehat{H}_{pq} + \widehat{\Omega}_u \widehat{H}_{pp} \widehat{\Omega}_s \tag{6.98a}$$

which admits also two equivalent representations written in terms of the introduced operators $\widehat{\Omega}_{s,u}$ and $\widehat{D}_{s,u}$

$$\widehat{H}_{us} = \left(\widehat{\Omega}_u - \widehat{\Omega}_s\right) \widehat{D}_s \,, \tag{6.98b}$$

$$\widehat{H}_{us} = \widehat{D}_u^\dagger \left(\widehat{\Omega}_s - \widehat{\Omega}_u\right) . \tag{6.98c}$$

In deriving these expressions we also have taken into account equality (6.96).[33]

Besides, it worthy of noting that according to Liouville's theorem (e.g., Landau and Lifshitz 1976; Arnold 1989) the phase space volume does not change during the motion of Hamiltonian systems. In our case it implies that the exterior form

$$dQ_s \wedge dQ_u = dx_s \wedge dv_s \wedge da_s \wedge dx_u \wedge dv_u \wedge da_u$$

should not change during the system motion, which is reduced to the condition

---

[32] The fact that the Hamiltonian $H(z)$ is equal to zero at the manifolds Ξ_s and Ξ_u is also demonstrated by Exp. (6.98b) or Exp. (6.98c) written for the case when both the components z_s and z_u formally belong to one of the two manifolds.

[33] Expressions (6.98b) and (6.98c) also directly demonstrate that the value of H_{ini} is conserved during the system motion. Indeed, from Exp. (6.97) we have

$$\begin{aligned}
\frac{dH_{\text{ini}}}{dt} &= \left\langle \frac{dQ_u}{dt} \middle| \widehat{H}_{us} \middle| Q_s \right\rangle + \left\langle Q_u \middle| \widehat{H}_{us} \middle| \frac{dQ_s}{dt} \right\rangle \\
&= \langle Q_u | \widehat{D}_u^\dagger \widehat{H}_{us} | Q_s \rangle + \langle Q_u | \widehat{H}_{us} \widehat{D}_s | Q_s \rangle \\
&= \langle Q_u | \widehat{D}_u^\dagger (\widehat{\Omega}_u - \widehat{\Omega}_s) \widehat{D}_s | Q_s \rangle + \langle Q_u | \widehat{D}_u^\dagger (\widehat{\Omega}_s - \widehat{\Omega}_u) \widehat{D}_s | Q_s \rangle = 0,
\end{aligned}$$

by virtue of Eq. (6.94) and Exps. (6.98b), (6.98b).

$$\mathrm{tr}\left(\widehat{D}_s + \widehat{D}_u\right) = 0 \tag{6.99}$$

the operators $\widehat{D}_{s,u}$ have to meet.

The obtained expressions enable us to relate the initial state of action strategy implementation to the initiation-induced Hamiltonian H_{ini}. Let at the initiation time t_{ini} the system state be specified by the position $Q^\dagger = \{x, v, t\}$ and the generalized momentum $P^\dagger = \{p_x, p_v, p_a\}$ in the Hamilton space $\mathbb{R}_H$. Then the partition of this state into the corresponding components Q_s and Q_u belonging to the stable and unstable manifolds Ξ_s and Ξ_u obeys the equalities

$$Q_s + Q_u = Q \quad \text{and} \quad \widehat{\Omega}_s Q_s + \widehat{\Omega}_u Q_u = P\,. \tag{6.100}$$

Resolving the two equation, e.g., with respect to Q_u we find

$$Q_u = \left(\widehat{\Omega}_u - \widehat{\Omega}_s\right)^{-1}\left(P - \widehat{\Omega}_s Q\right). \tag{6.101}$$

Within the initiation noise approximation, the system deviation from the stable manifold is treated as a small perturbation, so the inequality $Q_u \ll Q$ at the moment of initiation can be accepted. In this case, Exp. (6.98b) immediately gives

$$H_{\text{ini}} = \langle * |\, P - \widehat{\Omega}_s Q \widehat{D}_s \,| * \rangle\, Q. \tag{6.102}$$

It should be noted that Exp. (6.102) is actually Exp. (6.88) written in the vector form.

Expressions (6.101) and (6.102) describe the desired relationship between:

- the initial deviation of the system state from the stable manifold Ξ quantified in terms of the momentum difference $(P - \Omega_s Q)$,
- the magnitude of the system state component belonging to the unstable manifold, Exp. (6.101),
- the value of initiation-induced Hamiltonian H_{ini}, Exp. (6.102),

at the initiation time t_{ini}. In our account of skilled subject's actions we turn to noted quantities in order to describe:

- the imperfectness of the skilled subject's actions mainly determined by random and uncontrollable deviations of the momenta $\{p_x, p_v, p_a\}$ from the optimal values,
- the subject's evaluation of the current implementation of action strategy via the observed deviation of the object motion state $\{x, v, a\}$ from the desired goal point, which becomes evident as time goes on,
- the quality of action strategy implementation in terms of a certain aggregated measure based on the initiation-induced Hamiltonian.

Leaping ahead, we note that the desired measure of the proximity to the optimal implementation of action strategies is actually reduced to the initiation-induced Hamiltonian H_{ini} after its averaging over possible variations of the momentum difference $(P - \Omega_s Q)$. This averaging actually corresponds to the transition in our description

- **from** the level of external observer dealing with the trajectories of system motion in the Hamilton space $\mathbb{R}_H$ or, what is equivalent, in the Lagrange space $\mathbb{R}_L$ with two manifolds $\Xi_{s,u}$
- **to** the level of the body-in-the-self dealing with the space-time clouds of action strategy implementations as fuzzy regions each of them comprises
  - a cloud of action strategy initiation $\mathbb{C}_H^{\text{ini}}$ (Exp. 6.74) and
  - the related fragments of the stable manifold cloud $\mathbb{C}_\Xi^s$ (Exp. 6.86) representing all the acceptable action strategy implementations starting from $\mathbb{C}_H^{\text{ini}}$ and terminated inside the action point layer $\partial\mathbb{C}_\Xi^s$ (Exp. 6.87) or completed inside the goal cloud $\mathbb{C}_H^g$ (Exp. 6.85).

The space-time clouds of action strategy implementations are the mathematical description of the skill the subject acquired within long-term learning of how to respond and evaluate acceptable actions in governing the motion of controlled object. This skill, treated as a property of embodied cognition, takes into account the ability and limitations of perception, anticipation, and motor behavior.

6.2.5 *Spatial Measure of Proximity to Action Strategy Optimal Implementations*

The temporal criterion of proximity to the optimal implementation of action strategies identified with the proximity to the stable manifold Ξ_s (Sect. 6.2.3) prompts us focus on the jerk j in constructing the desired measure. The purpose of the present subsection is to construct this measure based on the results of the previous subsection and the statistical properties of the jerk.

The arguments for employing the jerk as main variable in quantifying the proximity to the stable manifold Ξ_s are as follows:

- The object motion phase variables $Q = \{x, v, a\}$ are the inner parameters of both the stable and unstable manifolds, so they cannot serve as quantities involved in differentiating these manifolds from each other.
- The momenta p_x, p_v are outside the subject's direct experience of object motion state within the experiential now, these variables describe the subject's skill in automatic (unconscious) implementations of action strategies.
- The jerk is a phase variable characterizing directly the experiential now, it is a component of the space $\mathbb{R}_{[4]}$ and is due to the tripartite structure of the experiential now (Sect. 2.6.6). Therefore the jerk is directly accessible to the mind.
- Due to the proportionality $p_a \propto j$ (Exp. 6.41), the jerk is actually a phase variable involved explicitly in the developed Hamiltonian description of the skilled subject's actions.
- According to (6.94) and (6.95) the inner dynamics of these stable and unstable manifolds differs only in the time derivative of the object acceleration, i.e., in the jerk j.

So the jerk does be a natural variable in quantifying the deviation of the object motion state from the stable manifold Ξ_s.

The uncertainty of subject's perception of the current jerk j as well as the jerk $j_s(x, v, a)$ corresponding to the inner dynamics of the stable manifold Ξ_s precludes us to from using the jerk as a single-value variable. In Sect. 6.2.3 we proposed the concept of action-point layer $\partial\mathbb{C}^s_\Xi$ separated from the stable manifold cloud $\mathbb{C}^s_\Xi$ by a distance exceeding the uncertainty in the subject's jerk perception. The subject is supposed to terminate the current action strategy implementation when he recognizes that the value of jerk j growing with time has reached the layer $\partial\mathbb{C}^s_\Xi$. It is essential that the subject started to recognize the deviation of the object motion state from the manifold Ξ_s earlier before the motion state has gotten the layer $\partial\mathbb{C}^s_\Xi$. It opens a gate toward statistical description of the jerk properties and using them in quantifying the deviation from Ξ_s.

Because the deviation from the stable manifold Ξ_s quantified in the jerk difference $\Delta j = j - j_s(x, v, a)$ is equiprobable into the region of positive and negative values, may use the value $\overline{\Delta j} = \langle(\Delta j)^2\rangle^{1/2}$ averaged in a certain way. Then the particular realization of Δj may be treated as a random variable whose distribution is localized in the region $|\Delta j| \lesssim \overline{\Delta j}$.

If the subject is able to estimate the time growth of the mean value $\overline{\Delta j}(t)$ adequately, it becomes possible to quantify probabilistically the proximity to the stable manifold turning to the temporal criterion (Sect. 6.2.3). However, the time increase of the variable $\overline{\Delta j}(t)$ is exponential, which makes its prediction a rather difficult task for the subject. At the same time the inner dynamics of the stable manifold is characterized by exponential decrease in the variables Q. As a result, the initiation-induced Hamiltonian H_{ini}, Exp. (6.88),

- relates the motion state components of the stable and unstable manifolds to each other, Exps. (6.101) and (6.102), such that
- its value is conserved during the time course of action strategy implementation, which is the general property of Hamiltonian systems (see also footnote 33).

The gist of our constructions is to relate the time increase in the jerk difference $\Delta j(t)$ to the position of the object state motion $Q(t)$ in the Lagrange space $\mathbb{R}_L$ via H_{ini}. In this case, the anticipation of the further increase in the jerk difference $\Delta j(t)$ can be much easier for the subject.

We take into account two sources of the random variations in the jerk difference Δj which contribute not only to the value of $\Delta j\big|_{\text{ini}}$ at the moment of the action strategy initiation but also to its further growth. They are:

- *The uncertainty of subject's perception and action strategy implementation under the same initial conditions*: Within the initiation noise approximation (Sect. 6.2.2), the scattering of the momenta p_x, p_v, and p_a near the stable manifold Ξ_s is the main factor responsible for the initial deviation of the object motion state from Ξ_s. Probabilistically this source of uncertainty is treated as random variables (see Exp. 6.101)

$$\Delta P \stackrel{\text{def}}{=} P - \widehat{\Omega}_s = \begin{cases} \Delta p_x = p_x - \pi^s_x(x, v, a) \\ \Delta p_v = p_v - \pi^s_v(x, v, a) \\ \Delta p_a = p_a - \pi^s_a(x, v, a) \end{cases} \tag{6.103}$$

distributed smoothly in the region

$$\mathbb{C}^{\text{ini}}_P \Rightarrow \frac{(\Delta p_x)^2}{\delta^2_{p_x}} + \frac{(\Delta p_v)^2}{\delta^2_{p_v}} + \frac{(\Delta p_a)^2}{\delta^2_{p_a}} \lesssim 1 \tag{6.104}$$

being the P-component part of the cloud $\mathbb{C}^{\text{ini}}_H$ of action strategy initiation in the Hamiltonian phase space $\mathbb{R}_H$ (Exp. 6.74).

- *Subject's categorization of different initial states of object motion as equivalent*: The subject's skill acquired during long-term learning concerns a wide variety of different situations related to the given *type* of goal-oriented behavior rather than a family of action strategies initiated in a small neighborhood of some point Q. So in evaluating the final results of action strategy implementations the subject has to aggregate them in some way reducing the dimensions of initial conditions. To be specific, let us consider that the subject categorizes the initial states of object motion according to the their distances to the action goal just rescaling the Q-component part of the cloud $\mathbb{C}^g_H$. In this case, the distance between the initial state and the goal is quantified by a variable s which corresponds to the spherical-like layer in the Lagrange space $\mathbb{R}_L$

$$\mathbb{L}^{\text{ini}}_Q(s) \Rightarrow \frac{(x - x_g)^2}{\delta^2_x} + \frac{v^2}{\delta^2_v} + \frac{a^2}{\delta^2_a} \sim s^2. \tag{6.105}$$

  So in studying the relationship between the jerk difference Δj and the initial state of object motion characterized by the distance s to the goal point, the position of the motion state inside the layer $\mathbb{L}^{\text{ini}}_Q(s)$ will be treated as a random variable.

The direct averaging over the two sources of "randomness" gives us the mean value of the initiation-induced Hamiltonian H_{ini}

$$\langle H^2_{\text{ini}} \rangle \sim \varkappa^2_{\text{eff}}\, s^2. \tag{6.106}$$

In deriving this expression we have used Exps. (6.75c), (6.83), and (6.102). The obtained estimate enables us to write the desired statistical relationship between the time increase in the jerk difference $\overline{\Delta j}(t)$ and the advance $s(t)$ in the action strategy implementation as

$$\overline{\Delta j}(t) \sim \delta_j \frac{s(t_{\text{ini}})}{s(t)}. \tag{6.107}$$

Concluding this subsection, we want to note the following aspects concerning the constructed *statistical* spatial measure (6.107) for the temporal criterion of proximity to the stable manifold Ξ_s.

- The relation between the estimates (6.106) and (6.107) shows that the subject focusing on the control over the jerk difference $\Delta j(t)$ controls implicitly the time variations in all the momenta p_x, p_v, and p_a intertwined with one another.
- The layer $\mathbb{L}_Q^{\text{ini}}(s)$ for $s \sim 1$ practically represents the goal cloud in the Lagrange space $\mathbb{R}_L$. So supposing that the skilled subject

  - typically returns the object motion state to this region via a single, automatic implementation of action strategies,
  - initiates the action strategy implementation when the jerk difference Δj gets the action-point layer characterized by the (fuzzy) threshold $\Delta j_\Xi \sim \theta_\Xi \delta_j$ (Sect. 6.2.3),

  from Ex. (6.107) we immediate obtain

$$s_{\text{ini}} = s_{\text{cont}} \sim \theta_\Xi \,. \tag{6.108}$$

  Here the value s_{cont} characterizes the *region of control* in the Lagrange space $\mathbb{R}_L$, i.e., the region within which the skilled subject is able to keep the object motion state for a long time in comparison with the duration of complex present.
- Expression (6.107) describes the subject's *anticipation* of action strategy implementation. A particular time course of action strategy implementation is outside the subject's control, which will be considered in Sect. 6.3 as the dynamics of action strategies.

6.2.6 Initial State of Action Strategy Space-Time Cloud: Cloud Density and Density Function of Hamiltonian Systems

The description of subject's perception of action strategy implementation comprises two related, nevertheless, individual problems. One concerns how the subject evaluates the plausible time course of action strategy implementation just after the initiation of a given action strategy $\mathcal{AS}$. The other is about the dynamics of this implementation when the subject

- perceives the current state of object motion—a particular realization of the motion state from all the expectable states—within the accuracy of its monitoring and
- again evaluates the further dynamics of the action strategy implementation starting, now, from the current state of motion.

The analysis of the action strategy dynamics will be presented in the next section (Sect. 6.3). The present section (Sect. 6.2) is focused on the former problems directly and partly concerns the latter one.

We consider the mathematical description of how the subject anticipates the time course of action strategy implementation based on the acquired skill. This skill implies that the subject has accumulated the information from the numerous implementations of action strategies whose multitude forms the cloud $\mathbb{C}^s_\Xi$ of the stable manifold Ξ_s combined with the action point layer $\partial\mathbb{C}^s_\Xi$ (Sect. 6.2.3).

The subject's anticipation of the controlled object motion in the connected future is described by the *forward temporal part* $\mathbb{C}^>_{\text{AS}}$ of the action strategy space-time cloud $\mathbb{C}_{\text{AS}}$.[34] Just after the action strategy initiation, the forward temporal part $\mathbb{C}^>_{\text{AS}}$ coincides with the action strategy space-time cloud $\mathbb{C}_{\text{AS}}$ as a whole. So we will call it the *initial state* $\mathbb{C}^{\text{ini}}_{\text{AS}}$ of the space-time cloud $\mathbb{C}_{\text{AS}}$.

The action strategy implementation between the initiation and interruption (termination) is considered to be automatic. So the initial state $\mathbb{C}^{\text{ini}}_{\text{AS}}$ of the space time cloud $\mathbb{C}_{\text{AS}}$ and its forward temporal part $\mathbb{C}^>_{\text{AS}}$ differ from each other in two aspects:

- the position of initial state of object motion and its current state in the Lagrange spaces $\mathbb{R}_L$ (Sect. 6.2.4),
- the corresponding values of the momenta $\{p_x, p_v, p_a\}$; for $\mathbb{C}^{\text{ini}}_{\text{AS}}$ the initial values of $\{p_x, p_v, p_a\}$ belong to the cloud of action strategy initiation $\mathbb{C}^{\text{ini}}_H$ (Exp. (6.74)), for $\mathbb{C}^>_{\text{AS}}$ in the general case their values result form the dynamics of action strategy implementation and may be located outside $\mathbb{C}^{\text{ini}}_H$.

In the present section we confine our consideration to the initial state $\mathbb{C}^{\text{ini}}_{\text{AS}}$ of the space time cloud $\mathbb{C}_{\text{AS}}$. The corresponding generalization to the forward temporal part $\mathbb{C}^>_{\text{AS}}$ will be considered in Sect. 6.3.

The subject is not able to recognize precisely the initial position of the trajectory

$$P^H_{\text{AS}}(t) = \{x(t),\ v(t),\ a(t),\ p_x(t),\ p_v(t),\ p_a(t)\}$$

being the current implementation of a given action strategy $\mathcal{AS}$ in the Hamilton space $\mathbb{R}_H$. So the subject has to regards the cloud of action strategy initiation $\mathbb{C}^{\text{ini}}_H$ as a whole indivisible entity "generating" a bundle of action strategy implementations. So, as zeroth approximation, we may represent the cloud $\mathbb{C}^{\text{ini}}_{\text{AS}}$ in terms of the ensemble of trajectories $\{P^H_{\text{AS}}\}$ whose distribution in the space-time continuum $\mathbb{R}_H \times t$ is described by the density function $\psi(t, Q, P)$. At the initiation time t_{ini} the density function $\psi(t, Q, P)$ represents the cloud $\mathbb{C}^{\text{ini}}_H$. This is exactly the first step in constructing the general concept of action strategy space-time cloud we presented in Sect. 3.3.4; the present section actually elaborates this concept to enable its mathematical description.

To develop the desired mathematical representation of action strategy space time clouds we need

- starting from the continuous description of the trajectory ensemble $\{P^H_{\text{AS}}\}$, to elucidate the required modification to allow for the uncertainty in the subject's perception of object motion;
- to clarify the basic features in generalizing the physical models of distributed systems to the cloud-type description of human evaluation of observed and imaginary processes and phenomena.

[34] We discussed the general notion of temporal parts in Sect. 1.2.3.

The present subsection is focused on the former issue.

Hamiltonian description of the $\{P^H_{\text{AS}}\}$-ensemble: Let us start from the Liouville equation for the density function $\psi(t, Q, P)$ of the Hamiltonian system (6.55) describing the conservation of the phase volume in the Hamilton space $\mathbb{R}_H$ (e.g., Liboff 2003, Chap. 1). Usually the Liouville equation is written in the form using the Poisson bracket

$$\frac{\partial \psi}{\partial t} + [\psi, H] \stackrel{\text{def}}{=} \frac{\partial \psi}{\partial t} + \frac{\partial \psi}{\partial Q}\frac{\partial H}{\partial P}\bigg|^{\text{T}} - \frac{\partial \psi}{\partial P}\frac{\partial H}{\partial Q}\bigg|^{\text{T}} = 0$$

we, however, turn to its equivalent form

$$\frac{\partial \psi}{\partial t} + \frac{\partial}{\partial Q}\left[\psi \frac{dQ}{dt}\right] + \frac{\partial}{\partial P}\left[\psi \frac{dP}{dt}\right] = 0\,, \tag{6.109}$$

where the used notions are understood as vectors and covectors of the Lagrange space $\mathbb{R}_L$

$$Q = \begin{bmatrix} x \\ v \\ a \end{bmatrix} \in \mathbb{R}_L\,, \qquad \frac{\partial}{\partial Q} = \left[\frac{\partial}{\partial x}\ \frac{\partial}{\partial v}\ \frac{\partial}{\partial a}\right],$$

$$P = \begin{bmatrix} p_x \\ p_v \\ p_a \end{bmatrix} \in \mathbb{R}_L\,, \qquad \frac{\partial}{\partial P} = \left[\frac{\partial}{\partial p_x}\ \frac{\partial}{\partial p_v}\ \frac{\partial}{\partial p_a}\right].$$

In Eq. (6.109) the time derivatives of the variables $\{Q, P\}$ are related to the system state by the Hamilton's equations

$$\frac{dQ}{dt} = \left[\frac{\partial H(Q, P)}{\partial P}\right]^{\text{T}} \quad \text{and} \quad \frac{dP}{dt} = -\left[\frac{\partial H(Q, P)}{\partial Q}\right]^{\text{T}}, \tag{6.110}$$

and, as in Sect. 6.2.4, we will use the Hamiltonian $H(Q, P)$ specified by Exp. (6.42) with $\gamma_s = 2$. The Liouville equation (6.109) describes the time evolution of the initial density $\psi(t, Q, P)_{t=t_{\text{ini}}}$ determining the number of instantiations of a given Hamiltonian system in the unit volume of its phase space.

Within the approximation of dual-manifold initiation, the Hamiltonian description of action strategies (Sect. 6.2.4) demonstrates the basic features underlying our description of the skilled subject's behavior. These results prompt us to represent the system dynamics in the form dealing with:

– the system position $Q = Q_s + Q_u$ in the Lagrange space $\mathbb{R}_L$ taking into account both the components of the system state expansion relative to the stable and unstable manifolds, Ξ_s and Ξ_u,

– the component of the system state belonging to the unstable manifold Ξ_u parameterized by the vector Q_u of the Lagrange space $\mathbb{R}_L$.

First, this approach focuses on the inner phase variables $Q = \{x, v, a\}$ characterizing the motion state of controlled object. Second, the variable Q_u, describing the deviation of action strategy implementation from the optimal one, is also brought into focus.

To do this we change the canonical variables $\{Q, P\}$ for the new variables $\{Q, Q_u\}$:

$$\{Q, P\} \Leftarrow \{Q, Q_u\} \quad \text{via} \quad P = \widehat{\Omega}_s Q + \big(\widehat{\Omega}_u - \widehat{\Omega}_s\big) Q_u \tag{6.111a}$$

by virtue of Exp. (6.100). The inverse relations are

$$\{Q, P\} \Rightarrow \{Q, Q_u\} \quad \text{via} \quad Q_u = \big(\widehat{\Omega}_u - \widehat{\Omega}_s\big)^{-1} \big(P - \widehat{\Omega}_s Q\big). \tag{6.111b}$$

In this case, the gradient written in the variables $\{Q, P\}$ should be replaced by the gradient written in the new variables $\{Q, Q_u\}$:

$$\left\{\frac{\partial}{\partial Q}, \frac{\partial}{\partial P}\right\}_{QP} \Leftarrow \left\{\frac{\partial}{\partial Q}, \frac{\partial}{\partial P}\right\}_{QQ_u},$$

which is reduced to the replacements

$$\frac{\partial}{\partial Q} \Leftarrow \frac{\partial}{\partial Q} - \frac{\partial}{\partial Q_u}\big(\widehat{\Omega}_u - \widehat{\Omega}_s\big)^{-1} \widehat{\Omega}_s\,, \tag{6.112a}$$

$$\frac{\partial}{\partial P} \Leftarrow \frac{\partial}{\partial Q_u}\big(\widehat{\Omega}_u - \widehat{\Omega}_s\big)^{-1}. \tag{6.112b}$$

Using the results of Sect. 6.2.4, namely, (*i*) Exp. (6.90) specifying the embedding of the stable and unstable manifolds $\Xi_{s,u}$ into the Hamilton space $\mathbb{R}_H$ and (*ii*) Exp. (6.94) governing the inner dynamics of these manifolds, we can represent the Hamilton's equations (6.110) as

$$\frac{dQ}{dt} = \widehat{D}_s Q + \big(\widehat{D}_u - \widehat{D}_s\big) Q_u\,, \tag{6.113a}$$

$$\frac{dP}{dt} = \widehat{\Omega}_s \widehat{D}_s Q + \big(\widehat{\Omega}_u \widehat{D}_u - \widehat{\Omega}_s \widehat{D}_s\big) Q_u\,. \tag{6.113b}$$

The derived expressions (6.112) and (6.113) enable use to rewrite the Liouville equation (6.109) as

$$\frac{\partial \psi}{\partial t} + \frac{\partial}{\partial Q}\left\{\left[\widehat{D}_s Q + \big(\widehat{D}_u - \widehat{D}_s\big) Q_u\right]\psi\right\} + \frac{\partial}{\partial Q_u}\left\{\big(\widehat{D}_u Q_u\big)\psi\right\} = 0\,. \tag{6.114a}$$

To facilitate understanding our further constructions, we represent this equation also in the detailed form

$$\frac{\partial\psi}{\partial t}+\frac{\partial}{\partial x}(v\psi)+\frac{\partial}{\partial v}(a\psi)+\frac{\partial}{\partial a}\big\{\big[j_s(x,v,a)+\Delta j(x_u,v_u,a_u)\big]\psi\big\}$$
$$+\frac{\partial}{\partial x_u}(v_u\psi)+\frac{\partial}{\partial v_u}(a_u\psi)+\frac{\partial}{\partial a_u}\big\{\big[j_u(x_u,v_u,a_u)\big]\psi\big\}=0\,, \tag{6.114b}$$

where, by virtue of Exp. (6.95), the linear functions

$$\begin{aligned} j_s(x,v,a) &= h_a(\omega^s_{ax}x+\omega^s_{av}v+\omega^s_{aa}a)\,,\\ j_u(x_u,v_u,a_u) &= h_a(\omega^u_{ax}x_u+\omega^s_{av}v_u+\omega^u_{aa}a_u)\,,\\ \Delta j(x_u,v_u,a_u) &= j_u(x_u,v_u,a_u)-j_s(x_u,v_u,a_u) \end{aligned} \tag{6.115}$$

describe the jerk $j = da/dt$ of the inner dynamics of the stable (Ξ_s) and unstable (Ξ_u) manifolds, $h_a = H_{p_a p_a}$, and the coefficients $\{\omega^{s,u}_{\dots}\}$ are the components of the matrices $\widehat{\Omega}_{s,u}$; their particular values are given by Exps. (6.291) and (6.292) in Appendix 6.7.3.

To avoid possible misunderstanding, it should be noted that replacement (6.111a) leads also to the renormalization of the density function $\psi(Q, P, t)$ caused by the corresponding rescaling of the phase space volume

$$\psi(Q,P,t)_{QP} \Leftarrow \left(\det\big[\widehat{\Omega}_u-\widehat{\Omega}_s\big]\right)^{-1}\psi(Q,P,t)_{QQ_u}.$$

However, in the following constructions this renormalization will be ignored because the related normalization of action-strategy space-time clouds to be introduced below is based on another formalism rather than the formalism of Newtonian mechanics. This new formalism deals with bilinear normalization of the space-time cloud $\mathbb{C}_{\text{AS}}$, which also is not essential in the context of the $\psi(Q, P, t)$-function renormalization. It is due the invariance of Eq. (6.114) with respect to the replacement $\psi \Rightarrow \psi^p$ for an arbitrary exponent $p > 0$.[35] So we may regard the function $\psi(t, Q, Q_u)$ as a certain prototype of the cloud density.

The choice of the canonical variables $\{Q, P\}$ and the corresponding Liouville equation (6.109) for constructing the cloud density is due to that this description

[35] This invariance stems directly from the fact that Eq. (6.114a) can be rewritten in the form related directly to the standard representation of the Liouville equation, i.e.,

$$\frac{\partial\psi}{\partial t}+\frac{\partial\psi}{\partial Q}\big[\widehat{D}_sQ+(\widehat{D}_u-\widehat{D}_s)Q_u\big]+\frac{\partial\psi}{\partial Q_u}(\widehat{D}_uQ_u)=0$$

due to the equality $\operatorname{tr}(\widehat{D}_u+\widehat{D}_s)=0$ (Exp. 6.99). So the replacement $\psi \Rightarrow F_\psi(\psi)$, where $F_\psi(\psi)$ is a monotonous function, merely leads to multiplication of this equation by the term $dF_\psi(\psi)/d\psi$.

- directly emphasizes the characteristic properties of Hamiltonian systems, on the one side, and deals with the phase space admitting interpretation as the standard Euclidean space, i.e., linear space with constant metric, on the other side[36];
- the inner phase variables of the object motion state $Q = \{x, v, a\}$ have clear meaning and conform to human ordinary experience in evaluating the motion of physical objects.

The change of variables $\{Q, P\} \Rightarrow Q, Q_u$ corresponds also to the phase space with constant metric and concerns only the components of the variables P and Q_u not accessible to the mind directly.

The density functions of the space-time cloud $\mathbb{C}^{\text{ini}}_{\text{AS}}$ and the ensemble of Hamiltonian systems: As noted above, the trajectory ensemble $\{P^H_{\text{AS}}\}$ is regarded as the prototype of the action strategy cloud $\mathbb{C}^{\text{ini}}_{\text{AS}}$. Turing to this interpretation, we deal with the space-time continuum $\{Q\} \times \{Q_u\} \times t$ of *another nature*. In this case, the space-time continuum $\{Q\} \times \{Q_u\} \times t$ belongs to the human temporality and only reflects physical reality. As a result, the status of time and space in this space-time continuum becomes actually the same. All the spatial and temporal elements can affect the "here-and-now" experience of the subject. In particular, distant spatial and temporal elements can cooperate or interfere in their "simultaneous" contribution to analyzed mental phenomena. It is due to the ownership of the self with respect to the human temporality (Sect. 3.2.3); below we will return to this issue.

In this case, the space-time cloud $\mathbb{C}^{\text{ini}}_{\text{AS}}$ (and $\mathbb{C}^{>}_{\text{AS}}$ too) describes how the subject mentally embeds the possible trajectories $\{P^H_{\text{AS}}\}$ into the Lagrange space $\mathbb{R}_L$ combined with some quantity characterizing their proximity to the stable manifold Ξ_s. In Sect. 3.3.4 within the general constructions we described action strategy clouds as some fuzzy regions in the space-time continuum $\mathbb{R}_{[4]} \times t$. For the problem in issue the notion of action strategy space-time clouds is detailed. Leaping ahead, we state that the cloud $\mathbb{C}^{\text{ini}}_{\text{AS}}$ ($\mathbb{C}^{>}_{\text{AS}}$) may be regarded as a fuzzy region in the space-time continuum $\mathbb{R}^H_{[4]} \times t$, where

$$\mathbb{R}^H_{[4]} = \{Q, j_u\} = \{x, v, a, \Delta j\} \tag{6.116}$$

is the representation of $\mathbb{R}_{[4]}$ in the Hamilton space as a certain subspace. Here the variable Δj quantifying the deviation from the stable manifold is not an ordinary inner phase variable of object motion. It is reflected in that the subject recognizes the system deviation from the stable manifold Ξ_s but is not able to relate this deviation to the current state of object motion $\{x, v, a\}$ even for large magnitudes of this deviation.

Put differently, the space-time cloud $\mathbb{C}^{\text{ini}}_{\text{AS}}$ ($\mathbb{C}^{>}_{\text{AS}}$) is the mental-mental embedding of the trajectories $\{P^H_{\text{AS}}\}$ as a whole multitude into the mental image of the Hamilton space $\mathbb{R}_H \times t$. So the cloud $\mathbb{C}^{\text{ini}}_{\text{AS}}$ ($\mathbb{C}^{>}_{\text{AS}}$) is the mathematical eidos of the action strategy

[36] The use of other types phase variables can make it necessary to deal with a phase space with metric depending on the position in this space. Under such conditions the corresponding density function combines the properties of Hamiltonian systems and the properties of this phase space, which causes additional difficulties in the desired generalization.

$\mathcal{AS}$. The cloud density[37] $\Phi(t, Q, \Delta j)$ representing the cloud $\mathbb{C}_{\text{AS}}^{\text{ini}}$ ($\mathbb{C}_{\text{AS}}^{>}$) inherits the properties of the density function $\psi(t, Q, Q_u)$.

In order to elucidate the relationship between

- the density $\Phi(t, Q, \Delta j)$ of space time cloud $\mathbb{C}_{\text{AS}}^{\text{ini}}$ (and $\mathbb{C}_{\text{AS}}^{>}$ too) and
- the density $\psi(t, Q, Q_u)$ of the Hamiltonian system ensemble

we accept the following four premises $\mathbf{P}_{ss}^{1}$–$\mathbf{P}_{ss}^{4}$ characterizing the properties of the skilled subject.

$\mathbf{P}_{ss}^{1}$: The subject does not recognize directly the splitting of the motion state $Q = Q_s + Q_u$ of controlled object into two its components belonging to the stable (Ξ_s) and unstable (Ξ_s) manifolds. He is able to operate only with their cumulative contribution.

$\mathbf{P}_{ss}^{2}$: The subject perceives the deviation of the object motion state Q from the stable manifold Ξ_s within some accuracy. We describe it as the subject's perception of the jerk difference Δj within the accuracy δ_j. Keeping in mind the previous premise, we suppose that the subject perceives the jerk difference as a certain uncontrollable and unpredictable factor. It implies that the dynamics of the momenta p_x and p_v is governed by the acquired skill and implemented unconsciously in the body-in-the-self.[38] Only the current value of the momentum p_a as the jerk $j \propto p_a$ is accessible to the mind without the feasibility of its conscious control.

Therefore, at the first step in constructing the governing equation for the cloud density $\Phi(t, Q, \Delta j)$, we integrate the Liouville equation (6.114) and, respectively, the density function $\psi(t, Q, Q_u)$, over the variables $Q_u = \{x_u, v_u, a_u\}$ within the replacement

$$\left\langle \Delta j_{us}(Q_u)\, \psi(t, Q, Q_u) \right\rangle_{Q_u} \quad \rightarrow \quad \overline{\psi}(t, Q) \Delta j(t)\,, \tag{6.117}$$

where the dynamics of the jerk difference $\Delta_j(t)$ is not controllable by the subject and determined by initiation noise,

$$\overline{\psi}(t, Q) = \left\langle \psi(t, Q, Q_u) \right\rangle_{Q_u} \quad \text{and} \quad \left\langle \ldots \right\rangle_{Q_u} = \int_{\mathbb{R}_L} dQ_u \left(\ldots \right).$$

[37] The cloud density $\Phi(t, Q, \Delta j)$ may be treated as a certain analogy to the membership function in the theory of fuzzy sets (e.g., Dubois and Prade 1980, Chap. 1). The main difference between them is that the cloud density $\Phi(t, Q, \Delta j)$ appears in binary operations with clouds like their overlapping rather mini-max relations.

[38] Such unconscious processes are exemplified by neural mechanisms of motor behavior described by the theory called the equilibrium-point hypothesis (see, e.g., Feldman et al. 2007; Feldman 2009; Feldman and Levin 2009; Latash 2008b, 2010, 2012, for a review). The body movements are considered to be governed by the muscle length-force relation. In these terms a particular equilibrium or quasi-equilibrium configuration of the body has to be determined by the equilibrium points of the individual length-force relations for all the muscles involved into its implementation. Changing these relations, the neural network affects their equilibrium points and, thereby, generates body movements.

Naturally, the terms of Eq. (6.114) related to the gradient $\partial/\partial Q_u$ vanish after the averaging.

$\mathbf{P}^3_{ss}$: The subject cannot differentiate neighboring points of the space-time continuum in their contribution to his current experience, we already discussed this issue in Sects. 2.5, 2.6 concerning the experiential now. In evaluating the contribution of past, current, and future events, the subject deals with their cloud-type properties rather than point-like characteristics similar to a position in space, an instant speed, the moment of time when a given event happened, etc. In other words, that what the subject treats as his current experience *consciously* indivisible into smaller fragments is the result of integration over

- a certain fuzzy neighborhood of the current spatial position—in our case it is the point $\{x_0, v_0, a_0, j_0\} \in \mathbb{R}_{[4]}$—whose size is determined by perceptual uncertainty of external stimuli,
- a certain fuzzy time interval around the current instant of time t_0 whose size is about the duration of experiential now τ_d.

To take into account this feature we describe the mathematical eidos $\mathbb{C}_\eta$ of the point $\eta = \{x, v, a, j, t\}$ belonging to the space-time continuum $\mathbb{R}_{[4]} \times t$ as a function

$$\Gamma_\eta(\eta') = \Gamma_\eta\left(\frac{t'-t}{\tau_d}, \frac{x'-x}{\delta_x}, \frac{v'-v}{\delta_v}, \frac{a'-a}{\delta_a}, \frac{j'-j}{\delta_j}\right).$$

This function meets the condition

$$\int_{\mathbb{R}_{[4]}\times t} d\eta' \Gamma_\eta(\eta') = 1\,, \tag{6.118}$$

meaning that on scales exceeding substantially $\{\tau_d, \delta_x, \delta_v, \delta_a, \delta_j\}$ it may be identified with the Dirac δ-function—the standard representation of point-like objects in the realm of functions. Besides, we will assume the function $\Gamma_\eta(\eta')$ to be symmetric with respect to its arguments,

$$\Gamma_\eta(\eta') = \Gamma_\eta(\eta'') \text{ for } \eta''_w - \eta_w = \eta_w - \eta'_w \text{ and the dimensions } w = x, v, a, j\,.$$

As far as the temporal dimension is concerned, the center of the function $\Gamma_\eta(\eta')$ is shifted along the time axis backward, which reflects the asymmetry of the experiential now. Indeed the immediate past exceeds the immediate now and the immediate future in duration (Sect. 2.6, see, e.g., Fig. 2.6). We describe this temporal shift by the parameter $\zeta_d \lesssim 0.5$ introduced via the expression

$$\int_{\mathbb{R}_{[4]}\times t} d\eta' \Gamma_\eta(\eta')(t'-t) = -\zeta_d\,\tau_d\,. \tag{6.119}$$

The value $\zeta_d \tau_d$ may be interpreted as the mean delay time in subject's perception of the object motion state.[39]

In mathematical terms the conversion

- from the point-type description of the trajectory ensemble $\{P^H_{\mathrm{AS}}\}$ based on the Liouville equation (6.114) *(the realm of the external observer)*
- to the desired cloud-type description of the initial state of action strategy space-time cloud $\mathbb{C}^{\mathrm{ini}}_{\mathrm{AS}}$ (and $\mathbb{C}^{>}_{\mathrm{AS}}$ too) *(the realm of the body-in-the-self)*

is implemented by applying the convolution operation to Eq. (6.114), where the function $\varGamma_\eta(\eta')$ plays the role of convolution kernel. The particular implementation of this transformation is based on (*i*) the identity

$$\int_{\mathbb{R}_{[4]}\times t} d\eta' \varGamma_\eta(\eta') \frac{\partial J_w(\eta')}{\partial \eta'_w} = \frac{\partial}{\partial \eta_w} \int_{\mathbb{R}_{[4]}\times t} d\eta' \varGamma_\eta(\eta') J_w(\eta')$$

for any smooth function $J_w(\eta)$ and (*ii*) the assumption that the density function $\psi(t, Q, Q_u)$ does not vary substantially on the spatial and temporal scales of the cloud $\mathbb{C}_\eta$. The latter enables us to expand the function $J_w(\eta')$ near the point η in a Taylor series and truncate it at the second term as the leading order in τ_d, δ_x, δ_v, and δ_a. It should be noted that in the analyzed case the uncertainty in time perception takes into account two factors. One of them is the integration of perceived information within the experiential now. The other is related to that in the cloud $\mathbb{C}^{\mathrm{ini}}_H$ different implementations of action strategy $\mathcal{AS}$ are initiated non-simultaneously.

In this way we, first, obtain the relationship between the density function $\overline{\psi}(t, Q)$ and the density $\varPhi(t, Q)$ of the action strategy space-time cloud $\mathbb{C}^{\mathrm{ini}}_{\mathrm{AS}}$,

$$\begin{aligned}\varPhi(t, Q) = \overline{\psi}(t, Q) &+ \frac{\tau_d^2}{2} \frac{\partial^2 \overline{\psi}(t, Q)}{\partial t^2} \\ &+ \frac{\delta_x^2}{2} \frac{\partial^2 \overline{\psi}(t, Q)}{\partial x^2} + \frac{\delta_v^2}{2} \frac{\partial^2 \overline{\psi}(t, Q)}{\partial v^2} + \frac{\delta_a^2}{2} \frac{\partial^2 \overline{\psi}(t, Q)}{\partial a^2}, \end{aligned} \tag{6.120}$$

where the identities

---

[39] It should be noted that in a wide variety of models developed for human control over unstable systems or human reaction to external stimuli this time shift is described as the *fixed* time delay τ in human reaction. These models usually deal with delay differential equations similar to

$$\left.\frac{dX}{dt}\right|_t = F(X_{t-\tau}, X_t),$$

where X is some system characteristics and the function $F(X, Y)$ describes human actions and physical regularities. However, justification of this approach, especially, to modeling complex phenomena causes by the delay-induced instability characterized by scales about τ is problematic. The matter is that any description of human reaction even confined to the experiential now has to deal with the tripartite structure of the experiential now. In comparison with the scale of τ this structure spans over a wide time interval containing the instants of the immediate past as well as the immediate future.

$$\int_{\mathbb{R}_{[4]}\times t} d\eta' \Gamma_\eta(\eta')(t' - t + \zeta_d \tau_d)^2 = \tau_d^2\,, \tag{6.121a}$$

$$\int_{\mathbb{R}_{[4]}\times t} d\eta' \Gamma_\eta(\eta')(w' - w)^2 = \delta_w^2 \quad \text{for } w = x, v, a. \tag{6.121b}$$

may be regarded as the definitions of the particular values of the scales τ_d and δ_w. Expression (6.120) describes the smoothing of the density function $\overline{\psi}(t, Q)$ caused by the uncertainty in perceiving the object motion state including the time moments.

Second, the Liouville equation written in form (6.114b) is reduced to

$$\left(\frac{\partial}{\partial t} + \widehat{\mathcal{L}}_{\Delta j}\right)\Phi\big[t - \zeta_d\tau_d, Q\,\big|\Delta j(t - \zeta_d\tau_d)\big] = 0 \tag{6.122}$$

where the operator $\widehat{\mathcal{L}}_{\Delta j}$ is specified by the expression

$$\begin{aligned}\widehat{\mathcal{L}}_{\Delta j} \overset{\text{def}}{=} &\frac{\partial}{\partial x}\left[v + \delta_v^2 \frac{\partial}{\partial v}\right] + \frac{\partial}{\partial v}\left[a + \delta_a^2 \frac{\partial}{\partial a}\right] \\ &+ \frac{\partial}{\partial a}\left[j_s\left(x + \delta_x^2 \frac{\partial}{\partial x},\, v + \delta_v^2 \frac{\partial}{\partial v},\, a + \delta_a^2 \frac{\partial}{\partial a}\right) + \Delta j(t)\right],\end{aligned} \tag{6.123}$$

by virtue of Exp. (6.115),

$$j_s\left(x + \delta_x^2 \frac{\partial}{\partial x},\, v + \delta_v^2 \frac{\partial}{\partial v},\, a + \delta_a^2 \frac{\partial}{\partial a}\right) = \sum_{w=x,v,a} h_a \omega_{aw}^s \left(w + \delta_w^2 \frac{\partial}{\partial w}\right), \tag{6.124}$$

and the quantity $\Delta(t)$ playing here the role of time-dependent parameter is treated as an uncontrollable variable with internal dynamics independent of the object motion state. Besides, time variations in the value of $\Delta j(t)$ are ignorable on the scales of the experiential now. Therefore in solving Eq. (6.122) the time delay $\zeta_d\tau_d$ in the cloud density $\Phi(t - \zeta_d\tau_d, Q)$ as well as in the time-dependence of $\Delta j(t - \zeta_d\tau_d, Q)$ may be omitted. Only in describing the subject's current actions it might be necessary to take into account that the subject reacts to delayed information about the system state.

To complete this description of the action strategy space-time cloud $\mathbb{C}_{\text{AS}}^{\text{ini}}$ we need to add a model for the subject's evaluation of the deviation from the stable manifold Ξ_s, which is quantified by the value Δj.

$\mathbf{P}_{ss}^4$: Being unable to relate the current state of motion $\{x, v, a\}$ to the deviation from the stable manifold Ξ_s, the subject treats the value Δj quantifying this deviation as a random variable that

- at the moment t_{ini} of instantiation takes some value related to the cloud of action strategy initiation $\mathbb{C}_H^{\text{ini}}$ (Exp. (6.74)) via the expression

$$\Delta j = (\widehat{D}_u - \widehat{D}_s)Q_u = h_a\big[p_a - \pi_a^s(x, v, a)\big]$$

by virtue of (6.95), (6.111b), (6.114a);
- grows in magnitude according to the exponential law $e^{\lambda_u t}$, where the increment λ_u stems from the approximation of the inner dynamics of the unstable manifold Ξ_u by growth of spherical region (in dimensionless units of the inner phase variables $\{x, v\, a\}$), i.e., $\lambda_u \sim \lambda_0, \lambda_r$ in units of $1/\tau_{cp}$ (see the eigenvalue equation (6.50));

The noted features prompt us to write the Fokker-Planck equation for the stochastic dynamics of the variables Δj described by the probability density function $\Upsilon(\Delta j, t)$ as (e.g., Gardiner 2009; Mahnke et al. 2009)

$$\left(\frac{\partial}{\partial t} - \widehat{\mathcal{L}}_{\text{FP}}\right)\Upsilon = 0\,, \tag{6.125}$$

where

$$\widehat{\mathcal{L}}_{\text{FP}} \stackrel{\text{def}}{=} \frac{\partial}{\partial \Delta j}\left[\frac{\delta_j^2}{\tau_d}\frac{\partial}{\partial \Delta j} - \frac{\lambda_u}{\tau_d \delta_j}\,\Delta j\right]. \tag{6.126}$$

Equation (6.125) implies that the dynamics of the variable $\Upsilon(\Delta j, t)$ is mainly governed by

- the noise component[40] on temporal scales about the duration τ_d of experiential now, which actually represents the action strategy initiation process and gives rise to the Δj-distribution localized in the neighborhood of the origin $\Delta j = 0$ of thickness about δ_j,
- the regular component[40] of the Δj-dynamics on larger scale, which is responsible for the exponential growth of the variable Δj.

Uniting equations (6.122) and (6.125), we obtain the final equation

$$\frac{\partial \Phi}{\partial t} = \widehat{\mathcal{H}}\,\Phi \tag{6.127}$$

for the function Φ being actually the next step prototype of the density of action strategy cloud which aggregates in itself also the distribution function $\Upsilon(t, \Delta j)$, i.e.,

$$\Phi(t, x, v, a, \Delta t) = \Phi(t, x, v, a \mid \Delta j)\Upsilon(t, \Delta j)\,. \tag{6.128}$$

In this product the variable Δj entering the first cofactor, $\Phi(t, x, v, a \mid \Delta j)$, play the role of parameter, so the right-hand side operator of Eq. (6.125) does not act on $\Phi(t, x, v, a \mid \Delta j)$, it acts only on the second cofactor in the product (6.128). The operator $\widehat{\mathcal{H}}$ is specified by the expression

[40] In the Fokker-Planck equation (6.125) the random and regular components of the Δj-dynamics are represented by the first and second terms on its right-hand side.

$$\widehat{\mathcal{H}} = \frac{\partial}{\partial \Delta j}\left[\frac{\delta_j^2}{\tau_d}\frac{\partial}{\partial \Delta j} - \frac{\lambda_u}{\tau_d \delta_j}\Delta j\right] - \frac{\partial}{\partial a}\left[j_s\left(x + \delta_x^2\frac{\partial}{\partial x},\, v + \delta_v^2\frac{\partial}{\partial v},\, a + \delta_a^2\frac{\partial}{\partial a}\right) + \Delta j\right] - \frac{\partial}{\partial v}\left[a + \delta_a^2\frac{\partial}{\partial a}\right] - \frac{\partial}{\partial x}\left[v + \delta_v^2\frac{\partial}{\partial v}\right]. \quad (6.129)$$

The presented discussion of premises $\mathbf{P}_{ss}^1$–$\mathbf{P}_{ss}^4$ has demonstrated the modifications in the conversion of the density function $\psi(t, Q, Q_u)$ to the cloud density $\Phi(t, Q, \Delta j)$ which are required to take into account the uncertainty of human perception. Equation (6.127) together with the operator $\widehat{\mathcal{H}}$ given by Exp. (6.129) is the final result of averaging the Liouville equation (6.109) over the scales of this uncertainty.

6.2.7 *Action Strategy Space-Time Cloud: Nonlocality of Human Temporality and Bilinear Normalization*

In the previous subsection we analyzed the transformation of the Liouville equation required for describing uncertainty in the subject's evaluation of motion state. The obtained results are necessary but not sufficient to finalize the conversion of the Liouville equation into the equation governing the action strategy space-time clouds. Going in this way, we face up to the challenge that

- the Liouville equation, describing the ensemble of particles treated as material points, belongs to the formalism of Newtonian mechanics, whereas
- the desired governing equation, describing the human perception of action strategies, does not meet the paradigm of Newtonian mechanics.

In other words, the transformation of the Liouville equation to the equation for action strategy clouds is accompanied by the change in the *type of formalism*. The present subsection is focused on elucidating this fundamental difference and clarifying how to tackle it in constructing the mathematical formalism for such human actions.

Point-type *vs* cloud-type formalism: Equations similar to the Liouville equation or the Fokker-Planck equation describe the statistical properties of physical particles or, what is the same, the properties of ensembles comprising a large number of such particles independent of one another. Within Newtonian mechanics these particles admit interpretation as material points whose motion state is characterized by specific values of spatial position $x(t)$, instant velocity $v(t)$ and acceleration $a(t)$ at each instant t of time.

Exactly the formalism dealing with point-like particles enables one to introduce the notion of their density $\psi(t, Q, P)$ met in Sect. 6.2.6. The density function gives the number $\psi(t, Q, P)dQdP$ of particles located inside the neighborhood of the point $\{Q, P\}$ of elementary volume $dQdP$ in the Hamiltonian phase space $\mathbb{R}_H = \big\{Q, P\big\}$. So for a given ensemble of independent particles any function F of its state is determined by the *additive* contribution of all its elements[41]

$$F(t) \stackrel{\text{def}}{=} \langle F(t, Q, P)\rangle = \int_{\mathbb{R}_H} dQdP\, F(Q, P)\psi(t, Q, P)\,.$$

In particular, the normalization of the density function $\psi(t, Q, P)$

$$\int_{\mathbb{R}_H} dQdP\, \psi(t, Q, P) = 1 \tag{6.130}$$

reflects conservation of the total number of particles at any instant t of time.

For the space-time clouds of action strategies as well as the space-time clouds representing the subject's perception of moving objects the situation is entirely distinct. It is so even when the controlled object is categorized by the external observer as a material point whose motion state is described by the point-like position in the space $\mathbb{R}_{[4]} = \big\{\{x, v, a, j\}\big\}$; we already discussed this issue in Sect. 3.3. The mind-in-the-self and the body-in-the-self do not deal with point-like entities like material points in operating with reality.

In Sect. 3.3.3 we considered the $\mathbb{C}_\text{I}$-type clouds—mathematical eidoi of mental entities—to cope with the mental images of physical objects including objects whose spatial dimensions are ignorable, i.e., material points. Generally, the cloud $\mathbb{C}_\text{mp}$ of a material point $\text{mp} = \{x, v, a, j\}$ is a finite-size fuzzy region (region with blurred boundaries) in the space-time continuum $\mathbb{R}_{[4]} \times t$. The cloud $\mathbb{C}_\text{mp}$ bears the information about the perception uncertainty in evaluating (*i*) the motion state of observed object and (*ii*) the corresponding instant of time.

In Sect. 5.6.1 we proposed a more detailed account making a distinction between the clouds in the mind-in-the-self and the corresponding clouds in the body-in-the-self. In the realm of the mind-in-the-self, the cloud $\mathbb{C}_\text{mp}$ of material point is just an elementary indivisible entity, the mind can operate with it only as a whole. In the realm of the body-in-the-self, the cloud $\mathbb{C}_\text{mp}$ may be conceived of as some function $\Phi(t, x, v, a, j)$ in the space-time continuum $\mathbb{R}_{[4]} \times t$ localized near the point $\text{mp} = \{x, v, a, j\}$. The characteristic scales τ_d, δ_x, δ_v, δ_a, and δ_j of this localization along the dimensions $\{t, x, v, a, j\}$, respectively, reflect the uncertainty in perceiving the corresponding quantities. Naturally, an internal representation of the function $\Phi(t, x, v, a, j)$ in the body-in-the-self can be encoded in some particular as a certain distributed neural pattern. This neural pattern can be compared with other neural patterns, merged with them, etc. However, the mind-in-the-self gets access only to the final result. Broadly speaking:

---

[41] In mathematics such systems are categorized as the L^1-functional spaces. In the case under consideration, this categorization emphasizes that the measure of a given ensemble is $dQdP\psi(Q, P)$ for calculating its other integral properties.

- the mind-in-the-self initiates operations with clouds $\mathbb{C}_{\text{mp}_1}$, $\mathbb{C}_{\text{mp}_2}$, ...in the body-in-the-self,
- the body-in-the-self implements these operations with the clouds $\{\mathbb{C}_{\text{mp}_i}\}$ as distributed neural patterns,
- the final result becomes accessible to the mind-in-the-self.

So, in particular, the space-time cloud $\mathbb{C}_{\text{mp}}$ of material point has be categorized as an entity belonging to the level of E-consciousness (see Sect. 3.2.5).

It should be emphasized that the cloud $\mathbb{C}_{\text{mp}}$ in the-mind-in-the-self and the cloud $\mathbb{C}_{\text{mp}}$ in the body-in-the-self are not two individual entities, they are just different aspects of one entity. Thereby the originally introduced notion of space-time clouds refers to both of them simultaneously. In this sense, the cloud $\mathbb{C}_{\text{mp}}$ should be categorized as an entity belonging to the self. Because the cloud $\mathbb{C}_{\text{mp}}$ represents the corresponding point of the space-time continuum $\mathbb{R}_{[4]} \times t$ as the mental and cognitive image of the real space-time continuum, $\mathbb{R}_{[4]} \times t$ is also an entity belonging to the self.

In the mathematical description of the mental operations with the clouds $\{\mathbb{C}_{\text{mp}}\}$ we will use their representations as some functions $\{\Phi(t, x, v, a, j)\}$. We suppose that the body-in-the-self operates with internal representation of another form. Nevertheless, the long-term adaptation to reality should make the two types of representations equivalent, at least, approximately; we discussed this issue and related ones in Sect. 3.2.4.

So, posing questions about describing the proximity of two states of moving object, their individual and cumulative efficiency, etc. within a common formalism, we inevitably turn to operators acting in the space-time continuum $\mathbb{R}_{[4]} \times t$. Keeping mind the Liouville equation as the prototype of the plausible equation governing the action strategy space-time cloud, we may confine our consideration to quasi-linear operators. Put differently, treating the concept of space-time cloud as the cornerstone in the desired description of human actions, we have to accept that:

- All the properties attributed to the motion state of controlled object as well as the properties characterizing the subject's evaluation of the object motion state are linear operators. For example, a numerical value F_{21} is attributed to an operator $\widehat{\mathcal{F}}$ via it mapping one cloud $\mathbb{C}_{\text{mp}_1}$ on another cloud $\mathbb{C}_{\text{mp}_2}$[42]:

$$F_{21} = \left\langle\!\left\langle \mathbb{C}_{\text{mp}_2} \right| \widehat{\mathcal{F}} \left| \mathbb{C}_{\text{mp}_1} \right\rangle\!\right\rangle \stackrel{\text{def}}{=} \int\limits_{\mathbb{R}_{[4]}\times t} d\mu_{[4]} dt \, \Phi_2^{\dagger}(t, x, v, a, j) \widehat{\mathcal{F}} \, \Phi_1(t, x, v, a, j),$$

  where the double angle brackets means integration over space and time, the symbol $d\mu_{[4]}$ stands for the elementary volume of the space $\mathbb{R}_{[4]}$, i.e., $d\mu_{[4]} = dx\, dv\, da\, dj$.

[42] In mathematics such systems are categorized as the L^2-functional spaces and also called the Hilbert spaces. The formalism of Hilbert spaces underlies quantum mechanics and physics of waves. In the case under consideration, it emphasizes the bilinear measure in calculating cloud properties to imply the use of the corresponding operators in the space-time continuum $\mathbb{R}_{[4]} \times t$.

The symbol of complex conjugate "†" has been added to reserve the possibility of using complex-value components of the functions $\{\Phi(t, x, v, a, j)\}$.

– Because any cloud $\mathbb{C}_{\rm mp}$ coincides with itself, the normalization condition

$$\langle\!\langle \mathbb{C}_{\rm mp} | \, | \mathbb{C}_{\rm mp} \rangle\!\rangle = 1 \tag{6.131}$$

must hold.

– The description of cloud dynamics may contain nonlinear operators reflecting the interaction between different clouds or, even, different parts of one cloud for clouds with complex structure.

As far as the space-time cloud $\mathbb{C}_{\rm AS}$ of action strategy is concerned, in the discussed aspects its properties are quite similar to that of the cloud $\mathbb{C}_{\rm mp}$ (for details see Sect. 3.3.4 devoted to the general construction of $\mathbb{C}_{\rm AS}$).

The action strategy space-time cloud $\mathbb{C}_{\rm AS}$ emerges via the fusion of many $\mathbb{C}_{\rm mp}$-clouds related to the multitude of action strategy implementations in the physical space. In the case under consideration this multitude it is the collection $\{P^H_{\rm AS}\}$ of trajectories representing the ensemble of Hamiltonian systems. Therefore the space-time cloud $\mathbb{C}_{\rm AS}$ is not elementary entity, it contains the forward ($\mathbb{C}^{>}_{\rm AS}$) and backward ($\mathbb{C}^{<}_{\rm AS}$) temporal parts belonging to the connected future and the connected past, respectively. The mind has individual access to the two parts of $\mathbb{C}_{\rm AS}$, so the space-time cloud $\mathbb{C}_{\rm AS}$ of action strategy must be categorized as an entity belonging the level of AM-consciousness.

In anticipating the results of current actions in the connected future the subject has to deal with the space-time cloud $\mathbb{C}^{>}_{\rm AS}$ as a whole indivisible entity. It is caused by that the subject is not able to predict the expected motion of controlled object within accuracy required for differentiating individual trajectories in the collection $\{P^H_{\rm AS}\}$. It should be emphasized that the subject can discriminate between two $\mathbb{C}_{\rm mp}$-clouds separated by spatial or temporal distance exceeding the uncertainty of his perception. The subject cannot attribute them to one trajectory $P^H_{\rm AS}$ and, for this reason, fuses them into one space-time cloud $\mathbb{C}^{>}_{\rm AS}$ having access to its different parts treated individually without understanding their dynamical connectivity.

As result, e.g., for describing how the subject evaluates the current action strategy based on the anticipation of the object further motion, we have to employ expressions similar to

$$\langle\!\langle \mathbb{C}^{>}_{\rm AS} | \, \widehat{\mathcal{P}}_s \, | \mathbb{C}^{>}_{\rm AS} \rangle\!\rangle \stackrel{\rm def}{=} \int\limits_{\mathbb{R}_{[4]}\times t} d\mu_{[4]} dt \, \Phi^{\dagger}_{\rm AS}(t, x, v, a, \Delta j) \widehat{\mathcal{P}}_s \, \Phi_{\rm AS}(t, x, v, a, \Delta j). \tag{6.132}$$

Here the function $\Phi_{\rm AS}(t, x, v, a, \Delta j)$ represents the forward temporal part $\mathbb{C}^{>}_{\rm AS}$ of the action strategy cloud $\mathbb{C}_{\rm AS}$ in the space-time continuum $\mathbb{R}_{[4]} \times t$, the operator $\widehat{\mathcal{P}}_s$ describes the proximity of action strategies to the stable manifold Ξ_s, and the superscript ">" at the integral symbol means the integration over the connected future of the complex present. The normalization of the space-time cloud $\mathbb{C}^{>}_{\rm AS}$

$$\langle\!\langle \mathbb{C}^{>}_{\text{AS}} | | \mathbb{C}^{>}_{\text{AS}} \rangle\!\rangle = \mathfrak{N}(T_{\text{AS}}) \sim \frac{T_{\text{AS}}}{\tau_d} \tag{6.133}$$

may depend only on the duration T_{AS} of the forward temporal part of the action strategy cloud $\mathbb{C}_{\text{AS}}$, however, as clarified below, the $\mathbb{C}^{>}_{\text{AS}}$-cloud normalization can be more complex in structure.

As we have seen, the Hamiltonian description of action strategy implementations and their cloud-type description differ basically in the mathematical background, which confronts us with the challenge that:

- On the one hand, we regard the multitude $\{P^{H}_{\text{AS}}\}$ of trajectories—representing action strategy $\mathcal{AS}$ within the Hamiltonian description—as the prototype of the space time cloud $\mathbb{C}_{\text{AS}}$. Thereby we intend to construct the equation governing for the $\mathbb{C}_{\text{AS}}$-cloud properties employing the corresponding Liouville equation. Naturally, this Liouville equation has to be averaged on scales reflecting the uncertainty in the subject's perception of object motion state, which can be done directly as shown in Sect. 6.2.6.
- On the other hand, the Hamiltonian description of subject' actions and the desired description of space-time clouds of action strategies belong to *different types* of mathematical formalism. The former one deals with point-like objects of Newtonian mechanics, whereas for the latter one the notion of clouds is one the main cornerstones. In particular, their difference is reflected in the normalization conditions (6.130), (6.131) being entirely distinct from each other in structure. Therefore the ordinary averaging of the Liouville equation cannot bridge the gap between the two types of formalism.[43]

Strong nonlocality of cloud dynamics and the self's ownership: The *principle of pre-reflective ownership of mental images* attributed to the self (Sect. 3.2.3) opens a gate toward overcoming the difference in the used two types of mathematical formalism noted above. The final step of overcoming this difference will be made in the next subsection (Sect. 6.2.8), below in this subsection we will focus on the basic premises underlying the introduction of the space (the Hilbert space) required for describing action strategies space-time clouds.

The concept of the self's ownership of the mental images implies that in comparing, e.g., two space-time clouds $\mathbb{C}_1$ and $\mathbb{C}_2$, the subject can try fitting them into each other to verify whether both the clouds are of the same form. In this case, their amplitudes, e.g., the maxima of the corresponding functions $\Phi_1(t, x, v, a, t)$

[43] It should pointed out that the cloud-type formalism known in mathematics as L^2-functional spaces or the Hilbert spaces underlies quantum mechanics and physics of wave. Speaking about quantum mechanics, the uncertainty in the spatial position of a quantum particle and its velocity is an intrinsic property of quantum objects. Ostrogradsky's formalism can be used for constructing the quantum description of systems with higher-order time-derivatives (e.g., Woodard 2007, 2009, 2015a, b; Smilga 2017; Raidal and Veermäe 2017; Ganz and Noui 2020; Aoki and Motohashi 2020; Motohashi and Suyama 2020). So, for the general reasons, the corresponding quantum version of the Liouville equitation may be attractive as the starting point in deriving the governing equation for human goal-oriented behavior, at least, within the complex present. The discussion of this approach is postponed to the end of the next subsection as an open problem.

and $\Phi_2(t, x, v, a, t)$ may be of minor importance. In mathematical terms this degree of freedom is reduced to the possibility of introducing additional multipliers $\{Z_i\}$ $(i = 1, 2)$ in the cloud density

$$\Phi_i(t, x, v, a, t) \Rightarrow Z_i \cdot \Phi_i(t, x, v, a, t) \tag{6.134}$$

such that the normalization condition (6.131) or (6.131) hold.[44] When for some reasons the clouds change their form, e.g., become wider, the multipliers $\{Z_i\}$ also vary to keep up the normalization condition.

The self's ownership of the space-time continuum $\mathbb{R}_{[4]} \times t$ as the mental image of the real space-time continuum endows the subject's evaluation of object motion state with an additional degree of freedom. Dealing with a space-time cloud $\mathbb{C} = \mathbb{C}_1 \bigcup \mathbb{C}_2$ of complex structure, e.g., containing two parts $\mathbb{C}_1$ and $\mathbb{C}_2$ the subject can change voluntarily their relative arrangement, form, and amplitudes smoothly or step-wise (on the scales of the immediate now). This change does not depend on the spatial and temporal distance between the clouds $\mathbb{C}_1$ and $\mathbb{C}_2$ in the human temporality, we already noted this degree of freedom in Sect. 6.2.6.

Put differently, the human temporality equipped with the space-time continuum $\mathbb{R}_{[4]} \times t$ admits "instant" interaction of its elements for any spatial and temporal distance between them. We will call this feature the *strong nonlocality of the human temporality*.[45] It should be emphasized that the strong nonlocality of the human temporality is a property belonging to the realm of the mind-in-the-self. So its specific implementation can depend on the subject's intentions and particular conditions characterizing a mental phenomenon in issue.

Bilinear normalization of action strategy clouds: The space-time cloud $\mathbb{C}_{\text{mp}}$ of material point is indivisible entity of the human temporality. Thereby such clouds can be compared only via their direct overlapping as whole entities and equality (6.131) is the only one way to introduce the normalization of the $\mathbb{C}_{\text{mp}}$-clouds. The situation changes when we deal with the action strategies clouds $\mathbb{C}_{\text{AS}}$, in particular, their

---

[44] We already discussed this feature in Sect. 5.4.1 concerning the cloud-type structure of effort mental measure.

[45] It is worthy of noting that according to the paradigm of modern physics, at the very basic level the interaction between its entities (elementary particles and quantum fields) is local. In particular, Einstein's theory of special relativity asserts that two events cannot be in a causal relation if light cannot pass the distance between them for the required time interval. During the last century there were a number of attempts to attack the locality principle or modify its understanding, the modern conception of quantum nonlocality is one of them (see, e.g., Grib and Rodrigues (Jr.) 1999; Popescu 2014; Tipler 2014; Cramer 2016, for review and discussion of related phenomena), which up to now is a matter of ongoing debates. The concept of local interaction may be referred to as a reason why the majority of models developed for micro- and meso-level phenomena turn to partial differential equations or integral differential equations with kernels decreasing as spatial distance increases. In particular, the Liouville equation as well as its modification constructed in Sect. 6.2.6 to allow for uncertainty in human perception belongs to the class of partial differential equations. In our case, for the human temporality spatial and temporal distance does not matter at all, at least, on the scales of the complex present.

forward temporal parts $\mathbb{C}^{>}_{\mathrm{AS}}$. Because the mind has access to different fragments of $\mathbb{C}^{>}_{\mathrm{AS}}$, the subject can compare them in various manners, which should be reflected in the mathematical description of their comparison, in particular, their normalization.

To clarify the possible approaches to describing the $\mathbb{C}^{>}_{\mathrm{AS}}$-cloud comparison, we turn to the detailed representation of the space-time cloud $\mathbb{C}_{\mathrm{AS}}$ of action strategy $\mathcal{AS}$ as a function $\Phi_{\mathrm{AS}}(t, x, v, a, \Delta j)$ localized in a fuzzy region—a region with blurred boundaries—of the space-time continuum

$$\mathbb{R}_{[4]} \times t = \big\{\{t, x, v, a, \Delta j\}\big\}.$$

The boundary blurring of this region is described by the gradual decrease of the function $\Phi_{\mathrm{AS}}(t, x, v, a, \Delta j)$.

In these terms, the common feature of the $\mathbb{C}^{>}_{\mathrm{AS}}$-cloud comparison, i.e., the relative proximity of two clouds $\mathbb{C}^{>}_{\mathrm{AS}_1}$ and $\mathbb{C}^{>}_{\mathrm{AS}_2}$ is the bi-linearity of the corresponding measure that generally can be written in the symbolic form and the coordinate-wise form as

$$\begin{aligned}\big\langle\!\!\big\langle \mathbb{C}^{>}_{\mathrm{AS}_1} \big| \widehat{\mathcal{B}} \big| \mathbb{C}^{>}_{\mathrm{AS}_2} \big\rangle\!\!\big\rangle &= \iint\limits_{\mathbb{R}_{[4]}\times t}^{>} dt\, dt'\, dQ\, dQ'\, d\Delta j\, d\Delta j' \\ &\times \Phi^{\dagger}_{\mathrm{AS}_1}(t', Q', \Delta j') \cdot \big\langle t', Q', \Delta j' \big| \widehat{\mathcal{B}} \big| t, Q, \Delta j \big\rangle \cdot \Phi_{\mathrm{AS}_2}(t, Q, \Delta j)\,, \end{aligned} \tag{6.135}$$

where, as before, $Q = \{x, v, a\}$ are the inner variables of object motion, and

$$\big\langle t', Q', \Delta j' \big| \widehat{\mathcal{B}} \big| t, Q, \Delta j \big\rangle$$

is the kernel of the comparison operator.

On the one hand, the subject recognizes the temporal order of $\mathbb{C}_{\mathrm{mp}}$-clouds separated by temporal distances exceeding or being about the duration τ_d of experiential now. On the other hand, the characteristic duration of the forward temporal part $\mathbb{C}^{>}_{\mathrm{AS}}$ is about the duration τ_{cp} of the connected future in the complex present. Therefore, accepting that $\tau_d \ll \tau_{cp}$, we may suppose that the subject can conceive of the cloud $\mathbb{C}^{>}_{\mathrm{AS}}$ as a temporally ordered set of layers of width about τ_d and analyze their properties individually. It implies that in comparing two $\mathbb{C}^{>}_{\mathrm{AS}}$-clouds the subject does this layer-by-layer. Mathematically we describe this type comparison with the kernel

$$\big\langle t', Q', \Delta j' \big| \widehat{\mathcal{B}} \big| t, Q, \Delta j \big\rangle \propto \frac{1}{\tau_d}\delta(t' - t)\,,$$

where $\delta(t' - t)$ is the Dirac δ-function and the coefficient $1/\tau_d$ effectively takes into account its blurring on scales of the experiential now.

Deviation from the optimal behavior, i.e., from the stable manifold Ξ_s is of prime importance in controlling the object motion, so the subject has to compare the $\mathbb{C}^{>}_{\mathrm{AS}}$-clouds in the dimension Δj as strictly as possible. We take into account this feature actually in the same way, namely, the kernel of comparison operator is approximated as

$$\left\langle t', Q', \Delta j' \middle| \widehat{\mathcal{B}} \middle| t, Q, \Delta j \right\rangle = \frac{1}{\tau_d \delta_j} \delta(t'-t)\delta(\Delta j' - \Delta j) \left\langle Q' \middle| \widehat{\mathcal{B}} \middle| Q \right\rangle . \quad (6.136)$$

Concerning the inner variables Q of the object motion state, there are various situations where the comparison kernel takes different forms. Keeping in mind the strong nonlocality of the human temporality we will consider two limit cases. First, it is the *complete comparison* of the $\mathbb{C}^{>}_{\text{AS}}$-clouds (*active comparison*), meaning the subject tries to compare the $\mathbb{C}^{>}_{\text{AS}}$-clouds as strictly as possible, which meets the kernel

$$\left\langle t', Q', \Delta j' \middle| \widehat{\mathcal{B}}_{\text{ac}} \middle| t, Q, \Delta j \right\rangle = \frac{1}{\tau_d \delta_j \delta_x \delta_v \delta_a} \delta(t'-t)\delta(\Delta j' - \Delta j)\delta(Q' - Q) . \quad (6.137)$$

In this case, the comparison measure (6.135) takes the standard form

$$\begin{aligned} \left\langle\!\left\langle \mathbb{C}^{>}_{\text{AS}_1} \middle| \widehat{\mathcal{B}}_{\text{ac}} \middle| \mathbb{C}^{>}_{\text{AS}_2} \right\rangle\!\right\rangle &= \left\langle\!\left\langle \mathbb{C}^{>}_{\text{AS}_1} \middle| \mathbb{C}^{>}_{\text{AS}_2} \right\rangle\!\right\rangle \\ &= \frac{1}{\tau_d \delta_j \delta_x \delta_v \delta_a} \int\limits_{\mathbb{R}_{[4]} \times t}^{>} dt\, dQ\, d\Delta j\, \Phi^{\dagger}_{\text{AS}_1}(t, Q, \Delta j) \cdot \Phi_{\text{AS}_2}(t, Q, \Delta j) . \end{aligned} \quad (6.138)$$

For the complete comparisons of action strategies via their forward temporal parts to be realized, the subject has to consciously focus on the current state of object motion and its prediction in the connected future. So we suppose that the limit of completer comparison of the $\mathbb{C}^{>}_{\text{AS}}$-clouds is the case when the subject recognizes that the current action strategy should be changed for another one. The corresponding normalization to be imposed on the space-time cloud $\mathbb{C}^{>}_{\text{AS}}$ is given by Exp. (6.133), namely,

$$\left\langle\!\left\langle \mathbb{C}^{>}_{\text{AS}} \middle| \mathbb{C}^{>}_{\text{AS}} \right\rangle\!\right\rangle \propto \int\limits_{\mathbb{R}_{[4]} \times t}^{>} dt\, dQ\, d\Delta j\, \Phi^{\dagger}_{\text{AS}}(t, Q, \Delta j) \cdot \Phi_{\text{AS}}(t, Q, \Delta j) \propto T_{\text{AS}} , \quad (6.139)$$

where the appropriate proportionality coefficient is will determined below. We have imposed on this normalization the condition of proportionality to the mean duration T_{AS} of action strategy implementation to enable its additivity in analyzing the action strategy with different temporal horizons.

Second, in the opposite limit, to be called the *automatic comparison* of action strategies (*passive comparison*), the subject does not pay attention to the details of object motion state Q at all because

- the motion state is assumed to belong to the stable manifold Ξ_s and
- the only that is necessary is to monitor the jerk difference Δj as a quantitative measure of the deviation from the stable manifold Ξ_s.

This situation is described with the comparison kernel

$$\left\langle t', Q', \Delta j' \middle| \widehat{\mathcal{B}}_{\text{ps}} \middle| t, Q, \Delta j \right\rangle = \frac{1}{\tau_d \delta_j} \delta(t'-t)\delta(\Delta j' - \Delta j) \quad (6.140)$$

and corresponds to the cloud comparison measure (6.135) specified by the expression

$$\langle\!\langle \mathbb{C}^{>}_{\mathrm{AS}_1} | \widehat{\mathcal{B}}_{\mathrm{ps}} | \mathbb{C}^{>}_{\mathrm{AS}_2} \rangle\!\rangle = \int\limits_{\Delta j \times t}^{>} dt\, dQ\, d\Delta j \\ \times \left[\int\limits_{\mathbb{R}_L} dQ\, \Phi^{\dagger}_{\mathrm{AS}_1}(t, Q, \Delta j) \right] \cdot \left[\int\limits_{\mathbb{R}_L} dQ\, \Phi_{\mathrm{AS}_2}(t, Q, \Delta j) \right]. \tag{6.141}$$

Mathematically, the mode of passive comparison can be reduced to the mode of active comparison within shortened phase space via transformations

$$\int_{\mathbb{R}_L} dQ\, \Phi_{\mathrm{AS}}(t, Q, \Delta j) \;\to\; \Phi_{\mathrm{AS}}(t, \Delta j), \tag{6.142a}$$

$$\mathbb{R}_{[4]} \times t \;\to\; \mathbb{R}_{[1]} \times t, \tag{6.142b}$$

$$\widehat{\mathcal{B}}^{[\mathbb{R}_{[4]} \times t]}_{\mathrm{pc}} \;\to\; \widehat{\mathcal{B}}^{[\mathbb{R}_{[1]} \times t]}_{\mathrm{ac}}, \tag{6.142c}$$

where $\mathbb{R}_{[1]} = \{\Delta j\}$. Given these transformations implemented, the normalization of the action strategy cloud $\mathbb{C}_{\mathrm{AS}}$ is again specified by Exp. (6.139) but within the shortened phase space $\mathbb{R}_{[1]} \times t$.

Transformations (6.142) can be employed in deriving the governing equation for the passive comparison from the governing equation for the active comparison via its integration over the Laplacian space $\mathbb{R}_L$.

It should be noted that the transformation (6.142a) on its own does not belong the formalism of the Hilbert spaces. In our case, the integral transformation (6.142a) is not just some operation with an entity of a given instantiation of the Hilbert space, it is accompanied by the other transformations (6.142b) and (6.142c) changing the basic metric of this Hilbert space. In other words, integrating the corresponding governing equation, we *switch to another instantiation of the Hilbert space.*

6.2.8 Action Strategy Space-Time Cloud: Governing Equation

This subsection is devoted to deriving the governing equation for the forward temporal part $\mathbb{C}^{>}_{\mathrm{AS}}$ of a given action strategy space-time cloud $\mathbb{C}_{\mathrm{AS}}$, including its initial state $\mathbb{C}^{\mathrm{ini}}_{\mathrm{AS}}$, based on:

- the modified Liouville equation allowing for the effects of the subject's perception uncertainty, Exp. (6.127),
- the general invariance of space-time clouds representing
  - the perceived states of object motion,
  - subject's individual actions with these states,

  – action strategy implementations

  with respect to rescaling (6.134) within the cloud mental evaluation and comparison,

- the accepted bilinear form (6.135) with its particular limits (6.135) and (6.141) describing the subject's comparison of space-time clouds.

The cases of active and passive comparison of the $\mathbb{C}^{>}_{\text{AS}}$-clouds will be analyzed within a common description, which becomes possible due to transformations (6.142).

The gist of our approach to deriving the governing equation for the space-time structure of the $\mathbb{C}^{>}_{\text{AS}}$-clouds is to combing together two sources of the variations in the $\mathbb{C}^{>}_{\text{AS}}$-cloud within the human temporarily. One of them is the dynamics of the Hamiltonian systems forming the multitude $\{P^{H}_{\text{AS}}\}$ of trajectories in the Lagrange space $\mathbb{R}_L = \{Q\}$ combined with the space $\mathbb{R}_{[1]} = \{\Delta j\}$ and described by Eq. (6.127). The other is the cloud rescaling (6.134) such that the cumulative variations of the space-time cloud $\mathbb{C}^{>}_{\text{AS}}$ do not change its normalization during "motion" in the human temporality.

Beforehand, to avoid possible misunderstanding, we want to note that the continuous variable t represents the temporal extent of the human temporality rather than physical time. How to reconcile the use of continuous variable t and the fact that the human temporality does not contain point-like instants of time as basic elements will be clarified below.

Let us write the equation governing for the cloud density Φ_{AS} as

$$\frac{\partial \Phi_{\text{AS}}}{\partial t} = \widehat{\mathcal{H}}\Phi_{\text{AS}} - E\Phi_{\text{AS}}\,, \tag{6.143}$$

where, in the case of active comparison of action strategy space-time clouds, the cloud density $\Phi_{\text{AS}}(t, Q\Delta j)$ in the list of its argument contains t, Q, and Δj, the operator $\widehat{\mathcal{H}}$ is given by Exp. (6.129), and the quantity E is independent of the variables Q and Δj. In the case of passive comparison, it is $\Phi_{\text{AS}}(t, \Delta j)$, the operator $\widehat{\mathcal{H}}$ is integrated over Q, and the quantity E is assume to be independent of Δj.

The condition

$$\langle \Phi_{\text{AS}}|\,|\Phi_{\text{AS}}\rangle = 1\,, \tag{6.144}$$

where the bilinear norm $\langle \cdot|\,|\cdot\rangle$ is understood appropriately, is imposed on the solution of Eq. (6.143). It should be noted that norm (6.144) contains only the spatial component of the operator (6.135) specifying the relative proximity of action strategy space-time clouds. Namely, in the case of active comparison of the $\mathbb{C}^{>}_{\text{AS}}$-clouds

$$\big\langle \Phi_{\text{AS}}\big|\Phi_{\text{AS}}\big\rangle \overset{\text{def}}{=} \int\limits_{\mathbb{R}_{[4]}}^{>} dQ\, d\Delta j\; \Phi^{\dagger}_{\text{AS}}(t, Q, \Delta j) \cdot \Phi_{\text{AS}}(t, Q, \Delta j) \tag{6.145a}$$

and in the case of passive comparison

$$\left\langle \Phi_{\rm AS} \middle| \Phi_{\rm AS} \right\rangle \overset{\rm def}{=} \int\limits_{\mathbb{R}_{[1]}}^{>} d\Delta j\, \Phi^{\dagger}_{\rm AS}(t, \Delta j) \cdot \Phi_{\rm AS}(t, \Delta j). \tag{6.145b}$$

In these expressions we omitted cofactors similar to one entering Exp. (6.138) or, what is the same, aggregated them into the function $\Phi_{\rm AS}$.

Equation (6.143) and normalization (6.144) immediate lead us to the expression specifying the value of E,

$$E = \langle \Phi_{\rm AS} | \widehat{\mathcal{H}}_s | \Phi_{\rm AS} \rangle \,, \tag{6.146}$$

where $\widehat{\mathcal{H}}_s$ is the Hermitian component of the operator $\widehat{\mathcal{H}}$, i.e.,

$$\widehat{\mathcal{H}}_s = \frac{1}{2}\big(\widehat{\mathcal{H}}^{\dagger} + \widehat{\mathcal{H}}\big)\,. \tag{6.147}$$

Below, for completeness, we present the particular expressions for the operator $\widehat{\mathcal{H}}$ and its Hermitian component $\widehat{\mathcal{H}}_s$. In the case of the $\mathbb{C}^{>}_{\rm AS}$-cloud active comparison, they are

$$\begin{aligned}
\widehat{\mathcal{H}} &= \frac{\partial}{\partial \Delta j}\left[\frac{\delta_j^2}{\tau_d}\frac{\partial}{\partial \Delta j} - \frac{\lambda_u}{\tau_d \delta_j}\Delta j\right] \\
&\quad - \frac{\partial}{\partial a}\left[j_s\left(x + \delta_x^2\frac{\partial}{\partial x},\ v + \delta_v^2\frac{\partial}{\partial v},\ a + \delta_a^2\frac{\partial}{\partial a}\right) + \Delta j\right] \\
&\quad - \frac{\partial}{\partial v}\left[a + \delta_a^2\frac{\partial}{\partial a}\right] - \frac{\partial}{\partial x}\left[v + \delta_v^2\frac{\partial}{\partial v}\right], \qquad\qquad (6.148a) \\
\widehat{\mathcal{H}}_s &= \frac{\delta_j^2}{\tau_d}\frac{\partial^2}{\partial \Delta j^2} - \frac{\lambda_u}{2\tau_d\delta_j} \\
&\quad - j_s\left(\delta_x^2\frac{\partial^2}{\partial a \partial x},\ \delta_v^2\frac{\partial^2}{\partial a \partial v},\ \frac{1}{2} + \delta_a^2\frac{\partial^2}{\partial a^2}\right) - \delta_a^2\frac{\partial^2}{\partial v \partial a} - \delta_v^2\frac{\partial^2}{\partial x \partial v}\,.
\end{aligned}$$

In the case of the $\mathbb{C}^{>}_{\rm AS}$-cloud passive comparison, they are

$$\widehat{\mathcal{H}} = \frac{\partial}{\partial \Delta j}\left[\frac{\delta_j^2}{\tau_d}\frac{\partial}{\partial \Delta j} - \frac{\lambda_u}{\tau_d \delta_j}\Delta j\right], \qquad \widehat{\mathcal{H}}_s = \frac{\delta_j^2}{\tau_d}\frac{\partial^2}{\partial \Delta j^2} - \frac{\lambda_u}{2\tau_d\delta_j}\,. \tag{6.148b}$$

Governing equation for the action strategy space-time clouds: The obtained results enable us to write the desired governing equation for the initial state $\mathbb{C}^{\rm ini}_{\rm AS}$ of the space-time cloud $\mathbb{C}_{\rm AS}$ in the form

$$\frac{\partial \Phi_{\rm AS}}{\partial t} - \widehat{\mathcal{H}}\,\Phi_{\rm AS} + \langle \Phi_{\rm AS} | \widehat{\mathcal{H}}_s | \Phi_{\rm AS} \rangle\, \Phi_{\rm AS} = 0\,. \tag{6.149}$$

At the moment of initiation $t_{\rm ini}$, the function $\Phi_{\rm AS}$ formally meets the condition

$$\Phi_{\mathrm{AS}}\Big|_{t=t_{\mathrm{ini}}} = \Phi_{\mathrm{AS}}^{\mathrm{ini}}\,, \tag{6.150}$$

where, in the case of the $\mathbb{C}_{\mathrm{AS}}^{>}$-cloud active comparison, the function $\Phi_{\mathrm{AS}}^{\mathrm{ini}}$ represents the cloud of action strategy initiation $\mathbb{C}_{\mathrm{ini}[4]}$, Exp. (6.67), i.e., the fuzzy neighborhood of the start point $\{x_s, v_s, a_s\}$ and the jerk $j_s(x_s, v_s, a_s)$ belonging to the stable manifold Ξ_s,

$$\mathbb{C}_{\mathrm{ini}[4]} = \frac{(x - x_s)^2}{\delta_x^2} + \frac{(v - v_s)^2}{\delta_v^2} + \frac{(a - a_s)^2}{\delta_a^2} + \frac{(j - j_s)^2}{\delta_j^2} \lesssim 1\,.$$

In the case of the $\mathbb{C}_{\mathrm{AS}}^{>}$-cloud passive comparison, the function $\Phi_{\mathrm{AS}}^{\mathrm{ini}}$ represents this region projected onto the Δj-axes, i.e.,

$$\mathbb{C}_{\mathrm{ini}[1]} = \frac{(j - j_s)^2}{\delta_j^2} \lesssim 1\,.$$

At a certain instant T_{AS} of time the subject terminates the action strategy implementation.

The governing equation (6.149) describes the *distribution* of the action strategy cloud $\mathbb{C}_{\mathrm{AS}}$ in the *human temporality* within the interval $[t_{\mathrm{ini}}, T_{\mathrm{AS}}]$ whose boundaries are blurred on scales about τ_d, i.e., the characteristic duration of experiential now. The forward temporal part $\mathbb{C}_{\mathrm{AS}}^{>}$ of the space-time cloud $\mathbb{C}_{\mathrm{AS}}$ is described by the function Φ_{AS} for $t > t_{\mathrm{cur}}$, where t_{cur} is current instant of time again blurred on scales of τ_d.

The normalization of the space-time cloud $\mathbb{C}_{\mathrm{AS}}$ specified by Exp. (6.139) for the $\mathbb{C}_{\mathrm{AS}}^{>}$-cloud active comparison is of the desired form,

$$\langle\!\langle \mathbb{C}_{\mathrm{AS}}^{>} | \mathbb{C}_{\mathrm{AS}}^{>} \rangle\!\rangle = \frac{1}{\tau_d \delta_j \delta_x \delta_v \delta_a} \int_{\mathbb{R}_{[4]}\times t}^{>} dt\, \langle \Phi_{\mathrm{AS}} | \Phi_{\mathrm{AS}} \rangle = \frac{(T_{\mathrm{AS}} - t_{\mathrm{cur}})}{\tau_d \delta_j \delta_x \delta_v \delta_a}\,, \tag{6.151}$$

for the $\mathbb{C}_{\mathrm{AS}}^{>}$-cloud passive comparison the multiplier $\delta_x \delta_v \delta_a$ in this given expression should be omitted.

We want to emphasize several aspects related to the obtained governing equation (6.149).

- On the one hand, there are no point-like elements such as spatial points or point-like instants of time in the human temporality. On the other hand, first, the governing equation (6.149) deals with the cloud density $\Phi_{\mathrm{AS}}(t, Q, \Delta j)$ (or $\Phi_{\mathrm{AS}}(t, \Delta j)$) as well as the operators $\widehat{\mathcal{H}}$, $\partial/\partial t$ whose definition is based on the notion of spatial points and time instants. Second, the initial and terminal conditions as well as the current time are also are directly related to the notion of point-like instants of time. To reconcile these features we want to note the following.
  - The present description of action-strategy cloud $\mathbb{C}_{\mathrm{AS}}$ is the mathematical representation of $\mathbb{C}_{\mathrm{AS}}$, it is a particular form of the mathematical eidos of the given

action strategy space time cloud (Sects. 3.3.3, 3.3.4). This type eidoi enable the mind to infer the characteristic properties of action strategy implementations based on logical manipulations.

- In realm of the body-in-the-self the forward temporal part $\mathbb{C}^{>}_{\text{AS}}$ of the action strategy $\mathcal{AS}$, in particular, its initial state $\mathbb{C}^{\text{ini}}_{\text{AS}}$, is encoded as some neural patterns. The adequacy of the relationship between the mathematical eidoi and the corresponding neural patterns is a result of long-term learning and skill acquisition. Actually, we discussed this issue in the context of reconciling the phenomenology and the cognition embodiment (Sect. 3.2.4). In Sect. 5.6.3 we also put forward the principle of cloud-fusion self-consistency (Sect. 5.6.3) emphasizing this adequacy explicitly. The given principle posits the self-consistency in the properties of (*i*) E-conscious fragments of the human temporality related to the experiential now and (*ii*) AM-conscious fragments related to the acquired skill.
- The neural pattern encoding the function $\Phi_{\text{AS}}(t, Q, \Delta j)$ or $\Phi_{\text{AS}}(t, \Delta j)$ in the body-in-the-self inevitably represents it only approximately. It stems from that no neural pattern containing many but finite number of elements can represent a continuous object. So when the mind-in-the-self gets access to this pattern and converts it into a mental image being a multitude of the material point space-time clouds $\mathbb{C}_{\text{mp}}$. So in the mind-in-the-self the mental image of the action strategy $\mathcal{AS}^{>}$ in the connected future can be treated as the union

$$\mathcal{AS}^{>} = \bigcup\nolimits_{\text{mp}} |\mathbb{C}_{\text{mp}}\rangle\langle\langle\mathbb{C}_{\text{mp}}|\mathbb{C}^{>}_{\text{AS}}\rangle\rangle. \tag{6.152}$$

  First, this representation describes the smoothing of spatial and temporal points caused by the human perception uncertainty. Second, it explains the cloud-type nature of the properties attributed to mental images in the mind, e.g., the cloud-type effort attributed to atomic action (Sect. 5.4.1). In estimating a certain property, the mind-in-the-self just deals with a collection of $\mathbb{C}_{\text{mp}}$-clouds rather than one of them.

- The constructed governing equation (6.149) is nonlinear in contrast to the underlying Liouville equation. It illustrates that nonlinearity of mental operations is an essential aspect in human evaluation of goal-oriented actions.
- In Sect. 6.2.5 we have obtained relationship (6.107) between the mean value of the jerk difference $\overline{\Delta j}(t)$ and the advance $s(t)$ in implementing the corresponding action strategy the Lagrange space $\mathbb{R}_L$. It enables us to pose a question about the possibility of representing or generalizing the obtained governing equation (6.149) such that the derived equation describe the Δj-component of action strategy space-time cloud in terms of the spatial variables $Q = \{x, v, a\}$ of the Lagrange space $\mathbb{R}_L$ rather than the temporal variable t.

Finalizing our discussion of the constructed description of the space-time structure of the action strategy clouds we want to note a possible novel turn in the developed theory.

Open problem: Quantum Ostrogradsky's formalism and action strategy space-time clouds: In footnote (43) we already pointed out that Ostrogradsky's formalism can be used for constructing the quantum description of systems with higher-order time-derivatives (e.g., Woodard 2007, 2009, 2015a, b; Smilga 2017; Raidal and Veermäe 2017; Ganz and Noui 2020; Aoki and Motohashi 2020; Motohashi and Suyama 2020). So in developing the cloud-type description of action strategies, it could possible to start from the quantum Liouville equation instead of the classical Liouville equation. One hand, this approach seems promising because:

- the mathematical formalism underlying quantum mechanics is originally of the required type—it is the Hilbert spaces equipped, by definition, with the bilinear metric,
- in quantum mechanics the uncertainty in the spatial position of a quantum particle and its velocity is an intrinsic property of quantum objects.

On the other hand, as we have illustrated in this subsection, the introduction of bilinear metric for the space-time cloud comparison is rooted in the basic properties of human mind and gives rise to the dissipative "dynamics" of space time clouds.

So it could be interesting to elucidate whether it is possible to merge (*i*) the bilinear formalism of quantum Ostrogradsky's approach with (*ii*) the formalism of action strategy bilinear comparison and the strong nonlocality of the human temporality. Going in this way, one might find

- a more self-consistent description of human uncertainty in evaluating goal-oriented actions,
- new phenomena in the emergence of the action strategies space-time clouds from the multitude of their possible implementations whose description is rooted in purely quantum phenomena like the Bose-Einstein condensation,
- a new way of introducing dissipation in quantum systems, which could give rise to a new class of open quantum systems.

It should be emphasized that the possibility of employing the quantum mechanics formalism for describing related problems in the control theory was demonstrated, e.g., by Contreras et al. (2017), Contreras and Peña (2019).

6.3 2D-Time Dynamics: Time & Temporality

In this section, we develop the description of action strategy evolution occurring in two temporal dimensions simultaneously. One of them is objective time being a property of physical reality, the other is the human temporality being a property of the human mind. We consider such processes to be characterized by 2D-time dynamics.

However, before passing directly to our constructions, let us elucidate the general aspects of 2D-time.

- In the present book we deal with physical (objective) time strictly following the paradigm of Newtonian mechanics. It concerns the laws governing the dynamics of physical objects from the standpoint of the external observer. In particular, in their form these laws are local in time, i.e., they operate with quantities determined at the current point-like instant of physical time.
- In describing the human temporality we take into account its two levels: the mind-in-the-self and the body-in-the-self.
  - All the entities belonging to the mind-in-the-self, on the one hand, are related to some events that happened in the real past, have happened just now, or are anticipated in the future. On the other hand, these entities exist in the mind and affect our current actions as well as one another.
  - In the realm of the mind-in-the-self there are no point-like instants, all its elements are finite-size clouds.
  - The realm of the body-in-the-self belongs to the human temporality as well as physical time. Therefore, the dynamics of processes in the body-in-the-self is determined by their states at the current point-like instant of time.
  - In the temporality the mind-in-the-self can initiate some processes in the body-in-the-self. These processes run and are completed according to the regularities of the body-in-the-self. After their completion the mind-in-the-self gains access to the resulting information. Processes initiated by the mind-in-the-self are interdependent via intentional relations in the mind. In physical reality processes run without interruption and their interdependence is of causal type.

The motion of a physical object controlled by the subject's goal-oriented behavior is reflected in both the temporal dimensions. In physical reality, the corresponding human actions are reduced to some additional forces exerted on the object among other physical forces. These forces, however, depend not only on the object motion state; they depend also on the mental image of object motion existing in the subject's mind. In turn, the mental image of object motion is affected by time changes of the object motion state in physical reality.

Thereby, we conceive of the object dynamics as interrelated dynamics of its two temporal components, each of them being governed by their own regularities. It is the gist of 2D-time dynamics.

6.3.1 Mechanism of 2D-Time Dynamics

In Sect. 4.2 we discussed the dynamics of human temporality from the general point of view. The pivot of this discussion is that the mind deals with two temporal dimensions. One of them—the human temporality—is its inner temporal dimension with various events belonging, in particular, to the past or future and forming an intentional network (Sects. 3.4.2–3.4.4). Within the complex present these events are united into action strategies, where intentional relations are specified by the orientation at attaining a particular goal. The other temporal dimension is physical time. The human temporality is embedded into physical time as a whole entity. So as physical time goes on, the human temporality also changes with respect to past and future events as well as the corresponding intentional networks spanning the past retained in memory and the imaginary future. This change of the human temporality induced by the passage of physical time is the essence of 2D-time dynamics.

Speaking about the complex present, we discriminate between two quite distinct dynamic processes. One of them is the automatic implementation of a selected action strategy to be referred as to the *passive phase* of 2D-time dynamics. The other is the deliberate change of one action strategy for another action strategy to be referred to as the *active phase* of 2D-time dynamics. Both of them should be described within the common conception. Indeed, as physical time goes on, in *implementing a given action strategy* $\mathcal{AS}$

- the experiential now "moves" along the selected implementation of $\mathcal{AS}$, i.e., the subject directly experiences the change in the motion state of controlled object and
- the subject continuously reevaluates his actions in the connected past as well as the action anticipated in the connected future.

It should be emphasized that In this case, the intentional relations uniting subject's actions into the given action strategy are fixed. In *changing action strategies*

- the subject, responding to the current state of object motion, changes the characteristics of action strategies sharply within the experiential now,
- this subject's action gives rise to the transformation of intentional relations spanning the complex present as a whole.

The former and latter dynamic processes clarify the essence of the *passive* and *active* phases of 2D-time dynamics of action strategy implementation, respectively.

In the following subsections we will consider the basic features of the two dynamic processes separately. However, before it we need to elaborate in more detail the notion of motion trajectory space-time cloud, which is the purpose of the present subsection. Previously, in Sect. 3.3.4, we introduced the space-time clouds of motion trajectory from the general point of view, here we have to concretize this notion to use it in describing action strategy implementations.

Initiated Implementation Cloud

In the previous section we developed the formalism for describing the subject's anticipation of action strategy implementation in the connected future. This anticipation is represented by the forward temporal part $\mathbb{C}^{>}_{\text{AS}}$ of an action strategy $\mathcal{AS}$. At the initiation of the action strategy $\mathcal{AS}$, it is just the space-time cloud $\mathbb{C}^{\text{ini}}_{\text{AS}}$ of its initial state which represents the multitude of all the probable implementations of $\mathcal{AS}$.

Therefore we also call the $\mathbb{C}^{\text{ini}}_{\text{AS}}$-cloud the space-time cloud of probable implementations. In this sense, the $\mathbb{C}^{\text{ini}}_{\text{AS}}$-cloud or, speaking more generally, the forward temporal part $\mathbb{C}^{>}_{\text{AS}}$ of the action strategy $\mathcal{AS}$ describes the deterministic aspects of the action strategy implementations.

As far as the particular implementation of $\mathcal{AS}$ realized at the moment of initiation is concerned, the subject is not able to control its choice because he cannot discriminate between different points inside the cloud $\mathbb{C}_{\text{ini}[4]}$ of the action strategy initiation, Exp. (6.67). Moreover, just after the initiation, the subject even does not recognize which $\mathcal{AS}$-implementation has been initiated. So, having started the *automatic* actions, the subject can only monitor the dynamics of the controlled object in evaluating their efficiency. When the object motion deviates essentially from the desired state, the subject terminates the currently executed implementation of action strategy and initiates a new action strategy.

Therefore, to describe the initiated implementation of the action strategy $\mathcal{AS}$ as it is perceived by the subject, we need an additional object—the space-time cloud $\mathbb{C}_P$ of object motion trajectory. Here we keep in mind the trajectory[46]

$$P_{\text{AS}} = \{x(t), v(t), a(t), j(t)\} \tag{6.153}$$

of object motion in the space-time continuum $\mathbb{R}_{[4]} \times t$ representing the initiated implementation of the action strategy $\mathcal{AS}$ in physical reality.

At the current instant of physical time t_{cur}, the fragment of the trajectory P_{AS}, corresponding to the experiential now, specifies the object motion state as it is experienced by the subject right now. This experience is described by the space-time cloud $\mathbb{C}^{\text{en}}_P$—a fuzzy region in the space-time continuum $\mathbb{R}_{[4]} \times t$ centered at the point of the trajectory $P_{\text{AS}}(t_{\text{cur}})$ and whose characteristic dimensions represent the uncertainties in the subject's monitoring of the quantities $\{x, v, a, j\}$ and the perception of time t. As physical time goes on from the initiation to the termination of the action strategy $\mathcal{AS}$, the cloud $\mathbb{C}^{\text{en}}_P$ "moves" in the space-time continuum $\mathbb{R}_{[4]} \times t$ forming the space-time cloud $\mathbb{C}_P$ of the trajectory P_{AS}. We call the $\mathbb{C}_P$-cloud *the space-time cloud of the initiated implementation* of the action strategy $\mathcal{AS}$.

Let us remind that the implementation of action strategies within the complex present is supposed to be based on subject's automatic actions. Therefore, after the initiation of the action strategy $\mathcal{AS}$ up to its termination, the trajectory P_{AS} may be treated as a fixed integral entity. It is a reason for dealing with the space-time cloud $\mathbb{C}_P$ of the trajectory P_{AS} rather than only with the space-time cloud $\mathbb{C}^{\text{en}}_P$ of the experiential now. Nevertheless, the fragment $\mathbb{C}^{\text{en}}_P$ of the $\mathbb{C}_P$-cloud is special. Because of the limit capacity of working memory and the subject's inability to predict in detail the initiated action strategy implementation, only the object motion state currently experienced by the subject and described by the $\mathbb{C}^{\text{en}}_P$-cloud is accessible to the mind. So in further constructions of the 2D-time dynamics exactly the $\mathbb{C}^{\text{en}}_P$-cloud takes central position. The role of the space-time cloud $\mathbb{C}_P$ of initiated implementation

[46] We use the symbol P to denote trajectories keeping in mind the word "path."

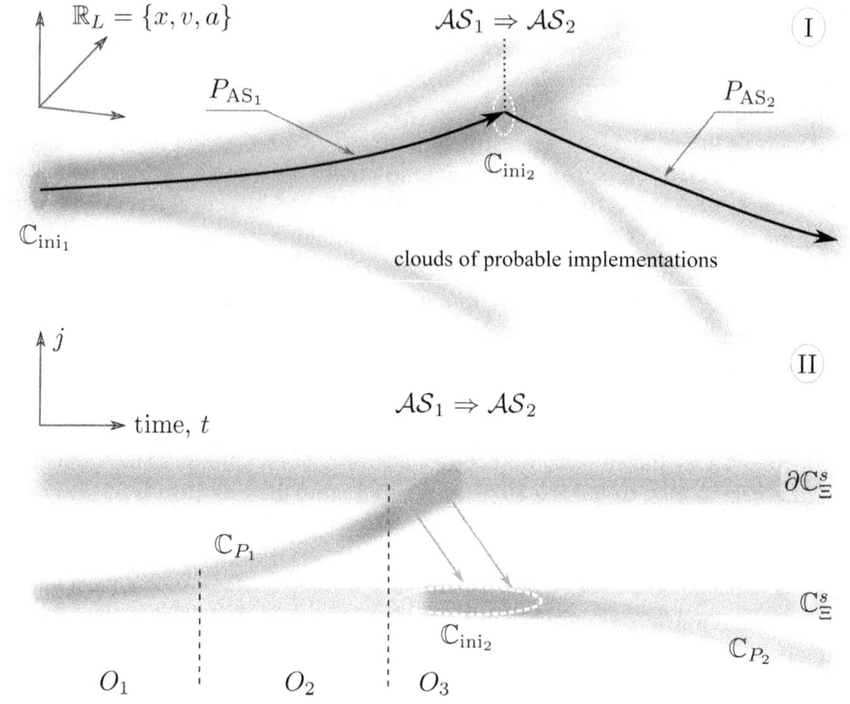

Fig. 6.2 Illustration of the space-time cloud $\mathbb{C}_P$ of initiated action strategy implementation based on the object motion trajectory P_{AS}. Fragment I depicts the motion trajectories P_{AS_1} and P_{AS_2} for two successive action strategies in the Lagrange space $\mathbb{R}_L$. The symbols $\mathbb{C}_{\text{ini}_{1,2}}$ denote the space-time clouds of action strategies $\mathcal{AS}_{1,2}$ initiation. The color stripes represent the space-time clouds $\mathbb{C}_{P_{1,2}}$ of initiated implementations in the connected past (blue) and the connected future (green). The experiential now (orange) corresponds to the $\mathcal{AS}_1 \Rightarrow \mathbb{AS}_2$-transition. The gray lines are the boundaries of the clouds of probable implementations just after the initiation of the action strategies $\mathcal{AS}_{1,2}$. Fragment II illustrates the transition between the space-time clouds $\mathbb{C}_{P_{1,2}}$ caused by the action strategy changes $\mathcal{AS}_1 \Rightarrow \mathcal{AS}_2$ on the phase plane $\{j, t\}$. The symbols $\mathbb{C}^s_{\Xi}$ and $\partial\mathbb{C}^s_{\Xi}$ denote the space-time cloud of the stable manifold Ξ_s and the action point layer, respectively (Sect. 6.2.3)

is reduced to determining the changes in the $\mathbb{C}^{\text{en}}_P$-cloud induced by the passage of physical time.

Figure 6.2 (fragment I) depicts two trajectories $P_{\text{AS}_{1,2}}$ representing the implementation of two successive action strategies $\mathcal{AS}_{1,2}$ in the Lagrange space $\mathbb{R}_L$ comprising the inner phase variables of the controlled object motion, i.e., the spatial position x, velocity v, and acceleration a of moving object. In this phase space the integral trajectory $P_{\text{AS}_1} \cup P_{\text{AS}_2}$ must be a continuous curve. The sharp cusp of the acceleration is related to the transition $\mathcal{AS}_1 \Rightarrow \mathcal{AS}_2$.

The color stripes denote the space-time clouds $\mathbb{C}_{P_{1,2}}$ of initiated implementations of action strategies $\mathcal{AS}_{1,2}$. The blue stripe around the trajectory P_{AS_1} stands for the space-time cloud $\mathbb{C}_{P_1}$ of the initiated implementation of $\mathcal{AS}_1$ and belong to the connected past. The green stripe around the trajectory P_{AS_2} stands for the space-time cloud $\mathbb{C}_{P_2}$ of action strategy implementation being currently under initiation and belonging to the connected future.

Figure 6.2 (fragment II) illustrates the details of the transition $\mathcal{AS}_1 \Rightarrow \mathcal{AS}_2$ between the action strategies in terms of transition between the space-time clouds $\mathbb{C}^{\mathrm{en}}_{P_{1,2}}$ on the plane $\{j, t\}$. Three characteristic cases can be singled out.

O_1: The cloud $\mathbb{C}_{P_1}$ overlaps substantially with the the stable manifold cloud $\mathbb{C}^s_{\Xi}$ (gray stripe). Under this condition, the subject just cannot recognize the deviation of the action strategy implementation from the optimal behavior represented by the stable manifold Ξ_s, i.e., the center of the $\mathbb{C}^s_{\Xi}$-cloud.

O_2: The deviation of the cloud $\mathbb{C}_{P_1}$ from the stable manifold cloud $\mathbb{C}^s_{\Xi}$ is remarkable but not too essential to make the correction of subject's action efficient. Put differently, the difference between the clouds $\mathbb{C}_{P_1}$ and $\mathbb{C}^s_{\Xi}$ in jerk is below a certain fuzzy threshold θ_{Ξ} separating the the stable manifold cloud $\mathbb{C}^s_{\Xi}$ and the action point layer $\partial\mathbb{C}^s_{\Xi}$ (Sect. 6.2.3).

O_3: The space-time cloud $\mathbb{C}_{P_1}$ deviates substantially from the stable manifold Ξ_s and reaches the action point layer $\partial\mathbb{C}^s_{\Xi}$ (red stripe). In this case, the subject via sharp change in the jerk j returns the object motion state into the stable manifold cloud $\mathbb{C}^s_{\Xi}$. The action point layer $\partial\mathbb{C}^s_{\Xi}$ is shown in red to accentuate that the subject not only recognizes the deviation of the object motion state from the optimal one but also considers this deviation essential and requiring urgent correction.

Anticipation-Implementation Cycle

The forward temporal part $\mathbb{C}^{>}_{\mathrm{AS}}$ of action strategy cloud $\mathbb{C}_{\mathrm{AS}}$ and the space-time cloud $\mathbb{C}_P$ of the initiated realization of the action strategy $\mathcal{AS}$ represent the opposite aspects in action strategy implementations.

The forward temporal part $\mathbb{C}^{>}_{\mathrm{AS}}$ represents the *predictable*, *controllable* aspects of the subject's actions. Namely, the $\mathbb{C}^{>}_{\mathrm{AS}}$-cloud as an entity of the human temporality describes the subject's anticipation of the controlled object dynamics within the connected future. Therefore the space-time cloud $\mathbb{C}^{>}_{\mathrm{AS}}$ should be categorized as (Sect. 5.6.1)

- an entity $\mathbb{C}^{>}_{\mathrm{AS}}$ belonging to the realm of the mind-in-the-self, i.e., the mental image of the action strategy $\mathcal{AS}$ in the connected future treated as a whole,[47] as well as;
- an entity $\Phi^{>}_{\mathrm{AS}}$ belonging to the realm of the body-in-the-self, i.e., the embodied union of two components:
  - a neural pattern admitting interpretation as a fuzzy space-time region in the space-time continuum $\mathbb{R}_{[4]} \times t$ and

[47] We use the same symbol $\mathbb{C}^{>}_{\mathrm{AS}}$ to denote the $\mathbb{C}^{>}_{\mathrm{AS}}$-cloud treated as a entity spanning the levels of the mind-in-the-self and the body-in-the-self as well as confined to its representation at the realm of the mind-in-the-self.

- a sequence of automatic subject's actions determining the implementation of $\mathcal{AS}$,

which emerge during the long-term learning process giving rise to the skill acquisition and, then, are enslaved by the mental image of $\mathbb{C}^{>}_{\text{AS}}$.

It should be pointed out that, in the given context, the noted items of the mind-in-the-self and the body-in-the-self have to be treated as two aspects of the same entity $\mathbb{C}^{>}_{\text{AS}}$ rather than two different entities. We will refer to the space-time cloud $\mathbb{C}^{>}_{\text{AS}}$ with the two aspects being accentuated, as a *top-down representation* of the subject's actions and the observed object in *the self*.[48] Figure 6.3 (fragment I) illustrates the top-down representation by the pair $\mathbb{C}^{>}_{\text{AS}} \mapsto \Phi^{>}_{\text{AS}}$, where, as noted above, the symbol $\Phi^{>}_{\text{AS}}$ denotes the embodied component of the $\mathbb{C}^{>}_{\text{AS}}$-cloud. The mapping $\mathbb{C}^{>}_{\text{AS}} \mapsto \Phi^{>}_{\text{AS}}$ is deterministic. Only at the final step of this top-down process the initiation of action strategy implementation in physical reality becomes random. This randomness is caused by the contribution of various factors, e.g., neural noise in the motor system, which are beyond the control not only of the mind-in-the-self but also of the body-in-the-self. In Fig. 6.3 this incompleteness of the transition from $\Phi^{>}_{\text{AS}}$ to P_{AS} is underlined by the red wave arrow.

As far as the space-time cloud $\mathbb{C}_P$ is concerned, the situation is opposite. The space-time cloud $\mathbb{C}_P$ emerges via two bottom-up transitions from the level of physical reality to the level of the body-in-the-self and, then, to the level of the mind-in-the-self. These transitions admitting the interpretation as a *bottom-up representation* of the observed object in *the self*[48 (p. 446)] are based on (Fig. 6.3, fragment I):

- The trajectory P_{AS}, Exp. (6.153), specifying the action strategy realization in physical reality. It is not accessible to the mind in the strict sense as well as cannot be recorded precisely in the body-in-the-self. The properties of the trajectory P_{AS} reflect
  - the acquired skill governing the subject's automatic actions and belonging to the body-in-the-self;
  - various uncontrolled processes that can be attributed to the body but not the body-in-the-self.

---

[48] Turning to the gist of Robinson's (1994) "two-perspective" view on Kant's triad "thing-in-itself–appearance–representation" (Sect. 1.1), the top-down representation may be categorized as the perspective on the "appearance–representation" union from the mind side. This union implies the governing role of the mind-in-the-self.

In turn, the bottom-up representation may be categorized as the prospective on the "appearance–thing-in-itself" union from the side of reality. This union admits the interpretation as the supervenience of the mental upon the physical. Within a simplified understanding, "a set of properties A supervenes upon another set B just in case no two things can differ with respect to A-properties without also differing with respect to their B-properties." (McLaughlin and Bennett 2018). For a detailed discussion of mental-physical supervenience a reader may be referred, e.g., to Kim (1993), Francescotti (2014), McLaughlin and Bennett (2018). The attractor-induced mental-physical supervenience, where the mental and the physical may be of the same basic level, is discussed by Lubashevsky (2017).

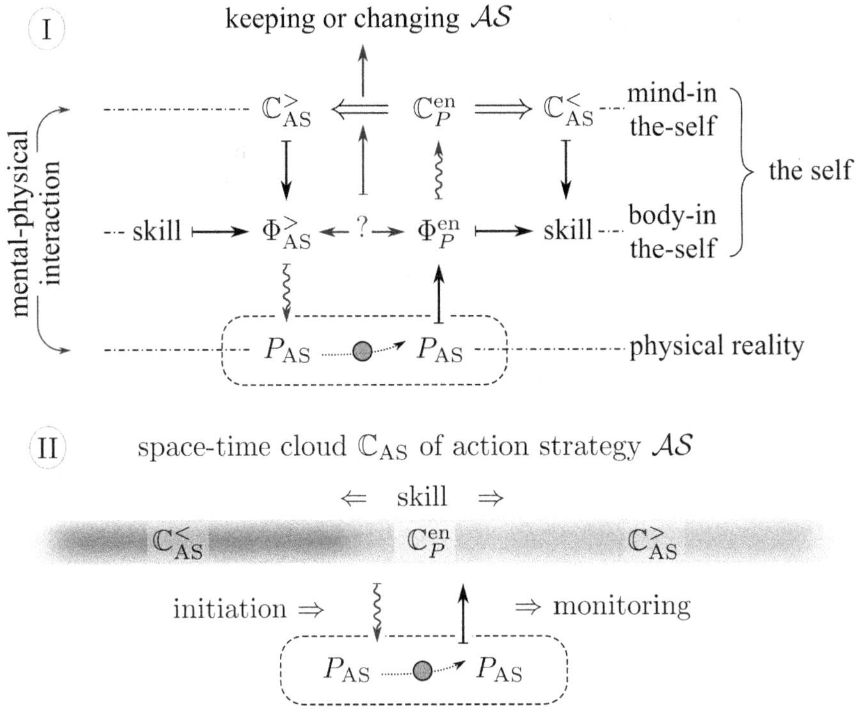

Fig. 6.3 Schematic illustration of the hierarchical mechanism governing the interaction between the physical object controlled by subject's actions and the mental image of action strategy implementation. Fragment I illustrates this interaction in detailed form, fragment II represents its essence

- The neurophysiological image Φ^{en}_{P} of the observed object arising in the body-in-the-self via sensory modalities. It represents the object motion state as it is experienced right now, i.e., within the experiential now. This image bears the information about the subject's current perception of the object position in the space-time continuum $\mathbb{R}_{[4]} \times t$. The transition $P_{AS} \mapsto \Phi^{en}_{P}$ is deterministic and the main characteristics of the object motion state in physical reality are preserved in the neural image Φ^{en}_{P} within their *blurring* caused by the limit capacity of sensory modalities.
- The space-time cloud $\mathbb{C}^{en}_{P}$ of the object motion state experienced right now. The mind-in-the-self gains only partial access the neurophysiological image Φ^{en}_{P} at the level of E-consciousness and, in this sense, is enslaved by the neurophysiological image Φ^{en}_{P}. In Fig. 6.3 the incompleteness of the transition from Φ^{en}_{P} to $\mathbb{C}^{en}_{P}$ is also underlined by the red wave arrow.

The given sequence of two bottom-up transitions $P_{AS} \to \Phi^{en}_{P} \to \mathbb{C}^{en}_{P}$, in particular, exemplifies the notion of the supervenience of the mental upon the physical widely discussed in philosophy of mind.[48]

The space-time cloud $\mathbb{C}_P$ is the mathematical eidos of the trajectory P_{AS}, i.e., the image of the mental-mental embedding of P_{AS} into the space-time continuum $\mathbb{R}_{[4]} \times t$ (Sects. 3.3.3 and 3.3.4). Nevertheless, the $\mathbb{C}_P$-cloud is not completely accessible to the mind. The matter is that, on the one hand, the sensory system continuously creates neurophysiological images $\{\Phi_P\}$ of the observed object representing its motion. On the other hand, because of the limit capacity of working memory this sequence of images $\{\Phi_P\}$ cannot be retained, at least, without loss of connectivity. In our account we suppose that the mind-in-the-self has partial access only to the neurophysiological image Φ_P^{en} created within the experiential now. The object motion at the previous moments of time, i.e., within the connected past is *reconstructed* by the mind-in-the-self involving also the body-in-the-self in the common, integral process. Put differently, based on the spaced-time cloud $\mathbb{C}_P^{\text{en}}$ and using the skill enabling him to anticipate the object motion in the connected future, the subject *reconstructs* the object motion in the connected past. This reconstruction, which is a top-done process belonging to the realm of the mind-in-the-self, gives rise to the backward temporal part $\mathbb{C}_{\text{AS}}^{<}$ of the space-time cloud $\mathbb{C}_{\text{AS}}$ of the action strategy $\mathcal{AS}$ (Fig. 6.3, fragment I).

The complex mental-physical interaction based on these top-down and bottom-up representations of the subject's actions and the perceived object motion is the mechanism governing the passive phase of 2D-time dynamics of action strategy implementation from its initiation to the initiation of the next action strategy. This mechanism called the *anticipation-implementation cycle* involves (Fig. 6.3):

- the top-down process $\mathbb{C}_{\text{AS}}^{>} \to \Phi_{\text{AS}}^{>} \to P_{\text{AS}}$ initiates a particular implementation of the action strategy $\mathcal{AS}$, the subject's skill affects this initiation;
- the continuous (in physical time) bottom-up process $P_{\text{AS}} \to \Phi_P^{\text{en}} \to \mathbb{C}_P^{\text{en}}$ governs the subject's monitoring of object motion;
- at the level of the body-in-the-self, the neural images of the anticipated object motion ($\Phi_{\text{AS}}^{>}$) and the currently perceived motion state (Φ_P^{en}) are compared with each other, the result of this comparison raised to the level of the mind-in-the-self is used to update the acquired skill and to make decision about keeping the current action strategy or terminating it;
- based on the information represented by the space-time cloud $\mathbb{C}_P^{\text{en}}$ the mind-in-the-self continuously modifies the forward and backward temporal parts of the action strategy space-time cloud $\mathbb{C}_{\text{AS}}$, which also modifies the neural image $\Phi_{\text{AS}}^{>}$ and the acquired skill.

Figure 6.3 (fragment II) illustrates the essence of this mental-physical interaction.

We want to remind once more that the space-time cloud $\mathbb{C}_P$ inaccessible as a whole to the mind plays the role of the source of the space-time cloud $\mathbb{C}_P^{\text{en}}$. This source is hidden from the mind-in-the-self and so, for the subject the details in the $\mathbb{C}_P^{\text{en}}$-appearance are only partly predictable. It is a reason to categorize the dynamics of the action strategy implementation as a process with hidden determinism.

6.3.2 Basic Elements of Passive Phase

In the previous subsection, we discussed the main aspect of the mechanism governing the passive and active phases of dynamic processes in subject's goal-oriented behavior united within the concept of 2D-time dynamics of action strategy.

In this subsection, we confine our consideration to the passive phase of 2D-time dynamics and the related elements of its mathematical description. Let us remind that the passive phase of 2D-time dynamics is related to the automatic process of the action strategy implementation from the initiation to the moment of its termination. In describing this dynamic process we take into account that in implementing an action strategy $\mathcal{AS}$ the subject continuously (Fig. 6.4):

- monitors the state of the object motion governed by his automatic actions and continuously evaluates the current values of the inner phase variables $\{x, v, a\}$, which are described by the space-time cloud $\mathbb{C}_P^{\text{en}}$;
- estimates the probable motion of controlled object in the connected future based on his skill acquired previously and the current perception of the motion state;
- reconstructs the object motion in the connected past using his skill and estimates the object initial position in the Lagrange space $\mathbb{R}_L = \{x, v, a\}$.

At terminating the current implementation of the action strategy $\mathcal{AS}$ the subject

- fuses the perceived effort of bodily executed actions and the mental effort of monitoring the object motion (Sect. 5.5),
- incorporates this fusion-induced effort with the effort accumulated previously and "retained" in his skill (body memory),
- attributes this aggregated effort to the action strategy $\mathcal{AS}$ as a whole entity or, speaking more strictly, attributes this effort to a certain set of action strategies. The member of this set are categorized by the position of their initiation clouds $\{\mathbb{C}_{\text{ini}}\}$ that are regarded by the subject as similar.

These actions are mainly implemented automatically by the body-in-the-self; in the mind-in-the-self they are represented as the subject's awareness about them like some kind anchors.

Let us outline the corresponding mathematical formalism describing the passive phase of 2D-time dynamics of action strategy.

Perceived dynamics of the current motion state: The space-time cloud $\mathbb{C}_P^{\text{en}}$ represents the subject's perception of the current state of object motion. Although being able to anticipate or remember the position of $\mathbb{C}_P^{\text{en}}$ at the nearest moments of time, the subject is supposed to focus only on the current state of object motion. The subject is unable to discriminate between different implementations of a given action strategy $\mathcal{AS}$ and additional effort required for anticipating and remembering the nearest states of object motion is unreasonable.

Therefore in describing the subject's actions during the automatic implementation of action strategy the space-time cloud $\mathbb{C}_P^{\text{en}}$ is treated as a given source of sensory

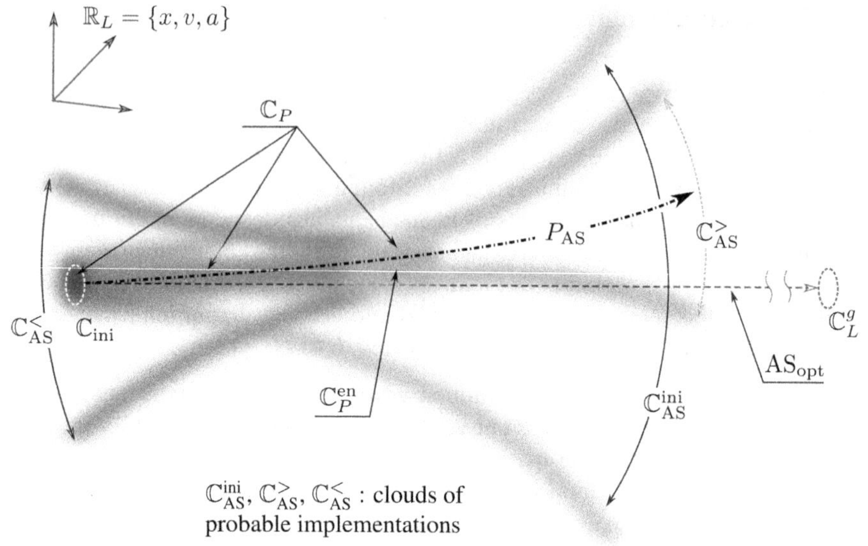

Fig. 6.4 Fragments of subject's actions during the automatic implementation of an action strategy $\mathcal{AS}$ aimed at getting the space-time cloud $\mathbb{C}^{g}_{L}$ of the desired goal. The dashed red line depicts the corresponding strictly optimal implementation $\mathrm{AS}_{\mathrm{opt}}$ of $\mathcal{AS}$ which is inaccessible to the mind-in-the-self as well as the body-in-the-self. The dash-dotted line P_{AS} represents the initiated implementation of action strategy in the Lagrange space $\mathbb{R}_L$. The blurred orange region $\mathbb{C}^{\mathrm{en}}_{P}$ represents the subject's evaluation of the current state of object motion within the experiential now. The neighborhood $\mathbb{C}_P$ of the path P_{AS} bounded by $\mathbb{C}^{\mathrm{en}}_{P}$ and the space-time cloud $\mathbb{C}_{\mathrm{ini}}$ of the $\mathcal{AS}$-initiation is a fragment to be analyzed in the next Sect. 6.3.3. Only the region $\mathbb{C}^{\mathrm{en}}_{P}$ is considered to be accessible to the mind. The blurred gray curves illustrate the boundary of the space-time cloud $\mathbb{C}^{\mathrm{ini}}_{\mathrm{AS}}$ representing the object anticipated motion just after the action strategy initiation. The blurred green (blue) curves illustrate the boundary of the forward (backward) temporal part $\mathbb{C}^{>}_{\mathrm{AS}}$ ($\mathbb{C}^{<}_{\mathrm{AS}}$) of the $\mathcal{AS}$-implementation

information about the object current motion. The subject just observes the object motion without trying to modify it *gradually*. For this reason we treat the automatic implementation of action strategies as the passive phase of subject's goal-oriented behavior. This passiveness does not mean that the subject bodily executes no actions, his motor behavior may be intensive but the subject just does not modify his actions *intentionally*. The active phase of subject's goal-oriented behavior will be considered in the next subsections.

As physical time goes on, the space-time cloud $\mathbb{C}^{\mathrm{en}}_{P}$ moves in the Lagrange space $\mathbb{R}_L$, which endows the subject with the perception of the passage of time in the case under consideration. Speaking about the human temporality confined to the complex present, motion of the cloud $\mathbb{C}^{\mathrm{en}}_{P}(t)$ is the pivot point relating the change of this temporality fragment to the flow of physical (external) time. The introduced previously term *hidden determinism* accentuates the $\mathbb{C}^{\mathrm{en}}_{P}$-motion to be, on the one side, not *consciously* controlled. On the other side, the $\mathbb{C}^{\mathrm{en}}_{P}$-motion representing the current implementation of the action strategy $\mathcal{AS}$ is fixed at the moment of the $\mathcal{AS}$-initiation.

Dynamics of the forward temporal part: Just after the initiation of action strategy implementation the subject evaluates the possible motion of controlled object taking into account all the probable implementations of the selected action strategy $\mathcal{AS}$. Its evaluation is represented by the space-time cloud $\mathbb{C}^{\text{ini}}_{\text{AS}}$ illustrated in Fig. 6.4 by blurred gray curves. For an action strategy *adequately* selected for reaching the desired goal, the space-time cloud $\mathbb{C}^{\text{ini}}_{\text{AS}}$ has to contain the path AS_{opt} in the Lagrange space $\mathbb{R}_L = \{x, v, a\}$ connecting the initial point to the target in the optimal way.

To recognize how close is the initiated implement AS of the action strategy $\mathcal{AS}$ to the optimal implementation AS_{opt} and when it deviates from AS_{opt} essentially, the subject

- continues to execute the corresponding automatic actions,
- reconstructs the probable implementations of the action strategy $\mathcal{AS}$ but now from the current state of object motion based on its monitoring.

In Fig. 6.4 this reconstruction of the $\mathcal{AS}$-implementations is represented by the region $\mathbb{C}^{>}_{\text{AS}}(t)$—the forward temporal part of action strategy implementation. As the space-time cloud $\mathbb{C}^{\text{en}}_{P}(t)$ moves along the path P_{AS} representing the current implementation AS of $\mathcal{AS}$, the discrepancy between AS and AS_{opt} becomes more and more pronounced and, finally, the subject starts to recognize that the action strategy should be modified.

Mathematically the dynamics of the forward temporal part $\mathbb{C}^{>}_{\text{AS}}$ in the space-time $\mathbb{R}_L \times t$ is described by the governing equation (6.149) dealing with the image $\Phi^{>}_{\text{AS}}$ of the space-time cloud $\mathbb{C}^{>}_{\text{AS}}$ in the body-in-the-self. This equation is subject to the initial condition (6.150) imposed on the cloud $\Phi^{>}_{\text{AS}}$ at the current instant of time $t = t_{\text{cur}}$ and where the right-hand side contains the image $\Phi^{\text{en}}_{P}(t_{\text{cur}})$ of the space-time cloud $\mathbb{C}^{\text{en}}_{P}(t_{\text{cur}})$ in the body-in-the-self.

It should be emphasized that although physical time t formally enters this equation, it is not in conflict with that the instants of physical time individually are inaccessible to the mind. Indeed, first, the nonlinear equation (6.149) is autonomous, i.e., physical time does not enter it explicitly. The corresponding initial condition just relates the current state of the cloud $\Phi^{>}_{\text{AS}}$ to the current state of object motion as it is perceived by the subject. Second, the mind deals with the space-time cloud $\Phi^{>}_{\text{AS}}(t)$ as a whole entity distributed in the complex present rather than some entity given a fixed instant of time.[49] Third, physical time can be removed from this consideration via replacing it by the distance (within the perception accuracy) to the goal position (Sect. 6.2.5).

Dynamics of the backward temporal part: In parallel to updating the anticipated motion of the controlled object, the subject reconstructs the object initial position in the Lagrange space $\mathbb{R}_L$. The two continuous mental processes take place during the automatic action strategy implementation simultaneously. When the subject decides to terminate the current action strategy implementation and to initiate a new action strategy, he

[49] See also the discussion of this issue started at Sect. 6.2.8.

- integrally evaluates the made effort,
- relates the terminated action strategy to the class of action strategies with equivalent positions of the initiation regions (and, thus, similar in the effort of implementation),
- incorporates the currently experienced effort into the effort attributed to the given class.

As elucidated in Appendix 6.7.4, the object motion governed by the Hamilton's equations (6.39), at least for Hamiltonian (6.42) (and respectively for Hamiltonian (6.44)), is reversible in the Lagrange space $\mathbb{R}_L$. It means that if at the terminal point of object motion along a certain trajectory P the motion direction changes for the opposite one, the object will return to the initial point moving along the same trajectory in $\mathbb{R}_L$. Put differently, the inverse motion of object along the trajectory P is described by the same governing equations within the replacement

$$\begin{aligned} \mathbb{R}_L : \quad & x \to x\,, \quad v \to -v\,, \quad a \to a\,, \\ \Xi_s \leftrightarrow \Xi_u : \quad & \text{in particular } p_a \to -p_a \ \ (j \to -j)\,. \end{aligned} \tag{6.154}$$

Therefore, we may suppose that the subject reconstructs the backward motion of the object in the same way as he estimates the forward motion of the object. He just anticipates the object motion starting from the space-time cloud $\mathbb{C}_P^{\text{en}}(t_{\text{cur}})$ moving along the manifold Ξ_s in the opposite direction (formally it is equivalent to $\Xi_s \to \Xi_u$). This reconstruction may be regarded as the *backward temporal part* $\mathbb{C}_{\text{AS}}^{<}(t_{\text{cur}})$ of the $\mathcal{AS}$-implementation.

As a result, mathematically the dynamics of the backward temporal part $\mathbb{C}_{\text{AS}}^{<}$ in the space-time $\mathbb{R}_L \times t$ is described by the governing equation (6.149) dealing with the image $\Phi_{\text{AS}}^{<}$ of the space-time cloud $\mathbb{C}_{\text{AS}}^{<}$ in the body-in-the-self. This equation is subject to the initial condition (6.150) imposed on the cloud $\Phi_{\text{AS}}^{<}$ at the current instant of time $t = t_{\text{cur}}$ and where the right-hand side contains the image $\Phi_P^{\text{en}}(t_{\text{cur}})$ of the space-time cloud $\mathbb{C}_P^{\text{en}}(t_{\text{cur}})$ in the body-in-the-self. The main difference from the case of the $\mathbb{C}_{\text{AS}}^{>}$-dynamics is that in Hamiltonian (6.148) the jerk dependence $j_s(x, v, a)$ specifying the stable branch Ξ_s of the zero-Hamiltonian manifold has to be replaced by the jerk dependence $j_u(x, v, a)$ for the unstable branch,

$$j_s(x, v, a) \to j_u(x, v, a)\,.$$

Beside the operator replacement h

$$\frac{\delta_j^2}{\tau_d}\frac{\partial}{\partial \Delta j} - \frac{\lambda_u}{\tau_d \delta_j}\Delta j \quad \to \quad \frac{\delta_j^2}{\tau_d}\frac{\partial}{\partial \Delta j} + \frac{\lambda_u}{\tau_d \delta_j}\Delta j$$

has to be also done. Moreover, for the backward temporal part $\mathbb{C}_{\text{AS}}^{<}$ its component related to the jerk difference Δj may be excluded from consideration.

Incorporation of the experienced effort into the skill-based effort: To complete this outline of the passive phase of 2D-time dynamics we should elucidate how the integral effort related to the terminated action strategy implementation is incorporated into the *skill-based effort*. The term *skill-based effort* is introduced to refer to the effort attributed to a class of similar action strategies characterized, on the average, by the same effort of their implementation.

The skill-based effort emerges via the initial accumulation during learning and skill acquisition on temporal scales exceeding the duration of complex present. Then the skill-based effort can change as a result of adaptation to graduate, continuous variations in the environment and the subject's physical and mental states. Besides, the skill-based effort

- fades in time without constant repetition of the corresponding actions and
- can contain random variations caused by that in incorporating the particular effort of the just now terminated action strategy its particular implementation is not consciously identified.

These issues will be analyzed in the next Sect. 6.3.3, here we focus on the effort incorporation on its own assuming the properties of the skill-based effort are given.[50]

In constructing the mathematical description of effort incorporation we turn to the concept of the self and the cloud-type entities with constituent components[51] belonging to the realms of the mind-in-the-self and the body-in-the-self (Sect. 4.3 and Sect. 5.6).

Let us assume that the effort of the currently finalized action strategy implementation and the skill-based effort are characterized by two effort clouds $\mathbb{E}_{\text{ex}}$ and $\mathbb{E}_{\text{sb}}$, respectively.[52] In the realm of the body-in-the-self these clouds are represented by two functions $\Phi_{\text{ex}}(e)$ and $\Phi_{\text{sb}}(e)$ meeting the normalization condition

$$\left\langle \Phi_{\text{ex}} \middle| \Phi_{\text{ex}} \right\rangle = 1 \quad \text{and} \quad \left\langle \Phi_{\text{sb}} \middle| \Phi_{\text{sb}} \right\rangle = 1\,. \tag{6.155}$$

These functions are supposed to have a special form $\Phi(e)$ and so can be specified by a few parameters. In the case under consideration, the function

$$\Phi(e) = \Phi(e \mid E_m, \Delta_m)$$

[50] We want to emphasize that this effort incorporation and the effort fusion analyzed in Sects. 5.5 and 5.6 are principally different processes. In the incorporation, the effort components are commensurable, i.e., there are common units in which these components can be quantified. In the fusion, the effort components as they are experienced by the subject are not commensurable.

[51] In this context, the constituent components should not be understood as different parts of a cloud-type entity admitting separation from each other. They may be treated as different aspects of one entity with internal integrity that are singled out in operating with this entity within the realms of the mind-in-the-self and the body-in-the-self.

[52] We introduced the notion of cognitive effort as a cloud in the Hermitian space of effort magnitudes and analyzed its general properties including the cloud comparison in Sect. 5.4.1.

is specified by the effort magnitude E_m where it attains its maximum and the characteristic width Δ_m quantifying uncertainty in the subject's perception or evaluation of a given type effort. For the cloud $\mathbb{E}_{\text{ex}}$ related to the subject's experience (perception) of action execution effort, the value of $\Delta_m = \Delta_m(E_m)$ is related to the value of E_m by the laws of psychophysics (Sect. 5.4.1). For the cloud $\mathbb{E}_{\text{sb}}$ representing the class of action strategies treated by the subject as equivalent in effort, the value of Δ_m should be larger. In this case, the width Δ_m represents the categorization of action strategies into different classes also according to the related effort of implementation. The difference in effort between these classes quantified by the value of Δ_m is an entity of AM-consciousness and, thereby, has to exceed, maybe essentially, the threshold of effort perception.

We treat the effort incorporation as the update process of the skill-based effort $\mathbb{E}^i_{\text{sb}} \to \mathbb{E}^{i+1}_{\text{sb}}$ and in the realm of the body-in-the-self we describe its initial stage by the linear expression

$$\Phi'_{\text{sb}} = (1-\varpi)\,\Phi^i_{\text{sb}} + \varpi\,\Phi_{\text{ex}}\,. \tag{6.156a}$$

Its coefficients meet the general requirement that the incorporation of an effort $\mathbb{E}$ into itself must not change anything. In Exp. (6.156) the quantity $\varpi \in [\varepsilon, 1]$ and the parameter $\varepsilon \ll 1$ characterize the state of embodied memory with respect to the given class of action strategies. After a relatively long-time time interval when no actions of this class were executed, the value $\varpi = 1$ and the skill-based effort is generated by the first implementation of an action strategy from the given class. When these actions were executed rather often in the course goal-oriented behavior, the value ϖ gets its limit $\varpi = \varepsilon$ determined by the capacity of embodied memory and the rate of its temporal decay. The dynamics of the quantity ϖ will be discussed below in Sect. 6.3.9.

Below we will consider the effort update features in the limit $\varpi \ll 1$, for simplicity. Following Sect. 5.6, we describe the final stage of the effort incorporation as the body-mind raising (5.78) of the result Φ'_{sb} from the level of the body-in-the-self to the level of the mind-in-the-self

$$\mathfrak{S}:\ \Psi'_{\text{sb}} \overset{\text{the self}}{\mapsto} \Phi^{i+1}_{\text{sb}} \quad \Leftarrow \quad \max_{E^{i+1}_m,\,\Delta^{i+1}_m} \left\langle \Phi^{i+1}_{\text{sb}} \middle| \Phi'_{\text{sb}} \right\rangle, \tag{6.156b}$$

modeled by the optimal square fitting. In the case $\varpi \ll 1$, the system (6.156) gives rise to the following coupled equations specifying the increments

$$\delta E_m = E^{i+1}_m - E^i_m\,,\quad \delta\Delta_m = \Delta^{i+1}_m - \Delta^i_m$$

of the parameters E_m, Δ_m in updating the skill-based effort:

$$\begin{aligned}
\left\langle \frac{\partial\Phi_{\text{sb}}}{\partial E_m} \middle| \frac{\partial\Phi_{\text{sb}}}{\partial E_m} \right\rangle \delta E_m + \left\langle \frac{\partial\Phi_{\text{sb}}}{\partial E_m} \middle| \frac{\partial\Phi_{\text{sb}}}{\partial \Delta_m} \right\rangle \delta\Delta_m &= \varpi \left\langle \frac{\partial\Phi_{\text{sb}}}{\partial E_m} \middle| \Phi_{\text{ex}} - \Phi_{\text{sb}} \right\rangle \\
\left\langle \frac{\partial\Phi_{\text{sb}}}{\partial \Delta_m} \middle| \frac{\partial\Phi_{\text{sb}}}{\partial E_m} \right\rangle \delta E_m + \left\langle \frac{\partial\Phi_{\text{sb}}}{\partial \Delta_m} \middle| \frac{\partial\Phi_{\text{sb}}}{\partial \Delta_m} \right\rangle \delta\Delta_m &= \varpi \left\langle \frac{\partial\Phi_{\text{sb}}}{\partial \Delta_m} \middle| \Phi_{\text{ex}} - \Phi_{\text{sb}} \right\rangle
\end{aligned} \tag{6.157}$$

The given equations are desired description of the effort incorporation in the main course of subject's goal-oriented actions.

Summarizing the aforesaid, the passive phase of 2D-time dynamics of action strategy can be conceived of as the dynamic relationship between

- the spaced-time cloud $\mathbb{C}_P^{\text{en}}$ describing the subject's perception of the current state of controlled object motion along the path P_{AS} in the Lagrange space $\mathbb{R}_L = \{x, v, a\}$,
- the forward temporal part $\mathbb{C}_{\text{AS}}^{>}$ of action strategy implementation representing the subject's anticipation of the object motion in the connected future;
- the backward temporal part $\mathbb{C}_{\text{AS}}^{>}$ of action strategy implementation representing the subject's reconstruction of the object motion in the connected past which is required for classifying the action strategy after terminating its implementation.

The temporal change of the $\mathbb{C}_P^{\text{en}}$-cloud position in $\mathbb{R}_L$ forces the subject to reconstruct continuously his anticipation and re-estimation of the action strategy initiation for its further classification. After terminating the current action strategy implementation, the subject incorporates the cumulative effort experienced during the course of given actions into the effort characterizing the corresponding set of action strategies. This skill-based effort is an essential aspect in selecting a next action strategy for implementation.

6.3.3 Action Strategy Termination

In the previous subsection, we discussed the passive phase of 2D-time dynamics of action strategy implementation treated also as the passive phase of subject's behavior. In this case, the subject's mind does not contribute to the dynamics of the controlled object. Indeed, the subject's automatic actions are determined by body movements initiated by some neural signals generated by the brain and the following implementation of the corresponding scenario encoded in muscles and the peripheral neural network.[53]

The situation changes drastically when the subject intentionally terminates the current action strategy $\mathcal{AS}$ and selects another one or, maybe, returns to the previous action strategy for a new implementation. We regard these intentional actions as the *active phase* of subject's behavior. This phase is not deterministic, its dynamics is reflected in changing the action strategies or their implementations (if the same action strategy is selected again).

The present subsection is focused on the action strategy termination which may be treated as an intermediate stage between the passive and active phases of 2D-time dynamics. So the action strategy termination has aspects of both the two phases. In

[53] In this statement we actually have turned to the gist of the equilibrium-point hypothesis. This hypothesis posits that the body movements are governed by the length-force relation for all the muscles involved in a given movement scenario. Changing these relations, the neural network affects their equilibrium points and, thereby, generates body movements (see Feldman et al. 2007; Feldman 2009; Feldman and Levin 2009; Latash 2008b, 2010, 2012, for a review).

the previous subsection, we considered the incorporation of the experienced effort into the skill-based effort arising in the course of the termination process at the level of body memory.

In this section, we consider the action strategy termination from the standpoint of the active phase of 2D-time dynamics and outline the basic elements of its mathematical description.

Two stages in terminating action strategy implementation: According to the temporal criterion of action strategy optimality (Sect. 6.2.3) the subject terminates the action strategy implementation when he (see Sect. 6.2.3):

(*i*) recognizes the deviation of the controlled object motion from the stable manifold Ξ_s (the stable branch of the zero-Hamiltonian manifold); which is based on the anticipation and mathematically described in terms of the overlap between the forward temporal part $\mathbb{C}^{>}_{\mathrm{AS}}$ of the action strategy implementation and the stable manifold cloud $\mathbb{C}^{s}_{\Xi}$ (event *i* in Fig. 6.5);

(*ii*) and, then, waits until the object motion state gets the action point layer $\partial\mathbb{C}^{s}_{\Xi}$ when the correction of the current action strategy implementation can be done efficiently[54]; which is described by the overlap between the space-time cloud $\mathbb{C}^{\mathrm{en}}_{P}$ representing the subject's perception of the current state of controlled object motion and the action point layer $\partial\mathbb{C}^{s}_{\Xi}$ (event *ii* in Fig. 6.5).

It should be emphasized that due to event *i* the subject is able to anticipate event *ii* in advance and so can compensate to some degree the delay in processing the perceived information which is about the duration of experiential now (Sect. 6.2.6). Figure 6.5 illustrates these actions. Event *i* represents the first stage in terminating the currently executed action strategy implementation when the subject recognizes the direction of the object motion deviation from the stable manifold Ξ_s on scales of its cloud $\mathbb{C}^{s}_{\Xi}$. Event *ii* represents the second stage when the current state of object motion admits its efficient correction. It is the case when the motion state gets the action point layer $\partial\mathbb{C}^{s}_{\Xi}$.

Formalism of cloud overlapping: To discuss in detail events *i* and *ii* noted above, we need to clarify the meaning of the term the overlap between the clouds $\mathbb{C}^{>}_{\mathrm{AS}}$ and $\mathbb{C}^{s}_{\Xi}$ as well as for the pair $\mathbb{C}^{\mathrm{en}}_{P}$ and $\partial\mathbb{C}^{s}_{\Xi}$. The matter is that the clouds $\mathbb{C}^{>}_{\mathrm{AS}}$, $\mathbb{C}^{\mathrm{en}}_{P}$, on the one hand, and the clouds $\mathbb{C}^{s}_{\Xi}$, $\partial\mathbb{C}^{s}_{\Xi}$, on the other hand, are essentially different in space-time structure.

In the realm of the body-in-the-self the space-time structure of the former pair $\mathbb{C}^{>}_{\mathrm{AS}}$, $\mathbb{C}^{\mathrm{en}}_{P}$ is represented by the two functions

$$\Phi^{>}_{\mathrm{AS}}(x, v, a, p_a, t)\,, \qquad \Phi^{\mathrm{en}}_{P}(x, v, a, p_a, t) \tag{6.158}$$

[54] Because of the uncertainty in subject's motor actions in addition to the perception uncertainty, correction of object motion just after recognizing its deviation from the desired state and returning it into the stable manifold cloud cannot be efficient.

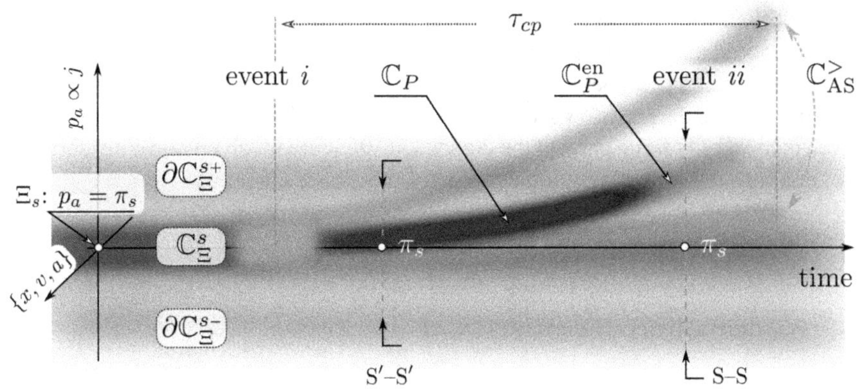

Fig. 6.5 Illustration of the two-stage mechanism via which the subject terminates the action strategy implementation. This figure combines together two events *i* and *ii* happened at different time moments. Event *i* is related to the moment when the subject recognizes that the anticipated motion of controlled object—the forward temporal part $\mathbb{C}_{\text{AS}}^{>}$ of action strategy implementation (region bounded by blurred green curves)—clearly deviates from the stable manifold Ξ_s in one (here upward) of the possible directions. Event *ii* is related to the moment when the space-time cloud $\mathbb{C}_P^{\text{en}}$ (the second blurred orange region) representing the subject's perception of the current motion state of controlled object gets one of the branches $\partial\mathbb{C}_\Xi^{s+}$, $\partial\mathbb{C}_\Xi^{s-}$ of the action point layer (two red horizontal stripes). The figure depicts the mutual arrangement of these clouds in the space-time $\{x, v, a\} \times p_a \times t$

localized in the space-time continuum $\{x, v, a, p_a\} \times t$ (actually in the space-time continuum $\mathbb{R}_{[4]} \times t$ because $p_a \propto j$). Their normalization is defined as follows

$$\begin{aligned}\left\langle \mathbb{C}_{\text{AS}}^{>} \middle| \mathbb{C}_{\text{AS}}^{>} \right\rangle &= \int_{-\infty}^{+\infty} \cdots \int dx\, dv\, da\, dp_a \big[\Phi_{\text{AS}}^{>}(x, v, a, p_a, t)\big]^2 = 1\,,\\ \left\langle \mathbb{C}_P^{\text{en}} \middle| \mathbb{C}_P^{\text{en}} \right\rangle &= \int_{-\infty}^{+\infty} \cdots \int dx\, dv\, da\, dp_a \big[\Phi_P^{\text{en}}(x, v, a, p_a, t)\big]^2 = 1\end{aligned} \tag{6.159}$$

according to the constructions presented in Sect. 6.2.7, where the time argument t is fixed.

By contrast, the space-time structure of the latter pair $\mathbb{C}_\Xi^s$, $\partial\mathbb{C}_\Xi^s$ is represented in the realm of the body-in-the-self by two functions

$$\Phi_\Xi^s(x, v, a, p_a, t) = \Omega_\Xi^s(p_a)\,, \qquad \Phi_\Xi^\partial(x, v, a, p_a, t) = \Omega_\Xi^\partial(p_a) \tag{6.160}$$

depending only on the argument p_a and, thus, are not localized in the unbounded space-time continuum $\{x, v, a\} \times t$. In Sect. 6.2.3 we introduced the cloud $\mathbb{C}_\Xi^s$ of stable manifold and the action point layer $\partial\mathbb{C}_\Xi^s$ as some density functions localized in the corresponding regions. Below we will extend the description of these clouds to the density functions $\Omega^s_\Xi(p_a)$ and $\Omega^\partial_\Xi(p_a)$ taking positive and negative values. Here we have noted that the given constructions of the cloud's overlap are not confined to the positive cloud densities.

Appealing to the self's ownership of mental entities (Sect. 3.2.3) and to that in the given context only finite parts of the clouds $\mathbb{C}^s_\Xi$, $\partial\mathbb{C}^s_\Xi$ are involved into consideration, we, first, use the following normalization of the clouds $\mathbb{C}^s_\Xi$, $\partial\mathbb{C}^s_\Xi$

$$\begin{aligned} \left\langle \mathbb{C}^s_\Xi \middle| \mathbb{C}^s_\Xi \right\rangle &\overset{\text{def}}{=} \left\langle \Omega^s_\Xi \middle| \Omega^s_\Xi \right\rangle = \int_{-\infty}^{\infty} dp_a \big[\Omega^s_\Xi(p_a)\big]^2 = 1\,, \\ \left\langle \partial\mathbb{C}^s_\Xi \middle| \partial\mathbb{C}^s_\Xi \right\rangle &\overset{\text{def}}{=} \left\langle \Omega^\partial_\Xi \middle| \Omega^\partial_\Xi \right\rangle = \int_{-\infty}^{\infty} dp_a \big[\Omega^\partial_\Xi(p_a)\big]^2 = 1\,. \end{aligned} \tag{6.161}$$

Second, we introduce the operator of embedding a space-time cloud $\mathbb{C}$ similar to $\mathbb{C}^>_{\text{AS}}$ and $\mathbb{C}^{\text{en}}_P$ into the cloud $\mathbb{C}^s_\Xi$ of the stable manifold and the action point layer $\partial\mathbb{C}^s_\Xi$ as follows:

$$\uparrow\!\downarrow^w \mathbb{C}: \qquad \Phi(x, v, a, p_a, t) \overset{w}{\hookrightarrow} \Omega^w_\Xi(p_a)\phi^w(x, v, a, t)\,, \tag{6.162a}$$

where the function $\Phi(x, v, a, p_a, t)$ represents the cloud $\mathbb{C}$ in the realm of the body-in-the-self,

$$\phi^w(x, v, a, t) = \int_{-\infty}^{\infty} dp_a \Omega^w_\Xi(p_a)\Phi(x, v, a, p_a, t)\,, \tag{6.162b}$$

and the index $w = s,\ \partial$ points to the clouds $\mathbb{C}^s_\Xi$ and $\partial\mathbb{C}^s_\Xi$, respectively. In the bra-ket notions the operator of embedding (6.162a) can be also written as, e.g., for the stable manifold cloud

$$\left|\uparrow\!\downarrow^s \mathbb{C}\right\rangle = \left|\mathbb{C}^s_\Xi\right\rangle \left\langle \mathbb{C}^s_\Xi \middle| \overline{\underline{\mathbb{C}}} \right\rangle, \tag{6.162c}$$

where the lines above and below the symbol $\mathbb{C}$ mean that the arguments $\{x, v, a, t\}$ of the function $\Phi(x, v, a, p_a, t)$ are fixed. In this sense the function $\phi^w(x, v, a, t)$, Exp. (6.162b), may be referred to as the partial overlap between the space-time cloud $\mathbb{C}$ and the stable manifold cloud $\mathbb{C}^s_\Xi$ or the action point layer $\partial\mathbb{C}^s_\Xi$.

Finally, we define the overlap between a space-time cloud $\mathbb{C} = \mathbb{C}^>_{\text{AS}}, \mathbb{C}^{\text{en}}_P$ and one of the clouds $\mathbb{C}^s_\Xi$, $\partial\mathbb{C}^s_\Xi$, e.g., $\mathbb{C}^s_\Xi$ as

$$\left\langle \mathbb{C} \middle| \uparrow\downarrow^{s} \mathbb{C} \right\rangle = \int_{-\infty}^{+\infty} \cdots \int dx\, dv\, da\; dp_a \Phi(x, v, a, p_a, t) \Omega_{\Xi}^{s}(p_a) \phi^{s}(x, v, a, t)$$

$$= \int_{-\infty}^{+\infty} \cdots \int dx\, dv\, da\, dp_a\, dp_a' \Phi(x, v, a, p_a, t) \Omega_{\Xi}^{s}(p_a) \Omega_{\Xi}^{s}(p_a') \Phi(x, v, a, p_a', t)$$

$$= \int_{-\infty}^{+\infty} \cdots \int dx\, dv\, da \left[\phi^{s}(x, v, a, t)\right]^2 = \left\langle \overline{\underline{\mathbb{C}|\mathbb{C}_{\Xi}^{s}\rangle\langle\mathbb{C}_{\Xi}^{s}|\mathbb{C}}} \right\rangle. \qquad (6.163)$$

by virtue of (6.162b). It should be emphasized that the given definition meets the intuitive understanding of the overlapping. Namely, we can write

$$\Delta C \stackrel{\text{def}}{=} \int_{-\infty}^{+\infty} \cdots \int dx\, dv\, da\, dp_a \left[\Phi(x, v, a, p_a, t) - \Omega_{\Xi}^{w}(p_a) \phi^{s}(x, v, a, t)\right]^2$$

$$= 1 - \int_{-\infty}^{+\infty} \cdots \int dx\, dv\, da \left[\phi^{s}(x, v, a, t)\right]^2 = 1 - \left\langle \mathbb{C} \middle| \uparrow\downarrow^{w} \mathbb{C} \right\rangle. \qquad (6.164)$$

In deriving this equality Exps. (6.159), (6.161), (6.162b), and (6.163) have been taken into account. Whence it follows that the value $\left\langle \mathbb{C} \middle| \uparrow\downarrow^{w} \mathbb{C} \right\rangle$ can be used to quantify the degree of the overlap between space-time clouds similar to $\mathbb{C}_{\text{AS}}^{>}$, $\mathbb{C}_{P}^{\text{en}}$ and the cloud $\mathbb{C}_{\Xi}^{s}$ of the stable manifold Ξ_s and the action point layer $\partial\mathbb{C}_{\Xi}^{s}$. In this case, the value $\left\langle \mathbb{C} \middle| \uparrow\downarrow^{w} \mathbb{C} \right\rangle$ admits interpretation as a measure in quantifying the proximity of a given fragment of action strategy implementation to the stable manifold Ξ_s. Indeed, due to (6.163) and (6.164), the value $\left\langle \mathbb{C} \middle| \uparrow\downarrow^{w} \mathbb{C} \right\rangle$ is

- positive even if space-time clouds are characterized by negative densities;
- less or equal to unity,
- increasing as the function $\Phi(x, v, a, p_a, t)$ representing the analyzed space-time cloud $\mathbb{C}$ in the body-in-the-self tends to the form $\Omega_{\Xi}^{w}(p_a)\phi(x, v, a, t)$ representing its embedding into the cloud $\mathbb{C}_{\Xi}^{s}$ or $\partial\mathbb{C}_{\Xi}^{s}$.

Probabilities of terminating action strategy implementation: The obtained results, mainly, Exp. (6.163) enable us to construct the general expressions for

- the probability $\mathcal{P}_i(t)$ of the subject identifying the direction of the deviation of object motion from the stable manifold and preparing himself for the efficient change of action strategy implementation (event *i*) at a given instant of time t;
- the probability $\mathcal{P}_{ii}(t)$ of terminating the currently executed action strategy implementation followed by selecting a new one (event *ii*) at a given instant of time t.

We regard the cloud overlaps (changing as physical time t goes on)

$$r^s = \left\langle \mathbb{C}^{>}_{\text{AS}} \middle| \uparrow\!\downarrow^s \mathbb{C}^{>}_{\text{AS}} \right\rangle(t)\,, \qquad r^\partial = \left\langle \mathbb{C}^{\text{en}}_P \middle| \uparrow\!\downarrow^\partial \mathbb{C}^{\text{en}}_P \right\rangle(t)\,, \tag{6.165}$$

as the quantities that determine the rates R_i and R_{ii} of the subject recognizing events i and ii when the arrangement of the space-time clouds $\mathbb{C}^{>}_{\text{AS}}$, $\mathbb{C}^{\text{en}}_P$ with respect to the stable manifold Ξ_s is appropriate. The quantities $R_i dt$ and $R_{ii} dt$ mathematically describe the probability densities that under the current conditions the subject makes decision about events i and ii within the time interval dt. It immediately gives rise to the expressions relating the probabilities $\mathcal{P}_i$ and $\mathcal{P}_{ii}$ to the rates R_i and R_{ii} as follows

$$\mathcal{P}_i(t) = R_i \exp\left\{-\int_0^t dt' R_i\right\}\,, \tag{6.166a}$$

$$\mathcal{P}_{ii}(t) = R_{ii} \exp\left\{-\int_0^t dt' R_{ii}\right\}\,. \tag{6.166b}$$

So to complete the description we need to elucidate how the rates R_i and R_{ii} are related to the subject's perception and evaluation of the executed action strategy implementation.

In Sect. 5.4.1 we discussed the cloud-type measure of cognitive effort, in particular, its confinement (Sect. 5.4.1) justified by experiments presented in Appendix 5.10. In our discussion about recognizing events i and ii, the confinement effect must be also reflected in perceiving the proximity of the object motion state to the stable manifold Ξ_s or the action point layer $\partial\mathbb{C}^s_\Xi$.

The action point layer $\partial\mathbb{C}^s_\Xi$: When the object motion state is rather far from the region $\partial\mathbb{C}^s_\Xi$, i.e., the formal overlap between the clouds $\mathbb{C}^{\text{en}}_P$ and $\partial\mathbb{C}^s_\Xi$ is small, i.e., $r^\partial \ll 1$, the subject is not able to recognized the value of $r^\partial \ll 1$. In this case, the subject has to treat the values r^∂ as equal to zero and consider the clouds $\mathbb{C}^{\text{en}}_P$ and $\partial\mathbb{C}^s_\Xi$ to be strictly disjoint. The given situation is described by the value $R_{ii} = 0$.

When the overlap r^∂ increases and exceeds some threshold r^∂_c, the subject becomes able to recognize the *possibility* of the cloud overlapping and the value R_{ii} should start to grow. Accepting the scale-free organization of the mind functioning we turn to the following ansatz

$$\mathcal{R}_{ii} = \begin{cases} 0, & \text{if } r^\partial \le r^\partial_c\,, \\ \dfrac{1}{\tau_d}\left(\dfrac{r^\partial}{r^\partial_c} - 1\right)^{\alpha_\partial}, & \text{if } r^\partial > r^\partial_c\,, \end{cases} \tag{6.167}$$

where the exponent $\alpha_\partial > 0$ is a model parameter and the time scale τ_d is the characteristic duration of experiential now—the time interval required for the subject to recognize the current state of the object motion. According to the experimental data

(Sect. 5.10) we may use the estimate $\alpha_\partial > 1$. We want to note that the existence of this strict threshold and uncertainty in human perception do not contradict each other. The matter is that this threshold is related to the transition between automatic and deliberate modes of decision-making, i.e., the transition between E-consciousness and AM-consciousness which is not controlled, at least explicitly, by the mind (see Sect. 5.10).

When for the subject both the options "no-overlapping" and "partial-overlapping" between the clouds $\mathbb{C}_P^{\text{en}}$ and $\partial\mathbb{C}_\Xi^s$ are available, which branch of the action point layer is involved in consideration is firmly determined. Put differently, the subject's categorization of the object motion state comprises two primary situations (Fig. 6.5): the space-time cloud $\mathbb{C}_P^{\text{en}}$ overlaps with

– the upper branch of the action point layer $\partial\mathbb{C}_\Xi^s$,
– the lower branch of the action point layer $\partial\mathbb{C}_\Xi^s$.

So the choice of the "no-overlapping" category is the result of rejecting both the primary categories rather than an additional option analyzed individually. Moreover, there no values of the overlap r_∂ when the subject chooses between the two primary categories simultaneously (we discuss this feature in Sect. 5.4.1).

To represent this type subject's evaluation of the mutual arrangement of the clouds $\mathbb{C}_P^{\text{en}}$ and $\partial\mathbb{C}_\Xi^s$ in more formal terms, let us generalize the description of the action point layer $\partial\mathbb{C}_\Xi^s$ as follows. In Sect. 6.2.3 we constructed the action point layer $\partial\mathbb{C}_\Xi^s$ treating its density as a positive quantity to quantify the degree of belonging to the corresponding fuzzy region in the Hamilton space $\mathbb{R}_H$. In the following sections we reduced the Hamilton space to the space-time $\{x, v, a\} \times p_a \times t$ of the human temporality in describing the space-time cloud of action strategy implementation. Now we modify the cloud density such that the upper branch $\partial\mathbb{C}_\Xi^{s+}$ of the action point layer (Fig. 6.5) is positive, whereas the lower branch $\partial\mathbb{C}_\Xi^{s-}$ is characterized by the negative density of the same magnitude. In other words, the cloud density $\Omega_\Xi^\partial(p_a)$ entering Exp. (6.160) is treated as an asymmetric function relative to the point π_s (the value of the momentum p_a belonging to the stable manifold for a given state of object motion, $p_a = \pi_s(x, v, a)$):

$$\Omega_\Xi^\partial(p_a) > 0 \quad \text{for } p_a > \pi_s \text{ and } \quad \Omega_\Xi^\partial(\pi_s - p_a) = -\Omega_\Xi^\partial(p_a - \pi_s)\,.$$

After the given generalization, we can use the partial overlap, Exp. (6.162b),

$$\phi^\partial = \left\langle \partial\mathbb{C}_\Xi^s \middle| \overline{\mathbb{C}_P^{\text{en}}} \right\rangle$$

whose sign singles out the branch of the $\partial\mathbb{C}_\Xi^s$-cloud involved in consideration and then the quantity $\left\langle \phi^\partial \middle| \phi^\partial \right\rangle$, Exp. (6.163), via relationship (6.167) determines the rate R_{ii} of the subject recognizing event ii. Figure 6.6 (right column) illustrates these constructions.

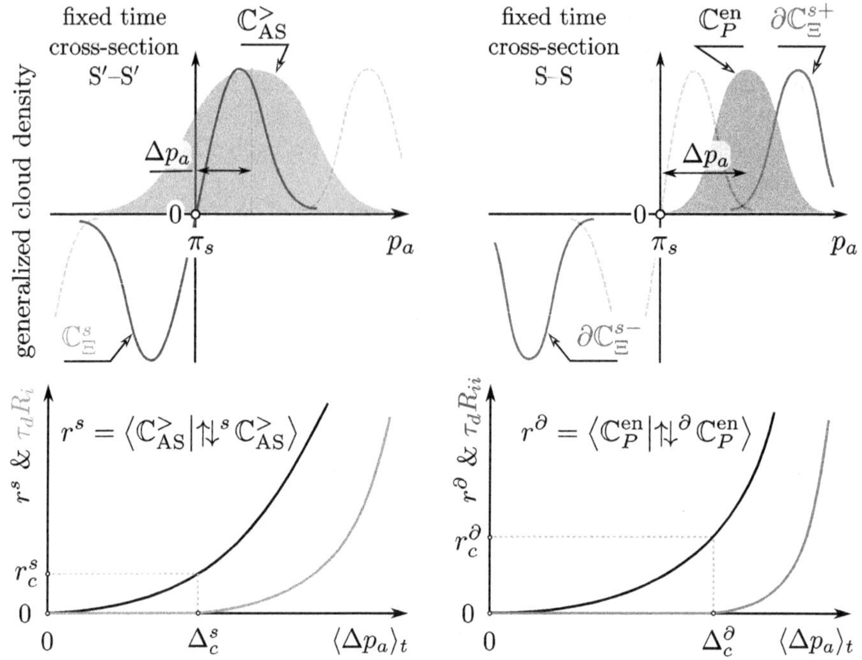

Fig. 6.6 The cloud overlapping and its characteristics. Upper row represents the arrangement of the space-time cloud components of $\mathcal{AS}$-implementation with respect to the stable manifold Ξ_s. Lower row represents the related quantities r^w, $\tau_d R^w$ ($w = r,\ \partial$) characterizing how the subject evaluates the proximity of the $\mathcal{AS}$-implementation to the manifold Ξ_s. Left column illustrates the first stage of action strategy termination. Right column illustrates its second stage. Left, upper plot depicts the overlapping between the forward temporal part $\mathbb{C}^{>}_{\text{AS}}$ of the $\mathcal{AS}$-implementation and the cloud $\mathbb{C}^{s}_{\Xi}$ of the stable manifold Ξ_s (at the time moment corresponding to the cross-section S′–S′ in Fig. 6.5). Here the value Δp_a quantifies the shift of $\mathbb{C}^{>}_{\text{AS}}$ from Ξ_s (specified by the momentum π_a). Left, lower plot illustrates (*i*) the value r^s of the overlap between $\mathbb{C}^{>}_{\text{AS}}$ and $\mathbb{C}^{s}_{\Xi}$ (black curve) and (*ii*) the subject's evaluation of this overlap (green curve) as functions of the mean shift $\langle \Delta p_a \rangle_t$ of $\mathbb{C}^{>}_{\text{AS}}$ from Ξ_s. Here the quantities Δ^s_c and r^s_c represent the threshold of subject's perception caused by the transition between E-consciousness and AM-consciousness. Right column represents similar features for the space-time cloud $\mathbb{C}^{\text{en}}_{P}$ and the two branches $\partial\mathbb{C}^{s+}_{\Xi}$, $\partial\mathbb{C}^{s-}_{\Xi}$ of the action point layer

The stable manifold cloud $\mathbb{C}^{s}_{\Xi}$: As far as the first stage of terminating action strategy implementation is concerned, its description is actually the same provided the density of the $\mathbb{C}^{s}_{\Xi}$-cloud is modified in a similar way. Namely, initially the cloud $\mathbb{C}^{s}_{\Xi}$ of the stable manifold Ξ_s is represented by the density $\Omega^{s0}_{\Xi}(p_a)$ being a positive-definite function of the momentum $p_a \propto j$ characterizing the accuracy in the subject's monitoring of the object jerk j. At the moment of action strategy initiation, the subject recognizes that finally the initiated implementation P of the selected action strategy $\mathcal{AS}$ will inevitably deviate from the optimal implementation AS_{opt} but he cannot predict the direction of this deviation. Mathematically it is reflected in that the initial space-time cloud $\mathbb{C}^{\text{ini}}_{\text{AS}}$ being much wider than the cloud $\mathbb{C}^{s}_{\Xi}$ is located such that the manifold $\varXi_s$ containing $\mathrm{AS}_{\mathrm{opt}}$ is in the center of $\mathbb{C}^{\mathrm{ini}}_{\mathrm{AS}}$ (see, e.g., Fig. 6.4). This situation matches the maximal overlap between the forward temporal part $\mathbb{C}^{>}_{\mathrm{AS}}$ of the $\mathcal{AS}$-implementation and the stable manifold cloud $\mathbb{C}^{s}_{\varXi}$. As time goes on, the direction of the P-deviation from the manifold $\varXi_s$ becomes more pronounced and the overlap between the clouds $\mathbb{C}^{>}_{\mathrm{AS}}$ and $\mathbb{C}^{s}_{\varXi}$ decreases. We want to relate the deviation of the forward temporal part $\mathbb{C}^{>}_{\mathrm{AS}}$ from the manifold $\varXi_s$ to the increase in the overlap between the clouds $\mathbb{C}^{>}_{\mathrm{AS}}$ and $\mathbb{C}^{s}_{\varXi}$. For this purpose we modify the concept of the cloud $\mathbb{C}^{s}_{\varXi}$ by allowing its density to take negative values. It is done via the replacement in Exp. (6.160)

$$\varOmega^{s0}_{\varXi}(p_a) \to \varOmega^{s}_{\varXi}(p_a) \propto -\frac{d\varOmega^{s0}(p_a)}{dp_a} \tag{6.168}$$

with the proportionality coefficient enabling normalization (6.161). In this way the cloud $\mathbb{C}^{s}_{\varXi}$ is generalized such that its density above the manifold $\varXi_s$ is positive, whereas below $\varXi_s$ is negative. Therefore, the partial overlap, Exp. (6.162b),

$$\phi^s = \langle \mathbb{C}^{s}_{\varXi} | \overline{\underline{\mathbb{C}^{>}_{\mathrm{AS}}}} \rangle$$

- takes zero-value at the moment of action strategy initiation,
- increases when the cloud $\mathbb{C}^{>}_{\mathrm{AS}}$ start to deviate from the manifold $\varXi_s$, and
- its sign indicates the direction of this deviation.

Figure 6.6 (left column) illustrates these aspects.

For the stable manifold cloud $\partial\mathbb{C}^{s}_{\varXi}$ we can repeat the constructions presented above for the action point layer $\partial\mathbb{C}^{s}_{\varXi}$. Namely, we accept that when the overlap r^s specified by Exp. (6.165) increases and exceeds some threshold r^s_c, the subject becomes able to recognize the *possibility* of the space-time cloud $\mathbb{C}^{>}_{\mathrm{AS}}$ deviating from the manifold $\varXi_s$. In this case, the value R_i should start to grow, which is described by the ansatz

$$\mathcal{R}_i = \begin{cases} 0, & \text{if } r^s \le r^s_c\,, \\ \dfrac{1}{\tau_d}\left(\dfrac{r^\partial}{r^\partial_c} - 1\right)^{\alpha_s}, & \text{if } r^s > r^s_c\,, \end{cases} \tag{6.169}$$

where the exponent $\alpha_s > 1$ is a model parameter.

The found expressions (6.167) and (6.169) specify the rates R_i and R_{ii} at which the subject recognizes the direction of the object motion deviation from the manifold $\varXi_s$ (event *i*) and the appropriateness of terminating the action strategy implementation for current state of object motion (event *ii*). Together with Exp. (6.166) it completes the description of terminating the action strategy implementation.

6.3.4 Action Strategy Efficiency

In implementing any action strategy, the subject faces up to finding a certain compromise between two opposing incentives. One of them is the minimization of the cognitive effort related to the action strategy implementation. The other is the displacement of the object toward the goal as close as feasible under the given conditions. Both the incentives impose restrictions on the effort of bodily executed actions and the effort related to the object motion monitoring. In Sects. 6.1, 6.2 we constructed the mathematical description of the former incentive underlying the passive phase of 2D-time dynamics of action strategy clouds. However, the Ostrogradsky instability has compelled us to describe the latter incentive in other terms than the action strategy implementation effort, namely, in the characteristics of action strategy termination analyzed in Sect. 6.3.3. So the purpose of the present subsection is to overcome the gap between the two descriptions and find a way to combine the two incentives within one functional—the action strategy efficiency.

At first, let us consider the action strategy efficiency in the *ideal case*. The hypothetical ideal subject is assumed to be able to minimize the effort functional (6.28) precisely (see Introduction to Sect. 6.2). It means that the subject is able to drive the object according to the optimal implementation $\mathrm{AS}_{\mathrm{opt}}$ of the selected action strategy $\mathcal{AS}$, i.e., along the optimal trajectory

$$P_{\mathrm{opt}} = \{x(t), v(t), a(t), j(t)\}_{\mathrm{opt}} \times t$$

in the space-time continuum $\mathbb{R}_{[4]} \times t$.

The value of the effort functional that it takes at the optimal trajectory P_{opt} may be treated as the zero-th approximation $\mathcal{E}^0_{\mathrm{AS}}$ of the action strategy cost $\mathcal{E}_{\mathrm{AS}}$. By virtue of Exp. (6.28), the value $\mathcal{E}^0_{\mathrm{AS}}$ can be written as

$$\mathcal{E}^0_{\mathrm{AS}} \stackrel{\mathrm{def}}{=} \mathcal{E}_{\mathrm{AS}}\{P_{\mathrm{opt}}\} = \int_{t_{\mathrm{ini}}}^{\infty} dt\, e^{-(t-t_{\mathrm{ini}})/\tau_{cp}} L\left(x, \frac{dx}{dt}, \frac{d^2x}{dt^2}, \frac{d^3x}{dt^3}\right)_{\mathrm{opt}}. \tag{6.170}$$

Here t_{ini} is the time moment of action strategy initiation, the analyzed form of the Lagrangian $L(x, v, a, j)$ is presented in Sect. 6.1.3, in particular by Exp. (6.26), and τ_{cp} is the characteristic duration of the complex present. The exponential factor $e^{-(t-t_{\mathrm{ini}})/\tau_{cp}}$ takes into account that the subject confines his anticipation of the object motion dynamics to the connected future only. It matches the accepted formulation of goal-oriented actions as the temporal boundary value problem (6.33) combined with the temporal criterion of proximity to $\mathrm{AS}_{\mathrm{opt}}$ (Sect. 6.2.3).

Expression (6.170) specifies the quantity $\mathcal{E}^0_{\mathrm{AS}}$ as a function of the action strategy $\mathcal{AS}$, i.e., as a certain function of its basic parameters including

- the position of the controlled object in the Lagrange space $\mathbb{R}_L = \{x, v, a\}$ at the moment of initiation with respect to the goal position and

– the accuracy in monitoring the variables $\{x(t), v(t), a(t), j(t)\}$ during the object motion.

Below we will regard the function $\mathcal{E}^0_{\text{AS}}(\mathcal{AS})$ is known beforehand for all the action strategies under consideration. The optimal trajectory leads directly to the desired goal, so in the ideal case the action strategy efficiency is reduced to the cumulative effort of action strategy implementation.

In the *real case*, the situation changes substantially because the real subject cannot drive the object in the optimal way as a consequence of the bounded capacity of human perception.

We turned to the concept of the skilled subject to describe the action strategy implementation (see Introduction to Sect. 6.2). Each action strategy $\mathcal{AS}$ can be implemented by the skilled subject in many ways. However, which one of these trajectories represents the action strategy within a given situation is beyond the subject's cognition. Almost all of these trajectories do not converge to the goal as time goes on. Moreover, they formally go to infinity, which is a result of the Ostrogradsky instability (Sect. 6.2.3). To overcome this problem we turned to the temporal criterion of proximity to the action strategy optimal implementation and constructed the action-point layer $\partial\mathbb{C}^s_\Xi$ surrounding the stable manifold $\mathbb{C}^s_\Xi$ (Sect. 6.2.3). An action strategy implementation AS initiated near the stable manifold $\mathbb{C}^s_\Xi$ inevitably deviates from $\mathbb{C}^s_\Xi$ with time. So when reaching the action-point layer $\partial\mathbb{C}^s_\Xi$ the action strategy implementation is terminated by the subject, which was described in the previous Sect. 6.3.3. According to the temporal criterion of proximity, the *acceptable* implementations of the action strategy $\mathcal{AS}$ should not reach the action-point layer $\partial\mathbb{C}^s_\Xi$ within the connected future of the complex present whose experiential now matches the $\mathcal{AS}$-initiation.

This approach poses a problem of describing the contribution of the noted confinement condition to the $\mathcal{AS}$-efficiency because it no more operates with the notion of atomic action effort, at least, explicitly.[55]

There are two ways to resolving this problem. First, it is (*i*) to use the results obtained in the previous Sect. 6.3.3 which describe the action strategy termination and, then, (*ii*) to average them over all the probable implementations of action strategy. Second, it is to turn to the subject's anticipation of further motion of controlled object within the connected future just after the $\mathcal{AS}$-initiation. The two approaches have to give the same result due to the self-consistency of skill acquisition on scales exceeding substantially the duration of complex present. It should be noted that this self-consistency is another facet of the principle of cloud-fusion self-consistency formulated in Sect. 5.6.3.

[55] In Sect. 6.1.2 we turned to the gist of cost-benefit tradeoffs to make comparable the cost of (*i*) action strategy implementation and (*ii*) attaining the goal within the complex present. Then, we constructed the functional quantifying the action strategy efficiency uniting both the factors. It is well justified when the goal is attained within the complex present after the action strategy initiation. Moreover, at that stage, it was sufficient for theory development. The temporal criterion of proximity to action strategy optimal implementation makes it necessary to generalize the notion of action strategy efficiency, which is done in the present section.

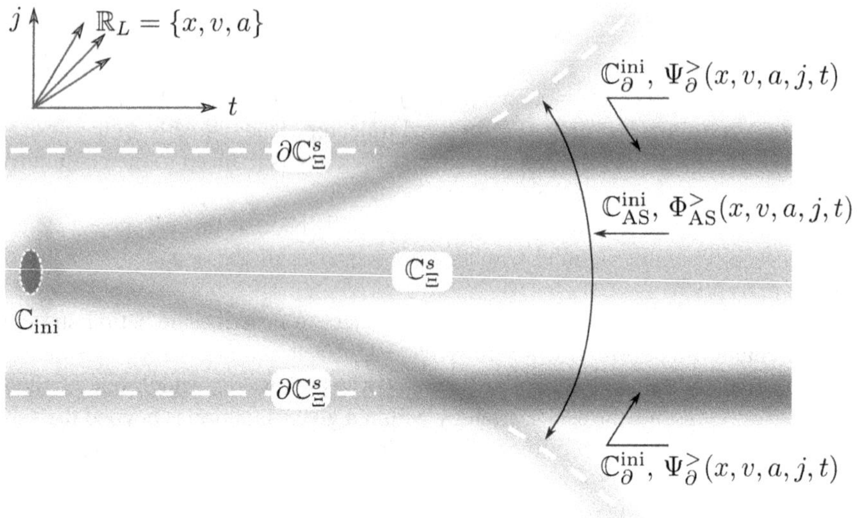

Fig. 6.7 The space-time cloud $\mathbb{C}^{\rm ini}_{\rm AS}$ of the action strategy $\mathcal{AS}$ just after its initiation in the case when the action point layer $\partial\mathbb{C}^s_{\Xi}$ is interpreted as the trap of the controlled object. The trapped motion states are represented by the space-time cloud $\mathbb{C}^{\rm ini}_{\partial}$ located inside the action point layer $\partial\mathbb{C}^s_{\Xi}$. Here $\mathbb{R}_L \times j \times t$ is the coordinate system of the human temporality and the distributions $\Phi^{>}_{\rm AS}(x, v, a, j, t)$, $\Psi^{>}_{\partial}(x, v, a, j, t)$ represent the space-time clouds $\mathbb{C}^{\rm ini}_{\rm AS}$, $\mathbb{C}^{\rm ini}_{\partial}$, respectively. The intensity of these clouds (distributions) is illustrated with the color (green, red) saturation. White dashed lines indicate the regions where the cloud intensity is ignorable, $\mathbb{C}_{\rm ini}$ is the cloud of the $\mathcal{AS}$-initiation, $\mathbb{C}^s_{\Xi}$ is the space-time cloud of the stable manifold Ξ

We will turn to the second way because which branch of the action point layer $\partial\mathbb{C}^s_{\Xi}$ is involved is not essential. Moreover, in this case, the formalism of forward temporal parts of action strategies averages over all the probable implementations of a given action strategy automatically. The gist of the proposed description is illustrated in Fig. 6.7.

Extended concept of action strategy space-time cloud: At the first step, we need to generalize the notion of the space-time cloud $\mathbb{C}_{\rm AS}$ of the action strategy $\mathcal{AS}$, mainly, its forward temporal part $\mathbb{C}^{\rm ini}_{\rm AS}$ that emerges just after the initiation. As previously, the cloud of action strategy $\mathcal{AS}$ represents the possible implementations of action strategy which describe the *motion* of the controlled object in the space-time $\mathbb{R}_{[4]} \times t$. The extended concept of the space-time cloud $\mathbb{C}_{\partial}$ as well as its forward temporal part $\mathbb{C}^{\rm ini}_{\partial}$ deals with the terminated implementations of $\mathcal{AS}$ which are treated as the object motion state trapped by the action-point layer $\partial\mathbb{C}^s_{\Xi}$. In Fig. 6.7 the intensity of these clouds is illustrated with color (green and red) saturation, white dashed lines denote the regions where the cloud intensity is ignorable.

In the realm of the body-in-the-self the clouds $\mathbb{C}^{\rm ini}_{\rm AS}$ and $\mathbb{C}^{\rm ini}_{\partial}$ are represented by the distributions

$$\Phi^{>}_{\mathrm{AS}}(x, v, a, j, t)\,, \quad \text{and} \quad \Psi^{>}_{\partial}(x, v, a, j, t) = \psi^{>}_{\partial}(x, v, a, t)\Omega^{\partial}_{\Xi}(j)$$

in the human temporality as illustrated below:

state	the mind-in-the-self		the body-in-the-self
moving:	$\mathbb{C}^{\mathrm{ini}}_{\mathrm{AS}}$	$\leftrightarrow$	$\Phi^{>}_{\mathrm{AS}}(x, v, a, j, t)$
trapped:	$\mathbb{C}^{\mathrm{ini}}_{\partial}$	$\leftrightarrow$	$\Psi^{>}_{\partial}(x, v, a, j, t)$

Here the function $\Omega^{\partial}_{\Xi}(j)$ represents structure of the action-point layer $\partial\mathbb{C}^{s}_{\Xi}$ and meets the normalization

$$\left\langle \mathbb{C}^{\partial}_{\Xi} \middle| \mathbb{C}^{\partial}_{\Xi} \right\rangle \overset{\mathrm{def}}{=} \left\langle \Omega^{\partial}_{\Xi} \middle| \Omega^{\partial}_{\Xi} \right\rangle = \int_{-\infty}^{\infty} dj \left[\Omega^{\partial}_{\Xi}(j)\right]^2 = 1\,. \tag{6.171}$$

The function $\Omega^{\partial}_{\Xi}(j)$ was introduced in the previous Sect. 6.3.3. Its form is depicted in Fig. 6.6 (right upper plot) within the replacement $j \leftrightarrow p_a$; in the analyzed case the sign of the function $\Omega^{\partial}_{\Xi}(j)$ is not essential.

In the mind the clouds $\mathbb{C}^{\mathrm{ini}}_{\mathrm{AS}}$ and $\mathbb{C}^{\mathrm{ini}}_{\partial}$ represent different aspects of the action strategy implementations and, so, are not mixed. It implies that in the body-in-the-self they are represented by the vector

$$V_{\mathrm{AS}} = \begin{Vmatrix} \Phi^{>}_{\mathrm{AS}}(x, v, a, j, t) \\ \Psi^{>}_{\partial}(x, v, a, j, t) \end{Vmatrix} \tag{6.172}$$

normalized by the condition

$$\left\langle \mathbb{C}^{\mathrm{ini}}_{\mathrm{AS}} \middle| \mathbb{C}^{\mathrm{ini}}_{\mathrm{AS}} \right\rangle + \left\langle \mathbb{C}^{\mathrm{ini}}_{\partial} \middle| \mathbb{C}^{\mathrm{ini}}_{\partial} \right\rangle = \int_{-\infty}^{+\infty} \!\!\cdots\! \int dx\, dv\, da\, dj \left[\Phi^{>}_{\mathrm{AS}}(x, v, a, j, t)\right]^2$$
$$+ \int_{-\infty}^{+\infty} \!\!\cdots\! \int dx\, dv\, da \left[\psi^{>}_{\partial}(x, v, a, t)\right]^2 = 1 \tag{6.173}$$

in the line with the general constructions of spaced-time action strategy clouds discussed in Sect. 6.2.7. In writing the normalization condition (6.173) we have taken into account identity (6.171).

Generalization of governing equation: At the second step of theory development we need to generalize the governing equation (6.149) to allow for the trapping by action-point layer $\partial\mathbb{C}^{s}_{\Xi}$. We accept that this generalization

- has to describe the effect of trapping the space-time cloud $\mathbb{C}^{\mathrm{ini}}_{\mathrm{AS}}$ by the action-point layer $\partial\mathbb{C}^{s}_{\Xi}$ in the form of a linear operator acting on the distribution function $\Phi^{>}_{\mathrm{AS}}(x, v, a, j, t)$. It is due to the irreversibility of the trapping in the framework of the internal 2D-time dynamics of action strategies;

– may deal with nonlinear operators acting on the distribution function $\Psi_{\partial}^{>}(x, v, a, j, t)$ because the trap effect gives rise to the additive increase in the intensity of the object trapped state;
– should describe the trap effect that on its own conserves the normalization condition (6.173),
– can be confined to the local dynamics of the distribution $\Psi_{\partial}^{>}(x, v, a, j, t)$ in the Lagrange space $\mathbb{R}_L = \{x, v, a\}$ because only the time moment when the action strategy implementation is "trapped" is essential.

Besides, let us quantify the rate of trapping via the embedding of the space-time cloud $\mathbb{C}_{\text{AS}}^{\text{ini}}$ into the action-point layer $\partial\mathbb{C}_{\Xi}^{s}$ according to the constructions presented in the previous Sect. 6.3.3, namely, Exps. (6.162). Put differently, the trap operator acting on the function $\Phi_{\text{AS}}^{>}(x, v, a, j, t)$ is written as

$$-\frac{1}{\tau_d}\cdot\left|\uparrow\!\downarrow^{\partial}\mathbb{C}_{\text{AS}}^{\text{ini}}\right\rangle = -\frac{1}{\tau_d}\cdot\left|\partial\mathbb{C}_{\Xi}^{s}\right\rangle\left\langle\partial\mathbb{C}_{\Xi}^{s}\,\overline{\underline{|\mathbb{C}}}\right\rangle$$
$$= -\frac{1}{\tau_d}\Omega_{\Xi}^{\partial}(j)\left\langle\Omega_{\Xi}^{\partial}\,\overline{\underline{|\Phi_{\text{AS}}^{>}}}\right\rangle = -\frac{1}{\tau_d}\Omega_{\Xi}^{\partial}(j)\int_{-\infty}^{\infty}dj\,\Omega_{\Xi}^{\partial}(j)\Phi_{\text{AS}}^{>}(x, v, a, j, t)\,,$$

where, as previously, τ_d is the characteristic duration of experiential now (the diachronic unit).

In this terms the generalized equation governing the distribution of vector (6.172) in the human temporality takes the form

$$\frac{\partial\Phi_{\text{AS}}^{>}}{\partial t} = \widehat{\mathcal{H}}\,\Phi_{\text{AS}}^{>} - \frac{1}{\tau_d}\Omega_{\Xi}^{\partial}(j)\left\langle\Omega_{\Xi}^{\partial}\,\overline{\underline{|\Phi_{\text{AS}}^{>}}}\right\rangle - E\Phi_{\text{AS}}^{>}\,, \tag{6.174a}$$

$$\frac{\partial\Psi_{\partial}^{>}}{\partial t} = \frac{1}{\tau_d\psi_{\partial}^{>}}\Omega_{\Xi}^{\partial}(j)\cdot\left\langle\overline{\underline{\Phi_{\text{AS}}^{>}}}|\Omega_{\Xi}^{\partial}\right\rangle\left\langle\Omega_{\Xi}^{\partial}\,\overline{\underline{|\Phi_{\text{AS}}^{>}}}\right\rangle - E\Psi_{\partial}^{>}\,. \tag{6.174b}$$

Here $\widehat{\mathcal{H}}$ is the same Hamiltonian that enters Eq. (6.149). By virtue of Exp. (6.171) Eq. (6.174b) after multiplicaiton by $\Psi_{\partial}^{>}$ and integration over j can be also rewritten as

$$\psi_{\partial}^{>}\frac{\partial\psi_{\partial}^{>}}{\partial t} = \frac{1}{\tau_d}\cdot\left\langle\overline{\underline{\Phi_{\text{AS}}^{>}}}|\Omega_{\Xi}^{\partial}\right\rangle\left\langle\Omega_{\Xi}^{\partial}\,\overline{\underline{|\Phi_{\text{AS}}^{>}}}\right\rangle - E\left[\psi_{\partial}^{>}\right]^2. \tag{6.174c}$$

Here as before the lines above and below the symbol $\Phi_{\text{AS}}^{>}$ mean that the arguments $\{x, v, a\}$ of the function $\Phi_{\text{AS}}^{>}(x, v, a, j, t)$ are fixed. Following the constructions of the cloud dynamics presented in Sect. 6.2.8 the value of E is found from the condition of conserving the normalization of the state vector, Exp. (6.173). Whence it follows that

$$E = \left\langle\Phi_{\text{AS}}^{>}\right|\widehat{\mathcal{H}}_s\left|\Phi_{\text{AS}}^{>}\right\rangle \tag{6.175}$$

where $\widehat{\mathcal{H}}_s$ is the Hermitian component of the operator $\widehat{\mathcal{H}}$, see Exp. (6.147). It should be noted that Exp. (6.175) gives the same result as Exp. (6.146) corresponding to the case when the presence of the action-point layer $\partial\mathbb{C}^{s}_{\Xi}$ is not taken into account. The governing equation (6.174) is subject to the "initial" condition reflecting the properties of the $\mathcal{AS}$-initiation. Namely,

$$\Phi^{>}_{\mathrm{AS}}\Big|_{t=t_{\mathrm{ini}}} = \Phi^{\mathrm{ini}}_{\mathrm{AS}}(x, v, a, j)\,, \qquad \Psi^{>}_{\partial}\Big|_{t=t_{\mathrm{ini}}} = 0\,, \tag{6.176}$$

where the function $\Phi^{\mathrm{ini}}_{\mathrm{AS}}(x, v, a, j)$ represents the cloud of action strategy initiation $\mathbb{C}_{\mathrm{ini}[4]}$, Exp. (6.67).

Efficiency of action strategy: To finalize our constructions we need to introduce a numerical criterion to quantify the efficiency of reaching the desired goal within a single instantiation of a given action strategy $\mathcal{AS}$. Qualitatively this criterion implies that, on the average, the $\mathcal{AS}$-implementations should not be terminated within the connected future after their initiations. The quantity to be constructed actually is a measure of the opposite criterion, it measures the degree that the $\mathcal{AS}$-implementations are terminated within the connected future. Therefore this quantity will be called the *inefficiency* of the action strategy $\mathcal{AS}$ in single instantiation and designated by the symbol $\mathcal{E}^{\partial}_{\mathrm{AS}}$.

Keeping in mind Exp. (6.170) let us specify the *single instantiation inefficiency* (or just *inefficiency*) of an action strategy $\mathcal{AS}$ via the expression

$$\begin{aligned}\mathcal{E}^{\partial}_{\mathrm{AS}} &= \frac{1}{\tau_{cp}}\int\limits_{t_{\mathrm{ini}}}^{\infty} dt\, e^{-(t-t_{\mathrm{ini}})/\tau_{cp}} \left\langle \Psi^{>}_{\partial} \middle| \Psi^{>}_{\partial} \right\rangle (t) \\ &= \frac{1}{\tau_{cp}}\int\limits_{t_{\mathrm{ini}}}^{\infty} dt\, e^{-(t-t_{\mathrm{ini}})/\tau_{cp}} \int\limits_{-\infty}^{+\infty}\!\!\cdots\!\int dx\, dv\, da \big[\psi^{>}_{\partial}(x, v, a, t)\big]^{2}.\end{aligned} \tag{6.177}$$

If a given action strategy is extremely inefficient, i.e., it is terminated practically immediately after initiation, the value $\langle \Psi^{>}_{\partial} | \Psi^{>}_{\partial} \rangle = 1$ almost everywhere in the connected future and the quantity gets its maximum $\mathcal{E}^{\partial,\overline{\lim}}_{\mathrm{AS}} = 1$. When the termination of a given action strategy within the connected future is improbable and so the value $\langle \Psi^{>}_{\partial} | \Psi^{>}_{\partial} \rangle \ll 1$ and highly efficient action strategies are characterized by $\mathcal{E}^{\partial,\underline{\lim}}_{\mathrm{AS}} = 0$.

Using the identity

$$\frac{1}{\tau_{cp}} e^{-(t-t_{\mathrm{ini}})/\tau_{cp}} = -\frac{d}{dt} e^{-(t-t_{\mathrm{ini}})/\tau_{cp}}\,,$$

the initial condition (6.176), the normalization (6.173), and the governing equation (6.174a) we rewrite Exp. (6.177) as

$$\mathcal{E}^{\partial}_{\rm AS} = 2\int\limits_{t_{\rm ini}}^{\infty} dt\, e^{-(t-t_{\rm ini})/\tau_{cp}} \left\{ \frac{1}{\tau_d} \left\langle \overline{\Phi^{>}_{\rm AS}|\Omega^{\partial}_{\Xi}\rangle\langle\Omega^{\partial}_{\Xi}|\Phi^{>}_{\rm AS}} \right\rangle \right.$$

$$\left. - \left\langle\Phi^{>}_{\rm AS}\right| \widehat{\mathcal{H}}_s \left|\Phi^{>}_{\rm AS}\right\rangle \left[1 - \left\langle \Phi^{>}_{\rm AS} \middle| \Phi^{>}_{\rm AS} \right\rangle\right]\right\}. \tag{6.178}$$

Here, according to definition (6.163),

$$\left\langle \overline{\Phi^{>}_{\rm AS}|\Omega^{\partial}_{\Xi}\rangle\langle\Omega^{\partial}_{\Xi}|\Phi^{>}_{\rm AS}} \right\rangle = \int\limits_{-\infty}^{+\infty} \cdots \int dx\, dv\, da\, dj\, dj'$$

$$\times\, \Phi^{>}_{\rm AS}(x, v, a, j, t)\Omega^{\partial}_{\Xi}(j)\Omega^{\partial}_{\Xi}(j')\Phi^{>}_{\rm AS}(x, v, a, j', t)\,.$$

Expression (6.178) finalizes our construction of the inefficiency of action strategy.

Equation (6.174a) combined with relation (6.175) form the closed description of the space-time cloud $\mathbb{C}^{\rm ini}_{\rm AS}$. So its solution specifies the value $\mathcal{E}^{\partial}_{\rm AS}$ as a function of the action strategy parameters including the coordinates of the $\mathcal{AS}$-initiation and the accuracy of monitoring. In this sense, (*i*) the zero-th approximation of the action strategy implementation cost, i.e., the cost of the action strategy optimal implementation $\mathcal{E}^{0}_{\rm AS}$ (Exp. 6.170) and (*ii*) the inefficiency $\mathcal{E}^{\partial}_{\rm AS}$ of getting the desired goal within a single instantiation of the action strategy $\mathcal{AS}$ (Exp. 6.178)

- possess actually the same structure in the human temporality as distributed in time characteristics of subject's actions,
- depend on the same characteristics of action strategies including the cognitive effort of bodily executed actions and the mental effort of the object motion monitoring (this characteristics determine the Hamiltonian $\widehat{H}$ and the mutual arrangement of the stable manifold Ξ_s and the action-point layer $\partial\mathbb{C}^{s}_{\Xi}$),
- characterize the action strategy efficiency in the same way, namely, the lower the value of $\mathcal{E}^{0}_{\rm AS}$ or $\mathcal{E}^{\partial}_{\rm AS}$, the higher efficiency of the corresponding action strategy.

Moreover, the value $\mathcal{E}^{\partial}_{\rm AS}$ may be regarded as an implicit measure quantifying the variations of the implementation cost functional (6.170) caused by deviation of action strategy implementation from the optimal one.

Phenomena and mechanisms underlying the construction of the quantities $\mathcal{E}^{0}_{\rm AS}$ or $\mathcal{E}^{\partial}_{\rm AS}$ seem to be principally different and incommensurable. Nevertheless the long-term learning and skill acquisition should endow the subject with the ability to compare the two criteria and fuse them in finding appropriate ways of reaching desired goals. Actually it is the gist of the principle of cloud-fusion self-consistency (Sect. 5.6.3). Therefore we suppose that the skilled subject can combine the two quantities into one measure

$$\mathcal{E}_{\rm AS}(\mathcal{AS}) = \mathcal{E}^{0}_{\rm AS}(\mathcal{AS}) + g_{\partial}\cdot\mathcal{E}^{\partial}_{\rm AS}(\mathcal{AS}) \tag{6.179}$$

to be called the *extended cost* of action strategy implementations. Here the coupling coefficient g_∂ having dimension of effort is the characteristics of the subject's skill. In the future analysis we will consider the extended cost to be known for each action strategy $\mathcal{AS}$. Its minimum properties determine the subject's preferences in selecting action strategies and, so, the extended cost $\mathcal{E}_{\text{AS}}(\mathcal{AS})$ may be treated as the free energy of action strategies.

6.3.5 *Neurophysiological Background of Action Strategy Selection*

Starting from this subsection we focus on the active phase of 2D-time dynamics of subject's goal-oriented behavior. Let us remind that the *active phase* of 2D-time dynamics implies the deliberate, intentional change of one action strategy for another action strategy (Sect. 6.3.1). The action strategy selection plays a crucial role in this process. Below we will outline the basic features of the corresponding formalism within the phenomenological description of human goal-oriented behavior in the complex present. However, before passing directly to the model description, we present some neurological data elucidating our constructions.

Selection of action strategies is a cognitive process during which the subject compares plausible action strategies with one another according to some criterion and selects one of them for implementation. Nowadays, selection as a psychological process has been investigated in a variety of contexts including perceptual categorization, action choice, decision-making, and attention; for a comprehensive review a reader may be referred, e.g., to Mysore and Kothari (2020).

Among the crucial components of this cognitive process we want to note selective attention required for involving possible options into processing (see, e.g., Knudsen 2007; Moore and Zirnsak 2017, for the underlying neural mechanisms). The possibility of dividing selective attention, in particular, the human ability to track four or more targets simultaneously is reviewed, e.g., by McMains and Somers (2004), Cavanagh and Alvarez (2005). Turning to economic choice behavior, Padoa-Schioppa (2011), Padoa-Schioppa and Conen (2017) summarize the neural mechanisms of comparing subjective values represented explicitly at the neuronal level and involved in decision-making. For a review of the neural correlates of perceptual categorization and related aspects in decision-making with respect to potentially ambiguous or noisy sensory stimuli, thresholds, the speed-accuracy tradeoff a reader may be referred, e.g., to Gold and Ding (2013), Mulder et al. (2014), Hanks and Summerfield (2017). The issue specific for competitive selection in the relation to executing the appropriate motor plan among competing alternatives are reviewed by, e.g., Cisek and Kalaska (2010), Freedman and Assad (2011), Murakami and Mainen (2015).

We want also to emphasize the accumulated evidence for that different forms of cognitive selection may share key computational principles as well as neural mechanisms, despite differences in selection tasks and the brain areas studied (Jiang and Kanwisher 2003a, b; Schumacher and Jiang 2003; Wang 2008; Cisek and Kalaska 2010; Freedman and Assad 2011; Padoa-Schioppa 2011; Mysore and Knudsen 2012; Padoa-Schioppa and Conen 2017; O'Connell et al. 2018; Mysore and Kothari 2020). This universality seems to have similar roots as the universality of inner psychophysics (Sect. 1.3.2) as well as the general concept of global workspace emphasizing the integrative nature of consciousness (Sect. 3.2.5). The noted commonality and modality independence argue for the feasibility of constructing mathematical models for the selection process turning to the general principles rather than particular details of sensory modalities.

As far as modeling perspective is concerned, there have been proposed two broad classes of models for cognitive selection (Mysore and Kothari 2020). One of them can be called the *accumulation-to-threshold* approach turning to the first passage time problem. The gist of this approach is that:

- the evidence for each option accumulates individually with the rate exhibiting random variations;
- when for one of options the accumulated evidence crosses a certain strict threshold for the first time, the selection process is terminated and
- this options becomes the "winner" of competitive selection.

For a review and discussion of various problems analyzed within the accumulation-to-threshold approach a reader may be referred, e.g., to Bogacz et al. (2006), Ratcliff and McKoon (2008), Simen (2012), Solway and Botvinick (2012), Zhang (2012), Ratcliff et al. (2016), Shinn et al. (2020).

The other class of models turns to the general concepts of nonlinear dynamical systems. In particular, for systems with a set of attractors this class of models is often referred to as the *winner-take-all* approach. The attractors represent different options of the selection process. When the input—the initial system state—falls into the basin of a given attractor, reaching this attractor is treated as the selection process result—its output. The relative volumes of these basins quantify the preference of the options. The presence of noise endows the selection process with probabilistic properties especially for the input located near the watershed in the basin's space. The description of brain functioning as a dynamical system within the paradigm of synergistics (e.g., Kelso 1995; Haken 1996; Kelso and Engstrøm 2006; Haken 2008; Kelso et al. 2013; Pillai and Jirsa 2017; Grossberg 2018) underlies such models. The formalism of stochastic dynamical systems in potential landscape can be also employed for describing neural network dynamics (Yan et al. 2013), in particular, decision-making (Yan et al. 2016) and the stability-flexibility-energy tradeoff in working memory (Yan and Wang 2020).

Taking into account our constructions, let us note dynamical systems of a special type making the gist of the winner-take-all approach more pronounced. Namely, it is the competitive Lotka-Volterra models (e.g., Murray 2002, for a general discussion) and similar models proposed for describing neural networks (e.g., Grossberg 1980, 1988; Haken 2008; Ashwin et al. 2016). The Lotka-Volterra models also admit the generalization describing more complex processes of competitive selection when the system, at first, evolves towards a certain attractor and, then, "jumps" between different clusters of its states, i.e., is characterized by the transient dynamics of neural processing (Rabinovich et al. 2008; Buonomano and Maass 2009; Rabinovich and Varona 2012; Pessoa 2019, for a review see also Ashwin et al. 2016). This continuous jumps can be caused by additive noise in heteroclinic networks (e.g., Kori and Kuramoto 2001; Ashwin and Postlethwaite 2013) or the *winnerless competition* (Rabinovich et al. 2001, 2006; Afraimovich et al. 2008, for a general discussion see also Rabinovich et al. 2015). The generalized Lotka-Volterra models also take into account several factors—e.g., cognition, emotion, and attention (Rabinovich and Muezzinoglu 2010; Afraimovich et al. 2011, also for a review Pessoa 2019)—contributing to the competitive selection in a similar way. Generally speaking the dynamics of such systems may be conceived of as permanent transitions between the saddle type stationary points and look like fast transitions between some regions in the corresponding phase space wherein the system resides for relatively long time intervals. In this sense the dynamics of winnerless competition resembles the behavior of the dynamical systems possessing Zaslavsky's type dynamical traps (Zaslavsky 1995, 2002, see also Zaslavsky 2005).

For comparison with our constructions, let us present in shortened form a generalized Lotka-Volterra model describing the winnerless competition affected by emotions (cf. Rabinovich and Muezzinoglu 2010):

$$\frac{dC_i}{dt} = C_i\Big[\sigma_i^c A_C - \sum_j \rho_{ij}^c C_j\Big] + C_i\,\eta_i^c(t)\,,$$

$$\frac{dE_i}{dt} = E_i\Big[\sigma_i^e A_E - \sum_j \rho_{ij}^e E_j\Big] + E_i\,\eta_i^e(t)\,,$$

$$\frac{dA_C}{dt} = A_C\Big[\zeta^c \sum_j C_j - \kappa_{cc} A_C - \kappa_{ce} A_E\Big],$$

$$\frac{dA_E}{dt} = A_E\Big[\zeta^e \sum_j E_j - \kappa_{ec} A_C - \kappa_{ee} A_E\Big],$$

where the non-negative state variables C_i and E_i represent the cognitive and emotional components of mode i of neural network functioning, the coefficients $\{c_i^{c,e}\}$, $\{\rho_{ij}^{c,e}\}$, $\{\zeta^{c,e}\}$, $\{\kappa_{\cdot,\cdot}\}$ can depend on the system state, $\{\eta_i^{c,e}\}$ are the noise components of neural network functioning, and A_C and A_E are the attentional components quantifying the computational resources allocated to cognitive and emotional channels of information processing. In selective processes the contribution of the attention dynamics and attention fluctuations can be also essential (Fisher 2017; Rangelov and Mattingley 2020). This type models are also employed in describing neural processes governing motor behavior (Perdikis, Huys and Jirsa 2011; Perdikis, Raoul and Viktor 2011; Huys et al. 2014).

For comparison of the two approaches, i.e.,

– the accumulation-to-threshold models and

– the winner-take-all models combined with winnerless models

with respect to their common features and the underlying mechanisms a reader may be addressed to Busemeyer et al. (2019), Mysore and Kothari (2020).

Conscious-unconscious aspects of decision-making: Another issue we need to discuss briefly before passing directly to our constructions is the conscious aspects of the cognitive selection process. Libet et al. (1983), Libet (2002, 2004) conducted experiments that argue for the existence of "preparatory" brain activity shortly before a subject starts actions seeming to be spontaneous. Libet interpreted this phenomenon as that the brain on its own "makes decision" what to do before the subject *consciously* selects an option. These results impacted upon the modern debates about the free will problem. In particular, Wegner (2004, 2018) surveys a verity of such findings including Libet's experiments and argues that the experience of conscious willing is an illusion. The free will problem does not belong to the scope of the present book and for the corresponding discussion a reader may be referred to Baumeister and Monroe (2014), Clarke and Capes (2017), O'Connor and Franklin (2018), McKenna and Coates (2019). Following Bode et al. (2014) we focus our discussion on the conscious-unconscious aspects and underlying neural mechanisms of human decision-making within cognitive selection rather than on resolving the philosophical "free will" debate.

Turning to the arguments by Bargh et al. (2001) as well as the dual-system theory applied to describing phenomenal consciousness and intentional actions by Marvan and Polák (2017), Schlosser (2019), we actually accept that the action strategy selection is *implemented* automatically (unconsciously). Put differently, the action strategy selection is supposed to belong to the realm of the body-in-the-self. The result of this selection is just given to the mind-in-the-self. So the action strategy selection has to be categorized as a process belonging to the level of E-consciousness (Sect. 3.2.5). In the line with Haggard (2008, 2019), Bonicalzi and Haggard (2019), Balaguer (2019) we assume that the mind-in-the-self is responsible only for setting up the possible options for selection, which implies also focusing attention on them. In this context we want to note also a model proposed by Bignetti (2014) where, in his terminology, voluntary actions are decided and performed by the subject's unconscious mind.

Reinforcement learning: To make goal-oriented actions efficient the subject has to acquire the skill of selecting the relevant actions for a given state of controlled system. He does this via learning and accumulating the information from the previous actions under the same conditions or similar ones. It turned out that neural mechanisms governing this learning admit interpretation in terms of models of learning developed in computer science (Montague et al. 1996; Schultz et al. 1997; Schultz 1998; Doya 1999). In particular, the skill acquisition in goal-oriented actions is usually modeled employing various reinforcement learning algorithms (see, e.g., Sutton and Barto 2018, for introduction); for a review about the details of the relationship between the corresponding mathematical formalism and the underlying neural mechanisms a reader may be referred, e.g., to Daw et al. (2005), Dayan and Daw (2008), Dayan and Berridge (2014), Daw (2018), O'Doherty et al. (2015, 2017).

In modeling human behavior two classes of reinforcement learning algorithms are typically singled out, the model-free and the model-based reinforcement learning (e.g., O'Doherty et al. 2017; Daw 2018; Sutton and Barto 2018, for detail). To allow for the aspects of goal-oriented behavior Daw et al. (2005) turn to the gist of the model-based reinforcement learning. This learning process deals with internal models of the decision problem, actions, rewards, the cognitive states, and the transitions between them caused by the executed actions. The model-free reinforcement learning mechanism assumes that rewards are attributed only to actions on their own and does not track the corresponding cognitive states of the subject as well as his anticipation.

However, in the problem we deal with the two types of reinforcement learning look rather similar. The main difference between them is hidden in that an action strategy bears the information about the current motion state of controlled object, i.e., it describes the transition between two states: $\mathcal{AS} \mapsto S_t \to S_{t'}$. In this sense the efficiency $\mathcal{E}_{\text{AS}}$, i.e., the preference of the selected action strategy $\mathcal{AS}$ depends on the internal characteristics of the action strategy implementation as well as the external characteristics of its initiation. So the contribution of models underling the subject's behavior and the dynamics of controlled object should determine the relationship between the quantities $\{\mathcal{E}_{\text{AS}}\}$ for action strategies $\{\mathcal{AS}\}$ close to one another in the external and internal characteristics. The other details of the two types of reinforcement learning algorithms are similar. The subject anticipation is assumed to be confined to the complex present and the action strategy efficiency is recognized via accumulation of the subject's experience at the previous time moments.

Within the discussed features of model-based contribution to the action strategy selection, a simple model of reinforcement learning can be illustrated as follows. The preference $\mathcal{E}_{\text{AS}}(S)$ of a given action strategy $\mathcal{AS}$ depends on the system current state $S_i = S$. It also reflects the trade-off between the control benefit of getting the state S_t and some cost of action strategy implementation (e.g., effort). The subject infers this preference via the trial-end-error learning. Each time moment t when the action strategy $\mathcal{AS}$ is implemented, the information about the preference $\mathcal{E}_{\text{AS}}(S', t)$ of the neighboring action strategies is updated:

$$\mathcal{E}_{\text{AS}}(S', t + \tau) = \left(1 - \frac{\epsilon_M}{\tau}\right) \mathcal{E}_{\text{AS}}(S', t) + T_{S'S} \cdot R_{\text{AS}}(S, t)\,.$$

Here the time step τ is the characteristic duration of action strategy implementations, the parameter $0 < \epsilon_M < 1$ quantifies the memory decay, the reword $R_{\text{AS}}(S, t)$ represents the subject's experience, and the function $T_{S'S}$ describes the mutual proximity of the two action strategies $\mathcal{AS}$ and $\mathcal{AS}'$. The probability $\mathcal{P}_{\text{AS}}$ of selecting the action strategies $\mathcal{AS}$ for implementation is typically approximated by the *soft-max* ansatz (Sutton and Barto 2018)

$$\mathcal{P}_{\mathrm{AS}} = \exp\{\beta\mathcal{E}_{\mathrm{AS}}(S,t)\}\left[\sum_{\mathcal{AS}'}\exp\{\beta\mathcal{E}_{\mathrm{AS'}}(S,t)\}\right]^{-1},$$

where the parameter β quantifies the uncertainty in the preference evaluation and the experienced reword. It should be noted that the soft-max ansatz meets the assumption that in selecting action strategies the subject take into account only the difference in the efficiency of action strategies rather than their absolute values. See also (Gläscher et al. 2010; Kool et al. 2016, 2017) for a detailed description of the human reinforcement learning combining the model-free and model-based algorithms.

Keeping in mind our cloud-type approach to modeling human goal-oriented actions, we want to note the modern branch of the reinforcement learning dealing with the notions of quantum mechanics (e.g., Dong et al. 2008; Fakhari et al. 2013; Bukov et al. 2018; Li et al. 2020). The basic idea is that the current mental state of the subject is the superposition of possible actions, in our case, action strategies $\{\mathcal{AS}\}$

$$|S\rangle = \sum_{\mathcal{AS}} \psi_{S,\mathcal{AS}}\,|\mathrm{AS}\rangle\,.$$

The probability of selecting a given action strategy $\mathcal{AS}$ for implementation is given by the coefficient $|\psi_{S,\mathcal{AS}}|^2$. The dynamics of the subject's choice preferences is described a Hermitian operator acting on the wave-function $|S\rangle$ which depends on the current choice and the corresponding preference function. This operator is typically described within two possible models proposed by Shor (1994) and Grover (1997, 1996).

Finalizing the description of the neurophysiological background, we want to accentuate a number of aspects significant for our constructions.

- Action strategies as mental entities bear the properties (e.g., spatial position, monitoring accuracy) specifying their efficiency.
- Attention playing the role of computational resources can be divided between action strategies.

Due to the cognitive selection of possible action strategies in its *implementation* belonging to the realm of the body-in-the-self:

- This selection process admits a description in terms of (*i*) distributions representing the action strategies in the body-in-the-self and (*ii*) mathematical operations with these distributions (Sects. 5.6.1, 5.6.2).
- The implementation of the action strategy comparison and reallocation of attention is automatic (unconscious). The result of selection is raised to the level of the mind-in-the-self and belongs to E-consciousness (Sect. 3.2.5).

As far as the skill acquisition based on reinforcement learning is concerned, the subject's evaluation of action strategies, at least within the first approximation, is governed by the fusion of

– the efficiency of the action strategy that has been terminated right now,

– the aggregated efficiency accumulated within the previous implementations of action strategies.

Therefore, the cloud-type description of action strategies proposed in the given book enables one to model (*i*) the action strategy selection and (*ii*) the reinforcement learning where the uncertainty of human perception plays a crucial role.

6.3.6 Types of Active Phase

In describing the passive phase of subject's goal-oriented behavior, we focused on the 2D-time dynamics of the space-time cloud $\mathbb{C}_{\mathrm{AS}}$ representing all the probable implementations of a given action strategy $\mathcal{AS}$ in the *complex present*. In analyzing the active phase of the subject's goal-oriented behavior, we meet another problem. The subject has to select one of several action strategies suitable for implementation and to initiate it within the *experiential now*. Strictly speaking, we should draw a distinction between two different types of active phase.

– Active phase of the *corrective type* is the case when there is only one action strategy $\mathcal{AS}$ for implementation and the subject just initiates it.
– Active phase of the *selective type* is the case when there are two action strategies or more, $\{\mathcal{AS}_k\}$ $(k = 1, 2, \ldots)$, and the subject needs to select one of them for initiation.

Leaping ahead, we want to note that in mathematical description the action strategy selection resembles the competitive selection discussed in Sect. 6.3.5 but is governed by an entirely distinct mechanism to be considered below. Moreover, this mechanism enables us to describe the action strategy correction and the action strategy selection within common frameworks.

However, before passing directly to the desired description of active phase, we need to elucidate the basic common features and the difference of the two types of active phase. Their comparative analysis is based on the results obtained for the passive phase of the 2D-times dynamics of action strategy implementation. The present section is devoted to this issue.

Active phase of the corrective type, in particular, is the case in the situation considered in Sect. 6.2 within describing the passive phase of the 2D-time cloud dynamics. To elucidate this relationship let us note following characteristic features of the developed formalism for the 2D-time dynamics of the space-time cloud $\mathbb{C}_{\mathrm{AS}}$ in the complex present.

- According to its construction (Sects. 6.1 and 6.2), the Hamiltonian $\widehat{\mathcal{H}}$ entering the governing equation (6.149) of the $\mathbb{C}_{\mathrm{AS}}$-cloud dynamics explicitly contains the coordinates of the final goal of subject's actions. So, at least in principle, the subject considers it is possible to get the desired goal directly within one implementation AS of the action strategy $\mathcal{AS}$. If he fails to do this, the subject initiates a new implementation AS$'$ of $\mathcal{AS}$ rather than changes the action strategy.

- In describing the passive phase of subject's behavior we confined our constructions to a certain proximity to the goal; it was sufficient for elucidating the basic properties of the $\mathbb{C}_{\text{AS}}$-cloud dynamics. As a result the stable branch Ξ_s of the zero-Hamiltonian manifold is reduced to a linear subspace of the Hamilton space $\mathbb{R}_H$. Thereby, within this approximation there is a one-to-one map between

  - the current state $\{x, v, a\}$ of object motion—the point in the Lagrange space $\mathbb{R}_L = \{x, v, a\}$—where the previous $\mathcal{AS}$-implementation has been terminated and
  - the point $\sigma_s(x, v, a)$ on the stable manifold Ξ_s in whose neighborhood $\mathbb{C}_H^{\text{ini}}$ the next implementation of the action strategy $\mathcal{AS}$ is initiated (Sect. 6.2.2).
  The position of the point $\sigma_s(x, v, a)$ on the stable manifold Ξ_s specifies the value of jerk $j_\Xi(x, v, a)$ near which the skilled subject initiates the next $\mathcal{AS}$-implementation. In studying the active phase of subject's behavior we may confine our consideration to the value $j_\Xi(x, v, a)$.

- The unique optimal implementation AS$_{\text{opt}}$ of the action strategy $\mathcal{AS}$ lying on the stable manifold Ξ_s joins the point $\sigma_s(x, v, a)$ related to the object motion state at the instant of the $\mathcal{AS}$-initiation to the the desired goal.

So, in the analyzed case, the initiation of the next $\mathcal{AS}$-implementation may be conceived of as a step-wise *correction* of the current $\mathcal{AS}$-implementation.

The noted features prompt us to characterize the corrective type active phase by the following two properties the implementation of the action strategy $\mathcal{AS}$ to be initiated.

- *Action strategy uniqueness*. For each motion state $S = \{x, v, a\}$ of controlled object within a given region of the Lagrange space $\mathbb{R}_L$ there are only one value $j_\Xi(x, v, a) = j_\Xi(S)$ of jerk and only one optimal implementation AS$_{\text{opt}}$ of the action strategy $\mathcal{AS}$ which joins the point $\{S, j_\Xi(S)\}$ to the final goal in the space $\mathbb{R}_{[4]} = \{x, v, a, j\}$.
- *Action strategy continuity*. For two object motion states S and S' the corresponding action strategy clouds $\mathbb{C}_{\text{AS}}(S)$ and $\mathbb{C}_{\text{AS}}(S')$ of the action strategy $\mathcal{AS}$ tend to each other in property when $S \to S'$.

These properties allow us to treat the function

$$\Xi_s^j(x, v, a) \stackrel{\text{def}}{=} j_\Xi(x, v, a) \tag{6.180}$$

as a certain projection of the stable manifold $\Xi_s \subset \mathbb{R}_H$ onto the space $\mathbb{R}_{[4]}$. In the case under consideration, the function (manifold) $\Xi_s^j(x, v, a)$ as a whole entity may be regarded as the representation of all the possible optimal implementations of the action strategy $\mathcal{AS}$ in the space $\mathbb{R}_{[4]}$. We will call $\Xi_s^j(x, v, a)$ the *manifold of action strategy initiation* (or just *initiation manifold*) in the space $\mathbb{R}_{[4]}$.

Examples of systems with the selective type active phase: Below we list several particular examples characterizing human goal-oriented behavior where the selective type active phase of the 2D-time dynamics of action strategy space-time cloud should play a crucial role.

- Balancing a pendulum with overdamped dynamics (e.g., Zgonnikov et al. 2014) illustrates systems governed by human actions where the control parameter is the object acceleration a rather than the jerk j. In this case, when the pendulum is located in the vicinity of its upright position, the subject can
  - either try to drive the pendulum toward the desired position, which possibly will necessitate higher effort of monitoring the pendulum motion within the appropriate accuracy,
  - or do nothing and wait until the pendulum deviates from the upright position for a relatively large distance such that returning it to a small neighborhood of the desired position will require less effort.

  In this case, the general goal of subject's actions—to keep the pendulum near the upright position being unstable without subject's control—is implemented via the two action strategies which are selected alternately.
- In driving a car, e.g., in traffic flow on highway, the driver regularly makes decision about
  - decreasing the headway distance to the car ahead,
  - increasing this headway distance,
  - changing the lane for overtaking the car ahead,
  - not affecting the current car arrangement and wait until the situation changes and the relevant maneuver becomes clear.

  All these actions may be treated as possible action strategies competing with one another in reaching the desired goal—to drive the car at the speed convenient for the driver (see also Sect. 4.2.3).
- Choosing between two options, e.g., "light gray" and "dark gray" in classifying a visualized gray shade in computer-based experiments illustrates the experimental setup in studying the change of mind in decision making (Sect. 4.2.2). In this case, the made choice and its demonstration are separated in time by two types of subject's actions, e.g., moving the mouse courser to the upper left or upper right corner of the screen. These subject's actions are characterized by that after one of the possible actions has been initiated, the active phase is continued in parallel with the passive phase. It can give rise to the change of mind, i.e., the selection of the opposite action during the mouse motion and changing the mouse motion direction.

The noted examples are generalized within a hypothetical situation shown in Fig. 6.8 and underlying the following description of the selective type active phase.

Active phase of the selective type is illustrated by dynamics of a nonlinear physical system controlled by actions of the skilled subject. These actions are aimed at keeping

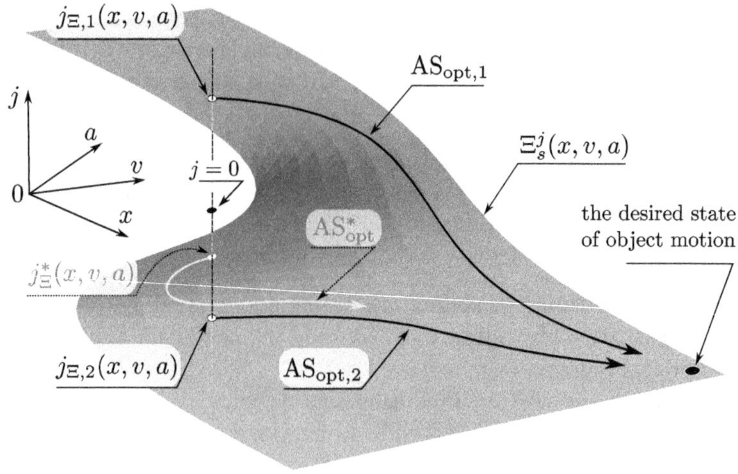

Fig. 6.8 Illustration of a nonlinear system where the subject's goal-oriented behavior is characterized by the selective type active phase. This type active phase is described by a manifold (surface) $\Xi^j_s(x, v, a)$ representing the initiation and optimal implementation of action strategies leading to the desired state of object motion starting from the current state $\{x, v, a\}$ of object motion. The surface $\Xi^j_s(x, v, a)$ has a fold in the space $\{x, v, a, j\}$ which makes the action strategy initiation ambiguous. Here the values $j_{\Xi(1,2)}(x, v, a)$ of the object jerk represent the initiation of two possible action strategies whose optimal implementations are shown by the lines $\mathrm{AS}_{\mathrm{opt}(1,2)}$ lying on the manifold Ξ_s. The jerk $j^*_\Xi(x, v, a)$ illustrates the point of the manifold $\Xi^j_s(x, v, a)$ irrelevant to the action strategy initiation because of evident inefficiency of the corresponding action strategy illustrated by its optimal implementation $\mathrm{AS}^*_{\mathrm{opt}}$. The object motion state pointed with $j = 0$ represents the object motion state when the subject initiates new action strategy implementation

a point-like object near some unstable equilibrium (Fig. 6.8). The stable branch Ξ_s of the zero-Hamiltonian manifold, which represents the subject's optimal behavior, takes the form a folded surface (manifold) $\Xi^j_s(x, v, a)$ after its embedding into the space $\mathbb{R}_{[4]} = \{x, v, a, j\}$,

$$\Xi_s \overset{\mathbb{R}_{[4]}}{\hookrightarrow} \Xi^j_s(x, v, a)\,. \tag{6.181}$$

Turning to the previous terminology, the folded surface $\Xi^j_s(x, v, a)$ will be also referred to as the *manifold of action strategy initiation* (or just *initiation manifold*) in the space $\mathbb{R}_{[4]}$.

Due to the fold, there arises a region of the object motion states $\{x, v, a\} \subset \mathbb{R}_L$ for which the action strategy initiation is ambiguous. The latter means that for any motion state $\{x, v, a\}$ of controlled object from this region, there are several values $j_{\Xi,k}(x, v, a)$ $(k = 1, 2, \ldots)$ of the jerk representing points on the initiation manifold $\Xi^j_s(x, v, a)$ that are projected on the object motion state $\{x, v, a\}$. Some of these jerk values, e.g., $j^*_\Xi(x, v, a)$ in Fig. 6.8, seem to be irrelevant to initiating action strategies because of the evident inefficiency of the related action strategies. Such jerk values should be omitted from this collection. The others, e.g., $j_{\Xi(1,2)}(x, v, a)$ shown in Fig. 6.8 together with the optimal implementations $\mathrm{AS}_{\mathrm{opt}(1,2)}$ of the corresponding action strategies, represent the plausible action strategies among which the subject selects one for implementation.

In this way we come to the following description of the selective type active phase of action strategy implementation based on the properties listed below.

- *Multiplicity of action strategy initiation*. For each motion state $S = \{x, v, a\}$ of controlled object within a given region $\mathfrak{S}$ of the Lagrange space $\mathbb{R}_L$, there are several jerk values $\{j_{\Xi,k}(x, v, a)\} = \{j_{\Xi,k}(S)\}$ $(k = 1, 2, \ldots)$ and the corresponding optimal implementations $\mathrm{AS}_{\mathrm{opt},k}$ joining the points $\{S, j_{\Xi,k}(S)\}$ to the final goal in the space $\mathbb{R}_{[4]} = \{x, v, a, j\}$ individually.
- *Multiplicity of selection options*. The noted set $\{j_{\Xi,k}(S)\}$ of jerk values represent the collection $\{\mathcal{AS}_k\}$ of action strategies that are similar in their efficiency. So in selecting one of them for implementation, the subject takes in into account all of them before making a final decision. It allows us to regard the set of values $\{j_{\Xi,k}(S)\}$ and the corresponding set of actions strategies $\{\mathcal{AS}_k\}$ as the available options of the selection processes. None of these options can be omitted completely and for any motion state S at least two options determine the subject's choice.
- *Perceptual discreteness of selection options*. The jerk values $\{j_{\Xi,k}(S)\}$, i.e., the selection options are separated in the space $\mathbb{R}_j$ by distances exceeding substantially the uncertainty δ_j in the subject's perception of the object jerk j. In other words, for any $k \neq k'$ and $S, S' \in \mathfrak{S}$ the difference $|j_{\Xi,k}(S) - j_{\Xi,k'}(S')| \gg \delta_j$. This condition, first, enables us to suppose that the subject is able to compare plausible action strategies based on their individual properties.[56] Second, in this case, not only the values $\{j_{\Xi,k}(S)\}$ but also the initiation clouds $\{\mathbb{C}_{\mathrm{ini},k}\}$ in the space $\mathbb{R}_j$ are perceived by the subject as separated entities.

Below the corrective and selective types of active phase in the 2D-time dynamics of action strategies are analyzed separately. Naturally, there can be some intermediate state of the active phase bearing the properties of both the corrective and selective types of active phase, which, however, requires an individual consideration and is not included in the present book.

6.3.7 Dynamics of Corrective Type Active Phase

The previous implementation of the action strategy $\mathcal{AS}$ in question is terminated when the subject has recognized the current value j of object jerk to deviate substantially from the optimal value $j_{\Xi}(x, v, a)$ for the current state $\{x, v, a\}$ of object motion. To initiate the next $\mathcal{AS}$-implementation the subject needs to identify the

[56] When for two action strategies $\mathcal{AS}_1$ and $\mathcal{AS}_2$ the inequality $|j_{\Xi,1} - j_{\Xi,2}| \lesssim \delta_j$ is the case, we actually meet the problem of action strategy fusion into one entity, which is outside the scope of the present book. The same concerns the emergence of two new action strategies via the old one splitting.

acceptable neighborhood of the value $j_\Xi(x, v, a)$ and to start object motion with some jerk from this neighborhood.

The noted identification of the jerk values relevant to efficient initiation of action strategy is a mental process that is started and completed within the experiential now. Therefore we describe it as evolution of a certain cloud $\mathfrak{C}$ in the space $\mathbb{R}_j = \{j\}$ and ignore the corresponding time variations in the inner phase variables x, v, a of object motion.

Put differently, the corrective type active phase is treated as temporal transformation of the cloud $\mathfrak{C} \subset \mathbb{R}_j$ within the experiential now:

$$\mathfrak{C}_0 \Longrightarrow \mathfrak{C}_\Xi \tag{6.182}$$

- *from* the state $\mathfrak{C}_0$ representing no subject's preference to any value j of the object jerk inside a rather wide region containing the jerk at terming the previous $\mathcal{AS}$-implementation as well as the value $j_\Xi(x, v, a)$ (Sect. 6.2.3).
- *to* the state $\mathfrak{C}_\Xi$ characterizing the subject preference to the value j_Ξ within the accuracy δ_j quantifying the subject's perception under a given conditions including the degree of attention focused on the correction process.

According to neurophysiological data (Sect. 6.3.5), the active phase of subject's behavior should be a complex process containing deliberate as well as automatic fragments. So, at first, we need to elucidate this type active phase from the standpoint of the mind-body interrelation.

Mind-body contribution to the active phase dynamics: Because the active phase of subject's behavior is confined to the experiential now, in the realm of the mind-in-the-self transformation (6.182) takes the form of an instantaneous transition. In the realm of the body-in-the-self this transition in detail should be described as a continuous dynamic processes governed by an unconscious mechanism (see Sect. 5.6.1 for the underlying concepts as well as Sect. 6.3.5 for neurophysiological evidence).

For this reason we introduce the *cloud function* $\Psi(j, t)$ (or just the function Ψ) which, generally speaking, is the image of the cloud $\mathfrak{C}$ in the body-in-the-self. The initial and terminal states $\Psi_0(j), \Psi_\Xi(j)$ of the function $\Psi(j, t)$ are the representations of the clouds $\mathfrak{C}_0, \mathfrak{C}_\Xi$ in the body-in-the-self and only they are accessible to the mind.

This mind-body description of the corrective type active phase in the action strategy dynamics is illustrated by the diagram (the used notions have been introduced in Sect. 5.6.1):

$$\begin{array}{lccc} \text{the mind-in-the-self:} & \mathfrak{C}_0 & & \mathfrak{C}_\Xi \\ & {\scriptstyle\text{the self}}\big\downarrow & & \big\uparrow{\scriptstyle\text{the self}} \\ \text{the body-in-the-self:} & \Psi_0(j) & \xrightarrow[\text{transformation}]{\text{unconscioius}} & \Psi_\Xi(j) \end{array} \tag{6.183}$$

which the gist of the mind-body realization of transition (6.182).

Governing equation of active phase dynamics: Turning to the gist of the approach to describing cloud dynamics summarized in Sect. 6.2.8, in particular, the governing equation (6.149), let us represent the dynamics of the cloud function $\Psi(j,t)$ within the following model

$$\tau_\Psi \frac{\partial \Psi}{\partial t} = \widehat{\mathcal{H}}\,\Psi - \langle \Psi |\, \widehat{\mathcal{H}}\, | \Psi \rangle\, \Psi\,. \tag{6.184}$$

Here $\tau_\Psi \lesssim \tau_d$ is the time scale characterizing the action strategy correction which should be competed within the experiential of duration about τ_d and the Hamiltonian $\widehat{\mathcal{H}}$ being a Hermitian operator is specified by the expression

$$\widehat{\mathcal{H}} = \delta_j^2 \frac{\partial^2}{\partial j^2} + \frac{A\mathcal{E}_\Xi}{\cosh^2\left\{\left[A^{1/2}/(\mathcal{E}_\Xi^{1/2}\delta_j)\right]\cdot(j-j_\Xi)\right\}}$$
$$\approx \delta_j^2 \frac{\partial^2}{\partial j^2} - \frac{A^2(j-j_\Xi)^2}{\delta_j^2} + A\mathcal{E}_\Xi\,, \tag{6.185}$$

where the given approximation holds for

$$|j-j_\Xi| \ll j_{\text{pa}} \overset{\text{def}}{=} \frac{\mathcal{E}_\Xi^{1/2}\delta_j}{A^{1/2}}\,, \tag{6.186}$$

the dimensionless quantity $\mathcal{E}_\Xi \gg 1$ characterizes the efficiency of the action strategy $\mathcal{AS}$ corresponding to the jerk j_Ξ and, thereby, will be referred to as the dimensionless efficiency of $\mathcal{AS}$, the value A is the degree of attention focused on the $\mathcal{AS}$-initiation. This model captures the basic features of the correction process followed by the action strategy initiation.[57]

[57] To elucidate this statement let us, first, explain how the dimensionless efficiency $\mathcal{E}_\Xi$ is related to the extended cost $\mathcal{E}_{\text{AS}}$ (Exp. 6.179) of the given action strategy $\mathcal{AS}$.

- We treat the quantity $-\mathcal{E}_{\text{AS}}$ as a certain representation of the action strategy efficiency because it corresponds to the intuitive understanding of this notion. Namely, the higher the efficiency, the more preferable the action strategy.
- The subject's perception of action strategy efficiency is characterized by some fuzzy threshold $\delta\mathcal{E}$ such that the subject can reliably order two action strategies according to their preference only if the difference in their efficiency exceeds substantially $\delta\mathcal{E}$. We use the value $\delta\mathcal{E}$ as the measurement unit, which enables us to deal with the dimensionless form of action strategy efficiency.
- For a particular implementation of the action strategy $\mathcal{AS}$ that is initiated with the jerk j, its efficiency has to decrease as the difference $|j-j_\Xi|$ increases. When this difference become essential, the probability of initiating the corresponding implementations is very low and, moreover, should not depend on the difference $|j-j_\Xi|$. Indeed, such initiation is just a result of subject's erroneous actions. Formally we can attribute to these action strategy "implementations" a certain efficiency $\mathcal{E}_0$ much less than $\mathcal{E}_\Xi$.
- Taking into account that the action strategy efficiency is determined up to a constant, we may set $\mathcal{E}_0 = 0$ and quantify the efficiency of the action strategy $\mathcal{AS}$ relatively to this level.

As far as the form of Hamiltonian (6.185) is concerned, we have related it with the action strategy efficiency and taken into account the invariance of the governing equation (6.184) with respect

Due to the accepted inequality $\mathcal{E}_\Xi \gg 1$, we may confine our consideration of the cloud function dynamics governed by Eq. (6.184) to the parabolic approximation of the Hamiltonian $\widehat{\mathcal{H}}$ specified by expression (6.185). In this case, the eigenfunctions $\{\psi_n(j)\}$ ($\left\langle \psi_n \middle| \psi_n \right\rangle = 1$) with the eigenvalues $\{E_n\}$ of the Hamiltonian $\widehat{\mathcal{H}}$ are given by the quantum theory of the one-dimensional oscillator (e.g., Flügge 1971, vol. I, problem 30). Namely, for $n = 0, 1, 2, \dots$

$$\psi_n(j) = A^{1/4}\,\psi_n^{(A=1)}\left(\sqrt{A}\,\frac{j - j_\Xi}{\delta_j}\right), \tag{6.187a}$$

$$E_n = A\big[\mathcal{E}_\Xi - (2n+1)\big] = A E_n^{(A=1)}\,. \tag{6.187b}$$

As far as the proposed dependence of Hamiltonian (6.185) on the attention degree A is concerned, the following can be regarded as arguments for it. The dependence of ansatz (6.185) as well as the dependence of eigenfunctions (6.187a) on the attention degree A give rise to the same fuzzy threshold δ_j^A in the jerk perception which scales with the attention degree as

$$\delta_j^A \sim \delta_j\, A^{-1/2}.$$

This scaling is of the same form as Weber's law under partial attention (5.40) and fits the experimental estimates of Weber's law exponent β_M (Sect. 5.4.4). The linear dependence of eigenvalues (6.187b) on the attention degree meets the intuitive understanding of the attention effect on processing the perceived information. If the attention degree A quantifies the amount of computational resources allocated to processing the information about the action strategy $\mathcal{AS}$, then the rate $\partial\Psi/\partial t$ of the cloud function change in time, i.e., the eigenvalue of the Hamiltonian $\widehat{\mathcal{H}}$ should be proportional to A.

Besides, taking into account the inequality $\delta_j^A \ll j_{\text{pa}}$ for $\mathcal{E}_\Xi \gg 1$, we may confine our analysis of the cloud function dynamics to the parabolic approximation (6.185) of the Hamiltonian $\widehat{\mathcal{H}}$ which holds within the limit (6.186).

Using the expansion of the cloud function $\Psi(j,t)$ in the bases comprising the eigenfunctions $\{\psi_n\}$

$$\Psi(j,t) = \sum_n c_n(t)\psi_n(j)\,, \tag{6.188}$$

the governing equation (6.184) can be reduced to the generalized Lotka-Volterra model (see Sect. 6.3.5)

to the replacement $\widehat{\mathcal{H}} \to \widehat{\mathcal{H}} + C$, where C is an arbitrary quantity independent of the jerk j. The chosen ansatz (6.185) is the modified Pöschl-Teller potential well (Flügge 1971, vol. I, problem 39), which is a particular example admitting a strict solution to the analyzed problem also for $\mathcal{E}_\Xi \sim 1$. The proposed dependence of Hamiltonian (6.185) on the degree of attention focused on the action strategy $\mathcal{AS}$ is explained in the text following the given footnote.

$$\tau_\Psi \frac{dc_n}{dt} = c_n \left(E_n - \sum_{n'} E_{n'} c_{n'}^2 \right) \tag{6.189}$$

for $n = 0, 1, 2, \ldots$, which, as it must be, conserves the cloud function normalization

$$\left\langle \Psi \middle| \Psi \right\rangle = \sum_n c_n(t) = 1\,.$$

Via expansion (6.188) the initial values of the coefficients $\{c_n^0\}$ are specified by the initial form $\Psi_0(j)$ of the cloud function, the terminal values of the coefficients $\{c_n\}$ as time $t \to \infty$ are

$$c_0^\infty = 1 \quad \text{and} \quad c_n^\infty = 0 \quad \text{for } n \geq 1.$$

In other words, the given Lotka-Volterra model describes the winner-take-all "competition" (Sect. 6.3.5) between the eigenfunctions $\{\psi_n\}$. The winner is the basic eigenfunction ψ_0 possessing the maximal eigenvalue $E_0 = A(\mathcal{E}_\Xi - 1)$.

Initiation of new action strategy implementation: The noted terminal state of the system dynamics corresponds to the formal limit $t \to \infty$. The correction process takes a finite time interval. To model the termination of the correction process and the initiation of the next action strategy implementation, we turn to the gist of our description of action strategy termination (Sect. 6.3.3).

The subject perception of the action strategy efficiency is characterized in the given model (6.184) by the fuzzy threshold about unity. So when the difference between the current value $\langle \Psi | \widehat{\mathcal{H}} | \Psi \rangle_t$ and its limit value $\langle \Psi | \widehat{\mathcal{H}} | \Psi \rangle_\infty$ becomes remarkably less than unity, the subject cannot draw a distinction between the current and formally terminal states. In other words, when the difference

$$\Delta\mathcal{E}(t) = \langle \Psi | \widehat{\mathcal{H}} | \Psi \rangle_\infty - \langle \Psi | \widehat{\mathcal{H}} | \Psi \rangle_t = 2 \sum_{n=1}^{\infty} n c_n^2(t) \approx 2\big[1 - c_0^2(t)\big], \tag{6.190}$$

admitting the noted approximation when only two eigenfunctions ψ_0 and ψ_1 are taken into account, becomes less than or compatible with unity, the unconscious process of correction is terminated with probability rate

$$R_a(t) \sim \frac{1}{\tau_\Psi} \big[\Delta\mathcal{E}(t)\big]^{-\beta_{\text{ini}}}\,, \tag{6.191}$$

where $\beta_{\text{ini}} > 0$ is some exponent being a model parameter. The correction process is terminated at time t with the probability density

$$\mathcal{P}_a(t) \approx R_a(t) \exp\left\{ - \int_{-\infty}^{t} dt' R_a(t') \right\}. \tag{6.192}$$

For details a reader may be referred to Sect. 6.3.3 and expressions (6.166).

Finally, the cloud function $\Psi(t)$ is raised from the level of the body-in-the-self to the level of the mind-in-the-self (diagram (6.183)) and becomes accessible to the mind in the form of the cloud $\mathfrak{C}_{\Xi}$ (diagram 6.183).

6.3.8 Dynamics of Selective Type Active Phase

The selective and corrective types of active phase in the subject's behavior are similar in many aspects. The main difference between them is that in the case of the selective type active phase:

- There are several action strategies $\{\mathcal{AS}_k\}$ characterized by their positions $\{j_{\Xi,k}\}$ in the space $\mathbb{R}_j$ and the different values $\{\mathcal{E}_{\Xi,k}\}$ of their efficiency. These action strategies "compete" with one another for implementation via the common resource represented by the cloud function $\Psi(j,t)$ (the second term in the governing equation 6.184).
- Attention as the computational resource is divided between these action strategies and the degree A_k of attention allocated to a given action strategy $\mathcal{AS}_k$ determines the rate of information processing about the $\mathcal{AS}_k$-properties.
- Changes in the system state can induce the attention reallocation between the action strategies. Besides, the attention degree should exhibit random fluctuations, endowing the cloud dynamics with probabilistic properties. These fluctuations are considered to be the main source of noise in the cloud dynamics.

The listed features are the characteristic properties of the selective type of active phase. However, they can be taken into account in the mathematical terms employed in the description of the corrective type active phase after a certain generalization. It should be emphasized that this generalization allows for phenomena not existing in the corrective type active phase.

Generalized Hamiltonian: We turn to the following modification of the Hamiltonian $\widehat{\mathcal{H}}$ (6.185) to take into account the characteristic properties of the selective type active phase:

$$\begin{aligned}\widehat{\mathcal{H}} =& \delta_j^2 \frac{\partial^2}{\partial j^2} + \sum_k \frac{A_k \, \mathcal{E}_{\Xi,k}}{\cosh^2\left\{\left[\overline{A}_k^{1/2}/(\mathcal{E}_{\Xi,k}^{1/2}\delta_j)\right]\cdot(j - j_{\Xi,k})\right\}} \\ \approx& \bigcup_{|j-j_{\Xi,k}|\ll j_{\mathrm{pa,k}}} \left[\delta_j^2 \frac{\partial^2}{\partial j^2} - \frac{\overline{A}_k^2 (j - j_{\Xi,k})^2}{\delta_j^2} + \left(\overline{A}_k + \widetilde{A}_k\right)\mathcal{E}_{\Xi,k}\right].\end{aligned} \tag{6.193}$$

Here the distances between neighboring elements in the set $\{j_{\Xi,k}\}$ are assumed to exceed substantially the perception threshold δ_j of the jerk, i.e.,

$$|j_{\Xi,k} - j_{\Xi,k'}| \gg \delta_j \tag{6.194}$$

for any pair $k' \neq k$, the degree $A_k = \overline{A}_k + \widetilde{A}_k$ of attention allocated to the action strategy $\mathcal{AS}_k$ $(k = 1, 2, \ldots)$ is regarded as a dynamical variable $\overline{A}_k$ with a noise component $\widetilde{A}_k$, and the degrees $\{A_k\}$ of attention divided between the action strategies $\{\mathcal{AS}_k\}$ meet the equality

$$\sum_k \overline{A}_k = 1. \tag{6.195}$$

It should be noted that the given generalization of the Hamiltonian $\widehat{\mathcal{H}}$ remains Hermitian. Effects like a subject's bias, e.g., in favor of positive values of the jerk in the selection preference, which require the introduction of skew-Hermitian components in the Hamiltonian $\widehat{\mathcal{H}}$ are ignored.

The individual eigenfunctions $\{\psi_{n(k)}(j, \overline{A}_k)\}$ and the corresponding eigenvalues $\{E_{n(k)}\}$ of a given action strategy $\mathcal{AS}_k$ are described by Exp. (6.187) which contains the degree of attention $\overline{A}_k$ allocated to the action strategy $\mathcal{AS}_k$ and averaged over its random fluctuations. Within approximation (6.193) the effect of attention noise on the action strategy selection is described by the terms $\{\widetilde{A}_k \mathcal{E}_{\Xi,k}\}$, meaning this effect to be reduced mainly to noise in comparing different action strategies in their preference.

We want to note that the developed description of action strategy selection using the cloud function formalism proposes a novel mechanism of transitions between "potential wells" specified by the Hamiltonian $-\widehat{\mathcal{H}}$. Indeed:

- This description deals with the interaction between different action strategies $\{\mathcal{AS}_k\}$ represented by the individual collections of eigenfunctions $\{\psi_{n(k)}(j)\}$ of quantum oscillators formally corresponding to the "potential wells" specified by the Hamiltonian $-\widehat{\mathcal{H}}$ within approximation (6.193).
- As seen below, the intensity of bidirectional transitions between two "potential wells" in the framework of the cloud function dynamics depends essentially on the proximity of the well minima. This proximity is quantified by the difference between the values of action strategy efficiency, actually, the difference $|E_{0k} - E_{0k'}|$.
- These transitions are caused neither by the cloud function tunneling (similar to the quantum-mechanical tunneling) nor by overcoming a potential barrier (similar to the escape from the potential well in classical physics). They are duo to the strong nonlocality of the human temporality (Sect. 6.2.7).[58]

The remaining part of this subjection is devoted to describing the dynamics of selective type active phase in terms of the cloud function expansion in the basis of the eigenfunction $\{\psi_{n(k)}(j, \overline{A}_k)\}$. First, we will derive a governing equation for the

[58] The effect of cloud function tunneling can become essential when the difference between the action strategy jerks j_Ξ and j'_Ξ is comparable with the jerk perception threshold δ_j. However, this effect seems to be responsible mainly for the splitting of one action strategy into two action strategies as an emergent phenomenon.

coefficients of this expansion, second, we will construct a model for the dynamics of attention. According to neurophysiological data discussed in Sect. 6.3.5, attention reallocated between different options during competitive selection in human decision-making is an intrinsic component of competitive selection.

Governing equation of cloud function dynamics: In describing the dynamics of the cloud function $\Psi(j,t)$, how the attention degrees $\{\overline{A}_k(t)\}$ vary in time is assumed to be known. Besides, by virtue of (6.194), the dynamics of the cloud function $\Psi(j,t)$ near the action strategies $\{\mathcal{AS}_k\}$—the inverted "potential wells" represented by the Hamiltonian approximation (6.193) and illustrated in Fig. 6.9—can be treated separately. In this case, the cloud function $\Psi(j,t)$ admits a representation as the linear combination of the eigenfunctions $\{\psi_{n(k)}(j,\overline{A}_k)\}$ of levels $n = 0, 1, 2, \ldots$ attributed to the action strategy $\mathcal{AS}_k$ which are treated individually (Fig. 6.9):

$$\Psi(j,t) = \sum_{n,k} c_{n(k)}(t)\, \psi_{n(k)}\big[j, \overline{A}_k(t)\big] \tag{6.196a}$$

or, rewriting this expansion in the bra-ket notions,

$$|\Psi\rangle = \sum_{n,k} c_{n(k)}(t)\, |n,k\rangle\,. \tag{6.196b}$$

The time dependence of the expansion coefficients $\{c_{n(k)}(t)\}$ represents the dynamics of the cloud function $\Psi(j,t)$ in addition to the time variations in the eigenfunctions $\{\psi_{n(k)}(j,\overline{A}_k)\}$ caused by the time dependence of the attention degrees $\{\overline{A}_k(t)\}$.

In these terms the governing equation (6.184) is reduced to the collection of dynamical equations governing the time variations of the coefficients $\{c_{n(k)}\}$

$$\begin{aligned}
\tau_\Psi \frac{dc_{n(k)}}{dt} = & c_{n(k)}\Big(E_{n(k)} - \sum_{n',k'} E_{n'(k')}\, c^2_{n'(k')}\Big) \\
& + \frac{1}{4\overline{A}_k}\frac{\overline{A}_k}{dt} \sum_{n'} \big\langle n,k\big|\big(b_k^\dagger b_k^\dagger - b_k b_k\big)\big|n',k\big\rangle\, c_{n'(k)} \\
& + c_{n(k)}\Big(\widetilde{A}_k(t)\mathcal{E}_{\Xi,k} - \sum_{k'} \widetilde{A}_{k'}(t)\mathcal{E}_{\Xi,k'} \sum_{n'} c^2_{n'(k')}\Big).
\end{aligned} \tag{6.197}$$

Here $\{E_{n(k)}(\overline{A}_k)\}$ are the eigenvalues of the eigenfunctions $\{\psi_{n(k)}\}$ given by Exp. (6.187b). The eigenvalue $E_{0(k)} = \overline{A}_k(\mathcal{E}_{\Xi,k} - 1) \approx \overline{A}_k\mathcal{E}_{\Xi,k}$ admits interpretation as the *attention-dependent efficiency* of the action strategy $\mathcal{AS}_k$ affecting its comparison with the other action strategies from the set $\{\mathcal{AS}_{k'}\}$ within subject's attention distributed between all the action strategies. Besides we have introduced the lowering b_k and raising $b_k^\dagger$ operators acting only on the eigenfunctions of the action strategy $\mathcal{AS}_k$ as (cf. Feynman 2018/1972)

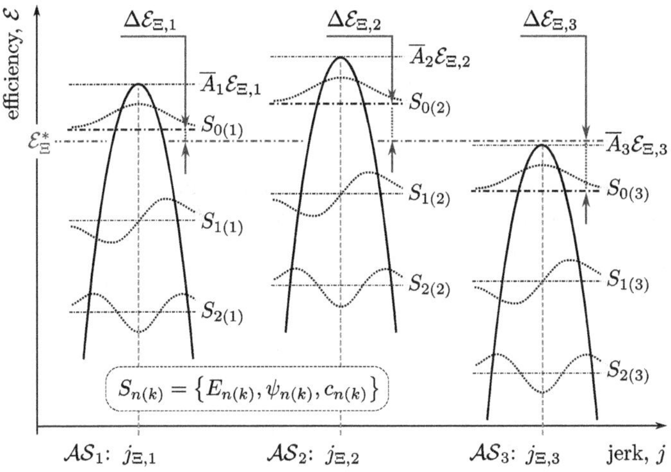

Fig. 6.9 The characteristic elements forming the description of the action strategies selection in terms of the cloud function dynamics. The shown system consists of three action strategies $\mathcal{AS}_k$ ($k = 1, 2, 3$) characterized by the jerk values $j_{\Xi,k}$ representing the initial position of the action strategy optimal implementations in the space of the object jerk j. Each action strategy $\mathcal{AS}_k$ treated separately is characterized by Hamiltonian (6.193) containing the only one corresponding term of the sum and, respectively, the collection of the eigenfunctions $\{\psi_{nk}(j)\}$ (dotted lines) and their eigenvalues $\{E_{n(k)}\}$ (horizontal dash-dotted lines) belonging to the levels $n = 0, 1, 2, \ldots$. The contribution of each level to the dynamics of the cloud function is specified by its triplet $S_{n(k)} = \{E_{n(k)}, \psi_{n(k)}, c_{n(k)}\}$, where the coefficient $c_{n(k)}$ determines the contribution of this level to the cloud function, see Exp. (6.196). The line red horizontal line of the efficiency level $\mathcal{E}^*_\Xi$ shows the reference level chosen for quantifying the action strategy efficiency

$$b_k = \frac{1}{\sqrt{2}}\left[\sqrt{A_k}\,\frac{(j - j_{\Xi,k})}{\delta_j} + \frac{1}{\sqrt{A_k}}\,\delta_j\frac{\partial}{\partial j}\right], \tag{6.198a}$$

$$b_k^\dagger = \frac{1}{\sqrt{2}}\left[\sqrt{A_k}\,\frac{(j - j_{\Xi,k})}{\delta_j} - \frac{1}{\sqrt{A_k}}\,\delta_j\frac{\partial}{\partial j}\right] \tag{6.198b}$$

meeting the commutation relations

$$\left[b_k, b_{k'}\right] = 0\,, \qquad \left[b_k^\dagger, b_{k'}^\dagger\right] = 0\,, \qquad \left[b_k, b_{k'}^\dagger\right] = \delta_{kk'}\,, \tag{6.199}$$

where $\delta_{kk'}$ is the Kronecker delta.

The governing equation (6.184) and, correspondingly, the governing equation (6.197) are invariant under an arbitrary common shift C of action strategy efficiency, i.e., under the transformation

$$\widehat{\mathcal{H}} \to \widehat{\mathcal{H}}' + C(t)\,. \tag{6.200}$$

Here the efficiency shift $C(t)$ may be a function of time t but not of the jerk j. For convenience, we choose the shift $C(t)$ equal to the attention-dependent efficiency averaged over the action strategies $\{\mathcal{AS}_k\}$, i.e., set

$$C(t) = \mathcal{E}^*_{\Xi}(t) = \langle E_{0(k)}(t) \rangle_{\text{AS}} \stackrel{\text{def}}{=} \frac{1}{N_{\text{AS}}} \sum_k E_{0(k)} \approx \frac{1}{N_{\text{AS}}} \sum_k \overline{A}_k \mathcal{E}_{\Xi,k} , \qquad (6.201)$$

where N_{AS} is the number of action strategies involved into the selection process. In this case, the Hamiltonian transformation (6.200) is reduced to measuring the eigenvalues $\{E_{n(k)}\}$ of the action strategies $\{\mathcal{AS}_k\}$ relatively to the level $\mathcal{E}^*_{\Xi}$. In particular, the base level eigenvalues $\{E_{0(k)}\}$ of the action strategies $\{\mathcal{AS}_k\}$ can be set equal to (Fig. 6.9):

$$\begin{aligned} E_{0(k)} &= \Delta\mathcal{E}_{\Xi,k}(t) = A_k(t)\mathcal{E}_{\Xi,k} - \mathcal{E}^*_{\Xi}(t) \\ &= \frac{1}{N_{\text{AS}}} \sum_{k'} \left[\overline{A}_k(t)(\mathcal{E}_{\Xi,k} - 1) - \overline{A}_{k'}(t)(\mathcal{E}_{\Xi,k'} - 1) \right]. \end{aligned} \qquad (6.202)$$

Exactly these quantities and their time dependence determine how the difference in efficiency of the action strategies is reflected in their selection.

Governing equation of attention dynamics: For describing the attention reallocation between the action strategies $\{\mathcal{AS}_k\}$ during their selection we want propose a principle posting that:

- The neural mechanics underlying this attention reallocation should be aimed at shortening the selection process to minimize the time of decision making as it possible.
- The criterion of this minimization cannot depend essentially on a system *particular* state, e.g., the number of action strategies involved in consideration, their positions $\{j_{\Xi,k}\}$ in the space $\mathbb{R}_j$, the accuracy δ_j of jerk perception under given conditions, etc.
- Because the action strategy selection is realized as an unconscious automatic process it, like physical processes in the inanimate world, has to be governed by regularities determined at the current instant of time. The same concerns quantities these regularities deal with.

The given principle posits that the criterion of optimizing the attention reallocation between the "competing" action strategies must be formulated at the general level using the quantities applicable to describing various particular situations. These quantities should characterize the system state and its instantaneous variations at a given moment of time. The generality of the proposed principle stems from the long-term adaptation of underlying neural mechanism to the efficient functioning under various conditions.

At a given instant of time the dynamics of attention reallocation as a regular continuous process can be quantified by the time variations $\{d\overline{A}_k/dt\}$ of the attention degrees $\{\overline{A}_k\}$ averaged over its random fluctuations. So, the desired principle should determine the time derivatives $\{d\overline{A}_k/dt\}$ treated as dynamical variables $\{d\overline{A}_k/dt\}$ locally independent of the current value of the attention degrees $\{\overline{A}_k\}$.

Let us introduce two functions characterizing the current state of (*i*) the action strategy selection and (*ii*) the attention reallocation between action strategies as individual processes with internal integrity.

One of the two functions is the integral dimensionless rate of the selection advance at the current moment.

$$\begin{aligned}
\mathcal{D}_{\text{AS}}(\{\overline{A}_k\}) &= \frac{1}{2}\tau_\Psi^2 \left\langle \frac{\partial \Psi}{\partial t} \middle| \frac{\partial \Psi}{\partial t} \right\rangle_r \\
&= \frac{1}{2}\left\langle \Psi \middle| \left(\widehat{\mathcal{H}} - \langle \Psi | \widehat{\mathcal{H}} | \Psi \rangle\right)^2 \middle| \Psi \right\rangle_r = \frac{1}{2}\langle \Psi | \widehat{\mathcal{H}}^2 | \Psi \rangle_r - \frac{1}{2}\langle \Psi | \widehat{\mathcal{H}} | \Psi \rangle_r^2 \\
&= \frac{1}{2}\sum_{n,k} c_{n(k)}^2 E_{n(k)}^2 - \frac{1}{2}\left(\sum_{n,k} c_{n(k)}^2 E_{n(k)}\right)^2, \qquad (6.203)
\end{aligned}$$

where the subscript r means that all the sources of noise in the cloud function dynamics governed by Eq. (6.184) are omitted. As seen from the governing equation (6.197), the time variations in the expansion coefficients $\{c_{n(k)}\}$, representing the individual dynamics of the cloud function Ψ, depend in their magnitude and sign on particular details of the selection process. On the contrary, the eigenvalues $\{E_{n(k)}(A_k)\}$ depend on the attention degrees $\{\overline{A}_k\}$ in a regular way, namely, linearly (Exp. 6.187b). So, studying how the value of $\mathcal{D}_{\text{AS}}$ changes with small variations $\{\delta\overline{A}_k\}$ in the attention degrees $\{\overline{A}_k\}$, we consider the coefficients $\{c_{n(k)}\}$ to be given constants and take into account only the induced changes of the eigenvalues $\{E_{n(k)}(A_k)\}$. In particular, the rate of time change in the value of $\mathcal{D}_{\text{AS}}$ caused solely by the time variations $\{\delta\overline{A}_k\}$ in the attention degrees is written as

$$\left.\frac{d}{dt}\right|_A \mathcal{D}_{\text{AS}} = \sum_k \frac{d\overline{A}_k}{dt}\frac{\partial \mathcal{D}_{\text{AS}}}{\partial \overline{A}_k}, \qquad (6.204)$$

where, by virtue of Exp. (6.203) together with Exp. (6.187b),

$$\mathcal{D}'_{\text{AS},k} \stackrel{\text{def}}{=} \frac{\partial \mathcal{D}_{\text{AS}}}{\partial \overline{A}_k} = \sum_n c_{n(k)}^2 \frac{\partial E_{n(k)}}{\partial \overline{A}_k}\left(E_{n(k)} - \sum_{n',k'} c_{n'(k')}^2 E_{n'(k')}\right). \qquad (6.205)$$

and the partial derivative $\partial E_{n(k)}/\partial \overline{A}_k = \mathcal{E}_{\Xi,k} - (2n+1) \approx \mathcal{E}_{\Xi,k}$ due to the accepted inequality $\mathcal{E}_{\Xi,k} \gg 1$.

The second function forming the mathematical bases of attention dynamics describes the rate of attention reallocation on its own. This integral rate as a function of time variations $\{d\overline{A}_k/dt\}$ in the attention degrees $\{\overline{A}_k\}$ is quantified by the value

$$\mathcal{R}_{\text{AS}}\left(\left\{\frac{d\overline{A}_k}{dt}, \overline{A}_k\right\}\right) = \frac{1}{2}\sum_k \tau_A^2 \frac{1}{\overline{A}_k}\left(\frac{d\overline{A}_k}{dt}\right)^2 + \frac{1}{2\epsilon_A}\tau_A^2\left(\sum_k \frac{d\overline{A}_k}{dt}\right)^2. \qquad (6.206)$$

Here the parameter $\tau_A \gtrsim \tau_\Psi$ is the characteristic time scale of attention dynamics and, in principle, can be influenced by the mind-in-the-self. The structure of the summands in the former sum reflects that the attention degrees $\{\overline{A}_k\}$ must remain positive quantities in the course of the attention reallocation. The latter sum with the formal parameter $\epsilon_A \to +0$ takes into account the attention conservation (6.195).

In describing the attention reallocation, we treat the integral rate $\mathcal{R}_{\rm AS}$, Exp. (6.206), as some analogy to Rayleigh's dissipation function in Lagrange mechanics (e.g., Goldstein et al. 2014). Therefore we posits the attention dynamics to be governed by the equation for $k = 1, 2, \dots$

$$\frac{1}{\tau_A}\frac{\partial \mathcal{R}_{\rm AS}}{\partial(d_t\overline{A}_k)} = \frac{\partial \mathcal{D}_{\rm AS}}{\partial \overline{A}_k}, \tag{6.207}$$

where we have used the notion $d_t\overline{A}_k = d\overline{A}_k/dt$. In particular, within the limit $\epsilon_A \to +0$, Eq. (6.207) leads to the equality

$$2\mathcal{R}_{\rm AS} = \tau_A \left.\frac{d}{dt}\right|_A \mathcal{D}_{\rm AS} \tag{6.208}$$

which admits interpretation as the acceleration of the action strategy selection caused by the attention reallocation and limited by the rate of this reallocation process.

Combining Exps. (6.205) and (6.206) with Eq. (6.207) we can represent the desired equation governing the attention reallocation between the action strategies $\{\mathcal{AS}_k\}$ during the course of their selection as

$$\tau_A \frac{dA_k}{dt} = A_k\left(\mathcal{D}'_{{\rm AS},k} - \sum_{k'} A_{k'}\mathcal{D}'_{{\rm AS},k'}\right) + A_k\xi_{A,k}(t), \tag{6.209}$$

where the random variations in the attention degree $\{A_k\}$ are taken into account via the noise $\xi_{A,k}(t)$ with some amplitude.[59] It should be pointed out that the obtained equation is of the Lotka-Volterra type, which meets the models of attention dynamics proposed previously for competitive selective (Sect. 6.3.5). In particular, Exp. (6.209) gives rise to the following equation governing the total amount of attention

$$A = \sum_k A_k,$$

i.e., computational resource allocated to the given selection process

---

[59] If the noise $\xi_{A,k}(t)$ is regarded as white noise, then the term $A_k\xi_{A,k}(t)$ should be treated within the Stratonovich interpretation to avoid the noise-induced decrease or increase of the attention degree A_k, which is an artifact for the analyzed system (e.g., Mahnke et al. 2009, for the corresponding properties of stochastic processes with nonlinear noise).

$$\tau_A \frac{dA}{dt} = \left(1 - A\right) \sum_k A_k \mathcal{D}'_{\text{AS},k'} + \sum_k A_k \xi_{A,k}(t) \,.$$

This equation describes stable fluctuations in the total amount of attention A near the mean value $\overline{A} = 1$, which enables us to consider the random variations $\{\widetilde{A}_k\}$ of individual components mutually independent.

By virtue of (6.187b) and the accepted inequality $\mathcal{E}_{\Xi,k} \gg 1$ for $k = 1, 2 \ldots$ the term $\mathcal{D}'_{\text{AS},k}$ coupling the dynamics of attention allocation to the dynamics of action strategy selection can be approximated as

$$\mathcal{D}'_{\text{AS},k} \approx \mathcal{E}_{\Xi,k} \sum_n c^2_{n(k)} \left(E_{n(k)} - \sum_{n',k'} c^2_{n'(k')} E_{n'(k')} \right) \tag{6.210}$$

because the contribution of the eigenfunctions $\{\psi_{n(k)}(j)\}$ with $n \gg 1$ to the selection process is ignorable. If in the governing equation (6.197) for the cloud function dynamics we take into account only the first term on the right-hand side, which can be justified on short time scales, then the quantity $\mathcal{D}'_{\text{AS},k}$ admits the estimation

$$\mathcal{D}'_{\text{AS},k} \approx \frac{\tau_A}{2} \mathcal{E}_{\Xi,k} \frac{d\rho_k}{dt} \tag{6.211}$$

where the quantity

$$\rho_k(t) = \sum_n c^2_{n(k)}(t) \tag{6.212}$$

may be treated the weight attributed to the action strategy $\mathcal{AS}_k$ at the current instant t of the selection process. Therefore the coupling term $\mathcal{D}'_{\text{AS},k}$ may be interpreted as the individual current "attractiveness" of the action strategy $\mathcal{AS}_k$ in the course of selection process for focusing attention on this action strategy. Indeed, it is the product of two factors:

- the efficiency $\mathcal{E}_{\Xi,k}$ of the given action strategy; the higher the efficiency, the more reasonable its choice is;
- the growth rate $\tau_A d\rho_k/dt$ of the action strategy weight; the higher the growth rate at the current instant of time, the more plausible is the choice of the given action strategy and the more amount of computational resources is necessary to clarify the action strategy property (when $\tau_A d\rho_k/dt \ll 1$ the action strategy either is not relevant to the selection under a given conditions or its properties are already well clarified).

The second term in parenthesis in Eq. (6.209) represents the action strategy "competition."

The system of the governing equation (6.197) for the action strategy selection process and the governing equation (6.209) for the attention reallocation in the course of this process form the desired description of the dynamics of the selective type active phase of subject's behavior.

Initiation of new action strategy implementation is the moment when the selection process is categorized as completed, the active phase in the subject's behavior is terminated, and the action strategy $\mathcal{AS}^*$ $(k = k^*)$ treated as the selected winner is initiated. We can single out two typical scenarios of completing the selection process, let us discuss them individually.

Quasi-deterministic action strategy selection. It is the case when there is one action strategy seeming preferable because its efficiency is perceptibly higher than that of the other action strategies. Under this condition the total amount of attentional resources is initially divided, e.g., uniformly between the involved action strategies $\{\mathcal{AS}_k\}$. After that, the selection process takes place without further reallocation of attention. This scenario of active phase implementation is described by the dynamics of the cloud function Ψ with the Hamiltonian $\widehat{\mathcal{H}}$ not depending on time explicitly nor via the attention degrees $\{A_k\}$.

In the give case, the action strategy selection is characterized by the winner-take-all behavior. Indeed, from the governing equation (6.184) we have

$$\tau_\Psi \frac{d}{dt} \langle\Psi|\,\widehat{\mathcal{H}}\,|\Psi\rangle = 2\Big\langle\Psi\Big|\left(\widehat{\mathcal{H}} - \langle\Psi|\,\widehat{\mathcal{H}}\,|\Psi\rangle\right)^2\Big|\Psi\Big\rangle + \Big\langle\Psi\Big|\frac{\partial\widehat{\mathcal{H}}}{\partial t}\Big|\Psi\Big\rangle. \tag{6.213}$$

So, when the Hamiltonian $\widehat{\mathcal{H}}$ does not depend on time t explicitly, the function

$$\langle\Psi|\,\widehat{\mathcal{H}}\,|\Psi\rangle = \sum_{n,k} E_{n(k)} c^2_{n(k)} \le \max_k\{E_{0(k)}\} = E_{0(k_m)} = A_{k_m}(\mathcal{E}_{\Xi,k_m} - 1)$$

bounded from above must grow monotonically until it attains the maximum. This maximum corresponds to the eigenfunction $\psi_{0(k_m)}$, the expansion coefficient $c_{0(k_m)} = 1$ and the others coefficients $c_{n(k)} = 0$. Put differently, in the case of the Hamiltonian $\widehat{\mathcal{H}}$ independent of time, the dynamics of the cloud function Ψ is globally stable and its stability is guaranteed by the existence of Lyapunov's function given by $\langle\Psi|\,\widehat{\mathcal{H}}\,|\Psi\rangle$. The action strategy $\mathcal{AS}_{k_m}$ becomes the winner.

The finite moment of time, when the selection process is terminated and the selected action strategy becomes accessible to the mind, is determined by the uncertainty in evaluating the action strategy efficiency. This situation practically coincides with the case considered in Sect. 6.3.7 provided the difference between the basic eigenvalue $E_{9(k_m)} = A_{k_m}(\mathcal{E}_{\Xi,k_m} - 1)$ of the winner and the corresponding eigenvalues of the others action strategies meet the inequality

$$E_{0(k_m)} - E_{0(k')} \gtrsim 1 \quad \text{for } k = 1, 2, \ldots \text{ and } k \neq k_m. \tag{6.214}$$

Exactly inequality (6.214) is the mathematical representation of perceptible preference of the action strategy $\mathcal{AS}_{k_m}$.

As a rare event, within the given scenario of selection process another action strategy rather than the action strategy $\mathcal{AS}^*$ with maximal efficiency can be selected. This event can be caused by a nonuniform division of attention such that the degree A_{k^*}

of attention allocated to processing information about $\mathcal{AS}^*$ turns out to be relatively small. Random fluctuations $\widetilde{A}_k(t)$ in attentional resources is another plausible cause, which, however, requires the description with the time-dependent Hamiltonian $\widehat{\mathcal{H}}(t)$.

Probabilistic (unpredictable) action strategy selection. It is the case when there are two or more action strategies $\{\mathcal{AS}_k\}$ $(k = 1, 2, \ldots)$ whose efficiency values $\{\mathcal{E}_{\Xi,k}\}$ are rather close to one another, i.e., the formal inequality

$$|\mathcal{E}_{\Xi,k} - \mathcal{E}_{\Xi,k'}| \ll 1 \tag{6.215}$$

holds for any pair of $\{k, k'\}$. The contribution of other action strategies to the selection process is ignorable. Under this condition, within a relatively short time interval of duration about τ_Ψ only the eigenfunctions $\{\psi_{0(k)}\}$ in the expansion of the cloud function Ψ in the basis of the eigenfunctions $\{\psi_{n(k)}\}$ (Exp. 6.196) "survive." Thereby, for the weights $\{\rho_k\}$ (Exp. 6.212), quantifying the individual contributions of the action strategies $\{\mathcal{AS}_k\}$ to the cloud function Ψ, the governing equation (6.197) can be approximately reduced to

$$\frac{\tau_\Psi}{2}\frac{d\rho_k}{dt} = \rho_k\left\{\left[\overline{A}_k + \widetilde{A}_k(t)\right]\mathcal{E}_{\Xi,k} - \sum_{k'}\left[\overline{A}_{k'} + \widetilde{A}_{k'}(t)\right]\mathcal{E}_{\Xi,k'}\,\rho_{k'}\right\}. \tag{6.216}$$

Because of the proximity of the action strategies $\{\mathcal{AS}_k\}$ in efficiency, the influence of random fluctuations $\{\widetilde{A}_k(t)\}$ in the attentional degrees becomes essential. If, in addition, the attention division between the involved action strategies is fixed at the moment of the active phase initiation, i.e., the attention degrees $\{\overline{A}_k\}$ do not change in time, the dynamics of selection process becomes substantially irregular. This irregularity can be reflected in random variations of the quantity

$$\langle\Psi|\,\widehat{\mathcal{H}}\,|\Psi\rangle \approx \sum_k \overline{A}_k\mathcal{E}_{\Xi,k}\rho_k(t)$$

on scales about or exceeding unity within a time interval about τ_Ψ. It is due to $\mathcal{E}_{\Xi,k} \gg 1$ for $k = 1, 2\ldots.$ and possible, not too essential difference in the attention degrees $\{\overline{A}_k\}$ at the moment of active phase initiation. It should be emphasized that in analyzing a solution, e.g., to the governing equation (6.216) we may measure the attention-dependent efficiency with respect to any arbitrary chosen level, in particular, quantify this efficiency as it is specified by Exp. (6.202). However, in turning to the function $\langle\Psi|\,\widehat{\mathcal{H}}\,|\Psi\rangle$ to describe the completeness of selection process, the action strategy efficiency must be measured in the absolute units given by the Hamiltonian representation (6.193).

As in the case of the corrective type active phase (Sect. 6.3.7), here we turn to the mechanism of terminating the selection process based on a certain stability of the value $\langle\Psi|\,\widehat{\mathcal{H}}\,|\Psi\rangle$ of scales of τ_Ψ. Therefore the noise-induced strong irregular variations in $\langle\Psi|\,\widehat{\mathcal{H}}\,|\Psi\rangle$ do not allow to compete the active phase implementation.

In this case, the mind-in-the-self can initiate the dynamics of attention reallocation, which is described by the governing equation (6.209) on time scales $\tau_A \gtrsim \tau_\Psi$. On these scales the mind can access to dynamics of the selection process, which endow it with some properties characterizing deliberate decision-making. It seems that this type transition between the automatic and deliberate modes of color categorization was found in our experiments reviewed in Appendix 5.10.

The form of the term $\mathcal{D}'_{\mathrm{AS},k}$ (Exp. 6.211) governing the attention dynamics argues that the attention reallocation should give rise to cyclic variations in the action strategy weights $\{\rho_k(t)\}$ on scales of τ_A. We want to point our that such variations may be categorized in terms of the winnerless competitive selection noted in Sect. 6.3.5. These variations on scales τ_A can be treated as quasi-stable values of $\langle\Psi|\widehat{\mathcal{H}}|\Psi\rangle$ and, thereby, quasi-stable states of selection process. In its turn, this quasi-stability, treated as the evidence accumulation for the corresponding action strategy, can induce terminating the selection process in the way described in Sect. 6.3.7 and initiating the implementation of the action strategy currently classified as the winner. Naturally, these variations in the selective process caused by attention dynamics are also rather irregular, which endows the selection process with probabilistic properties, which is reflected in the name of this action strategy scenario.

A detailed mathematical analysis of the probabilistic action strategy selection, seeming to be rather rich in properties, is beyond the goal of the present book. Let us list only some phenomena that admit the description within the proposed formalism.

- *The change of mind in decision-making*. We already discussed this phenomenon in Sect. 4.2.2 as well as in Sect. 6.3.6, so here we note only its relation to the proposed mechanism of probabilistic action strategy selection. Let us assume that the action strategy selection and the implementation of selected action strategies take place in parallel. In this case, after initiating the selection process accompanied by the attention dynamics exhibits cyclic variations on time scales about τ_A such that one or another action strategy mainly contributes to the cloud function Ψ within a time interval about τ_Ψ. Each of such quasi-stationary fragments of the τ_Ψ-dynamics may be regarded the accumulated evidence for selecting and initiating the corresponding action strategy. However, when in time about τ_A the selection process singles out another action strategy the subject can terminate the previous one and initiation the new action strategy. This action strategy change is caused by the internal dynamics of the ongoing selection process rather than the feature of the current implementation of the selected action strategies. Therefore such human actions may be characterized as the change of mind.
- *Biases in human choice*. There are a number of phenomena like the status-quo bias (Simon 1956; Samuelson and Zeckhauser 1988; Tversky and Shafir 1992; Ritov and Baron 1992, for review see also Eidelman and Crandall 2014; Dean et al. 2017), novelty seeking (Deci and Ryan 1985, 2000; Cloninger 2004; Liao et al. 2011; Fayn et al. 2015), inertia in decision-making (Akaishi et al. 2014; Alós-Ferrer et al. 2016; Jung et al. 2018; Jung and Dorner 2018; Power and Alison 2019) exhibiting one common feature. The preference of options in a sequence of repeated decision-making events depends, positively or negatively, on how many

times the subject previously selected a given option. Naturally, the own properties of options contribute to their preferences too.

The cumulative contribution of selection history to the magnitude of this effect can be taken into account within the paradigm of reinforcement learning to be considered in the next subjection. Here we want to note a plausible generalization of the proposed model for the action strategy efficiency allowing for the nonlinear self-interaction—self-attraction or self-repulsion—of the cloud function Ψ. This self-interaction affects the action strategy efficiency and, thereby, plays the role of the "core" these effects. Namely, a plausible modification of the Hamiltonian $\widehat{\mathcal{H}}$ specified by Exp. (6.193) is

$$\mathcal{E}_{\Xi,k} \quad \Rightarrow \quad \mathcal{E}_{\Xi,k}\big(1 + g_{\text{si}}\delta_j\,\Psi^2\big),$$

where the coefficient g_{si}, positive or negative, can change due to the learning process or, even, be a dynamical variable. This modification may be regarded as some analogy to the nonlinear Schrödinger equation (e.g., Catherine Sulem 1999; Fibich 2015; Copie et al. 2020, for a review).[60] The cloud function self-interaction can also underlie the description of the observed hysteresis in human perception of external stimuli (Sect. 1.3.2).

- *Cyclic variations in human perception and behavior.* Coupled dynamics of the cloud function Ψ and attention degrees $\{A_k\}$ may be regarded as a plausible mesolevel mechanism of the effect called bistable perception. Rubin's face/vase picture alternate perceived as the central white vase on black background and two black face profiles on white background is a paradigmatic example of bistable perception. This phenomenon was investigated during the last decades from various standpoints (e.g., Sterzer and Rees 2009; Wang, Arteaga and He 2013; Aydın et al. 2011; Weilnhammer et al. 2017; Gage and Baars 2018; Cao et al. 2018; Martínez and Parra 2018; Meilikhov and Farzetdinova 2019, for a review).

  Our description of probabilistic action strategy selection can also be used for modeling self-organized criticality in human behavior. When the subject cyclically shifts the focus of attention between different options of his actions, he can lose from time to time the control over the unstable system governed by his actions. It endows the system with complex multiscale dynamics. For discussion of this issue a reader may be referred to, e.g., Van Orden et al. (2003), Wagenmakers et al. (2005), Ramos et al. (2011), Ihlen and Vereijken (2013), Ton and Daffertshofer (2016), Cocchi et al. (2017), Tagliazucchi (2017), Palva and Palva (2018) and, also, collections edited by He et al. (2013), Massobrio et al. (2015).

[60] For describing the self-interaction of the cloud function Ψ without the directly introduced action strategy components of the Hamiltonian $\widehat{\mathcal{H}}$ (Exp. 6.193), an additional description of the attention distribution, e.g., in the space $\mathcal{R}_j$ can be necessary.

6.3.9 Fusion-Based Reinforcement Learning

Finalizing our description of action strategy dynamics we want to outline the mechanism via which the subject accumulates the information about the properties of actions strategies, in particular, their efficiency. The efficiency of action strategies, as it is evaluated by the subject, plays a crucial role in the active phase of subject's behavior. The skilled subject is supposed to able to adequately estimate the efficiency of action strategies determining his goal-oriented behavior. This skill is a result of many events when the subject realized given action strategies, evaluated them based on his experience with respect to the required effort and attaining the desired goal, and aggregating the information in bodily memory. We will refer to this process as to the reinforcement learning of efficient implementation of action strategies. The reinforcement learning is an ongoing process. It is related to the necessary of adaption to possible variations in the subject' state and the state of environment. To describe this reinforcement learning we need to consider some aspects of action strategy dynamics outside the complex present.

Strictly speaking, in the present subsection we consider the learning process on its own. Nevertheless, we use the term reinforcement learning because the aspects of reinforcement becomes explicit in coupling the learning dynamics with the subject's action dynamics.

In our account the reinforcement learning consists of two basic aspects. *First*, it is the fusion of

- the efficiency $\mathbb{E}^{\text{en}}_{\text{AS}}$ experienced during the action strategy implementation terminated just now and
- the efficiency $\mathbb{E}^{\text{sb}}_{\text{AS}}$ attributed to the action strategy $\mathcal{AS}$ based on the acquired skill.

This fusion is implemented within the experiential now. We will call this efficiency fusion the *experience-skill fusion* of action strategy efficiency.

Second, it is the long-term dynamics of parameters specifying the details of experience-skill fusion. These parameters change on temporal scales exceeding substantially the duration of complex present. Their dynamics reflects the accumulation of information gained. We will call this aspect the *long-term dynamics of experience-skill fusion*.

Let us consider the two aforementioned aspects separately.

Experience-skill fusion. Taking into account the active phase dynamics discussed previously, we consider the experience-skill fusion within the following frameworks.

- In analyzing efficiency of his actions, the subject units action strategies similar in property into one set $\mathfrak{A} = \{\mathcal{AS}_\alpha\}$ and treats it as a certain entity with internal integrity. The similarity of these action strategies is due to they belonging to one branch of the initiation manifold $\Xi^j_s(x, v, a)$ (Fig. 6.8). Therefore, the difference of the action strategies $\{\mathcal{AS}_\alpha\}$ from one another is caused by the variety of the initial states $\{x, v, a\}$ of object motion in the Lagrange space $\mathbb{R}_L$;

– Just after terminating an action strategy $\mathcal{AS}_\alpha$, the subject evaluates its efficiency as it is experienced by him right now. In our account, this experience-based efficiency is represented as a cloud $\widetilde{\mathbb{E}}_\alpha \subset \mathbb{R}_E$ in the 1D-space of efficiency magnitudes $\mathbb{R}_E = \{e\}$. The experience-based efficiency $\widetilde{\mathbb{E}}_\alpha$ can change between successive implementations of one action strategy $\mathcal{AS}_\alpha$. We consider these changes in $\widetilde{\mathbb{E}}_\alpha$ random because each time a particular implementation of the action strategy $\mathcal{AS}_\alpha$ is different and the subject cannot discriminate it from the other probable implementations of $\mathcal{AS}_\alpha$.
– The subject attributes the skill-based efficiency $\mathbb{E}_\mathfrak{A}$ of action strategies to the set $\mathfrak{A}$ as a whole entity. Therefore, in our account the skill-based efficiency $\mathbb{E}_\mathfrak{A}$ is a cloud not only in the space $\mathbb{R}_E$ of efficiency magnitudes but also in the Laplace space $\mathbb{R}_L$ of the object motion states $\{x, v, a\}$.

There are, at least, two reasons for the given interpretation of skill-based efficiency. One of them is the limited capacity of working memory as well as unconscious bodily memory. It makes keeping the individual information about each of the action strategies $\{\mathcal{AS}_\alpha\}$ for a long time problematic. It should be emphasized that the skill-based efficiency is retained in the long-term bodily memory, which is the essence of skill acquisition. The experience-based efficiency is kept in sensory memory only within the experiential now until the process of its fusion with the skill-based efficiency is completed.

The other reason is that the learning process of the "infinite" number of different action strategies based on the "finite" number of their implementations seems to be merely infeasible.

The noted features of the experience-skill fusion demonstrate that the entities $\{\widetilde{\mathbb{E}}_\alpha\}$ and $\mathbb{E}_\mathfrak{A}$ involved into the fusion process should be categorized as clouds belonging to the Cartesian product of the space $\mathbb{R}_E = \{e\}$ of efficiency magnitudes and the Lagrange space $\mathbb{R}_L = \{x, v, a\}$, i.e.,

$$\{\widetilde{\mathbb{E}}_\alpha\},\ \mathbb{E}_\mathfrak{A} \subset \mathbb{R}_E \times \mathbb{R}_L\,.$$

Let us elucidate the structural details of these clouds.

Concerning the experience-based efficiency $\widetilde{\mathbb{E}}_\alpha$ of the just now terminated action strategy $\mathcal{AS}_\alpha$, the subject's perception of this efficiency is characterized by some cloud function $\widetilde{\Phi}(e \mid \widetilde{E}_\alpha, \Delta_p)$ of the efficiency magnitude e. This cloud function is parameterized by the magnitude $\widetilde{E}_\alpha$, where the function attains its maximum, and the thickness Δ_p quantifying the uncertainty in subject's perception of action strategy efficiency (see Sect. 5.4.1 for details). The magnitude $\widetilde{E}_\alpha$ exhibits random variations in the sequence of the $\mathcal{AS}_\alpha$-implementations, by contrast, the quantity Δ_p may be treated as a fixed value given beforehand.

The properties of the experience-based efficiency $\widetilde{\mathbb{E}}_\alpha$ as a cloud in the Lagrange space $\mathbb{R}_L$ are related to that the $\mathcal{AS}_\alpha$-implementation is also characterized by the initial state $\{x_\alpha, v_\alpha, a_\alpha\}$ of object motion. The particular values of these state variables are accessible to the mind only in the form of the initiation cloud $\mathbb{C}^\alpha_\text{ini} \subset \mathbb{R}_L$. Blur of the initiation cloud $\mathbb{C}^\alpha_\text{ini}$ is quantified by the uncertainties $\{\delta_x, \delta_v, \delta_a\}$ in the subject's perception of the state variables. We describe the initiation cloud $\mathbb{C}_{\text{ini}}$ in the Lagrange space $\mathbb{R}_L$ using the membership function

$$\Omega_\alpha(x, v, a) = \Omega_{\text{ini}}\left[\frac{(x - x_\alpha)}{\delta_x}, \frac{(v - v_\alpha)}{\delta_v}, \frac{(a - a_\alpha)}{\delta_a}\right]. \tag{6.217}$$

The particular form the membership function Ω_{ini} may be chosen for the sake of convenience, e.g.,

$$\Omega_{\text{ini}}(u_x, u_v, u_a) = \exp\left[-\frac{1}{2}\left(u_x^2 + u_v^2 + u_a^2\right)\right], \tag{6.218}$$

where $u_x = (x - x_\alpha)/\delta_x$, the other variables u_v and u_a are constructed in a similar way.

Therefore, in the realm of the body-in-the-self, the experience-based efficiency $\widetilde{\mathbb{E}}_\alpha$ is represented by the cloud function

$$\widetilde{\Psi}_\alpha(e, x, v, a) = Z_\alpha\, \widetilde{\Phi}(e \mid \widetilde{E}_\alpha, \Delta_p)\Omega_\alpha(x, v, a)\,. \tag{6.219}$$

Here

$$\Pi_\alpha = \left\{E_\alpha, \Delta_p, x_\alpha, v_\alpha, a_\alpha\right\} \tag{6.220}$$

is the set of its parameters whose strict values are inaccessible to the mind. The normalization coefficient Z_α is determined by the condition

$$\left\langle \widetilde{\Psi}_\alpha \middle| \widetilde{\Psi}_\alpha \right\rangle = 1 \tag{6.221}$$

imposed on the cloud function $\widetilde{\Psi}_\alpha$.

As far as the skill-based efficiency $\mathbb{E}_{\mathfrak{A}}$ is concerned, in the realm of the body-in-the-self it is represented by the cloud function

$$\Psi_{\mathfrak{A}}(e, x, v, a) = Z_{\mathfrak{A}}\, \Phi_{\mathfrak{A}}\big[e \big| \, E_{\mathfrak{A}}(x, v, a), \Delta_{\mathfrak{A}}\big]\Omega_{\mathfrak{A}}(x, v, a)\,. \tag{6.222}$$

with the normalization condition

$$\left\langle \Psi_{\mathfrak{A}} \middle| \Psi_{\mathfrak{A}} \right\rangle = 1 \tag{6.223}$$

specifying the coefficient $Z_{\mathfrak{A}}$. Here the former cofactor,

$$\Phi_{\mathfrak{A}}[e|\, E_{\mathfrak{A}}(x, v, a), \Delta_{\mathfrak{A}}],$$

describes the distribution of the efficiency magnitude e which is characterized by two quantities. One of them is the mean efficiency $E_{\mathfrak{A}}(x, v, a)$ of action strategy implementations for the initial state of object motion located at a given point $\{x, v, a\}$ of the Lagrange space $\mathbb{R}_L$. We will call the function $E_{\mathfrak{A}}(x, v, a)$ the *efficiency field* of the action strategy set $\mathfrak{A}$. The other is the thickness $\Delta_{\mathfrak{A}}$ of the cloud $\mathbb{E}_{\mathfrak{A}}$ along the e-axis. The value of $\Delta_{\mathfrak{A}}$ has to be much larger than the uncertainty Δ_p characterizing the subject's perception of efficiency magnitude, $\Delta_{\mathfrak{A}} \gg \Delta_p$. It is due to the quantity $\Delta_{\mathfrak{A}}$ aggregating in itself not only the perception uncertainty but also variations in the efficiency of different particular implementations of action strategies $\{\mathcal{AS}_\alpha\}$.

The latter cofactor—the cloud function $\Omega_{\mathfrak{A}}(x, v, a)$—is a membership function representing the fuzzy region in the Laplace space $\mathbb{R}_L$ where the cloud $\mathbb{E}_{\mathfrak{A}}$ is localized, i.e.,

$$\mathbb{O}_{\mathfrak{A}} = \left\{ \max_{x',v',a'} \left[E_{\mathfrak{A}}(x', v', a')\right] - E_{\mathfrak{A}}(x, v, a) \lesssim 1 \right\} \subset \mathbb{R}_L \,. \tag{6.224}$$

This region is characterized by that when the motion state of controlled object enters it, the subject makes decision about terminating the previously initiated action strategies and regards action strategies from the set $\mathfrak{A}$ as plausible candidates for the next implementation. Inside the region $\mathbb{O}_{\mathfrak{A}}$ the efficiency field $E_{\mathfrak{A}}(x, v, a)$ admits the approximation

$$E_{\mathfrak{A}}(x, v, a) = E_m - \langle \mathbf{s}| \, \mathbf{E}_{\mathfrak{A}} \, |\mathbf{s}\rangle \ , \tag{6.225}$$

where the maximum of the efficiency field

$$E_m = \max_{x,v,a} \left[E_{\mathfrak{A}}(x, v, a)\right],$$

is attained at the point $\{x_m, v_m, a_m\}$ of the Laplace space, $\mathbf{s}$ is the dimensionless vector

$$\mathbf{s} = \left\{ \frac{(x - x_m)}{\delta_x}, \frac{(v - v_m)}{\delta_v}, \frac{(a - a_m)}{\delta_a} \right\}$$

of this space, and the matrix $\mathbf{E}_{\mathfrak{A}}$ represents a certain positive-definite quadratic form. The components of the matrix $\mathbf{E}_{\mathfrak{A}} = \|e_{ij}\| \; (i, j = x, v, a)$ have to be small values,

$$|e_{ij}| \ll 1$$

because the dimensions of the region $\mathbb{O}_{\mathfrak{A}}$ must exceed substantially the uncertainties $\{\delta_x, \delta_v, \delta_a\}$ in the subject's perception of the inner phase variables. In these terms the membership function $\Omega_{\mathfrak{A}}(x, v, a)$ can be written as

$$\Omega_{\mathfrak{A}}(x, v, a) = \exp\left\{ -\frac{1}{2} \langle \mathbf{s}| \, \mathbf{E}_{\mathfrak{A}} \, |\mathbf{s}\rangle \right\}, \tag{6.226}$$

which is the direct analogy to the membership function (6.218).

Summarizing this description of the skill-based efficiency $\mathbb{E}_{\mathfrak{A}}$, we see that its representation in the body-in-the-self by the cloud function $\Psi_{\mathfrak{A}}(e, x, v, a)$ (Exp. 6.222) is specified by the set of parameters

$$\Pi_{\mathfrak{A}} = \left\{E_m, x_m, v_m, a_m, \Delta_{\mathfrak{A}}, \|e_{ij}\|\right\}. \tag{6.227}$$

To finalize the description of the experience-skill fusion, we turn to the formalism developed in Sect. 6.3.2 for a similar process of incorporating the experience-based effort into the skill-based effort. The experience-skill fusion of

- the efficiency $\widetilde{\mathbb{E}}_\alpha$ the subject experiences just after terminating the action strategy $\mathcal{AS}_\alpha$ and
- the skill-based efficiency $\mathbb{E}_{\mathfrak{A}}$ attributed to the set $\mathfrak{A} = \{\mathcal{AS}_{\alpha'}\}$ of action strategies similar in property

is implemented via updating the skill-based efficiency symbolically written as

$$\mathbb{E}_{\mathfrak{A}} \overset{\widetilde{\mathbb{E}}_\alpha}{\Longrightarrow} \mathbb{E}'_{\mathfrak{A}} \tag{6.228}$$

and comprising two steps:

- In the realm of the body-in-the-self, the fusion process is described by the linear sum (cf. Eq. 6.156a)

$$\Psi'_{\mathfrak{A}}(e, x, v, a) \propto \Phi_{\mathfrak{A}}\big[e\big|\, E_{\mathfrak{A}}(x, v, a), \Delta_{\mathfrak{A}}\big]\Omega_{\mathfrak{A}}(x, v, a) +$$
$$\varpi_{\mathfrak{A}}\left\{\widetilde{\Phi}(e \mid \widetilde{E}_\alpha, \Delta_p) - \Phi_{\mathfrak{A}}\big[e\big|\, E_{\mathfrak{A}}(x, v, a), \Delta_{\mathfrak{A}}\big]\right\}\Omega_\alpha(x, v, a)\Omega_{\mathfrak{A}}(x, v, a), \tag{6.229}$$

  where $\varpi_{\mathfrak{A}} \in [\varepsilon, 1]$ is the parameter of fusion determining the relative weights of the two components and attributed to the action strategy set $\mathfrak{A}$ as a whole entity. Expression (6.229) implies that the experience-based efficiency and the skill-based efficiency are compared in the initiation region $\mathbb{C}^\alpha_{\text{ini}}$ in the Lagrange space $\mathbb{R}_L$.
- In the realm of the mind-in-the-self, the skill-based efficiency $\mathbb{E}'_{\mathfrak{A}}$ of the action strategy set $\mathfrak{A}$ emerging via this fusion process is determined by raising the cloud function $\Psi'_{\mathfrak{A}}(e, x, v, a)$ from the level of the body-in-the-self to the level of the mind-in-the-self.[61] Mathematically, this raising is reduced to the renormalization of the coefficients

$$\Pi_{\mathfrak{A}} \overset{\Psi'_{\mathfrak{A}}}{\Longrightarrow} \Pi'_{\mathfrak{A}} \tag{6.230}$$

  governed practically by equations (6.157) within their simple generalization taking into account the increased number of parameters.

The transformation of the parameters $\Pi_{\mathfrak{A}}$ governed by Exp. (6.230) is the heart of the experience-skill fusion on scales of the experiential now. It is worthy of noting that the transformation of the skill-based efficiency governed by Exps. (6.229) and

[61] The notion of raising a cloud function Ψ from the level of the body-in-the-self to the level of the mind-in-the-self is elaborated in Sect. 5.6. This raising determines the relationship between cloud type entities of the mind-in-the-self and operations with these entities via the body-in-the-self.

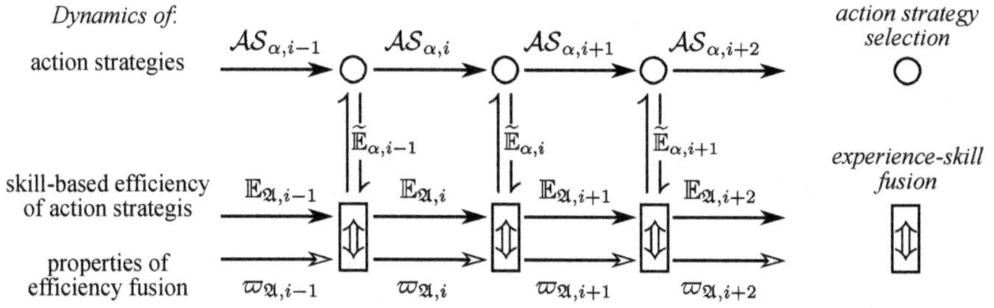

Fig. 6.10 The basic elements of fusion-based reinforcement learning, schematic illustration. The set $\{\mathcal{AS}_{\alpha,n}\}$ represents the time sequence of action strategies initiated and implemented by the subject during his goal-oriented actions. Correspondingly, the set $\{\mathbb{E}_{\mathfrak{A},n}, \varpi_{\mathfrak{A},n}\}$ represents the related sequence of the skill-based efficiency and the properties of the experience-skill fusion. Horizontal arrows shows the temporal order of the evens combining the action strategy selection (empty circles) and the experience-skill fusion (empty rectangular). Vertical harpoon-like arrows ($\upharpoonleft\downharpoonright$) illustrate (*i*) the effect of the skill-based efficiency of action strategies on their selection, (*ii*) the contribution of currently terminated action strategies to the experience-skill fusion via their experience-base efficiency $\{\widetilde{\mathbb{E}}_{\alpha,n}\}$. Vertical bidirectional arrows ($\Updownarrow$), illustrate the internal interaction between the components of the experience-skill fusion

(6.230) meets the predictive coding paradigm (Sect. 2.6.4). Namely, the cognitive states of the subject change only when the anticipation and the experience do not coincide with each other.

Long-term dynamics of experience-skill fusion. At the very beginning of skill acquisition the subject, trying to drive the controlled object toward the desired goal, collects information about possible actions. As his skill is acquired, the subject becomes able to perform more complex actions arranged into action strategies. We suppose that starting from this stage, the subject becomes familiar with the properties of relevant action strategies and can approximately estimate their efficiency. This state of subject's skill is the starting point in our constructions of the fusion-based reinforcement learning on long-term scales.

We suppose that the subject with partial skill can initially evaluate the boundaries of efficiency for the relevant action strategies. Put differently, the subject is assumed to *anticipate* certain efficiency $\{\mathbb{E}_{\mathfrak{A}}\}$ of the action strategy sets $\{\mathfrak{A}\}$ to be involved in his further actions. In particular, it concerns the sets $\{\Pi_{\mathfrak{A}}\}$ (Exp. 6.227) of the main parameters and their rough estimates. After starting his actions, the subject becomes able to compare the experience-base efficiency of trial action strategies and the efficiency attributed to the corresponding action strategy sets and, then, to fuse them. The action strategy efficiency emerging via this fusion may be regarded as the skill-based efficiency whose evolution is determined by the skill acquisition.

The events individually representing the action strategy initiation and implementation combined with the action strategy selection and the experience-skill fusion are illustrated in Fig. 6.10. The temporal chain of these events with

- the internal interaction between their constituent components on scales of the complex present and
- the influence of these events on one another on long-term scales

form the process being the heart of fusion-based reinforcement learning.

The upper line in Fig. 6.10 represents the dynamics of action strategies the subject selects and implements; this process was described in the previous subsections. In particular, it concerns the action strategy selection affected directly by the skill-based efficiency, which is illustrated by upward harpoon-like arrows.

The two lower lines represent dynamics of the skill-based efficiency of action strategies and the parameter $\varpi_{\mathfrak{A}}$ describing the details in the experience-skill fusion. A single event of fusing the experience-based efficiency together with the skill-based efficiency has been considered above in this subsection. The two-step transformation (6.228) describes this fusion on scales of the experiential now. Temporal chain of such events (three of them are shown in Fig. 6.10) describes the long-term dynamics of the action strategy skill-based efficiency, in particular, the efficiency field $E_{\mathfrak{A}}(x, v, a \mid t)$ of the main types $\{\mathfrak{A}\}$ of action strategies. This dynamic process may be regarded as the subject's skill acquisition governed by the fusion-induced reinforcement learning.

The fusion parameter $\varpi_{\mathfrak{A}}$ specifying the relative weights of the experience-based efficiency and the skill-based efficiency in their fusion also changes during the learning process. At the first approximation, we describe the $\varpi_{\mathfrak{A}}$-dependence on the current state of learning process by the equation written for temporal scales exceeding the duration τ_s of experiential now

$$\tau_{\varpi}\frac{d\varpi_{\mathfrak{A}}}{dt} = (1 - \varpi_{\mathfrak{A}}) - \varpi_{\mathfrak{A}}\,\wp_{\mathfrak{A}}\,\tau_s\delta(t - t_{\mathfrak{A},k}) \tag{6.231}$$

subject to the formal initial condition $\varpi_{\mathfrak{A}}\big|_{t=0} = 1$. Here the time scale $\tau_{\varpi} \gg \tau_d$ describes the decay of bodily memory, i.e., the characteristic time scale of skill dynamics, $\delta(t)$ is the Dirac δ-function, $\{t_{\mathfrak{A},k}\}$ is the sequence of time moments when the subject terminates an action strategy $\mathcal{AS}_\alpha \in \mathfrak{A}$ from the set $\mathfrak{A}$ and fuses the experience-based efficiency $\mathbb{E}_\alpha$ and the skill-based efficiency $\mathbb{E}_{\mathfrak{A}}$. The quantity $\wp_{\mathfrak{A}}$ takes into account the amount of information contained in the experience-based efficiency relative to amount of information contained the skill-based efficiency, which must be a crucial factor in the experience-skill fusion. For this reason we related the quantity $\wp_{\mathfrak{A}}$ to the ratio of the relative volumes of the two efficiency components in the space $\mathbb{R}_E \times \mathbb{R}_L$, namely,

$$\wp_{\mathfrak{A}} = \Big[\delta_x\delta_v\delta_a \det\big(\mathbf{E}_{\mathfrak{A}}\big)\Big]^{-\beta_L}\left(\frac{\Delta_{\mathfrak{A}}}{\Delta_p}\right)^{\beta_E}. \tag{6.232}$$

Here the values $\{\delta_x, \delta_v, \delta_a\}$ quantify uncertainty in the subject's perception of the inner phase variables $\{x, v, a\}$ of the controlled object motion, the determinant of the matrix $\mathbf{E}_{\mathfrak{A}}$ quantifies the inverse volume of the region $\mathbb{O}_{\mathfrak{A}}$ in the Laplace space

$\mathbb{R}_L$ attributed to the action strategy set $\widetilde{\mathfrak{A}}$ (cf., Exp. 6.226). The quantities Δ_p and $\Delta_{\mathfrak{A}}$ characterize the width of the clouds $\mathbb{E}_\alpha$ and $\mathbb{E}_{\mathfrak{A}}$ with respect to the efficiency magnitudes. The exponents β_L, $\beta_E > 0$ are the model parameters. As follows from Eq. (6.231), for the skilled subject the parameter attains its quasi-equilibrium value

$$\varpi_\infty = \varepsilon \sim \frac{\tau_d}{\wp_{\mathfrak{A}} \tau_s} \ll 1.$$

It should be noted that the quantity $\wp_{\mathfrak{A}}$, in principle, can change during the course of learning because of variations in the parameters of the skill-based efficiency. *Concluding* our account of fusion-based reinforcement learning, we note that the proposed description combines within one scheme all the main characteristic fragments in the dynamics of action strategy clouds. Namely,

- the automatic implementation of action strategies and their termination when the motion state of controlled object deviates essentially from the desired goal;
- the fusion of local properties of just now terminated action strategy with the subject's long-term skill;
- the selection of an action strategy for the next implementation.

It should be emphasized that these fragments are characterized by time scales changing from the duration of experiential now to the duration of complex present. The fusion-based reinforcement learning unites them into a process with internal integrity and characterized by time scales exceeding substantially the duration of complex present. The subject's skill acquisition is an additional aspect of the given learning process.

Beside we want to add the following comments.

- The outlined formalism of fusion-based reinforcement learning may be regarded as a gate toward constructing the description of human goal-oriented behavior beyond the complex present. Naturally, it can require the development of the theory of human memory and anticipation taking into account multiscale functioning of the mind.
- The learning process illustrated in Fig. 6.10 deals with the cloud functions $\{\Psi_{\mathfrak{A}}(e, x, v, a)\}$ (Exp. 6.222) representing the skill-based efficiency $\{\mathbb{E}_{\mathfrak{A}}\}$ of the main action strategy sets $\{\mathfrak{A}\}$ in the realm of the body-in-the-self. In this sense the proposed formalism may be treated as some analogy to the quantum reinforcement learning discussed in Sect. 6.3.5.
- The long-term dynamics of the subject actions combined with the skill acquisition may be treated a sequence alternating active and passive phases of subject's actions (Fig. 6.10). The passive phase is related to automatic implementation of selected action strategies within the complex present. The action strategy implementation is a random process with respect the object motion in the Laplace space and its duration whose mean duration is about τ_d—the characteristic duration of complex present. The active phase is related to conscious selection of action strategy for further implementation and is also a random process including its duration about the characteristic duration $\tau_s \ll \tau_d$ of the experiential now. In

this sense, the considered fusion-based reinforcement learning may be regarded as some analogy of the continuous-time random walks in the space of action strategies.[62]

At this place we have finalized our mathematical description of the human temporality.

6.4 Conclusion

The present chapter is the final part in our constructions of mathematical formalism for human goal-oriented actions within the complex present and the proposed concept of action strategy cloud is the gist of this formalism. The developed description of goal-oriented actions comprises several constituent components. We summarize them separately because each of them represents a special aspect of such actions.

Temporal extension of the complex present. For describing action strategies spanning the complex present (i.e., the connected past, the experiential now, and the connected future) a special formalism had to be developed. As a result:

- The novel formalism of dynamical systems with higher-order time derivatives has been developed. In its basis, this formalism takes into consideration the tripartite structure of the experiential now and proposes a new interpretation of Ostrogradsky's mechanics. In particular, this formalism
  - is based on the action-strategy Lagrangian quantifying the implementation cost of action strategies treated as trajectories of object motion in the space $\mathbb{R}_{[4]} = \{x, v, a, j\}$. In constructing this Lagrangian we used the results obtained in Chap. 5 devoted to physics of experiential now;
  - singles our two types of phase variables: the inner phase variables $\{x, v, a\}$ of the object motion and the control phase variable j—the object jerk— the subject is able to change voluntarily[63];
  - introduces the strictly optimal implementation of action strategy described by a temporal boundary value problem, which may be regarded as a novel class of problems in mathematical physics.
- The Hamiltonian description of the strictly optimal implementation of action strategies for the ideal subject has been developed. This formalism is based on the constructed higher-order Lagrangian of action strategies and enabled us to introduce

[62] For an introductory review of the concept of continuous-time random walks and its generalization a reader may be refereed, e.g., to Kleinhans and Friedrich (2007), Masoliver and Lindenberg (2017), Kutner and Masoliver (2017) and collections edited by Klages et al. (2008), Paul and Baschnagel (2013).

[63] When the dynamics of the controlled object is overdamped, the inner phase variables are $\{x, v\}$, whereas the control variable is the acceleration a.

- o the notion of the Lagrange space as a subspace of the complete Hamilton space and
- o the notion of the zero-Hamiltonian manifold consisting of stable and unstable branches such that

  - all the strictly optimal implementations of action strategies lie on the stable branch, whereas
  - the unstable branch formally represents the system dynamics deviating from the optimal behavior.

Fuzziness of the complex present. To allow for uncertainty in human perception, cognitive evaluation, and anticipation, the cloud type formalism for action strategies has been developed. It takes into account (*i*) the temporal extension of action strategies as cognitive entities and (*ii*) uncertainty in the mental-mental embedding. The proposed formalism comprises several new concepts:

- The concept of skilled subject has been proposed to allow for the uncertainty in human perception and evaluation as well as the bounded capacity of human anticipation It assumes that the subject with acquired skill is able to follow the optimal action strategy only approximately.
- The Hamiltonian formalism of goal-oriented actions has been developed within the skilled subject approximation. This formalism is not confined to the optimal action strategy implementations and describes also the deviation of skilled subject's actions from the optimal behavior.
- The temporal criterion of action strategy proximity to the optimal implementation has been constructed to overcome the problem of Ostrogradsky instability,. The pivot of this construction is expansion of the object motion into its projections on the stable and unstable branches of the zero-Hamiltonian manifold.
- The cloud type description of action strategy implementation has been developed using the zero-Hamiltonian manifold and the Liouville equation. This description deals with the space-time cloud treated as a certain spatio-temporal distribution over the human temporality confined to the complex present.
- A governing equation for the spatio-temporal structure of the cloud function of action strategy has been developed. It describes the formal dynamics of the cloud function in the corresponding Hilbert space. This governing equation may be treated as some analogy to the Schrödinger equation in imaginary time.

2D-time dynamics of the complex present. Evolution of action strategies embedded into the complex present as physical time goes on has been described in terms of 2D-time dynamics. This 2D-time dynamics implies that action strategies are attributed with properties (*i*) having temporal extension in the human temporality and (*ii*) changing with physical time. The developed description of the cloud 2D-time dynamics includes the following:

- The concept of anticipation-implementation cycle has been put forward as the governing mechanism of the 2D-time dynamics of action strategy clouds. This mechanism reflects the interplay between the real motion of the controlled object and the mental image of action strategy.

- The concept of the action strategy efficiency which takes into account
  - cognitive effort of action strategy implementation and
  - the effectiveness in keeping the object motion near the optimal implementation of this strategy

  has been constructed.
- Two principally different phases have been singled out in the 2D-time dynamics of action strategy clouds. They are:
  - the passive phase, that represents automatic implementation of the selected action strategy within the complex present. For this phase the constructed formalism describes individually
    - the 2D-time dynamics of the forward and backward temporal parts of the action strategy cloud,
    - the action strategy termination as a probabilistic process when the forward temporal part and the action-point layer start to overlap with each other,
    - the incorporation of the experienced effort into the skill-based effort based on the interplay between the mind-in-the-self and the body-in-the-self;
  - the active phase, that represents the selection of an action strategy and its initiation; this process is completed within the experiential now. Two types of active phase have been singled out:
    - the corrective type, corresponding to the initiation of a new implementation of the same action strategy,
    - the selective type, related to the selection of an action strategy among several different action strategies and then its initiation.
- The formalism developed for both the types of active phase may be regarded as a certain integration of the cloud-type dynamics and the gist of the Lotka-Volterra models. The probabilistic properties of the action strategy selection have been demonstrated to be governed by a new type transitions between "potential wells" caused by the strong nonlocality of the human temporality.

The gist of the complex present long-term dynamics. The fusion-based reinforcement learning has been elaborated based on the concept of experience-skill fusion. This learning mechanism unites the characteristic fragments of the action strategy dynamics into a process with internal integrity and characterized by time scales exceeding substantially the duration of complex present. The subject's skill acquisition is a component of this learning process.

The formalism developed for the 2D-time dynamics of action strategy clouds in the framework of the complex present is not confined to particular problems underlying its construction. It sheds light on the basic properties of the human temporality and opens a gate towards the novel direction in physics—physics of the human mind.

Appendix B

Appendices to Chapter 6: Higher-Order Systems

The present appendix is devoted to the dynamics of systems with higher-order time-derivatives. The formalism to be developed below takes the central position in describing human goal-oriented behavior when the desired goal to be reached in the future affects the subject's current actions. This formalism is rooted in Lagrangian mechanics of classical systems and uses the gist of Ostrogradsky's description of systems with higher-order time-derivatives (Ostrogradsky 1850) within the new interpretation (Sect. 6.1).

6.5 Lagrangian Description

In this section we focus on the dynamics of some system whose states can be *completely* specified as a point in the phase space

$$\mathbb{R}_{[n]} = \left\{x, \dot{x}, \ddot{x}, \dddot{x}, \ldots, x^{(n-1)}\right\} , \tag{6.233}$$

where the spatial coordinate x may contain several components

$$x = \{x_1, x_2, \ldots, x_m\}$$

representing the spatial dimensions of the phase space $\mathbb{R}_{[n]}$. Below this aspect is taken into account implicitly supposing that the operators like

$$\frac{\partial}{\partial x} = \left\{\frac{\partial}{\partial x_1}, \frac{\partial}{\partial x_2}, \ldots, \frac{\partial}{\partial x_m}\right\}$$

imply the corresponding component-wise differentiation. The completeness of the phase space $\mathbb{R}_{[n]}$ is understood as the closure of the system local description. Namely, any property that can be attributed to the given system at the current instant of time is assumed to admit a description as a function of the point of $\mathbb{R}_{[n]}$. Put differently, no higher time derivative $x^{(r)}$ with $r \geq n$ can be treated as an independent quantity completing the system description based on the phase space $\mathbb{R}_{[n]}$.

6.5.1 *Higher-Order Lagrangian*

Within the description of the complex present, for evaluating a path $\{x(t)\}$ of hypothetical system motion we accept the model of additive cumulative contribution from different time instants. It is reduced to the following *action functional*

$$\mathcal{E}\{x(t)\} = \int_{t_-}^{t_+} dt\, L\left(x, \dot{x}, \ddot{x}, \dddot{x}, \dots, x^{(n-1)}\right), \tag{6.234}$$

where the Lagrangian $L(\dots)$ is a given function determined in the phase space $\mathbb{R}_{[n]}$. In principle the Lagrangian $L(\dots)$ may depend directly on time

$$L\left(x, \dot{x}, \ddot{x}, \dddot{x}, \dots, x^{(n-1)} \mid t\right). \tag{6.235}$$

Below we confine our analysis to the *principle of least action*. Within this principle, the system dynamics is supposed to meet the minimum of action functional. In other words, the real trajectory of system motion within the time interval $t_- < t < t_+$ obeys the condition

$$\{x(t)\}_{\text{real}} \Rightarrow \min_{\{x(t)\}} \mathcal{E}\{x(t)\}, \tag{6.236}$$

which means that the system motion is described by the extremals of functional (6.234).

6.5.2 Higher-Order Euler-Lagrange Equation

For internal instants of time $t \in (t_-, t_+)$ the extremals of $\mathcal{E}\{x(t)\}$ obeys the condition

$$\min_{\{x(t)\}} \mathcal{E}\{x(t)\} \Rightarrow \int_{t_-}^{t_+} dt\, \delta L\left(x, \dot{x}, \ddot{x}, \dddot{x}, \dots, x^{(n-1)}\right) = 0, \tag{6.237}$$

provided there are no perturbations in the trajectory $\{x(t)\}$ at the boundary points $t = t_-, t_+$, i.e., for $\delta x(t) = x(t) - x_{\text{real}}(t)$ the boundary values $\delta x(t_-) = \delta x(t_+) = 0$. In this case the equality

$$\int_{t_-}^{t_+} dt\, \delta L\left(x, \dot{x}, \ddot{x}, \dddot{x}, \dots, x^{(n-1)}\right) = \sum_{r=0}^{n-1} \int_{t_-}^{t_+} dt\, \frac{\partial L}{\partial x^{(r)}}\, \delta x^{(r)}$$

holds and by virtue of integration by parts

$$\int_{t_-}^{t_+} dt\, \frac{\partial L}{\partial x^{(r)}} \delta x^{(r)} = \int_{t_-}^{t_+} dt\, (-1)^r \left\{\left[\frac{d}{dt}\right]^r \frac{\partial L}{\partial x^{(r)}}\right\} \delta x\,. \tag{6.238}$$

Thereby, condition (6.237) is reduced to the governing equation

$$\frac{\partial L}{\partial x} - \frac{d}{dt}\frac{\partial L}{\partial \dot{x}} + \left[\frac{d}{dt}\right]^2 \frac{\partial L}{\partial \ddot{x}} - \ldots + (-1)^{n-1}\left[\frac{d}{dt}\right]^{n-1} \frac{\partial L}{\partial x^{(n-1)}} = 0 \quad (6.239a)$$

or in the shortened form

$$\sum_{r=0}^{n-1}(-1)^r \left[\frac{d}{dt}\right]^r \frac{\partial L}{\partial x^{(r)}} = 0\,. \quad (6.239b)$$

because it has to be fulfilled for any variation $\delta x(t)$ at the internal points $t \in (t_-, t_+)$. Equation (6.239) is often called the *higher-order Euler-Lagrange equation.*

When the perturbation $\delta x = x(t) - x_{\text{real}}(t)$ does not disappear at the boundary points $t = t_-,\ t_+$, the variation $\delta\mathcal{E}$ of the action functional (6.234) contains also the boundary variations

$$\begin{aligned}\Delta\mathcal{E} &= \sum_{r=1}^{n-1}\sum_{k=0}^{r-1}(-1)^k \left\{\left[\frac{d}{dt}\right]^k \frac{\partial L}{\partial x^{(r)}}\right\}\left\{\left[\frac{d}{dt}\right]^{r-1-k} \delta x\right\}\Bigg|_{t=t_-}^{t=t_+} \\ &= \sum_{r=1}^{n-1}\sum_{p=0}^{r-1}(-1)^{r-1-p} \left\{\left[\frac{d}{dt}\right]^{r-1-p} \frac{\partial L}{\partial x^{(r)}}\right\}\left\{\left[\frac{d}{dt}\right]^{p} \delta x\right\}\Bigg|_{t=t_-}^{t=t_+} \\ &= \sum_{p=0}^{n-2}\left(\sum_{r=p+1}^{n-1}(-1)^{r-1-p} \left\{\left[\frac{d}{dt}\right]^{r-1-p} \frac{\partial L}{\partial x^{(r)}}\right\}\right)\left\{\left[\frac{d}{dt}\right]^{p} \delta x\right\}\Bigg|_{t=t_-}^{t=t_+}\end{aligned} \quad (6.240)$$

which follows from integration by parts of the left-hand side of Exp. (6.238). In studying the problem of action functional minimality with free boundaries in the phase space $\mathbb{R}_{[n]}$ value (6.240) must be set equal to zero.

6.5.3 *Temporal Boundary Value Problem*

In the general case Eq. (6.239) is of $2(n-1)$ order, i.e., it contains $2n-2$ "free" variables, $\{x, \dot{x}, \ddot{x}, \ldots, x^{(2n-3)}\}$, which together determine the highest order time derivative $x^{(2n-2)}$. All the higher order time derivatives $x^{(r)}$ (for $r > 2n-2$) can be calculated based on the obtained relationship.

It should be noted that the governing equation (6.239) specifying the extremals of the action functional (6.234) contains $n-2$ additional variables, $\{x^{(n)}, x^{(n+1)}, \ldots, x^{(2n-3)}\}$ which do not belong to the phase space $\mathbb{R}_{[n]}$. Therefore, finding the extremals of functional (6.234) based on the higher-order Euler-Lagrange equation (6.239) cannot be categorized as the initial value problem. Indeed, the accepted concept of the phase space $\mathbb{R}_{[n]}$ (Exp. 6.233) implies that only the variables $\{x, \dot{x}, \ddot{x}, \ldots, x^{(n-1)}\}$ can be related to some physical properties of a given system which enable one to treat these variables as independent quantities in ascribing to them some particular values. The remaining $n-2$ "free" variables should be determined turning to another principle rather than the reference to the system physical state at the initial instant of time.[64]

The theory of human goal-oriented behavior opens a gate toward a novel solution to this puzzle. In order to describe physical regularities of system motion together with human intentions, the phase space $\mathbb{R}_{[n]}$ should be composed of two types of variables:

- the phase variables describing the system physical states and which cannot be changed explicitly by the subject based on his intentions; we call them the *inner phase variables*,
- the phase variables describing directly the subject's intentional actions and which the subject can change explicitly, and in this say, affects implicitly the dynamics of the inner phase variables; we call this type variables the *control phase variables*.

We regard the quantities $\{x, \dot{x}, \ddot{x}, \ldots, x^{(n-2)}\}$ as the inner phase variables of system dynamics, whereas the highest-order time-derivative $x^{(n-1)}$ belonging to the space $\mathbb{R}_{[n]}$ is treated as the control phase variable.

In this terms the higher-order Euler-Lagrange equation (6.239) may be conceived of as a temporal boundary value problem. The inner phase variables fixed at the initial and terminal instants of time, t_- and t_+,

$$S_- = \begin{bmatrix} x \\ \dot{x} \\ \ddot{x} \\ \cdots \\ x^{(n-2)} \end{bmatrix}_{t=t_-}, \qquad S_+ = \begin{bmatrix} x \\ \dot{x} \\ \ddot{x} \\ \cdots \\ x^{(n-2)} \end{bmatrix}_{t=t_+} \tag{6.241}$$

specify the physical state S_- of the given system at the initial moment, when the subject starts his actions, and the desired state S_+ being the goal of subject's action that should be attained at the terminal moment. Within this interpretation, the total number of variables that should be determined beforehand is $2n-2$. Therefore, in the general case, the higher-order Euler-Lagrange equation (6.239) subject to the *temporal boundary conditions* (6.241) specifies completely the desired extremal of the action functional (6.234). It is the essence of the temporal boundary value problem when the goal to be attained in future influences the current subject's actions.

We want to emphasize that the temporal boundary conditions (6.241) do not contain the boundary values of the control phase variables $x^{(n-1)}$. It is justified by that the control variable admits variations including step-wise ones directly caused the subject's actions and, so, cannot be included into the boundary conditions. Indeed, the values

[64] The problem of determining the variables $\{x^{(n)}, x^{(n+1)}, \ldots, x^{(2n-3)}\}$ for systems with higher-order time-derivatives is know under the name the Ostrogradsky' ghost; for details and the corresponding references see Sect. 6.1.6.

$$x_{-}^{(n-1)} = \lim_{t \to t_{-}+0} x^{(n-1)}(t) \quad \text{and} \quad x_{+}^{(n-1)} = \lim_{t \to t_{+}-0} x^{(n-1)}(t)$$

are determined by subject's actions when he starts to implement the desired system motion or terminates it rather than the system states just before the actions or after their termination. By contrast, the inner phase variables $\{x, \dot{x}, \ddot{x}, \dots, x^{(n-2)}\}$ can exhibit only continuous time variations. Thereby their boundary values (6.241) must be related to the corresponding physical states of system motion.

6.5.4 Action Functional in Extended Phase Space

For constructing the Hamiltonian description of systems with higher-order time-derivatives condition (6.237) should be rewritten in the form enabling us to introduce Hamiltonian function and its canonical variables in the standard way. Following Plyushchay (1988), Pons (1989), Motohashi and Suyama (2015) we turn to the technique of Lagrange multipliers. Namely let us introduce the formal variables treated as mutually independent quantities

$$q_r = x^{(r)} \equiv \frac{d^r x}{dt^r} \quad \text{for } r = 0, 1, 2, \dots, n-1, \qquad q_0 = x \tag{6.242}$$

and the collection $\{\lambda_r\}_{r=0}^{r=n-2}$ of $(n-1)$ additional variables—the Lagrange multipliers. Then Lagrangian (6.234) is treated as a function of the arguments $\{q_r\}$

$$L\left(q_0, q_1, q_2, \dots, q_{n-1} \mid t\right) \tag{6.243}$$

and the action functional is generalized (extended) to

$$\mathcal{E}_{\Lambda}\{q(t), \lambda(t)\} = \int_{t_-}^{t_+} dt \left\{ L\left(\{q_r\}_{r=0}^{r=n-1}\right) + \sum_{r=0}^{n-2} \lambda_r \left[\frac{dq_r}{dt} - q_{r+1}\right] \right\}. \tag{6.244}$$

The second term on the right-hand side of (6.244) restores the relationship of time derivatives $\{x^{(r)}\}$ after setting the variational derivatives of functional (6.244) with respect to the Lagrange multipliers equal to zero, $\delta\mathcal{E}_{\Lambda}/\delta\lambda_r = 0$ for all $r = 0, 1, 2, \dots, n-1$.

The crucial feature of the generalized action functional (6.244) is that it contains only the first-order time derivatives of its arguments. In this case the standard variation technique for Lagrangians with first-order time derivatives leads to the following equations[65]

[65] Actually these equations stem directly from expressions obtained in Sect. 6.5.2 for systems with first-order time derivatives, i.e., $n = 2$.

$$\frac{\delta \mathcal{E}_\Lambda}{\delta \lambda_r} = \frac{dq_r}{dt} - q_{r+1} = 0\,, \qquad (r = 0, 1, 2, \ldots, n-2) \tag{6.245a}$$

$$\frac{\delta \mathcal{E}_\Lambda}{\delta q_{n-1}} = \frac{\partial L}{\partial q_{n-1}} - \lambda_{n-2} = 0\,, \tag{6.245b}$$

for the variables $\{\lambda_r\}$ and q_{n-1} whose time derivatives do not enter the list of the Lagrangian arguments and for the other variables

$$\frac{\delta \mathcal{E}_\Lambda}{\delta q_r} = \frac{\partial L}{\partial q_r} - \lambda_{r-1} - \frac{d\lambda_r}{dt} = 0\,, \qquad (r = 1, 2, \ldots, n-2) \tag{6.245c}$$

$$\frac{\delta \mathcal{E}_\Lambda}{\delta q_0} = \frac{\partial L}{\partial q_0} - \frac{d\lambda_0}{dt} = 0\,. \tag{6.245d}$$

governing the extremals of functional (6.244). In obtaining Eqs. (6.245c) and (6.245d) we have taken into account the equality for $r = 0, 1, 2, \ldots, n-2$

$$\frac{\delta \mathcal{E}_\Lambda}{\delta \dot{q}_r} = \lambda_r\,.$$

In the introduced terms the variation of functional (6.244) at the boundary points $t = t_-, t_+$ is given by the expression

$$\Delta \mathcal{E} = \sum_{r=1}^{n-2} \lambda_r\, \delta q_r \Big|_{t=t_-}^{t=t_+}\,. \tag{6.246}$$

Comparing the two approaches to describing the extremals of action functional for systems with higher-order time derivatives we may say the following. First, Eqs. (6.245a) restore the meaning of variables $\{q_r\}$ is the collection of time derivatives. Second, using Eqs. (6.245b) and (6.245c) within the iteration starting from $r = n-2$ and going down in r we obtain for $p = 0, 1, 2, \ldots, (n-2)$

$$\lambda_p = \sum_{r=p+1}^{n-1} (-1)^{r-1-p} \left[\frac{d}{dt}\right]^{r-1-p} \frac{\partial L}{\partial q_r}\,. \tag{6.247}$$

Then combining Eqs. (6.245d) and (6.247) for $r = 0$ we get the governing equation (6.239). Third, Exps. (6.246) and (6.247) immediately lead to (6.240).

These results demonstrate, as it must be, the equivalence of the two approaches to describing the desired extremals.

6.6 Hamiltonian Description

The present section is devoted to constructing Hamiltonian mechanics for systems with higher-order time-derivatives based on the higher-order Lagrangian introduced in Sect. 6.5.4.

Employing Ostrogradsky's technique (e.g., Nakamura and Hamamoto 1996; Woodard 2015b) we turn to the least action principle for the extended action functional (6.244). Namely, the starting point is the extended Lagrangian

$$L_\Lambda\Big(\{q_r\}_{r=0}^{r=n-1},\{\lambda_r\}_{r=0}^{r=n-2}\mid t\Big) = L\Big(\{q_r\}_{r=0}^{r=n-1}\mid t\Big)+\sum_{r=0}^{n-2}\lambda_r\left[\frac{dq_r}{dt}-q_{r+1}\right]. \tag{6.248}$$

Then, we introduce the new phase variables $\{p_k\}_{k=0}^{k=n-2}$ to be called the *generalized momenta*, that are related to the time derivatives $\{\dot{q}_k\}_{k=0}^{k=n-2}$ of the variables $\{q_k\}_{k=0}^{k=n-2}$ as

$$p_k=\frac{\partial L_\Lambda}{\partial \dot{q}_k}. \tag{6.249}$$

The variables $\{\lambda_k\}_{k=0}^{k=n-2}$ and q_{n-1} whose derivatives are not included in the extended Lagrangian are subjected to the condition of local minimality, i.e.,

$$\frac{\partial L_\Lambda}{\partial \lambda_k}=0 \quad (k=0,1,2,n-2)\,, \qquad\qquad \frac{\partial L_\Lambda}{\partial q_{n-1}}=0\,, \tag{6.250a}$$

which is another form of Eqs. (6.245a) and (6.245c). The variables $\{q_k\}_{k=0}^{k=n-2}$ whose time derivatives $\{\dot{q}_k\}_{k=0}^{k=n-2}$ enter the extended Lagrangian (6.248) obey the equations

$$\frac{\partial L_\Lambda}{\partial q_k}-\frac{d}{dt}\frac{\partial L_\Lambda}{\partial \dot{q}_k}, \quad k=0,1,2,\ldots,n-2\,, \tag{6.250b}$$

which is also another form of Eqs. (6.245c) and (6.245d). Equations (6.250) are no more than the standard Euler-Lagrange equations for the extended Lagrangian (6.248).

The key idea of introducing the new phase variables (6.249) is to base the system description upon the variables $\{q_k, p_k\}$ only. For the first step, we note that, according to Eq. (6.249),

$$p_k=\lambda_k \qquad (k=0,1,2,\ldots,n-2)\,. \tag{6.251}$$

Then Eqs. (6.250b) i.e., Eqs. (6.245c) and (6.245d) give rise to

$$\frac{dp_0}{dt} = \frac{\partial L}{\partial q_0}, \tag{6.252a}$$

$$\frac{dp_k}{dt} = \frac{\partial L}{\partial q_k} - p_{k-1} \quad \text{for } k = 1, 2, \ldots, n-2\,. \tag{6.252b}$$

Now everything is ready to make the next step.

6.6.1 *Ostrogradsky's Hamiltonian*

The desired Hamiltonian, to be also referred to as Ostrogradsky's Hamiltonian,

$$H\Big(\{q_r\}_{r=0}^{r=n-2}, \{p_r\}_{r=0}^{r=n-2} \mid t\Big)$$

as a function of the variables $\{q_k, p_k\}$ including $k = 0, 1, 2, \ldots, n-2$ is introduced as

$$H\Big(\{q_r\}_{r=0}^{r=n-2}, \{p_r\}_{r=0}^{r=n-2} \mid t\Big) = \sum_{k=0}^{n-2} p_k \frac{dq_k}{dt} - L_\Lambda\Big(\{q_r\}_{r=0}^{r=n-1}, \{p_r\}_{r=0}^{r=n-2} \mid t\Big), \tag{6.253}$$

where the quantities

$$q_{n-1}\,, \quad \left\{\frac{dq_k}{dt}\right\}_{k=0}^{k=n-2} \quad \Leftarrow \quad \{q_k, p_k\}_{k=0}^{k=n-2} \tag{6.254}$$

have to be treated as some functions of the canonical variables $\{q_k, p_k\}$ for $k = 0, 1, 2, \ldots, n-2$. We call the collection of variables

$$\{q, p\} \overset{\text{def}}{=} \{q_k, p_k\}_{k=0}^{k=n-2} \tag{6.255}$$

the *canonical variables* of the Ostrogradsky's Hamiltonian $H(q, p)$.

The extended Lagrangian (6.248) together with equality (6.251) enable us to rewrite (6.253) as

$$H(q, p \mid t) \equiv H(q_0, q_1, \ldots, q_{n-2}, p_0, p_1, \ldots, p_{n-2} \mid t) = p_{n-2}q_{n-1} + \sum_{k=0}^{n-3} p_k q_{k+1} - L(q_0, q_1, \ldots, q_{n-2}, q_{n-1} \mid t) \tag{6.256}$$

with the variable q_{n-1} related to p_{n-1} and $q_0, q_1, \ldots, q_{n-2}$ via the equation

$$p_{n-2} = \frac{\partial L(q_0, q_1, \ldots, q_{n-2}, q_{n-1} \mid t)}{\partial q_{n-1}} \tag{6.257}$$

by virtue of the first equality of pair (6.250a) and equality (6.251). To find the quantity

$$q_{n-1} = q_{n-1}\big(q_0, q_1, \ldots, q_{n-2}, p_{n-2} \mid t\big)$$

as a function of canonical variables $\{q, p\}$ Eq. (6.257) has to be resolved with respect to q_{n-1}.

6.6.2 *Hamilton's Equations*

Now let us demonstrate that the Hamiltonian function (6.256) constructed in the given way does meet the standard Hamilton's equations for systems with first-order time derivatives.

First, we turn to the following chain of equalities written for the Hamiltonian function (6.256) and the indices $k = 0, 1, 2, \ldots, n-3$

$$\frac{\partial H}{\partial p_k} = q_{k+1} = \frac{dq_k}{dt}$$

and for $k = n-2$

$$\frac{\partial H}{\partial p_{n-2}} = q_{n-1} + \left(p_{n-2} - \frac{\partial L}{\partial q_{n-1}}\right)\frac{\partial q_{n-1}}{\partial p_{n-2}} = q_{n-1} = \frac{dq_{n-2}}{dt}$$

by virtue of (6.245a) and (6.257). Second, for $k = 0$

$$\frac{\partial H}{\partial q_0} = \left(p_{n-2} - \frac{\partial L}{\partial q_{n-1}}\right)\frac{\partial q_{n-1}}{\partial q_0} - \frac{\partial L}{\partial q_0} = -\frac{\partial L}{\partial q_0} = -\frac{dp_0}{dt}$$

and for $k = 1, 2, \ldots, n-2$

$$\frac{\partial H}{\partial q_k} = \left(p_{n-2} - \frac{\partial L}{\partial q_{n-1}}\right)\frac{\partial q_{n-1}}{\partial q_k} + p_{k-1} - \frac{\partial L}{\partial q_k} = p_{k-1} - \frac{\partial L}{\partial q_k} = -\frac{dp_k}{dt}$$

by virtue of (6.252).

In this way we have gotten the classic Hamilton's equations for the canonical variable $\{q, p\}$:

$$\frac{dq_k}{dt} = \frac{\partial H(q, p \mid t)}{\partial p_k}, \tag{6.258a}$$

$$\frac{dp_k}{dt} = -\frac{\partial H(q, p \mid t)}{\partial q_k}, \tag{6.258b}$$

with the general feature of Hamiltonian dynamics

$$\frac{dH}{dt} = \frac{\partial H}{\partial t} \tag{6.259}$$

admitting the interpretation of the Hamiltonian $H(p, q)$ independent of time as a first integral of system motion, i.e., $dH/dt = 0$.

As far as the relationship between the momenta $\{p_k\}$ and the time derivatives of spatial coordinates $\{x, \dot{x}, \ddot{x}, \ldots\}$ are concerned, we can turn to Exps. (6.252b) and (6.257). Namely,

$$\begin{aligned} p_{n-2} &= \frac{\partial L}{\partial x^{(n-1)}}, \\ p_{n-3} &= \frac{\partial L}{\partial x^{(n-2)}} - \left[\frac{d}{dt}\right]\frac{\partial L}{\partial x^{(n-1)}} \\ p_{n-4} &= \frac{\partial L}{\partial x^{(n-3)}} - \left[\frac{d}{dt}\right]\frac{\partial L}{\partial x^{(n-2)}} + \left[\frac{d}{dt}\right]^2\frac{\partial L}{\partial x^{(n-1)}} \\ &\ldots \end{aligned} \tag{6.260a}$$

or, in the shortened form, for $k = 0, 1, 2, 3, \ldots, n-2$

$$p_k = \sum_{r=k+1}^{n-1} (-1)^{r-k-1} \left[\frac{d}{dt}\right]^{r-k-1} \frac{\partial L}{\partial x^{(r)}} \tag{6.260b}$$

which corresponds to Exp. (6.247). In particular, in terms of momenta, the boundary variations of the initial action functional (6.234) as follows from Exp. (6.240) can be represented as

$$\Delta\mathcal{E} = \sum_{k=0}^{n-2} p_k\, \delta x^{(k)} \Big|_{t=t_-}^{t=t_+}. \tag{6.261}$$

6.7 Dynamics of Linear Hamiltonian Systems in the Hamilton Space $\mathbb{R}_H$

In the present section we consider a special case of Hamiltonian systems described by Eqs. (6.258) for the Ostrogradsky's Hamiltonian constructed in Sect. 6.6.1. It is the case that was studied in Sect. 6.1 and is characterized by the following conditions.

- A system under consideration is described by Ostrogradsky's Hamiltonian $H(q, p)$ being of a certain symmetric quadratic form of the canonical variables $\{q, p\}$. In this case, Hamilton's equations (6.258) become linear. Hamiltonian (6.42) with $\gamma_s = 2$ will be referred to as a particular example.
- The system phase space $\mathbb{R}_{[n]}$ is
  - four-dimensional with respect to the time-derivatives, i.e.,

$$\mathbb{R}_{[n]} = \{x, \dot{x}, \ddot{x}, \dddot{x}\};$$

  - one-dimensional with respect to its spatial structure, i.e., the spatial position x of the system, its velocity $v = \dot{x}$, acceleration $a = \ddot{x}$, and jerk $j = \dddot{x}$ are 1D quantities.

  For the given phase space $\mathbb{R}_{[n]}$, the space $\mathbb{R}_H = \{q, p\}$ of the canonical variables—the *Hamilton space* $\mathbb{R}_H$—may be conceived of as a linear space of vectors $\{x, v, a, p_x, p_v, p_a\}$.

Even under these conditions, the analysis of Hamiltonian dynamics requires intricate mathematical manipulations, so we have confined our consideration to the given case to avoid unnecessary over-complication.

For such systems Hamilton's equation (6.55) can be rewritten in the form of a linear dynamical system in the six-dimensional Hamilton space $\mathbb{R}_H$

$$\frac{dz}{dt} = \mathbf{J}\frac{\partial H(z)}{\partial z} = \mathbf{J}\mathbf{H}z\,, \tag{6.262}$$

where the real vector z and matrix (operator) $\mathbf{J}$ have the following structures

$$z = \begin{bmatrix} x \\ v \\ a \\ p_x \\ p_v \\ p_a \end{bmatrix}, \qquad \mathbf{J} = \left[\begin{array}{ccc|ccc} 0 & 0 & 0 & 1 & 0 & 0 \\ 0 & 0 & 0 & 0 & 1 & 0 \\ 0 & 0 & 0 & 0 & 0 & 1 \\ \hline -1 & 0 & 0 & 0 & 0 & 0 \\ 0 & -1 & 0 & 0 & 0 & 0 \\ 0 & 0 & -1 & 0 & 0 & 0 \end{array}\right]. \tag{6.263}$$

Here the Hamiltonian $H(z)$ is regarded as a quadratic form

$$H(z) = \frac{1}{2}\langle z|\mathbf{H}|z\rangle = \frac{1}{2}z^{\dagger}\mathbf{H}z \tag{6.264}$$

with the matrix

$$\mathbf{H} = \left\| H_{ij} = \frac{\partial^2 H(z)}{\partial z_i \partial z_j} \right\|$$

and the symbol "†" stands for the *conjugate* transpose of the corresponding matrix or vector because below we will deal with complex eigenvectors of the operator **JH**, in particular, $|z\rangle^\dagger = \langle z|$.

We have turned to this general representation of the analyzed problem to figure out its properties that do not depend on particular details of employed models.

6.7.1 J-Orthogonality and Hamiltonian

For the matrix **J** the following equalities

$$\mathbf{J}^\dagger = \mathbf{J}^{-1} = -\mathbf{J} \tag{6.265}$$

hold, thereby,

$$\mathbf{J} \cdot (\mathbf{JH}) = -\mathbf{H}\,, \qquad (\mathbf{JH})^\dagger \cdot \mathbf{J} = \mathbf{H}\,. \tag{6.266}$$

Because the matrix **JH** represents the operator governing the dynamics of the vectors $\{z\}$ in the Hamilton space $\mathbb{R}_H$, let us consider the eigenvectors $\{z_\alpha := |\alpha\rangle\}$ of this operator

$$\mathbf{JH}\,|\alpha\rangle = \lambda_\alpha\,|\alpha\rangle\,, \tag{6.267}$$

where λ_α is the eigenvalue of the eigenvector $|\alpha\rangle$ and α is their index. The matrix **JH** is not Hermitian, so its eigenvectors and their eigenvalues may be complex-value entities.

In the case under consideration, **H** is a real-value matrix. Therefore, if $|\lambda\rangle$ is an eigenvector with the eigenvalue λ, then the operator **JH** possesses the collection of four eigenvectors (e.g., Meyer and Offin 2017)

$$|\lambda\rangle\,, \quad \left|\overline{\lambda}\right\rangle, \quad |-\lambda\rangle\,, \quad \left|-\overline{\lambda}\right\rangle.$$

When the eigenvalue λ is a real number, a purely imaginary number, or equal to zero this collection is reduced. Keeping in mind the problem in issue, we will assume that all the eigenvalues have non-zero real parts and the eigenvectors form a basis of the Hamilton space $\mathbb{R}_H$.

Let $|1\rangle$ and $|2\rangle$ be two eigenvectors with eigenvalues λ_1 and λ_2. Then, by virtue of (6.266) and (6.267),

$$\langle 1|\,\mathbf{H}\,|2\rangle = -\lambda_2\,\langle 1|\,\mathbf{J}\,|2\rangle\,, \tag{6.268a}$$

$$\langle 1|\,\mathbf{H}\,|2\rangle = \overline{\lambda}_1\,\langle 1|\,\mathbf{J}\,|2\rangle\,. \tag{6.268b}$$

Whence it follows that:

$$\langle 1|\,\mathbf{H}\,|2\rangle = 0 \text{ and } \langle 1|\,\mathbf{J}\,|2\rangle = 0, \qquad \text{if } \overline{\lambda}_1 + \lambda_2 \neq 0, \tag{6.269}$$

$$\langle 1|\,\mathbf{H}\,|2\rangle = -\lambda_2 \langle 1|\,\mathbf{J}\,|2\rangle\,, \qquad \text{if } \overline{\lambda}_1 + \lambda_2 = 0. \tag{6.270}$$

This result with respect to the symplectic inner product $\langle 1|\,\mathbf{J}\,|2\rangle$—the J-orthogonality—is well-known and presented in almost all textbooks on the theory of linear Hamiltonian systems. In out constructions its relationship with the Hamiltonian $H(z)$ via $\langle 1|\,\mathbf{H}\,|2\rangle$ is also crucial.

6.7.2 Lagrange Splitting and Dual-Expansion

Let us decompose the Hamilton space into two Lagrange subspaces

$$\mathbb{R}_H = \Xi_s \bigoplus \Xi_u$$

to be called the stable and unstable manifolds, Ξ_s and Ξ_u. By definition, the stable manifold Ξ_s is a linear real-value span of all the eigenvectors whose eigenvalues have negative real parts. Respectively, the unstable manifold Ξ_u is a linear real-value span of all the eigenvectors whose eigenvalues have positive real parts. To describe the Lagrange splitting, we deal with, let us accept that the set of eigenvectors comprises:

– a pair D of two real eigenvectors with eigenvalues $\lambda_0 > 0$ and $-\lambda_0$

$$|u\rangle\,, \qquad |s\rangle\,,$$

or, in other notations,

$$|\lambda_0\rangle\,, \qquad |-\lambda_0\rangle\,;$$

– a group T of four eigenvectors possessing eigenvalues $\lambda_r + i\lambda_i$, $\lambda_r - i\lambda_i$, $-\lambda_r + i\lambda_i$, $-\lambda_r - i\lambda_i$ with $\lambda_r > 0$ and $\lambda_i > 0$,

$$|u+\rangle\,, \qquad |u-\rangle\,, \qquad |s+\rangle\,, \qquad |s-\rangle\,,$$

or, in other notations,

$$|\lambda_r + i\lambda_i\rangle\,, \qquad |\lambda_r - i\lambda_i\rangle\,, \qquad |-\lambda_r + i\lambda_i\rangle\,, \qquad |-\lambda_r - i\lambda_i\rangle\,.$$

According to Exp. (6.269) only the combinations

$$\begin{aligned}\langle u|\,\mathbf{J}\,|s\rangle &= -\overline{\langle s|\,\mathbf{J}\,|u\rangle} \neq 0\,,\\ \langle u+|\,\mathbf{J}\,|s+\rangle &= -\overline{\langle s+|\,\mathbf{J}\,|u+\rangle} \neq 0\,,\\ \langle u-|\,\mathbf{J}\,|s-\rangle &= -\overline{\langle s-|\,\mathbf{J}\,|u-\rangle} \neq 0\,,\end{aligned} \tag{6.271}$$

for all the other combinations the symplectic inner product is equal to zero.

Within the basis composed of the given eigenvalues the manifolds Ξ_s and Ξ_u can be parameterized using

- real-value coefficients ξ^s, ξ^u for the pair D and
- complex-value coefficients ζ^s, $\overline{\zeta^s}$, ζ^u, $\overline{\zeta^u}$ for the four-eigenvector group T

of the vector expansion with respect to this basis. Namely,

$$\Xi_s\{\xi^s, \zeta^s\} = \xi^s\,|s\rangle + \zeta^s\,|s+\rangle + \overline{\zeta^s}\,|s-\rangle\,, \tag{6.272a}$$

$$\Xi_u\{\xi^u, \zeta^u\} = \xi^u\,|u\rangle + \zeta^u\,|u+\rangle + \overline{\zeta^u}\,|u-\rangle\,. \tag{6.272b}$$

The variables $\{\xi, \zeta\}$ play the role of local coordinates on the two manifolds.

By virtue of (6.271), for a vector $|z\rangle$ of the general position in the Hamilton space $\mathbb{R}_H$, its expansion relative to the manifolds Ξ_s and Ξ_u can be specified as follows:

$$\xi^s = \frac{\langle u|\,\mathbf{J}\,|z\rangle}{\langle u|\,\mathbf{J}\,|s\rangle}\,, \qquad \xi^u = \frac{\langle s|\,\mathbf{J}\,|z\rangle}{\langle s|\,\mathbf{J}\,|u\rangle}\,, \tag{6.273a}$$

$$\zeta^s = \frac{\langle u+|\,\mathbf{J}\,|z\rangle}{\langle u+|\,\mathbf{J}\,|s+\rangle}\,, \qquad \zeta^u = \frac{\langle s+|\,\mathbf{J}\,|z\rangle}{\langle s+|\,\mathbf{J}\,|u+\rangle}\,. \tag{6.273b}$$

The values of $\overline{\zeta^s}$ and $\overline{\zeta^u}$ can be obtained by just finding the complex conjugate of ζ^s and ζ^u, respectively, or via expressions similar of Exp. (6.273a) and Exp. (6.273b) using the corresponding eigenvectors. It is due to equality (6.265) and the identities

$$\overline{|s+\rangle} = |s-\rangle\,, \quad \overline{|u+\rangle} = |u-\rangle\,.$$

To simplify the representation of intermediate manipulations, let us rewrite Exp. (6.272) in the shortened form

$$\Xi_s\{\eta^s\} = \sum_\alpha \eta^s_\alpha\,|\alpha, s\rangle\,, \qquad \Xi_u\{\eta^u\} = \sum_\alpha \eta^u_\alpha\,|\alpha, u\rangle\,, \tag{6.274}$$

where the notations $|{*}\rangle\ \alpha, (s, u)$ and $\eta_\alpha^{(s,u)}$ represent the eigenvectors $|{*}\rangle\ (s, u)$, $|{*}\rangle\ (s, u)\pm$ and the variables $\xi^{(s,u)}$, $\zeta^{(s,u)}$, $\overline{\zeta^{(s,u)}}$. Respectively, Exp. (6.273) becomes

$$\eta^s_\alpha = \frac{\langle \alpha, u|\,\mathbf{J}\,|z\rangle}{\langle \alpha, u|\,\mathbf{J}\,|\alpha, s\rangle}\,, \qquad \eta^u_\alpha = \frac{\langle \alpha, s|\,\mathbf{J}\,|z\rangle}{\langle \alpha, s|\,\mathbf{J}\,|\alpha, u\rangle}\,, \tag{6.275}$$

the other combinations are equal to zero. The symbol λ_α denotes the eigenvalue of the eigenvector $|{*}\rangle\ \alpha, u$. Besides, by virtue of (6.269) and (6.270), all the elements

$\langle\ldots|\,\mathbf{H}\,|\ldots\rangle = 0$ except for

$$\begin{aligned}\langle\alpha, s|\,\mathbf{H}\,|\alpha, u\rangle &= -\lambda_\alpha \langle\alpha, s|\,\mathbf{J}\,|\alpha, u\rangle\,,\\ \langle\alpha, u|\,\mathbf{H}\,|\alpha, s\rangle &= \;\;\overline{\lambda}_\alpha \langle\alpha, u|\,\mathbf{J}\,|\alpha, s\rangle\,.\end{aligned} \tag{6.276}$$

Using the expansion of the vector $z \in \mathbb{R}_H$ with respect to these eigenvectors and taking into account expressions (6.275) and identities (6.271), (6.276), Hamiltonian (6.264) can be rewritten as

$$\begin{aligned}H(z) &= \frac{1}{2}\langle z|\,\mathbf{H}\,|z\rangle = \frac{1}{2}\sum_\alpha \Big[\overline{\eta^u_\alpha}\eta^s_\alpha \langle\alpha, u|\,\mathbf{H}\,|\alpha, s\rangle + \overline{\eta^s_\alpha}\eta^u_\alpha \langle\alpha, s|\,\mathbf{H}\,|\alpha, u\rangle\Big]\\ &= \frac{1}{2}\sum_\alpha \left\{\frac{\langle z|\,\mathbf{J}\,|\alpha, s\rangle\,\overline{\lambda}_\alpha\,\langle\alpha, u|\,\mathbf{J}\,|z\rangle}{\langle\alpha, u|\,\mathbf{J}\,|\alpha, s\rangle} - \frac{\langle z|\,\mathbf{J}\,|\alpha, u\rangle\,\lambda_\alpha\,\langle\alpha, s|\,\mathbf{J}\,|z\rangle}{\langle\alpha, s|\,\mathbf{J}\,|\alpha, u\rangle}\right\}.\end{aligned} \tag{6.277}$$

Whence it follows, in particular, that the stable and unstable manifolds Ξ_s, Ξ_u correspond to zero-value of the Hamiltonian $H(z)$, which is a reason to call the union $\Xi_s \bigcup \Xi_u$ the zero-Hamiltonian manifold, where Ξ_s and Ξ_u are its stable and unstable branches. Thereby the value, the Hamiltonian $H(z)$ takes at a given state z of system motion, can be used for quantifying the system deviation from the zero-Hamiltonian manifold.

By way of example, we illustrate the obtained results for Hamiltonian (6.42) with $\gamma_s = 2$ written in the dimensionless form, i.e., for the dimensionless phase variables x, v, a, p_x, p_v, p_a, and dimensionless time t. Besides, the final state of system motion, which attaining is the goal of subject's action, is placed at the origin, i.e., we set $x_g = 0$. In these terms Hamiltonian (6.42) takes the form

$$\begin{aligned}H(p_x, p_v, p_a, x, v, a) = p_x v + p_v a + \frac{1}{2\epsilon_j} p_a^2\\ -\frac{\epsilon_x}{2}x^2 - \frac{\epsilon_v}{2}v^2 - \frac{\epsilon_a}{2}a^2 - \frac{\widetilde{g}}{2}\big(f_a a + f_v v - f_x x\big)^2,\end{aligned} \tag{6.278}$$

where the dimensionless parameters specified by Exp. (6.51) meet the estimates $\widetilde{g} \sim k_m \sim k_v \sim k_x \sim 1$. The corresponding $\mathbf{H}$-matrix is

$$\mathbf{H} = \left[\begin{array}{ccc|ccc} & & & 0 & 0 & 0\\ & -\mathbf{L} & & 1 & 0 & 0\\ & & & 0 & 1 & 0\\ \hline 0 & 1 & 0 & 0 & 0 & 0\\ 0 & 0 & 1 & 0 & 0 & 0\\ 0 & 0 & 0 & 0 & 0 & 1/\epsilon_j\end{array}\right], \tag{6.279}$$

where the Lagrangian matrix

Table 6.1 Classification of eigenvectors according to their eigenvectors

Unstable manifold Ξ_u		Stable manifold Ξ_s	
Eigenvector	Eigenvalue	Eigenvector	Eigenvalue
$z_0^u = \lvert u\rangle$	$\lambda_0^u = \lambda_0$	$z_0^s = \lvert s\rangle$	$\lambda_0^s = -\lambda_0$
$z_+^u = \lvert u+\rangle$	$\lambda_+^u = \lambda_c = \lambda_r + i\lambda_i$	$z_+^s = \lvert s+\rangle$	$\lambda_+^s = -\overline{\lambda}_c = -\lambda_r + i\lambda_i$
$z_-^u = \lvert u-\rangle$	$\lambda_-^u = \overline{\lambda}_c = \lambda_r - i\lambda_i$	$z_-^s = \lvert s-\rangle$	$\lambda_-^s = -\lambda_c = -\lambda_r - i\lambda_i$

$$\mathbf{L} = \begin{bmatrix} L_{xx} & L_{xv} & L_{xa} \\ L_{vx} & L_{vv} & L_{va} \\ L_{ax} & L_{av} & L_{aa} \end{bmatrix} = \begin{bmatrix} (\epsilon_x + \widetilde{g}f_x^2) & -\widetilde{g}f_x f_v & -\widetilde{g}f_x f_a \\ -\widetilde{g}f_x f_v & (\epsilon_v + \widetilde{g}f_v^2) & \widetilde{g}f_v f_a \\ -\widetilde{g}f_x f_a & \widetilde{g}f_v f_a & (\epsilon_a + \widetilde{g}f_a^2) \end{bmatrix}. \tag{6.280}$$

For the given **JH**-operator, its eigenvector $|\lambda\rangle = \{x, v, a, p_x, p_v, p_a\}^{\mathrm{T}}$ with an eigenvalue λ is specified by the expressions

$$\begin{aligned} v &= \lambda x\,, \qquad a = \lambda^2 x\,, \qquad p_a = \epsilon_j \lambda^3 x\,, \\ \lambda p_x &= \big[L_{xx} + L_{xv}\lambda + L_{xa}\lambda^2\big]x \\ \lambda p_v &= \big[L_{vx} + L_{vv}\lambda + L_{va}\lambda^2\big]x - p_x\,, \\ \lambda p_a &= \big[L_{ax} + L_{av}\lambda + L_{aa}\lambda^2\big]x - p_v\,. \end{aligned} \tag{6.281}$$

The resulting eigenvalue equation is of the form

$$L_{xx} - \big(L_{vv} - 2L_{xa}\big)\lambda^2 + L_{aa}\lambda^4 - \epsilon_j\lambda^6 = 0\,, \tag{6.282}$$

which coincides with the eigenvalue equation (6.50) for $\gamma_s = 2$.

As stated above, this system is supposed to possess six eigenvectors divided into two groups illustrated by Table 6.1. The eigenvectors belonging to different rows of Table 6.1 are J-orthogonal to one another.

It should be noted that the Lagrangian matrix components L_{xv}, L_{vx}, L_{va}, L_{av} do not enter the eigenvalue equation (6.282) and the components L_{vv}, $L_{xa} = L_{ax}$ enter it as the difference $(L_{vv} - 2L_{xa})$ only. Thereby, the extremals of the corresponding action functional (6.234), described by Hamilton's equation (6.262) and giving a solution to the temporal boundary value problem (6.241), are independent of the particular values of the components $L_{xv} = L_{vx}$ and $L_{va} = L_{av}$. The same concerns the Lagrangian matrix components L_{vv} and $L_{xa} = L_{ax}$ provided their difference is fixed.

If we construct the corresponding Lagrangian and the Ostrogradsky's Hamiltonian based on introducing the eigenvalues λ_0 and λ_c at the first step, then only the components L_{xx}, L_{aa}, and the difference $(L_{vv} - 2L_{xa})$ are fixed,

$$
\begin{aligned}
L_{xx} &= \epsilon_j \lambda_0^2 |\lambda_c|^4, \\
L_{aa} &= \epsilon_j \big[\lambda_0^2 + 2(\lambda_r^2 - \lambda_i^2)\big], \\
L_{vv} - 2L_{xa} &= \epsilon_j \big[|\lambda_c|^4 + 2\lambda_0^2(\lambda_r^2 - \lambda_i^2)\big].
\end{aligned}
\tag{6.283}
$$

The other components of the Lagrangian matrix are "free parameter." This "freedom" demonstrates the existence of a certain family of Lagrangians describing the same set of strictly optimal action strategies. Therefore, in constructing the action-strategy Lagrangian $L(x, v, a, j)$ for subject's goal-oriented behavior (Sect. 6.1.3) we may turn to additional arguments concerning non-ideal subject's actions, which is posed, in particular, in Sect. 6.2.2 within the discussion of canonical transformations. Here we want to note that exactly the momenta p_x and p_v reflect the properties of non-ideal subject's behavior, in particular, via their individual dependence on the "free" components $L_{xv} = L_{vx}$ and $L_{va} = L_{av}$.

6.7.3 *Inner Dynamics of Zero-Hamiltonian Manifold*

The spatial position x of the analyzed system, its velocity v and acceleration a are treated as the *inner phase variable*, i.e., phase variables that are not *directly* controllable by the subject. Put differently, the subject cannot change the inner phase variables explicitly, he can affect directly only the jerk j of system motion and, in this way, change the inner variables indirectly. Within the used Hamiltonian description, the subject's intentions are taken into account by the momenta $\{p_x, p_v, p_a\}$, whereas the variables $\{x, v, a\}$ describe the physical states of the system governed by the subject's actions. So, we may consider that the variables $\{x, v, a\}$ represent the system motion state, whereas the variables $\{p_x, p_v, p_a\}$ describe the subject's intentional actions.

To take into account these features explicitly, let us introduce the Lagrange space $\mathbb{R}_L$ being the linear space of the inner phase variables, i.e.,

$$
\mathbb{R}_L = \Big\{\big[x\,,\, v\,,\, a\big]^{\mathrm{T}}\Big\}. \tag{6.284}
$$

The Lagrange space $\mathbb{R}_L$ is a subspace of the Hamilton space $\mathbb{R}_H$ and the linear subspace of the vectors $\{p_x, p_v, p_a\}$ may be treated as the complement $\mathbb{R}_L^{\dagger}$ of the Lagrange space $\mathbb{R}_L$, i.e.,

$$
\mathbb{R}_L^{\dagger} = \Big\{\big[p_x\,,\, p_v\,,\, p_a\big]^{\mathrm{T}}\Big\} \quad \text{and} \quad \mathbb{R}_L^{\dagger} \bigoplus \mathbb{R}_L = \mathbb{R}_H\,.
$$

Dealing with the stable and unstable manifolds, Ξ_s and Ξ_u, we refer to this complement to categorize the deviation of a given system state from the two manifolds.

To simplify referencing to the Lagrange space $\mathbb{R}_L$, we set the x-component of the eigenvectors $|u, s\rangle$, $|u, s, \pm\rangle$ (Table 6.1) equal to unity. In this case, the relationship between the system state z in the Hamilton space $\mathbb{R}_H$ and the components of its expansion relative to the stable and unstable manifolds Ξ_s and Ξ_u, Exp. (6.272), can be written as

$$\mathbf{x}_s = \mathbf{S}_s\, \boldsymbol{\zeta}_s \quad \text{and} \quad \mathbf{x}_u = \mathbf{S}_u\, \boldsymbol{\zeta}_u\,, \tag{6.285}$$

where the vectors $\boldsymbol{\zeta}_{s,u}$ and the matrices (operators) $\mathbf{S}_{s,u}$

$$\boldsymbol{\zeta}_{s,u} = \begin{bmatrix} \xi^{s,u} \\ \zeta^{s,u} \\ \overline{\zeta^{s,u}} \end{bmatrix}, \quad \mathbf{S}_s = \begin{bmatrix} 1 & 1 & 1 \\ -\lambda_0 & -\overline{\lambda}_c & -\lambda_c \\ \lambda_0^2 & \overline{\lambda}_c^{2} & \lambda_c^2 \end{bmatrix}, \quad \mathbf{S}_u = \begin{bmatrix} 1 & 1 & 1 \\ \lambda_0 & \lambda_c & \overline{\lambda}_c \\ \lambda_0^2 & \lambda_c^2 & \overline{\lambda}_c^{2} \end{bmatrix} \tag{6.286}$$

are treated as entities of the Lagrange space $\mathbb{R}_L$. The vectors $\mathbf{x}_{s,u}$ belonging to the Lagrange space $\mathbb{R}_L$ admit the interpretation as the binary projection of the system state z—the vector of the Hilbert space $\mathbb{R}_H$. At the first step, the vector z is projected onto the manifolds Ξ_s and Ξ_u, i.e., expanded relative to the bases of the two manifolds as it is described by Exp. (6.275). At the second step, the results are projected onto the Lagrange space $\mathbb{R}_L$. This binary projection is illustrated by the diagram:

$$\begin{array}{ccccc} \Xi_s & \longleftarrow & z(t) \in \mathbb{R}_H & \longrightarrow & \Xi_u \\ \downarrow & & & & \downarrow \\ \mathbf{x}_s(t) & =\!=\!=\!= & \in\ \mathbb{R}_L\ \ni & =\!=\!=\!= & \mathbf{x}_u(t) \end{array} \tag{6.287}$$

As time goes on, the time dependent cofactors $\{e^{\lambda t}\}$ of the eigenvectors $\{|\lambda\rangle\}$ induce time variations in the vectors $\mathbf{x}_{s,u}$ stemming from time variations in the system state $z(t)$ which are governed by the Hamilton's equation (6.262). By virtue of their construction (diagram 6.287), the motion of the vectors $\mathbf{x}_{s,u}$ in the Lagrange space $\mathbb{R}_L$ may be conceived of as the motion of the shadows on the Lagrange space $\mathbb{R}$ that are made individually by the system state projections onto the stable and unstable manifolds $\Xi_{s,u}$. In particular, the equations

$$\frac{d\mathbf{x}_s}{dt} = \mathbf{D}_s\, \mathbf{x}_s \quad \text{and} \quad \frac{d\mathbf{x}_u}{dt} = \mathbf{D}_u\, \mathbf{x}_u\,, \tag{6.288}$$

where the matrices (operators) $\mathbf{D}_{s,u}$ to be constructed below, describe the dynamics of the system state projections onto the manifolds $\Xi_{s,u}$. We call it the *inner dynamics* of the zero-Hamiltonian manifold or, dealing with equations (6.288) individually, the *inner dynamics* of its stable and unstable branches (manifolds) $\Xi_{s,u}$. If initially the system state z belongs to one of the branches $\Xi_{s,u}$, then during the further motion it remains on this branch. Thereby, the corresponding equation of inner dynamics completely describes the system motion.

To construct the matrices $\mathbf{D}_{s,u}$ we need the inverses of the matrices $\mathbf{S}_{s,u}$. The direct calculations give us

$$\mathbf{S}_s^{-1} = \frac{1}{D}\begin{bmatrix} 2\lambda_i(\lambda_i^2+\lambda_r^2) & 4\lambda_i\lambda_r & 2\lambda_i \\ -i\lambda_0\lambda_c(\lambda_0-\lambda_c) & -i(\lambda_0^2-\lambda_c^2) & -i(\lambda_0-\lambda_c) \\ i\lambda_0\overline{\lambda}_c(\lambda_0-\overline{\lambda}_c) & i(\lambda_0^2-\overline{\lambda}_c^2) & i(\lambda_0-\overline{\lambda}_c) \end{bmatrix},$$

$$\mathbf{S}_u^{-1} = \frac{1}{D}\begin{bmatrix} 2\lambda_i(\lambda_i^2+\lambda_r^2) & -4\lambda_i\lambda_r & 2\lambda_i \\ i\lambda_0\overline{\lambda}_c(\lambda_0-\overline{\lambda}_c) & -i(\lambda_0^2-\overline{\lambda}_c^2) & i(\lambda_0-\overline{\lambda}_c) \\ -i\lambda_0\lambda_c(\lambda_0-\lambda_c) & i(\lambda_0^2-\lambda_c^2) & -i(\lambda_0-\lambda_c) \end{bmatrix},$$

where the coefficient

$$D = 2\lambda_i\Big[(\lambda_0-\lambda_r)^2+\lambda_i^2\Big].$$

By virtue of Exp. (6.285), the inner dynamics of the manifolds $\varXi_s$ and $\varXi_u$ is specified by the equations

$$\frac{d\mathbf{x}_s}{dt} = \mathbf{S}_s\begin{bmatrix} -\lambda_0 & 0 & 0 \\ 0 & -\overline{\lambda}_c & 0 \\ 0 & 0 & -\lambda_c \end{bmatrix}\mathbf{S}_s^{-1}\,\mathbf{x}_s\,, \qquad \frac{d\mathbf{x}_u}{dt} = \mathbf{S}_u\begin{bmatrix} \lambda_0 & 0 & 0 \\ 0 & \lambda_c & 0 \\ 0 & 0 & \overline{\lambda}_c \end{bmatrix}\mathbf{S}_u^{-1}\,\mathbf{x}_u\,.$$

Whence we get the desired expressions for the operators of inner dynamics

$$\mathbf{D}_s = \begin{bmatrix} 0 & 1 & 0 \\ 0 & 0 & 1 \\ -\omega_{ax} & -\omega_{av} & -\omega_{aa} \end{bmatrix}, \qquad \mathbf{D}_u = \begin{bmatrix} 0 & 1 & 0 \\ 0 & 0 & 1 \\ \omega_{ax} & -\omega_{av} & \omega_{aa} \end{bmatrix}, \tag{6.289}$$

where the coefficients $\{\omega_{\dots}\}$ are given by the expressions

$$\omega_{ax} = \lambda_0|\lambda_c|^2, \qquad \omega_{av} = (2\lambda_0\lambda_r + |\lambda_c|^2), \qquad \omega_{aa} = (\lambda_0+2\lambda_r). \tag{6.290}$$

The upper two rows of the operators $\mathbf{D}_{s,u}$ represent just kinematic relations between the inner phase variables $\{x, v, a\}$, the lower row is the essence of the inner dynamics relating the time derivative $\dot{a} = j$ of the system acceleration, i.e., the system jerk to the current values of $\{x, v, a\}$.

By virtue of (6.281) and (6.283), the generalized momenta $\{p_x, p_v, p_a\}$ corresponding to the vectors $\mathbf{x}_{s,u} = \{x^{s,u}, v^{s,u}, a^{s,u}\}$ of the Lagrange space and denoted as $\boldsymbol{\pi}_{s,u} = \{\pi_x^{s,u}, \pi_v^{s,u}, \pi_a^{s,u}\}$, respectively, are given by the expressions

$$\boldsymbol{\pi}_s = \boldsymbol{\varOmega}_s\,\mathbf{x}_s \quad \text{and} \quad \boldsymbol{\pi}_u = \boldsymbol{\varOmega}_u\,\mathbf{x}_u\,, \tag{6.291}$$

where the matrices $\boldsymbol{\varOmega}_{s,u}$, mapping the Lagrange space $\mathbb{R}_L$ onto its complement $\mathbb{R}_L^{\dagger}$, are specified as

$$\boldsymbol{\Omega}_s = \epsilon_j \begin{bmatrix} -\frac{\tilde{g}}{\epsilon_j} f_x f_v - \omega_{ax}\omega_{av}; & -\frac{\tilde{g}}{\epsilon_j} f_x f_a - \omega_{ax}\omega_{aa}; & -\omega_{ax} \\ -\frac{\tilde{g}}{\epsilon_j} f_x f_a - \omega_{ax}\omega_{aa}; & \frac{\tilde{g}}{\epsilon_j} f_v f_a - 2\lambda_r(\lambda_0^2 + \omega_{av}); & -\omega_{av} \\ -\omega_{ax} & -\omega_{av} & -\omega_{aa} \end{bmatrix} \tag{6.292a}$$

$$\boldsymbol{\Omega}_u = \epsilon_j \begin{bmatrix} -\frac{\tilde{g}}{\epsilon_j} f_x f_v + \omega_{ax}\omega_{av}; & -\frac{\tilde{g}}{\epsilon_j} f_x f_a - \omega_{ax}\omega_{aa}; & \omega_{ax} \\ -\frac{\tilde{g}}{\epsilon_j} f_x f_a - \omega_{ax}\omega_{aa}; & \frac{\tilde{g}}{\epsilon_j} f_v f_a + 2\lambda_r(\lambda_0^2 + \omega_{av}); & -\omega_{av} \\ \omega_{ax} & -\omega_{av} & \omega_{aa} \end{bmatrix} \tag{6.292b}$$

Equations (6.288) together with the mapping (6.291) specify the inner dynamics of the stable and unstable manifolds $\Xi_{s,u}$ in the Hamilton space $\mathbb{R}_H$.

6.7.4 *Dual Time-Reversibility of Inner Dynamics*

The inner dynamics of the stable and unstable manifolds Ξ_s and Ξ_u, which is described by Eq. (6.288) with the operators $\mathbf{D}_{s,u}$ given by Exp. (6.289), demonstrates the *dual time-reversibility of the inner dynamics* of the zero-Hamiltonian manifold. Let, e.g., for given values of the inner phase variables $\mathbf{x} = \{x, v, a\}$ the corresponding state $z_s = \{x, v, a, \pi_x^s, \pi_v^s, \pi_a^s\}$ belong to the stable manifold Ξ_s. Then, the time inversion $t \to -t$ converts the inner dynamics of the stable manifold Ξ_s into the inner dynamics of the unstable manifold Ξ_u. The corresponding system state is $z_u = \{x, -v, a, \pi_x^u, \pi_v^u, -\pi_a^s\}$ and belongs to the unstable manifold.

In the general form the dual time-reversibility can be represented as the follows transformations:

stable manifold Ξ_s		unstable manifold Ξ_u	$t \longmapsto -t$	stable manifold Ξ_s		unstable manifold Ξ_u
x_s	$\longleftrightarrow$	x_u		v_s	$\longleftrightarrow$	$-v_u$
a_s	$\longleftrightarrow$	a_u		π_a^s	$\longleftrightarrow$	$-\pi_a^u$

The conversion of the momenta $\{\pi_x^s, \pi_v^s\} \longleftrightarrow \{\pi_x^u, \pi_v^u\}$ is more complicated; according to the mapping (6.291), some parts of these moments change the sign; the others do not do this.

References

Afraimovich, V., Tristan, I., Huerta, R., & Rabinovich, M. I. (2008). Winnerless competition principle and prediction of the transient dynamics in a Lotka–Volterra model. *Chaos: An Interdisciplinary Journal of Nonlinear Science, 18*(4), 043103 (9 pages).

Afraimovich, V., Young, T., Muezzinoglu, M. K., & Rabinovich, M. I. (2011). Nonlinear dynamics of emotion-cognition interaction: When emotion does not destroy cognition? *Bulletin of Mathematical Biology, 73*(2), 266–284.

Akaishi, R., Umeda, K., Nagase, A., & Sakai, K. (2014). Autonomous mechanism of internal choice estimate underlies decision inertia. *Neuron, 81*(1), 195–206.

Allen, C., & Neal, J. (2020). Teleological notions in biology. In E. N. Zalta (Ed.), *The Stanford Encyclopedia of Philosophy* (spring 2020 ed.). Metaphysics Research Lab, Stanford University.

Alós-Ferrer, C., Hügelschäfer, S., & Li, J. (2016). Inertia and decision making. *Frontiers in Psychology, 7*, Article 169 (9 pages).

Andrzejewski, K., Gonera, J., Machalski, P., & Maślanka, P. (2010). Modified Hamiltonian formalism for higher-derivative theories. *Physical Review D, 82*(4), 045008 (12 pages).

Andrzejewski, K. (2014). Hamiltonian formalisms and symmetries of the pais–uhlenbeck oscillator. *Nuclear Physics B, 889*, 333–350.

Aoki, K., & Motohashi, H. (2020). Ghost from constraints: a generalization of Ostrogradsky theorem. *Journal of Cosmology and Astroparticle Physics, 2020*(8), 026 (36 pages).

Arnold, V. I. (1989). *Mathematical methods of classical mechanics* (2nd ed.). Berlin: Springer. Translated from Russian by K. Vogtmann and A. Weinstein.

Asai, Y., Tasaka, Y., Nomura, K., Nomura, T., Casadio, M., & Morasso, P. (2009). A model of postural control in quiet standing: Robust compensation of delay-induced instability using intermittent activation of feedback control. *PLoS One, 4*(7), e6169 (14 pages).

Asai, Y., Tateyama, S., & Nomura, T. (2013). Learning an intermittent control strategy for postural balancing using an EMG-based human-computer interface. *PLoS ONE, 8*(5), e62956 (19 pages).

Ashwin, P., Coombes, S., & Nicks, R. (2016). Mathematical frameworks for oscillatory network dynamics in neuroscience. *The Journal of Mathematical Neuroscience, 6*(1), Article 2 (91 pages).

Ashwin, P., & Postlethwaite, C. (2013). On designing heteroclinic networks from graphs. *Physica D: Nonlinear Phenomena, 265*, 26–39.

Aydın, M., Herzog, M. H., & Öğmen, H. (2011). Attention modulates spatio-temporal grouping. *Vision Research, 51*(4), 435–446.

Balaguer, M. (2019). Free will, determinism, and epiphenomenalism. *Frontiers in Psychology, 9*, Article 2623 (14 pages).

Bargh, J. A., Gollwitzer, P. M., Lee-Chai, A., Barndollar, K., & Trötschel, R. (2001). The automated will: Nonconscious activation and pursuit of behavioral goals. *Journal of Personality and Social Psychology, 81*(6), 1014–1027.

Baumeister, R. F., & Monroe, A. E. (2014). Recent research on free will: Conceptualizations, beliefs, and processes. In J. M. Olson & M. P. Zanna (Eds.), *Advances in experimental social psychology* (pp. 1–52). Waltham, MA: Academic Press, Elsevier Inc.

Bellman, R. (1957). *Dynamic programming*. Princeton, NJ: Princeton University Press.

Bertsekas, D. P. (2017). *Dynamic programming and optimal control: Volume I* (4th ed.). Belmont, Mass: Athena Scientific.

Bignetti, E. (2014). The functional role of free-will illusion in cognition: "The Bignetti Model." *Cognitive Systems Research, 31–32*, 45–60.

Birrell, J., & Wehr, J. (2018). Homogenization of dissipative, noisy, Hamiltonian dynamics. *Stochastic Processes and their Applications, 128*(7), 2367–2403.

Bode, S., Murawski, C., Soon, C. S., Bode, P., Stahl, J., & Smith, P. L. (2014). Demystifying "free will": The role of contextual information and evidence accumulation for predictive brain activity. *Neuroscience & Biobehavioral Reviews, 47*, 636–645.

Bogacz, R., Brown, E., Moehlis, J., Holmes, P., & Cohen, J. D. (2006). The physics of optimal decision making: A formal analysis of models of performance in two-alternative forced-choice tasks. *Psychological Review, 113*(4), 700–765.

Bonicalzi, S., & Haggard, P. (2019). From freedom from to freedom to: New perspectives on intentional action. *Frontiers in Psychology, 10*, Article 1193 (14 pages).

Bottaro, A., Yasutake, Y., Nomura, T., Casadio, M., & Morasso, P. (2008). Bounded stability of the quiet standing posture: An intermittent control model. *Human Movement Science, 27*(3), 473–495.

Bukov, M., Day, A. G. R., Sels, D., Weinberg, P., Polkovnikov, A., & Mehta, P. (2018). Reinforcement learning in different phases of quantum control. *Physical Review X, 8*(3), 031086 (15 pages).

Buonomano, D. V., & Maass, W. (2009). State-dependent computations: Spatiotemporal processing in cortical networks. *Nature Reviews Neuroscience, 10*, 113.

Busemeyer, J. R., Gluth, S., Rieskamp, J., & Turner, B. M. (2019). Cognitive and neural bases of multi-attribute, multi-alternative, value-based decisions. *Trends in Cognitive Sciences, 23*(3), 251–263.

Cao, T., Wang, L., Sun, Z., Engel, S. A., & He, S. (2018). The independent and shared mechanisms of intrinsic brain dynamics: Insights from bistable perception. *Frontiers in Psychology, 9*, Article 589 (11 pages).

Catherine Sulem, P.-L.S. (1999). *The nonlinear Schrödinger equation: Self-focussing and wave collapse*. New York, NY: Springer-Verlag GmbH.

Cavanagh, P., & Alvarez, G. A. (2005). Tracking multiple targets with multifocal attention. *Trends in Cognitive Sciences, 9*(7), 349–354.

Chen, T., Fasiello, M., Lim, E. A., & Tolley, A. J. (2013). Higher derivative theories with constraints: exorcising ostrogradski's ghost. *Journal of Cosmology and Astroparticle Physics, 2013*(2), 042 (18 pages).

Chetrite, R., & Gawędzki, K. (2008). Fluctuation relations for diffusion processes. *Communications in Mathematical Physics, 282*(2), 469–518.

Cisek, P., & Kalaska, J. F. (2010). Neural mechanisms for interacting with a world full of action choices. *Annual Review of Neuroscience, 33*(1), 269–298.

Clarke, R., & Capes, J. (2017). Incompatibilist (nondeterministic) theories of free will. In E. N. Zalta (Ed.), *The Stanford encyclopedia of philosophy* (fall 2020 ed.). Metaphysics Research Lab, Stanford University.

Cloninger, C. (2004). *Feeling good: The science of well-being*. New York, NY: Oxford University Press.

Cocchi, L., Gollo, L. L., Zalesky, A., & Breakspear, M. (2017). Criticality in the brain: A synthesis of neurobiology, models and cognition. *Progress in Neurobiology, 158*, 132–152.

Contreras, M., Pellicer, R., & Villena, M. (2017). Dynamic optimization and its relation to classical and quantum constrained systems. *Physica A: Statistical Mechanics and its Applications, 479*, 12–25.

Contreras, M., & Peña, J. P. (2019). The quantum dark side of the optimal control theory. *Physica A: Statistical Mechanics and its Applications, 515*, 450–473.

Copie, F., Randoux, S., & Suret, P. (2020). The physics of the one-dimensional nonlinear Schrödinger equation in fiber optics: Rogue waves, modulation instability and self-focusing phenomena. *Reviews in Physics, 5*, 100037 (17 pages).

Cramer, J. G. (2016). *The quantum handshake: Entanglement, nonlocality and transactions*. Cham, Switzerland: Springer International Publishing.

Cresson, J., & Darses, S. (2007). Stochastic embedding of dynamical systems. *Journal of Mathematical Physics, 48*(7), 072703.

Cruz, M., Gómez-Cortés, R., Molgado, A., & Rojas, E. (2016). Hamiltonian analysis for linearly acceleration-dependent Lagrangians. *Journal of Mathematical Physics, 57*(6), 062903 (21 pages).

Dai, J. (2020). Stability and consistent interactions in higher derivative matter field theories. *The European Physical Journal Plus, 135*(7), 555 (22 pages).

Daw, N. D. (2018). Are we of two minds? *Nature Neuroscience, 21*(11), 1497–1499.
Daw, N. D., Niv, Y., & Dayan, P. (2005). Uncertainty-based competition between prefrontal and dorsolateral striatal systems for behavioral control. *Nature Neuroscience, 8*(12), 1704–1711.
Dayan, P., & Berridge, K. C. (2014). Model-based and model-free Pavlovian reward learning: Revaluation, revision, and revelation. *Cognitive, Affective, & Behavioral Neuroscience, 14*(2), 473–492.
Dayan, P., & Daw, N. D. (2008). Decision theory, reinforcement learning, and the brain. *Cognitive, Affective, & Behavioral Neuroscience, 8*(4), 429–453.
Dean, M., Kıbrıs, O., & Masatlioglu, Y. (2017). Limited attention and status quo bias. *Journal of Economic Theory, 169*, 93–127.
Deci, E. L., & Ryan, R. M. (2000). The "what" and "why" of goal pursuits: Human needs and the self-determination of behavior. *Psychological Inquiry, 11*(4), 227–268.
Deci, E. L., & Ryan, R. M. (1985). *Intrinsic motivation and self-determination in human behavior*. New York, NY: Springer Science+Business Media.
Deng, J., Anton, C., & Wong, Y. S. (2014). High-order symplectic schemes for stochastic Hamiltonian systems. *Communications in Computational Physics, 16*(1), 169–200.
Di Paolo, E. A. (2005). Autopoiesis, adaptivity, teleology, agency. *Phenomenology and the Cognitive Sciences, 4*(4), 429–452.
Dong, D., Chen, C., Li, H., & Tarn, T.-J. (2008). Quantum reinforcement learning. *IEEE Transactions on Systems, Man, and Cybernetics, Part B (Cybernetics), 38*(5), 1207–1220.
Doya, K. (1999). What are the computations of the cerebellum, the basal ganglia and the cerebral cortex? *Neural Networks, 12*(7–8), 961–974.
Droege, P. (2009). Now or never: How consciousness represents time. *Consciousness and Cognition, 18*(1), 78–90.
Dubois, D., & Prade, H. (1980). *Fuzzy sets and systems: Theory and applications*. New York, NY: Academic Press Inc.
Eidelman, S., & Crandall, C. S. (2014). The intuitive traditionalist: How biases for existence and longevity promote the status quo. In J. M. Olson & M. P. Zanna (Eds.), *Advances in experimental social psychology* (Vol. 50, pp. 53–104). Amsterdam: Elsevier Inc.
Fakhari, P., Rajagopal, K., Balakrishnan, S. N., & Busemeyer, J. R. (2013). Quantum inspired reinforcement learning in changing environment. *New Mathematics and Natural Computation, 9*(3), 273–294.
Fayn, K., MacCann, C., Tiliopoulos, N., & Silvia, P. J. (2015). Aesthetic emotions and aesthetic people: Openness predicts sensitivity to novelty in the experiences of interest and pleasure. *Frontiers in Psychology, 6*, Article 1877 (11 pages).
Feldman, A. G. (2009). Origin and advances of the equilibrium-point hypothesis. In D. Sternad (Ed.), *Progress in motor control: A multidisciplinary perspective* (pp. 637–643). New York: Springer Science+Business Media, LLC.
Feldman, A. G., & Levin, M. F. (2009). The equilibrium-point hypothesis - past, present and future. In D. Sternad (Ed.), *Progress in motor control: A multidisciplinary perspective* (pp. 699–726). New York: Springer Science+Business Media, LLC.
Feldman, A. G., Goussev, V., Sangole, A., & Levin, M. F. (2007). Threshold position control and the principle of minimal interaction in motor actions. In P. Cisek, T. Drew, & J. F. Kalaska (Eds.), *Computational neuroscience: Theoretical insights into brain function*, Vol. 165 of Progress in brain research (pp. 267–281). Amsterdam, The Netherlands: Elsevier.
Feynman, P. R. (2018/1972). *Statistical mechanics: A set of lectures*. Boca Raton, FL: CRC Press, Taylor & Francis Group. First published 1972 by Westview Press.
Fibich, G. (2015). *The nonlinear Schrödinger equation*. Cham, Switzerland: Springer International Publishing.
Fisher, G. (2017). An attentional drift diffusion model over binary-attribute choice. *Cognition, 168*, 34–45.
Flügge, S. (1971). *Practical quantum mechanics*. Berlin: Springer.
Francescotti, R. (2014). *Physicalism and the Mind*. Dordrecht, Netherlands: Springer.

Franklin, G. F., Powell, J. D., & Emami-Naeini, A. (2018). *Feedback control of dynamic systems* (8th ed.). Upper Saddle River, NJ: Pearson Higher Education Inc.
Freedman, D. J., & Assad, J. A. (2011). A proposed common neural mechanism for categorization and perceptual decisions. *Nature Neuroscience, 14*(2), 143–146.
Gage, N. M., & Baars, B. J. (2018). *Fundamentals of cognitive neuroscience: A beginner's guide* (2nd ed.). London, UK: Academic Press, Elsevier. Chap. 4, The Art of Seeing.
Ganz, A., & Noui, K. (2020). Reconsidering the Ostrogradsky theorem: Higher-derivatives Lagrangians, ghosts and degeneracy, arXiv:2007.01063 [hep-th].
Gardiner, C. (2009). *Stochastic methods: A handbook for the natural and social sciences* (4th ed.). Berlin: Springer.
Gläscher, J., Daw, N., Dayan, P., & O'Doherty, J. P. (2010). States versus rewards: Dissociable neural prediction error signals underlying model-based and model-free reinforcement learning. *Neuron, 66*(4), 585–595.
Gold, J. I., & Ding, L. (2013). How mechanisms of perceptual decision-making affect the psychometric function. *Progress in Neurobiology, 103,* 98–114.
Goldstein, H., Poole, C. P., & Safko, J. L. (2014). *Classical Mechanics* (3rd ed.). Harlow: Pearson Education Limited.
Grib, A. A., & Rodrigues (Jr.), W. A. (1999). *Nonlocality in quantum physics*. New York: Springer Science+Business Media, LLC.
Grossberg, S. (2018). Desirability, availability, credit assignment, category learning, and attention: Cognitive-emotional and working memory dynamics of orbitofrontal, ventrolateral, and dorsolateral prefrontal cortices. *Brain and Neuroscience Advances, 2,* 239821281877217 (50 pages).
Grossberg, S. (1980). Biological competition: Decision rules, pattern formation, and oscillations. *Proceedings of the National Academy of Sciences, 77*(4), 2338–2342.
Grossberg, S. (1988). Nonlinear neural networks: Principles, mechanisms, and architectures. *Neural Networks, 1*(1), 17–61.
Grover, L. K. (1996). A fast quantum mechanical algorithm for database search. In G. L. Miller (ed.), *Proceedings of the Twenty-Eighth Annual ACM Symposium on Theory of Computing - STOC'96* (pp. 212–219). New York, NY: Association for Computing Machinery.
Grover, L. K. (1997). Quantum mechanics helps in searching for a needle in a haystack. *Physical Review Letters, 79*(2), 325–328.
Haggard, P. (2008). Human volition: Towards a neuroscience of will. *Nature Reviews Neuroscience, 9*(12), 934–946.
Haggard, P. (2019). The neurocognitive bases of human volition. *Annual Review of Psychology, 70*(1), 9–28.
Haken, H. (1996). *Principles of brain functioning: A synergetic approach to brain activity, behavior and cognition*. Berlin: Springer.
Haken, H. (2008). *Brain dynamics: An introduction to models and simulations* (2nd ed.). Berlin: Springer.
Hamamoto, S., & Nakamura, M. (2000). Path-integral measures in higher-derivative gravitational theories. *Progress of Theoretical Physics, 104*(3), 691–702.
Hanks, T. D., & Summerfield, C. (2017). Perceptual decision making in rodents, monkeys, and humans. *Neuron, 93*(1), 15–31.
He, B. J., Daffertshofer, A., & Boonstra, T. W. (Eds.). (2013). *Scale-free dynamics and critical phenomena in cortical activity*. Frontiers in Physiology. Lausanne: Frontiers Media SA.
Huys, R., Perdikis, D., & Jirsa, V. K. (2014). Functional architectures and structured flows on manifolds: A dynamical framework for motor behavior. *Psychological Review, 121*(3), 302–336.
Ihlen, E. A. F., & Vereijken, B. (2013). Multifractal formalisms of human behavior. *Human Movement Science, 32*(4), 633–651.
IUPAC. (1993). *Compendium of chemical terminology*, 2nd (the "green book") edn. Oxford: Blackwell Scientific Publications. Prepared for publication by I. Mills, T. Cvitaš, K. Homann, N. Kallay, and K. Kuchitsu.

Jiang, Y., & Kanwisher, N. (2003). Common neural substrates for response selection across modalities and mapping paradigms. *Journal of Cognitive Neuroscience, 15*(8), 1080–1094.

Jiang, Y., & Kanwisher, N. (2003). Common neural mechanisms for response selection and perceptual processing. *Journal of Cognitive Neuroscience, 15*(8), 1095–1110.

Jung, D., & Dorner, V. (2018). Decision inertia and arousal: Using neurois to analyze bio-physiological correlates of decision inertia in a dual-choice paradigm. In F. D. Davis, R. Riedl, J. vom Brocke, P.-M. Léger & A. B. Randolph (Eds.), *Information systems and neuroscience (Lecture notes in information systems and organisation, Vol. 25)* (pp. 159–166). Cham: Springer International Publishing AG.

Jung, D., Stäbler, J., & Weinhardt, C. (2018). Investigating cognitive foundations of inertia in decision-making: Discussion paper heikamaxy 2018. *Technical report*, Karlsruher Institut für Technologie (KIT).

Kelso, J. A. S. (2009a). Coordination dynamics. In R. A. Meyers (Ed.), *Encyclopedia of complexity and systems science* (pp. 1537–1565). New York, NY: Springer Science+Business Media, LLC.

Kelso, J. A. S. (2009b). Synergies: Atoms of brain and behavior. In D. Sternad (Ed.), *Progress in motor control: A multidisciplinary perspective* (pp. 83–91). Boston, MA: Springer Science+Business Media, LLC.

Kelso, J. A. S. (1995). *Dynamic patterns: The self-organization of brain and behavior*. Cambridge, MA: The MIT Press.

Kelso, J. A. S. (2012). Multistability and metastability: Understanding dynamic coordination in the brain. *Philosophical Transactions of the Royal Society of London B: Biological Sciences, 367*(1591), 906–918.

Kelso, J. A. S., Dumas, G., & Tognoli, E. (2013). Outline of a general theory of behavior and brain coordination. *Neural Networks, 37,* 120–131.

Kelso, J. A. S., & Engstrøm, D. A. (2006). *The complementary nature*. Cambridge, MA: The MIT Press.

Kim, J. (1993). *Supervenience and mind: Selected philosophical essays*. Cambridge: Cambridge University Press.

Klages, R., Radons, G., & Sokolov, I. M. (Eds.). (2008). *Anomalous transport: Foundations and applications*. KGaA, Weinheim: Wiley-VCH Verlag GmbH & Co.

Kleinhans, D., & Friedrich, R. (2007). Continuous-time random walks: Simulation of continuous trajectories. *Physical Review E, 76*(6), 061102 (6 pages).

Knudsen, E. I. (2007). Fundamental components of attention. *Annual Review of Neuroscience, 30*(1), 57–78.

Kool, W., Cushman, F. A., & Gershman, S. J. (2016). When does model-based control pay off?, *PLOS Computational Biology, 12*(8), e1005090 (34 pages).

Kool, W., Gershman, S. J., & Cushman, F. A. (2017). Cost-benefit arbitration between multiple reinforcement-learning systems. *Psychological Science, 28*(9), 1321–1333.

Kori, H., & Kuramoto, Y. (2001). Slow switching in globally coupled oscillators: Robustness and occurrence through delayed coupling. *Physical Review E, 63*(4), 046214 (10 pages).

Kutner, R., & Masoliver, J. (2017). The continuous time random walk, still trendy: Fifty-year history, state of art and outlook. *The European Physical Journal B, 90*(3), 50 (13 pages).

Lanczos, C. (1986). *The variational principles of mechanics* (4th ed.). Toronto: The University of Toronto Press.

Landau, L. D., & Lifshitz, E. M. (1976). *Mechanics* (3rd ed., Vol. 1). Burlington, MA: Elsevier Butterworth-Heinemann. Course of Theoretical Physics.

Latash, M. L. (2012). *Fundamentals of motor control*. London: Academic Press, Elsevier Inc.

Latash, M. L. (2008). *Synergy*. Oxford: Oxford University Press.

Latash, M. L. (2010). Motor synergies and the equilibrium-point hypothesis. *Motor Control, 14*(3), 294–322.

Lázaro-Camí, J.-A., & Ortega, J.-P. (2008). Stochastic Hamiltonian dynamical systems. *Reports on Mathematical Physics, 61*(1), 65–122.

Liao, H.-I., Yeh, S.-L., & Shimojo, S. (2011). Novelty vs. familiarity principles in preference decisions: Task-context of past experience matters. *Frontiers in Psychology, 2*, Article 43 (8 pages).

Libet, B. (2002). Do we have free will? In R. Kane (Ed.), *The Oxford handbook of free will* (1st ed., pp. 551–564). New York: Oxford University Press.

Libet, B. (2004). *Mind time: The temporal factor in consciousness*. Cambridge, MA: Harvard University Press.

Libet, B., Gleason, C. A., Wright, E. W., & Pearl, D. K. (1983). Time of conscious intention to act in relation to onset of cerebral activity (readiness-potential): The unconscious initiation of a freely voluntary act. *Brain, 106*(3), 623–642.

Liboff, R. L. (2003). *Kinetic theory: Classical, quantum, and relativistic descriptions* (3rd ed.). New York, NY: Springer.

Li, J.-A., Dong, D., Wei, Z., Liu, Y., Pan, Y., Nori, F., & Zhang, X. (2020). Quantum reinforcement learning during human decision-making. *Nature Human Behaviour, 4*(3), 294–307.

Lubashevsky, I. (2016). Human fuzzy rationality as a novel mechanism of emergent phenomena. In C. H. Skiadas & C. Skiadas (Eds.), *Handbook of applications of chaos theory* (pp. 827–878). London: CRC Press, Taylor & Francis Group.

Lubashevsky, I., & Morimura, K. (2019). Physics of mind and car-following problem. In B. S. Kerner (Ed.), *Complex dynamics of traffic management, Encyclopedia of complexity and systems science series* (pp. 559–592). New York, NY: Springer Science+Business Media, LLC.

Lubashevsky, I., Wagner, P., & Mahnke, R. (2003). Rational-driver approximation in car-following theory. *Physical Review E, 68*(5), 056109 (15 pages).

Lubashevsky, I. (2017). *Physics of the human mind*. Cham: Springer International Publishing AG.

Mahnke, R., Kaupužs, J., & Lubashevsky, I. (2009). *Physics of stochastic processes: How randomness acts in time*. KGaA, Weinheim: Wiley-VCH Verlag GmbH & Co.

Martínez, G. A. R., & Parra, H. C. (2018). Bistable perception: Neural bases and usefulness in psychological research. *International Journal of Psychological Research, 11*(2), 63–76.

Marvan, T., & Polák, M. (2017). Unitary and dual models of phenomenal consciousness. *Consciousness and Cognition, 56*, 1–12.

Masoliver, J., & Lindenberg, K. (2017). Continuous time persistent random walk: A review and some generalizations. *The European Physical Journal B, 90*(6), 107 (13 pages).

Massobrio, P., de Arcangelis, L., Pasquale, V., Jensen, H. J., & Plenz, D. (Eds.). (2015). *Criticality as a signature of healthy neural systems: Multi-scale experimental and computational studies*. Frontiers in Systems Neuroscience. Lausanne: Frontiers Media SA.

Masterov, I. (2016). An alternative Hamiltonian formulation for the Pais–Uhlenbeck oscillator. *Nuclear Physics B, 902*, 95–114.

McKenna, M., & Coates, D. J. (2019). Compatibilism. In E. N. Zalta (Ed.), *The Stanford encyclopedia of philosophy* (winter 2020 ed.). Metaphysics Research Lab, Stanford University.

McLaughlin, B., & Bennett, K. (2018). Supervenience. In E. N. Zalta (Ed.), *The Stanford encyclopedia of philosophy* (winter 2018 ed.). Metaphysics Research Lab, Stanford University. http://plato.stanford.edu/archives/win2018/entries/supervenience/.

McMains, S. A., & Somers, D. C. (2004). Multiple spotlights of attentional selection in human visual cortex. *Neuron, 42*(4), 677–686.

Meilikhov, E. Z., & Farzetdinova, R. M. (2019). Bistable perception of ambiguous images: simple Arrhenius model. *Cognitive Neurodynamics, 13*(6), 613–621.

Meyer, K. R., & Offin, D. C. (2017). *Introduction to Hamiltonian dynamical systems and the N-body problem* (3rd ed.). Cham, Switzerland: Springer International Publishing AG.

Milstein, G. N., Repin, Y. M., & Tretyakov, M. V. (2002a). Numerical methods for stochastic systems preserving symplectic structure. *SIAM Journal on Numerical Analysis, 40*(4), 1583–1604.

Milstein, G. N., Repin, Y. M., & Tretyakov, M. V. (2002b). Symplectic integration of Hamiltonian systems with additive noise. *SIAM Journal on Numerical Analysis, 39*(6), 2066–2088.

Montague, P. R., Dayan, P., & Sejnowski, T. J. (1996). A framework for mesencephalic dopamine systems based on predictive Hebbian learning. *The Journal of Neuroscience, 16*(5), 1936–1947.

Moore, T., & Zirnsak, M. (2017). Neural mechanisms of selective visual attention. *Annual Review of Psychology, 68*(1), 47–72.

Morozov, A. Y. (2008). Hamiltonian formalism in the presence of higher derivatives. *Theoretical and Mathematical Physics, 157*(2), 1542–1549.

Motohashi, H., & Suyama, T. (2015). Third order equations of motion and the Ostrogradsky instability. *Physical Review D, 91*(8), 085009.

Motohashi, H., & Suyama, T. (2020). Quantum Ostrogradsky theorem. *Journal of High Energy Physics, 2020*(9), 32 (10 pages).

Motohashi, H., Noui, K., Suyama, T., Yamaguchi, M., & Langlois, D. (2016). Healthy degenerate theories with higher derivatives. *Journal of Cosmology and Astroparticle Physics, 2016*(7), 033 (28 pages).

Motohashi, H., Suyama, T., & Yamaguchi, M. (2018a). Ghost-free theories with arbitrary higher-order time derivatives. *Journal of High Energy Physics, 2018*(6), 133 (28 pages).

Motohashi, H., Suyama, T., & Yamaguchi, M. (2018b). Ghost-free theory with third-order time derivatives. *Journal of the Physical Society of Japan, 87*(6), 063401 (4 pages).

Mukherjee, P. (2010). Poincaré gauge theory from higher derivative matter Lagrangians, *Classical and Quantum Gravity, 27*(21), 215008 (11 pages).

Mulder, M. J., van Maanen, L., & Forstmann, B. U. (2014). Perceptual decision neurosciences – a model-based review. *Neuroscience, 277,* 872–884.

Murakami, M., & Mainen, Z. F. (2015). Preparing and selecting actions with neural populations: Toward cortical circuit mechanisms. *Current Opinion in Neurobiology, 33,* 40–46.

Murray, J. D. (2002). *Mathematical biology: I. An introduction* (3rd ed.). Berlin: Springer.

Mysore, S. P., & Kothari, N. B. (2020). Mechanisms of competitive selection: A canonical neural circuit framework. *eLife, 9,* e51473 (45 pages).

Mysore, S. P., & Knudsen, E. I. (2012). Reciprocal inhibition of inhibition: A circuit motif for flexible categorization in stimulus selection. *Neuron, 73*(1), 193–205.

Nakamura, T., & Hamamoto, S. (1996). Higher derivatives and canonical formalisms. *Progress of Theoretical Physics, 95*(3), 469–484.

O'Connell, R. G., Shadlen, M. N., Wong-Lin, K., & Kelly, S. P. (2018). Bridging neural and computational viewpoints on perceptual decision-making. *Trends in Neurosciences, 41*(11), 838–852.

O'Connor, T., & Franklin, C. (2018). Free will. In E. N. Zalta (Ed.), *The Stanford encyclopedia of philosophy* (fall 2020 ed.). Metaphysics Research Lab, Stanford University.

O'Doherty, J. P., Cockburn, J., & Pauli, W. M. (2017). Learning, reward, and decision making. *Annual Review of Psychology, 68*(1), 73–100.

O'Doherty, J. P., Lee, S. W., & McNamee, D. (2015). The structure of reinforcement-learning mechanisms in the human brain. *Current Opinion in Behavioral Sciences, 1,* 94–100.

Ostrogradsky, M. V. (1850). Mémoires sur les équations différentielles relatives au problème des isopérimètres. *Mem. Acad. St. Petersbourg V, 1*(4), 385–517.

Padoa-Schioppa, C. (2011). Neurobiology of economic choice: A good-based model. *Annual Review of Neuroscience, 34*(1), 333–359.

Padoa-Schioppa, C., & Conen, K. E. (2017). Orbitofrontal cortex: A neural circuit for economic decisions. *Neuron, 96*(4), 736–754.

Palva, S., & Palva, J. M. (2018). Roles of brain criticality and multiscale oscillations in temporal predictions for sensorimotor processing. *Trends in Neurosciences, 41*(10), 729–743.

Paul, W., & Baschnagel, J. (2013). *Stochastic processes: From physics to finance* (2nd ed.). Switzerland: Springer International Publishing.

Perdikis, D., Huys, R., & Jirsa, V. K. (2011). Time scale hierarchies in the functional organization of complex behaviors. *PLoS Computational Biology, 7*(9), e1002198, 18 pages.

Perdikis, D., Raoul, H., & Viktor, J. (2011). Complex processes from dynamical architectures with time-scale hierarchy. *PLoS ONE, 6*(2), 1–12.

Pessoa, L. (2019). Neural dynamics of emotion and cognition: From trajectories to underlying neural geometry. *Neural Networks, 120,* 158–166.

Pillai, A. S., & Jirsa, V. K. (2017). Symmetry breaking in space-time hierarchies shapes brain dynamics and behavior. *Neuron, 94*(5), 1010–1026.

Plyushchay, M. S. (1988). Canonical quantization and mass spectrum of relativistic particle analogue of relativistic string with rigidity. *Modern Physics Letters A, 3*(13), 1299–1308.

Pons, J. M. (1989). Ostrogradski's theorem for higher-order singular Lagrangians. *Letters in Mathematical Physics, 17*(3), 181–189.

Popescu, S. (2014). Nonlocality beyond quantum mechanics. *Nature Physics, 10*(4), 264–270.

Power, N., & Alison, L. (2019). Decision inertia in critical incidents. *European Psychologist, 24*(3), 209–218.

Rabinovich, M. I., Huerta, R., & Varona, P. (2006). Heteroclinic synchronization: Ultrasubharmonic locking. *Physical Review Letters, 96*(1), 014101 (4 pages).

Rabinovich, M., Volkovskii, A., Lecanda, P., Huerta, R., Abarbanel, H. D. I., & Laurent, G. (2001). Dynamical encoding by networks of competing neuron groups: Winnerless competition. *Physical Review Letters, 87*(6), 068102 (4 pages).

Rabinovich, M., Huerta, R., & Laurent, G. (2008). Transient dynamics for neural processing. *Science, 321*(5885), 48–50.

Rabinovich, M. I., & Muezzinoglu, M. K. (2010). Nonlinear dynamics of the brain: Emotion and cognition. *Physics-Uspekhi, 53*(4), 357–372.

Rabinovich, M. I., Simmons, A. N., & Varona, P. (2015). Dynamical bridge between brain and mind. *Trends in Cognitive Sciences, 19*(8), 453–461.

Rabinovich, M. I., & Varona, P. (2012). Transient brain dynamics. In M. I. Rabinovich, K. J. Friston, & P. Varona (Eds.), *Principles of brain dynamics: Global state interactions* (pp. 71–92). Cambridge, MA: MIT Press.

Raidal, M., & Veermäe, H. (2017). On the quantisation of complex higher derivative theories and avoiding the Ostrogradsky ghost. *Nuclear Physics B, 916,* 607–626.

Ramm, E. (2011). Principles of least action and of least constraint. *GAMM-Mitteilungen, 34*(2), 164–182.

Ramos, R. T., Sassi, R. B., & Piqueira, J. R. C. (2011). Self-organized criticality and the predictability of human behavior. *New Ideas in Psychology, 29*(1), 38–48.

Rangelov, D., & Mattingley, J. B. (2020). Evidence accumulation during perceptual decision-making is sensitive to the dynamics of attentional selection. *NeuroImage, 220*, 117093 (12 pages).

Ratcliff, R., & McKoon, G. (2008). The diffusion decision model: Theory and data for two-choice decision tasks. *Neural Computation, 20*(4), 873–922.

Ratcliff, R., Smith, P. L., Brown, S. D., & McKoon, G. (2016). Diffusion decision model: Current issues and history. *Trends in Cognitive Sciences, 20*(4), 260–281.

Ritov, I., & Baron, J. (1992). Status-quo and omission biases. *Journal of Risk and Uncertainty, 5*(1), 49–61.

Robinson, H. (1994). Two perspectives on kant's appearances and things in themselves. *Journal of the History of Philosophy, 32*(3), 411–441.

Samuelson, W., & Zeckhauser, R. (1988). Status quo bias in decision making. *Journal of Risk and Uncertainty, 1*(1), 7–59.

Schlosser, M. (2019). Dual-system theory and the role of consciousness in intentional action. In B. Feltz, M. Missal, & A. Sims (Eds.), *Free will, causality, and neuroscience* (pp. 35–56). Leiden, The Netherlands: Brill Rodopi.

Scholz, J. P., & Schöner, G. (1999). The uncontrolled manifold concept: Identifying control variables for a functional task. *Experimental Brain Research, 126*(3), 289–306.

Schultz, W. (1998). Predictive reward signal of dopamine neurons. *Journal of Neurophysiology, 80*(1), 1–27.

Schultz, W., Dayan, P., & Montague, P. R. (1997). A neural substrate of prediction and reward. *Science, 275*(5306), 1593–1599.

Schumacher, E. H., & Jiang, Y. (2003). Neural mechanisms for response selection: Representation specific or modality independent? *Journal of Cognitive Neuroscience, 15*(8), 1077–1079.

Shinn, M., Lam, N. H., & Murray, J. D. (2020). A flexible framework for simulating and fitting generalized drift-diffusion models. *eLife, 9*, e56938 (27 pages).

Shor, P. W. (1994). Algorithms for quantum computation: Discrete logarithms and factoring. In *Proceedings 35th Annual Symposium on Foundations of Computer Science* (pp. 124–134). Los Alamitos, CA: IEEE Computer Society Press.

Simen, P. (2012). Evidence accumulator or decision threshold – which cortical mechanism are we observing?, *Frontiers in Psychology, 3*, Article 183 (14 pages).

Simon, H. (1956). Rational choice and the structure of the environment. *Psychological Review, 63*(2), 129–138.

Smilga, A. (2017). Classical and quantum dynamics of higher-derivative systems. *International Journal of Modern Physics A, 32*(33), 1730025 (30 pages).

Solway, A., & Botvinick, M. M. (2012). Goal-directed decision making as probabilistic inference: A computational framework and potential neural correlates. *Psychological Review, 119*(1), 120–154.

Stephen, N. G. (2008). On the Ostrogradski instability for higher-order derivative theories and a pseudo-mechanical energy. *Journal of Sound and Vibration, 310*(3), 729–739.

Sterzer, P., & Rees, G. (2009). Bistable perception and consciousness. In W. P. Banks (Ed.), *Encyclopedia of consciousness* (Vol. 1, pp. 93–106). Oxford, UK: Academic Press.

Sutton, R. S., & Barto, A. G. (2018). *Reinforcement learning: An introduction* (2nd ed.). Cambridge, MA: The MIT Press.

Suzuki, Y., Morimoto, H., Kiyono, K., Morasso, P. G., & Nomura, T. (2016). Dynamic Determinants of the Uncontrolled Manifold during Human Quiet Stance. *Frontiers in Human Neuroscience, 10*(Article 618), 20 pages.

Suzuki, Y., Nomura, T., Casadio, M., & Morasso, P. (2012). Intermittent control with ankle, hip, and mixed strategies during quiet standing: A theoretical proposal based on a double inverted pendulum model. *Journal of Theoretical Biology, 310*, 55–79.

Swanson, N. (2019). On the ostrogradski instability; or, why physics really uses second derivatives. *The British Journal for the Philosophy of Science*.

Tagliazucchi, E. (2017). The signatures of conscious access and its phenomenology are consistent with large-scale brain communication at criticality. *Consciousness and Cognition, 55*, 136–147.

Tipler, F. J. (2014). Quantum nonlocality does not exist. *Proceedings of the National Academy of Sciences, 111*(31), 11281–11286.

Todosiev, E. P. (1963). *The action point model of the driver-vehicle system*, Ph.D. thesis, The Ohio State University. (Ph.D. Dissertation, Ohio State University, 1963).

Todosiev, E. P., & Barbosa, L. C. (1963/64). A proposed model for the driver–vehicle system. *Traffic Engineering, 34*, 17—20.

Ton, R., & Daffertshofer, A. (2016). Model selection for identifying power-law scaling. *NeuroImage, 136*, 215–226.

Tversky, A., & Shafir, E. (1992). Choice under conflict: The dynamics of deferred decision. *Psychological Science, 3*(6), 358–361.

van der Schaft, A., & Jeltsema, D. (2014). Port-hamiltonian systems theory: An introductory overview. *Foundations and Trends® in Systems and Control, 1*(2-3), 173–378.

Van Orden, G. C., Holden, J. G., & Turvey, M. T. (2003). Self-organization of cognitive performance. *Journal of Experimental Psychology: General, 132*(3), 331–350.

Vladimirov, I. G., & Petersen, I. R. (2018). Dissipative linear stochastic Hamiltonian systems. In *2018 Australian & New Zealand Control Conference (ANZCC)* (pp. 227–232). Melbourne, VIC: IEEE.

Vujanovic, B. D., & Jones, S. E. (1989). *Variational methods in nonconservative phenomena*. San Diego, CA: Academic Press.

Wagenmakers, E.-J., Farrell, S., & Ratcliff, R. (2005). Human cognition and a pile of sand: A discussion on serial correlations and self-organized criticality. *Journal of Experimental Psychology: General, 134*(1), 108–116.

Wang, X.-J. (2008). Decision making in recurrent neuronal circuits. *Neuron, 60*(2), 215–234.

Wang, M., Arteaga, D., & He, B. J. (2013). Brain mechanisms for simple perception and bistable perception. *Proceedings of the National Academy of Sciences, 110*(35), E3350–E3359.
Wang, L., Hong, J., Scherer, R., & Bai, F. (2009). Dynamics and variational integrators of stochastic Hamiltonian systems. *International Journal of Numerical Analysis and Modeling, 6*(4), 586–602.
Wegner, D. M. (2018). *The illusion of conscious will* (new edition ed.). Cambridge, MA: The MIT Press. With Foreword by Daniel Gilbert and Introduction by Thalia Wheatley.
Wegner, D. M. (2004). Précis of the illusion of conscious will. *Behavioral and Brain Sciences, 27*(5), 649–659.
Weilnhammer, V., Stuke, H., Hesselmann, G., Sterzer, P., & Schmack, K. (2017). A predictive coding account of bistable perception - a model-based fMRI study. *PLOS Computational Biology, 13*(5), e1005536 (21 pages).
Woodard, R. (2007). Avoiding Dark Energy with 1/R Modifications of Gravity. In L. Papantonopoulos (Ed.), *The invisible universe: Dark matter and dark energy*, Vol. 720 of Lecture notes in physics (pp. 403–433). Berlin, Heidelberg: Springer.
Woodard, R. P. (2009). How far are we from the quantum theory of gravity?, *Reports on Progress in Physics, 72*(12), 126002 (42 pages).
Woodard, R. P. (2015a). Ostrogradsky's theorem on Hamiltonian instability. *Scholarpedia, 10*(8), 32243. revision #186559.
Woodard, R. P. (2015b). The theorem of Ostrogradsky, arXiv:1506.02210v2 [hep-th].
Yan, H., & Wang, J. (2020). Non-equilibrium landscape and flux reveal the stability-flexibility-energy tradeoff in working memory. *PLOS Computational Biology, 16*(10), e1008209.
Yan, H., Zhang, K., & Wang, J. (2016). Physical mechanism of mind changes and tradeoffs among speed, accuracy, and energy cost in brain decision making: Landscape, flux, and path perspectives. *Chinese Physics B, 25*(7), 078702 (21 pages).
Yan, H., Zhao, L., Hu, L., Wang, X., Wang, E., & Wang, J. (2013). Nonequilibrium landscape theory of neural networks. *Proceedings of the National Academy of Sciences, 110*(45), E4185–E4194.
Yoshikawa, N., Suzuki, Y., Kiyono, K., & Nomura, T. (2016). Intermittent feedback-control strategy for stabilizing inverted pendulum on manually controlled cart as analogy to human stick balancing. *Frontiers in Computational Neuroscience, 10*(Article 34), 19 pages.
Zaslavsky, G. M. (2002). Dynamical traps. *Physica D: Nonlinear Phenomena, 168–169*, 292–304. VII Latin American Workshop on Nonlinear Phenomena.
Zaslavsky, G. M. (1995). From Hamiltonian chaos to Maxwell's demon. *Chaos, 5*(4), 653–661.
Zaslavsky, G. M. (2005). *Hamiltonian Chaos and fractional dynamics*. Oxford, NY: Oxford University Press.
Zgonnikov, A., Lubashevsky, I., Kanemoto, S., Miyazawa, T., & Suzuki, T. (2014). To react or not to react? intrinsic stochasticity of human control in virtual stick balancing. *Journal of The Royal Society Interface, 11*(99), 20140636 (13 pages).
Zhang, J. (2012). The effects of evidence bounds on decision-making: Theoretical and empirical developments. *Frontiers in Psychology, 3*, Article 263 (19 pages).

Part III
Instead of Epilogue: Beyond Complex Present

Chapter 7
The Phenomenological Self in Physics of the Human Mind

In this rather brief discussion, we want to consider the notion of *the self* taking a central position in our account of the human temporality. In the present chapter, we want to merge various facets of this notion we dealt with in constructing our theory.

The notion of "the self" takes a central position in philosophy of mind, psychology, psychiatry, neuroscience. Our concept does not contradict the available accounts but has other roots. Our understanding of the self has been shaped by mathematical constructions, which enabled us to highlight new aspects in the self.

Before passing directly to our account, let us outline the understanding of the self met in literature.

7.1 The Modern Concepts of the Self

Nowadays, the concepts of the self and related issues are a matter of debate in many disciplines and a reader may be referred to Dainton (2008), Hardcastle (2008), Metzinger (2009), Gallagher (2013), Gallagher and Daly (2018), Friston (2018), Newen (2018), Woźniak (2018), Gallagher and Zahavi (2019), Smith (2020), Schütz et al. (2020) as well as collections edited by Qin et al. (2014), Gallagher (2011) as an introduction to the modern state of this question. In this subsection, following Dainton (2008), Gallagher and Daly (2018), Smith (2020), Schütz et al. (2020), we briefly note the concepts of the self that are related to our account.

An answer to the question of what the self is, divides concepts of the self into several classes. Broadly speaking, materialistic accounts reduce the self to neural processes; dualistic accounts accept that the mind and the brain are complementary components mutually irreducible; phenomenological accounts leaving aside the

I. Lubashevsky and N. Plavinska, *Physics of the Human Temporality*,
Understanding Complex Systems, https://doi.org/10.1007/978-3-030-82612-3_7

ontological aspects focus on the first-person point of view on the self properties. Let us acquaint with some of them.

No-self. Metzinger (2003, 2011, 2013, 2014) in his "no-self" account rejects all the dualistic conceptions of the self and reduces the notion of self to neuro-dynamics. In this context we want to note also the concept of the narrative self (Dennett 1991; Schechtman 2007, 2011; Hardcastle 2008; Muñoz-Corcuera 2016). It posits that the only self is a complex entity integrating our self-understanding, decision making, actions, etc. within the narrative identity (e.g., Parfit 2011; Merrill and Fivush 2016; McLean 2017, for are view).

Dualism. The concepts of the self reflect the basic mind-body problem in philosophy of mind. For example, Karl Popper and John Eccles, within a dualistic view, regarded the self as an autonomous entity that exercises a superior interpretative and controlling role upon the neural events (Popper and Eccles 1977; Eccles 1989, 1994). Their concept of the self motivated a debates between the philosopher Mario Bunge and the neuroscientist Donald MacKay (Bunge 1979, 1977; MacKay 1978, 1979, 1980) that clarified the watershed between materialism and dualism in understanding the self.[1] The debates stimulated by Bunge and MacKay are still ongoing and face up to the problem of using incomparable notions in describing the self and analyzing the brain functioning (LeDoux 2002; Northoff et al. 2006; Apps and Tsakiris 2014).

The self and the brain. During the last two decades, there has been a great deal of identifying specific parts of the brain related to various aspects of self; for a review a reader may be addressed to Vogeley and Gallagher (2011), Gallagher and Daly (2018), Gallup et al. (2011), Tsakiris (2011) and collection edited by Qin et al. (2014). Based on the available data (Gillihan and Farah 2005, see also Craik et al. 1999; LeDoux 2002; Legrand and Ruby 2009) drawn the conclusion that there is no specialized brain area on their own responsible for generating "the self." This led Vogeley and Gallagher (2011) to idea that the self is both everywhere and nowhere in the brain.

> Selves are experiential, ecological, and agentive; they are often engaged in reflective evaluations and judgments; they are capable of various forms of self-recognition, self-related cognition, self-narrative, and self-specific perception and movement. In many of these activities, selves are more 'in-the-world' than 'in-the-brain', and they are in-the-world *as-subject* more so than *as-object*. (Vogeley and Gallagher 2011, p. 129)

---

[1] Mackay (1978, p. 605) "take[s] the concept of conscious human agency as primary, and recognize[s] an irreducible duality in the two kinds of correlated data we have about it: (a) the data of our conscious experience, and (b) the data obtained from observation of our brain-workings. By distinguishing (as a communication engineer does) between two levels of determination—the *energetic* and the *informational*—we can then see how even if the chain-mesh of physical causality were unbroken at the energetic level, the *form* of our bodily actions can still be determined by our thinking and deciding. The essential condition is that our conscious form-determining activity must be *embodied* in, rather than running alongside, the brain activity that physically determines (at the energetic level) what our bodies do."

The *pattern theory of self* (Gallagher 2013) with its generalization (Gallagher and Daly 2018) to including dynamical relations in the self-pattern captures both the variety of involved factors and that the self is more "in-the-world" than "in-the-brain." This theory posits that the self comprises as a number of characteristic factors including embodiment, experience, inter-subjectivity, psychological/cognitive, reflective, narrative factors. All these factors are dynamically interrelated and can be reflectively reiterated in the narrative component of the self.

Embodiment of the self. Embodiment is generally understood as the concept that human mental processes are deeply rooted in and shaped by how the human body interacts with the world, for a review of the embodiment a reader may be referred to Wilson (2002), Varela et al. (2016) and collection edited by Shapiro (2014). In particular, Gallagher (2005) emphasized that embodiment precedes the emergence of our knowledge and, so, is not accessible to the mind. Therefore, how the body shapes the mind at the basic level has to be unconscious.

In the context of the self, the concept of embodiment is reflected in the notion of embodied self (e.g., Gallagher 2005; Bermúdez 2011; Cassam 2011). Below we will touch upon the embodied self once more, here, we want to point out, first, the account of embodied self by Newen (2018). He, turning to the dynamical pattern theory of self, relates self-consciousness to the integration of activated features of the self, which remain anchored in the embodied self. This integration can be realized without conscious experiences being involved, thereby the embodied self should be, at least partly, unconscious. Second, Schaefer and Northoff (2017) taking into account the unconscious aspects of embodiment, introduce the notions of the conscious self and the unconscious embodied self. They posit that the conscious self gains access to the external world via the unconscious embodied self. However, the neural mechanism governing the interaction of the conscious self with the unconscious self remains unclear.

Phenomenological aspects. The self-as-subject representing the first-person view on mental phenomena is a complex entity characterized by many aspects. Turning to Gallagher and Zahavi (2012, 2019), Gallagher (2012), let us note those which are related to our account of the self.

• *Pre-reflective self-consciousness.* It is (*i*) a non-reflective awareness that we have before we do anything reflecting on our experience and (*ii*) an implicit and first-order awareness. An *explicit, reflective self-consciousness* being a higher order awareness is possible only because there is a pre-reflective self-consciousness that is an on-going and more primary kind of self-awareness. Evidence for the pre-reflective self-consciousness was found, in particular, in studying the behavior of newborns (e.g., Rochat 2011) and autistic children (e.g., Hobson 2011).

• *Pre-reflective bodily self-awareness.* This aspect of the self-as-subject may be regarded as the pre-reflective self-consciousness embodied and embedded in the world. Pre-reflective body-awareness accompanies and shapes every spatial and temporal experience. The introduction of the pre-reflective body-awareness is due to the first-person point of view on the world is always determined by the arrangement of the subject's body and surrounding objects. For a detailed analysis of this type body-awareness a reader may be addressed to Legrand (2011), Henry and Thompson (2011).

• *Sense of self-ownership* meaning the pre-reflective experience or sense that I am the subject who is executing voluntary actions, having thoughts, undergoing an experience.

• *Sense of self-agency* meaning the pre-reflective experience or sense that I am the author of these voluntary actions and thoughts, i.e., I cause and can control them.

Nowadays, the pre-reflective aspects of the self are debated from various standpoints and a reader may be referred to Bermúdez (2011, 2017), Tsakiris (2011, 2015), Zahavi (2014, 2018, 2020), Dainton and Snowdon (2016), Billon (2017), Buhrmann and Di Paolo (2017), Guillot (2016), Howell and Thompson (2017), de Vignemont (2017), Mylopoulos and Shepherd (2020) for introductory review. In particular, Gallagher (2017) defends the phenomenological account of the self and its constituent concepts arguing that pre-reflective self-awareness and the sense of ownership are features intrinsic to experience and bodily actions.

Temporal structure of self-agency. Actions underlying the self-agency unfold in time. So agentive experiences should be conceptualized as entities retaining their identity, even though their states and relations may change. In particular, human agency is characterized by the interplay of automatic and conscious processes and experiences of agency plays the role in their integration. For a discussion and review of these issues a reader may be addressed to Pacherie (2011, 2015, 2018), Krisst et al. (2015), Mylopoulos and Pacherie (2017), Pacherie and Mylopoulos (2020), Mylopoulos and Shepherd (2020) and collection edited by Morsella and Poehlman (2014). In this context we want to note approaches to describing conscious aspects in bodily actions called *passive frame theory* (Morsella et al. 2016) and its generalization (Morsella et al. 2020) as well as *attention schema theory* (Graziano and Webb 2015). In addition, we want to point out the account of interplay between conscious and unconscious volition by Slors (2019). Considering actions caused by intentions, he makes a distinction between consciously formed intentions and intentions we become conscious of, with only the former being relevant to the notion of consciously initiated actions. Nevertheless, both of them are of complex temporal structure, where consciously intentions play the role of structuring causes, whereas the action triggering is unconscious.

As far as the self-agency and the brain functioning are concerned, we want to note the "free-energy" account of the self by Apps and Tsakiris (2014). It describes the self-awareness based on the interplay between top-down and bottom-up processes with predictive coding paradigm. Besides, the concept of the consciousness state space by Berkovich-Ohana and Glicksohn (2014) proposes a unifying model for consciousness, the self, and the brain functioning.

Concerning the temporal structure of the self, we may note also the problem of personal identity (e.g., Olson 2019; Drummond 2020, for a general discussion). In particular, Dainton (2008), Campbell (2011) discuss the personal identity over time from the standpoint of the first-person view.

The self is the basic entity in philosophical accounts of non-Cartesian dualism defended, e.g., by Lowe (2008, 2009, 2010, 2013), see also the collection edited by Lavazza and Robinson (2014) and the review by Robinson (2020).

As clear from this brief outline, the concept of the self, being in the process of developing, is stimulated by new advances in philosophy, psychology, and neuroscience. These disciplines and their branches highlight the different aspects in studying the self. They illustrate the multi-facet structure of the self and the fruitfulness of its analysis from various standpoints. Below, we want to formulate our account of the self, which, on the one hand, is consistent with the noted concepts of the self. On the other hand, principally different premises underlie it. The internal logic of mathematical construction has shaped our account of the self we followed in our description of human goal-oriented behavior.

7.2 The Self and Its Holistic Properties

The concept of the self, we intend to discuss in this section, arises out of the developed mathematical description of the human temporality. Exactly the underlying mathematical constructions with their inner logic have shaped our understanding of the self as a whole entity and also its particular aspects. We want to emphasize that our description of the self should not be treated as an elaborated conception; here, we present our current perspective on the self. Our account of the self to be presented in this section is a certain schema rather than rigorous construction.

In our constructions, we draw a distinction between matter (physical reality) and the mind. Up to now, there is no ontological definition of matter that would satisfy the majority of scholars. As far as the mind is concerned, there is no consensus even about its status and we also do not undertake to define what the mind is. We consider the absence of such definitions to indicate that the notions of matter and the mind belong to the very fundamental level of description and cannot be represented in terms of other basic notions. However, we can point out characteristic features of matter and the mind that are compatible with each other. Turning to matter, we single out the notions of point and time as such characteristics. For the mind, they are the notions of cloud and temporality. Let us consider them in more detail.

Point and Cloud. In physics, one of its basic notions is the point of the space-time continuum. In particular, in Newtonian mechanics, the idea of point is reflected in the concept of material point—the structureless physical object having no dimensions—and its motion in space. In classical field physics, the set of spatial points related to one instant of time represents fields as distributions in the space. In quantum mechanics and quantum field physics physical objects (particles and fields) become partly indivisible spatial distributions.

In the mathematical description of the human mind, the notion of cloud related to the mental image of the space-time continuum is one of the basic elements. This cloud may be regarded as some analogy to the point in physics. However, the mental cloud on its own, i.e., as an object the mind deals with directly, does not consist of points. The cloud represents how physical objects or events are reflected in the mind, which endows the cloud with spatio-temporal properties. In particular, it can bear the properties of the past, present, and future. For this reason, we call this mental entity the space-time cloud. In the realm of the mind space-time clouds may be indivisible, at least, their division cannot be arbitrary. However, in the embodying, i.e., embedding the mind into the body, a cloud converts into some pattern of neural signals admitting interpretation as a spatio-temporal distribution in the space-time continuum. In this way, the mind embodiment provides the frameworks where the space-time clouds can be compared, divided, merged, etc.

Time and Temporality. In physical reality, time united with space takes the central position. Time can be conceived as a set of point-like instants, leaving aside the aspects of Einstein's theories of relativity. The evolution of any physical system is described in terms of system states and their dynamics. The system state is considered to be specified by a collection of properties that are attributed to its elements. These properties are introduced within the system consideration confined to the current instant of time. For example, for an ensemble of physical particles, these properties are the current positions of particles and their instantaneous velocities. The time rate of variations in system states is determined by the system current state. Whence it follows that physical systems may be described in terms of presentism where only the current instant of time is essential. The states the system occupied in the past and the states the system will get in the future do not contribute to the current dynamics and are recognized only by the mind.

In the mind, the human temporality takes the central place instead of time. The temporality possesses internal integrity and time extension. This extension includes in itself simultaneously the past retained in memory, the complex present (current experience), and the imaginary future. Space-time clouds are elements of the human temporality and the mind operates with them as whole entities. The points of the space-time continuum do not belong to the realm of the mind. For the mind, the notion of mental state can be also introduced but in this case, the temporality as a whole (or, at least, its fragment relevant to the case in issue) must be also incorporated. In this way, the past and the future become active elements and contribute substantially to the mind dynamics.

The human temporality and physical time are closely interwoven. To describe the dynamics of mental states we need to take into account both the temporal dimensions, which is the essence of 2D-time dynamics. The integrity of 2D-time dynamics, i.e., the inseparability of the two temporal dimensions from each other, is due to that the self unites physical reality and the mind.

The Self and its Holistic Nature. Up to now, no exhaustive definition has been proposed for the self as well as no exhaustive definitions have been proposed for matter and the mind. Therefore, turning to the logic used in comparing matter and the mind, let us single out the features of the self which are crucial for our constructions. These features are united in that they underlie the relationship between physical objects and their mental images and reflect various details of this relationship. The union of these features is the heart of our account of the self.

First, we posit that the self inherits the basic properties of the mind including its temporality, which endows the self with temporal extension. Therefore, we attribute to the self its own temporal dimension to be called the *self-temporality*. It enables us to take into account not only the mental aspect of human goal-oriented actions but also their implementation in physical reality due to the belonging of the self to both the realm of the mind and physical reality.

Second, the embodiment and embedment of the mind endows the self also with properties characterizing physical objects. In particular, it enables us to introduce for the self the notion of state to be called the *self-state*. The self-state is introduced in a way similar to constructing phase states for physical systems.[2] Namely, the self-state is specified by:

- the physiological state of the body, including the states of the central and peripheral nervous systems, different organs of the body, the sensory modalities with their own neural networks, etc. These unconscious bodily states either affect the subject's current experience or are involved into the unconscious processing of perceived information;
- the psychological states such as emotions, feelings, habits which are currently experienced by the subject but which are not involved in the deliberate analysis of past or anticipated events.

The characteristic feature of the self-state is that the self-state at the next instant of physical time is determined by the self-state at the current instant provided the difference between these instants is about the duration of experiential now. In particular, it enables us to describe the dynamics of self-states in terms of differential equations, which may be regarded as a certain version of presentism.[3]

In this fashion, from the side of the mind, the self is endowed with temporal extension, whereas, from the side of the body, the self is endowed with the dynamics meeting the paradigm of physics. Therefore, speaking about the mathematical description of the self we need to take into account its two types of aspects. They are fundamentally different in property, on the one hand, complement each other, on the other hand. Namely, these aspects are:

- teleological aspects concerning the existence of various goals in the self-temporality which the subject intends to attain in future and makes special efforts for this in the present,

[2] It should be pointed out that the introduction of the self-temporality, taking into account subject's actions aimed at attaining goals lying in the future, enables us to introduce the self-state which deals with the bodily and mental properties related *only* to the subject's current experience.

[3] Our understanding of the concept of presentism—actually, thick presentism—and its relationship with the formalism of dynamical systems is elucidated in the book by Lubashevsky (2017).

– local-in-time aspects concerning the dynamics of the self-states which are governed by regularities admitting a mathematical description that deals with quantities whose introduction is based only on the current instant of time.

The contributions of the teleological and local-in-time aspects into the properties of the self are not independent in spite of the essential difference between time scales characterizing these aspects. The mutual dependence of these contributions is of asymmetric dynamic form.

Before passing directly to a detailed discussion of different aspects of the self, let us outline its status with respect to the mind and the body which is the pivot in our phenomenological account of the self.

- The self with respect to the mind and the body is an individual, partly independent entity. It implies that the self overlaps with the mind and the body in property but is not reduced to any of them.
- The self is an entity with internal integrity and its overlapping with the mind, on the one side, and the body, on the other side, may be treated as a gradual change of the corresponding aspects rather than a mechanistic composition of different components.
- The self implements the interaction between the mind and the body (formally including also the environment) playing the role of the "bridge" between them. This interaction becomes possible due to that the aspects of the mind and the body coexist in the self simultaneously; we will elucidate this conception below in more detail.

Finalizing this section, we want to emphasize that the self is an individual entity that can be analyzed separately from the mind and matter. It is so because all the relevant mental and bodily properties are already attributed to the self. The existence of the mind and matter is a necessary condition for the existence of the self.[4] It is the reason for categorizing the nature of the self as well as its properties as holistic. In the self, any property belonging to the mental or bodily level reflects implicitly the features of the other level.

Brief digression: Two kinds of holism

The term *holism* was coined by Jan C. Smuts in 1926. In his book "Holism and Evolution" (Smuts 1927, 1-st edition published in 1926) he defined holism as a general concept about the intricate interrelationship between the whole and the parts. He writes

> the whole and the parts ... reciprocally influence and determine each other, and appear more or less to merge their individual characters: the whole is in the parts and the parts are in the whole, and this synthesis of whole and parts is reflected in the holistic character of the functions of the parts as well as of the whole. (Smuts 1927, p. 88)

[4] Leaving aside ontological issues, we want to note that phenomenology of the human mind implies the first-person view to be treated as an individual dimension with its own properties and dynamics in studying human behavior in addition to physical reality. The existence of some entity belonging to this dimension is understood exactly in the phenomenological frameworks.

The modern accounts of holism, positing the priority of a whole over its parts, may be divided into two big classes that may be referred to as *holism of the whole* and *holism of the parts*; each of them is reflected in Smuts' definition.

- *Holism of the whole* implying that the whole is greater than the sum of its parts. Namely, a system or organism being a coherent, unified whole may have properties (as some entity or phenomenon) in addition to those of its parts. So the understanding of the parts does not necessarily provide an understanding of the whole (e.g., VandenBos 2015, *Holism*, p. 498).
- *Holism of the parts* turns to the idea that the parts of the whole are in intimate interconnection and they cannot have some type-properties or, even, exist independently of the whole, or cannot be understood without reference to the whole (e.g., Stevenson 2010, *Holism*, p. 836).

In particular, the former kind of holism is usually related to the theory of emergent phenomena and opposed to reductionism in philosophy of science (e.g., Esfeld 2001, 2009; Weber and Esfeld 2009; Healey 2016). The latter kind of holism is exemplified by social holism which is opposed to atomism (e.g., Pettit 1998; Esfeld 2015). Social holism maintains that without being involved in social relations with people, a person cannot possess the capacity for thinking and acting like a human. For historical background and a detailed comparison of holism and atomism, a reader may be addressed to Esfeld (2015) and collections edited by Ikäheimo and Laitinen (2011), Zahle and Collin (2014). Temporal aspects of social holism are considered, e.g., Zheng (2016).

The concept of diachronic holism for the irreducibility of the mental to the physical as well as the personal identity problem has been proposed by Slors (2001).

7.3 Two Facets of the Self

Before continuing the discussion of our account of the self, let us explain how it is related to the concept of the human temporality developed in this book.

First, for describing subject's goal-oriented actions as some entities in the mind, it was necessary to understand how their properties are related to the physical characteristics of the controlled object motion. These characteristics are perceived by the subject via sensory modalities and become accessible to the mind. In other words, the mental images of subject's actions are rooted in the subject's experience of physical reality. This experience arises via the sensory modalities of the subject's body and is reflected in features accessible to the mind but is not consciously analyzable, we discussed this issue in Sects. 3.2.5 and 3.3.1. In this way, we came to the recognition that the self should have a facet related to the level of consciousness regarded as E-consciousness. In our mathematical description of action strategies—purely mental entities—this facet of the self is related to the phase space of subject's actions or, speaking more strictly, their mental images. The basic elements of this phase space are mental images embodied in neural processes; so the given facet represents the self from the side of the mind.

How the mind deals with elements of this phase space admits interpretation in terms of purely mental operations and should be categorized as AM-consciousness. AM-consciousness exemplifies elements of the mind outside the self.

Second, another facet of the self appears in describing how the mind implements various operations (comparison, fusion, fission, etc.) within E-consciousness. At the level of E-consciousness, the object motion states are represented by their mental images—the space-time clouds of the phase state which have no structure required for such operations. At the level of the body, the neural patterns related to these images possess the structure required for implementing the operations that are initiated by the mind. In the mathematical description developed for subject's goal-oriented actions, this facet is reflected in the equations governing the dynamics of space-time clouds treated as spaced-time distributions (Sects. 6.2 and 6.3). On the one hand, these equations govern the local-in-time dynamics of these spaced-time distributions—which reflects the presentism of the physical system. On the other hand, these equations obey regularities partly shaped by the mind. Thereby, the given facet represents the self from the side of the human body.

Naturally, there are physiological processes, e.g., biochemical processes in the body cells, that are not affected by the mind. These processes exemplify matter (the body) outside the self.

Metaphorically, we can compare this self with *alexandrite*—the gemstone exhibiting green and red colors depending on viewing direction in partially polarised light and also changing color in artificial light compared to daylight. In our account, the self is an entity with internal integrity. However, in its analysis from different perspectives, different aspects become pronounced.

7.3.1 Mind-Body Supervenience of the Self

First of all, we want to note that below to adhere to the terminology of the mind-body problem, we will use the term the body in the sense of matter. In other words, the notion of the body is understood in a wide sense including (*i*) the human body itself with all its parts, organs, systems, (*ii*) cells and biochemical processes in them, (*iii*) physical stimuli perceived by sensory modalities, and, finally, components of physical reality reflected in some way in the functioning of the human body.

Confining our description to the phenomenological approach, we accept the self (an individual entity) to supervene upon the mind and the body, which will be referred to as the mind-body supervenience of the self. Generally, the supervenience of A upon B is understood as

> A set of properties A supervenes upon another set B just in case no two things can differ with respect to A-properties without also differing with respect to their B-properties. In slogan form, "there cannot be an A-difference without a B-difference" (McLaughlin and Bennett 2018).

There is a wide variety of supervenience whose versions are singled out by features in focus, for an introductory review a reader may be addressed to Kim (2005), Leuenberger (2008), Shagrir (2012), McLaughlin and Bennett (2018) and a collection edited by Savellos and Yalçin (1995). In our constructions, we turn to the supervenience which is grounded on mathematical constructions. In terms of necessity, the given type of supervenience can be reformulated as

> A property B necessitates a property A if it is necessary that when B is instantiated then A is instantiated (cf. Bliss and Trogdon 2014).

In our account, we regard this proposition as equivalent to the supervenience of A upon B; in general, it does not hold (McLaughlin and Bennett 2018).

According to our concept of the self, there are two subvenient bases: the mind and the body. So we say the self to supervene upon the mind and the body simultaneously in the sense that

> there is a set $\{S\}$ of properties of the self such that for a given property S to be instantiated, some properties M_s and B_s have to be instantiated in the mind and the body, respectively.

The properties $\{S\}$ play essential roles in the self. Depending on a particular feature of the self, which is the focus, the roles of the subvenient properties M_s and B_s may be different, in particular, being in the dominant-subordinate relations with respect to each other in their contribution to the self. Nonetheless, both the properties M_s and B_s must be presented simultaneously.

It should be emphasized that within the phenomenological approach to describing mental phenomena the mind and the body are two individual subvenient bases. However, outside this approach, the mental supervenes upon the physical. Therefore our account deals with two routes in relating properties of the self to properties of the body: one leads to the body directly, the other does this via the mind. To elucidate the reasons behind the concept of mind-body supervenience of the self characterized by this overdetermination, we turn to the concept of *grounding*. For a general discussion of the grounding, a reader may be referred to Rosen (2010), Trogdon (2013), Leuenberger (2013, 2014), Bliss and Trogdon (2014), Schaffer (2016), Bennett (2017), Rydéhn (2018) and collections edited by Correia and Schnieder (2012), Raven (2020). For the relationship between supervenience and grounding under particular conditions a reader may be referred, e.g., to Leuenberger (2013, 2014), Bliss and Trogdon (2014).

There are various approaches to treating the notion of grounding. In our constructions, we use the notion of grounding that links the concept of the self to the mathematical description of human goal-oriented behavior within the phenomenological approach. It should be noted the notion of grounding that is focused on (metaphysical) explanation is elaborated, in particular, by Fine (2002, 2013b), deRosset

(2013), Bliss and Trogdon (2014), Raven (2015b), Litland (2017), Glazier (2020). In this context, we want to note also the concept of *building* aimed at explaining how some phenomena are "built" from more fundamental phenomena (Bennett 2017).

Steiner (1978), Bliss and Trogdon (2014), Raven (2015b) posed a question about using mathematical constructions as the ground. In our account of the mind-body supervenience of the self we have turned exactly to this idea. Namely, the logic of the developed mathematical constructions was the ground for turning to the relevant philosophical conceptions as well as accepting and making hypotheses about psychological and physiological data and phenomena. The self-consistency of these constructions may be regarded as an argument for them.

Brief digression: We may be criticized for using the formalism, developed within the phenomenological description of mental phenomena, in grounding the self. Indeed, phenomenology is usually conceived of as a discipline dealing with entities introduced at meso- or macro-level which aggregate fundamental micro-level elements. The properties of these meso/macro-level entities represent *cooperative* behavior of micro-level elements and, as a result, the individual features of micro-level elements are lost. Within such interpretation, a phenomenological description cannot be fundamental and used as a ground.

However, accounts of reality that reject phenomenology and deal only with the fundamental elements meet also a number of challenging problems (Schaffer 2003, 2010, 2012; Sider 2011; Jenkins 2013; Fine 2013a). To overcome these problems Raven (2015a) put forward the concept he called *fundamentality without foundations*. According to his concept, an entity is considered to be fundamental when it has some *ineliminable* properties, i.e., properties that are not grounded in the properties of the other entities. In this case, complex entities may be also fundamental. For further discussion of the concepts of fundamentality a reader may be addressed to Bennett (2017), Leuenberger (2020).

In our account of human goal-oriented behavior, the temporal extension of events into the future is an ineliminable feature. This ineliminability is our argument for treating the developed description as fundamental and using it as the ground of the mind-body supervenience of the self.

Within our account of the self, we draw a distinction between mental operations (e.g., mathematical operations with mental images) and bodily processes (e.g., neurological, physiological, biochemical processes). This distinction is due to the entirely different formalisms describing mental operations and bodily processes. Indeed, although mental operations supervene upon bodily processes, their logic is grounded in their own regularities. Thereby the formalism of mental operations does not require referring to bodily processes. As far as bodily processes are concerned, they are governed by the regularities of the physical world, especially the governing regularities of complex systems. Thereby the formalism of bodily processes may have nothing in common with the formalism describing operations with mental images.

The self is the entity where the mental and the physical overlap with each other. In other words, all the features of the self are grounded simultaneously in the mental operations and physical processes, i.e., their mathematical description deals with mental phenomena, on the one hand, and their bodily implementations, on the other hand.

The mental and the physical, overlapping in the self, contribute dissimilarly to its properties. Depending on which one is in focus, we speak about different aspects of the self, namely, the mind-in-the-self or the body-in-the-self. Let us consider these aspects individually.

7.3.2 The Mind-in-the-Self

We are not able to give a definition of the mind-in-the-self, but can point out its characteristic property—E-consciousness. Exactly this property makes it possible to single out the mind-in-the-self from the human mind. Let us discuss this issue in more detail.

Dealing with the mind-in-the-self, we focus on specific mental entities, their emergence, evaluation, transformations, etc. Perception of external physical stimuli (sound, color, etc.) and their categorization exemplifies these entities. Each of them emerges in the mind via complex, unconscious processing of perceived information by the body. Only the final result of the information processing becomes accessible to the mind. On the one hand, It means that the mind is aware of these entities and able to categorize them according to some criteria (e.g., strong-weak or the membership to a certain class). On the other hand, these entities are recognized as just given to the mind and do not undergo conscious analysis. We categorize such entities as elements of consciousness related to the experiential now and called E-consciousness (Sect. 3.2.5). In particular, the pre-reflective self-consciousness and bodily self-awareness, as well as the sense of self-ownership and self-agency discussed in Sect. 7.1, may be treated as E-conscious features of the self.

As far as mental operations with these entities are concerned, the mind-in-the-self only initiates these operations, but they are implemented in the body-in-the-self. For example, in comparing two similar sounds in loudness the mind does not implement any logic operations, it just becomes aware that one of the sounds is louder. Therefore, the implementation of mental operations within the self should be regarded as unconscious (automatic) processes. The mind is able only to influence them, e.g., via attention allocation but is not involved in their implementation.

Entities of the mind-in-the-self have the corresponding representations in the body-in-the-self. Such representations are some neural patterns in the body-in-the-self. The results of operations with these patterns in the body-in-the-self are "returned" to the mind. We call this "return" the body-mind raising of a neural pattern from the level of the body-in-the-self to the level of the mind-in-the-self (Sect. 5.6.1).

It is worthy of noting that the presented description of the mental operations within the self argues that the human mind is irreducible to the mind-in-self. Indeed, there are various (e.g., algebraic) manipulations with mental entities that are grounded in their own logic and whose mathematical description does not require referring to the human body. These operations are implemented solely within the human mind, which is not possible for the mind-in-the-self. To characterize these operations we introduced the notion of AM-consciousness (consciousness completely accessible to the mind, Sect. 3.2.5).

Finalizing the present subsection, we want to note that the mind-in-the-self admits an interpretation as the embedding of the human mind into the human body. For this reason, the mind-in-the-self inherits from the human mind its temporal dimension including the connected past, the experiential now, and the connected future—the basic elements of the complex present. Therefore, the formalism describing the mind-in-the-self should operate with nonlocal-in-time entities.

7.3.3 The Body-in-the-Self

Speaking about the mind-in-the-self, we focused on its entities; dealing with the body-in-the-self we focus on processes. These processes are directly related to the implementation of mental operations with entities belonging to the realm of the mind-in-the-self.

Naturally, many other processes occur in the body but are irrelevant to the body-in-the-self; various biochemical processes inside living cells exemplify them. It allows us to posit that the notion of the human body is irreducible to the notion of the body-in-the-self, however, the body-in-the-self belongs to the human body. This relationship enables us to regard the human body as a complex, hierarchical system, where the body-in-the-self is its part.

According to the modern paradigm of physics, a characteristic feature of fundamental laws is the temporal locality of governing equations. In particular, this feature is reflected in the mathematical description of biochemical processes in living cells and their constituent elements.

For describing processes and entities at higher levels the quasi-adiabatic approximation is usually employed. This approximation is based on the concept of collective variables—a set of quantities introduced at higher levels. These collective variables are assumed to enslave the lower level quantities. For a detailed description of the quasi-adiabatic approximation and the notion of collective variables applied to describing human behavior, a reader may be referred to Kelso (2000, 2009a, b, 2014), Fuchs and Kelso (2009), Latash (2008, 2012). Within the quasi-adiabatic approximation, the temporal locality of governing equations for collective variables is conserved. Naturally, this temporal locality should be considered within the time scale of the immediate now—the fragment of the experiential now characterizing the unconscious processing of currently perceived information before it becomes accessible to the mind (Sects. 2.5 and 2.6.3). The duration of the immediate now bounds from above various delays in the dynamics of individual neural processes.

Therefore, in our account of the body-in-the self, we accept that the equations governing dynamics of neural patterns related to the entities of the mind-in-the-self meet the temporal locality.[5] As for the spatial structure of these equations, their form depends on the complex organization of the human body at each of its levels. For this reason, we suppose that the spatial structure of these equations may differ substantially from the form of the governing equations belonging to the bottom level. It enables us to employ the top-level regularities that are of a form stemming from the expected mind-body interaction even they have no analogy at the bottom level. In particular, these regularities have to reflect directly the current organization of the brain's functioning.

Speaking about the self within the phenomenological approach, it is necessary to clarify how the mind-in-the-self influences the body-in-the-self. We should note that the widely used argument against the effect of the mind on the body is the thesis that "every lower-level physical effect has a purely lower-level physical cause." Nowadays, however, advances in studying complex systems put forward arguments for equal, complementary contributions of bottom-up and top-down principles (Gibb 2015, 2019; Cat 2017; O'Connor 2020). This union of bottom-up and top-down interaction is necessary to explain and describe many emergent phenomena found in numerous scientific domains, from sociology to fundamental physics (Ellis et al. 2011), including the mind-body problem (e.g., Macdonald and Macdonald 2019; Robb 2019; Wong 2019).

Concerning this issue of *how* the mind-in-the-self affects the body-in-the-self, we adopt a standpoint reconciling the local-in-time dynamics of neural processes within the body-in-the-self and temporal extension of the mind-in-the-self. Actually, we follow Sperry (1969, 1980) in assuming that the mind, as an external factor, shapes the architecture of the neural process flow, whereas its dynamics is governed by local-in-time regulations. This concept is rooted in Sperry's assertion that

> conscious phenomena as emergent functional properties of brain processing exert an active control role as causal determinants in shaping the flow patterns of cerebral excitation. Once generated from neural events, the higher order mental patterns and programs have their own subjective qualities and progress, operate and interact by their own causal laws and principles which are different from and cannot be reduced to those of neurophysiology, ... The mental entities transcend the physiological just as the physiological transcends the molecular, the molecular, the atomic, and subatomic, etc. The mental forces do not violate, disturb or intervene in neuronal activity but they do supervene. Interaction is

[5] There are physical systems where the quasi-adiabatic approximation does not hold. By way of example, we note systems with scale-free memory when at any level all temporal scales contribute to the system dynamics. The mathematical formalism developed for describing such systems is based on differential equations with fractional time derivatives (e.g., Uchaikin 2013a, b, for are view of various system with scale-free memory). However, the "memory" of such systems is not the same as human memory, in particular, it is based on the current state of a given system mathematically described in terms of the system dynamics at previous moments of time. Put differently, the dynamics of such systems remains to be of the local-in-time type and the "memory" arises as a result of reducing the system dimensions. By contrast, human memory about the past is of another nature, in particular, it can be modified at the present because it did not cease to exist in the mind.

mutually reciprocal between the neural and mental levels in the nested brain hierarchies. (Sperry 1980, p. 201)

For a further discussion of this concept a reader may be referred, e.g., to Seth (2018), Marks (2019), Robb (2019).

Summarizing the aforementioned, let us note the basic features of our account of the body-in-the-self.

- The body-in-the-self is only a part of the human body.
- The equations governing the dynamics of neural processes in the body-in-the-self can be considered to be of the local-in-time type on scales of the immediate now.
- The body-in-the-self is influenced by the mind-in-the-self which affects the architecture of neural processes.

Finalizing the present subsection, we want to note that the body-in-the-self admits an interpretation as the embedding of the human body into the human mind, which is possible within the phenomenological approach. For this reason, the body-in-the-self is endowed with the local-in-time dynamics and temporal extension of its regularities.

The mind, the body, and the self (involving the mind-in-the-self and the body-in-the-self) are the entities mutually irreducible within the phenomenological approach. Each of them is individually involved in goal-oriented actions in various ways, which explains the diversity of human goal-oriented actions and their properties.

7.4 Evolution of the Self

The evolution of the self is closely intertwined with the 2D-time dynamics of human goal-oriented behavior the present book is devoted to. The self in its nature implies that it is simultaneously and equipollently embedded into two temporal dimensions—physical time and the human temporality.

The essence of this 2D-time dynamics can be represented as follows:

- The mind-in-the-self is embedded in the human temporality and, thereby, possesses temporal extension including the past and imaginary (anticipated) future. The mind-in-the-self operates with various space-time clouds (fragments of temporality) and is able to recognize their variations as physical time goes on. The entities of the mind-in-the-self are accessible to the mind. The mind is able to operate with these entities based on its own logic not requiring the reference to the body.
- The body-in-the-self is embedded into the flow of physical time and can be described from the standpoint of thick presentism. On the one hand, the body-in-the-self deals with neural patterns representing entities of the mind-in-the-self. The dynamics of these neural patterns is of the local-in-time type. On the other hand, these neural patterns reflect bodily actions accessible to the mind-in-the-self via sensory modalities, which may be regarded as features inherited from the body.

In the self these temporal dimensions are tightly interwoven because of the influence of the mind-in-the-self and the body-in-the-self on each other. Namely,

- the body-in-the-self provides access for the mind-in-the-self to the information about results of completed actions initiated by the mind-in-the-self,
- the main course of neural processes underlying the body-in-the-self dynamics is inaccessible to the mind-in-the-self,
- the mind-in-the-self can influence the regularities of the local-in-time dynamics of the body-in-the-self by affecting the architecture of neural processes during a *short physical time*.

Therefore, due to the two facet nature of the self, the 2D-time dynamics of human goal-oriented behavior may be regarded as a process

- which is reducible to the local-in-time dynamics[6] of neural processes (which is the characteristic feature of physical systems)
- whose regularities, however, can be affected, i.e., modified within short time intervals by the mind-in-the-self.

This type 2D-time dynamics admits an interpretation as a process reducible to the physical with bifurcations reducible to the mental. However, at the bifurcation points, the reducibility to the physical does not hold.

The 2D-time dynamics underlie the evolution of the self. Changes occurring in the mind-in-the-self as well as in the body-in-the-self are aggregated in memory and the skill. As a result, decision-making and action execution are based not only on the response to the current situation but also on the gained experience.

Experience changes the memory and the skill; these changes are different in nature but are correlated with each other. Because the experience is always individual, decisions and reactions to various situations emerging via the evolution of the self differ from person to person. For this reason, we, first, attribute to the self a new feature—the personality. Second, the temporal extension of the self can be extended to long-term scales. In particular, the two features—personality and long-term temporal extension—can be used for explaining personal identity over time.

The human temporality is the temporal dimension that is able to present the human personality as a whole entity with its history and future plans at each point-like instant of physical time. It prompts us to regard the human temporality not just as a container of events and intentional relations but as a certain mental medium that possesses its own properties and exceeds substantially the complex present.

[6] Speaking strictly, the dynamics of neural processes is based on local-in-time governing equations admitting some time delay between different regions of the neural network. These delay times are bounded from above by the duration of the immediate now—one of the basic synchronic units forming the experiential now. In our discussion, we understand the temporal locality of neural processes as the locality on scales on the immediate now.

References

Apps, M. A., & Tsakiris, M. (2014). The free-energy self: A predictive coding account of self-recognition. *Neuroscience & Biobehavioral Reviews, 41,* 85–97.

Bennett, K. (2017). *Making things up*. Oxford, UK: Oxford University Press.

Berkovich-Ohana, A., & Glicksohn, J. (2014). The consciousness state space (CSS)—a unifying model for consciousness and self. *Frontiers in Psychology,5*, Article 341 (19 pp).

Bermúdez, J. L. (2011). Bodily awareness and self-consciousness. In S. Gallagher (Ed.), *The Oxford handbook of the self* (pp. 157–179). Oxford, NY: Oxford University Press.

Bermúdez, J. L. (2017). Ownership and the space of the body. In F. de Vignemont & A. J. T. Alsmith (Eds.), *The subject's matter: Self-consciousness and the body* (pp. 117–144). Cambridge, Massachusetts: The MIT Press.

Billon, A. (2017). Mineness first: Three challenges to the recent theories of the sense of bodily ownership. In F. de Vignemont & A. J. T. Alsmith (Eds.), *The subject's matter: Self-consciousness and the body* (pp. 190–216). Cambridge, Massachusetts: The MIT Press.

Bliss, R., & Trogdon, K. (2014). Metaphysical grounding. In E. N. Zalta (Ed.), *The Stanford encyclopedia of philosophy* (Winter 2016 ed.). Stanford: Metaphysics Research Lab, Stanford University.

Buhrmann, T., & Di Paolo, E. (2017). The sense of agency - a phenomenological consequence of enacting sensorimotor schemes. *Phenomenology and the Cognitive Sciences, 16*(2), 207–236.

Bunge, M. (1977). Emergence and the mind. *Neuroscience, 2*(4), 501–509.

Bunge, M. (1979). The mind-body problem, information theory and Christian dogma. *Neuroscience, 4*(3), 453–454.

Campbell, J. (2011). Personal identity. In S. Gallagher (Ed.), *The Oxford handbook of the self* (pp. 339–351). Oxford, NY: Oxford University Press.

Cassam, Q. (2011). The embodied self. In S. Gallagher (Ed.), *The Oxford handbook of the self* (pp. 137–156). Oxford, NY: Oxford University Press.

Cat, J. (2017). The unity of science. In E. N. Zalta (Ed.), *The Stanford encyclopedia of philosophy* (Fall 2017 ed.). Stanford: Metaphysics Research Lab, Stanford University.

Correia, F., & Schnieder, B. (Eds.). (2012). *Metaphysical grounding: Understanding the structure of reality*. Cambridge, UK: Cambridge University Press.

Craik, F. I. M., Moroz, T. M., Moscovitch, M., Stuss, D. T., Winocur, G., Tulving, E., & Kapur, S. (1999). In search of the self: A positron emission tomography study. *Psychological Science, 10*(1), 26–34.

Dainton, B. (2008). *The phenomenal self*. New York: Oxford University Press Inc.

Dainton, B., & Snowdon, P. (2016). I—the sense of self. *Aristotelian Society Supplementary, 90*(1), 113–143.

de Vignemont, F. (2017). Agency and bodily ownership: The bodyguard hypothesis. In F. de Vignemont & A. J. T. Alsmith (Eds.), *The subject's matter: Self-consciousness and the body* (pp. 217–236). Cambridge, Massachusetts: The MIT Press.

Dennett, D. C. (1991). *Consciousness explained*. New York, NY: Back Bay Books, Little, Brown and Company.

deRosset, L. (2013). Grounding explanations. *Philosopher's Imprint, 13*(7), 1–26.

Drummond, J. J. (2020). Self-identity and personal identity. *Phenomenology and the Cognitive Sciences*. Published online August 7, 2020.

Eccles, J. C. (1989). *Evolution of the brain: Creation of the self.* London: Routledge.

Eccles, J. C. (1994). *How the SELF controls its BRAIN*. Berlin: Springer.

Ellis, G. F. R., Noble, D., & O'Connor, T. (2011). Top-down causation: An integrating theme within and across the sciences? *Interface Focus, 2*(1), 1–3.

Esfeld, M. (2001). *Holism in philosophy of mind and philosophy of physics*. Dordrecht: Kluwer Academic Publishers.

Esfeld, M. (2009). Philosophical holism. In G. H. Hadorn (Ed.), *Unity of knowledge (in Transdisciplinary research for sustainability), encyclopedia of life support systems* (Vol. 1, pp. 110–127). Oxford, UK: EOLLS Publishers.

Esfeld, M. (2015). Atomism and holism: Philosophical aspects. In J. D. Wright (Ed.), *International encyclopedia of the social & behavioral sciences* (2nd ed., Vol. 2, pp. 131–135). Netherlands: Elsevier, Amsterdam.
Fine, K. (2002). The question of realism. In A. Bottani, M. Carrara, & P. Giaretta (Eds.), *Individuals, essence and identity: Themes of analytic metaphysics* (pp. 3–48). Dordrecht, The Netherlands: Springer Science+Business Media.
Fine, K. (2013a). Fundamental truth and fundamental terms. *Philosophy and Phenomenological Research,87*(3), 725–732.
Fine, K. (2013b). Guide to ground. In F. Correia & B. Schnieder (Eds.), *Metaphysical grounding: Understanding the structure of reality* (pp. 37–80). Cambridge, UK: Cambridge University Press.
Friston, K. (2018). Am i self-conscious? (or does self-organization entail self-consciousness?). *Frontiers in Psychology,9*, Article 579.
Fuchs, A., & Kelso, J. A. S. (2009). Movement coordination. In R. A. Meyers (Ed.), *Encyclopedia of complexity and systems science* (pp. 5718–5736). New York, NY: Springer Science+Business Media, LLC.
Gallagher, S. (2005). *How the body shapes the mind*. Oxford: Oxford University Press.
Gallagher, S. (Ed.). (2011). *The Oxford handbook of the self*. Oxford: Oxford University Press.
Gallagher, S. (2012). *Phenomenology*. New York, NY: Palgrave Macmillan.
Gallagher, S. (2013). A pattern theory of self. *Frontiers in Human Neuroscience,7*, Article 443 (7 pp).
Gallagher, S. (2017). Self-defense: Deflecting deflationary and eliminativist critiques of the sense of ownership. *Frontiers in Psychology,8*, Article 1612 (10 pp).
Gallagher, S., & Daly, A. (2018). Dynamical relations in the self-pattern. *Frontiers in Psychology,9*, Article 664 (13 pp).
Gallagher, S., & Zahavi, D. (2012). *The phenomenological mind* (2nd ed.). London: Routledge, Taylor & Francis Group.
Gallagher, S., & Zahavi, D. (2019). Phenomenological approaches to self-consciousness. In E. N. Zalta (Ed.), *The Stanford encyclopedia of philosophy* (Summer 2019 ed.). Stanford: Metaphysics Research Lab, Stanford University.
Gallup, G. G., Jr., Anderson, J. R., & Platek, S. M. (2011). Self-recognition. In S. Gallagher (Ed.), *The Oxford handbook of the self* (pp. 80–110). Oxford, NY: Oxford University Press.
Gibb, S. (2015). The causal closure principle. *The Philosophical Quarterly, 65*(261), 626–647.
Gibb, S. (2019). The causal closure principle. In S. Gibb, R. Hendry, & T. Lancaster (Eds.), *The Routledge handbook of emergence* (pp. 111–120). London: Routledge, Taylor & Francis Group.
Gillihan, S. J., & Farah, M. J. (2005). Is self special? A critical review of evidence from experimental psychology and cognitive neuroscience. *Psychological Bulletin,131*(1), 76–97.
Glazier, M. (2020). Explanation. In M. J. Raven (Ed.), *The Routledge handbook of metaphysical grounding* (pp. 121–132). New York, NY: Routledge, The Taylor & Francis Group.
Graziano, M. S. A., & Webb, T. W. (2015). The attention schema theory: A mechanistic account of subjective awareness. *Frontiers in Psychology,06*, Article 500.
Guillot, M. (2016). I me mine: On a confusion concerning the subjective character of experience. *Review of Philosophy and Psychology, 8*(1), 23–53.
Hardcastle, V. G. (2008). *Constructing the self*. Amsterdam: John Benjamins Publishing Company.
Healey, R. (2016). Holism and nonseparability in physics. In E. N. Zalta (Ed.), *The Stanford encyclopedia of philosophy* (Spring 2016 ed.). Stanford: Metaphysics Research Lab, Stanford University.
Henry, A., & Thompson, E. (2011). Witnessing from here: Self-awareness from a bodily versus embodied perspective. In S. Gallagher (Ed.), *The Oxford handbook of the self* (pp. 228–249). Oxford, NY: Oxford University Press.
Hobson, R. P. (2011). Autism and the self. In S. Gallagher (Ed.), *The Oxford handbook of the self* (pp. 571–591). Oxford, NY: Oxford University Press.
Howell, R. J., & Thompson, B. (2017). Phenomenally mine: In search of the subjective character of consciousness. *Review of Philosophy and Psychology, 8*(1), 103–127.

Ikäheimo, H., & Laitinen, A. (Eds.). (2011). *Recognition and social ontology*. Leiden, The Netherlands: Brill, Hotei Publishing.

Jenkins, C. I. (2013). Explanation and fundamentality. In B. Schnieder, M. Hoeltje, & A. Steinberg (Eds.), *Varieties of dependence: Ontological dependence, grounding, supervenience, response-dependence* (pp. 211–241). Munich: Philosophia-Verlag GmbH.

Kelso, J. A. S. (2000). Principles of dynamic pattern formation and change for a science of human behavior. In L. R. Bergman, R. B. Cairns, L.-G. Nilsson, & L. Nystedt (Eds.), *Developmental science and the holistic approach* (pp. 63–83). Mahwah, NJ: Lawrence Erlbaum Associates, Inc.

Kelso, J. A. S. (2009a). Coordination dynamics. In R. A. Meyers (Ed.), *Encyclopedia of complexity and systems science* (pp. 1537–1565). New York, NY: Springer Science+Business Media, LLC.

Kelso, J. A. S. (2009b). Synergies: Atoms of brain and behavior. In D. Sternad (Ed.), *Progress in motor control: A multidisciplinary perspective* (pp. 83–91). Boston, MA: Springer Science+Business Media, LLC.

Kelso, J. A. S. (2014). The dynamic brain in action: Coordinative structures, criticality, and coordination dynamics. In D. Plenz & E. Niebu (Eds.), *Criticality in neural systems* (pp. 67–104). Weinheim: Wiley-VCH Verlag GmbH & Co. KGaA.

Kim, J. (2005). Supervenience, emergence, realization, reduction. In M. J. Loux & D. W. Zimmerman (Eds.), *The Oxford handbook of metaphysics* (pp. 556–584). Oxford: Oxford University Press.

Krisst, L., Montemayor, C., & Morsella, E. (2015). Unconscious and conscious component processes. In P. Haggard & B. Eitam (Eds.), *The sense of agency: Social cognition and social neuroscience* (pp. 25–62). Oxford, UK: Oxford University Press.

Latash, M. L. (2008). *Synergy*. Oxford: Oxford University Press.

Latash, M. L. (2012). *Fundamentals of motor control*. London: Academic, Elsevier Inc.

Lavazza, A., & Robinson, H. (Eds.). (2014). *Contemporary dualism: A defense*. New York: Routledge Taylor & Francis Group.

LeDoux, J. (2002). *Synaptic self: How our brains become who we are*. New York, NY: Viking Penguin.

Legrand, D. (2011). Phenomenological dimensions of bodily self-consciousness. In S. Gallagher (Ed.), *The Oxford handbook of the self* (pp. 204–227). Oxford, NY: Oxford University Press.

Legrand, D., & Ruby, P. (2009). What is self-specific? Theoretical investigation and critical review of neuroimaging results. *Psychological Review, 116*(1), 252–282.

Leuenberger, S. (2008). Supervenience in metaphysics. *Philosophy Compass, 3*(4), 749–762.

Leuenberger, S. (2013). From grounding to supervenience? *Erkenntnis, 79*(1), 227–240.

Leuenberger, S. (2014). Grounding and necessity. *Inquiry, 57*(2), 151–174.

Leuenberger, S. (2020). The fundamental: Ungrounded or all-grounding? *Philosophical Studies, 177*(9), 2647–2669.

Litland, J. E. (2017). Grounding ground. In K. Bennett & D. W. Zimmerman (Eds.), *Oxford studies in metaphysics* (Vol. 10, pp. 279–316). Oxford, UK: Oxford University Press.

Lowe, E. J. (2008). *Personal agency: The metaphysics of mind and action*. Oxford: Oxford University Press.

Lowe, E. J. (2009). Dualism. In A. Beckermann, B. P. McLaughlin, & S. Walter (Eds.), *The Oxford handbook of philosophy of mind* (pp. 66–84). Oxford: Oxford University Press.

Lowe, E. J. (2010). Substance dualism: A non-cartesian approach. In R. C. Koons & G. Bealer (Eds.), *The waning of materialism* (pp. 439–461). Oxford, NY: Oxford University Press.

Lowe, E. J. (2013). Branching time and temporal unity. In F. Correia & A. Iacona (Eds.), *Around the tree: Semantic and metaphysical issues concerning branching and the open future* (pp. 73–80). Dordrecht: Springer Science+Business Media.

Lubashevsky, I. (2017). *Physics of the human mind*. Cham: Springer International Publishing AG.

Macdonald, C., & Macdonald, G. (2019). Emergence and non-reductive physicalism. In S. Gibb, R. Hendry, & T. Lancaster (Eds.), *The Routledge handbook of emergence* (pp. 187–194). London: Routledge, Taylor & Francis Group.

MacKay, D. M. (1978). Selves and brains. *Neuroscience, 3*(7), 599–606.

MacKay, D. M. (1979). Reply to Bunge. *Neuroscience, 4*(3), 454.

MacKay, D. M. (1980). The interdependence of mind and brain. *Neuroscience, 5*(7), 1389–1391.

Marks, D. F. (2019). I am conscious, therefore, i am: Imagery, affect, action, and a general theory of behavior. *Brain Sciences,9*(5), 107 (27 pp).

McLaughlin, B., & Bennett, K. (2018). Supervenience. In E. N. Zalta (Ed.), *The Stanford encyclopedia ofphilosophy* (Winter 2018 ed.). Stanford: Metaphysics Research Lab, Stanford University. http://plato.stanford.edu/archives/win2018/entries/supervenience/.

McLean, K. C. (2017). And the story evolves. In J. Specht (Ed.), *Personality development across the lifespan* (pp. 325–338). London: Academic, Elsevier.

Merrill, N., & Fivush, R. (2016). Intergenerational narratives and identity across development. *Developmental Review, 40,* 72–92.

Metzinger, T. (2003). *Being no one: The self-model theory of subjectivity*. Cambridge, Massachusetts: The MIT Press.

Metzinger, T. (2009). *The ego tunnel: The science of the mind and the myth of the self.* New York: Basic Books.

Metzinger, T. (2011). The no-self alternative. In S. Gallagher (Ed.), *The Oxford handbook of the self* (pp. 279–296). Oxford, NY: Oxford University Press.

Metzinger, T. (2013). The myth of cognitive agency: Subpersonal thinking as a cyclically recurring loss of mental autonomy. *Frontiers in Psychology,4*, Article 931 (19 pp).

Metzinger, T. (2014). How does the brain encode epistemic reliability? Perceptual presence, phenomenal transparency, and counterfactual richness. *Cognitive Neuroscience, 5*(2), 122–124.

Morsella, E., & Poehlman, T. A. (Eds.). (2014). *Consciousness and action control*. Frontiers in psychology. Frontiers Media SA.

Morsella, E., Godwin, C. A., Jantz, T. K., Krieger, S. C., & Gazzaley, A. (2016). Homing in on consciousness in the nervous system: An action-based synthesis. *Behavioral and Brain Sciences,39*, e168.

Morsella, E., Velasquez, A. G., Yankulova, J. K., Li, Y., Wong, C. Y., & Lambert, D. (2020). Motor cognition: The role of sentience in perception and action. *Kinesiology Review, 9*(3), 261–274.

Muñoz-Corcuera, A. (2016). Persons, characters and the meaning of 'Narrative'. In R. Winkler (Ed.), *Identity and difference: Contemporary debates on the self* (pp. 37–61). Cham, Switzerland: Palgrave Macmillan.

Mylopoulos, M., & Pacherie, E. (2017). Intentions and motor representations: The interface challenge. *Review of Philosophy and Psychology, 8*(2), 317–336.

Mylopoulos, M., & Shepherd, J. (2020). The experience of agency. In U. Kriegel (Ed.), *The Oxford handbook of the philosophy of consciousness* (pp. 164–187). Oxford, UK: Oxford University Press.

Newen, A. (2018). The embodied self, the pattern theory of self, and the predictive mind. *Frontiers in Psychology,9*, Article 2270 (14 pp).

Northoff, G., Heinzel, A., de Greck, M., Bermpohl, F., Dobrowolny, H., & Panksepp, J. (2006). Self-referential processing in our brain—a meta-analysis of imaging studies on the self. *NeuroImage, 31*(1), 440–457.

O'Connor, T. (2020). Emergent properties. In E. N. Zalta (Ed.), *The Stanford encyclopedia of philosophy* (Fall 2020 ed.). Stanford: Metaphysics Research Lab, Stanford University.

Olson, E. T. (2019). Personal identity. In E. N. Zalta (Ed.), *The Stanford encyclopedia of philosophy* (Spring 2021 ed.). Stanford: Metaphysics Research Lab, Stanford University.

Pacherie, E. (2011). Self-agency. In S. Gallagher (Ed.), *The Oxford handbook of the self* (pp. 442–464). Oxford, NY: Oxford University Press.

Pacherie, E. (2015). Time to act: The dynamics of agentive experiences. In P. Haggard & B. Eitam (Eds.), *The sense of agency: Social cognition and social neuroscience* (pp. 3–24). Oxford, UK: Oxford University Press.

Pacherie, E. (2018). Motor intentionality. In A. Newen, L. D. Bruin, & S. Gallagher (Eds.), *The Oxford handbook of 4E cognition* (pp. 369–387). Oxford, UK: Oxford University Press.

Pacherie, E., & Mylopoulos, M. (2020). Beyond automaticity: The psychological complexity of skill. *Topoi*. Published online August 9, 2020.
Parfit, D. (2011). The unimportance of identity. In S. Gallagher (Ed.), *The Oxford handbook of the self* (pp. 419–441). Oxford, NY: Oxford University Press.
Pettit, P. (1998). Defining and defending social holism. *Philosophical Explorations, 1*(3), 169–184.
Popper, K. R., & Eccles, J. C. (1977). *The self and its brain*. Berlin: Springer.
Qin, P., Duncan, N. W., & Northoff, G. (Eds.). (2014). *Why and how is the self related to the brain midline regions?* Frontiers in human neuroscience. Lausanne: Frontiers Media SA.
Raven, M. J. (2015a). Fundamentality without foundations. *Philosophy and Phenomenological Research,93*(3), 607–626.
Raven, M. J. (2015b). Ground. *Philosophy Compass,10*(5), 322–333.
Raven, M. J. (Ed.). (2020). *The Routledge handbook of metaphysical grounding*. New York, NY: Routledge, The Taylor & Francis Group.
Robb, D. (2019). Emergent mental causation. In S. Gibb, R. Hendry, & T. Lancaster (Eds.), *The Routledge handbook of emergence* (pp. 195–205). London: Routledge, Taylor & Francis Group.
Robinson, H. (2020). Dualism. In E. N. Zalta (Ed.), *The Stanford encyclopedia of philosophy* (Fall 2020). Stanford: Metaphysics Research Lab, Stanford University.
Rochat, P. (2011). What is it like to be a newborn? In S. Gallagher (Ed.), *The Oxford handbook of the self* (pp. 57–79). Oxford, NY: Oxford University Press.
Rosen, G. (2010). Metaphysical dependence: Grounding and reduction. In A. H. Bob Hale (Ed.), *Modality: Metaphysics, logic, and epistemology* (pp. 109–135). Oxford, UK: Oxford University Press.
Rydéhn, H. (2018). Grounding and ontological dependence. *Synthese*. Published online on 30 May 2018.
Savellos, E. E., & Yalçin, U. D. (Eds.). (1995). *Supervenience: New essays*. Cambridge, UK: Cambridge University Press.
Schaefer, M., & Northoff, G. (2017). Who am i: The conscious and the unconscious self. *Frontiers in Human Neuroscience,11*, Article 126 (5 pp).
Schaffer, J. (2003). Is there a fundamental level? *Noûs, 37*(3), 498–517.
Schaffer, J. (2010). Monism: The priority of the whole. *Philosophical Review, 119*(1), 31–76.
Schaffer, J. (2012). Grounding, transitivity, and contrastivity. In F. Correia & B. Schnieder (Eds.), *Metaphysical grounding: Understanding the structure of reality* (pp. 122–138). Cambridge, UK: Cambridge University Press.
Schaffer, J. (2016). Grounding in the image of causation. *Philosophical Studies, 173*(1), 49–100.
Schechtman, M. (2007). Stories, lives, and basic survival: A refinement and defense of the narrative view. *Royal Institute of Philosophy Supplement, 60,* 155–178.
Schechtman, M. (2011). The narrative self. In S. Gallagher (Ed.), *The Oxford handbook of the self* (pp. 394–416). Oxford, NY: Oxford University Press.
Schütz, A., Fehn, T., & Baumeister, R. F. (2020). Self. In V. Zeigler-Hill & T. K. Shackelford (Eds.), *Encyclopedia of personality and individual differences* (pp. 4628–4637). Cham, Switzerland: Springer Nature Switzerland AG.
Seth, A. K. (2018). Consciousness: The last 50 years (and the next). *Brain and Neuroscience Advances, 2,* 1–6.
Shagrir, O. (2012). Concepts of supervenience revisited. *Erkenntnis, 78*(2), 469–485.
Shapiro, L. (Ed.). (2014). *The Routledge handbook of embodied cognition*. London: Routledge, Taylor & Francis Group.
Sider, T. (2011). *Writing the book of the world*. Oxford, UK: Oxford University Press.
Slors, M. (2001). *The diachronic mind: An essay on personal identity, psychological continuity and the mind-body problem*. Dordrecht, The Netherlands: Kluwer Academic Publishers.
Slors, M. (2019). Two distinctions that help to chart the interplay between conscious and unconscious volition. *Frontiers in Psychology,10*, Article 552.
Smith, J. (2020). Self-consciousness. In E. N. Zalta (Ed.), *The Stanford encyclopedia of philosophy* (Summer 2020 ed.). Stanford: Metaphysics Research Lab, Stanford University.

Smuts, J. C. (1927). *Holism and evolution* (2nd ed.). London: The Macmillan and Co., Ltd.
Sperry, R. W. (1969). A modified concept of consciousness. *Psychological Review, 76*(6), 532–536.
Sperry, R. W. (1980). Mind-brain interaction: Mentalism, yes; dualism, no. *Neuroscience, 5*(2), 195–206.
Steiner, M. (1978). Mathematical explanation. *Philosophical Studies, 34*(2), 135–151.
Stevenson, A. (Ed.). (2010). *Oxford dictionary of English* (3rd ed.). Oxford, New York: Oxford University Press.
Trogdon, K. (2013). An introduction to grounding. In B. Schnieder, M. Hoeltje, & A. Steinberg (Eds.), *Varieties of dependence: Ontological dependence, grounding, supervenience, response-dependence* (pp. 97–122). Munich: Philosophia-Verlag GmbH.
Tsakiris, M. (2011). The sense of body ownership. In S. Gallagher (Ed.), *The Oxford handbook of the self* (pp. 180–203). Oxford, NY: Oxford University Press.
Tsakiris, M. (2015). The relations between agency and body ownership. In P. Haggard & B. Eitam (Eds.), *The sense of agency: Social cognition and social neuroscience* (pp. 235–256). Oxford, UK: Oxford University Press.
Uchaikin, V. V. (2013a). *Fractional derivatives for physicists and engineers: Volume I. Background and theory*. Berlin: Higher Education Press, Beijing and Springer.
Uchaikin, V. V. (2013b). *Fractional derivatives for physicists and engineers: Volume II. Applications*. Berlin: Higher Education Press, Beijing and Springer.
VandenBos, G. R. (Ed.). (2015). *APA dictionary ofpsychology* (2nd ed.). Washington, DC: American Psychological Association.
Varela, F. J., Thompson, E., & Rosch, E. (2016). *The embodied mind: Cognitive science and human experience* (Revised ed.). Cambridge, MA: The MIT Press. New foreword by Jon Kabat-Zinn and new introductions by Evan Thompson and Eleanor Rosch.
Vogeley, K., & Gallagher, S. (2011). Self in the brain. In S. Gallagher (Ed.), *The Oxford handbook of the self* (pp. 111–136). Oxford, NY: Oxford University Press.
Weber, M., & Esfeld, M. (2009). Holism in the sciences. In G. H. Hadorn (Ed.), *Unity of knowledge (in Transdisciplinary research for sustainability), encyclopedia of life support systems* (Vol. 1, pp. 91–109). Oxford, UK: EOLLS Publishers.
Wilson, M. (2002). Six views of embodied cognition. *Psychonomic Bulletin & Review,9*(4), 625–636.
Wong, H. Y. (2019). Emergent dualism in the philosophy of mind. In S. Gibb, R. Hendry, & T. Lancaster (Eds.), *The Routledge handbook of emergence* (pp. 179–186). London: Routledge, Taylor & Francis Group.
Woźniak, M. (2018). "i" and "me": The self in the context of consciousness. *Frontiers in Psychology,9*, Article 1656 (14 pp).
Zahavi, D. (2014). *Self and other: Exploring subjectivity, empathy, and shame*. Oxford, UK: Oxford University Press.
Zahavi, D. (2018). Consciousness, self-consciousness, selfhood: A reply to some critics. *Review of Philosophy and Psychology, 9*(3), 703–718.
Zahavi, D. (2020). Consciousness and selfhood: Getting clearer on for-me-ness and mineness. In U. Kriegel (Ed.), *The Oxford handbook of the philosophy of consciousness* (pp. 635–653). Oxford, UK: Oxford University Press.
Zahle, J., & Collin, F. (Eds.). (2014). *Rethinking the individualism-holism debate: Essays in the philosophy of social science*. Cham, Switzerland: Springer International Publishing.
Zheng, Y. (2016). The Swampman puzzle and diachronic holism. *The Philosophical Forum, 47*(2), 171–193.

Post Scriptum

Natalie Plavinska, my wife, for a long time refused to put her name on the cover of the present book, despite the fact that for many years she has been my co-author in research on human behavior and her contribution to the present book is crucial.

In fact, it was she who inspired me to focus on developing a mathematical formalism aimed at modeling the basic features of the human mind as an individual problem. It not only impacted my scientific interests but also opened up new horizons in understanding the human being; I am very grateful to her for that.

Our first paper was published more than ten years ago. The previous book, "The Physics of the Human Mind," was also written together. Actually, she thought out the idea of the first book and its structure. In the writing, our discussions, text corrections, and discussions again were endless. Without her permanent participation, the first book would never have been written. At that time, I asked her to include her name in the list of authors, but she refused because, according to her, it was not easy to indicate her contribution, and besides, she is not a physicist. Now I have persuaded her to become the coauthor explicitly because her contribution to the present book has become clear.

The present book is again devoted to physics, now, of the human temporality. The mathematical formalism developed in it—which is my contribution—is only one side of the story. The general ideas, Natalie Plavinska proposed, underlie the formalism, the others emerged at the intermediate stages. Naturally, her permanent participation in "discussing-writing-reading-rewriting-discussing again' was essential as previously.

I want to note the essential concepts and ideas proposed by Natalie Plavinska, which were unexpected for me and now are interwoven into the fabric of the book.

- The idea and the structure of this book, as well as the previous one, belongs to her.
- Although the notions of *complex present* and *space-time cloud* were introduced previously in the book "Physics of the Human Mind," in the present book, their content is principally different due to the interpretation she proposed.

I. Lubashevsky and N. Plavinska, *Physics of the Human Temporality, Understanding Complex Systems*, https://doi.org/10.1007/978-3-030-82612-3

  - Now we deal with the hierarchical structure of complex present, where each level has the individual tripartite structure. Natalie put forward this idea from the beginning, but I recognized its importance only after the first two chapters were written. As a result, we needed to revise the written chapters. This hierarchical structure has become the heart of the constructed theory.
  - Her interpretation of the space-time cloud as the basic element of the human temporality containing *no* spatial and temporal points was unclear for me for a long time. Initially, I conceived of the space-time cloud as some fuzzy region in the space-time continuum. Our discussions for hours and hours gave rise to the concept of the mind-in-the-self and the body-in-the-self, where her interpretation of the space-time cloud and my initial understanding of this entity were reconciled. It opened a gate to describing conscious (deliberate) and unconscious (automatic) human actions within a common formalism.

- Her understanding of that the human temporality is integrated by relations of another type than cause-effect relationship has become the key to describing the impact of the imaginary future on the current actions. It enabled me to categorize such relations as *intentional* and recognize that the properties of events united by intentional relations should be attributed to the corresponding relational network (or the relational chain) rather than to the events separately.
- Her novel, original concept of endless cloud cycle in the mind underlying the developed theory of effort-fusion is an example of solving theoretical puzzles having no solutions from inside. This idea has underpinned the principle of cloud-fusion self-consistency.
- She proposed the concept of *tunnel* taking the central position in the two-dimensional time description of human goal-oriented behavior.
- Her conception of the *self* raises the particular problems discussed in the present book to a higher level. This concept opens for us a new perspective on the human perception of surrounding objective reality. In fact, the concept of the self accentuates the role of human temporality in the fusion of short-term fragments of the human now into the mental image of reality forming an entity with internal integrity.

Naturally, there are many other particular examples, where her and my ideas are so intertwined that now we just cannot separate them individually.

I am very glad that my coauthor, Natalie Plavinska, has finally dared to come out of hiding.

Ihor Lubashevsky

Index

I. Lubashevsky and N. Plavinska, *Physics of the Human Temporality, Understanding Complex Systems*, https://doi.org/10.1007/978-3-030-82612-3

CPSIA information can be obtained
at www.ICGtesting.com
Printed in the USA
LVHW080416251021
701439LV00002B/11

9 783030 82611